The Mechanics of Robot Grasping

In this comprehensive textbook about robot grasping, you will find an integrated look at the major concepts and technical results in robot grasp mechanics. A large body of prior research – including key theories, graphical techniques, and insights on robot hand designs – is organized into a systematic review, using common notation and a common analytical framework.

With introductory and advanced chapters that support senior undergraduate- and graduate-level robotics courses, this book provides a full introduction to robot grasping principles that are needed to model and analyze multi-finger robot grasps. This textbook also serves as a valuable reference for robotics researchers and practicing robot engineers. Each chapter contains many worked-out examples, exercises with full solutions, and figures that highlight new concepts and help the reader master the use of the theories and equations presented.

Elon Rimon is a professor in the Department of Mechanical Engineering at the Technion – Israel Institute of Technology. He has also been a visiting associate faculty member at the California Institute of Technology. Professor Rimon was a finalist for the best paper awards at the IEEE International Conference on Robotics and Automation and the Workshop on Algorithmic Foundations of Robotics, and he was awarded best paper presentation at the Robotics Science and Systems Conference.

Joel Burdick is the Richard L. and Dorothy M. Hayman Professor of Mechanical Engineering and Bioengineering at the California Institute of Technology. Professor Burdick has received the NSF Presidential Young Investigator Award, the Office of Naval Research Young Investigator Award, and the Feynman Fellowship. He has been a finalist for the best paper awards at the IEEE International Conference on Robotics and Automation, and he received the Popular Mechanics Breakthrough Award.

"*The Mechanics of Robot Grasping*, by two of the world's leading experts, fills an important gap in the literature by providing the first comprehensive survey of the mathematical tools needed to model the physics of robot grasping. The book uses configuration space to consistently characterize equilibrium, immobilizing and caging grasps, and clearly conveys important points such as the distinction between first-order and second-order form closure. The book also contains new material on the effects of gravity, compliance, and hand mechanism design. Grasping remains a Grand Challenge for robots, and this book provides the solid foundation for progress for students and researchers in the years ahead."

Ken Goldberg,
University of California–Berkeley

The Mechanics of Robot Grasping

ELON RIMON
Technion – Israel Institute of Technology

JOEL BURDICK
Caltech – California Institute of Technology

CAMBRIDGE
UNIVERSITY PRESS

CAMBRIDGE
UNIVERSITY PRESS

University Printing House, Cambridge CB2 8BS, United Kingdom

One Liberty Plaza, 20th Floor, New York, NY 10006, USA

477 Williamstown Road, Port Melbourne, VIC 3207, Australia

314–321, 3rd Floor, Plot 3, Splendor Forum, Jasola District Centre, New Delhi – 110025, India

79 Anson Road, #06–04/06, Singapore 079906

Cambridge University Press is part of the University of Cambridge.

It furthers the University's mission by disseminating knowledge in the pursuit of education, learning, and research at the highest international levels of excellence.

www.cambridge.org
Information on this title: www.cambridge.org/9781108427906
DOI: 10.1017/9781108552011

First published 2019

Printed in the United Kingdom by TJ International Ltd. Padstow Cornwall

A catalogue record for this publication is available from the British Library.

Library of Congress Cataloging-in-Publication Data
Names: Rimon, Elon, author. | Burdick, Joel, author.
Title: The mechanics of robot grasping / Elon Rimon, Joel Burdick.
Description: Cambridge, United Kingdom ; New York, NY : Cambridge University Press, 2019. | Includes bibliographical references and index.
Identifiers: LCCN 2018059144 | ISBN 9781108427906 (hardback : alk. paper)
Subjects: LCSH: Robot hands–Textbooks. | Mechanical movements–Textbooks. | Robotics–Textbooks.
Classification: LCC TJ211.43 .R56 2019 | DDC 629.8/933–dc23
LC record available at https://lccn.loc.gov/2018059144

ISBN 978-1-108-42790-6 Hardback

Contents

1 Introduction and Overview

1.1 Motivation and Background

Robots are already widely used in manufacturing, where they assemble products, handle machining operations, transfer parts on factory floors or package orders for shipping. Robots are now being assigned to assist humans in a much broader range of tasks, such as medical procedures, elderly care, agricultural work, disaster response, remediation of toxic waste and, somewhat unfortunately, battlefield operations. In all of these current and future applications, robots must *grasp and manipulate* objects in their environment in order to realize their assigned task goals. These robotic operations rely upon the use of *multi-finger robot hands* to securely grasp the task-related objects and realize the task goals.

This book is devoted to the kinematics and mechanics principles that govern the behavior of multi-finger robot hands. Robot hands typically make multiple contacts with physical objects during grasping and manipulation tasks. However, the principles described in this book are relevant to a much wider range of robotic applications. To see the common issues in these seemingly different applications, let us briefly describe each one of these areas.

Multi-finger robot grasping: Figure 1.1 shows an example of a multi-finger robot hand that can grasp and manipulate a wide variety of objects. This hand was designed by Salisbury at Stanford University in the 1980s. As seen in Figure 1.1, a *grasp* is used to fix the object position securely within the robot hand, which itself is typically attached to a robot arm. Properly designed robot hands can implement a wide variety of grasps – from *precision grasps*, where only the fingertips touch the grasped object (Figure 1.1), to *power grasps*, where the robot hand makes additional midfinger and palm contacts with the grasped object in order to provide highly secure grasps (Figure 1.2). Once grasped, the object can be securely transported or manipulated as part of a more complex robotic task.

In the context of robot grasping, the finger joints serve two purposes. First, the mechanical torques or forces generated at each finger joint control the contact forces between the fingers and the grasped object to maintain a secure grasp in the presence of external disturbances. Second, they allow the finger mechanisms to be positioned so that the robot hand can grasp a wide variety of objects with a range of contact arrangements. Beyond basic grasping, the finger joints allow manipulation of a grasped object within the robot hand.

Figure 1.1 A three-finger robot hand demonstrating *precision grasps.* Original image ©Hank Morgan/Science Source, (Left). Original image ©RGB Ventures / SuperStock / Alamy Stock Photo, (Right).

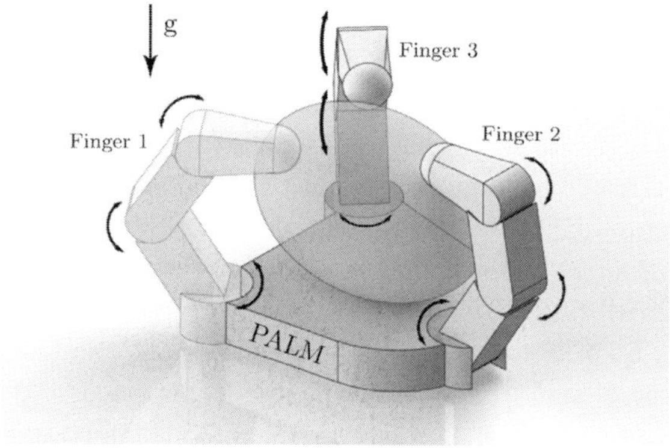

Figure 1.2 A three-finger robot hand demonstrating a *power grasp.*

The quality of a grasp is largely dictated by the location of the finger contacts on the grasped object surface. While the finger joints can actively change the contact forces applied on the grasped object, for purposes of grasp analysis, one can conceptually replace the finger mechanisms by *finger bodies* that apply equivalent forces at the contacts. Alternatively, one can take the conceptual viewpoint that the fingertips, if held rigidly by the robot hand, provide constraints on the grasped object's motion. Thus, as depicted in Figure 1.3, the analysis of a multi-finger grasp can be simplified to the study of a central *grasped object*, $\mathcal{B}$, in contact with surrounding *finger bodies*, $\mathcal{O}_1, \mathcal{O}_2, \ldots, \mathcal{O}_k$. Parts I, II and III of the book focus on this simplified grasp system. Building on these foundations, Part IV considers the kinematics and mechanics principles that govern the full robot hand during grasping.

Workpiece fixturing: Similar situations that involve multiple contacts arise in workpiece fixturing. Figure 1.4(a) shows a *modular fixture* that consists of fixuring elements, or *fixels,* that restrain a workpiece object. Fixtures are commonly used to securely hold workpieces during manufacturing and assembly processes. While some fixturing systems include actuation to actively control some of the clamping forces, most fixturing

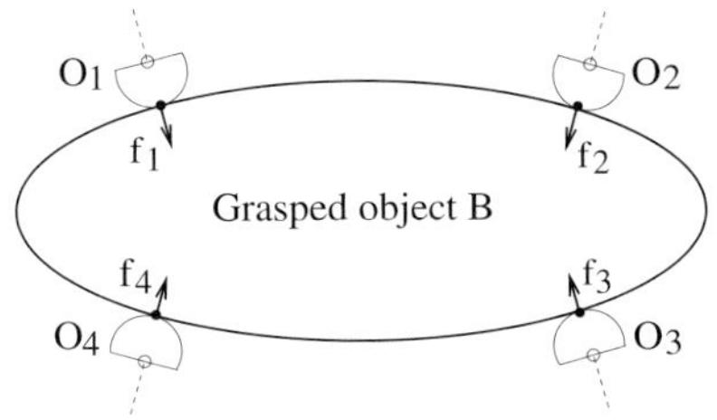

Figure 1.3 A multi-finger grasp can be abstracted as a system of finger bodies $\mathcal{O}_1, \mathcal{O}_2, \ldots, \mathcal{O}_k$ interacting with a grasped object $\mathcal{B}$.

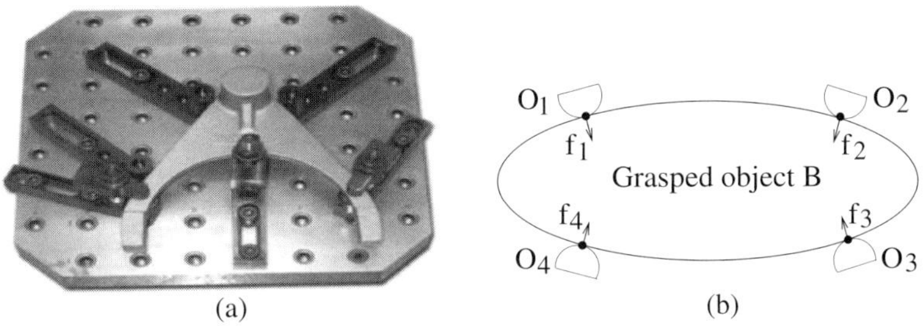

Figure 1.4 (a) A modular fixture in which a curved workpiece is held by multiple fixels. (b) Abstraction of the fixture system as a central workpiece in contact with multiple rigid bodies. Credit: Advanced Machine & Engineering (www.ame.com)

systems are unactuated. The contact forces between the fixels and the workpiece are typically generated during preloading of the fixtures, or arise from reactions of the fixels and workpiece materials as they are stressed and strained by machining or assembly operations. As depicted in Figure 1.4(b), the analysis of fixtures can also be idealized as a central workpiece in contact with multiple fixel bodies.

Quasistatic legged robot locomotion: The physical arrangements seen in robot grasps can also be applied to the study of some key issues in quasistatic legged robot locomotion. Figure 1.5(a) shows the Adaptive Suspension Vehicle, a 7,000-pound (about 3,000-kilogram) six-legged walking machine built at Ohio State University in the 1980s. This amazing vehicle could walk over uneven and muddy terrain while towing a load of 2,000 pounds (about 1,000 kilograms). A smaller four-legged robot developed at NASA JPL for disaster recovery missions is shown in Figure 1.5(b). Legged robot postures are analogous to robot grasps in the following ways. The terrain supporting the legged robot is analogous to the terrain formed by a robot hand supporting an object against gravity via multiple contacts. Legged robots are said to walk with *quasistatic gait* when the mechanism's center of mass lies within the *support region* of the leg contacts on the terrain.[1] If all the leg joints were to be instantaneously fixed, the robot's rigidified posture must be *statically stable* to ensure stable gait. Thus, as suggested in

[1] When walking on a flat horizontal terrain, the support region is the convex hull of the contact points. On uneven terrain, the support region assumes a more complex shape fully described in this book.

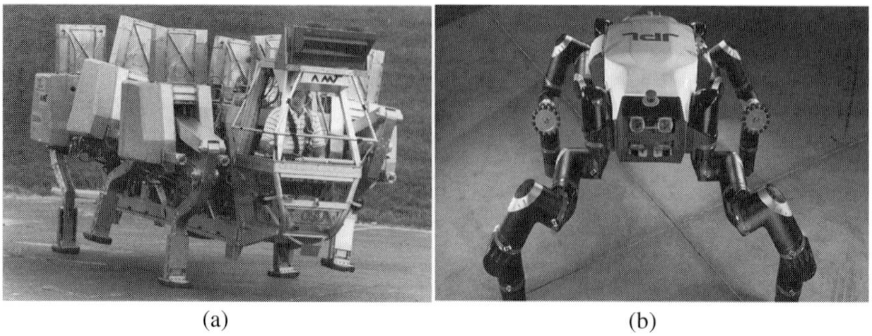

Figure 1.5 (a) The six-legged Adaptive Suspension Vehicle built in the 1980s. (b) A four-legged robot developed in the 2010s for disaster recovery missions. Picture of Ohio State University's Adaptive Suspension Vehicle, obtained from www.theoldrobots.com.

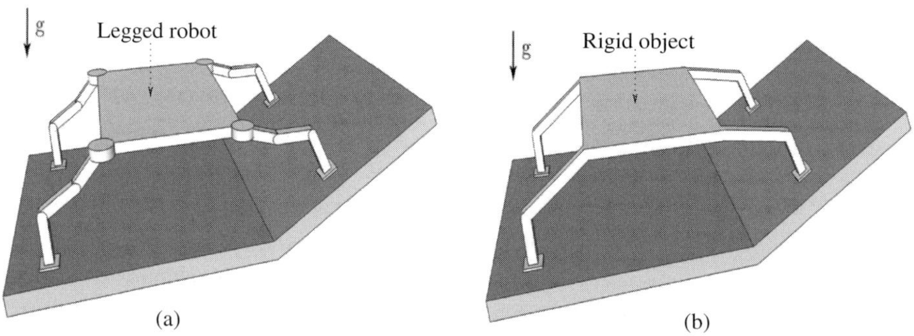

Figure 1.6 (a) A conceptual four-legged robot posture. (b) An abstraction of the legged robot posture as a rigidified body supported by multiple contacts on uneven terrain.

Figure 1.6, the question of legged robot posture stability can be reduced to the stability of a rigid object supported against gravity by a robot hand.

Multi agent transport systems: A team of ground-based mobile robots is tasked with transporting a heavy object to a new location. Each robot may only push against the object on its periphery and therefore can be thought of as a fingertip body that can apply transport forces to the object through contact or provide motion constraints for the object while it is pushed by other team members. A team of such robots is analogous to multiple finger bodies performing a manipulation task. Large objects can also be transported by a team of flying robots. In such transport schemes, a large object is suspended from cables attached to each agent. The cables provide *unidirectional* force constraints. That is, each flying robot can only *pull* on its cable to provide needed transport and maneuvering forces. In an analogous manner, the fingers of a robot hand provide *unidirectional* force constraints, as they can only *push* against the grasped object surface.

These examples show that the category of robot grasping includes diverse problems that can all be reduced to the study of multiple contacting bodies. While the principles described in this book apply equally well to all of these areas, the book describes these principles in the context of *robot grasping*.

1.2 Purpose of This Book

The first goal of this book is to assemble historical and recent results on robot grasping into one volume. There are many compelling reasons to do so. While several issues in robot grasping clearly merit further research, the basic kinematics and mechanics principles underlying the processes of grasping, fixturing and quasistatic legged locomotion have become well understood. Since the publication of the first book dedicated to grasp mechanics, *Robot Hands and the Mechanics of Manipulation* [1] published in 1985, real progress has been made on basic geometric modeling of these processes. Many robotics texts – such as [2–5] – discuss some important aspects of robot grasping. However, they do not provide a complete survey of the topics that we feel should be collected into one comprehensive volume, which can serve both as an educational and as a reference text.

A second goal of this book is to formulate the diversity of issues in robot grasping into a common language. We focus on the *configuration space* (c-space) formulation of multi-finger robotic grasps, as it provides an intuitively appealing viewpoint on many issues in grasp mechanics in a manner which is consistent with mainstream robot motion planning theory. This synergy of kinematics, mechanics and c-space should simplify future efforts in the development of algorithms for robot grasping systems. A third goal of this book is to highlight the impact of higher-order kinematics and mechanics effects such as surface curvature on robot grasping performance. Such effects, which were often ignored in the early robot grasping literature, can be significant in many practical circumstances.

We have purposefully limited the scope of this book to the kinematics and mechanics principles underlying robot grasping. We do not present or analyze grasp planning algorithms. While such algorithms are essential to maximize the utility of robot grasping systems, we purposely choose to avoid a survey of such algorithms. However, because grasping mechanisms must incorporate the basic principles of grasp mechanics, successful planning algorithms, even one based on machine learning, will be based on the theories presented in this book. Other than an advanced chapter on robot hand *manipulability,* the book does not study the mechanics of manipulation (the process of repositioning parts using a robot hand and possibly a robot arm). While the book touches upon a small set of issues concerning legged robots, it does not survey the field of legged robot locomotion. Instead, three chapters discuss the *static stability* of objects supported against gravity by a robot hand. This material appeared only as research papers and can be directly applied to legged robot locomotion on uneven terrains.

Thus, the goal of this book is to present material in the context of robot grasping, with the belief that this material can serve as a foundation for students, researchers and practitioners interested in developing new algorithms for all related robotics tasks.

1.3 How to Use This Book

The book is organized so that it can serve different purposes. First, it could serve as a textbook for those portions of an introductory graduate robotics course that touch upon robot grasping. Consequently, we have added exercises and some worked-out examples to help students master the key material through the problem-solving process. Note that exercises marked with an asterisk (*) are advanced problems, often bordering upon basic research questions. Second, the book can serve as a useful reference for researchers and practitioners. Bibliographical notes at the end of each chapter serve as a jumping-off point for further investigation. Moreover, we have included detailed proofs of some of the more critical and foundational results. To develop these analyses, we rely upon a few mathematical tools, such as *non-smooth analysis* and *stratified Morse theory,* which are not standard in the robotics literature. We therefore provide self-contained reviews of these methods in the book appendices so that their use can be readily followed in the main body of the text and so that their effectiveness can be appreciated.

The book is organized into *four parts,* whose content and purpose are summarized as follows.

- **Part I – Basic Geometry of Robot Grasping:** Chapters 2, 3, 4 and 5 review the essential geometric and kinematic concepts behind the *configuration space* of multiple contacting rigid bodies. For students who are newly starting their study of grasp mechanics, Part I is *essential background reading.* Practitioners already familiar with the field can skim most of these chapters to take note of the notational and kinematic conventions used in this book. However, it should be noted that Chapter 2 presents a novel formulation of the geometry of *c-space obstacles* using non-smooth distance functions. This approach provides a more coherent and straightforward framework for the analysis of multiple contacting bodies. Chapter 3 introduces a graphical representation of the tangent space to a c-space obstacle. This representation, which is not standard in the robotics literature, plays an important role in subsequent chapters. Chapter 4 describes *equilibrium grasps,* where the grasped object is held stationary within the robot hand. Equilibrium grasps are necessary for maintaining *secure grasps,* whose study forms the bulk of Parts II and III of the book. Chapter 5 subsequently describes a comprehensive catalog of the equilibrium grasps in terms of *line geometry.* This line geometry formulation has not previously appeared in a textbook, and some equilibrium grasps in the catalog have not appeared previously in any publication. Thus, even experienced readers may want to devote some attention to these portions of Part I.
- **Part II – Frictionless Rigid-Body Grasps and Stances:** Chapters 6–11 focus on multi-finger grasps under *frictionless* contact conditions. While no grasp contacts are truly frictionless, nor are any physical objects truly rigid, such models constitute useful idealizations. First, they serve as a conservative approximation to many real robot grasping applications. Second, friction has no bearing on the notion of *immobilizing grasps* considered in Chapters 7–9, nor does it play any role in the synthesis of *caging grasps* considered in Chapter 10.

For students, Part II introduces alternative notions of *secure grasps*. Individual chapters are devoted to the different types of secure grasps associated with frictionless contacts: *first-order* immobilization in Chapter 7, *second-order* immobilization in Chapter 8 and then *caging grasps* in Chapter 10. This part of the book stresses how the geometric and kinematic ideas of Part I can be used to analyze these types of grasps and serves as a useful bridge between the purely geometric ideas of Part I and the mechanics-oriented methods of Part III. In particular, the notion of *second-order immobilization* has only appeared in the research literature. Similarly, no text has previously organized the material on caging grasps into a coherent whole.

Chapter 9 considers *minimalistic robot grasps*. The chapter shows how to use mobility analysis to bound the number of fingers that are necessary to grasp different classes of objects in a secure immobilizing grasp. This material is not essential reading for a first introduction to the subject of robot grasping. Practitioners should note that this material collects together results which appeared only in the research literature. The final chapter of Part II considers the gravitational stability of objects supported by a robot hand via frictionless contacts. This chapter is not essential for beginning students. Experienced practitioners should note that this subject relies upon second-order mobility analysis and the use of *stratified Morse theory*, which is a unique and powerful tool for the analysis of multi-contact grasps. Neither of these analyses has previously appeared in a textbook.

- **Part III – Frictional Rigid-Body Grasps and Stances:** Chapters 12–15 study multi-finger grasps under *frictional* contact conditions. Chapter 12 introduces the notion of secure *wrench resistant* grasps. Both students and practitioners may be interested in the portions of Chapter 12 that formulate the wrench resistance property as a linear matrix inequality and relate this same property to the notion of small-time local controllability. Chapters 14 and 15 consider the gravitational stability of hand-supported stances when friction is present at the contacts. These chapters are not essential reading for beginning students. However, experienced practitioners should note that these chapters are highly relevant for legged robot locomotion on uneven terrains.

  Chapter 13 describes *grasp quality functions* that assess the value of alternative grasps. This chapter summarizes the key concepts behind most grasp quality functions that appeared in the literature. One of these key concepts concerns the behavior a grasp quality function under different choices of reference frames. The analysis of *frame invariance* that appears in Chapter 13 should interest both students and practitioners.
- **Part IV – Robot Hand Mechanisms:** The previous chapters abstracted the finger mechanisms to *finger bodies* that can generate fingertip motions and contact forces as necessary. Chapters 16, 17, and 18 consider the full robot hand during grasping processes. Chapter 16 considers how the kinematic structure of a robot hand affects key properties of the entire grasp system. The chapter studies the division of forces within a grasping hand into distinct linear subspaces and

considers how these different subspaces manifest as different types of wrench resistant grasps at the fingertips. Practitioners should note that these include *active* wrench resistant grasps, which correspond to precision grasps, and *structurally dependent* wrench resistant grasps, which correspond to power grasps.

Chapter 17 studies the property of *grasp manipulability,* defined as the ability of a robot hand to impart arbitrary local motions to the grasped object. This chapter forms an important link between basic grasp mechanics principles and the ability of a robot hand to locally manipulate objects while maintaining the grasp. Chapter 18 looks at the influence of compliance in grasping mechanisms. The chapter introduces the *grasp stiffness matrix*, which describes how the grasped object reacts to disturbances – assuming that the hand's joints behave in a compliant manner. The grasp stiffness matrix determines the stability of compliant grasps and provides information on the basin of attraction surrounding the grasp. Practitioners should note that robot hand control algorithms often implement proportional–integral–derivative (PID) control of the hand mechanism joints during grasping. Hence, this chapter describes the behavior and stability properties of objects that are grasped in controlled hand mechanisms.

- **Appendices:** Appendix A summarizes some useful tools of *non-smooth analysis.* These tools are central in modeling the *c-space obstacles* induced by grasping fingers and are used to explain the theory of immobilizing grasps. Appendix B summarizes the subject of *stratified Morse theory.* This theory offers analytical tools that characterize extremum points on *stratified sets.* One finds these sets in the configuration space formed by a robot hand that interacts with the grasped object. Stratified sets also appear as the sets of wrenches that can be affected on a grasped object by multiple fingers. Hence, all readers may wish to read these short appendices in order to appreciate the elegant and effective tools offered by these theories.

We hope that this organizational approach provides a text that is relatively self-contained for readers who are new to this field and that this format allows experienced practitioners to quickly pinpoint topics of interest.

1.4 Suggested Reading

As part of an introductory robotics course, Part I, Chapters 6 and 7 and Chapter 12 of Part III would give students a working knowledge of key issues in robot grasp modeling and analysis. Armed with this knowledge, students could pursue a self-study of this book's other chapters. A longer introduction would additionally include Chapter 8, on contact curvature effects, and possibly Chapter 10, on caging grasps. For courses that also include some study of legged robot locomotion, Chapters 14 and 15 of Part III would give students thorough insight into the stability of legged robot postures on uneven terrains. The remaining chapters primarily serve as advanced topics or reference material. The portions of the text that are more apt to be used as part of a course also have an increased number of exercises and worked-out examples.

Prerequisites: Obviously, it is difficult to write a book that can be both accessible and engaging to readers with a vast range of backgrounds. In writing this book, we have assumed that the reader has a basic understanding of matrix analysis and advanced calculus, which are reviewed when needed. We also assume working knowledge of rigid-body kinematics and the kinematics of serial chain robotic linkages. Both topics are covered in robotics texts such as [4] and [6].

1.5 A Brief History of Robot Grasp Mechanics

What follows is an abbreviated and incomplete history of the relevant developments in robot grasp mechanics, including connections to the contents of this book. Rather than give a detailed list of references, this review emphasizes some of the main trends and their influence on the organization of this book. This section can be skipped upon a first reading of the book.

1.5.1 The Early Influences

The theory underlying all robot grasping processes is built upon three classical foundations: kinematics, mechanics and mechanism theory. Before summarizing more modern developments, let us review some of the classical developments that underlie the principles and techniques described in this book.

All practical grasping systems involve multiple contacting bodies. To a very good approximation, in many practical situations, one can analyze multi-finger grasps using quasistatic rigid-body approach. The subject of *statics* has been around for centuries. However, Louis Poinsot's 1803 book *Théorie Nouvelle de la Rotation des Corps* can be identified as one of the first to geometrize rigid-body statics and mechanics principles. Poinsot observed that any system of forces acting on a rigid body can be resolved into a single force and a couple acting about the force axis. In the context of robot grasping, a system of contact forces applied to a grasped object can be replaced by an equivalent *net wrench.*[2] Coupled with the use of *line geometry,* these ideas are the basis of the equilibrium grasps described in Chapters 4 and 5.

The detailed mechanics of the contact between each finger and the grasped object can have significant influence upon the properties of the grasp. A *contact model* describes the types of forces that can be transmitted through a finger–object contact. Throughout the modern history of grasp analysis, contact models have included the effects of surface friction. Charles Augustin de Coulomb introduced the concept that friction and cohesion can have an influence on problems of statics. In his *Théorie des Machines Simples* (1781), Coulomb developed what we now call the *Coulomb friction model.* The

[2] The term *wrench* was introduced later by Ball [7] in his treatise on screw theory.

Coulomb friction model may be criticized as too simplistic and inaccurate, since it does not capture many tribological effects. However, experience has shown that it provides a surprisingly effective approximation for grasp analysis, and consequently it has been widely used in the study of robot grasping. The rigid-body contact models are reviewed in Chapter 4. Moreover, the influence of mechanics on the analysis of grasps drives the partition of the book into multiple parts.

In many ways, the theory of grasp mechanics starts with Reuleaux's 1875 book, *The Kinematics of Machinery* [8]. This book introduced a systematic theory for categorizing and analyzing mechanisms as well as multi-contact systems. In this work, Reuleaux established the basic concept of *force closure,* which is introduced in Chapter 6 and fully analyzed in Chapter 12. Intuitively, a grasp is said to be *force closure* if any force and torque applied to the grasped object can be resisted by feasible finger forces at the contacts. The force closure concept has been a central issue in robot grasping. This book will use the term *wrench resistant grasp,* as it encompasses a broader notion of a grasp's ability to resist forces and torques applied to the grasped object.

A grasp is said to be in *form closure* if the contacting fingers provide sufficient rigid-body constraints on the grasped object motion so that the object is fully restrained by the grasping fingers. This book will use the term *immobilizing grasp* rather than form closure, as it more accurately describes the notion of complete restraint based on rigid-body constraints. In 1900, Somoff provided the first systematic analysis of immobilization [9]. This book provides a thorough analysis of rigid-body immobilization in Chapters 7 and 8. A key contribution of this book is the extension of classical form closure concepts to include surface curvature effects, leading to *second-order immobilization.* Chapter 9 shows that by including curvature effects at the finger contacts, a robot hand can immobilize objects with fewer contacts than is predicted by these classical models.

The next historical developments concern techniques to represent and analyze grasp kinematics. Two different schools of analysis emerged independently in the late nineteenth and early twentieth centuries. In the mid-1870s, Lie developed his notion of infinitesimal groups, which we know today as Lie algebra. In 1900, Ball published *A Treatise on the Theory of Screws* [7], while in 1924 von Mises introduced the related concept of *motors* [10]. These techniques became what we now call *line geometry.* The engineering kinematics community largely relied upon line geometry until the 1980s, when Lie group approaches were introduced by control theory researchers working in the newly energized field of robotics. Because of the prevalence of line geometry notions in the robot grasping literature, we review the relevant aspects of line geometry where needed. More generally, this book uses standard notions of *differential geometry,* such as tangent and cotangent vectors to the grasped object configuration space manifold, to formulate all topics of grasp analysis.

In the late 1970s Lozano-Perez, Mason and Taylor introduced the *configuration space* (c-space) approach for robot motion planning. All of the analyses in this book are developed under the influential c-space framework in a manner that is compatible with robot grasp planning concepts [11–13]. This formulation comes with a rich body of analytical methods from differential geometry that can be leveraged to the analysis of robot grasps. This book reviews all concepts associated with the c-space framework in a self-contained manner.

1.5.2 The Autonomous Robot Hands Era

Significant research of immediate relevance to robot grasping did not emerge until the 1970s. The nascent research in the 1970s was driven by three different lines of practical activity in the 1950s and 1960s: industrial robotics, telerobotics and human prosthetics. When the first industrial robots were installed in a General Motors factory in 1961, these early factory robots relied upon special-purpose grippers to accomplish their grasping and part manipulation tasks. It was an obvious next step for robot engineers and researchers to imagine general-purpose dextrous hands modeled after the human hand. The hope was that anthropomorphic robot hands could eliminate the need to develop special-purpose tooling and robot grippers, thereby enabling robots to be easily reprogrammed for different assembly operations. An example of such an early investigation is the Waseda arm of the late 1960s, which included a four-degrees-of-freedom gripper. Similarly, the rise of nuclear power generators in the 1950s motivated significant work in teleoperated robot arms. Telerobots acted as mechanical proxies, thereby removing humans from highly radioactive environments. While simple parallel jaw grippers sufficed for most teleoperation tasks, some tasks demanded dextrous end effectors that could mimic human hand motions. Additionally, electronic and battery technology had sufficiently advanced to the point where dextrous prosthetic hands were worthy of serious research endeavors. A notable early work in this area is the Tomovic prosthetic hand of 1965.

Early robotics researchers were challenged by these practical demands to develop more comprehensive approaches to the analysis and synthesis of multi-finger robot grasps. In the late 1970s, Lakshminarayana [14] updated Somoff's classical form closure theory, or *first-order immobilization.* Subsequent work during the 1980s and early 1990s was dominated by the force closure, or *wrench resistance,* approach to grasp analysis, instead of the form closure approach. This book tries to present a balanced view between the force closure and form closure approaches to grasp analysis. From the viewpoint of geometric analysis, form closure is actually a more natural framework for rigid-body grasp analysis and certainly provides the highest level of grasp security.

In the late 1970s and through the 1980s, Roth and his students at Stanford University systematically analyzed many fundamental issues in robot grasping and dextrous manipulation. Much of their work was based on classical *line geometry* [15]. In his PhD thesis [16], Salisbury formalized the common rigid-body *contact models* which are reviewed in Chapter 4. He also gave one of the first general formulations of *grasp stiffness* in multi-finger robot hands [17], which underpins Chapter 18. Cai [18] subsequently determined how the point of contact between two bodies moves as a function of the bodies' relative motions and their surface curvatures. Simultaneously, Montana, in his PhD work at Harvard [19], developed what we now call the *Montana contact equations* to describe this same effect. Montana relied upon differential geometry to obtain results analogous to those of Cai. The influence of surface curvature on the mechanics of grasping is a common theme in many chapters of this book. Our presentation tends more toward Montana's geometric formulation, as this approach is more consistent with the configuration space framework.

The early 1980s also saw the first fully functional robot hand prototypes. Salisbury completed the JPL/Stanford hand in 1982 (see Figure 1.1), while Jacobsen, Hollerbach and coworkers completed the Utah/MIT hand in 1985. These hands represented a new level of technical sophistication and performance. While these devices were still somewhat impractical, their successful demonstration of basic grasping operations motivated a new generation of researchers to tackle grasping and dextrous manipulation issues. A recurrent theme of this book is how to transfer the theory of robot grasping into insights that can lead to successful robot hand designs. We will see in Chapter 9 and the bulk of Part IV that three-finger robot hands are *minimalistic,* provided that the hand's palm (or alternatively, an additional supporting finger) is used occasionally to achieve highly secure grasps.

The field of robot grasping is continuously expanding with new applications and designs. For example, *cable grasps* can hold and manipulate objects just like multi-finger robot hands. In a similar manner, *suction devices* used in manufacturing as well as *adhesives* can be integrated into heterogeneous robot hands. *Jamming grippers* use a phase transition in granular materials (caused by the withdrawal of air between the granules when a vacuum is applied to the granular material) to lock an object within a conformal envelope, thereby securing the object within the grasp. Advances in 3-D printing technology allows *flexure joints* to be rapidly assembled into compliant robot hands. Moreover, advances in soft materials have enabled a new generation of *soft hands* that can safely interact with biological materials that are themselves soft and easily damaged. While the physical actuation or grasping procedure is different in each of these novel grasping devices, the kinematics and mechanics principles underlying the grasping process in these diverse devices have many commonalities. We believe that the principles described in this book should be relevant to all future innovations in the broad and ever expanding field of robot grasping and manipulation.

References

[1] M. T. Mason and J. K. Salisbury, *Robots Hands and the Mechanics of Manipulation*. MIT Press, 1985.

[2] R. M. Murray, Z. Li and S. S. Sastry, *A Mathematical Introduction to Robotic Manipulation*. CRC Press, 1994.

[3] M. T. Mason, *Mechanics of Robotic Manipulation*. MIT Press, 2001.

[4] K. M. Lynch and F. C. Park, *Modern Robotics: Mechanics, Planning, and Control*. Cambridge University Press, 2017.

[5] J. M. Selig, *Geometrical Fundamentals of Robotics*. Springer-Verlag, 2003.

[6] M. W. Spong, S. Hutchinson and M. Vidyasager, *Robot Dynamics and Control.* 2nd Edition. John Wiley, 2004.
[7] R. S. Ball, *The Theory of Screws.* Cambridge University Press, 1900.
[8] F. Reuleaux, *The Kinematics of Machinery.* Macmillan, 1876, republished by Dover, 1963.
[9] P. Somoff, "Ueber Gebiete von Schraubengeschwindigkeiten eines starren Koerpers bei verschiedener Zahl von Stuetzflaechen" ["On the regions of rotational velocities for a rigid body with various numbers of supporting surfaces"], *Zeitschrift fuer Mathematik und Physik*, vol. 45, pp. 245–306, 1900.
[10] R. von Mises, "Motorrechnung, ein neues hilfsmittel in der mechanik (Motor calculus: a new theoretical device for mechanics)," *Zeitschrift fur Mathematic and Mechanik*, vol. 4, no. 2, pp. 155–181, 1924.
[11] J. C. Latombe, *Robot Motion Planning.* Kluwer Academic Publishers, 1990.
[12] S. M. LaValle, *Planning Algorithms.* Cambridge University Press, 2006.
[13] H. Choset, K. M. Lynch, S. Hutchinson, G. Kantor, W. Burgard, L. E. Kavraki and S. Thrun, *Principles of Robot Motion.* MIT Press, 2005.
[14] K. Lakshminarayana, "Mechanics of form closure," ASME, Tech. Rep. 78-DET-32, 1978.
[15] M. S. Ohwovoriole and B. Roth, "An extension of screw theory," *Journal of Mechanical Design*, vol. 103, pp. 725–735, 1981.
[16] J. K. Salisbury, "Kinematic and force analysis of articulated hands," PhD thesis, Dept. of Mechanical Engineering, Stanford University, 1982.
[17] J. K. Salisbury and J. J. Craig, "Analysis of multi-fingered hands: Force control and kinematic issues," *International Journal of Robotics Research*, vol. 1, no. 1, pp. 4–17, 1982.
[18] C. C. Cai and B. Roth, "On the spatial motion of a rigid body with point contact," in *IEEE International Conference on Robotics and Automation*, Raleigh, NC, pp. 686–695, 1987.
[19] D. J. Montana, "The kinematics of contact and grasp," *International Journal of Robotics Research*, vol. 7, no. 3, pp. 17–25, 1988.

Part I

Basic Geometry of the Grasping Process

2 Rigid-Body Configuration Space

This chapter considers a freely moving rigid body, $\mathcal{B}$, surrounded by stationary rigid bodies $\mathcal{O}_1, \ldots, \mathcal{O}_k$. The body $\mathcal{B}$ represents an object being grasped by a robot hand. The surrounding bodies represent fingertips, or segments of a robot hand supporting an object $\mathcal{B}$ against gravity. The chapter introduces the notion of rigid-body configuration space, or *c-space*, which is essential for analyzing the mobility and stability of the object $\mathcal{B}$ with respect to the surrounding finger bodies. The chapter begins with a parametrization of $\mathcal{B}$'s c-space in terms of Euclidean coordinates, effectively transforming c-space from an abstract manifold to a familiar Euclidean space. Configuration space obstacles, or *c-obstacles*, are then introduced and several of their properties are described. The chapter proceeds to describe the first- and second-order geometric properties of the c-obstacles, as this geometry plays a key role in subsequent chapters.

2.1 The Notion of Configuration Space

The object $\mathcal{B}$ is assumed to be a *rigid body* that moves freely in Euclidean space $\mathbb{R}^n$, where $n = 2$ or 3. Rigidity implies that the distance between the object's points remains fixed as $\mathcal{B}$ moves in $\mathbb{R}^n$. The object's *configuration* specifies the position of its points in $\mathbb{R}^n$. The parametrization of $\mathcal{B}$'s configurations requires a selection of two frames, depicted in Figure 2.1. The first is a fixed *world frame*, denoted $\mathcal{F}_W$, which establishes a coordinate system for the physical space $\mathbb{R}^n$. The second is a *body frame*, denoted $\mathcal{F}_B$, which is rigidly attached to $\mathcal{B}$. The configuration of $\mathcal{B}$ is specified by a vector $d \in \mathbb{R}^n$

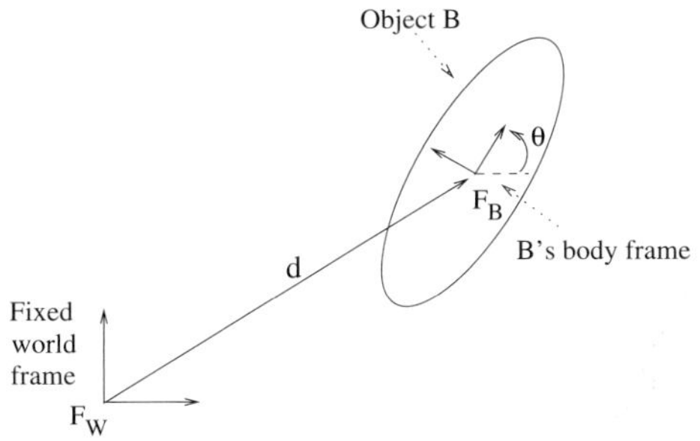

Figure 2.1 The basic setup for the c-space representation of a rigid object $\mathcal{B}$.

describing the position of $\mathcal{F}_B$'s origin with respect to $\mathcal{F}_W$, and an *orientation matrix*, $R \in \mathbb{R}^{n \times n}$, whose columns describe the orientation of $\mathcal{F}_B$'s axes with respect to those of $\mathcal{F}_W$. The orientation matrices form a group under matrix multiplication, which is defined as follows (see Exercises).

DEFINITION 2.1 (**orientation matrices**) *The $n \times n$* **orientation matrices** *form the special orthogonal group:*

$$SO(n) = \left\{ R \in \mathbb{R}^{n \times n} : R^T R = I \quad \textit{and} \quad \det(R) = 1 \right\},$$

where I is the $n \times n$ identity matrix.

The orientation matrices possess two important properties. First, every orientation matrix acts as a rotation on vectors $v \in \mathbb{R}^n$. That is, R preserves the norm of v: $\|Rv\| = (v^T R^T R v^T)^{1/2} = \|v\|$ for $v \in \mathbb{R}^n$. Second, $SO(n)$ is a smooth manifold of dimension $\frac{1}{2}n(n-1)$ in the space of $n \times n$ matrices.[1] In particular, $SO(2)$ is a compact *one-dimensional* manifold in the space of 2×2 matrices, while $SO(3)$ is a compact *three-dimensional* manifold in the space of 3×3 matrices. The practical meaning of this topological property is that one needs a single scalar, $\theta \in \mathbb{R}$, to parametrize $SO(2)$, and three scalars, $\theta = (\theta_1, \theta_2, \theta_3) \in \mathbb{R}^3$, to parametrize $SO(3)$.

Manifold structure of $SO(3)$: The manifold $SO(3)$ is topologically equivalent to a canonical three-dimensional manifold, the *projective space RP^3*. One way to construct RP^3 is to take the unit ball centered at the origin of $\mathbb{R}^3$ and identify antipodal points on its bounding sphere. The manifold RP^3 is path connected, compact and orientable. ○

The configuration space, or *c-space*, of the freely moving object $\mathcal{B}$ consists of pairs (d, R) as stated in the following definition.

DEFINITION 2.2 (**configuration space**) *The* **c-space** *of an n-dimensional rigid body $\mathcal{B}$ is the smooth manifold $\mathcal{C} = \mathbb{R}^n \times SO(n)$, consisting of pairs (d, R) such that $d \in \mathbb{R}^n$ and $R \in SO(n)$, where $n = 2$ or 3.*

The dimension of $\mathcal{C}$ is the sum: $m = n + \frac{1}{2}n(n-1) = \frac{1}{2}n(n+1)$, giving $m = 3$ when $\mathcal{B}$ is two-dimensional (2-D) and $m = 6$ when $\mathcal{B}$ is three-dimensional (3-D). Every position and orientation of $\mathcal{B}$ is represented by a *point* in $\mathcal{C}$, while every continuous motion of $\mathcal{B}$ is represented by a *curve* in $\mathcal{C}$. However, practical analysis requires coordinates for the c-space manifold. Therefore, we shall introduce *global* coordinates for $\mathcal{C}$ in terms of a Euclidean space $\mathbb{R}^m$, with some periodicity rules for the coordinates representing the orientation matrices.

First consider the coordinates for the orientation matrices, which form a *matrix group*. A standard means for parametrizing such groups is by *exponential coordinates*:

$$R(\theta) = e^{[\theta \times]} \qquad \theta \in \mathbb{R}^{\frac{1}{2}n(n-1)},$$

[1] An m-dimensional manifold is a hypersurface, $\mathcal{M}$, such that at each point $p \in \mathcal{M}$ the manifold can be locally represented by Euclidean coordinates $\mathbb{R}^m$.

where the matrix exponential is defined by the series: $e^A = I + A + \frac{1}{2!}A^2 + \ldots$, and $[\theta\times]$ is the following skew-symmetric matrix.[2] In the case of $SO(2)$, the parameter $\theta \in \mathbb{R}$ represents the orientation of $\mathcal{F}_B$ relative to $\mathcal{F}_W$ in $\mathbb{R}^2$. The 2×2 matrix $[\theta\times]$ is given by

$$[\theta\times] = \begin{bmatrix} 0 & -\theta \\ \theta & 0 \end{bmatrix} \qquad \theta \in \mathbb{R}.$$

In the case of $SO(3)$, the parameters $\theta = (\theta_1, \theta_2, \theta_3) \in \mathbb{R}^3$ represent the orientation of $\mathcal{F}_B$ relative to $\mathcal{F}_W$ in $\mathbb{R}^3$. The 3×3 matrix $[\theta\times]$ is given by

$$[\theta\times] = \begin{bmatrix} 0 & -\theta_3 & \theta_2 \\ \theta_3 & 0 & -\theta_1 \\ -\theta_2 & \theta_1 & 0 \end{bmatrix} \qquad \theta = (\theta_1, \theta_2, \theta_3) \in \mathbb{R}^3.$$

The skew symmetric matrix $[\theta\times]$ acts as a *cross-product* on vectors $v \in \mathbb{R}^3$: $[\theta\times]v = \theta \times v$ for $v \in \mathbb{R}^3$. Hence, it is called the *cross-product matrix.* When the skew symmetric matrices are substituted into the exponential series, one obtains the following parametrization of $SO(n)$.

THEOREM 2.1 **(exponential coordinates for $SO(n)$)** *The 2×2 orientation matrices are globally parametrized by $\theta \in \mathbb{R}$, according to the formula:*

$$R(\theta) = \begin{bmatrix} \cos\theta & -\sin\theta \\ \sin\theta & \cos\theta \end{bmatrix} \qquad \theta \in \mathbb{R},$$

where θ is measured using the right-hand rule (Figure 2.1). The 3×3 orientation matrices are globally parametrized by $\theta = (\theta_1, \theta_2, \theta_3) \in \mathbb{R}^3$ according to Rodrigues' formula:

$$R(\theta) = I + \sin(\|\theta\|)[\hat{\theta}\times] + \big(1 - \cos(\|\theta\|)\big)[\hat{\theta}\times]^2 \qquad \theta \in \mathbb{R}^3$$

where I is the 3×3 identity matrix, $[\hat{\theta}\times]$ is a 3×3 cross-product matrix, and $\hat{\theta} = \theta/\|\theta\|$.

The parametrization of $SO(2)$ in terms of $\theta \in \mathbb{R}$ is periodic in 2π, with each 2π interval parametrizing the entire manifold $SO(2)$ (Figure 2.2). In Rodrigues' formula, the unit vector $\hat{\theta}$ represents the axis of rotation of $R(\theta)$, while the scalar $\|\theta\|$ corresponds to the angle of rotation about this axis according to the right-hand rule. The parametrization of $SO(3)$ in terms of $\theta \in \mathbb{R}^3$ satisfies the following periodicity rule. The origin $\theta = \vec{0}$ is mapped to the identity orientation matrix I. Similarly, all concentric spheres of radius $\|\theta\| = 2\pi, 4\pi, \ldots$ are mapped to I. Each pair of antipodal points on the sphere of radius $\|\theta\| = \pi$ is mapped to the same matrix R, since $R(\pi\hat{\theta}) = R(-\pi\hat{\theta})$ for all $\hat{\theta}$ (similarly, antipodal points on the spheres of radius $\|\theta\| = 3\pi, 5\pi, \ldots$ are identified). Since $\hat{\theta}$ can have any direction in $\mathbb{R}^3$, the manifold $SO(3)$ is fully parametrized by the closed ball of radius π centered at the origin, with antipodal points on its bounding sphere identified. Note that this rule matches the definition of RP^3, thus providing a constructive proof that $SO(3)$ is topologically equivalent to RP^3.

The pair (d, θ) provides a *global* parametrization of the c-space manifold $\mathcal{C} = \mathbb{R}^n \times SO(n)$ in terms of c-space coordinates, $q \in \mathbb{R}^m$, as stated in the following definition.

[2] The skew-symmetric matrices $[\theta\times]$ form the *Lie algebra* of the matrix group $SO(n)$.

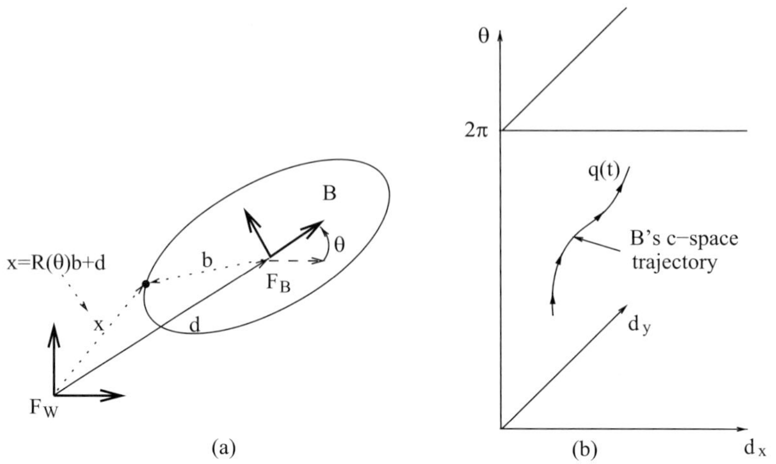

Figure 2.2 (a) The c-space coordinates $q=(d_x,d_y,\theta)$ of a 2-D object $\mathcal{B}$, where θ is periodic in 2π. (b) A physical motion of $\mathcal{B}$ is represented by a c-space trajectory in $\mathbb{R}^m$.

DEFINITION 2.3 **(c-space coordinates)** *When $\mathcal{B}$ is a 2-D body, its* **c-space coordinates** *are $q=(d,\theta)\in\mathbb{R}^3$, where $d=(d_x,d_y)\in\mathbb{R}^2$ and $\theta\in\mathbb{R}$. When $\mathcal{B}$ is a 3-D body, its* **c-space coordinates** *are $q=(d,\theta)\in\mathbb{R}^6$, where $d=(d_x,d_y,d_z)\in\mathbb{R}^3$ and $\theta=(\theta_1,\theta_2,\theta_3)\in\mathbb{R}^3$.*

The c-space coordinates allow us to model the physical motions of $\mathcal{B}$ as trajectories of a *point* in $\mathbb{R}^m$, where $m=3$ for a 2-D object and $m=6$ for a 3-D object. For simplicity we will refer to $\mathbb{R}^m$ as the *c-space* of $\mathcal{B}$. Before we discuss how the finger bodies appear as forbidden regions in $\mathcal{B}$'s c-space, let us review the key notion of *rigid-body transformation.*

DEFINITION 2.4 **(rigid-body transformation)** *When a rigid body $\mathcal{B}$ is located at a configuration q, the position of its point $b\in\mathcal{B}$ expressed in $\mathcal{F}_B$ relative to $\mathcal{F}_W$ is given by the* **rigid-body transformation**, *$X(q,b):\mathbb{R}^m\times\mathcal{B}\rightarrow\mathbb{R}^n$, according to the formula* (Figure 2.2(a)):

$$X(q,b)=R(\theta)b+d \qquad q=(d,\theta)\in\mathbb{R}^m,\ b\in\mathcal{B}$$

where $m=3$ in 2-D and $m=6$ in 3-D.

We will occasionally use the notation $X_b(q)$ to specify the rigid-body transformation with the point $b\in\mathcal{B}$ held fixed on $\mathcal{B}$. In this case, $X_b(q)$ gives the world position of the fixed point b as a function of $\mathcal{B}$'s configuration q.

2.2 Configuration Space Obstacles

Rigid bodies cannot interpenetrate in physical space. Hence, when a rigid object $\mathcal{B}$ is surrounded by stationary rigid finger bodies $\mathcal{O}_1,\ldots,\mathcal{O}_k$, the finger bodies form obstacles that constrain the object's possible motions. The finger bodies induce the following forbidden regions in $\mathcal{B}$'s c-space, called *c-obstacles.*

DEFINITION 2.5 (**c-obstacle**) *Let $\mathcal{B}(q)$ be the set of points in $\mathbb{R}^n$ occupied by $\mathcal{B}$ at a configuration q. The* **c-obstacle** *induced by a stationary finger body $\mathcal{O}$, denoted $\mathcal{CO}$, is the set of configurations at which $\mathcal{B}(q)$ intersects $\mathcal{O}$:*

$$\mathcal{CO} = \left\{ q \in \mathbb{R}^m : \mathcal{B}(q) \cap \mathcal{O} \neq \emptyset \right\}$$

where $m = 3$ for a 2-D object $\mathcal{B}$ and $m = 6$ for a 3-D object $\mathcal{B}$.

When $\mathcal{B}$ is a full body with nonempty interior, its c-obstacle $\mathcal{CO}$ occupies an m-dimensional set in $\mathcal{B}$'s c-space $\mathbb{R}^m$, even when $\mathcal{O}$ is a point obstacle. The c-obstacle boundary, denoted $\mathcal{S}$, forms a piecewise smooth $(m - 1)$-dimensional manifold whose points satisfy the following property.

LEMMA 2.2 (**c-obstacle boundary**) *The* **c-obstacle boundary** *consists of configurations q at which $\mathcal{B}(q)$ touches $\mathcal{O}$ strictly from the outside:*

$$\mathcal{S} = \{ q \in \mathbb{R}^m : \mathcal{B}(q) \cap \mathcal{O} \neq \emptyset \ \textit{and} \ \mathsf{int}(\mathcal{B}(q)) \cap \mathsf{int}(\mathcal{O}) = \emptyset \},$$

where int *denotes set interior.*

In the case of 2-D bodies, one can conceptually construct the c-obstacle surface as follows. First fix the orientation of $\mathcal{B}$'s reference frame to a particular θ. Then slide $\mathcal{B}$ along the perimeter of $\mathcal{O}$ with fixed orientation, making sure that $\mathcal{B}$ maintains continuous contact with $\mathcal{O}$. The trace of $\mathcal{B}$'s frame origin during this circumnavigation forms a closed curve. The resulting curve is the fixed-θ slice of $\mathcal{S}$. When this process is repeated for all θ, the resulting stack of loops forms the c-obstacle boundary.

Example: Figure 2.3 shows an elliptical object $\mathcal{B}$ moving in a planar environment in the presence of a stationary disc finger $\mathcal{O}$. The c-obstacle $\mathcal{CO}$ is depicted in Figure 2.3(a) for a choice of $\mathcal{F}_B$'s origin at the ellipse's center. The c-obstacle forms a spiraling stack of 2-D ovals. Each oval is formed by sliding the object $\mathcal{B}$ about $\mathcal{O}$

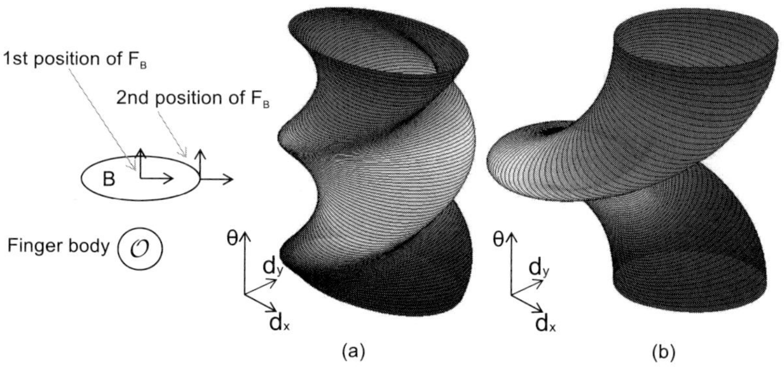

Figure 2.3 The c-obstacle surface induced by a stationary disc representing a finger body, shown for two choices of $\mathcal{B}$'s reference frame: (a) at the ellipse's center, and (b) at the ellipse's tip.

with fixed orientation, and the spiraling ovals represent the 2π periodicity of the θ axis. The same c-obstacle is depicted in Figure 2.3(b) for a choice of $\mathcal{F}_B$'s origin at the tip of the ellipse's major axis. While the c-obstacle geometric shape (surface normal and curvature) has changed, it is topologically equivalent to the c-obstacle depicted in Figure 2.3(a). This observation holds under all choices of the reference frames $\mathcal{F}_W$ and $\mathcal{F}_B$. ○

A detailed discussion of the c-obstacles can be found in on robot motion planning texts; see Bibliographical Notes. The following are some key properties of the c-obstacles.

1. **Connectivity and compactness propagate.** When $\mathcal{B}$ is compact and connected, a compact and connected finger body $\mathcal{O}$ induces a compact and connected c-obstacle $\mathcal{CO}$.
2. **Union propagates.** When $\mathcal{B}$ is a union of two sets, $\mathcal{B} = \mathcal{B}_1 \cup \mathcal{B}_2$, the c-obstacle induced by a finger body $\mathcal{O}$ is a union of the c-obstacles corresponding to the two sets, $\mathcal{CO} = \mathcal{CO}_1 \cup \mathcal{CO}_2$.
3. **Convexity propagates.** A set $\mathcal{A} \subseteq \mathbb{R}^n$ is said to be *convex* when every pair of points in $\mathcal{A}$ can be connected by a line segment lying in $\mathcal{A}$. When $\mathcal{B}$ and $\mathcal{O}$ are convex bodies, each fixed-orientation slice of $\mathcal{CO}$ forms a convex set.
4. **Polygonality propagates.** When $\mathcal{B}$ and $\mathcal{O}$ are polygonal 2-D bodies, each fixed-orientation slice of $\mathcal{CO}$ forms a polygonal set. When $\mathcal{B}$ and $\mathcal{O}$ are polyhedral 3-D bodies, each fixed-orientation slice of $\mathcal{CO}$ forms a polyhedral set.

Parametrization of c-obstacle boundary: Explicit parametrization of the c-obstacle boundary is available when $\mathcal{B}$ and $\mathcal{O}$ are convex bodies. This example considers a 2-D smooth convex object $\mathcal{B}$ and a disc finger $\mathcal{O}$ centered at x_0 of radius r. Let $\beta(s)$ for $s \in \mathbb{R}$ be a counterclockwise parametrization of $\mathcal{B}$'s perimeter in its body frame $\mathcal{F}_B$, such that the tangent $\beta'(s)$ is a unit vector. Note that $J\beta'(s)$ is the unit *outward* normal to $\mathcal{B}$ in $\mathcal{F}_B$, where $J = \begin{bmatrix} 0 & 1 \\ -1 & 0 \end{bmatrix}$. When $\mathcal{B}$ touches $\mathcal{O}$ at a point $x(s,q) = R(\theta)\beta(s) + d$ where $q = (d, \theta)$, the two bodies share collinear contact normals:

$$\frac{1}{r}(x - x_0) = -R(\theta)J\beta'(s). \tag{2.1}$$

It follows from Eq. (2.1) that $x(s,q) = x_0 - rR(\theta)J\beta'(s)$. Substituting for x in the rigid-body transformation, then solving for d in terms of s and θ gives

$$d(s,\theta) = x(s,\theta) - R(\theta)\beta(s) = x_0 - R(\theta)\big(\beta(s) + rJ\beta'(s)\big).$$

The function $\varphi(s,\theta): \mathbb{R}^2 \to \mathbb{R}^3$ given by

$$\varphi(s,\theta) = \begin{pmatrix} d(s,\theta) \\ \theta \end{pmatrix} \qquad d(s,\theta) = x_0 - R(\theta)\big(\beta(s) + rJ\beta'(s)\big) \tag{2.2}$$

parametrizes the c-obstacle boundary in terms of s and θ. The curves depicted on the c-obstacle surfaces in Figure 2.3 are fixed-θ curves of the parametrization (2.2). ○

The c-obstacle boundary, $\mathcal{S}$, inherits smoothness properties from those of $\mathcal{B}$ and $\mathcal{O}$. When both bodies are smooth and convex, $\mathcal{S}$ forms a smooth $(m-1)$-dimensional manifold in $\mathbb{R}^m$. More generally, $\mathcal{S}$ is locally smooth at any configuration $q \in \mathcal{S}$ at which $\mathcal{B}(q)$ touches $\mathcal{O}$ at a single point, such that the two bodies possess smooth boundaries at the contact point. When the two bodies are piecewise smooth, for instance when $\mathcal{B}$ and $\mathcal{O}$ are polygons, $\mathcal{S}$ becomes piecewise smooth. In the latter case $\mathcal{S}$ consists of smooth $(m-1)$-dimensional "patches," meeting along lower-dimensional manifolds. For instance, when $\mathcal{B}$ is a convex polygon and $\mathcal{O}$ is a disc, $\mathcal{S}$ consists of surface patches generated by an edge of $\mathcal{B}$ sliding on $\mathcal{O}$, and surface patches generated by a vertex of $\mathcal{B}$ sliding on $\mathcal{O}$. Similar observations hold for the five-dimensional boundary of $\mathcal{CO}$ in the 3-D case.

Example – Smoothness of c-obstacle boundary: Consider the previous example of a 2-D convex object $\mathcal{B}$ and a disc finger $\mathcal{O}$. Using the boundary parametrization of Eq. (2.2), let us verify that the c-obstacle boundary forms a smooth surface in $\mathcal{B}$'s c-space. It suffices to show that the tangent vectors $\frac{\partial}{\partial s}\varphi(s,\theta)$ and $\frac{\partial}{\partial \theta}\varphi(s,\theta)$ are linearly independent and thus span a well defined tangent plane at every point $q = \varphi(s,\theta) \in \mathcal{S}$. The tangent vector $\frac{\partial}{\partial s}\varphi(s,\theta)$ is given by

$$\frac{\partial}{\partial s}\varphi(s,\theta) = \begin{pmatrix} -R(\theta)\big(\beta'(s) + rJ\beta''(s)\big) \\ 0 \end{pmatrix} = -(1 + r\kappa_{\mathcal{B}}(s)) \begin{pmatrix} R(\theta)\beta'(s) \\ 0 \end{pmatrix},$$

where we used the fact that $J\beta''(s)$ is collinear with $\beta'(s)$, and that $\kappa_{\mathcal{B}}(s) = \beta'(s) \cdot J\beta''(s)$ is the *curvature* of $\mathcal{B}$ at $\beta(s)$ (more details on curvature are provided in Section 2.4). The tangent vector $\frac{\partial}{\partial \theta}\varphi(s,\theta)$ is given by

$$\frac{\partial}{\partial \theta}\varphi(s,\theta) = \begin{pmatrix} JR(\theta)\big(\beta(s) + rJ\beta'(s)\big) \\ 1 \end{pmatrix} = \begin{pmatrix} JR(\theta)\beta(s) \\ 1 \end{pmatrix} - r\begin{pmatrix} R(\theta)\beta'(s) \\ 0 \end{pmatrix},$$

where we used the identities $R'(\theta) = -JR(\theta) = -R(\theta)J$ and $J^2 = -I$. Since $\kappa_{\mathcal{B}}(s) > 0$ for a convex object, $\frac{\partial}{\partial s}\varphi(s,\theta)$ and $\frac{\partial}{\partial \theta}\varphi(s,\theta)$ are linearly independent and the c-obstacle boundary forms a smooth surface in c-space (see Figure 2.3). ○

2.3 The C-Obstacle Normal

When a rigid object $\mathcal{B}$ is contacted by stationary rigid finger bodies $\mathcal{O}_1, \ldots, \mathcal{O}_k$, the object's configuration q lies at the intersection of the finger c-obstacle boundaries. We will see in Part II of the book that the free motions of $\mathcal{B}$ are determined in this case by the first and second-order geometry of the c-obstacle boundaries – i.e., by their normals and curvatures at q. This section derives a formula for the *c-obstacle normal,* while the next section derives a formula for the *c-obstacle curvature.*

We shall assume that the object $\mathcal{B}$ touches a stationary body $\mathcal{O}$ at a *single point,* such that the two bodies have locally smooth boundaries at the contact point. Under this assumption, the c-obstacle boundary is locally smooth and has a well-defined normal.[3]

[3] The c-obstacle normal is well defined even when one of the contacting bodies is non-smooth at the contact point, provided that the other body has a smooth boundary at this point.

In order to compute the c-obstacle normal, consider the following inter-body distance function.

DEFINITION 2.6 **(C-obstacle distance function)** *Let $\mathcal{O}$ be a stationary body in $\mathbb{R}^n$, and let $\mathsf{dst}(x,\mathcal{O}):\mathbb{R}^n \to \mathbb{R}$ be the minimal distance of a point x from $\mathcal{O}$: $\mathsf{dst}(x,\mathcal{O}) = \min_{y\in\mathcal{O}}\{\|x-y\|\}$. The* **c-obstacle function,** *$o(q):\mathbb{R}^m \to \mathbb{R}$, is the minimal distance between $\mathcal{B}(q)$ and $\mathcal{O}$:*

$$o(q) = \min_{x\in\mathcal{B}(q)} \{\mathsf{dst}(x,\mathcal{O})\},$$

where $\mathcal{B}(q)$ is the set of points $\mathbb{R}^n$ occupied by $\mathcal{B}$ at a configuration q.

The c-obstacle distance function $o(q)$ is identically zero in the interior of the c-obstacle $\mathcal{CO}$ and is strictly positive outside $\mathcal{CO}$. Hence $\mathcal{CO} = \{q \in \mathbb{R}^m : o(q) \le 0\}$. If $o(q)$ would have been differentiable at $q \in \mathcal{S}$, its gradient $\nabla o(q)$ would be collinear with the c-obstacle outward normal at q. But $o(q)$ is identically zero inside $\mathcal{CO}$ and monotonically increasing away from $\mathcal{CO}$, implying that it is *non-differentiable* on the c-obstacle boundary $\mathcal{S}$. However, $o(q)$ is Lipschitz continuous and can be analyzed with tools that resemble the classical ones (Lipschitz continuity and other relevant aspects of non-smooth analysis are reviewed in Appendix A). Lipschitz continuous functions are piecewise smooth, and they possess a *generalized gradient* at points were the function is non-differentiable. The *generalized gradient* of f at x, denoted $\partial f(x)$, is the convex combination of the gradients $\nabla f(y)$ at points y that approach x from all sides (see Appendix A). In particular, $\partial f(x)$ reduces to the usual gradient $\nabla f(x)$ at points where f is differentiable.

Let us compute the generalized gradient of $o(q)$ and see how it determines the c-obstacle normal. To emphasize that only q is a free variable in $o(q)$, we write:

$$o(q) = \min_{b\in\mathcal{B}} \left\{\mathsf{dst}\big(X_b(q),\mathcal{O}\big)\right\},$$

where $X_b(q)$ is the rigid-body transformation specified in Definition 2.4. According to Property (3) in Appendix A, $o(q)$ is Lipschitz continuous when its constituent functions, $\mathsf{dst}\big(X_b(q),\mathcal{O}\big)$ for $b \in \mathcal{B}$, are Lipschitz continuous. The rigid-body transformation $X_b(q)$ is smooth and therefore Lipschitz continuous in q. The minimal distance function, $\mathsf{dst}(x,\mathcal{O})$, is shown in Appendix A to be Lipschitz continuous in x. Since Lipschitz continuity is preserved under function composition, the functions $\mathsf{dst}(X_b(q),\mathcal{O})$ are Lipschitz continuous. Hence, $o(q)$ is Lipschitz continuous and, therefore, possesses a *generalized gradient,* $\partial o(q)$, which determines the c-obstacle normal. As discussed in Appendix A, $\partial o(q)$ is the convex combination of the generalized gradient of the functions $\mathsf{dst}\big(X_b(q),\mathcal{O}\big)$ that attain the minimum distance at q. When $\mathcal{B}(q)$ touches $\mathcal{O}$ at a *single point,* one obtains the following corollary.

COROLLARY 2.3 *Let $\mathcal{B}(q)$ contact a stationary body $\mathcal{O}$ at a single point $b_0 \in \mathcal{B}$. Then $o(q)$ is attained by a single function, $o(q) = \mathsf{dst}(X_{b_0}(q),\mathcal{O})$, and the generalized gradient of $o(q)$ is given by $\partial o(q) = \partial\mathsf{dst}(X_{b_0}(q),\mathcal{O})$.*

The computation of $\partial o(q)$ thus reduces to the computation of $\partial \mathsf{dst}(X_{b_0}(q), \mathcal{O})$. The function $\mathsf{dst}(X_{b_0}(q), \mathcal{O})$ is a composition of $\mathsf{dst}(x, \mathcal{O})$ with $x(q) = X_{b_0}(q)$. The generalized gradient of such a composition can be computed using a *generalized chain rule,* described in Appendix A. It specifies that $\partial \mathsf{dst}\big(X_{b_0}(q), \mathcal{O}\big) = \partial \mathsf{dst}(x_0, \mathcal{O}) \cdot DX_{b_0}(q)$, where $x_0 = X_{b_0}(q)$ is the position of the contact point b_0 in the world frame $\mathcal{F}_W$, and $DX_{b_0}(q)$ is the Jacobian matrix of $X_{b_0}(q)$. The resulting formula for $\partial o(q)$ is as follows.

PROPOSITION 2.4 (**generalized gradient of $o(q)$**) *The generalized gradient of $o(q)$ at $q \in \mathcal{S}$ is a line segment based at q:*

$$\partial o(q) = s \cdot \begin{pmatrix} n(x_0) \\ R(\theta)b_0 \times n(x_0) \end{pmatrix} \qquad 0 \leq s \leq 1, \tag{2.3}$$

where $n(x_0)$ is the unit normal pointing out of $\mathcal{O}$ and into $\mathcal{B}$ at the contact point x_0.

Proof: Based on the generalized chain rule:

$$\partial o(q) = \partial \mathsf{dst}\big(X_{b_0}(q), \mathcal{O}\big) = \partial \mathsf{dst}(x_0, \mathcal{O}) \cdot DX_{b_0}(q). \tag{2.4}$$

The Jacobian $DX_{b_0}(q)$ is given by (see Exercises): $DX_{b_0}(q) = \big[I \quad -R(\theta)b_0 \times\big]$. As shown in Appendix A, the generalized gradient of $\mathsf{dst}(x_0, \mathcal{O})$ is a line segment based at x_0:

$$\partial \mathsf{dst}(x_0, \mathcal{O}) = s \cdot n(x_0) \qquad 0 \leq s \leq 1,$$

where $n(x_0)$ is $\mathcal{O}$'s outward unit normal at x_0. Substituting for $DX_{b_0}(q)$ and $\partial \mathsf{dst}(x_0, \mathcal{O})$ in Eq. (2.4), then taking the transpose so that $\partial o(q)$ would become a column vector, gives Formula (2.3) for $\partial o(q)$. □

The generalized gradient of $o(q)$ forms a line segment based at $q \in \mathcal{S}$. This line segment points *outward* with respect to $\mathcal{CO}$, since $o(q)$ increases away from $\mathcal{CO}$. The c-obstacle normal is simply the *direction* of this line segment, as summarized in the following theorem, which uses the notation b instead of b_0 and x instead of x_0.

THEOREM 2.5 (**c-obstacle normal**) *Let a freely moving rigid body $\mathcal{B}$ located at q contact a stationary rigid body $\mathcal{O}$. The* **c-obstacle outward normal** *at $q \in \mathcal{S}$, denoted $\eta(q)$, is given by*

$$\eta(q) = DX_b^T(q)n(x) = \begin{pmatrix} n(x) \\ R(\theta)b \times n(x) \end{pmatrix}, \tag{2.5}$$

where b is $\mathcal{B}$'s contact point with $\mathcal{O}$ expressed in $\mathcal{F}_B$, $x = X_b(q)$ is the contact point expressed in $\mathcal{F}_W$, and $n(x)$ is the unit contact normal at x, pointing outward with respect to $\mathcal{O}$ at x.[4]

Proof: Consider a c-space path $\alpha(t)$ that lies in $\mathcal{S}$, such that $\alpha(0) = q$. Since $\mathcal{B}$ maintains continuous contact with $\mathcal{O}$ along α, $o(\alpha(t)) = 0$ along this path. By the generalized chain rule, $\frac{d}{dt} o(\alpha(t)) = \partial o(\alpha(t)) \cdot \frac{d}{dt}\alpha(t) = 0$ for $t \in \mathbb{R}$. Since $\frac{d}{dt}\big|_{t=0} \alpha(t)$ is an arbitrary

[4] In the 2-D case, $Rb \times n = (Rb)^T Jn$, where $J = \begin{bmatrix} 0 & 1 \\ -1 & 0 \end{bmatrix}$.

tangent vector to $\mathcal{S}$ at q, the line segment $\partial o(q)$ is perpendicular to the tangent space $T_q\mathcal{S}$. This line segment points outward with respect to $\mathcal{CO}$. The vector $\eta(q)$ specified in Eq. (2.5), obtained by substituting $s=1$ in $\partial o(q)$, therefore forms the c-obstacle outward normal at q. □

We will see in the next chapter that the c-obstacle normal $\eta(q)$ can be interpreted as the generalized force, or *wrench*, generated by a unit-magnitude normal force acting on $\mathcal{B}$ at x. The vanishing of the product: $\partial o(\alpha(t)) \cdot \frac{d}{dt}\alpha(t)=0$, reflects the physical fact that a normal contact force does no work along any contact preserving motion of $\mathcal{B}$ as it slides along the boundary of the stationary body $\mathcal{O}$.

Example: Consider the parametrization $\varphi(s,\theta)$ of the c-obstacle boundary $\mathcal{S}$, obtained in Eq. (2.2). We already verified that the tangent vectors $\frac{\partial}{\partial s}\varphi(s,\theta)$ and $\frac{\partial}{\partial\theta}\varphi(s,\theta)$ span the tangent plane to $\mathcal{S}$ at $q=\varphi(s,\theta)$. The cross-product of the two tangent vectors should therefore be collinear with $\eta(q)$. A straightforward calculation gives

$$\frac{\partial}{\partial s}\varphi(s,\theta) \times \frac{\partial}{\partial\theta}\varphi(s,\theta) = \begin{pmatrix} -JR(\theta)\beta'(s) \\ (JR(\theta)\beta'(s)) \cdot JR(\theta)\beta(s) \end{pmatrix} = \begin{pmatrix} n(x) \\ (R(\theta)b)^T J n(x) \end{pmatrix},$$

where $b=\beta(s)$, and $n(x)=-JR(\theta)\beta'(s)$ (since $JR(\theta)=R(\theta)J$, and $-J\beta'(s)$ is $\mathcal{B}$'s inward unit normal at b expressed in $\mathcal{F}_B$). The c-obstacle normal computed from $\varphi(s,\theta)$ thus matches formula (2.5) for $\eta(q)$. ○

2.4 The C-Obstacle Curvature

As mentioned in the previous section, when a rigid object $\mathcal{B}$ is held by stationary rigid finger bodies $\mathcal{O}_1,\ldots,\mathcal{O}_k$, the free motions of $\mathcal{B}$ are determined by the first- and second-order geometry of the finger c-obstacles. The first-order geometry corresponds to the c-obstacle normal; the second-order geometry corresponds to the c-obstacle curvature, which is studied in this section.

We will develop the *c-obstacle curvature formula* for a freely moving 2-D object $\mathcal{B}$ and a stationary 2-D body $\mathcal{O}$. The 3-D version of the c-obstacle curvature formula is summarized at the end of this section. In the 2-D case, the c-obstacle boundary forms a surface in $\mathbb{R}^3$. The curvature of this surface depends on the curvature of the contacting bodies for which we need to introduce notation. First consider the stationary body $\mathcal{O}$. As before, n denotes the outward unit normal to $\mathcal{O}$ at the contact point x. Let $x(t)$ parametrize the boundary of $\mathcal{O}$, such that $x(0)=x$ and $\frac{d}{dt}\big|_{t=0}x(t)=\dot{x}$. The *curvature* of $\mathcal{O}$ at x, denoted $\kappa_{\mathcal{O}}(x)$, is the scalar measuring the change of the unit normal $n(x)$ along $x(t)$:

$$\frac{d}{dt}\bigg|_{t=0} n(x(t)) = \kappa_{\mathcal{O}}(x)\dot{x}.$$

Note that the change in $n(x)$ is tangent to $\mathcal{O}$'s boundary at x, since $\|n(x)\|=1$ along $x(t)$. The sign of $\kappa_{\mathcal{O}}(x)$ is positive when $\mathcal{O}$ is convex at x, negative when $\mathcal{O}$ is concave at x and zero when $\mathcal{O}$ is flat at x. The *radius of curvature* of $\mathcal{O}$ at x, denoted $r_{\mathcal{O}}(x)$, is the reciprocal of the curvature, $r_{\mathcal{O}}(x) = 1/\kappa_{\mathcal{O}}(x)$. The *circle of curvature* of $\mathcal{O}$ at x is the circle of radius $|r_{\mathcal{O}}(x)|$ tangent to $\mathcal{O}$ at x. It forms the boundary's second-order approximation at x. The curvature of $\mathcal{B}$ is similarly defined with respect to its body frame $\mathcal{F}_B$. The curvature of $\mathcal{B}$ at a boundary point b is the signed scalar $\kappa_{\mathcal{B}}(b)$, and its radius of curvature is $r_{\mathcal{B}}(b) = 1/\kappa_{\mathcal{B}}(b)$. In the following discussion, $\kappa_{\mathcal{B}}$ and $\kappa_{\mathcal{O}}$ will be used for $\kappa_{\mathcal{B}}(b)$ and $\kappa_{\mathcal{O}}(x)$.

The curvature of the c-obstacle boundary is defined as follows. Denote by $T_q\mathcal{S}$ the tangent plane to the c-obstacle boundary $\mathcal{S}$ at q, and recall that $\eta(q)$ denotes the c-obstacle outward normal at $q \in \mathcal{S}$. Let $\hat{\eta}(q) = \eta(q)/\|\eta(q)\|$ be the unit-magnitude outward normal, and let $q(t)$ be a c-space path that lies in $\mathcal{S}$, such that $q(0) = q$ and $\frac{d}{dt}\big|_{t=0} q(t) = \dot{q}$. The *curvature form* of $\mathcal{S}$ at q, denoted $\kappa(q, \dot{q})$, measures the change in $\hat{\eta}(q)$ along tangent directions $\dot{q} \in T_q\mathcal{S}$:

$$\kappa(q, \dot{q}) = \dot{q} \cdot \frac{d}{dt}\bigg|_{t=0} \hat{\eta}(q(t)) = \dot{q} \cdot D\hat{\eta}(q)\dot{q} \qquad \dot{q} \in T_q\mathcal{S}. \tag{2.6}$$

The curvature $\kappa(q, \dot{q})$ is a *quadratic form* that acts on tangent vectors $\dot{q} \in T_q\mathcal{S}$. Since $T_q\mathcal{S}$ is a two-dimensional subspace of the ambient tangent space $T_q\mathbb{R}^3$, $D\hat{\eta}(q)$ acts as a 2×2 symmetric matrix on $T_q\mathcal{S}$. The eigenvalues and eigenvectors of $D\hat{\eta}(q)$ are the *principal curvatures* and *principal directions of curvature* of $\mathcal{S}$ at q. The principal curvatures are analogous to the curvature of a planar curve. That is, the principal curvatures can be used to construct a quadratic surface tangent to $\mathcal{S}$ at q, which forms the second-order approximation of $\mathcal{S}$ at q.

We now turn to the task of computing the c-obstacle curvature. When the object $\mathcal{B}$ moves along a c-space path $q(t)$ that lies in $\mathcal{S}$, it maintains continuous contact with the stationary body $\mathcal{O}$. Let $x(t)$ be the contact point of $\mathcal{B}$ with $\mathcal{O}$ along $q(t)$, expressed in the world frame $\mathcal{F}_W$. Since $\eta(q) = DX_b^T(q)n(x)$ according to Theorem 2.5, the derivative of $\hat{\eta}(q)$ along $q(t)$ involves the contact point velocity, $\dot{x}(t)$, along the c-space path $q(t)$. The contact point velocity depends on the curvatures of $\mathcal{B}$ and $\mathcal{O}$, as stated in the following proposition.

PROPOSITION 2.6 **(contact point velocity)** *Let $q(t)$ be a c-space path that lies in $\mathcal{S}$, and let $x(t)$ be $\mathcal{B}$'s contact point with $\mathcal{O}$ along $q(t)$, expressed in the world frame $\mathcal{F}_W$. The* **contact point velocity** *along $q(t)$ is given by*

$$\dot{x}(t) = \tfrac{\kappa_{\mathcal{B}}}{\kappa_{\mathcal{B}} + \kappa_{\mathcal{O}}} \begin{bmatrix} I & -J R(\theta) b_c \end{bmatrix} \dot{q}(t), \tag{2.7}$$

where $\kappa_{\mathcal{B}}$ and $\kappa_{\mathcal{O}}$ are the curvatures of $\mathcal{B}$ and $\mathcal{O}$ at $x(t)$, b_c is $\mathcal{B}$'s center of curvature at $x(t)$ expressed in $\mathcal{F}_B$, I is a 2×2 identity matrix and $J = \left[\begin{smallmatrix} 0 & 1 \\ -1 & 0 \end{smallmatrix}\right]$.

The proof of the proposition appears in the chapter's appendix. The object's *center of curvature*, b_c, is the center of $\mathcal{B}$'s circle of curvature at x. The denominator, $\kappa_{\mathcal{B}} + \kappa_{\mathcal{O}}$, is positive semi-definite. For instance, when a concave body $\mathcal{O}$ touches a convex object $\mathcal{B}$ at x, $r_{\mathcal{B}} \leq |r_{\mathcal{O}}|$; otherwise the two bodies would interpenetrate. In this case, $|\kappa_{\mathcal{O}}| \leq \kappa_{\mathcal{B}}$ and, indeed, $\kappa_{\mathcal{B}} + \kappa_{\mathcal{O}} \geq 0$. The quantity $\kappa_{\mathcal{B}} + \kappa_{\mathcal{O}}$ is strictly positive when the bodies' second-order approximations maintain point contact at x. Since we assume a single point contact, we may as well assume that $\kappa_{\mathcal{B}} + \kappa_{\mathcal{O}} > 0$.

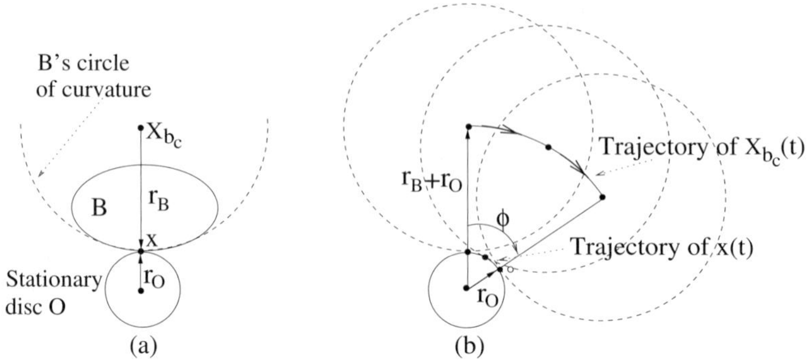

Figure 2.4 (a) The contacting bodies are replaced by their circles of curvature at x. (b) The $\mathcal{B}$-circle executes a contact preserving motion along the boundary of the stationary disc $\mathcal{O}$.

Geometric interpretation of the contact point velocity formula: Let $X_{b_c}(q)$ be the position of $\mathcal{B}$'s center of curvature at x, expressed in the world frame $\mathcal{F}_W$. Then $X_{b_c}(q) = R(\theta)b_c + d$. Assuming that b_c is held fixed on $\mathcal{B}$, $\dot{X}_{b_c} = \begin{bmatrix} I & -JR(\theta)b_c \end{bmatrix} \dot{q}$. Eq. (2.7) can thus be written as $\dot{x} = \frac{\kappa_{\mathcal{B}}}{\kappa_{\mathcal{B}}+\kappa_{\mathcal{O}}} \dot{X}_{b_c}$. In order to justify this formula, replace the object $\mathcal{B}$ by its circle of curvature at x, and assume that $\mathcal{O}$ is a stationary disc (Figure 2.4(a)). As the $\mathcal{B}$-circle executes a contact preserving motion along $\mathcal{O}$ (Figure 2.4(b)), the $\mathcal{B}$-circle's center, X_{b_c}, moves along a circular arc of radius $|r_{\mathcal{B}}+r_{\mathcal{O}}|$. The circles' contact point, x, moves along a concentric circular arc of radius $|r_{\mathcal{O}}|$ during this motion. Since x and X_{b_c} lie on a common radius vector emanating from $\mathcal{O}$'s center, the two points move with identical angular velocities about $\mathcal{O}$'s center. Moreover, the two points move in the same direction when $r_{\mathcal{B}} \geq 0$. Assuming this case, let $\dot{\phi}$ be the common angular velocity of the two points (Figure 2.4(b)). Then $\dot{x} = |r_{\mathcal{O}}|\dot{\phi}$ while $\dot{X}_{b_c} = |r_{\mathcal{B}}+r_{\mathcal{O}}|\dot{\phi}$. Equating $\dot{\phi}$ in both expressions gives $\dot{x} = \frac{|r_{\mathcal{O}}|}{|r_{\mathcal{B}}+r_{\mathcal{O}}|} \dot{X}_{b_c}$. Finally, $\frac{r_{\mathcal{O}}}{r_{\mathcal{B}}+r_{\mathcal{O}}} = \frac{\kappa_{\mathcal{B}}}{\kappa_{\mathcal{B}}+\kappa_{\mathcal{O}}}$, giving the contact point velocity formula of Eq. (2.7). ○

Based on the contact point velocity formula, the c-obstacle curvature form is as follows.

THEOREM 2.7 **(c-obstacle curvature form in 2-D)** *Let $\mathcal{CO}$ be a c-obstacle associated with 2-D bodies $\mathcal{B}$ and $\mathcal{O}$. The* **curvature form** *of the c-obstacle boundary at $q \in \mathcal{S}$ is given by*

$$\kappa(q,\dot{q}) = \frac{1}{\|\eta(q)\|} \cdot \frac{1}{\kappa_{\mathcal{B}}+\kappa_{\mathcal{O}}} \dot{q}^T \times \begin{bmatrix} \kappa_{\mathcal{B}}\kappa_{\mathcal{O}} I & -\kappa_{\mathcal{B}}\kappa_{\mathcal{O}} JRb_c \\ -\kappa_{\mathcal{B}}\kappa_{\mathcal{O}}(JRb_c)^T & (\kappa_{\mathcal{O}}Rb - n(x))^T(\kappa_{\mathcal{B}}Rb + n(x)) \end{bmatrix} \dot{q} \qquad \dot{q} \in T_q\mathcal{S},$$

where $\eta(q)$ is the c-obstacle outward normal at q, $\kappa_{\mathcal{B}}$ and $\kappa_{\mathcal{O}}$ are the curvatures of $\mathcal{B}$ and $\mathcal{O}$ at the contact point $x = X_b(q)$, $n(x)$ is $\mathcal{B}$'s inward unit normal at x, and b_c is $\mathcal{B}$'s center of curvature at x expressed in $\mathcal{F}_B$; I is a 2×2 identity matrix and $J = \begin{bmatrix} 0 & 1 \\ -1 & 0 \end{bmatrix}$.

The proof of the theorem appears in the chapter's appendix. The key step in the proof is based on the contact point velocity formula. Consider a c-space path $q(t)$ that lies in $\mathcal{S}$, such that $q(0) = q$ and $\frac{d}{dt}\big|_{t=0} q(t) = \dot{q}$. According to Theorem 2.5, the c-obstacle

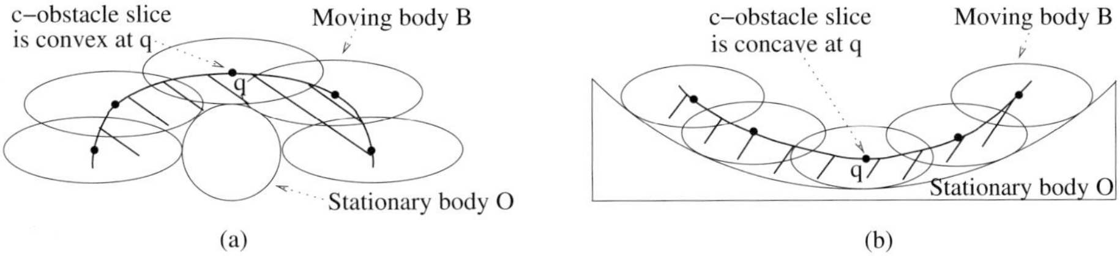

Figure 2.5 The c-obstacle slice is (a) convex when $\mathcal{B}$ and $\mathcal{O}$ are convex at x, and (b) concave when one of the two bodies is concave at x.

normal is given by $\eta(q) = DX_b^T(q)n(x)$, where $n(x)$ is $\mathcal{B}$'s inward unit normal at x. The curvature of $\mathcal{S}$ along $\dot{q}$ is determined by the derivative:

$$\frac{d}{dt}\bigg|_{t=0} \eta(q(t)) = DX_b^T(q)\, \frac{d}{dt}\bigg|_{t=0} n(x(t)) + \left(\frac{d}{dt}\bigg|_{t=0} DX_b^T(q(t)) \right) n(x). \qquad (2.8)$$

Since $\mathcal{B}$ maintains continuous contact with the stationary body $\mathcal{O}$ along $q(t)$, the contact point $x(t)$ moves along $\mathcal{O}$'s boundary during this motion. Hence $\frac{d}{dt}\big|_{t=0} n(x(t)) = \kappa_{\mathcal{O}}\dot{x}$ in the first summand of Eq. (2.8). Substituting $\dot{x} = \frac{\kappa_{\mathcal{B}}}{\kappa_{\mathcal{B}}+\kappa_{\mathcal{O}}}[I \;\; -JRb_c]\dot{q}$ according to Formula (2.7), then substituting the Jacobian formula $DX_b = [I \;-JRb]$ (see Exercises), gives

$$DX_b^T(q)\, \frac{d}{dt}\bigg|_{t=0} n(x(t)) = \frac{\kappa_{\mathcal{B}}\kappa_{\mathcal{O}}}{\kappa_{\mathcal{B}}+\kappa_{\mathcal{O}}} \begin{bmatrix} I \\ (-JRb)^T \end{bmatrix} [I \;\; -JRb_c]\,\dot{q}$$

$$= \frac{\kappa_{\mathcal{B}}\kappa_{\mathcal{O}}}{\kappa_{\mathcal{B}}+\kappa_{\mathcal{O}}} \begin{bmatrix} I & -JRb_c \\ (-JRb)^T & b \cdot b_c \end{bmatrix} \dot{q},$$

where we used the identities $R^T R = I$ and $J^T J = I$. The remainder of the proof appears in the chapter's appendix.

Curvature of c-obstacle slices: Consider the curvature of the fixed-θ slices of the c-obstacle boundary $\mathcal{S}$, denoted $\mathcal{S}|_\theta$. Each slice $\mathcal{S}|_\theta$ is a planar curve embedded in a fixed-θ plane in c-space $\mathbb{R}^3$. The vector tangent to $\mathcal{S}|_\theta$ at $q = (d,\theta)$ is $\dot{q} = (\dot{d}, 0)$, such that $\dot{d}$ is orthogonal to $n(x)$. This tangent vector corresponds to instantaneous translation of $\mathcal{B}$ along the tangent to $\mathcal{O}$'s boundary at x. Based on Theorem 2.7, the c-obstacle curvature along $\dot{q} = (\dot{d}, 0)$ is given by

$$\kappa\big(q, (\dot{d}, 0)\big) = \frac{\kappa_{\mathcal{B}}\kappa_{\mathcal{O}}}{\kappa_{\mathcal{B}} + \kappa_{\mathcal{O}}} \|\dot{d}\|^2.$$

The coefficient preceding $\|\dot{d}\|^2$ is the curvature of the c-obstacle slice $\mathcal{S}|_\theta$ at q. Its reciprocal is the slice's radius of curvature at q:

$$\left(\frac{\kappa_{\mathcal{B}}\kappa_{\mathcal{O}}}{\kappa_{\mathcal{B}} + \kappa_{\mathcal{O}}} \right)^{-1} = r_{\mathcal{B}} + r_{\mathcal{O}},$$

where $r_{\mathcal{B}}$ and $r_{\mathcal{O}}$ are the radii of curvature of $\mathcal{B}$ and $\mathcal{O}$ at x. We see that the radius of curvature of $\mathcal{S}|_\theta$ is the *algebraic sum* of the contacting bodies' radii of curvature. In particular, $\mathcal{S}|_\theta$ is convex at q when the bodies are convex at x (Figure 2.5(a)), while $\mathcal{S}|_\theta$ is concave at q when one of the two bodies is concave at x (Figure 2.5(b)). ◦

C-obstacle curvature in 3-D: The c-obstacle curvature in the 3-D case depends on the same geometric data as in the 2-D case, with the bodies' surface curvatures replacing the curvatures $\kappa_{\mathcal{B}}$ and $\kappa_{\mathcal{O}}$. Let $\mathcal{S}$ be the five-dimensional boundary of $\mathcal{CO}$, let $q \in \mathcal{S}$, and let $x = X(q,b)$ be $\mathcal{B}$'s contact point with $\mathcal{O}$. We assume that $\mathcal{S}$ is locally smooth at q, and denote by $T_q\mathcal{S}$ the five-dimensional tangent space of $\mathcal{S}$ at q. Let $q(t)$ be a c-space path that lies in $\mathcal{S}$, such that $q(0) = q$ and $\frac{d}{dt}\big|_{t=0} q(t) = \dot{q}$. The *curvature form* of $\mathcal{S}$ at q is given by $\kappa(q,\dot{q}) = \dot{q} \cdot \frac{d}{dt}\big|_{t=0} \hat{\eta}(q(t))$, where $\dot{q} \in T_q\mathcal{S}$ and $\hat{\eta}(q(t))$ is the unit outward normal to $\mathcal{S}$ along $q(t)$. The surface curvatures of $\mathcal{B}$ and $\mathcal{O}$ at x are determined by the linear maps $L_{\mathcal{B}}$ and $L_{\mathcal{O}}$. These linear maps act on the tangent plane of the respective surface to yield the change in the surface normal along a given tangent direction. The *curvature form* of $\mathcal{S}$ at q is given by (see Bibliographical Notes):

$$\kappa(q,\dot{q}) = \frac{1}{\|\eta(q)\|}\dot{q}^T \left(\begin{bmatrix} I & -[Rb\times] \\ O & [n(x)\times] \end{bmatrix}^T \begin{bmatrix} L_{\mathcal{B}}[L_{\mathcal{B}}+L_{\mathcal{O}}]^{-1}L_{\mathcal{O}} & -L_{\mathcal{O}}[L_{\mathcal{B}}+L_{\mathcal{O}}]^{-1} \\ -[L_{\mathcal{B}}+L_{\mathcal{O}}]^{-1}L_{\mathcal{O}} & -[L_{\mathcal{B}}+L_{\mathcal{O}}]^{-1} \end{bmatrix} \begin{bmatrix} I & -[Rb\times] \\ O & [n(x)\times] \end{bmatrix} + \begin{bmatrix} O & O \\ O & -\left([Rb\times]^T[n(x)\times]\right)_s \end{bmatrix} \right) \dot{q} \qquad \dot{q} \in T_q\mathcal{S},$$

where $\eta(q)$ is the c-obstacle outward normal at $q \in \mathcal{S}$, $n(x)$ is $\mathcal{O}$'s outward unit normal at x, I is a 3×3 identity matrix, O is a 3×3 matrix of zeroes and $(A)_s = \frac{1}{2}(A^T + A)$. Two comments are in order here. First, $L_{\mathcal{O}} + L_{\mathcal{B}} \geq 0$, otherwise the two bodies would interpenetrate at the contact. In particular, $L_{\mathcal{O}} + L_{\mathcal{B}} > 0$ in the generic case where the second-order approximations to the contacting surfaces of $\mathcal{B}$ and $\mathcal{O}$ maintain point contact at x. Second, the tangent vector $\dot{q} = (Rb \times n(x), n(x)) \in T_q\mathcal{S}$ is an eigenvector with zero eigenvalue of the matrix associated with the curvature form. This tangent vector corresponds to instantaneous rotation of $\mathcal{B}$ about its contact normal with $\mathcal{O}$. Hence, the c-obstacle $\mathcal{CO}$ always possesses *zero curvature* along instantaneous rotation of $\mathcal{B}$ about its contact normal with $\mathcal{O}$.

Bibliographical Notes

The notion of *configuration space* originated during a collaboration between two MIT doctoral students, assigned to develop one of the first robotic assembly stations [1–3]. While discussing the automation of the peg-in-a-hole insertion task, they observed that the best insertion approach would be to slide the peg along the horizontal surface at an oblique angle rather than at the vertical angle required for insertion (Figure 2.6). Once the oblique peg wedges itself into the hole, a rotational motion aligns the peg with the hole while completing the insertion task. This approach makes perfect sense when the c-obstacles induced by the hole's two sides are considered. The θ slice of the c-obstacles at a vertical angle contains only a *thin segment* of collision-free configurations (Figure 2.6(a)). In contrast, the θ slice of the c-obstacles at any oblique angle contains a full notch of collision-free configurations (Figure 2.6(b)). This observation led to the formulation of c-space as a framework for planning the motions of bodies in contact, a framework that has served virtually all motion planning algorithms. Some

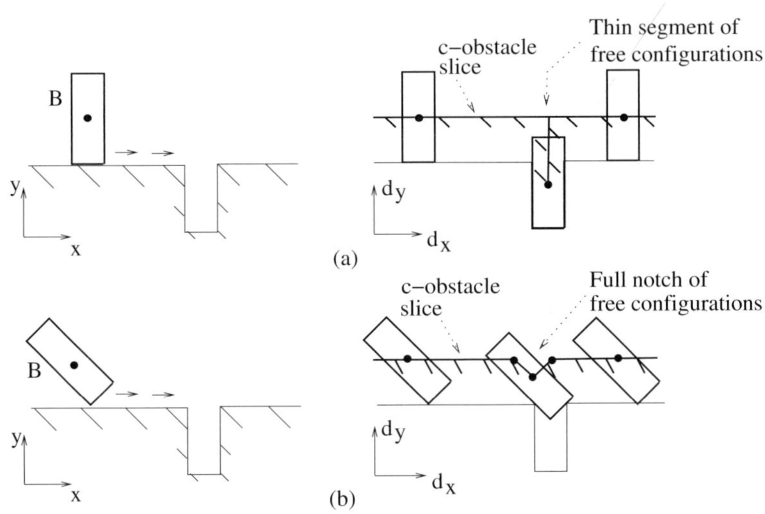

Figure 2.6 (a) At a vertical angle the peg's c-space contains only a thin segment of free configurations inside the hole. (b) At an oblique angle the peg's c-space contains a full notch of free configurations inside the hole.

recommended texts on robot motion planning are by Canny [4], Latombe [5], Lavalle [6] and, most recently, Lynch and Park [7].

The formulas for the c-obstacle normal and the c-obstacle curvature form are based on Rimon and Burdick [8, 9]. These papers contain the detailed derivation of the c-obstacle curvature form associated with 3-D bodies, which has been summarized here without any formal derivation. The contact point velocity formula of Proposition 2.6, which plays an important role in the derivation of the c-obstacle curvature form was derived by Montana [10].

Appendix: Details of Proofs

This appendix contains a derivation of the c-obstacle curvature formula in the 2-D case. We begin with the contact point velocity formula.

Proposition 2.6 *Let $q(t)$ be a c-space path that lies in $\mathcal{S}$, and let $x(t)$ be $\mathcal{B}$'s contact point with $\mathcal{O}$ along $q(t)$, expressed in the world frame $\mathcal{F}_W$. The* **contact point velocity** *along $q(t)$ is given by*

$$\dot{x}(t) = \frac{\kappa_{\mathcal{B}}}{\kappa_{\mathcal{B}}+\kappa_{\mathcal{O}}} \begin{bmatrix} I & -J R(\theta) b_c \end{bmatrix} \dot{q}(t), \tag{2.9}$$

where $\kappa_{\mathcal{B}}$ and $\kappa_{\mathcal{O}}$ are the curvatures of $\mathcal{B}$ and $\mathcal{O}$ at $x(t)$, b_c is $\mathcal{B}$'s center of curvature at $x(t)$ expressed in $\mathcal{F}_B$, I is a 2×2 identity matrix, and $J = \begin{bmatrix} 0 & 1 \\ -1 & 0 \end{bmatrix}$.

Proof: When the object $\mathcal{B}$ moves along a c-space path $q(t) = (d(t), \theta(t))$ which lies in $\mathcal{S}$, the contact point position is given by

$$x(t) = X(q(t), b(t)) = R(\theta(t))b(t) + d(t).$$

Taking the time derivative of both sides: $\dot{x} = DX_b(q)\dot{q} + R(\theta)\dot{b}$. In order to obtain an expression for $\dot{x}$ as a function of $\dot{q}$, we need a second equation relating $(\dot{x}, \dot{b})$ to $\dot{q}$. Since $\mathcal{B}$ maintains continuous contact with $\mathcal{O}$ along $q(t)$, $\mathcal{O}$'s *outward* unit normal at x, $n(x)$, must match $\mathcal{B}$'s *inward* unit normal at x. Denote by $\bar{n}(b)$ the outward unit normal of $\mathcal{B}$ at b, expressed in $\mathcal{F}_B$. Then $-R(\theta)\bar{n}(b)$ is the direction of $\mathcal{B}$'s *inward* unit normal at x in the world frame $\mathcal{F}_W$, and, therefore, $n(x(t)) = -R(\theta(t))\bar{n}(b(t))$ along $q(t)$. Taking the time derivative of both sides gives

$$\tfrac{d}{dt}n(x(t)) = JR(\theta)\bar{n}(b)\dot{\theta} - R(\theta)\tfrac{d}{dt}\bar{n}(b(t)),$$

where we used the formula $\dot{R}(\theta) = -JR(\theta)\dot{\theta}$. The curvature of $\mathcal{B}$ satisfies the relation $\frac{d}{dt}\bar{n}(b(t)) = \kappa_{\mathcal{B}}\dot{b}$. The curvature of $\mathcal{O}$ satisfies the relation $\frac{d}{dt}n(x(t)) = \kappa_{\mathcal{O}}\dot{x}$. Therefore,

$$\kappa_{\mathcal{O}}\dot{x} = JR(\theta)\bar{n}(b)\dot{\theta} - \kappa_{\mathcal{B}}R(\theta)\dot{b}.$$

Substituting $R(\theta)\dot{b} = \dot{x} - DX_b(q)\dot{q}$ in the latter equation gives

$$(\kappa_{\mathcal{B}} + \kappa_{\mathcal{O}})\dot{x} = \kappa_{\mathcal{B}}DX_b(q)\dot{q} + JR(\theta)\bar{n}(b)\dot{\theta} \qquad \dot{q} = (\dot{d}, \dot{\theta}).$$

Substituting $DX_b(q)\dot{q} = \dot{d} - JRb\dot{\theta}$, then pulling $\kappa_{\mathcal{B}}$ as a common factor gives

$$(\kappa_{\mathcal{B}} + \kappa_{\mathcal{O}})\dot{x} = \kappa_{\mathcal{B}}\left\{\dot{d} - JR\big(b - r_{\mathcal{B}}\bar{n}(b)\big)\dot{\theta}\right\} \qquad r_{\mathcal{B}} = 1/\kappa_{\mathcal{B}}.$$

The term $b - r_{\mathcal{B}}\bar{n}(b)$ is the position of $\mathcal{B}$'s center of curvature in $\mathcal{F}_B$: $b_c = b - r_{\mathcal{B}}\bar{n}(b)$. This gives $\dot{x} = \frac{\kappa_{\mathcal{B}}}{\kappa_{\mathcal{B}}+\kappa_{\mathcal{O}}}\left[I \;\; -JR(\theta)b_c\right]\dot{q}$, where $\dot{q} = (\dot{d}, \dot{\theta})$. □

The following theorem specifies the c-obstacle curvature formula in the 2-D case.

Theorem 2.7 *Let $\mathcal{CO}$ be a c-obstacle associated with 2-D bodies $\mathcal{B}$ and $\mathcal{O}$. The* **curvature form** *of the c-obstacle boundary at $q \in \mathcal{S}$ is given by*

$$\kappa(q, \dot{q}) = \frac{1}{\|\eta(q)\|} \cdot \frac{1}{\kappa_{\mathcal{B}}+\kappa_{\mathcal{O}}}\, \dot{q}^T \begin{bmatrix} \kappa_{\mathcal{B}}\kappa_{\mathcal{O}} I & -\kappa_{\mathcal{B}}\kappa_{\mathcal{O}} JRb_c \\ -\kappa_{\mathcal{B}}\kappa_{\mathcal{O}}(JRb_c)^T & (\kappa_{\mathcal{O}}Rb - n(x))^T(\kappa_{\mathcal{B}}Rb + n(x)) \end{bmatrix} \dot{q}$$
$$\dot{q} \in T_q\mathcal{S},$$

where $\eta(q)$ is the c-obstacle outward normal at q, $\kappa_{\mathcal{B}}$ and $\kappa_{\mathcal{O}}$ are the curvatures of $\mathcal{B}$ and $\mathcal{O}$ at the contact point $x = X_b(q)$, $n(x)$ is $\mathcal{B}$'s inward unit normal at x, and b_c is $\mathcal{B}$'s center of curvature at x expressed in $\mathcal{F}_B$; I is a 2×2 identity matrix and $J = \left[\begin{smallmatrix} 0 & 1 \\ -1 & 0 \end{smallmatrix}\right]$.

Proof: Let $q(t)$ be a c-space curve that lies in $\mathcal{S}$, such that $q(0) = q$ and $\frac{d}{dt}\big|_{t=0} q(t) = \dot{q}$. Based on the definition of $\kappa(q, \dot{q})$, we have to compute the derivative:

$$\left.\frac{d}{dt}\right|_{t=0} \hat{\eta}(q(t)) = \left.\frac{d}{dt}\right|_{t=0} \frac{1}{\|\eta(q(t))\|}\eta(q(t)) = \frac{1}{\|\eta(q)\|}\left[I - \hat{\eta}(q)\hat{\eta}(q)^T\right] \left.\frac{d}{dt}\right|_{t=0} \eta(q(t)).$$

Since $\left[I - \hat{\eta}(q)\hat{\eta}(q)^T\right]\dot{q} = \dot{q}$ on $T_q\mathcal{S}$, the curvature form can be equivalently written as

$$\kappa(q,\dot{q}) = \frac{1}{\|\eta(q)\|}\dot{q}\cdot\frac{d}{dt}\Big|_{t=0}\eta(q(t)) \qquad \dot{q}\in T_q\mathcal{S}.$$

The c-obstacle outward normal is given by $\eta(q) = DX_b^T(q)n(x)$, where $DX_b(q) = [I \;\; -JRb]$ and $n(x)$ is $\mathcal{O}$'s outward unit normal at x (Theorem 2.5). Thus, we have to compute the derivative:

$$\frac{d}{dt}\Big|_{t=0}\eta(q(t)) = DX_b^T(q)\,\frac{d}{dt}\Big|_{t=0}n(x(t)) + \left(\frac{d}{dt}\Big|_{t=0}DX_b^T(q(t))\right)n(x).$$

Since $\mathcal{B}$ maintains continuous contact with the stationary body $\mathcal{O}$ along $q(t)$, the contact point $x(t)$ moves along $\mathcal{O}$'s boundary. Hence $\frac{d}{dt}\big|_{t=0}n(x(t)) = \kappa_{\mathcal{O}}\dot{x}$ in the first summand. Substituting $\dot{x} = \frac{\kappa_{\mathcal{B}}}{\kappa_{\mathcal{B}}+\kappa_{\mathcal{O}}}[I \;\; -JRb_c]\dot{q}$ according to Proposition 2.6 gives

$$\begin{aligned} DX_b^T(q)\,\frac{d}{dt}\Big|_{t=0}n(x(t)) &= \tfrac{\kappa_{\mathcal{B}}\kappa_{\mathcal{O}}}{\kappa_{\mathcal{B}}+\kappa_{\mathcal{O}}}\begin{bmatrix} I \\ (-JRb)^T\end{bmatrix}[I \;\; -JRb_c]\dot{q} \\ &= \tfrac{\kappa_{\mathcal{B}}\kappa_{\mathcal{O}}}{\kappa_{\mathcal{B}}+\kappa_{\mathcal{O}}}\begin{bmatrix} I & -JRb_c \\ (-JRb)^T & b\cdot b_c\end{bmatrix}\dot{q}, \end{aligned}$$

where we used the identities $R^TR = I$ and $J^TJ = I$. In the second summand, $\frac{d}{dt}\big|_{t=0} DX_b^T(q) = \left[O \;\; -J\frac{d}{dt}\big|_{t=0}(Rb)\right]^T$, where O is a 2×2 matrix of zeroes. Since $\mathcal{B}$ maintains continuous contact with $\mathcal{O}$ along $q(t)$, the contact point satisfies the equation: $x(t) = R(\theta(t))b(t) + d(t)$. Taking the time derivative of both sides: $\dot{x} = \frac{d}{dt}(Rb) + \dot{d}$. Substituting $\dot{x} = \frac{\kappa_{\mathcal{B}}}{\kappa_{\mathcal{B}}+\kappa_{\mathcal{O}}}[I \;\; -JRb_c]\dot{q}$ according to Proposition 2.6 gives

$$\frac{d}{dt}(Rb) = \tfrac{\kappa_{\mathcal{B}}}{\kappa_{\mathcal{B}}+\kappa_{\mathcal{O}}}[I \;\; -JRb_c]\dot{q} - \dot{d} = \tfrac{-1}{\kappa_{\mathcal{B}}+\kappa_{\mathcal{O}}}[\kappa_{\mathcal{O}}I \;\; \kappa_{\mathcal{B}}JRb_c]\dot{q} \qquad \dot{q} = (\dot{d},\dot{\theta}).$$

The second summand thus has the form:

$$\begin{aligned} \left(\tfrac{d}{dt}\Big|_{t=0}DX_b^T(q(t))\right)n(x) &= \tfrac{1}{\kappa_{\mathcal{B}}+\kappa_{\mathcal{O}}}\begin{bmatrix} O \\ n^T(x)J\,[\kappa_{\mathcal{O}}I \;\; \kappa_{\mathcal{B}}JRb_c]\dot{q}\end{bmatrix} \\ &= \tfrac{1}{\kappa_{\mathcal{B}}+\kappa_{\mathcal{O}}}\begin{bmatrix} O & \vec{0} \\ \kappa_{\mathcal{O}}n^T(x)J & -\kappa_{\mathcal{B}}n^T(x)Rb_c\end{bmatrix}\dot{q}, \end{aligned}$$

where we used the identity $J^2 = -I$. Substituting for the two summands in the derivative $\frac{d}{dt}\big|_{t=0}\eta(q(t))$ gives

$$\begin{aligned} \frac{d}{dt}\Big|_{t=0}\eta(q(t)) &= \tfrac{1}{\kappa_{\mathcal{B}}+\kappa_{\mathcal{O}}}\begin{bmatrix} O & \vec{0} \\ \kappa_{\mathcal{O}}n^T(x)J & -\kappa_{\mathcal{B}}n^T(x)Rb_c\end{bmatrix}\dot{q} \\ &\quad + \tfrac{\kappa_{\mathcal{B}}\kappa_{\mathcal{O}}}{\kappa_{\mathcal{B}}+\kappa_{\mathcal{O}}}\begin{bmatrix} I & -JRb_c \\ (-JRb)^T & b\cdot b_c\end{bmatrix}\dot{q} \\ &= \tfrac{1}{\kappa_{\mathcal{B}}+\kappa_{\mathcal{O}}}\begin{bmatrix} \kappa_{\mathcal{B}}\kappa_{\mathcal{O}}I & -\kappa_{\mathcal{B}}\kappa_{\mathcal{O}}JRb_c \\ \kappa_{\mathcal{O}}n^T(x)J + \kappa_{\mathcal{B}}\kappa_{\mathcal{O}}(-JRb)^T & -\kappa_{\mathcal{B}}n^T(x)Rb_c + \kappa_{\mathcal{B}}\kappa_{\mathcal{O}}b\cdot b_c\end{bmatrix}\dot{q}. \end{aligned}$$

The expression on the lower left simplifies as follows. Denote by $\bar{n}(b)$ the outward unit normal to $\mathcal{B}$ at b, expressed in $\mathcal{F}_{\mathcal{B}}$. Then $n(x) = -R(\theta)\bar{n}(b)$, and, therefore,

$$\kappa_{\mathcal{O}} n^T(x) J + \kappa_{\mathcal{B}}\kappa_{\mathcal{O}}(-JRb)^T = \kappa_{\mathcal{B}}\kappa_{\mathcal{O}}\big(-r_{\mathcal{B}}\bar{n}(b) + b\big)^T R^T J = \kappa_{\mathcal{B}}\kappa_{\mathcal{O}}(-JRb_c)^T,$$

where $b_c = b - r_{\mathcal{B}}\bar{n}(b)$ is $\mathcal{B}$'s center of curvature at x. The expression on the lower right simplifies as follows:

$$-\kappa_{\mathcal{B}} n^T(x) R b_c + \kappa_{\mathcal{B}}\kappa_{\mathcal{O}} b \cdot b_c = (\kappa_{\mathcal{O}} Rb - n(x))^T(\kappa_{\mathcal{B}} R b_c) = (\kappa_{\mathcal{O}} Rb - n(x))^T(\kappa_{\mathcal{B}} Rb + n(x)),$$

where we substituted $\kappa_{\mathcal{B}} R b_c = \kappa_{\mathcal{B}} R(b - r_{\mathcal{B}}\bar{n}(b)) = \kappa_{\mathcal{B}} Rb + n(x)$. Substituting the simplified terms in the derivative $\frac{d}{dt}\big|_{t=0}\eta(q(t))$ gives

$$\frac{d}{dt}\bigg|_{t=0}\eta(q(t)) = \frac{1}{\kappa_{\mathcal{B}}+\kappa_{\mathcal{O}}}\begin{bmatrix} \kappa_{\mathcal{B}}\kappa_{\mathcal{O}} I & -\kappa_{\mathcal{B}}\kappa_{\mathcal{O}} J R b_c \\ -\kappa_{\mathcal{B}}\kappa_{\mathcal{O}}(JRb_c)^T & (\kappa_{\mathcal{O}} Rb - n(x))^T(\kappa_{\mathcal{B}} Rb + n(x)) \end{bmatrix}\dot{q}. \tag{2.10}$$

Pre-multiplying both sides of Eq. (2.10) by $1/\|\eta(q)\|$ and by the row vector $\dot{q}$ gives the c-obstacle curvature form. □

Exercises

Section 2.1

Exercise 2.1: Justify the definition of $SO(3)$ by the conditions $R^T R = I$ and $\det(R) = 1$.

Solution: Write $R = [c_1\, c_2\, c_3]$. The condition $R^T R = I$ ensures that $\|c_i\| = 1$ while $c_i \cdot c_j = 0$ for $1 \le i, j \le 3$, implying that the columns of R describe an orthonormal triplet. The condition $\det[c_1\, c_2\, c_3] = 1$ is equivalent to the condition $(c_1 \times c_2) \cdot c_3 = 1$, implying that the columns of R form a right-handed triplet. ○

Exercise 2.2: The matrix group $SO(3)$ forms a three-dimensional manifold topologically equivalent to the projective space RP^3. The points of RP^3 correspond to lines passing through the origin in $\mathbb{R}^4$. Explain why RP^3 is topologically equivalent to a three-dimensional unit ball centered at the origin of $\mathbb{R}^3$, with antipodal points on its bounding sphere identified.

Solution: The collection of lines passing through the origin in $\mathbb{R}^4$ can be identified with antipodal pairs of points on the unit three-dimensional sphere, S^3, embedded in $\mathbb{R}^4$.[5] The latter set can be identified with unique points on the upper hemisphere of S^3, together with antipodal pairs of points on the "equator" of S^3, which is the unit sphere S^2. The upper hemisphere with its equator S^2 is topologically equivalent to the unit three-dimensional ball embedded in $\mathbb{R}^3$, with antipodal points on its bounding unit sphere identified. Note that Rodrigues' formula parametrizes $SO(3)$ in terms of a radius π ball and its bounding sphere. ○

[5] When one uses *quaternions*, the global parametrization of $SO(3)$ is in terms of S^3 embedded in $\mathbb{R}^4$.

Exercise 2.3: When the matrix group $SO(3)$ is viewed as a manifold, it contains two classes of loops: those that can be contracted to a point and those that cannot be contracted to a point within $SO(3)$ (such a manifold is *not* simply connected). Identify the non-shrinkable loops in $SO(3)$, using the topological model obtained in the previous exercise.

Solution: Rodrigues' formula parametrizes $SO(3)$ in terms of a radius π ball with identified antipodal points on its bounding sphere. Consider a loop that starts at the origin, moves to the radius-π sphere, and then wraps through the antipodal point back to the origin. An attempt to contract this loop within the radius π ball would break it. ○

Exercise 2.4: Verify that $R(\theta)$ in Rodrigues' formula satisfies the conditions $R^T R = I$ and $\det(R) = 1$.

Exercise 2.5: Using Rodrigues' formula, verify that $\theta = (\theta_1, \theta_2, \theta_3) \in \mathbb{R}^3$ is an eigenvector of $R(\theta) \in SO(3)$, and $v \cdot (R(\theta)v) = \cos(\|\theta\|)$ for any unit vector $v \in \mathbb{R}^3$ orthogonal to θ.

Exercise 2.6: Show that Rodrigues' formula gives the 2×2 orientation matrices when $\hat{\theta} = (0, 0, 1)$.

Exercise 2.7*: Verify that the rigid-body transformation, $X(q, b) = R(\theta)b + d$, is the general form of distance and orientation preserving embedding of a rigid body $\mathcal{B}$ in $\mathbb{R}^3$.

Solution: This basic property is discussed in geometry texts, such as Rees' *Notes on Geometry* [11]. ○

Section 2.2

Exercise 2.8: Explain the characterization of the c-obstacle boundary specified in Lemma 2.2.

Exercise 2.9: Prove that when $\mathcal{B}$ and $\mathcal{O}$ are path-connected bodies, the c-obstacle $\mathcal{CO}$ is also path connected.

Solution: A graphical proof of this property appears in Latombe [5](Proposition 2.6).

Exercise 2.10: A real-valued function forms a *convex function* when its epigraph (the set of points on or above the graph of the function) forms a convex set. Prove that each θ-slice of the c-obstacle associated with a convex object $\mathcal{B}$ and a convex stationary body $\mathcal{O}$ is convex, based on the fact that $\text{dst}(x, \mathcal{O})$ is a convex function when $\mathcal{O}$ is convex.

Exercise 2.11: Let $\mathcal{B}$ be a 2-D smooth convex body and $\mathcal{O}$ a stationary disc of radius r and center x_0. Prove that the boundary of $\mathcal{CO}$ is parametrized by the formula: $\varphi(s, \theta) = (d(s, \theta), \theta)$, where $d(s, \theta) = x_0 - R(\theta)\big(\beta(s) + rJ\beta'(s)\big)$.

Solution: Let θ_0 be a particular orientation of $\mathcal{B}$. When $\mathcal{B}$ traces $\mathcal{O}$'s perimeter at a fixed orientation θ_0, the curve traced by $\mathcal{B}$'s contact point in $\mathcal{F}_W$ is $x(s) = R(\theta_0)b(s) + d(s)$ for $s \in \mathbb{R}$. The c-obstacle boundary is the curve traced by $\mathcal{B}$'s origin during this motion:

$d(s) = x(s) - R(\theta_0)b(s)$. Since the contact normals of $\mathcal{O}$ and $\mathcal{B}$ are collinear at $x(s)$, $\mathcal{O}$'s center point satisfies the equation, $x_0 = x(s) + r R(\theta_0) J \beta'(s)$, since $J\beta'(s)$ points into $\mathcal{O}$. Substituting for $x(s)$ in the expression for $d(s)$ gives $d(s) = x_0 - r R(\theta_0) J \beta'(s) - R(\theta_0)b(s) = x_0 - R(\theta_0)\big(b(s) + r J \beta'(s)\big)$. ○

Exercise 2.12: Show that the boundary of the c-obstacle $\mathcal{CO}$ associated with 2-D smooth convex bodies $\mathcal{B}$ and $\mathcal{O}$ forms a single smooth surface in $\mathcal{B}$'s c-space.

Exercise 2.13: When $\mathcal{B}$ is a polygon and $\mathcal{O}$ a stationary disc, the boundary of $\mathcal{CO}$ forms a piecewise smooth surface in $\mathcal{B}$'s c-space. What types of two-dimensional patches form the c-obstacle boundary? Write the (s, θ) parametrization of the patch generated by an edge of $\mathcal{B}$.

Solution: There are two types of smooth patches on the c-obstacle surface. An edge patch, generated by an edge of $\mathcal{B}$ sliding on $\mathcal{O}$, and a vertex patch, generated by a vertex of $\mathcal{B}$ sliding on $\mathcal{O}$. Let $\mathcal{O}$ have a radius r and center x_0. Consider now an edge of $\mathcal{B}$ having endpoints b_1 and b_2 and length L. Let $v = (b_2 - b_1)/L$ be the edge's direction. Then $\beta(s) = b_1 + sv$ for $0 \le s \le L$ parametrizes the edge in $\mathcal{F}_B$. Since the edge can touch $\mathcal{O}$ from the outside at any orientation θ, the parameter θ varies freely in $\mathbb{R}$. Following the solution approach of Exercise 2.11, $d(s, \theta) = x_0 - R(\theta)(b_1 + sv + rJv)$ for $0 \le s \le L$ and $\theta \in \mathbb{R}$. Note that $d(s, \theta)$ is linear in s, implying that the patch $\varphi(s, \theta) = (d(s, \theta), \theta)$ forms a *ruled surface* in this case. ○

Exercise 2.14*: Use the fact that Lipschitz continuous functions are piecewise smooth to conclude that the c-obstacle boundary is a piecewise smooth surface.

Section 2.3

Exercise 2.15: Verify the formula for the 3-D Jacobian of the rigid-body transformation, $DX_b(q) = \frac{d}{dq} X_b(q)$, which appears in the proof of Proposition 2.4. Obtain the 2-D Jacobian as a special case of the 3-D formula.

Exercise 2.16: Consider the c-obstacle normal, $\eta(q)$, at a configuration q at which $\mathcal{F}_B$'s origin lies along the contact normal. Verify that the tangent plane to the c-obstacle boundary, $T_q\mathcal{S}$, is vertical in this case, implying that instantaneous rotations of $\mathcal{B}$ about $\mathcal{F}_B$'s origin are tangent to $\mathcal{S}$ at q.

Section 2.4

Exercise 2.17: Prove that each fixed-θ slice of the c-obstacle boundary, $\mathcal{S}|_\theta$, is convex at $q = (d, \theta)$ when the contacting bodies are convex at x (Figure 2.5(a)).

Solution: Since $r_\mathcal{B}(x) \ge 0$ and $r_\mathcal{O}(x) \ge 0$ for convex bodies, $r_\mathcal{B}(x) + r_\mathcal{O}(x) \ge 0$, implying that $\mathcal{S}|_\theta$ is convex at $q = (d, \theta)$. ○

Exercise 2.18: Prove that each fixed-θ slice of the c-obstacle boundary, $\mathcal{S}|_\theta$, is concave at $q = (d, \theta)$ when one of the contacting bodies is concave at x (Figure 2.5(b)).

Solution: Suppose that the stationary body $\mathcal{O}$ is concave while the moving body $\mathcal{B}$ is convex at the contact point x. Then $r_{\mathcal{B}}(x) \geq 0$ while $r_{\mathcal{O}}(x) < 0$. Since $|r_{\mathcal{O}}(x)| > r_{\mathcal{B}}(x)$ (otherwise the bodies would interpenetrate), $r_{\mathcal{B}}(x) + r_{\mathcal{O}}(x) < 0$, implying that $\mathcal{S}|_{\theta}$ is concave at $q = (d, \theta)$. ○

References

[1] H. Inoue, "Force feedback in precise assembly tasks," Dept. of CS, MIT, Artificial Intelligence Laboratory, Technical Report AIM 308, 1974.

[2] T. Lozano-Pérez, "The design of a mechanical assembly system," Dept. of CS, MIT, Artificial Intelligence Laboratory, Technical Report AI-TR 397, 1976.

[3] T. Lozano-Pérez, "Spatial planning: A configuration space approach," *IEEE Transactions on Computers*, vol. 32, no. 2, pp. 108–120, 1983.

[4] J. F. Canny, *The Complexity of Robot Motion Planning*. MIT Press, 1988.

[5] J. C. Latombe, *Robot Motion Planning*. Kluwer Academic Publishers, 1990.

[6] S. M. LaValle, *Planning Algorithms*. Cambridge University Press, 2006.

[7] K. M. Lynch and F. C. Park, *Modern Robotics: Mechanics, Planning, and Control*. Cambridge University Press, 2017.

[8] E. Rimon and J. W. Burdick, "A configuration space analysis of bodies in contact – (i): 1st order mobility," *Mechanisms and Machine Theory*, vol. 30, no. 6, pp. 897–912, 1995.

[9] E. Rimon and J. W. Burdick, "A configuration space analysis of bodies in contact – (ii): 2nd order mobility," *Mechanisms and Machine Theory*, vol. 30, no. 6, pp. 913–928, 1995.

[10] D. J. Montana, "The kinematics of contact and grasp," *International Journal of Robotics Research*, vol. 7, no. 3, pp. 17–25, 1988.

[11] E. G. Rees, *Notes on Geometry*. Springer-Verlag, 1983.

3 Configuration Space Tangent and Cotangent Vectors

The rigid-body configuration space provides a geometric framework for describing the constraints imposed on the object's motions by the grasping fingers. Important components of this framework is the representation of the object's velocities as c-space *tangent vectors* and the representation of the finger contact forces as c-space *cotangent vectors.* This chapter consists of three parts. Section 3.1 describes the relation of the c-space tangent vectors to the rigid-body linear and angular velocities. Based on the virtual work principle, Section 3.2 describes how the finger contact forces are represented as cotangent vectors in the grasped object c-space. Section 3.3 introduces the *line geometry* of the rigid-body tangent and cotangent vectors. Using this theory, we obtain a graphical depiction of the c-obstacle tangent space, a representation that will prove useful in subsequent chapters.

3.1 C-Space Tangent Vectors

When a rigid body $\mathcal{B}$ moves along a c-space trajectory, $q(t)$, the *tangent vector* $\dot{q}(t) = \frac{d}{dt}q(t)$ represents instantaneous motion of $\mathcal{B}$ in the physical environment. In the 2-D case, $\dot{q}(t) = (\dot{d}(t), \dot{\theta}(t))$ is simply $\mathcal{B}$'s linear and rotational velocity with respect to the world frame $\mathcal{F}_W$. In the 3-D case $\dot{q}(t) = (\dot{d}(t), \dot{\theta}(t))$, but only $\dot{d}(t)$ retains the interpretation of being $\mathcal{B}$'s linear velocity with respect to $\mathcal{F}_W$. In order to assign intuitive meaning to $\dot{\theta}(t)$, we describe under what condition it can be interpreted as $\mathcal{B}$'s angular velocity vector.

Let us first summarize the notion of a rigid-body angular velocity vector. Let $R(t)$ be a parametrized curve of 3×3 orientation matrices in the manifold $SO(3)$, such that $R(0) = R_0$. The tangent to $R(t)$ at R_0 is given by $\dot{R} = \frac{d}{dt}\big|_{t=0} R(t)$. The matrices $R \in SO(3)$ satisfy the identity $RR^T = I$. Hence, the derivative $\dot{R}$ satisfies the identity: $\dot{R}R_0^T + R_0\dot{R}^T = 0$ at $t = 0$. It follows $\dot{R}R_0^T$ forms a skew-symmetric matrix: $(\dot{R}R_0^T)^T = -\dot{R}R_0^T$. The *angular velocity vector,* denoted $\omega \in \mathbb{R}^3$, parametrizes the skew-symmetric matrices $\dot{R}R_0^T$ as cross-product matrices: $\dot{R}R_0^T = [\omega\times]$ for $\omega \in \mathbb{R}^3$. Based on this definition, the derivative of $R(t)$ at $t = 0$ has the form:

$$\frac{d}{dt}\bigg|_{t=0} R(t) = [\omega\times]R_0 \qquad \omega \in \mathbb{R}^3. \tag{3.1}$$

It follows from Eq. (3.1) that $\dot{R} = \frac{d}{dt}\big|_{t=0} R(t)$ represents instantaneous rotation of $\mathcal{B}$ about an axis collinear with ω which passes through $\mathcal{B}$'s frame origin, with the norm $\|\omega\|$ being $\mathcal{B}$'s rotational rate about this axis.

In order to interpret the c-space tangent vectors, $\dot{\theta} \in \mathbb{R}^3$, as angular velocity vectors, we shall assume that $\mathcal{B}$ has a distinguished configuration (d_0, R_0), which will later be its nominal equilibrium grasp configuration. Let the orientation matrices $SO(3)$ be parametrized by *exponential coordinates,* centered at R_0 (see Chapter 2):

$$R(\theta) = \exp\big([\theta\times]\big) R_0 \qquad \theta \in \mathbb{R}^3.$$

The c-space coordinates are still $(d, \theta) \in \mathbb{R}^3 \times \mathbb{R}^3$, but now $\mathcal{B}$'s nominal orientation, R_0, is parametrized by $\theta = \vec{0}$. The following lemma asserts that the tangent vectors $\dot{\theta}$ are $\mathcal{B}$'s angular velocity vectors at the orientation R_0.

LEMMA 3.1 **(angular velocity vector)** *Let $SO(3)$ be parametrized by $R(\theta) = \exp([\theta\times])R_0$ for $\theta \in \mathbb{R}^3$. Let $\theta(t)$ be chosen such that $\theta(0) = \vec{0}$ and $\frac{d}{dt}\big|_{t=0} \theta(t) = \dot{\theta}$. Then*

$$\frac{d}{dt}\bigg|_{t=0} R(\theta(t)) = [\dot{\theta}\times]R_0 = [\omega\times]R_0,$$

where ω is $\mathcal{B}$'s angular velocity vector at R_0.

Proof: Using the expansion $\exp([\theta\times]) = \Big(I + [\theta\times] + \frac{1}{2}[\theta\times]^2 + \cdots\Big)$, the derivative of $R(\theta)$ at $\theta = \vec{0}$ is: $\frac{d}{dt}\big|_{t=0} R(\theta(t)) = \big([\dot{\theta}\times] + \frac{1}{2}\big([\dot{\theta}\times][\theta\times] + [\theta\times][\dot{\theta}\times]\big) + \cdots\big)R_0 = [\dot{\theta}\times]R_0$. Observing that $\dot{\theta}$ plays the role of ω in Eq. (3.1), we conclude that $\dot{\theta} = \omega$. □

Lemma 3.1 implies that the 3×3 skew-symmetric matrices, $[\omega\times]$ for $\omega \in \mathbb{R}^3$, span the three-dimensional tangent space of $SO(3)$ at R_0. The skew-symmetric matrices form a vector space with respect to matrix addition and scalar multiplication. They also possess a *Lie algebra* structure that is used in the kinematic modeling of serial chains (see Bibliographical Notes).

For the sake of uniformity, let us also parametrize the orientation matrices $SO(2)$ by exponential coordinates centered at R_0:

$$R(\theta) = \exp(\theta \bar{J})R_0 \qquad \theta \in \mathbb{R},\ \bar{J} = \begin{bmatrix} 0 & -1 \\ 1 & 0 \end{bmatrix}.$$

The c-space coordinates in the 2-D case are $(d, \theta) \in \mathbb{R}^2 \times \mathbb{R}$, with the scalar $\theta = 0$ parametrizing the orientation R_0. The rotational velocity $\dot{\theta}$ at $\theta = 0$ is the angular velocity of $\mathcal{B}$ about an axis perpendicular to the planar environment. We denote this angular velocity by the same symbol ω, where the distinction between $\omega \in \mathbb{R}$ (2-D case) and $\omega \in \mathbb{R}^3$ (3-D case) will be clear from the context. The derivative of $R(\theta)$ at $\theta = 0$ satisfies a formula analogous to Eq. (3.1):

$$\frac{d}{dt}\bigg|_{t=0} R(\theta(t)) = \omega \bar{J} R_0 \qquad \omega \in \mathbb{R},\ \bar{J} = \begin{bmatrix} 0 & -1 \\ 1 & 0 \end{bmatrix}. \tag{3.2}$$

Note that $\bar{J}R_0 = R_0\bar{J}$ in the preceding formula, since the matrices of $SO(2)$ commute and R_0 and $\bar{J}$ both belong to $SO(2)$.

Let $\mathcal{B}$ be located at a c-space point $q_0 \in \mathbb{R}^m$, which corresponds to the nominal configuration (d_0, R_0). The *c-space tangent vectors* at q_0 are pairs, $\dot{q} = (v, \omega)$, where $v = \dot{d}$ is $\mathcal{B}$'s linear velocity and $\omega = \dot{\theta}$ is $\mathcal{B}$'s angular velocity vector (rotational rate in the 2-D case). The ambient *tangent space* of $\mathbb{R}^m$ at q_0 is the m-dimensional vector space, $T_{q_0}\mathbb{R}^m$, spanned by all tangent vectors $\dot{q}$ based at q_0. The *tangent space* to the c-obstacle boundary, $T_{q_0}\mathcal{S}$, is the $(m-1)$-dimensional subspace of all vectors $\dot{q} \in T_{q_0}\mathbb{R}^m$ tangent to $\mathcal{S}$ at q_0. The tangent space $T_{q_0}\mathcal{S}$ consists of instantaneous motions of $\mathcal{B}$ along which $\mathcal{B}$'s contact point, $x = X_b(q)$, moves tangentially with respect to the stationary body $\mathcal{O}$ (see Exercises).

Instantaneous rolling example: Consider the c-obstacle associated with a convex object $\mathcal{B}$ and a stationary disc $\mathcal{O}$ of radius r. The parametrization of the c-obstacle boundary, $\varphi(s,\theta)$, was specified in Eq. (2.2) of Chapter 2. We already verified that the tangent vectors:

$$\frac{\partial}{\partial s}\varphi(s,\theta) = \begin{pmatrix} -R(\theta)\beta'(s) \\ 0 \end{pmatrix} \quad \text{and} \quad \frac{\partial}{\partial \theta}\varphi(s,\theta) = \begin{pmatrix} -JR(\theta)\big(\beta(s) + rJ\beta'(s)\big) \\ 1 \end{pmatrix}$$

span the tangent plane $T_q\mathcal{S}$ at $q = \varphi(s,\theta)$, where we omitted a scalar factor in $\frac{\partial}{\partial s}\varphi(s,\theta)$ and $J = \begin{bmatrix} 0 & 1 \\ -1 & 0 \end{bmatrix}$ in $\frac{\partial}{\partial \theta}\varphi(s,\theta)$. In these expressions, $\beta(s)$ is the contact point position expressed in $\mathcal{F}_B$, and $\beta'(s)$ is the unit tangent to $\mathcal{B}$'s boundary. Retaining $\frac{\partial}{\partial s}\varphi(s,\theta)$ and taking the sum $r\frac{\partial}{\partial s}\varphi(s,\theta) + \frac{\partial}{\partial \theta}\varphi(s,\theta)$ gives the basis vectors for $T_q\mathcal{S}$:

$$\dot{q}_1 = \begin{pmatrix} -Jn(x) \\ 0 \end{pmatrix} \quad \text{and} \quad \dot{q}_2 = \begin{pmatrix} JR(\theta)b \\ 1 \end{pmatrix},$$

where we substituted $b = \beta(s)$ and $n(x) = -R(\theta)J\beta'(s)$. The tangent vector $\dot{q}_1$ represents instantaneous translation of $\mathcal{B}$ such that $\dot{x}$ is tangent to $\mathcal{O}$ at x. Since $DX_b(q) = [I \ \ -JRb]$, the tangent vector $\dot{q}_2$ satisfies $\dot{x} = DX_b(q)\dot{q}_2 = [I \ \ -JRb]\dot{q}_2 = \vec{0}$, implying that $\dot{q}_2$ represents *instantaneous rolling* of $\mathcal{B}$ on the stationary body $\mathcal{O}$. ○

We conclude this section with a formula for the Jacobian of the rigid-body transformation, $DX_b(q)$, which first appeared in Chapter 2. It will play an important role in the representation of forces as cotangent vectors and is repeated here with formal derivation.

LEMMA 3.2 (**Jacobian of $X_b(q)$**) *Let $x = X_b(q)$ be the rigid-body transformation such that b is held fixed on $\mathcal{B}$. Let q_0 be the nominal configuration described earlier. In the 2-D case, the* **Jacobian** *$DX_b(q)$ forms the 2×3 matrix:*

$$DX_b(q_0) = [I \ \ -JR_0 b],$$

where I is a 2×2 identity matrix, $J = \begin{bmatrix} 0 & 1 \\ -1 & 0 \end{bmatrix}$, and R_0 is $\mathcal{B}$'s orientation matrix at q_0. In the 3-D case, the **Jacobian** *$DX_b(q)$ forms the 3×6 matrix:*

$$DX_b(q_0) = [I \ \ -[R_0 b\times]],$$

where I is a 3×3 identity matrix and R_0 is $\mathcal{B}$'s orientation matrix at q_0.

Proof: Let $q(t)$ be a c-space curve such that $q(0) = q_0$ and $\dot{q}(0) = \dot{q}$. The rigid-body transformation along this curve is $X_b(q(t)) = R(\theta(t))b + d(t)$, where $q(t) = (d(t), \theta(t))$. Using the chain rule: $\frac{d}{dt}\big|_{t=0} X_b(q(t)) = DX_b(q)\dot{q} = \dot{R}b + \dot{d}$. Using Eq. (3.2) for $\dot{R}$ in the 2-D case gives

$$\frac{d}{dt}\bigg|_{t=0} X_b(q(t)) = \dot{R}b + \dot{d} = -\omega J R_0 b + v = [I \quad -JR_0 b]\begin{pmatrix} v \\ \omega \end{pmatrix} \qquad \dot{q} = (v, \omega),$$

implying that $DX_b(q) = [I \quad -JR_0 b]$ at q_0. Using Eq. (3.1) for $\dot{R}$ in the 3-D case gives

$$\begin{aligned} \frac{d}{dt}\bigg|_{t=0} X_b(q(t)) &= \dot{R}b + \dot{d} = [\omega\times]R_0 b + v = -[R_0 b\times]\omega + v \\ &= [I \quad -[R_0 b\times]]\begin{pmatrix} v \\ \omega \end{pmatrix} \qquad \dot{q} = (v, \omega), \end{aligned}$$

implying that $DX_b(q_0) = [I \quad -[R_0 b\times]]$ at q_0. □

3.2 C-Space Cotangent Vectors

The c-space *cotangent vectors* represent the action of physical forces on $\mathcal{B}$, based on the following principle. Let a force f act on $\mathcal{B}$ at a fixed body point b, such that $x = X_b(q)$ is the position of the force action point in the world frame $\mathcal{F}_W$. In our setting, f can be a contact force generated by a finger body $\mathcal{O}$, or it can be a gravitational force acting at $\mathcal{B}$'s center of mass. When $\mathcal{B}$ moves along a c-space trajectory $q(t)$, the point b moves along a trajectory $x(t) = X_b(q(t))$. Now let us think of b as a point mass attached to $\mathcal{B}$. The *instantaneous work* done by f on the point mass b (i.e., the change in the point-mass mechanical energy) is given by the *inner product:* $f \cdot \dot{x}(t)$. Since the point mass is rigidly attached to $\mathcal{B}$, the force f must induce an identical change in $\mathcal{B}$'s mechanical energy. This physical fact is the basis for the representation of forces as c-space covectors, called *wrenches* (see Bibliographical Notes).

Let us first summarize the notion of a covector in $\mathbb{R}^m$. The ambient *cotangent space* of $\mathbb{R}^m$ at q_0, $T^*_{q_0}\mathbb{R}^m$, consists of all real-valued functions $h : T_{q_0}\mathbb{R}^m \to \mathbb{R}$ that act linearly on tangent vectors $\dot{q} \in T_{q_0}\mathbb{R}^m$. The elements of $T^*_{q_0}\mathbb{R}^m$ are *cotangent vectors.* A cotangent vector can be represented by a particular tangent vector that acts on tangent vectors $\dot{q} \in T_{q_0}\mathbb{R}^m$ as follows. Let $e_1, \ldots, e_m$ be the standard basis for $T_{q_0}\mathbb{R}^m$, induced from the standard basis for the ambient vector space $\mathbb{R}^m$. The components of h with respect to this basis are the scalars $h(e_1), \ldots, h(e_m)$. Now let $u_1, \ldots, u_m$ be the components of $\dot{q}$ with respect to the same basis, so that $\dot{q} = \sum_{i=1}^m u_i e_i$. By the linearity of h:

$$\begin{aligned} h(\dot{q}) &= \sum_{i=1}^m u_i h(e_i) = \big(h(e_1), \ldots, h(e_m)\big) \cdot \big(u_1, \ldots, u_m\big) \\ &= \big(h(e_1), \ldots, h(e_m)\big)\begin{pmatrix} u_1 \\ \vdots \\ u_m \end{pmatrix} \qquad \dot{q} \in T_{q_0}\mathbb{R}^m, \end{aligned}$$

where '·' denotes the Euclidean inner product in $T_{q0}\mathbb{R}^m$. The action of h on $\dot{q}$ can thus be represented as an *inner product* of the fixed row vector $(h(e_1),\ldots,h(e_m))$ with variable column vectors $(u_1,\ldots,u_m)$. In the following discussion, we shall treat cotangent vectors as fixed tangent vectors appearing on the left side of the Euclidean inner product.

The virtual work principle: Let a rigid body $\mathcal{B}$ move along a c-space trajectory $q(t)$ such that $q(0)=q_0$ and $\dot{q}=\dot{q}(0)$. Let a force f act on $\mathcal{B}$ at a point $x=X_b(q(t))$ during this motion. The cotangent vector representing the action of f on $\mathcal{B}$ is denoted $\boldsymbol{w}$. The formula for $\boldsymbol{w}$ is based on the following *virtual work principle.* Along any instantaneous motion $\dot{q}\in T_{q_0}\mathbb{R}^m$ of $\mathcal{B}$, the instantaneous work done by $\boldsymbol{w}$ on $\mathcal{B}$ *must be equal* to the instantaneous work done by f on the point mass x along $\dot{x}$. That is, the inner products must satisfy the equality: $f\cdot\dot{x}=\boldsymbol{w}\cdot\dot{q}$, where $\boldsymbol{w}$ is yet unknown. Since $\dot{x}=DX_b(q_0)\dot{q}$ by the chain rule, we obtain

$$\boldsymbol{w}\cdot\dot{q}=f\cdot DX_b(q_0)\dot{q}=f^T(x)DX_b(q_0)\dot{q} \qquad \text{for all } \dot{q}\in T_{q_0}\mathbb{R}^m.$$

Adopting the convention that $\boldsymbol{w}$ is written as a column vector, the linear function corresponding to $\boldsymbol{w}$ satisfies the formula:

$$\boldsymbol{w}=DX_b^T(q_0)f \qquad \text{where } f \text{ acts on } \mathcal{B} \text{ at } x=X_b(q_0).$$

In rigid-body grasps, cotangent vectors are *wrenches* that act on $\mathcal{B}$. When a force f acts on $\mathcal{B}$, the wrench induced by f is obtained by substituting for $DX_b(q_0)$ according to Lemma 3.2, giving the following formula.

Wrench formula: *Let a force f act on a rigid body $\mathcal{B}$ at a point $x=X_b(q_0)$, where q_0 is $\mathcal{B}$'s configuration. The* **wrench** *generated on $\mathcal{B}$ by f is given by*

$$\boldsymbol{w}=\begin{pmatrix} f \\ \tau \end{pmatrix}=\begin{cases} \begin{pmatrix} f \\ f\cdot\bar{J}R_0b \end{pmatrix} & \text{2-D case} \\ \begin{pmatrix} f \\ R_0b\times f \end{pmatrix} & \text{3-D case} \end{cases}$$

where R_0 is $\mathcal{B}$'s orientation at q_0, and $\bar{J}=\begin{bmatrix} 0 & -1 \\ 1 & 0 \end{bmatrix}$ in the 2-D case.

The *wrenches* that act on $\mathcal{B}$ will be written as pairs, $\boldsymbol{w}=(f,\tau)$, where f and τ are the force and torque components. In the 2-D case, $\tau\in\mathbb{R}$ is the moment that acts on $\mathcal{B}$ about an axis perpendicular to the planar environment. In the 3-D case, $\tau\in\mathbb{R}^3$ satisfies the classical formula: $\tau=\rho\times f$, where $\rho=R_0b$ is the vector from $\mathcal{B}$'s frame origin to the force's action point, expressed in $\mathcal{F}_W$. The distinction between $\tau\in\mathbb{R}$ (2-D case) and $\tau\in\mathbb{R}^3$ (3-D case) will be clear from the context. When a rigid object $\mathcal{B}$ is contacted by multiple rigid finger bodies $\mathcal{O}_1,\ldots,\mathcal{O}_k$, the net wrench affected on $\mathcal{B}$ is the *sum* of the wrenches generated by the individual contact forces. The net wrench is illustrated in the following example.

Example: Consider a drill that acts on a fixtured workpiece $\mathcal{B}$. The drill applies a normal penetrating force, f_n, along its axis together with a matched pair of tangential forces, $\pm f_t$, at its two tips, called *flutes*. The drill's contact forces generate a net wrench on $\mathcal{B}$

as follows. Let $\mathcal{B}$'s frame origin be located on the drill's axis, such that the drill's flutes are located at $-b$ and b in $\mathcal{F}_B$. The net wrench affected on $\mathcal{B}$ by the drill's contact forces is the sum:

$$\boldsymbol{w} = \begin{pmatrix} \frac{1}{2}f_n - f_t \\ R_0(-b) \times (\frac{1}{2}f_n - f_t) \end{pmatrix} + \begin{pmatrix} \frac{1}{2}f_n + f_t \\ R_0 b \times (\frac{1}{2}f_n + f_t) \end{pmatrix} = \begin{pmatrix} f_n \\ 2R_0 b \times f_t \end{pmatrix},$$

which is no longer a single-force wrench. The net wrench on $\mathcal{B}$ consists of a force which acts along the drill's axis, coupled with a torque which acts about the drill's axis. We will see in the next section that every wrench can be described in this screw-like manner. ○

Coordinate transformation of tangent and cotangent vectors: Let $q \in \mathbb{R}^m$ and $\bar{q} \in \bar{\mathbb{R}}^m$ be c-space parametrizations associated with two choices of world and body frames, such that q_0 and $\bar{q}_0$ are $\mathcal{B}$'s nominal configuration in the two parametrizations. It can be verified that q and $\bar{q}$ are related by a coordinate transformation (i.e., a *diffeomorphism*) $q = F(\bar{q})$, such that $q_0 = F(\bar{q}_0)$ (see Exercises). The c-space trajectories are related by the transformation $q(t) = F(\bar{q}(t))$. Hence, by the chain rule $\dot{q}(t) = [DF(\bar{q}_0)]\dot{\bar{q}}(t)$, where DF is the $m \times m$ Jacobian matrix of F at $\bar{q}_0$. The Jacobian $DF(\bar{q}_0)$ thus maps tangent vectors in $T_{\bar{q}_0}\bar{\mathbb{R}}^m$ to tangent vectors in $T_{q_0}\mathbb{R}^m$:

$$\dot{q} = DF(\bar{q}_0)\dot{\bar{q}} \qquad \dot{q} \in T_{q_0}\mathbb{R}^m,\ \dot{\bar{q}} \in T_{\bar{q}_0}\bar{\mathbb{R}}^m. \tag{3.3}$$

Now let $\boldsymbol{w}$ be a covector acting on tangent vectors $\dot{q} \in T_{q_0}\mathbb{R}^m$. Using the transformation rule of Eq. (3.3):

$$\boldsymbol{w} \cdot \dot{q} = \boldsymbol{w}^T\left(DF(\bar{q}_0)\dot{\bar{q}}\right) = \left(\boldsymbol{w}^T DF(\bar{q}_0)\right)\dot{\bar{q}} \qquad \dot{q} \in T_{q_0}\mathbb{R}^m, \dot{\bar{q}} \in T_{\bar{q}_0}\bar{\mathbb{R}}^m.$$

It follows that the covector $\bar{\boldsymbol{w}}^T = \boldsymbol{w}^T DF(\bar{q}_0) \in T^*_{\bar{q}_0}\bar{\mathbb{R}}^m$ corresponds to the covector $\boldsymbol{w} \in T^*_{q_0}\mathbb{R}^m$. Writing $\boldsymbol{w}$ and $\bar{\boldsymbol{w}}$ as column vectors, the *covector transformation rule* has the form:

$$\bar{\boldsymbol{w}} = DF^T(\bar{q}_0)\boldsymbol{w} \qquad \boldsymbol{w} \in T^*_{q_0}\mathbb{R}^m,\ \bar{\boldsymbol{w}} \in T^*_{\bar{q}_0}\bar{\mathbb{R}}^m.$$

The Jacobian $DF(\bar{q}_0)$ thus maps tangent vectors *forward* from $T_{\bar{q}_0}\bar{\mathbb{R}}^m$ to $T_{q_0}\mathbb{R}^m$, while the transposed Jacobian $DF^T(\bar{q}_0)$ maps cotangent vectors *backward* from $T^*_{q_0}\mathbb{R}^m$ to $T^*_{\bar{q}_0}\bar{\mathbb{R}}^m$. ○

3.3 Line Geometry of Tangent and Cotangent Vectors

Line geometry provides an intuitive means for depicting the c-space tangent and cotangent vectors as *directed lines* in physical space. This section describes three useful facts concerning line geometry. First, every tangent vector $\dot{q} = (v, \omega)$ can be represented as an *instantaneous screw motion* about a spatial axis in $\mathbb{R}^3$. Second, every wrench $\boldsymbol{w} = (f, \tau)$ can be represented by a *screw-like application of force and torque* about a spatial axis in $\mathbb{R}^3$. Third, the inner product $\boldsymbol{w} \cdot \dot{q}$ can be expressed as a geometric relation

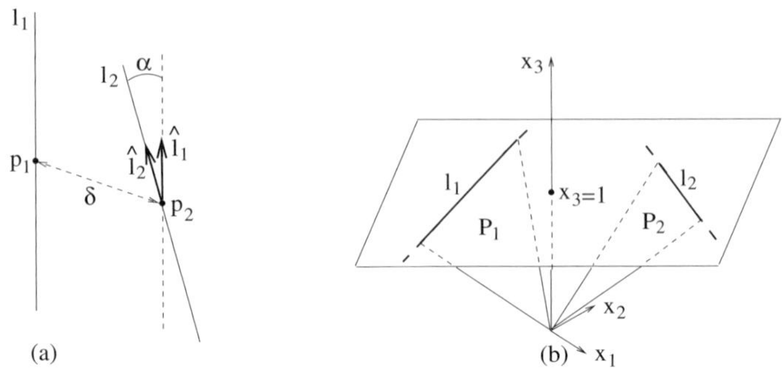

Figure 3.1 (a) The parameters specifying the relative position and orientation of l_1 and l_2. (b) The analogous correspondence between embedded planar lines l_1 and l_2 and planes P_1 and P_2 that pass through the origin of $\mathbb{R}^3$.

between the directed lines representing $\boldsymbol{w}$ and $\dot{q}$. Line geometry will then be used to graphically depict the c-obstacle tangent space.

The directed spatial lines in $\mathbb{R}^3$ are parametrized by the following *Plücker coordinates.*

DEFINITION 3.1 *The* **Plücker coordinates** *of directed lines in* $\mathbb{R}^3$ *are vectors* $l = (\hat{l}, p \times \hat{l}) \in \mathbb{R}^6$, *where* p *is any point on the line and* $\hat{l}$ *is the unit direction of the line* (Figure 3.1(a)).

We will use the symbol l both for the physical line in $\mathbb{R}^3$ and for its Plücker coordinates in $\mathbb{R}^6$. Note that any point on l can serve as a reference point, since $(p + s\hat{l}) \times \hat{l} = p \times \hat{l}$ for $s \in \mathbb{R}$.

Manifold structure of the spatial lines: The Plücker coordinates of all directed lines in $\mathbb{R}^3$ span a smooth four-dimensional manifold in $\mathbb{R}^6$. This means that the directed lines depend on *four* parameters, which are the line direction $\hat{l}$ and the line base point p. To explain this topological fact, consider the unit sphere in $\mathbb{R}^3$, denoted S^2. The collection of all directed lines in $\mathbb{R}^3$ is topologically equivalent to a canonical four-dimensional manifold, the *tangent bundle* of the unit sphere, TS^2, which is simply the union of the tangent planes at all points of S^2. The four-dimensional manifold of directed lines in $\mathbb{R}^3$ is topologically equivalent to TS^2: every point on S^2 specifies a particular direction $\hat{l}$, such that the sphere's tangent plane at $\hat{l}$ specifies the collection of base points, p, of all spatial lines having this particular direction. ○

An important relation between two spatial lines is the following *reciprocal product* of the lines' Plücker coordinates.

DEFINITION 3.2 *The* **reciprocal product** *of two lines* $l_1 = (\hat{l}_1, p_1 \times \hat{l}_1)$ *and* $l_2 = (\hat{l}_2, p_2 \times \hat{l}_2)$ *is the "swapped" inner product of the lines' Plücker coordinates:*

$$(\hat{l}_1, p_1 \times \hat{l}_1) \cdot (p_2 \times \hat{l}_2, \hat{l}_2) = (\hat{l}_1, p_1 \times \hat{l}_1) \begin{bmatrix} O & I \\ I & O \end{bmatrix} \begin{pmatrix} \hat{l}_2 \\ p_2 \times \hat{l}_2 \end{pmatrix},$$

where O *is a* 3×3 *zero matrix and* I *is a* 3×3 *identity matrix.*

The reciprocal product of l_1 and l_2 is determined by the relative position and orientation of the two lines (Figure 3.1(a)). Let $\delta \geq 0$ be the minimal distance between l_1 and l_2, and let $0 \leq \alpha \leq \pi$ be the angle between l_1 and l_2, measured about the lines' common normal (Figure 3.1(a)). The reciprocal product is next specified in terms of δ and α (see Exercises).

PROPOSITION 3.3 **(reciprocal product formula)** *The* **reciprocal product** *of two spatial lines l_1 and l_2 satisfies the formula:*

$$(\hat{l}_1, p_1 \times \hat{l}_1) \begin{bmatrix} O & I \\ I & O \end{bmatrix} \begin{pmatrix} \hat{l}_2 \\ p_2 \times \hat{l}_2 \end{pmatrix} = -\sigma \cdot \delta \sin \alpha \tag{3.4}$$

where δ and α are the minimal distance and angle between the two lines, and $\sigma \in \{-1, +1\}$ according to the sign of the expression $(p_2 - p_1) \cdot (\hat{l}_1 \times \hat{l}_2)$.

Based on Formula (3.4), the reciprocal product of two spatial lines is *zero* when $\delta \sin \alpha = 0$. If $\delta = 0$ in this expression, l_1 and l_2 intersect at a common point in $\mathbb{R}^3$. Otherwise $\sin \alpha = 0$, which means that l_1 and l_2 are parallel and hence intersect at infinity. This key property is summarized in the following corollary.

COROLLARY 3.4 **(reciprocal lines)** *The reciprocal product of two spatial lines is* **zero** *iff the two lines intersect in $\mathbb{R}^3$, with the understanding that parallel lines intersect at infinity.*

Remark: The geometric intuition behind Corollary 3.4 is as follows. Let (x_1, x_2, x_3, x_4) denote the coordinates of $\mathbb{R}^4$. The lines l_1 and l_2 can be embedded in the hyperplane $x_4 = 1$ of $\mathbb{R}^4$. Every embedded line l determines a unique two-dimensional plane passing through the origin of $\mathbb{R}^4$. (An analogous situation arises when two planar lines are embedded in the $x_3 = 1$ plane of $\mathbb{R}^3$. In this case, every embedded line determines a plane passing through the origin of $\mathbb{R}^3$, as shown in Figure 3.1(b).) Now let $l_1 = (\hat{l}_1, p_1 \times \hat{l}_1)$ and $l_2 = (\hat{l}_2, p_2 \times \hat{l}_2)$ be two spatial lines, and let $\mathcal{P}_1$ and $\mathcal{P}_2$ be the planes associated with the embedded lines in $\mathbb{R}^4$. The two lines intersect at a common point in $\mathbb{R}^3$ iff the two planes $\mathcal{P}_1$ and $\mathcal{P}_2$ intersect along a line passing through the origin of $\mathbb{R}^4$. The planes $\mathcal{P}_1$ and $\mathcal{P}_2$ intersect along a line in $\mathbb{R}^4$ iff their basis vectors are linearly dependent. Each embedded line, $l_i = (\hat{l}_i, p_i \times \hat{l}_i)$, passes through the point $(p_i, 1) \in \mathbb{R}^4$ along the direction $(\hat{l}_i, 0) \in \mathbb{R}^4$. It follows that the pair $\{(p_i, 1), (\hat{l}_i, 0)\}$ forms a basis for $\mathcal{P}_i$ $(i = 1, 2)$. The linear dependence of $\mathcal{P}_1$ and $\mathcal{P}_2$ is thus equivalent to the condition $\det \begin{bmatrix} p_1 & \hat{l}_1 & p_2 & \hat{l}_2 \\ 1 & 0 & 1 & 0 \end{bmatrix} = 0$, which is the zero reciprocal product condition. ○

Our next step would be to represent the c-space tangent vectors as instantaneous screw-like motions about a spatial axis in $\mathbb{R}^3$, called *instantaneous twists*. The following statement is the classical *Mozzi–Chasles theorem*.

THEOREM 3.5 **(instantaneous twist)** *Every c-space tangent vector $\dot{q} = (v, \omega) \in T_{q_0}\mathbb{R}^m$ can be written as instantaneous rotation of $\mathcal{B}$ about the line $l = (\hat{\omega}, p \times \hat{\omega})$, coupled with instantaneous translation of $\mathcal{B}$ along this line:*

$$\begin{pmatrix} v \\ \omega \end{pmatrix} = \begin{pmatrix} p \times \omega \\ \omega \end{pmatrix} + \begin{pmatrix} z\omega \\ \vec{0} \end{pmatrix} = \|\omega\| \begin{pmatrix} p \times \hat{\omega} \\ \hat{\omega} \end{pmatrix} + \begin{pmatrix} z\omega \\ \vec{0} \end{pmatrix}, \tag{3.5}$$

where $p = \omega \times v/\|\omega\|^2$, $z = v \cdot \omega/\|\omega\|^2$ *and* $\hat{\omega} = \omega/\|\omega\|$.

Proof: The linear velocity v can be expressed as the sum: $v = (v \cdot \hat{\omega})\hat{\omega} + (I - \hat{\omega}\hat{\omega}^T)v$. Using the identity $[\hat{\omega}\times]^2 = \hat{\omega}\hat{\omega}^T - I$, $v = (v \cdot \hat{\omega})\hat{\omega} - [\hat{\omega}\times]^2 v = (v \cdot \hat{\omega})\hat{\omega} - \hat{\omega} \times (\hat{\omega} \times v)$. Substituting $z = (v \cdot \hat{\omega})/\|\omega\| = (v \cdot \omega)/\|\omega\|^2$ and $p = (\hat{\omega} \times v)/\|\omega\| = (\omega \times v)/\|\omega\|^2$ gives the result. □

Note that the Plücker coordinates of the line representing an instantaneous rotation axis, $l = (\hat{\omega}, p \times \hat{\omega})$, are swapped in Eq. (3.5). The parameter z is called the *pitch* of the instantaneous twist. The pitch parameter specifies the amount of translation per unit rotation along the twist axis l. In particular, $z = 0$ corresponds to pure instantaneous rotation of $\mathcal{B}$ about l, while $z = \infty$ corresponds to pure instantaneous translation of $\mathcal{B}$ along the line l.

The wrenches that act on $\mathcal{B}$ can be similarly represented as a screw-like application of force and torque about a spatial axis in $\mathbb{R}^3$, called *wrench screws*. The following statement is the classical *Poinsot theorem*.

THEOREM 3.6 (**wrench screw**) *Every c-space cotangent vector* $\boldsymbol{w} = (f, \tau) \in T^*_{q_0}\mathbb{R}^m$ *can be written as a force acting along the line* $l = (\hat{f}, p \times \hat{f})$*, together with a torque acting about this line:*

$$\begin{pmatrix} f \\ \tau \end{pmatrix} = \begin{pmatrix} f \\ p \times f \end{pmatrix} + \begin{pmatrix} \vec{0} \\ zf \end{pmatrix} = \|f\| \begin{pmatrix} \hat{f} \\ p \times \hat{f} \end{pmatrix} + \begin{pmatrix} \vec{0} \\ zf \end{pmatrix} \tag{3.6}$$

where $p = f \times \tau/\|f\|^2$, $z = f \cdot \tau/\|f\|^2$ *and* $\hat{f} = f/\|f\|$.

Proof: The proof is similar to the proof of Mozzi–Chasles theorem. The torque τ can be expressed as the sum: $\tau = (\tau \cdot \hat{f})\hat{f} + (I - \hat{f}\hat{f}^T)\tau$. Using the identity $[\hat{f}\times]^2 = \hat{f}\hat{f}^T - I$, $\tau = (\tau \cdot \hat{f})\hat{f} - [\hat{f}\times]^2\tau = (\tau \cdot \hat{f})\hat{f} - \hat{f} \times (\hat{f} \times \tau)$. Substituting $z = (f \cdot \tau)/\|f\|^2$ and $p = (f \times \tau)/\|f\|^2$ gives the result. □

The parameter z is the *pitch* of the wrench screw. The pitch specifies the amount of torque applied by the wrench about l per unit force applied along this line. In particular, a zero-pitch wrench screw corresponds to a wrench generated by a pure force f acting along l. Having obtained a representation of tangent and cotangent vectors as screw lines, the inner product $\boldsymbol{w} \cdot \dot{q}$ can be expressed in terms of these screw lines as follows.

LEMMA 3.7 (**line geometry formula for $\boldsymbol{w} \cdot \dot{q}$**) *Let a wrench* $\boldsymbol{w} = (f, \tau) \in T^*_{q_0}\mathbb{R}^m$ *have a screw axis* l_1 *and pitch* z_1*. Let a tangent vector* $\dot{q} = (v, \omega) \in T_{q_0}\mathbb{R}^m$ *have a screw axis* l_2 *and pitch* z_2*. The inner product* $\boldsymbol{w} \cdot \dot{q}$ *satisfies the geometric formula:*

$$\boldsymbol{w} \cdot \dot{q} = \|f\|\|\omega\|\big(-\sigma \cdot \delta \sin\alpha + (z_1 + z_2)\cos\alpha\big), \tag{3.7}$$

where δ *is the minimum distance between the lines* l_1 *and* l_2*,* α *is the angle between* l_1 *and* l_2*, and* $\sigma \in \{-1, +1\}$ *is the sign specified in the reciprocal product formula.*

Proof: Substituting $\boldsymbol{w} = (f, p_1 \times f) + (\vec{0}, z_1 f)$ and $\dot{q} = (p_2 \times \omega, \omega) + (z_2\omega, \vec{0})$ in $\boldsymbol{w} \cdot \dot{q}$ gives:

$$\begin{aligned}\boldsymbol{w} \cdot \dot{q} &= (f, p_1 \times f)\begin{pmatrix} p_2 \times \omega \\ \omega \end{pmatrix} + (f, p_1 \times f)\begin{pmatrix} z_2\omega \\ \vec{0} \end{pmatrix} + (\vec{0}, z_1 f)\begin{pmatrix} p_2 \times \omega \\ \omega \end{pmatrix} \\ &= \|f\|\|\omega\| \left((\hat{f}, p_1 \times \hat{f})\begin{pmatrix} p_2 \times \hat{\omega} \\ \hat{\omega} \end{pmatrix} + (z_1 + z_2)\hat{f} \cdot \hat{\omega} \right),\end{aligned}$$

where $\hat{f} = f/\|f\|$ and $\hat{\omega} = \omega/\|\omega\|$. The first summand is equal to $-\sigma \cdot \delta \sin\alpha$ according to Proposition 3.3. The second summand is equal to $(z_1 + z_2)\cos\alpha$. □

Recall that the *reciprocal product* of the screw lines associated with $\boldsymbol{w}$ and $\dot{q}$ is equivalent to the inner product $\boldsymbol{w} \cdot \dot{q}$. When the reciprocal product of the screw lines associated with $\boldsymbol{w}$ and $\dot{q}$ is *zero*, we have that $\boldsymbol{w} \cdot \dot{q} = 0$. In this case, the wrench $\boldsymbol{w}$ *cannot impede* the instantaneous motion of $\mathcal{B}$ along $\dot{q}$. This fact will provide physical intuition for the mobility of an object $\mathcal{B}$ with respect to surrounding finger bodies $\mathcal{O}_1, \ldots, \mathcal{O}_k$ in subsequent chapters. We conclude this section with application of the reciprocal product to the depiction of the c-obstacle tangent space.

Graphical depiction of the c-obstacle tangent space: Let $\boldsymbol{w}$ be a wrench generated by a force aligned with $\mathcal{B}$'s inward contact normal at x. The tangent vectors satisfying $\boldsymbol{w} \cdot \dot{q} = 0$ span the c-obstacle tangent space at q_0. Formally speaking, the c-obstacle tangent space is given by $T_{q_0}\mathcal{S} = \{\dot{q} \in T_{q_0}\mathbb{R}^m : \eta(q_0) \cdot \dot{q} = 0\}$, where $\eta(q_0)$ is the c-obstacle normal at q_0 (Chapter 2). As discussed in the previous section, $\eta(q_0)$ can be interpreted as the wrench generated by a unit-magnitude normal force acting on $\mathcal{B}$ at x, so that $\boldsymbol{w} = \eta(q_0)$. Hence $T_{q_0}\mathcal{S}$ consists of tangent vectors satisfying $\boldsymbol{w} \cdot \dot{q} = 0$, where $\boldsymbol{w} = \eta(q_0)$. Since the wrench screw of a pure force has zero pitch, $z_1 = 0$ for the wrench screw of $\eta(q_0)$. Assuming that the frames $\mathcal{F}_W$ and $\mathcal{F}_B$ share a common origin at q_0, the screw axis of the wrench $\boldsymbol{w} = \eta(q_0)$ coincides with the contact normal line at x. Substituting $\|f\| = 1$ and $z_1 = 0$ in formula (3.7) for $\boldsymbol{w} \cdot \dot{q}$ gives

$$T_{q_0}\mathcal{S} = \left\{\dot{q} \in T_{q_0}\mathbb{R}^m : \sigma \cdot \|\omega\|\delta \sin(\alpha) = \|\omega\| z_2 \cos(\alpha)\right\}, \tag{3.8}$$

where (δ, α, z_2) parametrize the instantaneous twists, $\dot{q} \in T_{q_0}\mathcal{S}$, expressed relative to the contact normal line at x. Since $\|\omega\|$ appears on both sides of Eq. (3.8), an instantaneous screw motion that satisfies Eq. (3.8) does so for all $\|\omega\| \in \mathbb{R}$.

Formula (3.8) can be used to depict the c-obstacle tangent space in the 2-D case. Embed the (x, y) plane as the $(x, y, 0)$ plane in $\mathbb{R}^3$, and denote by $\boldsymbol{e} = (0, 0, 1)$ the unit upward direction in $\mathbb{R}^3$. The linear velocity of $\mathcal{B}$ is the vector $(v, 0)$, while the angular velocity vector of $\mathcal{B}$ is $\omega\boldsymbol{e}$, where $\omega \in \mathbb{R}$ is $\mathcal{B}$'s rotational rate in the (x, y) plane. The instantaneous twists of $\mathcal{B}$ have zero pitch, since $z_2 = \omega((v, 0) \cdot \boldsymbol{e}) = 0$. All tangent vectors in the ambient tangent space $T_{q_0}\mathbb{R}^3$ are therefore instantaneous rotations about points in the (x, y) plane. To determine which instantaneous rotations correspond to $T_{q_0}\mathcal{S}$, substitute $z_2 = 0$ and $\sin(\alpha) = 1$ in Eq.(3.8):

$$T_{q_0}\mathcal{S} = \{\dot{q} \in T_{q_0}\mathbb{R}^3 : \delta = 0\},$$

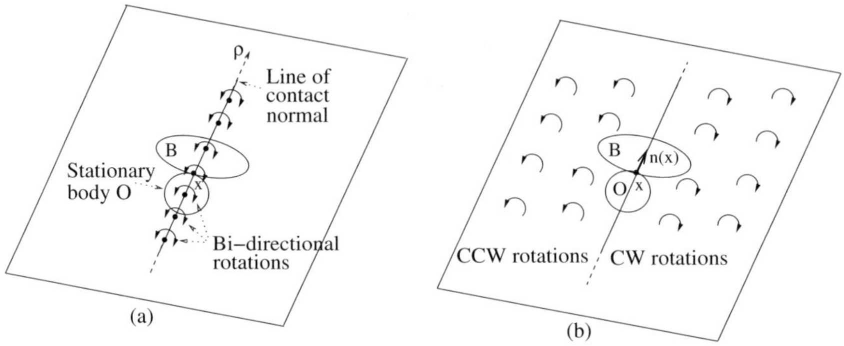

Figure 3.2 (a) $T_{q_0}\mathcal{S}$ consists of instantaneous rotations of $\mathcal{B}$ about all points on the contact normal line. (b) Tangent vectors pointing away from $\mathcal{CO}$ at q_0 are CCW rotations on the left half-plane and CW rotations in the right half-plane, measured with respect to $n(x)$.

where we cancelled out the sign parameter $\sigma = \pm 1$. It follows that $T_{q_0}\mathcal{S}$ consists of instantaneous rotations of $\mathcal{B}$ about all points along the contact normal line (Figure 3.2(a)):

$$T_{q_0}\mathcal{S} = \left\{ \dot{q} = \|\omega\| \begin{pmatrix} p \times \boldsymbol{e} \\ \boldsymbol{e} \end{pmatrix} : \omega \in \mathbb{R},\ p = x + \rho n(x) \text{ for } -\infty \leq \rho \leq \infty \right\},$$

where p parametrizes the contact normal line in terms of a scalar ρ. The resulting characterization of $T_{q_0}\mathcal{S}$ is depicted in Figure 3.2(a). Note that instantaneous rotations of $\mathcal{B}$ about points at infinity represent instantaneous translations of $\mathcal{B}$ along the tangent at the contact point x. Also note that $T_{q_0}\mathcal{S}$ is parametrized by two scalars, ρ and $\|\omega\|$, which is consistent with the fact that $T_{q_0}\mathcal{S}$ is a tangent plane in the 2-D case.

The graphical technique can be extended to represent the half-space of tangent vectors bounded by the c-obstacle tangent plane (Figure 3.2(b)). The half-space of $T_{q_0}\mathbb{R}^m$ bounded by the tangent plane $T_{q_0}\mathcal{S}$ and pointing *away* from the c-obstacle $\mathcal{CO}$ at q_0 consists of all tangent vectors $\dot{q} \in T_{q_0}\mathbb{R}^m$ satisfying the inequality $\eta(q_0) \cdot \dot{q} \geq 0$. In the 2-D case, these are instantaneous *clockwise rotations* about points on the right side of the contact normal line and instantaneous *counterclockwise rotations* about points on the left side of the contact normal line (Figure 3.2(b)). Note that instantaneous rotations about points on the contact normal line are bi-directional, as these rotations correspond to tangent vectors in $T_{q_0}\mathcal{S}$.

Bibliographical Notes

The composite structure of a manifold and its tangent spaces, called the *tangent bundle,* is described in texts such as Do Carmo [1] and Thorpe [2]. Each tangent space possesses a dual *cotangent space.* This notion can be explored in the context of Euclidean vector spaces using texts such as Fleming [3]. The representation of the action of a covector on vectors in terms of the Euclidean inner product is called *lowering or raising indices.*

The extension of these notions to manifolds is described in texts such as Do Carmo [1] and Guilleman and Pollack [4]. Cotangent vectors are alternatively called *differential 1-forms* on the manifold. Advanced texts on *geometric mechanics,* such as [5] and [6], discuss the tangent and cotangent bundles of configuration space manifolds.

Line geometry developed throughout the eighteenth, nineteenth and twentieth centuries. Starting with the works of Mozzi [7] (1763) and Chasles [8] (1830) on the representation of rigid-body velocities as instantaneous helicoidal motions, and the works of Poinsot [9] (1834) and von Mises [10] (1924) on the representation of rigid-body wrenches as screw-like action. The graphical representation of the instantaneous motions of a rigid object $\mathcal{B}$ in contact with a stationary body is due to Reuleaux [11] (1876). The terms *wrenches* and *twists* were introduced by Ball [12] (1900), in his treatise on screw lines and their reciprocal product. Freudenstein applied line geometry tools to mechanism analysis during the first half of the twentieth century. Line geometry is the main tool used to analyze robotic linkages, as discussed by Roth in "Screws, Motors, and Wrenches that Cannot be Bought in a Hardware Store" [13]. For instance, Merlet [14] and Simaan and Shoham [15] use line geometry to characterize parallel-chain mechanisms. Recommended texts on line geometry in mechanism theory are by McCarthy [16] and Selig [17].

Exercises

Section 3.1

Exercise 3.1: The tangent space to $\mathcal{CO}$'s boundary at q_0 is given by $T_{q_0}\mathcal{S} = \{\dot{q} \in T_{q_0}\mathbb{R}^m : \eta(q_0)\cdot\dot{q} = 0\}$, where $\eta(q_0)$ is the c-obstacle outward normal at q_0. Verify that $T_{q_0}\mathcal{S}$ is the collection of instantaneous motions along which $\mathcal{B}$'s contact point moves tangentially with respect to the stationary body $\mathcal{O}$.

Solution: Since $\eta(q_0) = DX_b(q_0)^T n(x)$ and $\dot{x} = DX_b(q_0)\dot{q}$ by the chain rule, $T_{q_0}\mathcal{S} = \{\dot{q} : \eta(q_0)\cdot\dot{q} = n(x)\cdot\dot{x} = 0\}$. ○

Section 3.2

Exercise 3.2: Let a force f act on $\mathcal{B}$ at a point x. Interpret f as a covector acting on the velocities $\dot{x}$ of a point mass at x. Interpret the wrench formula as a transformation from $T_x^*\mathbb{R}^n$ to $T_{q_0}^*\mathbb{R}^m$.

Exercise 3.3*: Let $q \in \mathbb{R}^m$ and $\bar{q} \in \overline{I\!R}^m$ be two c-space parametrizations associated with two choices of world and body frames, related by fixed rigid-body transformations (d_W, R_W) and (d_B, R_B). Derive the coordinate transformation $q = F(\bar{q})$ between the two parametrizations.

Exercise 3.4: Verify that under the coordinate transformation $q = F(\bar{q})$ associated with two choices of world and body frames, $\bar{\boldsymbol{w}}$ represents the same linear function as $\boldsymbol{w}$.

Solution: Based on the tangent and cotangent vector transformation rules, let $\dot{q} = DF(\bar{q}_0)\dot{\bar{q}}$ and $\bar{w} = DF^T(\bar{q}_0)w$. Hence, $w \cdot \dot{q} = w^T\left(DF(\bar{q}_0)\dot{\bar{q}}\right) = \left(w^T DF(\bar{q}_0)\right)\dot{\bar{q}} = \bar{w} \cdot \dot{\bar{q}}$. ○

Exercise 3.5*: Show that the virtual work principle can be used to obtain the c-obstacle normal, $\eta(q) = DX_b(q)^T n(x)$, where $DX_b(q)$ is the Jacobian of the rigid-body transformation.

Solution: Let the body $\mathcal{B}$ move while maintaining contact with a stationary body $\mathcal{O}$, and let q be any configuration of $\mathcal{B}$ along this motion. In the case of a *frictionless* contact, the force acting on $\mathcal{B}$ is collinear with $\mathcal{B}$'s inward unit normal at x, $f = \|f\|n(x)$. Since $\mathcal{B}$ maintains continuous contact with $\mathcal{O}$, the contact point velocity is tangent to $\mathcal{O}$ at x, implying that $f \cdot \dot{x} = \|f\|(n(x) \cdot \dot{x}) = 0$ during this motion. It follows from the virtual work principle that $w \cdot \dot{q} = 0$. Since this argument holds for *all* $\dot{q} \in T_q\mathcal{S}$, the wrench $w = \|f\|DX_b(q)^T n(x)$, seen as a fixed tangent vector at q_0, is orthogonal to the tangent space $T_q\mathcal{S}$. Indeed, the c-obstacle normal formula, which was derived based on purely geometric considerations, is given by $\eta(q) = DX_b(q)^T n(x)$. ○

Exercise 3.6: Let $(\mathcal{F}_W, \mathcal{F}_B)$ and $(\bar{\mathcal{F}}_W, \bar{\mathcal{F}}_B)$ be two choices of world and body frames. Let $q \in \mathbb{R}^m$ and $\bar{q} \in \bar{\mathbb{R}}^m$ be the c-space coordinates associated with the two choices of frames, chosen such that q_0 and $\bar{q}_0$ are $\mathcal{B}$'s nominal configuration in the two c-space parametrizations. The wrench formula is $w = DX_b(q_0)^T f$ and $\bar{w} = DX_{\bar{b}}(\bar{q}_0)^T \bar{f}$ with respect to the two choices of frames, where $DX_b(q)$ is the Jacobian of the rigid-body transformation. Verify that the wrench formula is *independent* on the choice of world and body frames.

Solution: Since $w \cdot \dot{q} = f^T DX_b(q_0)\dot{q} = f \cdot \dot{x}$ and $\bar{w} \cdot \dot{\bar{q}} = \bar{f}^T DX_{\bar{b}}(\bar{q}_0)\dot{\bar{q}} = \bar{f} \cdot \dot{\bar{x}}$, the wrench formula is frame invariant if $f \cdot \dot{x} = \bar{f} \cdot \dot{\bar{x}}$. Let $\mathcal{F}_W$ and $\bar{\mathcal{F}}_W$ be related by a fixed translation and rotation (d_W, R_W), so that points in the two frames are related by the transformation $x = R_W\bar{x} + d_W$. The Jacobian of this transformation is the constant matrix R_W, and by the chain rule, $\dot{x} = R_W\dot{\bar{x}}$. Similarly, f and $\bar{f}$ are related by the transformation $f = R_W\bar{f}$. Thus, $f \cdot \dot{x} = \bar{f}^T R_W^T R_W\dot{\bar{x}} = \bar{f} \cdot \dot{\bar{x}}$, implying that the wrench formula is independent on the choice of frames. ○

Exercise 3.7: A drill applies on a rigid object $\mathcal{B}$ contact forces $f_1 = \frac{1}{2}f_n - f_t$ and $f_2 = \frac{1}{2}f_n + f_t$, at points located at distances $\pm b$ from the drill's axis. Explain why the net wrench applied by the drill on $\mathcal{B}$ is *not* a simple single-force wrench.

Section 3.3

Exercise 3.8: The *Plücker coordinates* of a directed line l in $\mathbb{R}^3$ is the vector $(\hat{l}, p \times \hat{l}) \in \mathbb{R}^6$. Provide a geometric interpretation for the term $p \times \hat{l}$.

Solution: The term $p \times \hat{l}$, traditionally called the line's "arm," is the vector orthogonal to l and having its tip on l. ○

Exercise 3.9: Prove the *reciprocal product formula* specified in Proposition 3.3.

Solution: By definition of the reciprocal product, $(\hat{l}_1, p_1 \times \hat{l}_1) \cdot (p_2 \times \hat{l}_2, \hat{l}_2) = \hat{l}_1 \cdot (p_2 \times \hat{l}_2) + (p_1 \times \hat{l}_1) \cdot \hat{l}_2$. Since $\hat{l}_1 \cdot (p_2 \times \hat{l}_2) = -p_2 \cdot (\hat{l}_1 \times \hat{l}_2)$ and $(p_1 \times \hat{l}_1) \cdot \hat{l}_2 = p_1 \cdot (\hat{l}_1 \times \hat{l}_2)$ by the triple scalar product identity, $(\hat{l}_1, p_1 \times \hat{l}_1) \cdot (p_2 \times \hat{l}_2, \hat{l}_2) = -(p_2 - p_1) \cdot (\hat{l}_1 \times \hat{l}_2)$. In the latter inner product, $\hat{l}_1 \times \hat{l}_2$ is orthogonal to l_1 and l_2. Since p_1 and p_2 may freely vary along l_1 and l_2 without affecting the inner product, we may assume that the segment $p_2 - p_1$ is collinear with $\hat{l}_1 \times \hat{l}_2$ so that $\|p_2 - p_1\| = \delta$. Based on this argument, $-(p_2 - p_1) \cdot (\hat{l}_1 \times \hat{l}_2) = -\sigma \|p_2 - p_1\| \|\hat{l}_1 \times \hat{l}_2\| = -\sigma \cdot \delta \sin(\alpha)$, where $\sigma = \pm 1$ is the sign of $(p_2 - p_1) \cdot (\hat{l}_1 \times \hat{l}_2)$ and α is the angle between $\hat{l}_1$ and $\hat{l}_2$. ○

Exercise 3.10: The collection of all directed lines in $\mathbb{R}^3$ forms a smooth four-dimensional manifold in $\mathbb{R}^6$. Justify this statement by locally parametrizing the spatial lines in terms of four scalar parameters.

Solution: Since the collection of spatial lines forms a four-dimensional manifold, every line l is surrounded by a neighborhood of lines parametrized by four scalar parameters. Let l have a direction $\hat{l} \in S^2$ (S^2 being the unit sphere), and let $\mathcal{P}$ be the plane orthogonal to $\hat{l}$ through the origin of $\mathbb{R}^3$. The local neighborhood of l in the four-dimensional manifold can be parametrized by the lines' intersection point with $\mathcal{P}$ in the vicinity of the origin, together with a local neighborhood of directions surrounding $\hat{l} \in S^2$. ○

Exercise 3.11: The *Plücker coordinates* of l are given by $(\hat{l}, p \times \hat{l}) \in \mathbb{R}^6$, where p is a point on l and $\hat{l}$ is the unit direction of l. Since the collection of spatial lines forms a four-dimensional manifold, the Plücker coordinates contain redundancies. Identify the two scalar constraints that are always satisfied by the *Plücker coordinates* in $\mathbb{R}^6$.

Solution: Let $(x_1, \ldots, x_6)$ be the coordinates of $\mathbb{R}^6$. The first three coordinates represent the line's unit-magnitude direction on the unit sphere. Hence, $x_1^2 + x_2^2 + x_3^2 = 1$. Since $\hat{l} \cdot (p \times \hat{l}) = 0$, the second scalar constraint is given by $x_1 x_4 + x_2 x_5 + x_3 x_6 = 0$. ○

Exercise 3.12: Let l_1, l_2, l_3 be three planar lines having 2-D *Plücker coordinates* $l_i = (\hat{l}_i, p_i \times \hat{l}_i)$ for $i = 1, 2, 3$. Prove that the three lines intersect at a common point in $\mathbb{R}^2$ iff

$$\det \begin{bmatrix} \hat{l}_1 & \hat{l}_2 & \hat{l}_3 \\ p_1 \times \hat{l}_1 & p_2 \times \hat{l}_2 & p_3 \times \hat{l}_3 \end{bmatrix} = 0.$$

Solution: Let the three lines be embedded in the $x_3 = 1$ plane of $\mathbb{R}^3$, where (x_1, x_2, x_3) are the coordinates of $\mathbb{R}^3$. Let $\mathcal{P}_1, \mathcal{P}_2, \mathcal{P}_3$ be the planes through the origin of $\mathbb{R}^3$ determined by the three lines (Figure 3.1(b)). The three lines intersect at a common point in $\mathbb{R}^2$ iff $\mathcal{P}_1, \mathcal{P}_2, \mathcal{P}_3$ intersect along a common line in $\mathbb{R}^3$. The latter condition is equivalent to the requirement that the planes' normals be linearly dependent. The pair $\{(p_i, 1), (\hat{l}_i, 0)\}$ is a basis for $\mathcal{P}_i$ $(i = 1, 2, 3)$. Hence the normal to $\mathcal{P}_i$ is given by $(p_i, 1) \times (\hat{l}_i, 0) = (J\hat{l}_i, p_i \times \hat{l}_i)$ $(i = 1, 2, 3)$, where $J = \begin{bmatrix} 0 & 1 \\ -1 & 0 \end{bmatrix}$. The three normals are linearly dependant iff

$$\det \begin{bmatrix} J\hat{l}_1 & J\hat{l}_2 & J\hat{l}_3 \\ p_1 \times \hat{l}_1 & p_2 \times \hat{l}_2 & p_3 \times \hat{l}_3 \end{bmatrix} = -\det \begin{bmatrix} \hat{l}_1 & \hat{l}_2 & \hat{l}_3 \\ p_1 \times \hat{l}_1 & p_2 \times \hat{l}_2 & p_3 \times \hat{l}_3 \end{bmatrix} = 0.$$

○

Exercise 3.13: Determine the screw axis l and the pitch z of the wrench generated by a pure force f acting on $\mathcal{B}$ at a point $x = X_b(q_0)$. Under what condition on the frames $\mathcal{F}_W$ and $\mathcal{F}_B$ the screw line l can be interpreted as the force line through x?

Solution: The wrench generated by a pure force f acting on $\mathcal{B}$ at point $x = X_b(q_0)$ is given by $\boldsymbol{w} = (f, R_0 b \times f)$. Hence its screw axis is the line $l = (\hat{f}, R_0 b \times \hat{f})$. Since $x = R_0 b + d_0$, the frames $\mathcal{F}_W$ and $\mathcal{F}_B$ must share a common origin at q_0. In this case, $d_0 = \vec{0}$, and the wrench can be written as $\boldsymbol{w} = (f, x \times f)$. The screw axis of this wrench is the line $l = (\hat{f}, x \times \hat{f})$, which passes through the contact point x along the force direction $\hat{f}$. ○

Exercise 3.14: Consider the half-space of tangent vectors pointing *away* from $\mathcal{CO}$ at q_0, given by $\{\dot{q} \in T_{q_0}\mathbb{R}^m : \eta(q_0) \cdot \dot{q} \geq 0\}$. Justify the characterization of this half-space as *clockwise rotations* of $\mathcal{B}$ about points on the right side of the contact normal line, and *counterclockwise rotations* of $\mathcal{B}$ about points on the left side of the contact normal line (Figure 3.2(b)).

Solution: Based on Lemma 3.7, $\eta(q_0) \cdot \dot{q} = -\sigma \cdot \|\omega\|\delta$, where σ is the sign of the inner product $(p_2 - p_1) \cdot (\hat{l}_1 \times \hat{l}_2)$. In our case $p_1 = x$, $\hat{l}_1 = (n(x), 0)$, and $\hat{l}_2 = \mathrm{sgn}(\omega)\boldsymbol{e}$. Hence, $\hat{l}_1 \times \hat{l}_2 = \mathrm{sgn}(\omega)(t(x), 0)$, where $t(x)$ is the unit tangent at x such that $\{(t(x), 0), (n(x), 0), \boldsymbol{e}\}$ is a right-handed triplet. The inequality $\eta(q_0) \cdot \dot{q} \geq 0$ is thus equivalent to the inequality $-\mathrm{sgn}(\omega)(p_2 - x) \cdot t(x) \geq 0$. The latter inequality is linear in p_2 and therefore defines two half planes according to the sign of ω. When $\mathrm{sgn}(\omega) = +1$, the instantaneous rotations are counterclockwise, and in this case p_2 lies in the half plane $(p_2 - x) \cdot t(x) \leq 0$. When $\mathrm{sgn}(\omega) = -1$, the instantaneous rotations are clockwise, and in this case p_2 lies in the half plane $(p_2 - x) \cdot t(x) \geq 0$. ○

References

[1] M. P. do Carmo, *Differential Geometry of Curves and Surfaces*. Prentice-Hall, 1976.

[2] J. A. Thorpe, *Elementary Topics in Differential Geometry*. Springer-Verlag, 1979.

[3] W. Fleming, *Functions of Several Variables*. Springer-Verlag, 1987.

[4] V. Guillemin and A. Pollack, *Differential Topology*. Prentice-Hall, 1974.

[5] J. Marsden, *Elementary Classical Analysis*. Freeman, 1974.

[6] A. M. Bloch, *Nonholonomic Mechanics and Control*. Springer-Verlag, 2003.

[7] G. Mozzi, *Discorso Matematico sopra il Rotamento Momentaneo dei Corpi*. Stamperia di Donato Campo, 1763.

[8] R. Chasles, "Note sur les propriéteés géneérales du système de deux corps semblables entreux," *Bulletin des Sciences Mathématique, Astronomiques, Physiques et Chemiques*, vol. 14, 1830.

[9] L. Poinsot, *Théorie Nouvelle de la Rotation des Corps*. Bachelier, 1834.

[10] R. von Mises, "Motorrechnung, ein neues hilfsmittel in der mechanik" ("Motor calculus, a new theoretical device for mechanics"), *Zeitschrift fur Mathematic und Mechanik*, vol. 4, no. 2, pp. 155–181, 1924.

[11] F. Reuleaux, *The Kinematics of Machinery*. Macmillan, 1876, republished by Dover, 1963.

[12] R. S. Ball, *A Treatise on the Theory of Screws*. Cambridge University Press, 1900.

[13] B. Roth, "Screws, motors, and wrenches that cannot be bought in a hardware store," in *Robotics Research*, in Brady and Paul, eds., MIT Press, 679–693, 1984.

[14] J. P. Merlet, "Singular configurations of parallel manipulators and grassmann geometry," in *Geometry and Robotics, Lecture Notes in Computer Science*, vol. 391, Springer-Verlag, pp. 194–212, 1988.

[15] N. Simaan and M. Shoham, "Singularity analysis of a class of composite serial in-parallel robots," *IEEE Trans. on Robotics and Automation*, vol. 7, no. 3, pp. 301–311, 2001.

[16] J. M. McCarthy, *An Introduction to Theoretical Kinematics*. MIT Press, 1990.

[17] J. M. Selig, *Geometrical Fundamentals of Robotics*. Springer-Verlag, 2003.

4 Rigid-Body Equilibrium Grasps

Any robotic grasping system ought to fulfill three requirements. First, the fingers must hold the object at an equilibrium grasp. Second, the grasp should be stable with respect to small external disturbances. Third, the grasp should withstand finite-size disturbance sets serving as a model for the intended application. This chapter focuses on the equilibrium grasp requirement. The chapter begins with a description of the common rigid-body *contact models*. Each contact model is associated with a particular collection of forces that can be transmitted at a rigid-body contact. The chapter next focuses on the multi-contact setting where finger bodies $\mathcal{O}_1, \ldots, \mathcal{O}_k$ hold an object $\mathcal{B}$ at an equilibrium grasp. The chapter introduces the *grasp map,* which gives the net wrenches that can be affected on $\mathcal{B}$ by varying the finger forces at the contacts. The chapter then formulates the equilibrium grasp condition: a k-contact arrangement forms a *feasible equilibrium grasp* if some finger force combination satisfying the contact model constraints can affect a zero net wrench on the object $\mathcal{B}$. The chapter subsequently discusses the role of *internal grasp forces* in generating equilibrium grasps. Finally, a *moment labeling* technique for depicting the net wrenches that can be affected on an object $\mathcal{B}$ through multiple contacts is described.

4.1 Rigid-Body Contact Models

A *contact model* describes the possible forces that can be transmitted between a finger body and the grasped object when the two bodies are in contact. A *rigid-body contact model* describes the possible physical interaction when the two bodies are assumed to be perfectly rigid. This section introduces the most commonly used rigid-body contact models. While no real objects are truly rigid, the rigid-body contact models provide an excellent idealization in many robot grasping applications. Note that fingertips made of compliant material require a consideration of contact deformations. Such contact deformations can be lumped into a soft-point contact model.

In the following description of the contact models, a finger body $\mathcal{O}_i$ is assumed to be in *point contact* with an object $\mathcal{B}$. The contact point location in the fixed world frame is denoted x_i. The object $\mathcal{B}$ is located at a configuration q_0, such that its frame $\mathcal{F}_B$ coincides with the world frame $\mathcal{F}_W$ at q_0 (Figure 4.1). Under this assumption, the *wrench* generated by a force f_i acting on $\mathcal{B}$ at x_i is simply $\boldsymbol{w}_i = (f_i, x_i \times f_i)$ (see Chapter 3). We will describe the inter-body forces using local reference frames attached

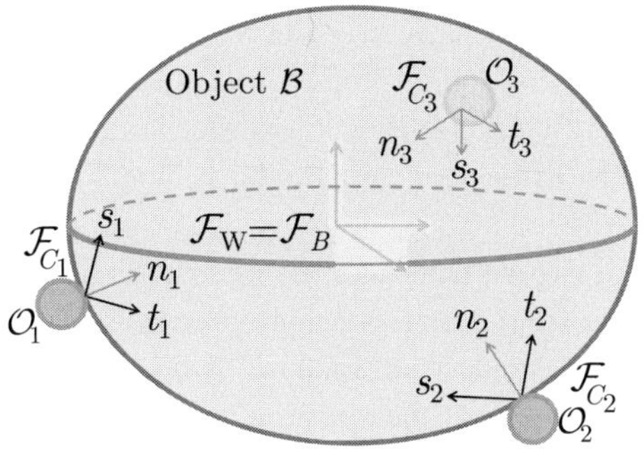

Figure 4.1 Depiction of the world frame $\mathcal{F}_W$, the object frame $\mathcal{F}_B$ and the contact frames $\mathcal{F}_{C_1}$, $\mathcal{F}_{C_2}$ and $\mathcal{F}_{C_3}$.

to $\mathcal{B}$ at the contacts. The ith contact frame, $\mathcal{F}_{C_i}$, is attached to $\mathcal{B}$ at x_i such that its z-axis is aligned with $\mathcal{B}$'s inward unit normal, denoted n_i. The remaining two axes are unit tangents to $\mathcal{B}$ at x_i, denoted s_i and t_i, such that $\{s_i, t_i, n_i\}$ forms a right-handed frame (Figure 4.1). The following are the most common rigid-body contact models.

1 The Frictionless Point Contact Model

The simplest contact model assumes that the contact is entirely frictionless. While no practical contact between two bodies is truly frictionless, this model serves as a conservative approximation in cases of low contact friction and variable surface traction. Additionally, we will see in Part II of the book that many problems in the analysis of frictionless grasps can be answered with purely geometric methods. Therefore, the study of frictionless grasps serves as a convenient starting point for the study of grasp mechanics.

At a frictionless contact, the inter-body force can only be sustained along the direction normal to the contacting bodies' surfaces. The contact force is thus given by $f_i = f_i^n n_i$, where f_i^n is a nonnegative scalar and n_i is $\mathcal{B}$'s inward unit normal at the contact point. Note that $f_i^n \geq 0$, since a rigid finger body can only apply *a unilateral pushing force* at the contact.[1] The collection of contact forces supported by a frictionless point contact is given by

$$C_i = \left\{ f_i \in \mathbb{R}^3 \,:\, f_i = f_i^n n_i,\ f_i^n \geq 0 \right\},$$

where $f_i \in \mathbb{R}^2$ in the case of 2-D grasps. Note that C_i forms a half line, or a one-dimensional cone, based at x_i and pointing along $\mathcal{B}$'s inward unit normal at this point.

2 The Coulomb Friction Point Contact Model

Any real bodies in contact experience friction, which is a force resisting the relative motion of the contacting bodies. The field of *tribology* studies the science and

[1] Grasping systems that incorporate suction devices as well as cables attached to the grasped object are able to realize *unilateral pulling forces* on $\mathcal{B}$.

technologies related to interacting surfaces in motion. Complex tribological models that predict the frictional forces between two bodies may include electrostatic effects, weak chemical bonding effects and mechanical interaction of microscopic asperities. These models, while accurate, are too cumbersome for practical grasp analysis.

Charles-Augustin Coulomb introduced a simple empirical friction model which was derived from experiments like the one depicted in Figure 4.2(a). Consider a block of mass m made from material A (say aluminum), resting under the influence of gravity on a flat surface made from material B (say steel). We assume that the materials have been cleaned of any surface residue and dirt, and that both materials are dry. Starting at time $t = 0$, the block is pulled with a force f_p parallel to the horizontal surface. Imagine that the pulling force is linearly increased from zero. Since the block is under the influence of gravity, a frictional reaction force, f_r, will arise from the interaction of the two contacting surfaces. Figure 4.2(b) shows a typical graph of f_r versus f_p that results from such experiments.

Initially the block remains at rest, even under the influence of an increasing pulling force. This implies that the frictional reaction force f_r counterbalances the pulling force f_p during the initial ramp-up of the applied force. However, as the increasing pulling force reaches a critical magnitude f_p^* (whose value depends on the block's mass and the choice of materials), the block begins to slide in the direction of f_p. Experiments with blocks made of the same material having different masses show that the ratio f_p^*/mg (the object mass multiplied by the gravitational constant) is roughly a constant. The latter ratio is termed the Coulomb *static coefficient of friction* μ_s. Once the block begins to slide, the magnitude of f_r drops slightly but remains constant even when f_p is further increased. The reaction force magnitude during sliding is modeled as the product of a constant dynamic coefficient of friction, μ_d, times $\mathsf{m}g$. In most materials, the difference between μ_s and μ_d is small and, to a reasonable approximation a *single coefficient of friction*, $\mu = \mu_s$, is used to model both effects. Moreover, our analysis of frictional rigid-body grasps will focus on the limits of frictional force that can be obtained when the finger bodies do not slide on the object's surface. Therefore, $\mu = \mu_s$ will form the basis of the frictional contact models used throughout this book.

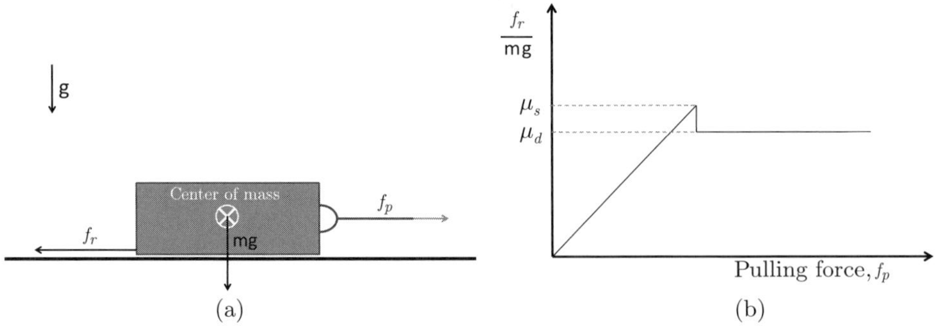

Figure 4.2 (a) A schematic diagram of Coulomb's friction experiment. (b) A typical plot of the normalized reaction force f_r versus the pulling force f_p, showing both static and dynamic friction effects. (Assume the small moment generated by f_p does not incur any tip over motion of the block.)

The coefficient of friction is a nonnegative parameter, $\mu \geq 0$, whose value varies across material types and increases with surface roughness. For metal-on-metal contacts, $0.1 \leq \mu \leq 0.5$. For teflon-on-teflon contacts, μ can be as low as 0.04. A contact between a rubber-coated fingertip and common materials such as plastic, metal or wood achieves high coefficient of friction in the range of $1.0 \leq \mu \leq 2.0$.

Let the *superscripts* (f_i^s, f_i^t, f_i^n) denote the tangential and normal components of the force f_i with respect to the ith contact frame: $(f_i^s, f_i^t) = (f_i \cdot s_i, f_i \cdot t_i)$ and $f_i^n = f_i \cdot n_i$. Under the static Coulomb friction law, the finger body can apply only a unilateral pushing force along $\mathcal{B}$'s inward contact normal, so that $f_i^n \geq 0$. The presence of friction additionally allows the finger body to independently apply tangential force components. However, if the magnitude of the tangential force becomes too large, the finger body will start to slide on the object surface. The Coulomb friction model states that the finger body will *not* slip on the object surface as long as $\| f_i^s s_i + f_i^t t_i \| < \mu_i f_i^n$, where μ_i is the coefficient of friction at the ith contact. Once the finger force violates this condition, the finger body will start to slide on the object surface, with the magnitude of the tangential reaction force satisfying the equality $\| f_i^s s_i + f_i^t t_i \| = \mu_i f_i^n$, such that the direction of the tangential reaction force opposes the finger's sliding direction.

The friction cone: Based on this discussion, the collection of forces supported by a frictional point contact is given in the 3-D case by

$$C_i = \left\{ f_i \in \mathbb{R}^3 \,:\, f_i \cdot n_i \geq 0,\ \sqrt{(f_i \cdot s_i)^2 + (f_i \cdot t_i)^2} \leq \mu_i f_i \cdot n_i \right\},$$

where $\{s_i, t_i, n_i\}$ are the axes of the ith contact frame and μ_i is the coefficient of friction at the ith contact. Since $\sqrt{x^2} = |x|$, the corresponding collection of forces in the 2-D case is given by

$$C_i = \left\{ f_i \in \mathbb{R}^2 \,:\, f_i \cdot n_i \geq 0,\ |f_i \cdot t_i| \leq \mu_i f_i \cdot n_i \right\}.$$

A set C in $\mathbb{R}^n$ forms a *convex cone* if for any vectors $v_1, v_2 \in C$, the linear combination $\lambda_1 v_1 + \lambda_2 v_2$ lies in C for all scalars $\lambda_1, \lambda_2 \geq 0$. The collection of forces C_i has a geometric interpretation in terms of a *friction cone* (Figure 4.3). Consider the convex

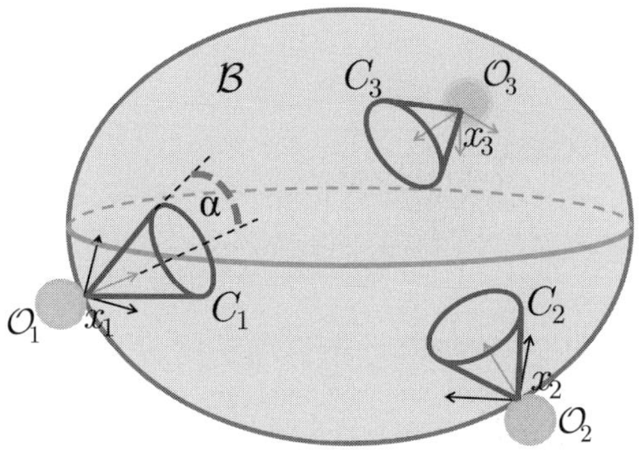

Figure 4.3 The friction cones associated with the frictional point contact model.

cone whose vertex is located at the contact point x_i. The cone's central axis is aligned with $\mathcal{B}$'s inward unit normal at x_i, and the cone's half-angle is $\alpha_i = \tan^{-1}(\mu_i)$. As long as the force applied by the finger lies inside this friction cone, the finger will not slip at the contact. The fingertip will remain *static* on the object surface, usually at a static equilibrium grasp that involves several fingers applying opposing contact forces on the object $\mathcal{B}$.[2]

3 The Soft-Point Contact Model

While this chapter focuses on rigid-body contact models, consider what happens when a fingertip that is made from compliant material comes into contact with a rigid object $\mathcal{B}$. The fingertip material will deform in the vicinity of the contact, resulting in a *contact area* rather than a point contact (Figure 4.4). In the case of 3-D grasps, the contact will be established along a two-dimensional region which can sustain some torsional forces of interaction about the contact normal. The *soft-point contact model* approximates these torsional forces within the rigid-body point contact framework. The potential effect of fingertip softness, adapted to the rigid-body point contact framework, adds to the frictional point contact model an independently modulated torque, τ_i^n, which acts about the contact normal (Figure 4.4). This torque is generated when a fingertip rotation about the contact normal is opposed by the integrative effect of Coulomb friction acting at each point of the contact area. To a rough approximation, the norm of this torque is bounded by $|\tau_i^n| \le \gamma_i f_i^n$, where $\gamma_i > 0$ is the *rotational coefficient of friction* at the ith contact. The collection of forces and torques supported by the *soft-point contact* model is thus given by

$$C_i = \left\{(f_i, \tau_i^n) \in \mathbb{R}^3 \times \mathbb{R} \,:\, f_i \cdot n_i \ge 0,\ \sqrt{(f_i \cdot s_i)^2 + (f_i \cdot t_i)^2} \le \mu_i f_i \cdot n_i,\ |\tau_i^n| \le \gamma_i f_i \cdot n_i\right\},$$

where $\{s_i, t_i, n_i\}$ are the axes of the ith contact frame, and μ_i and γ_i are the two friction coefficients at x_i. Note that C_i forms a *generalized friction cone* in the space $(f_i, \tau_i^n) \in \mathbb{R}^3 \times \mathbb{R}$ of force and torque components based at x_i.

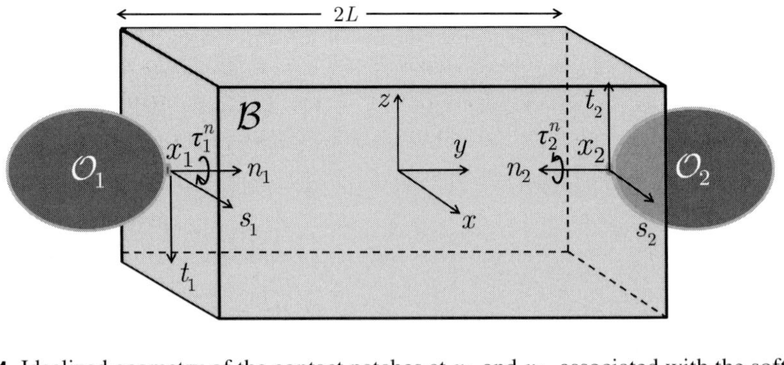

Figure 4.4 Idealized geometry of the contact patches at x_1 and x_2, associated with the soft-point contact model.

[2] A frictional fingertip can also *roll without sliding* on the object surface. However, we will assume sharp-tipped fingers whose contacts remain static on the object surface during such local rolling motions.

The rotational coefficient of friction: While μ_i is a unitless parameter, the relation $|\tau_i^n| \leq \gamma_i f_i^n$ requires that γ_i possess length units. An expression for γ_i can be obtained by assuming that the contact area forms a planar disc of radius R, centered at x_i and tangent to $\mathcal{B}$ (it is actually a planar ellipse). Assume that the normal inter-body force component, f_i^n, is evenly distributed over the contact area (it is actually maximal at x_i and decays to zero on the disc's boundary). The normal force at each point of the contact area is thus $f_n = f_i^n/(\pi R^2)$. Let the fingertip attempt to rotate about the contact normal, n_i, such that all points of the contact area are on the verge of slipping. Assuming that Coulomb friction holds at each point of the contact area, the torque generated at the individual points about n_i satisfies the equality: $|\tau(r)| = \mu_i r f_n$, where $r \in [0, R]$. When this torque is integrated over the contact area, the net torque affected on $\mathcal{B}$ through the contact area is $|\tau_i^n| = \frac{2}{3}\mu_i R f_i^n$. It follows that $\gamma_i = \frac{2}{3}\mu_i R$, which has length units. ○

4.2 The Grasp Map

This section studies the *net wrench* affected on $\mathcal{B}$ by contacting finger bodies $\mathcal{O}_1, \ldots, \mathcal{O}_k$. The ith finger wrench generated on $\mathcal{B}$, $\boldsymbol{w}_i = (f_i, \tau_i)$, has the form:

$$\boldsymbol{w}_i = \begin{pmatrix} f_i \\ x_i \times f_i \end{pmatrix} + \begin{pmatrix} \vec{0} \\ \tau_i^n n_i \end{pmatrix} = \begin{bmatrix} I \\ [x_i \times] \end{bmatrix} f_i + \begin{pmatrix} \vec{0} \\ n_i \end{pmatrix} \tau_i^n, \tag{4.1}$$

where the torque τ_i^n acts only under the soft-point contact model. In the 2-D case, the term $[x_i \times] = x_i^T J$ such that $J = \begin{bmatrix} 0 & 1 \\ -1 & 0 \end{bmatrix}$ forms a row vector in Eq. (4.1). The ith finger force and torque has an associated set of constraints, modeled by the generalized friction cone C_i.

When $\mathcal{B}$ is contacted by finger bodies $\mathcal{O}_1, \ldots, \mathcal{O}_k$, the *net wrench* on $\mathcal{B}$ is simply the sum of the finger wrenches described in Eq. (4.1). The *composite friction cone* for the grasp, defined as the product $C_1 \times \cdots \times C_k$, describes the set of all feasible forces and torques at the contacts.[3] The sum of finger wrenches described in Eq. (4.1) defines the *grasp map,* which maps the composite friction cone to the collection of net wrenches that can be affected on $\mathcal{B}$.

DEFINITION 4.1 (**grasp map**) *Let a rigid object $\mathcal{B}$ located at q_0 be contacted by rigid finger bodies $\mathcal{O}_1, \ldots, \mathcal{O}_k$. Let C_i be the cone of feasible forces and possibly torques at x_i. The rigid body* **grasp map** *is the linear map $\mathcal{L}_G : C_1 \times \cdots \times C_k \to T_{q_0}^* \mathbb{R}^m$ ($m = 3$ or 6) of the form:*

$$\boldsymbol{w} = \begin{bmatrix} I & \cdots & I \\ [x_1 \times] & & [x_k \times] \end{bmatrix} \begin{pmatrix} f_1 \\ \vdots \\ f_k \end{pmatrix} + \begin{bmatrix} \vec{0} & \cdots & \vec{0} \\ n_1 & & n_k \end{bmatrix} \begin{pmatrix} \tau_1^n \\ \vdots \\ \tau_k^n \end{pmatrix}, \tag{4.2}$$

where the first summand appears in all three contact models, while the second summand appears only under the soft-point contact model.

[3] The *product* of sets $\mathcal{S}_1, \ldots, \mathcal{S}_k$ in $\mathbb{R}^n$ is defined as $\mathcal{S}_1 \times \cdots \times \mathcal{S}_k = \{(s_1, \ldots, s_k) \in \mathbb{R}^{kn} : s_k \in \mathcal{S}_k \text{ for } i = 1 \ldots k\}$.

The cotangent space $T^*_{q_0}\mathbb{R}^m$ will be termed the object *wrench space* throughout the book. In practical analyzes throughout the book, we will use the *grasp matrix, $G(q_0)$*, which represents the grasp map in terms of the chosen contact frame bases. Let us obtain the grasp matrix for each of the three contact models. Let p denote the number of independent force and torque components supported by the contact model at each contact. For instance, $p=3$ under the 3-D frictional point contact model. With slight abuse of notation, let $f_i \in \mathbb{R}^p$ denote the vector of force as well as *torque* components at the ith contact. Let $\vec{f}=(f_1,\ldots,f_k)\in\mathbb{R}^{kp}$ denote the composite vector of all force and torque components of the grasp. The $m \times kp$ grasp matrix maps the composite vector $\vec{f}$ to the net wrench affecting $\mathcal{B}$, $\boldsymbol{w}=G(q_0)\vec{f}$. Under the frictionless point contact model, $f_i=f_i^n$ for $i=1\ldots k$, and G forms the $m \times k$ matrix:

$$G(q_0)=\begin{bmatrix} n_1 & \cdots & n_k \\ x_1 \times n_1 & & x_k \times n_k \end{bmatrix}.$$

The grasp matrix associated with frictionless contacts is fully determined by the contact locations and the contact normal directions. Under the frictional point contact model, say in the 3-D case, $f_i=(f_i^s,f_i^t,f_i^n)$ for $i=1\ldots k$, and G forms the $m \times 3k$ matrix:

$$G(q_0)=\begin{bmatrix} G_1(q_0) & \cdots & G_k(q_0) \end{bmatrix} \quad \text{where}$$

$$G_i(q_0)=\begin{bmatrix} s_i & t_i & n_i \\ x_i \times s_i & x_i \times t_i & x_i \times n_i \end{bmatrix} \qquad i=1\ldots k.$$

The grasp matrix associated with frictional contacts is fully determined by the contact locations and the contact frame orientations. Under the soft-point contact model, $f_i=(f_i^s,f_i^t,f_i^n,\tau_i^n)$ for $i=1\ldots k$, and G forms the $m \times 4k$ matrix:

$$G(q_0)=\begin{bmatrix} G_1(q_0) & \cdots & G_k(q_0) \end{bmatrix} \quad \text{where}$$

$$G_i(q_0)=\begin{bmatrix} s_i & t_i & n_i & \vec{0} \\ x_i \times s_i & x_i \times t_i & x_i \times n_i & n_i \end{bmatrix} \qquad i=1\ldots k.$$

Note that the force and torque components, $\vec{f}\in\mathbb{R}^{kp}$, are constrained by the contact model that is assumed to hold at the contacts: $f_i \cdot n_i \geq 0$, $\sqrt{(f_i\cdot s_i)^2+(f_i\cdot t_i)^2}\leq \mu_i f_i \cdot n_i$, and possibly $|\tau_i^n|\leq \gamma_i f_i \cdot n_i$ for $i=1\ldots k$.

Example: Figure 4.5 shows a two-finger grasp of an ellipse along its major axis, whose length is $2L$. Assuming frictional point contacts, each finger can independently modulate two force components so that $p=2$ at each contact. The grasp matrix is determined by the contact positions $x_1=(-L,0)$ and $x_2=(L,0)$, and by the contact frames $\{t_1,n_1\}=\left\{\begin{pmatrix}0\\-1\end{pmatrix},\begin{pmatrix}1\\0\end{pmatrix}\right\}$ and $\{t_2,n_2\}=\left\{\begin{pmatrix}0\\1\end{pmatrix},\begin{pmatrix}-1\\0\end{pmatrix}\right\}$. The *net wrench* affecting the ellipse results from application of the 3×4 grasp matrix on the four independent finger force components at the contacts:

$$\boldsymbol{w}=\begin{bmatrix} G_1 & G_2 \end{bmatrix}\vec{f}=\begin{bmatrix} 0 & 1 & 0 & -1 \\ -1 & 0 & 1 & 0 \\ L & 0 & L & 0 \end{bmatrix}\begin{pmatrix} f_1^t \\ f_1^n \\ f_2^t \\ f_2^n \end{pmatrix} \qquad f_i^n \geq 0,\ |f_i^t|\leq \mu_i f_i^n \ \ i=1,2.$$

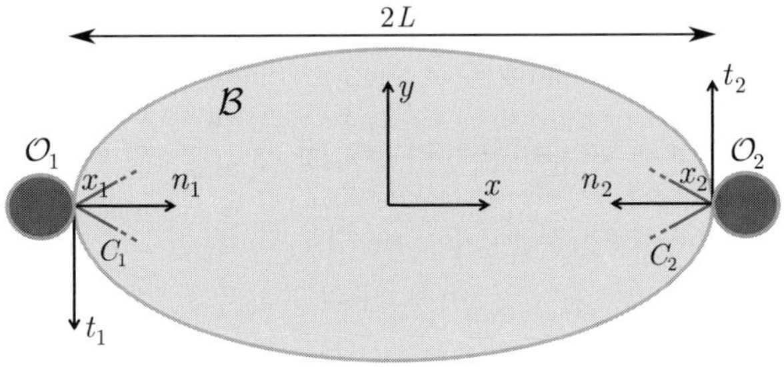

Figure 4.5 A frictional two-finger grasp of an ellipse. The 3×4 grasp matrix G has full row rank at this grasp.

Note that $G = [G_1\ G_2]$ has full row rank, implying that G forms a surjective map onto $\mathcal{B}$'s wrench space.[4] ○

Example: Figure 4.4 shows a two-finger grasp of a rectangular box $\mathcal{B}$ of size $2L$, whose contacts are governed by the soft-point contact model. Each of the two fingers can independently control the normal force component, the two tangential force components and the torque about the contact normal. Thus, $p = 4$ at each contact. The *net wrench* on $\mathcal{B}$ results from application of the 6×8 grasp matrix on the force and torque components:

$$w = \begin{bmatrix} G_1 & G_2 \end{bmatrix} \vec{f} = \begin{bmatrix} 1 & 0 & 0 & 0 & 1 & 0 & 0 & 0 \\ 0 & 0 & 1 & 0 & 0 & 0 & -1 & 0 \\ 0 & -1 & 0 & 0 & 0 & 1 & 0 & 0 \\ 0 & L & 0 & 0 & 0 & L & 0 & 0 \\ 0 & 0 & 0 & 1 & 0 & 0 & 0 & -1 \\ L & 0 & 0 & 0 & -L & 0 & 0 & 0 \end{bmatrix} \begin{pmatrix} f_1^s \\ f_1^t \\ f_1^n \\ \tau_1^n \\ f_2^s \\ f_2^t \\ f_2^n \\ \tau_2^n \end{pmatrix},$$

where $f_i^n \geq 0$, $\sqrt{(f_i^s)^2 + (f_i^t)^2} \leq \mu_i f_i^n$ and $|\tau_i^n| \leq \gamma_i f_i^n$ for $i = 1, 2$. The grasp matrix $G = [G_1\ G_2]$ has full row rank, implying that G forms a surjective map onto $\mathcal{B}$'s wrench space. ○

4.3 The Equilibrium Grasp Condition

This section characterizes the feasible equilibrium grasps of a rigid object $\mathcal{B}$ held by finger bodies $\mathcal{O}_1, \ldots, \mathcal{O}_k$. Consider the simplest scenario where the object $\mathcal{B}$ is influenced solely by the fingers, without any other external influences such as gravity. Subsequent chapters will consider equilibrium grasps in the presence of gravity. A k-contact arrangement forms a *feasible equilibrium grasp* when the fingers can apply forces and torques consistent with the contact model which is assumed to hold at the contacts,

[4] Let A be an $l \times r$ matrix with real entries. The matrix A has *full rank* when $\text{rank}(A) = \min\{l, r\}$, and *full row rank* when $l \leq r$ and $\text{rank}(A) = l$.

such that the net wrench affecting $\mathcal{B}$ is *zero.* A formal definition of an equilibrium grasp follows.

DEFINITION 4.2 (**equilibrium grasp**) *Let a rigid object $\mathcal{B}$ be contacted by rigid finger bodies $\mathcal{O}_1,\ldots,\mathcal{O}_k$ at points $x_1,\ldots,x_k$. The contact arrangement forms a* **feasible equilibrium grasp** *under the frictionless/frictional point contact models if there exist non-zero forces $f_1,\ldots,f_k$ satisfying the condition:*

$$\mathcal{L}_G(f_1,\ldots,f_k)=\begin{pmatrix} f_1 \\ x_1\times f_1\end{pmatrix}+\cdots+\begin{pmatrix} f_k \\ x_k\times f_k\end{pmatrix}=\vec{0}\qquad f_i\in C_i\quad i=1\ldots k \tag{4.3}$$

where $\mathcal{L}_G$ is the grasp map and C_i is the friction cone at x_i for $i=1\ldots k$.

The contact arrangement forms a **feasible equilibrium grasp** *under the soft-point contact model if there exist non-zero forces $f_1,\ldots,f_k$ and torques $\tau_1^n,\ldots,\tau_k^n$ satisfying the condition:*

$$\mathcal{L}_G(f_1,\ldots,f_k;\tau_1^n,\ldots,\tau_k^n)=\begin{pmatrix} f_1 \\ x_1\times f_1+\tau_1^n n_1\end{pmatrix}+\cdots+\begin{pmatrix} f_k \\ x_k\times f_k+\tau_k^n n_k\end{pmatrix}=\vec{0}$$

such that $(f_i,\tau_i^n)\in C_i$ for $i=1\ldots k$, where C_i is the generalized friction cone at x_i for $i=1\ldots k$.

An equilibrium grasp involves a balancing action of opposing forces and possibly torques on the object $\mathcal{B}$. Hence, all equilibrium grasps involve at least two finger contacts when there are no other external influences on $\mathcal{B}$. Equilibrium grasp feasibility requires that $\mathcal{L}_G$ map some non-zero combination of forces and torques to the zero net wrench. This observation highlights an important property of the equilibrium grasps. Recall that the *null space* of a matrix A is the linear subspace satisfying $Av=\vec{0}$. In order to maintain an equilibrium grasp, the grasp matrix of $\mathcal{L}_G$ must possess a nonempty null space, called the subspace of *internal grasp forces.* This subspace will be characterized in the next section.

A deeper insight can be obtained by studying the equilibrium grasps in $\mathcal{B}$'s wrench space. Consider those contact models where the fingers apply pure contact forces on $\mathcal{B}$. The collection of wrenches that can be generated by each finger is defined as follows.

DEFINITION 4.3 (**finger wrench cone**) *Let an object $\mathcal{B}$ located at q_0 be contacted by a finger body $\mathcal{O}_i$. The ith* **finger wrench cone,** *denoted $\mathcal{W}_i$, is the collection of wrenches that can be generated by the ith finger force: $\mathcal{W}_i=\{(f_i,x_i\times f_i)\in T^*_{q_0}\mathbb{R}^m\,:\,f_i\in C_i\}$.*

Each $\mathcal{W}_i$ forms a convex cone based at $\mathcal{B}$'s wrench space origin. For instance, at a planar frictional contact, $\mathcal{W}_i$ forms a two-dimensional sector based at $\mathcal{B}$'s wrench space origin. This sector is bounded by wrenches generated by forces aligned with the two edges of the friction cone C_i (Figure 4.6). When $\mathcal{B}$ is held by several finger bodies, the collection of net wrenches that can be affected on $\mathcal{B}$ forms the grasp's *net wrench cone,* defined as follows.

DEFINITION 4.4 (**net wrench cone**) *Let $\mathcal{B}$ be contacted by finger bodies $\mathcal{O}_1,\ldots,\mathcal{O}_k$. The* **net wrench cone** *is the sum $\mathcal{W}=\mathcal{W}_1+\cdots+\mathcal{W}_k$.*

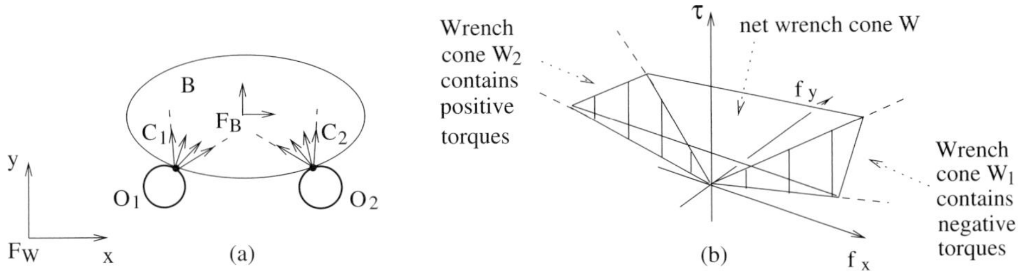

Figure 4.6 (a) A frictional two-contact arrangement in the horizontal (x, y) plane, without gravity. (b) The net wrench cone does not contains a linear subspace, implying that the object is *not* held at a feasible equilibrium grasp.

The sum of convex cones still forms a convex cone (see Exercises). Hence $\mathcal{W}$ forms a convex cone based at $\mathcal{B}$'s wrench space origin. At a feasible equilibrium grasp, the net wrench cone $\mathcal{W}$ contains a linear subspace passing through $\mathcal{B}$'s wrench space origin. This necessary and sufficient condition for equilibrium grasps is stated in the following proposition.

PROPOSITION 4.1 *A rigid object $\mathcal{B}$ is held by finger bodies $\mathcal{O}_1, \ldots, \mathcal{O}_k$ at a feasible equilibrium grasp iff the net wrench cone $\mathcal{W}$ contains a nontrivial* **linear subspace** *at the given grasp.*

Proof: First assume that $\mathcal{B}$ is held at a feasible equilibrium grasp. We may assume that the finger body $\mathcal{O}_1$ is active and hence applies a non-zero force on $\mathcal{B}$ at the equilibrium grasp. The finger wrench $\boldsymbol{w}_1 = (f_1, x_1 \times f_1)$ lies in $\mathcal{W}$. Hence, the equilibrium condition of Eq. (4.3) can be written in the form:

$$-\begin{pmatrix} f_1 \\ x_1 \times f_1 \end{pmatrix} = \begin{pmatrix} f_2 \\ x_2 \times f_2 \end{pmatrix} + \cdots + \begin{pmatrix} f_k \\ x_k \times f_k \end{pmatrix}. \tag{4.4}$$

It follows from Eq. (4.4) that the negated wrench, $-\boldsymbol{w}_1 = -(f_1, x_1 \times f_1)$, lies in the sum $\mathcal{W}_2 + \cdots + \mathcal{W}_k$, which implies that $-\boldsymbol{w}_1$ also lies in $\mathcal{W}$. Since $\boldsymbol{w}_1$ and $-\boldsymbol{w}_1$ both lie in $\mathcal{W}$, the half lines spanned by these wrenches also lie in $\mathcal{W}$. The union of the two half lines forms a one-dimensional subspace, which shows that $\mathcal{W}$ contains at least a one-dimensional linear subspace at a feasible equilibrium grasp.

Next assume that the net wrench cone $\mathcal{W}$ contains a nontrivial linear subspace. Let $\boldsymbol{w}_0 = \sum_{i=1}^{k}(f_i, x_i \times f_i)$ and $-\boldsymbol{w}_0 = \sum_{i=1}^{k}(f_i', x_i \times f_i')$ be two opposing wrenches generated by non-zero contact forces, which necessarily exist in the latter linear subspace. Note that f_i and f_i' are feasible contact forces that lie in the friction cone C_i for $i = 1 \ldots k$. The sum of the two wrenches gives a zero net wrench:

$$\sum_{i=1}^{k} \begin{pmatrix} f_i \\ x_i \times f_i \end{pmatrix} + \begin{pmatrix} f_i' \\ x_i \times f_i' \end{pmatrix} = \sum_{i=1}^{k} \begin{pmatrix} f_i + f_i' \\ x_i \times (f_i + f_i') \end{pmatrix} = \vec{0}.$$

Equilibrium grasp is feasible when the forces $f_i + f_i'$ $(i = 1 \ldots k)$ are not all zero and lie in their respective friction cones. Since each friction cone C_i forms a convex cone,

$f_i, f_i' \in C_i$ implies that $f_i + f_i'$ also lies in C_i. Since each C_i forms a *pointed cone,* $f_i + f_i' = \vec{0}$ only when f_i and f_i' are both zero forces. Since $-\boldsymbol{w}_0$ and $\boldsymbol{w}_0$ are generated by non-zero contact forces, their sums form non-zero contact forces. Hence, the forces $f_1 + f_1', \ldots, f_k + f_k'$ form a feasible equilibrium grasp of $\mathcal{B}$. □

The subspace property is illustrated with two examples.

Examples: Consider the two-contact arrangement depicted in Figure 4.6(a). Assuming *frictional* contacts, the net wrench cone of this grasp, $\mathcal{W} = \mathcal{W}_1 + \mathcal{W}_2$, forms a four-sided *polyhedral cone* in $\mathcal{B}$'s wrench space (Figure 4.6(b)). This contact arrangement does *not* form a feasible equilibrium grasp since the net wrench cone does not contain a linear subspace.

Next consider the three-contact arrangement depicted in Figure 4.7(a). Assuming *frictionless* contacts, each finger wrench cone forms a half line aligned with the finger c-obstacle outward normal at q_0. The net wrench cone of this grasp, $\mathcal{W} = \mathcal{W}_1 + \mathcal{W}_2 + \mathcal{W}_3$, forms a two-dimensional subspace passing through $\mathcal{B}$'s wrench space origin (Figure 4.7(b)). This three-finger grasp forms a feasible equilibrium grasp of $\mathcal{B}$. ○

We end this section with the notion of *essential finger contacts* at an equilibrium grasp.

DEFINITION 4.5 (**essential contact**) *Let an object $\mathcal{B}$ be held at a k-finger equilibrium grasp. A finger contact is said to be* **essential** *for the grasp if the corresponding finger must generate a non-zero contact force in order to maintain the equilibrium grasp.*

The notion of essential finger contact has a geometric interpretation in terms of *pointed cones.* A convex cone is said to be *pointed* when it has a sharp-tipped or strictly convex vertex. A finger contact is *essential* for maintaining the equilibrium grasp when the net wrench cone associated with all remaining finger contacts forms a pointed cone in $\mathcal{B}$'s wrench space. For instance, the removal of any finger contact from the three-finger grasp of Figure 4.7 results in a pointed sector in $\mathcal{B}$'s wrench space. Based on this insight, when all k contacts are essential for maintaining the equilibrium grasp, the net wrench cone $\mathcal{W} = \mathcal{W}_1 + \cdots + \mathcal{W}_k$ must form a $(k-1)$-dimensional linear subspace (assuming $k \leq m + 1$ contacts). For instance, it forms a two-dimensional linear subspace in the three-finger grasp of Figure 4.7.

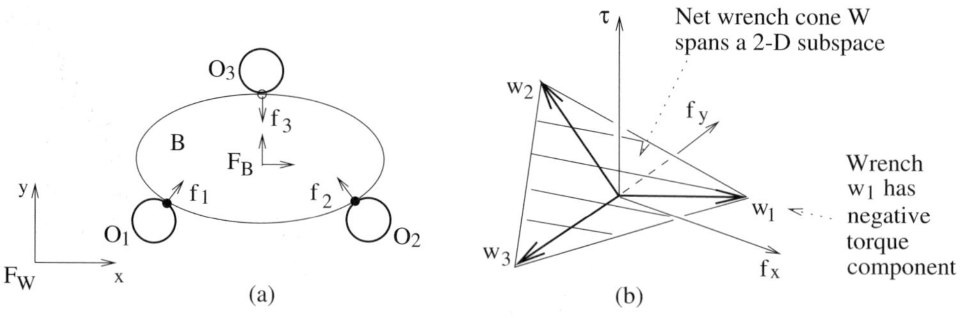

Figure 4.7 (a) Top view of a frictionless three-contact arrangement that forms a feasible equilibrium grasp. (b) The net wrench cone spans a two-dimensional subspace at this grasp.

4.4 The Internal Grasp Forces

We have seen that the grasp matrix, G, has a nonempty null space of contact forces and possibly torques at a feasible equilibrium grasp. That is, there exists a combination of non-zero contact forces and torques, $\vec{f} \in \mathbb{R}^{kp}$, such that $G\vec{f} = \vec{0}$. The forces and torques in the null space of G are called *internal grasp forces*, as these forces and torques are absorbed by the grasped object without disturbing the equilibrium grasp. Let us determine the dimension of the null space of G for the three contact models and then discuss some examples that provide an intuitive feel for these forces. The grasp matrix has dimensions $m \times kp$, where p is fixed by the contact model, $m = 3$ or $m = 6$ and k is the number of contacts.

In the case of *frictionless* contacts, each contact can modulate its normal force component so that $p = 1$ and G is $m \times k$. To determine the dimension of the subspace spanned by the internal grasp forces, consider the case of $k \leq m + 1$ contacts. That is, $2 \leq k \leq 4$ in 2-D grasps and $2 \leq k \leq 7$ in 3-D grasps. For these numbers of contacts, G generically has full rank. Since G possesses a nonempty null space at a feasible equilibrium grasp, it will generically possess a deficient rank at such grasps. The internal grasp forces therefore span a one-dimensional subspace at generic equilibrium grasps having $k \leq m + 1$ frictionless contacts. This subspace consists of a coordinated modulation of the grasp's *total preload,* defined as the totality of the force magnitudes, $f_T = \sum_{i=1}^{k} f_i^n$ (see Exercises). The dimension of the internal grasp force subspace increases for higher numbers of contacts. However, we will see in subsequent chapters that $k \leq m+1$ frictionless contacts are perfectly adequate for almost all rigid-body grasps.

In the case of *frictional* contacts, $p = 2$ in 2-D grasps and $p = 3$ in 3-D grasps. The grasp matrix G is thus $3 \times 2k$ in 2-D grasps and $6 \times 3k$ in 3-D grasps. At the frictional two-contact grasps, the internal grasp forces span a one-dimensional subspace. Much like the frictionless case, these forces can only modulate the total preload f_T. For higher numbers of frictional contacts, the internal grasp forces span a $(2k - 3)$-dimensional subspace in 2-D grasps and a $(3k - 6)$-dimensional subspace in 3-D grasps, as illustrated in the following example.

Example: In the example depicted in Figure 4.8, three disc fingers grasp a triangular object via frictional contacts. The grasp matrix G is 3×6 in this example, and its null space is three-dimensional. The forces (f_1, f_2, f_3) spanning the null space of G are given by

$$\begin{pmatrix} f_1 \\ f_2 \\ f_3 \end{pmatrix} \in \text{span}\left\{ \begin{pmatrix} x_2 - x_1 \\ -(x_2 - x_1) \\ \vec{0} \end{pmatrix}, \begin{pmatrix} \vec{0} \\ x_3 - x_2 \\ -(x_3 - x_2) \end{pmatrix}, \begin{pmatrix} x_1 - x_3 \\ \vec{0} \\ -(x_1 - x_3) \end{pmatrix} \right\}.$$

In contrast with two-finger grasps, the internal grasp forces do not merely modulate the equilibrium force magnitudes, but actually rotate the equilibrium force directions within the friction cones at the contacts. ○

Under the *soft*-point contact model, there exist internal grasp forces as well as internal squeeze torques about the contact normals. The grasp matrix is $6 \times 4k$, and its null space is a $(4k - 6)$-dimensional subspace for $k \geq 2$ contacts. Since $4k - 6 \geq 2$ for $k \geq 2$

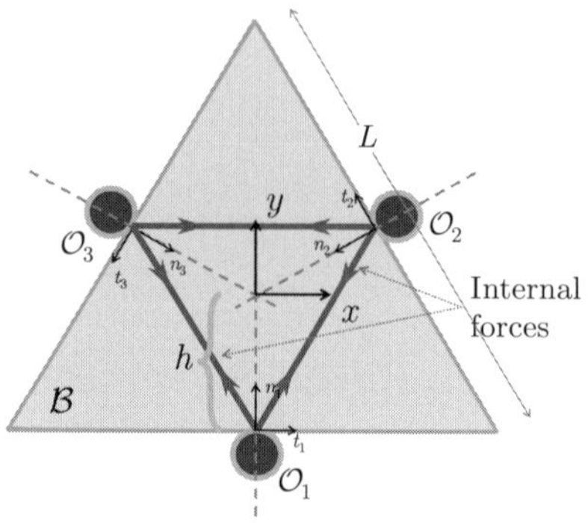

Figure 4.8 Top view of a frictional three-finger equilibrium grasp of a triangular object. The null space of G is three-dimensional. Its basis vectors are opposing forces aligned with the three contact point segments.

contacts, the internal forces and torques span at least a two-dimensional subspace, as illustrated in the following example.

Example: Consider the example of Figure 4.4, where a rectangular box $\mathcal{B}$ is grasped via opposing soft-point contacts. The grasp matrix is 6×8 in this example, and its null space forms a two-dimensional subspace. One basis vector of the null space is given by $(f_1^s, f_1^t, f_1^n, \tau_1^n, f_2^s, f_2^t, f_2^n, \tau_2^n) = (0, 0, 1, 0, 0, 0, 1, 0)$. It corresponds to opposing finger forces squeezing the object $\mathcal{B}$ along the line connecting the two contact points. The other basis vector of the null space is given by $(f_1^s, f_1^t, f_1^n, \tau_1^n,\ f_2^s, f_2^t, f_2^n, \tau_2^n) = (0, 0, 0, 1, 0, 0, 0, 1)$. This basis vector is feasible only under the soft-point contact model, as it consists of opposing finger torques squeezing $\mathcal{B}$ about the line connecting the two contact points. ○

Let us end this section with a physical interpretation of the internal grasp forces.

Structural rigidity interpretation of the internal grasp forces: The dimension of the subspace of internal grasp forces at a k-contact grasp is precisely the minimum number of rigid bars required to intconnect k point masses via rotational joints into a single rigid structure. First consider k point masses in $\mathbb{R}^2$. Two point masses require a single bar to form a rigid structure. Three point masses require three bars to form a rigid structure. Each additional point mass can be joined to the rigid structure using two bars, giving $2k - 3$ for the minimum number of bars required to connect k point masses into a rigid planar structure. This observation can be intuitively explained as follows. Consider a planar graph whose nodes are the k point masses and whose edges are rigid bars connected by *rotational joints* at the nodes. If the graph forms a rigid structure, the bars can absorb any combination of internal grasp forces acting at the k nodes. When any single bar is removed, the structure is incapable of absorbing at least one combination of internal grasp forces. The $2k - 3$ bars thus represent a *basis* for the subspace of internal grasp forces of the k-contact grasp.

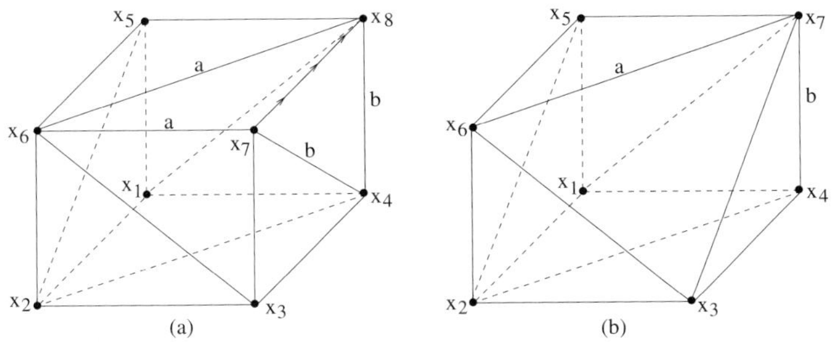

Figure 4.9 (a) A rigid structure connecting $k = 8$ point masses with $3k - 6 = 18$ rigid bars. (b) Its contaction to a rigid structure connecting $k = 7$ point masses having $3k - 6 = 15$ rigid bars.

Next consider k point masses in $\mathbb{R}^3$. One can construct a spatial graph whose nodes are the k point masses and whose edges are rigid bars connected by *spherical joints* at the nodes. That is, each bar ends with a ball that freely rotates inside a spherical socket attached to the point mass. Since each rigid bar has spherical joints at its endpoints, one cannot prevent self-rotation of the bar about its axis. Hence we will construct a spatial graph whose bars freely rotate about their axes, such that the graph as a whole forms a rigid structure. The construction is based on the fact that a triangle with spherical joints forms a rigid structure up to self-rotation of its bars. When k is even, split the k point masses into two sets of $k/2$ points; then embed these sets in two parallel planes (Figure 4.9(a)). Next connect the points of each subset into a rigid planar graph as discussed before. This task can be achieved with a total of $k - 3$ bars per planar graph. Next, arrange the two planar graphs on top of each other, and connect the vertically aligned points with $k/2$ vertical bars. Finally, add a diagonal bar within each vertical rectangle using $k/2$ additional bars. The resulting spatial graph is triangulated and has a total of $2(k - 3) + \frac{k}{2} + \frac{k}{2} = 3k - 6$ bars (Figure 4.9(a)). When k is odd, first construct the spatial graph for an even number of $k+1$ point masses, and then *contract* a pair of adjacent point masses and their connecting bar to a single point mass (x_7 and x_8 in Figure 4.9(b)). During the contraction, identify each pair of bars that connect the two original points with a common third point of the structure into a single bar. The resulting spatial graph is still triangulated and has $3k - 6$ bars. In both cases, the structure is rigid up to self-rotation of the bars and contains the minimum number of bars required for structural rigidity (see Exercises). ○

4.5 The Moment Labeling Technique

This section describes the *moment labeling technique,* which can be used to graphically depict the net wrenches that can be affected on $\mathcal{B}$ by contacting finger bodies $\mathcal{O}_1, \ldots, \mathcal{O}_k$. The technique is described for 2-D grasps, but the underlying principles also apply for 3-D grasps. We will use the technique to obtain additional information on the net wrench cone of a grasp, including graphical detection of equilibrium grasp feasibility.

The moment labeling technique is based on a duality between the sum of convex cones and the intersection of their *polar cones*. Let us first describe this duality and then apply it to our problem. The intersection of two convex cones clearly forms a convex cone. The sum of two convex cones, $W_1 + W_2 = \{w_1 + w_2 : w_1 \in W_1,\ w_2 \in W_2\}$, also forms a convex cone. Given a cone W_i, its *polar* cone W_i' consists of all vectors pointing *away* from W_i:

$$W_i' = \{v \in \mathbb{R}^m \ : w \cdot v \leq 0 \quad \text{for all } w \in W_i\}. \tag{4.5}$$

For instance, W_i depicted in Figure 4.10(a) forms a two-dimensional sector based at the origin of $\mathbb{R}^3$. Its polar cone W_i' forms the three-dimensional wedge depicted in Figure 4.10(a). The following lemma describes the duality between cones and their polar cones.

LEMMA 4.2 (**polar cone duality**) *Let W_1 and W_2 be convex cones based at the origin of $\mathbb{R}^m$, and let W_1' and W_2' be their polar cones in $\mathbb{R}^m$. The sum $W_1 + W_2$ is* **polar** *to the intersection $W_1' \cap W_2'$.*

Proof: Let v be a vector in the intersection of the polar cones, $v \in W_1' \cap W_2'$. By definition of polarity, $w_1 \cdot v \leq 0$ for all $w_1 \in W_1$, and $w_2 \cdot v \leq 0$ for all $w_2 \in W_2$. Hence v satisfies the inequality: $(w_1 + w_2) \cdot v \leq 0$ for all $w_1 + w_2 \in W_1 + W_2$, which implies that $W_1' \cap W_2'$ is polar to $W_1 + W_2$. □

Example: The duality property is illustrated in Figure 4.10(b). The figure shows a four-sided polyhedral cone, $\mathcal{W} = \mathcal{W}_1 + \mathcal{W}_2$, which is the sum of two sectors $\mathcal{W}_1$ and $\mathcal{W}_2$. The polar cone, $\mathcal{W}_1' \cap \mathcal{W}_2'$, forms the downward-pointing polyhedral cone depicted in Figure 4.10(b). It is the intersection of the three-dimensional wedges polar to the sectors $\mathcal{W}_1$ and $\mathcal{W}_2$. ○

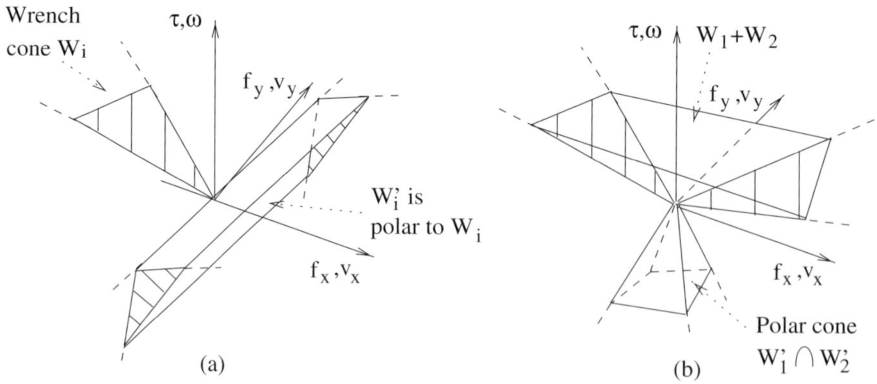

Figure 4.10 (a) A wrench cone $\mathcal{W}_i$ and its polar cone $\mathcal{W}_i'$. (b) The net wrench cone $\mathcal{W} = \mathcal{W}_1 + \mathcal{W}_2$ and its polar cone $\mathcal{W}' = \mathcal{W}_1' \cap \mathcal{W}_2'$.

Let us apply Lemma 4.2 to k-finger grasps. In 2-D grasps $\mathcal{B}$'s configuration space is parametrized by $q \in \mathbb{R}^3$ such that $T_{q_0}\mathbb{R}^3 \cong \mathbb{R}^3$ is the c-space tangent space at q_0. In the following discussion, $\mathcal{W}$ represents a cone of *wrenches* embedded in $T_{q_0}\mathbb{R}^3$, while its polar cone, $\mathcal{W}'$, represents a cone of *tangent vectors* in $T_{q_0}\mathbb{R}^3$. Note that the Euclidean inner product in Eq. (4.5) represents the action, or instantaneous work, of a wrench on tangent vectors.

Let $\mathcal{W}_i$ be the ith finger wrench cone, and let $\mathcal{W}'_i$ be its polar cone in $T_{q_0}\mathbb{R}^3$. Based on Lemma 4.2, the sum $\mathcal{W}_1+\cdots+\mathcal{W}_k$ is *polar* to the intersection $\mathcal{W}'_1\cap\cdots\cap\mathcal{W}'_k$. Therefore, our plan is to first obtain a graphical depiction of the polar cones $\mathcal{W}'_1,\ldots,\mathcal{W}'_k$, then take their intersection, and finally use duality to depict the sum $\mathcal{W}_1+\cdots+\mathcal{W}_k$. In frictional 2-D grasps, each finger wrench cone forms a two-dimensional sector: $\mathcal{W}_i = \{(f_i, x_i \times f_i) : f_i \in C_i\}$, where C_i is the physical friction cone at x_i (Figure 4.10(a)). The friction cone C_i can be written as a positive linear combination of its two edges:

$$C_i = \left\{s_1 f_i^L + s_2 f_i^R \; : \; s_1, s_2 \geq 0\right\},$$

where f_i^L and f_i^R are unit forces aligned with the two edges of C_i. It follows that $\mathcal{W}_i$ can be written as a positive linear combination of the wrenches generated by f_i^L and f_i^R:

$$\mathcal{W}_i = \left\{s_1 \begin{pmatrix} f_i^L \\ x_i \times f_i^L \end{pmatrix} + s_2 \begin{pmatrix} f_i^R \\ x_i \times f_i^R \end{pmatrix} \; : \; s_1, s_2 \geq 0\right\}.$$

Equivalently, $\mathcal{W}_i$ is the sum $\mathcal{W}_i = \mathcal{W}_i^L + \mathcal{W}_i^R$, where $\mathcal{W}_i^L$ and $\mathcal{W}_i^R$ are the half lines collinear with the wrenches $(f_i^L, x_i \times f_i^L)$ and $(f_i^R, x_i \times f_i^R)$. Using $(\mathcal{W}_i^L)'$ and $(\mathcal{W}_i^R)'$ to denote the polar cones of $\mathcal{W}_i^L$ and $\mathcal{W}_i^R$, let us next obtain the polar cone $\mathcal{W}'_i$, as $\mathcal{W}'_i = (\mathcal{W}_i^L)' \cap (\mathcal{W}_i^R)'$.

Let $\mathcal{W}_f$ be the half line of wrenches generated by a unit contact force f acting on $\mathcal{B}$ at x, given by $\mathcal{W}_f = \{s(f, x \times f) : s \geq 0\}$. The polar cone of $\mathcal{W}_f$ has the form: $\mathcal{W}'_f = \{\dot{q} \in T_{q_0}\mathbb{R}^3 : \begin{pmatrix} f \\ x \times f \end{pmatrix} \cdot \dot{q} \leq 0\}$. The polar cone occupies the half-space of $T_{q_0}\mathbb{R}^3$ passing through the origin, whose bounding plane is orthogonal to $(f, x \times f)$ and pointing away from this wrench. As depicted in Figure 4.11(a), the half-space occupied by $\mathcal{W}'_f$ consists of instantaneous *clockwise* rotations of $\mathcal{B}$ on the left side of the force line, *counterclockwise* instantaneous rotations on the right side of the force line, and *bi-directional* instantaneous rotations on the force line itself.

In our case, $\mathcal{W}'_i = (\mathcal{W}_i^L)' \cap (\mathcal{W}_i^R)'$ for $i = 1\ldots k$. When the intersection is performed on $\mathcal{B}$'s instantaneous rotations associated with $(\mathcal{W}_i^L)'$ and $(\mathcal{W}_i^R)'$, we obtain two labeled polygons: a polygon of counterclockwise rotations, denoted M_i^-, and a polygon of clockwise rotations, denoted M_i^+ (Figure 4.11(b)). Note that M_i^- corresponds to *positive* rotations while M_i^+ corresponds to *negative* rotations. The pair (M_i^-, M_i^+) represents the polar cone $\mathcal{W}'_i$. Also note that M_i^- and M_i^+ are constructed as intersections of convex polygons and therefore form convex and connected polygons in $\mathbb{R}^2$. Consider now the intersection of the k negative polygons, $M^- = \cap_{i=1}^k M_i^-$, and the intersection of the k positive polygons, $M^+ = \cap_{i=1}^k M_i^+$. If M^- or M^+ becomes empty during the intersection process, it is marked as an empty set. The pair (M^-, M^+) graphically represents the polar cone $\mathcal{W}' = \mathcal{W}'_1 \cap \cdots \cap \mathcal{W}'_k$.

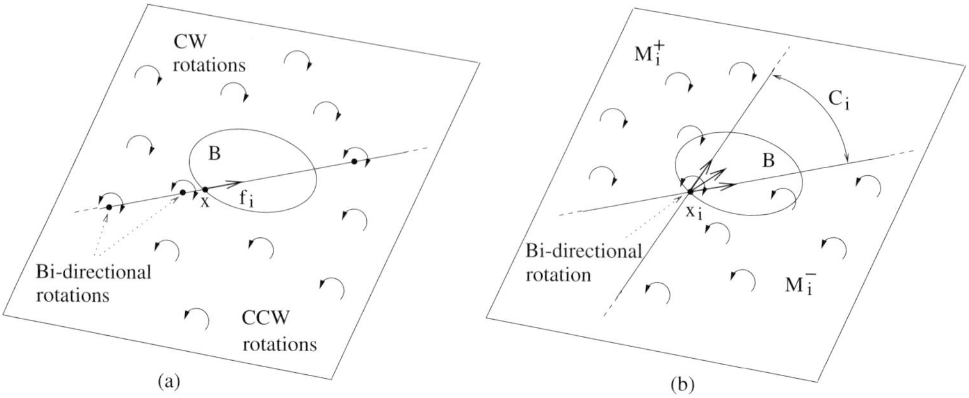

Figure 4.11 (a) The instantaneous rotations of $\mathcal{B}$ that represent the half-space of $T_{q_0}\mathbb{R}^3$ polar to the wrench $(f, x \times f)$. (b) The polygons M_i^- and M_i^+ represent the cone of tangent vectors $\mathcal{W}_i'$ polar to the wrench sector $\mathcal{W}_i$ in $T_{q_0}\mathbb{R}^3$.

Our final step is to depict the net wrench cone, $\mathcal{W} = \mathcal{W}_1 + \cdots + \mathcal{W}_k$, by identifying the meaning of polarity with respect to instantaneous rotations associated with the pair (M^-, M^+). According to Poinsot's theorem (see Chapter 3), any wrench that acts on $\mathcal{B}$ can be parametrized as a directed force line, $(f, p \times f)$, which represents a force f acting on $\mathcal{B}$ along a line passing through a point p in $\mathbb{R}^2$. The following lemma describes the collection of force lines polar to the instantaneous rotations of the pair (M^-, M^+).

LEMMA 4.3 **(moment labeling)** *Let M^- and M^+ be the counterclockwise and clockwise instantaneous rotation polygons, representing the polar cone $\mathcal{W}' = \mathcal{W}_1' \cap \cdots \cap \mathcal{W}_k'$. The net wrench cone, $\mathcal{W} = \mathcal{W}_1 + \cdots + \mathcal{W}_k$, consists of all force lines $(f, p \times f)$ satisfying the condition:*

$$\mathcal{W} = \left\{ \begin{pmatrix} f \\ p \times f \end{pmatrix} : (p - z) \times f \leq 0 \textit{ for all } z \in M^- \textit{ and } (p - z) \times f \geq 0 \textit{ for all } z \in M^+ \right\}.$$

The force lines of $\mathcal{W}$ thus generate **negative or zero** *moments about all points of M^-, and* **positive or zero** *moments about all points of M^+.*

A proof of Lemma 4.3 appears in the chapter's appendix. The following example illustrates the moment labeling technique.

Example: Figure 4.12 shows a frictional two-finger grasp of a rectangular object $\mathcal{B}$. The figure depicts the M^- and M^+ polygons for this grasp. The net wrench cone generated by the two frictional contacts, $\mathcal{W} = \mathcal{W}_1 + \mathcal{W}_2$, corresponds to all force lines passing between M^- and M^+, such that the lines generate negative or zero moments about *all* points of M^- and positive or zero moments about *all* points of M^+. The net wrench cone is bounded by planar facets (Figure 4.10(b)). These planar facets correspond to force lines passing through the vertices of M^- and M^+ (see Exercises). In this example, M^- and M^+ have a total of three vertices, and $\mathcal{W}$ is therefore bounded by three planar facets in $\mathcal{B}$'s wrench space. ○

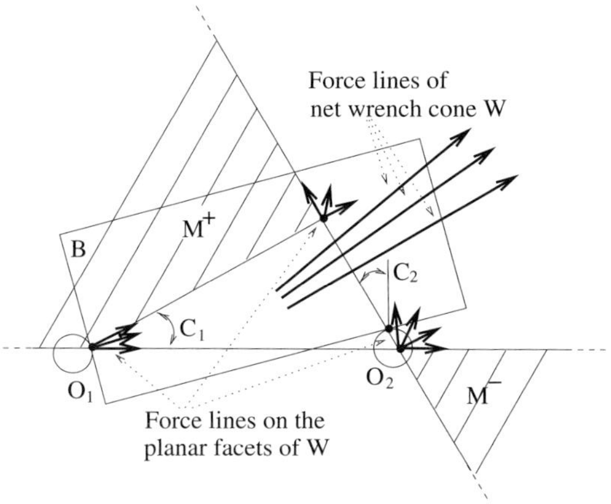

Figure 4.12 A 2-D frictional two-finger contact arrangement. The instantaneous rotation polygons, M^- and M^+, and the force lines spanning the net wrench cone $\mathcal{W} = \mathcal{W}_1 + \mathcal{W}_2$.

Detection of equilibrium grasps: When a contact arrangement forms a feasible equilibrium grasp of $\mathcal{B}$ the net wrench cone $\mathcal{W}$ contains at least a one-dimensional linear subspace (Proposition 4.1). Hence, two different contact force combinations can generate opposing forces that lie on a common line l in $\mathbb{R}^2$. Now choose a positive direction for l. Since the contacts can generate a net force line along l's positive direction, M^- must lie on the *left side* of l, while M^+ must lie on the *right side* of l. Since the contacts can also generate a net force line along l's negative direction, M^- and M^+ must also lie on the *opposite* sides of l. It follows that at an equilibrium grasp M^- and M^+ must form *subsets of some common line l in* $\mathbb{R}^2$. Since M^- and M^+ are convex and therefore connected sets, each of these sets can be an empty set, a single point or a single segment along the common line l. ○

Bibliographical Notes

The frictional point contact model is based on Coulomb's experimental measurements (1781). The friction cone interpretation of Coulomb's law is apparently due to Moseley (1839). The best source for rigid-body contact models in the robotics literature is Salisbury and Roth [1, 2]. One must keep in mind that Coulomb's law assumes *isotropic* finger contact forces. However, robot grasping systems may intentionally use *nonuniform* fingertip surface textures, which are able to realize *anisotropic* friction cones. As for the soft-point contact model, one should be aware that the generalized friction cone may not fully capture the true behavior of this type of contact. Howe, Kao and Cutkosky [3] suggested a more accurate model of the form:

$$C_i = \left\{ (f_i, \tau_i^n) \in \mathbb{R}^3 \times \mathbb{R} : f_i \cdot n_i \geq 0,\ \frac{1}{\mu_i}\big((f_i \cdot s_i)^2 + (f_i \cdot t_i)^2\big) + \frac{1}{\gamma_i}(\tau_i^n)^2 \leq (f_i \cdot n_i)^2 \right\},$$

where $\{s_i, t_i, n_i\}$ are the axes of the ith contact frame, and μ_i and γ_i are the two friction coefficients at x_i. This model ensures that under a given normal loading, the maximal allowed torque about n_i is coupled with the maximal allowed tangential force at the contact.

The linear algebraic modeling of multi-finger grasps in terms of the *grasp matrix* and its linear subspaces dates to Kerr and Roth [4, 5]. This reference as well as Yoshikawa and Nagai [6] discuss the role of the *internal grasp forces* in maintaining equilibrium grasps. Chapter 12 will discuss the role of these forces in maintaining *secure* grasps. Finally, the moment labeling technique is due to Mason [7, 8].

Appendix: Proof of Moment Labeling Lemma

Lemma 4.3 *Let M^- and M^+ be the counterclockwise and clockwise instantaneous rotation polygons, representing the polar cone $\mathcal{W}' = \mathcal{W}'_1 \cap \cdots \cap \mathcal{W}'_k$. The net wrench cone, $\mathcal{W} = \mathcal{W}_1 + \cdots + \mathcal{W}_k$, consists of all force lines $(f, p \times f)$ satisfying the condition:*

$$\mathcal{W} = \left\{ \begin{pmatrix} f \\ p \times f \end{pmatrix} : (p - z) \times f \leq 0 \text{ for all } z \in M^- \text{ and } (p - z) \times f \geq 0 \text{ for all } z \in M^+ \right\}.$$

The force lines of $\mathcal{W}$ thus generate **negative or zero** *moment about all points of M^-, and* **positive or zero** *moment about all points of M^+.*

Proof: Let the planar environment form the horizontal (x, y) plane in $\mathbb{R}^3$, and let $\boldsymbol{e} = (0, 0, 1)$ be the unit upward direction in $\mathbb{R}^3$. Under this embedding, $\mathcal{B}$'s angular velocity vector is given by $\omega \boldsymbol{e}$ for $\omega \in \mathbb{R}$. According to Theorem 3.5, the tangent vectors $\dot{q} = (v, \omega) \in T_{q_0}\mathbb{R}^3$ can be represented as instantaneous rotations of magnitude ω about points z in the (x, y) plane. The parametrization of $\dot{q} \in T_{q_0}\mathbb{R}^3$ in terms of z and ω is as follows. Let b be a fixed point on $\mathcal{B}$, described in the object frame $\mathcal{F}_B$. When $\mathcal{B}$ is located at $q_0 = (d_0, \theta_0)$, the position of b in the world frame $\mathcal{F}_W$ is given by $z = R(\theta_0)b + d_0$. Assume that $d_0 = \vec{0}$. When $\mathcal{B}$ moves along a c-space path $q(t)$ such that $\dot{q} = (v, \omega)$, the velocity of z in the (x, y) plane is given by $\dot{z} = \omega \boldsymbol{e} \times z + v$. The *center of rotation* associated with $\dot{q}$ is the point $z \in \mathbb{R}^2$ whose velocity vanishes: $\dot{z} = \omega \boldsymbol{e} \times z + v = \vec{0}$. Hence, $v = -\omega \boldsymbol{e} \times z$, and the desired parametrization of the tangent space has the form: $\dot{q} = \omega(z \times \boldsymbol{e}, \boldsymbol{e})$ for $(z, \omega) \in \mathbb{R}^2 \times \mathbb{R}$. Substituting for $\dot{q}$ in the polarity condition of Eq. (4.5) gives

$$\begin{pmatrix} f \\ p \times f \end{pmatrix} \cdot \dot{q} = \omega(f, p \times f) \begin{pmatrix} z \times \boldsymbol{e} \\ \boldsymbol{e} \end{pmatrix} = \omega((p - z) \times f)) \cdot \boldsymbol{e} \leq 0. \qquad (4.6)$$

There are two cases to consider. When $z \in M^-$, the object $\mathcal{B}$ executes a *counterclockwise* rotation about z. In this case, $\omega \geq 0$, and the polarity condition of Eq. (4.6) becomes $((p - z) \times f) \cdot \boldsymbol{e} \leq 0$. When $z \in M^+$, the object $\mathcal{B}$ executes a *clockwise* rotation about z. In this case, $\omega \leq 0$, and the polarity condition of Eq. (4.6) becomes $((p - z) \times f) \cdot \boldsymbol{e} \geq 0$. These two requirements describe the force lines of the net wrench cone $\mathcal{W}$. □

Exercises

Section 4.1

Exercise 4.1: Figure 4.13 shows a simple procedure to estimate the Coulomb friction coefficient. Place a block of mass m on an inclined slope that has angle α with respect to the horizontal direction. As the slope angle is slowly increased, the block begins to slide at a critical angle, α^*. Show that α^* is related to the Coulomb friction coefficient, μ, as follows:

$$\alpha^* = \tan^{-1}(\mu). \tag{4.7}$$

Solution: Gravity induces a force normal to the contact surface whose magnitude is $f_n = \mathsf{m}g\cos\alpha$. Similarly, gravity also acts on the block down the slope with a force of magnitude $f_p = \mathsf{m}g\sin\alpha$. If the block is stationary when the slope angle is α, the frictional reaction force, f_r, is equal to f_p, and the contact obeys the no-slip condition of the Coulomb friction law: $|f_r| < \mu f_n$. As the slope angle is slowly increased, the block will begin to slide at a critical angle, α^*. At this critical angle, $f_r = \mu f_n$. This implies that $\mathsf{m}g\sin\alpha^* = \mu\mathsf{m}g\cos\alpha^*$, which gives Eq. (4.7). ○

Exercise 4.2: A *line contact* occurs when a finger body touches an object $\mathcal{B}$ along a line segment. Assume a uniform Coulomb friction coefficient at all points of the line segment. Show that such a line contact is equivalent to a pair of distinct contact points placed at the line segment endpoints.

Solution: The simplest approach would be to discretize the line segment into closely spaced points $p_0, \ldots, p_N$ and discretize the contact force distribution into forces $f_0, \ldots, f_N$ that act at these points. The position of each point p_j can be expressed as a *convex combination* of the segment endpoints, p_0 and p_N:

$$p_j = s_{j1}p_0 + s_{j2}p_N \qquad 0 \le s_{j1}, s_{j2} \le 1, \; s_{j1} + s_{j2} = 1.$$

Using these convex combinations, the net wrench affected on $\mathcal{B}$ by the discretized line segment contact forces is the sum:

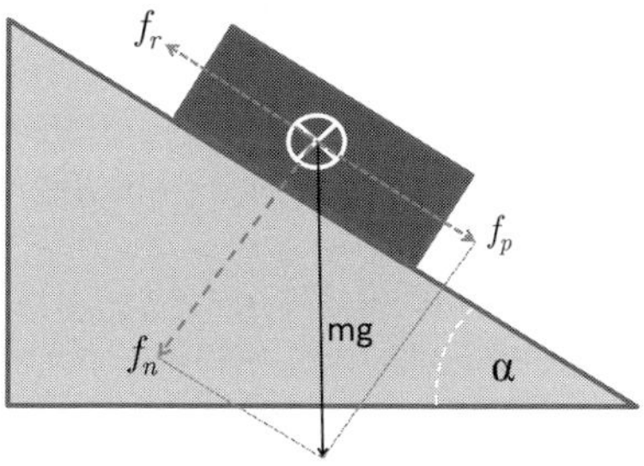

Figure 4.13 An alternative procedure to empirically determine the Coulomb friction coefficient.

$$w = \sum_{j=0}^{N} \begin{pmatrix} f_j \\ p_j \times f_j \end{pmatrix} = \sum_{j=0}^{N} \begin{pmatrix} f_j \\ (s_{j1}p_0 + s_{j2}p_N) \times f_j \end{pmatrix}$$

$$= \sum_{j=0}^{N} \left\{ \begin{pmatrix} s_{j1}f_j \\ p_0 \times s_{j1}f_j \end{pmatrix} + \begin{pmatrix} s_{j2}f_j \\ p_N \times s_{j2}f_j \end{pmatrix} \right\} = \begin{pmatrix} F_1 \\ p_0 \times F_1 \end{pmatrix} + \begin{pmatrix} F_2 \\ p_N \times F_2 \end{pmatrix},$$

where $F_1 = \sum_{j=0}^{N} s_{j1}f_j$ and $F_2 = \sum_{j=0}^{N} s_{j2}f_j$. Since all points of the line segment possess identical friction cones, the equivalent forces F_1 and F_2 lie in the friction cones at p_0 and p_N. ○

Exercise 4.3*: Can the result of the previous exercise be extended to a *plane contact,* where a polyhedral finger body maintains contact with an object $\mathcal{B}$ along a convex flat polygon?

Exercise 4.4: The two-finger grasp of the rectangular box depicted in Figure 4.4 is an example of a *wrench resistant grasp,* to be explored in subsequent chapters. At such grasps, the fingers can theoretically generate any desired net wrench on the object $\mathcal{B}$. List practical limitations to this ambitious statement.

Solution: From the object's perspective, its overall structural strength may limit the magnitude of the allowed finger forces. From the fingers' perspective, they are driven by mechanisms whose actuators are limited by practical considerations such as motor strength and power supply limitations. ○

Section 4.2

Exercise 4.5: Consider a planar two-finger grasp of a rigid object $\mathcal{B}$ via frictional point contacts. Under what condition the grasp map, $\mathcal{L}_G$, ceases to be *surjective* onto $\mathcal{B}$'s wrench space?

Solution: The 3×4 grasp map is given by

$$w = w_1 + w_2 = \begin{bmatrix} I & I \\ x_1^T J & x_2^T J \end{bmatrix} \begin{pmatrix} f_1 \\ f_2 \end{pmatrix} \qquad f_i \in C_i \;\; i = 1, 2$$

where I is a 2×2 identity matrix and $J = \begin{bmatrix} 0 & 1 \\ -1 & 0 \end{bmatrix}$. The upper two rows are spanned by $(1,0,1,0), (0,1,0,1) \in \mathbb{R}^4$. Hence, any linear combination of the upper two rows has the form $(a,b,a,b) \in \mathbb{R}^4$. The lower row can have this form only when $x_1 = x_2$, which is physically unrealizable for rigid finger bodies. Note that two *cables* can be attached through a common anchor point to a rigid platform $\mathcal{B}$, thus allowing for such special cases. ○

Section 4.3

Exercise 4.6: Given convex cones $\mathcal{W}_1, \ldots, \mathcal{W}_k$ based at the origin of $\mathbb{R}^m$, prove that their sum, $\mathcal{W}_1 + \cdots + \mathcal{W}_k$, also forms a convex cone based at the origin of $\mathbb{R}^m$.

Exercise 4.7: A planar object $\mathcal{B}$ is held at a parallel three-finger grasp. Verify that the grasp forms a feasible equilibrium when two forces are unidirectional and the third force has an opposing direction, such that the opposing force line lies in the strip bounded by the unidirectional force lines.

Solution: The finger forces must positively span the origin of $\mathbb{R}^2$ at an equilibrium grasp. Hence two forces must have the same direction, $\hat{f}$, while the third force must have an opposing direction, $-\hat{f}$. Let the two unidirectional forces act at on $\mathcal{B}$ at x_1 and x_2, and let the opposing force act on $\mathcal{B}$ at x_3. The force part of the equilibrium equation, $\lambda_1\hat{f} + \lambda_2\hat{f} - \lambda_3\hat{f} = \vec{0}$, implies that $\lambda_1 + \lambda_2 = \lambda_3$ at an equilibrium. Substituting for λ_3 in the torque part of the equilibrium equation gives $(\lambda_1 x_1 + \lambda_2 x_2) \times \hat{f} - (\lambda_1 + \lambda_2)x_3 \times \hat{f} = 0$. The latter equality holds true when x_3 satisfies the relation: $x_3 = \lambda_1/(\lambda_1 + \lambda_2)x_1 + \lambda_2/(\lambda_1 + \lambda_2)x_2$. Therefore, x_3 must lie in the infinite strip parallel to $\hat{f}$ and bounded by x_1 and x_2. ○

Section 4.4

Exercise 4.8: A rigid object $\mathcal{B}$ is held via $k \leq m + 1$ *frictionless* finger contacts at an equilibrium grasp ($m = 3$ or 6). Characterize the subspace of internal forces at this grasp.

Exercise 4.9: A rigid object $\mathcal{B}$ is held via two *frictional* finger contacts at an equilibrium grasp. Characterize the subspace of internal forces at this grasp.

Exercise 4.10: Consider the frictional three-finger grasp of the equilateral triangle depicted in Figure 4.8. Show that the grasp matrix G has full row rank at this grasp.

Solution: Assign a reference frame whose origin is located at the intersection point of the three contact normal lines, and whose orientation is parallel to that of the contact frame of the finger body $\mathcal{O}_1$. In this reference frame, the grasp matrix takes the form:

$$G = \begin{pmatrix} n_1 & t_1 & n_2 & t_2 & n_3 & t_3 \\ \vec{0} & -\frac{L}{2\sqrt{3}}(n_1 \times t_1) & \vec{0} & -\frac{L}{2\sqrt{3}}(n_2 \times t_2) & \vec{0} & -\frac{L}{2\sqrt{3}}(n_3 \times t_3) \end{pmatrix}$$

$$= \begin{pmatrix} 0 & 1 & -\frac{\sqrt{3}}{2} & -\frac{1}{2} & \frac{\sqrt{3}}{2} & -\frac{1}{2} \\ 1 & 0 & -\frac{1}{2} & -\frac{\sqrt{3}}{2} & -\frac{1}{2} & -\frac{\sqrt{3}}{2} \\ 0 & \frac{L}{2\sqrt{3}} & 0 & \frac{L}{2\sqrt{3}} & 0 & \frac{L}{2\sqrt{3}} \end{pmatrix}, \tag{4.8}$$

where the vector of contact force components is given by $\vec{f} = (f_1^n, f_1^t, f_2^n, f_2^t, f_3^n, f_3^t)$. Note that the first, second, and fourth columns of this matrix are linearly independent vectors, and thus the grasp matrix has full row rank of three. ○

Exercise 4.11: Consider the frictional three-finger grasp of Figure 4.8. Show that the three-dimensional null space of G is spanned by three basis vectors that physically correspond to opposing forces applied along each of the three lines connecting pairs of contact points. (This result generalizes to all frictional three-finger grasps in $\mathbb{R}^2$.)

Solution: Consider the application of equal forces, having magnitude $\|f\|$, along the line connecting the finger contact points x_1 and x_2. The finger contact force components in this case are: $\vec{f} = \|f\| \cdot (\frac{\sqrt{3}}{2}, -\frac{1}{2}, \frac{\sqrt{3}}{2}, \frac{1}{2}, 0, 0)$. Multiplication of the matrix G of Eq. (4.8) by $\vec{f}$ gives a zero net wrench on the triangular object $\mathcal{B}$, confirming that this is an internal force vector. By symmetry, the forces of equal magnitude along the lines connecting the other two possible pairs of contact points can also be shown to be squeeze forces. ○

Exercise 4.12: The internal grasp forces in the three-finger grasp of Figure 4.8 may not lie inside the friction cones at the contacts and, therefore, may not be feasible contact forces. Show that a null space vector consisting of only normal contact forces (all with the same magnitude) is feasible.

Solution: A force of magnitude $\|f\|$ applied along each of the three contact normals, $\vec{f} = \|f\| \cdot (1, 0, 1, 0, 1, 0)$ is a feasible combination of contact forces in the null space of the grasp matrix G in Eq. (4.8). ○

Exercise 4.13: We wish to interconnect k point masses in $\mathbb{R}^2$ by rigid bars via *rotational joints,* such that the structure will become a single rigid body. Verify that $2k - 3$ rigid bars suffice to interconnect k point masses into a single rigid structure.

Solution: When $k = 2$, a single bar connects two point masses into a rigid body. Hence, consider the case of $k \geq 3$ point masses in $\mathbb{R}^2$. Arrange the k points in a circle, and interconnect these points by a circle of k bars. In general, three rigid bars connected by three rotational joints form a rigid triangle. Hence, we can select one point mass on the circle and connect it with every other point mass that is not adjacent to it along the circle. The resulting triangulated structure forms a single rigid body, and the structure has a total of $k + (k - 3) = 2k - 3$ bars. ○

Exercise 4.14*: Under the conditions of the previous exercise, prove that $2k - 3$ is the *minimum* number of rigid bars required to connect k point masses into a single rigid structure.

Solution: A structure made of rigid bars connected by rotational joints has two types of degrees of freedom: *internal* degrees of freedom, associated with rotation of the structure's joints, and *external* degrees of freedom, associated with translation and rotation of the structure as a rigid body in $\mathbb{R}^2$. The total number of internal degrees of freedom can be computed using *Grübler's formula* as follows. Consider a planar structure consisting of n bars attached by rotational joints. The initially disconnected bars have $3n$ degrees of freedom in $\mathbb{R}^2$. Each attachment of two bars with a rotational joint reduces the total number of degrees of freedom by two. Since a planar structure has three external degrees of freedom, the total number of *internal* degrees of freedom is $3(n - 1) - 2j$, where j is the number of pairwise attachments of bars in the structure.

Let us next derive a relation between the number of joints, j, and the number of point masses, k. The structure forms a planar graph whose nodes are the k point masses. The *degree* of the ith node, d_i, is defined as the number of bars attached to the ith node.

The number of rotational joints at the ith node is $d_i - 1$. Hence, the total number of rotational joints is given by $j = \sum_{i=1}^{k}(d_i - 1) = \sum_{i=1}^{k} d_i - k$. Each bar is attached at its endpoints to two nodes. Hence, the sum of the node degrees satisfies the relation $\sum_{i=1}^{k} d_i = 2n$. Substituting $j = 2n - k$ in Grübler's formula gives: $3(n-1) - 2j = 3(n-1) - 2(2n-k) = 2k - n - 3$. The structure forms a rigid body when $2k - n - 3 \leq 0$. Any connection of k point masses into a rigid structure thus requires $n \geq 2k - 3$ bars. ○

Exercise 4.15: Consider the procedure for connecting k point masses in $\mathbb{R}^3$ into a rigid structure consisting of $3k - 6$ rigid bars connected by *spherical joints*. Verify that the structure forms a single rigid body up to self-rotation of the individual bars.

Solution: Since the rigid bars are connected by spherical joints, the structure has two types of *internal* degrees of freedom: self-rotations of the bars about their axes and rotations of the structure's joints while the bars are fixed with respect to their axes. The structure also has six *external* degrees of freedom associated with translation and rotation of the structure as a rigid body in $\mathbb{R}^3$. A spatial n-bar structure forms a rigid body when it has n internal degrees of freedom, as these degrees of freedom correspond to self-rotations of the n bars. The initially disconnected bars have $6n$ degrees of freedom. Each attachment of two bars with a spherical joint reduces the total number of degrees of freedom by three. Based on Grübler's formula, the structure's total number of *internal* degrees of freedom is $6(n-1) - 3j$, where j is the number of pairwise attachments of bars, or joints, in the structure. The procedure uses a total of $n = 3k - 6$ bars to connect the k point masses. Each of the planar graphs is constructed with $\frac{k}{2} + 2(\frac{k}{2} - 3) = \frac{3}{2}k - 6$ joints. The k vertical bars are attached with $2k$ additional joints. The total number of joints is thus $j = 2(\frac{3}{2}k - 6) + 2k = 5k - 12$. Substituting for j in Grübler's formula, then using the relation $3k = n + 6$, gives $6(n-1) - 3j = 6(n-1) - 3(5k-12) = 6(n-1) - 5(n+6) + 36 = n$. Since the structure consists of n bars that can freely rotate about their axes, it forms a rigid body up to self-rotations of its bars. ○

Exercise 4.16*: Prove that $3k - 6$ is the *minimum* number of bars required to connect k point masses via spherical joints into a rigid structure in $\mathbb{R}^3$.

Solution: A spatial structure of rigid bars connected by spherical joints has three types of degrees of freedom. The first type are *self rotations* of the bars about their axes. The second type are *internal* degrees of freedom associated with rotation of the structure's joints while the bars are kept fixed with respect to their axes. The third type are *external* degrees of freedom associated with translation and rotation of the structure as a rigid body in $\mathbb{R}^3$.

Consider a spatial structure consisting of n bars attached by spherical joints. The initially disconnected bars have $6n$ degrees of freedom. Each attachment of two bars with a spherical joint reduces the total number of degrees of freedom by three. Since a spatial structure has six external degrees of freedom and n self-rotation degrees of freedom, the total number of remaining *internal* degrees of freedom is $6(n-1) - n - 3j$, where j is the number of pairwise attachments of bars in the structure.

Let us next derive a relation between the number of joints, j, and the number of point masses, k. The structure forms a graph whose nodes are the k point masses. Recall that the *degree* of the ith node, d_i, is the number of bars attached to the ith node. The number of joints at the ith node is $d_i - 1$. Hence, the total number of spherical joints is given by $j = \sum_{i=1}^{k}(d_i - 1) = \sum_{i=1}^{k} d_i - k$. Each bar is attached at its endpoints to two nodes of the graph. Hence, the sum of the node degrees satisfies the relation $\sum_{i=1}^{k} d_i = 2n$. Substituting $j = 2n - k$ in Grübler's formula gives: $6(n-1) - n - 3j = 6(n-1) - n - 3(2n-k) = 3k - n - 6$. The structure forms a rigid body when $3k - n - 6 \leq 0$. Any connection of the k point masses into a rigid structure thus requires $n \geq 3k - 6$ bars. ○

Section 4.5

Exercise 4.17: Let $\mathcal{W}_f$ be the half line of wrenches spanned by the wrench $\boldsymbol{w}_i = (f, x \times f)$. Prove that the half-space *polar* to $\mathcal{W}_f$ consists of instantaneous clockwise rotations on the left side of the corresponding force line, instantaneous counterclockwise rotations on the right side of the force line and bi-directional rotations on the force line itself (Figure 4.11(a)).

Solution: Let z be $\mathcal{B}$'s instantaneous center of rotation associated with a tangent vector $\dot{q}$. Based on the proof of Lemma 4.3, the tangent vectors $\dot{q} = (v, \omega) \in T_{q_0}\mathbb{R}^3$ can be parametrized in terms of z and ω by the formula: $\dot{q} = \omega(z \times \boldsymbol{e}, \boldsymbol{e})$, where $(z, \omega) \in \mathbb{R}^2 \times \mathbb{R}$ and $\boldsymbol{e} = (0, 0, 1)$. Substituting for $\dot{q}$ in the polarity condition of Eq. (4.5) gives

$$\begin{pmatrix} f \\ x \times f \end{pmatrix} \cdot \dot{q} = \omega(f, x \times f) \begin{pmatrix} z \times \boldsymbol{e} \\ \boldsymbol{e} \end{pmatrix} = \omega((x - z) \times f)) \cdot \boldsymbol{e} \leq 0.$$

Since $\boldsymbol{e}$ is a unit vector perpendicular to the plane, the polarity condition is equivalent to the planar inequality $\omega(x - z) \times f \leq 0$. There are now two cases to consider. When $\omega \leq 0$, the object $\mathcal{B}$ executes a clockwise rotation about z. In this case, $(x - z) \times f \geq 0$, which means that z must lie on the left side of the force line. When $\omega \geq 0$, the object $\mathcal{B}$ executes a counterclockwise rotation about z. In this case, the inequality becomes $(x - z) \times f \leq 0$, implying that z must lie on the right side of the force line. ○

Exercise 4.18: A 2-D object $\mathcal{B}$ is held via two frictional contacts as depicted in Figure 4.12. The net wrench cone of such a grasp, $\mathcal{W} = \mathcal{W}_1 + \mathcal{W}_2$, forms a convex cone based at $\mathcal{B}$'s wrench space origin and bounded by planar facets. Prove that the planar facets of this cone correspond to force lines passing through the vertices of the polygons M^- and M^+.

Solution: Using line geometry, each planar facet of $\mathcal{W}$ lies in a two-dimensional subspace, which correspond to a *flat pencil* of force lines in $\mathbb{R}^2$ (see Chapter 5). Consider now a flat pencil based at x_0. When x_0 is located at a vertex of M^- or M^+, some perturbations of x_0 move it into the interior of either M^- or M^+, while other perturbations move it into the exterior of both polygons. It follows that such a pencil corresponds to a boundary facet of $\mathcal{W}_1 + \mathcal{W}_2$. Note that when x_0 is located at an interior edge point of

M^- or M^+, only a single force line of the pencil is feasible. This force line corresponds to a wrench that forms an *edge* of the net wrench cone $\mathcal{W}$. ○

Exercise 4.19: Consider the frictional two-finger grasp depicted in Figure 4.12. The net wrench cone, $\mathcal{W} = \mathcal{W}_1 + \mathcal{W}_2$, forms a convex cone bounded by three planar facets. Determine which positions of $\mathcal{O}_2$ along the edge of $\mathcal{B}$ transform the net wrench cone into a four-sided polyhedral cone.

Solution: When the contact point of $\mathcal{O}_2$ with $\mathcal{B}$ slides leftward outside the friction cone C_1 located at the opposing contact point, the polygons M^- and M^+ possess four vertices. Since the vertices of M^- and M^+ correspond to the planar facets of $\mathcal{W}$, in this case $\mathcal{W}$ is bounded by four planar facets. ○

Exercise 4.20: Figure 4.14(a) depicts a triangular object $\mathcal{B}$ that is held by three fingers via *frictionless* contacts. The lines underlying the three contact normals intersect at a common point x_0. Construct the polygons M^- and M^+ for this grasp; then determine if it forms a feasible equilibrium grasp. Describe the force lines corresponding to the net wrench cone of this grasp.

Solution: At a frictionless contact x_i, the regions M_i^- and M_i^+ form complementary half planes in $\mathbb{R}^2$, separated by the ith contact normal line. The intersection of these half planes gives $M^- = M_1^- \cap M_2^- \cap M_3^- = \{x_0\}$ and $M^+ = M_1^+ \cap M_2^+ \cap M_3^+ = \{x_0\}$. This is indeed a feasible equilibrium grasp. Since x_0 belongs to both polygons M^- and M^+, the force lines of the net wrench cone must generate zero moment about x_0. Based on the moment formula, $\tau = x_0 \times f$, the feasible force lines can have any direction but must pass through x_0. The resulting collection of force lines forms a *flat pencil* based at x_0. The net wrench cone of this grasp therefore forms a two-dimensional linear subspace in $\mathcal{B}$'s wrench space. ○

Exercise 4.21: Consider a slight shift of the finger $\mathcal{O}_1$ as shown in Figure 4.14(b). Repeat the previous exercise for the perturbed grasp arrangement.

Solution: Now $M^- = \emptyset$, while M^+ is a polygon with a nonempty interior. This contact arrangement is *not* a feasible equilibrium grasp, since no matter how we modulate the

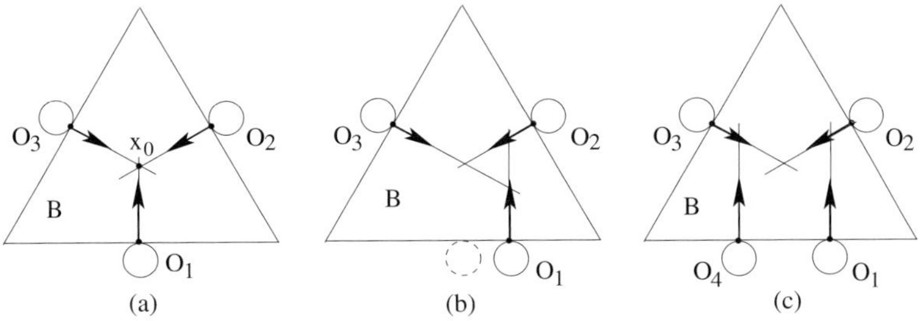

Figure 4.14 (a) A frictionless three-contact grasp with concurrent force lines. (b) A slightly perturbed frictionless three-contact grasp. (c) A frictionless four-contact grasp.

force magnitudes, these forces generate a strictly positive moment about the points of the polygon M^+. ◦

Exercise 4.22: Consider the four-finger grasp involving an additional finger, $\mathcal{O}_4$, depicted in Figure 4.14(c). Repeat the previous exercise for the four-contact arrangement.

Solution: When $\mathcal{O}_4$ is added to the grasp, the polygon M^+ becomes empty. Hence $M^- = M^+ = \emptyset$ at this grasp. This four-contact arrangement therefore forms a feasible equilibrium grasp. Since all force lines are feasible, the net wrench cone occupies the entire wrench space in this grasp. ◦

References

[1] J. K. Salisbury, "Kinematic and force analysis of articulated hands," PhD thesis, Dept. of Mechanical Engineering, Stanford University, 1982.

[2] J. K. Salisbury and B. Roth, "Kinematic and force analysis of articulated mechanical hands," *Journal of Mechanisms, Transmissions and Automation in Design*, vol. 105, no. 1, pp. 35–41, 1983.

[3] R. D. Howe, I. Kao and M. R. Cutkosky, "The sliding of robot fingers under combined torsion and shear loading," in *IEEE International Conference on Robotics and Automation*, pp. 103–105, 1988.

[4] J. R. Kerr, "An analysis of multi-fingered hands," PhD thesis, Dept. of Mechanical Engineering, Stanford University, 1984.

[5] J. R. Kerr and B. Roth, "Analysis of multi-fingered hands," *International Journal of Robotics Research*, vol. 4, no. 4, pp. 3–17, 1986.

[6] T. Yoshikawa and K. Nagai, "Manipulating and grasping forces in manipulation by multifingered robot hands," *IEEE Transactions on Robotics and Automation*, vol. 7, pp. 67–77, 1991.

[7] M. T. Mason, "Two graphical methods for planar contact problems," in *IEEE/RSJ Conference on Intelligent Robots and Systems (IROS)*, pp. 443–448, 1991.

[8] M. T. Mason, *Mechanics of Robotic Manipulation*. MIT Press, 2001.

5 A Catalog of Equilibrium Grasps

Multi-finger equilibrium grasps provide the means by which robot hands hold, transport, and manipulate objects. Equilibrium grasps thus form the most basic component of grasp mechanics. This chapter describes a catalog of the equilibrium grasps in two and three dimensions and in so doing increases the reader's familiarity with these grasp arrangements. The chapter shows how the equilibrium grasp condition of Chapter 4 can be interpreted as linear dependency of the finger force lines. The chapter then applies *line geometry* in order to graphically characterize the basic equilibrium grasps.

This chapter first describes the line geometry interpretation of the equilibrium grasps. Section 5.2 considers 2-D equilibrium grasps involving two, three and four contacts. Section 5.3 is more advanced and considers 3-D equilibrium grasps involving two, three, four and seven contacts. Section 5.4 considers grasps having higher numbers of contacts. The chapter also contains two appendices. Appendix I contains proofs of statements made throughout the chapter. Appendix II describes a useful property of the frictionless equilibrium grasps. Using *contact space* to parametrize the k-contact arrangements, the dimension of the set of feasible equilibrium grasps determines how many contacts can be freely perturbed on the grasped object surface without destroying equilibrium grasp feasibility. We will use this insight to construct equilibrium grasps in this chapter as well as in subsequent chapters.

5.1 Line Geometry of Equilibrium Grasps

We used line geometry to describe rigid-body tangent and cotangent vectors in Chapter 3. Here, line geometry will provide a graphical interpretation of the equilibrium grasps in two and three dimensions. Let a rigid object $\mathcal{B}$ be held by rigid finger bodies $\mathcal{O}_1, \ldots, \mathcal{O}_k$ at a configuration q_0, such that $\mathcal{B}$'s frame coincides with the world frame. Under this assumption, the *wrench* generated by a finger force f_i acting on $\mathcal{B}$ at x_i is given by $\boldsymbol{w}_i = (f_i, x_i \times f_i)$. The ith finger contact force will be parametrized by its magnitude and direction:

$$f_i = \lambda_i \hat{f}_i \qquad \lambda_i \geq 0,$$

where λ_i is the force's magnitude and $\hat{f}_i$ is the force's unit direction. A k-contact arrangement forms a *feasible equilibrium grasp* (in the absence of external influences such as gravity) when there exist feasible finger forces, $f_i \in C_i$ where C_i is the generalized friction at x_i, satisfying the condition:

$$\lambda_1 \begin{pmatrix} \hat{f}_1 \\ x_1 \times \hat{f}_1 \end{pmatrix} + \cdots + \lambda_k \begin{pmatrix} \hat{f}_k \\ x_k \times \hat{f}_k \end{pmatrix} = \vec{0} \qquad \lambda_1, \ldots, \lambda_k \geq 0 \qquad (5.1)$$

such that the coefficients $\lambda_1, \ldots, \lambda_k$ are not all zero. Equilibrium grasps under gravity can be included in this chapter's catalog of equilibrium grasps by interpreting gravity as a *virtual finger* that applies force at $\mathcal{B}$'s center of mass.

Recall from Chapter 3 that the *Plücker coordinates* of a directed spatial line, l, are given by the pair $(\hat{l}, p \times \hat{l}) \in \mathbb{R}^6$, where $\hat{l}$ is the unit direction of l and p is a point on l. Similarly, the *Plücker coordinates* of a directed planar line are given by the pair $(\hat{l}, p \times \hat{l}) \in \mathbb{R}^3$, where $\hat{l}$ is the unit direction of l, p is a point on l, and $p \times \hat{l} = p^T J \hat{l}$ such that $J = \begin{bmatrix} 0 & 1 \\ -1 & 0 \end{bmatrix}$.

DEFINITION 5.1 *The* **finger force line** *associated with a finger force f_i acting on $\mathcal{B}$ at x_i is the line passing through x_i along the direction $\hat{f}_i$, specified as $(\hat{f}_i, x_i \times \hat{f}_i)$ in Plücker coordinates.*

An important observation is that the Plücker coordinates of a finger force line can be interpreted as the *wrench* generated by a unit force $\hat{f}_i$ acting on $\mathcal{B}$ at x_i, $\mathbf{w}_i = (\hat{f}_i, x_i \times \hat{f}_i)$. It follows that the equilibrium grasp condition of Eq. (5.1) is equivalent to *linear dependency* of the finger force lines in *Plücker* coordinates, with the additional requirement that the coefficients $\lambda_1, \ldots, \lambda_k$ must be positive semi-definite.

Linear subspaces of finger force lines: The Plücker coordinates of all planar lines, $(\hat{l}, p \times \hat{l}) \in \mathbb{R}^3$, span a two-dimensional manifold in $\mathbb{R}^3$. The Plücker coordinates of all spatial lines, $(\hat{l}, p \times \hat{l}) \in \mathbb{R}^6$, span a four-dimensional manifold in $\mathbb{R}^6$ (see Chapter 3). On the other hand, the finger force lines satisfying the equilibrium grasp condition of Eq. (5.1) form linearly dependent vectors in the respective vector space. These vectors span a linear subspace when the coefficients $\lambda_1, \ldots, \lambda_k$ vary in $\mathbb{R}^k$. The intersection of the linear subspace with the respective manifold gives a collection of linearly dependent finger force lines. We will use the term *linear subspace of lines* for such collections of linearly dependent finger force lines. ○

The line geometric characterization of the equilibrium grasps is summarized in the next theorem. A set of vectors $v_1, \ldots, v_k \in \mathbb{R}^n$ is said to *positively span* the origin when there exist positive semi-definite scalars $\lambda_1, \ldots, \lambda_k$ that are not all zero, such that $\lambda_1 v_1 + \cdots + \lambda_k v_k = \vec{0}$.

THEOREM 5.1 (**line geometry of equilibrium grasps**) *Let a rigid object $\mathcal{B}$ be held by $k \geq 2$ fingers in $\mathbb{R}^n$, where $n = 2$ in 2-D grasps and $n = 3$ in 3-D grasps. The following two conditions are* **necessary** *for an equilibrium grasp:*

(i) *The finger force directions positively span the origin of $\mathbb{R}^n$.*
(ii) *The finger force lines are linearly dependent in their Plücker coordinates.*

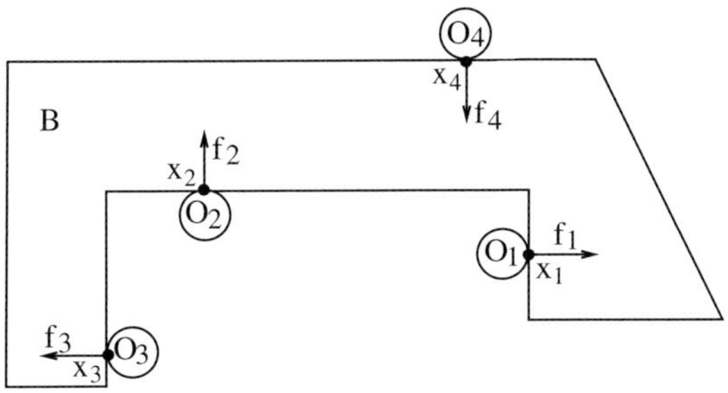

Figure 5.1 Top view of a four-finger grasp that satisfies the necessary conditions of Theorem 5.1 but is *not* a feasible equilibrium grasp.

Proof: Condition (i) is the force component of the equilibrium grasp condition of Eq. (5.1). Condition (ii) is the line geometric interpretation of the equilibrium grasp equation. The two conditions are thus necessary for equilibrium grasp feasibility. □

This chapter's catalog of equilibrium grasps will be largely based on Theorem 5.1. Before discussing the catalog, let us establish that the conditions of Theorem 5.1 are necessary and sufficient for equilibrium grasps involving a small number of contacts.

COROLLARY 5.2 *The conditions of Theorem 5.1 are* **necessary and sufficient** *for equilibrium grasp feasibility of generic 2-D grasps having* $2 \leq k \leq 3$ *contacts and generic 3-D grasps having* $2 \leq k \leq 4$ *contacts.*

The proof of Corollary 5.2 appears in Appendix I. The conditions of Theorem 5.1 are no longer sufficient for equilibrium grasps having a higher number of contacts, as illustrated in the following example.

Example: Figure 5.1 depicts a polygonal object $\mathcal{B}$ that is held by four disc fingers in a horizontal plane without gravity. The finger forces act along $\mathcal{B}$'s inward contact normals, and they positively span the origin of $\mathbb{R}^2$, as required by Condition (i) of Theorem 5.1. Since four planar lines are always linearly dependent in Plücker coordinates, the finger force lines also satisfy Condition (ii) of Theorem 5.1. We will see in Section 3 that at this particular grasp, the finger forces cannot generate zero net torque on $\mathcal{B}$ for any combination of finger force magnitudes. It follows that this contact arrangement is *not* a feasible equilibrium grasp, thus establishing that Conditions (i) and (ii) of Theorem 5.1 may not be sufficient for equilibrium grasp feasibility. ○

5.2 The Planar Equilibrium Grasps

The catalog of planar equilibrium grasps will include two-, three-, and four-finger grasps. Each of these grasps has an important role in certain situations as next described. For a grasp to be useful, the fingers must be able to hold the object $\mathcal{B}$ in a way that can resist external wrench disturbances that may act on $\mathcal{B}$. This notion of *wrench resistance* will be fully explored in Chapter 12. In the case of frictional contacts, two fingers can form wrench resistant grasps by varying their contact forces within the respective friction cones. A stronger notion of grasp security, termed *immobilizing grasps,* relies on the fingers' ability to prevent any motion of the object $\mathcal{B}$ based on rigid-body constraints. In the case of frictionless contacts, three fingers can immobilize almost all 2-D objects, except objects that have parallel edges. Four fingers can immobilize even objects that have parallel edges. Hence, we will fully describe the planar equilibrium grasps involving two, three, and four contacts.

The catalog of planar equilibrium grasps will be based on the following geometric characterization of the *linear subspaces* of directed lines in $\mathbb{R}^2$. A similar characterization of the linear subspaces associated with spatial lines will appear in the discussion of 3-D equilibrium grasps.

Linear Subspaces of Planar Lines

1. A single line in $\mathbb{R}^2$ spans a one-dimensional subspace in Plücker coordinates, consisting of the line itself.
2. Two lines in $\mathbb{R}^2$ span a two-dimensional subspace in Plücker coordinates, consisting of a *flat pencil,* defined as all lines in $\mathbb{R}^2$ that pass through the two lines' intersection point.
3. Three lines in $\mathbb{R}^2$ that do not intersect at a common point span the full three-dimensional space of Plücker coordinates of all planar lines.

When an object $\mathcal{B}$ is held at k-contact equilibrium grasp, the finger force lines must be linearly dependent in their Plücker coordinates according to Theorem 5.1. The finger force lines thus span a $(k-1)$-dimensional linear subspace in Plücker coordinates: a one-dimensional subspace at the two-finger equilibrium grasps, a two-dimensional subspace at the three-finger equilibrium grasps and the entire three-dimensional space at the four-finger equilibrium grasps. We now apply this characterization to the individual planar grasps for $k = 2, 3, 4$ contacts.

1 Planar Two-Finger Equilibrium Grasps

The two-finger equilibrium grasps enjoy a special place in industrial robotics, as these grasps are used by parallel-jaw grippers to pick and place parts during manufacturing operations. Based on line geometry, the force lines at a two-finger equilibrium grasp must lie in a common one-dimensional subspace in Plücker coordinates. A one-dimensional subspace corresponds to a single line in $\mathbb{R}^2$. The finger forces must therefore lie on a common line, which necessarily passes through the two contacts. The following lemma specializes the conditions of Theorem 5.1 to the case of two-finger grasps.

LEMMA 5.3 **(planar two-finger grasps)** *A 2-D object $\mathcal{B}$ is held at a feasible* **two-finger equilibrium grasp** *iff there exist feasible finger forces, $(f_1, f_2) \in C_1 \times C_2$, such that the two forces have opposing directions and lie on a common line passing through the two contacts.*

Note that the finger force magnitudes are equal at any two-finger equilibrium grasp. Lemma 5.3 is next applied to the cases of frictionless and frictional contacts.

1.1 Frictionless Two-Finger Equilibrium Grasps

Let us first clarify which finger forces can be realized at a frictionless contact. At a smooth boundary point of $\mathcal{B}$, the finger force acts along $\mathcal{B}$'s inward normal at this point. At a nonsmooth boundary point of $\mathcal{B}$, such as a polygon vertex, we need the notion of *generalized contact normal.* It is defined as the convex combination of the inward unit normals to the boundary curves adjacent to the vertex (Figure 5.2(a)).[1] At a non-smooth boundary point, any vector from the generalized contact normal can be realized as a physical finger force. The frictionless two-finger equilibria form *antipodal grasps,* according to the following definition.

DEFINITION 5.2 **(antipodal grasp)** *A two-finger grasp of a rigid object $\mathcal{B}$ forms an* **antipodal grasp** *when the two contact normals, or two vectors from the generalized contact normals, lie on the line passing through the contacts and have opposing directions.*

Every piecewise smooth object $\mathcal{B}$ possesses two special antipodal grasps, termed the *minimal* and *maximal* grasps. This property holds for 2-D as well as 3-D objects and is stated here for the case of planar grasps.

PROPOSITION 5.4 *Every piecewise smooth 2-D object $\mathcal{B}$ has at least two antipodal grasps along its outer boundary, the* **minimal** *and* **maximal** *equilibrium grasps* (Figure 5.2(a)).

C-space proof sketch: Consider the c-obstacles, $\mathcal{CO}_1$ and $\mathcal{CO}_2$, induced by the fingers $\mathcal{O}_1$ and $\mathcal{O}_2$ in $\mathcal{B}$'s c-space. Let the two fingers be initially located far apart; then move toward each other along some fixed line l in $\mathbb{R}^2$. As long as the inter-finger distance is sufficiently large, $\mathcal{B}$ can freely move between the two fingers

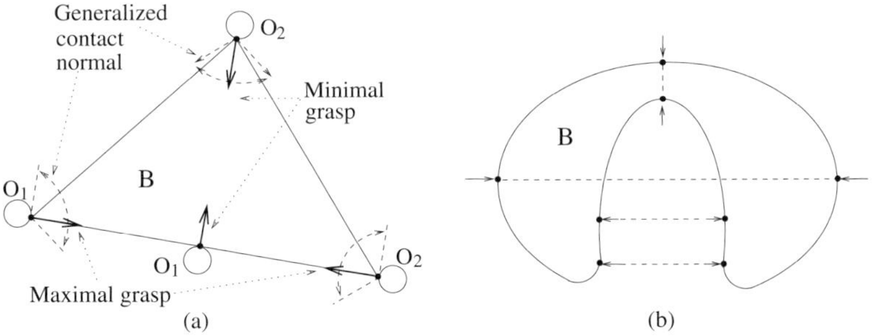

Figure 5.2 (a) Any piecewise smooth object has at least two antipodal grasps. (b) Generic piecewise smooth objects possess a discrete set of antipodal grasps.

[1] When the unit normals are outward pointing with respect to $\mathcal{B}$, the generalized contact normal is precisely the generalized gradient of the distance function, $\mathsf{dst}(x, \mathcal{B}(q_0))$, introduced in Appendix A.

from one side of l to the other. The finger c-obstacles, $\mathcal{CO}_1$ and $\mathcal{CO}_2$, are initially disjoint and approach each other along a fixed translation line while retaining their shape and size in $\mathcal{B}$'s c-space. When $\mathcal{B}$ can touch both fingers for the first time, it touches the two fingers along its *maximum-width segment,* which maximizes the inter-contact distance along $\mathcal{B}$'s boundary. At this instant, the surfaces of $\mathcal{CO}_1$ and $\mathcal{CO}_2$ touch for the first time, at the point q_1 depicted in Figure 5.3(a). Moreover, $\mathcal{CO}_1$ and $\mathcal{CO}_2$ share a common tangent plane and opposing normals at q_1. Each c-obstacle normal can be interpreted as the *wrench* generated by a unit normal force acting on $\mathcal{B}$ at x_i $(i=1,2)$. The maximum-width segment thus determines the maximal grasp of $\mathcal{B}$. As the inter-finger distance keeps decreasing, eventually $\mathcal{B}$ can touch both fingers along its *minimum-width segment.* At this instant, the opening between $\mathcal{CO}_1$ and $\mathcal{CO}_2$ shrinks to the single *puncture point,* q_2, depicted in Figure 5.3(b). Note that $\mathcal{CO}_1$ and $\mathcal{CO}_2$ again share a common tangent plane with opposing normals at q_2. The minimum-width segment thus determines the minimal grasp of $\mathcal{B}$. □

The existence of minimal and maximal equilibrium grasps can also be justified by studying the extrema of the inter-finger distance function, defined as $\sigma(x_1,x_2)=\|x_1-x_2\|$, where x_1 and x_2 are the finger contacts along the boundary of $\mathcal{B}$ (see Exercises). The following example illustrates the minimal and maximal grasps.

Example: Figure 5.2(a) depicts a triangular object, $\mathcal{B}$, together with the generalized contact normals at its vertices. The minimal grasp occurs at a vertex and an opposing edge of $\mathcal{B}$, while the maximal grasp occurs at two opposing vertices of $\mathcal{B}$. Note that objects such as an ellipse possess exactly two antipodal grasps (Figure 5.3). ○

Generic piecewise smooth objects (which include polygonal object without opposing parallel edges) possess only a *finite* set of antipodal grasps. This property holds for 2-D as well as 3-D grasps and is based on a general dimensionality result discussed in Appendix II. This property is next illustrated for the case of 2-D grasps.

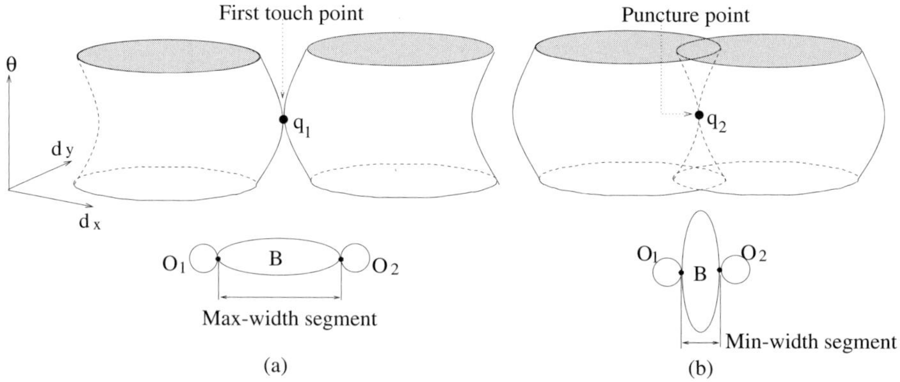

Figure 5.3 Two finger c-obstacles depicted at the maximal and minimal two-finger equilibrium grasps of an elliptical object.

Example: Figure 5.2(b) depicts a smooth object $\mathcal{B}$ having a deep concavity at its center. The object possesses four antipodal grasps. Two of these grasps are said to form *compressive grasps* (the finger forces act inward), while the other two grasps are said to form *expansive grasps* (the finger forces act outward within the object's concavity). ○

1.2 Frictional Two-Finger Equilibrium Grasps

Based on Lemma 5.3, the frictional two-finger equilibrium grasps occur at pairs of contacts whose friction cones contain opposing forces that lie on the line passing through the contacts. This condition generically holds along contiguous boundary segments rather than at discrete boundary points. The frictional two-finger equilibrium grasps are thus *robust* with respect to small finger placement errors. The following simple test determines if a candidate two-contact arrangement forms a feasible equilibrium grasp.

Graphical Test for Frictional Two-Finger Equilibrium Grasps

Let C_i^- denote the reflection of the friction cone C_i with respect to x_i $(i = 1,2)$.
If $x_1 \in C_2$ and $x_2 \in C_1$, or $x_1 \in C_2^-$ and $x_2 \in C_1^-$, a two-finger equilibrium grasp is feasible. Otherwise, a two-finger equilibrium grasp is not feasible.

Example: Consider the compressive equilibrium grasp depicted in Figure 5.4(a). The condition $x_1 \in C_2$ means that the vector pointing from x_1 to x_2 lies in C_1, while $x_2 \in C_1$ means that the opposite vector lies in C_2. Next consider the expansive equilibrium grasp depicted in Figure 5.4(b). The condition $x_1 \in C_2^-$ means that the vector pointing from x_1 to x_2 lies in C_1^-, while $x_2 \in C_1^-$ means that the opposite vector lies in C_2^-. ○

A final useful property is based on the *grasp matrix, G*, which is 3×4 in the case of two frictional contacts (see Chapter 4). Generically, G has full row rank and

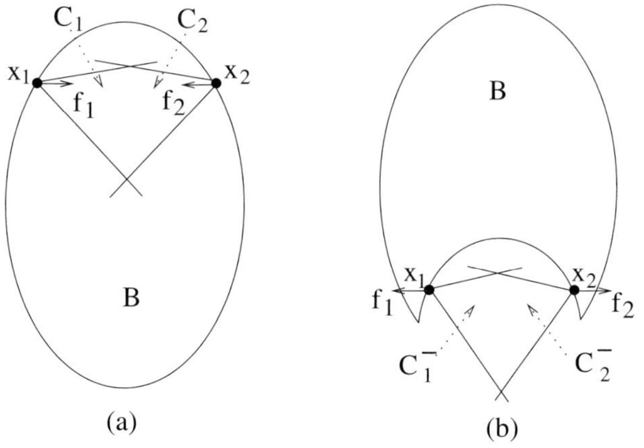

Figure 5.4 (a) A compressive equilibrium grasp satisfying the condition $(x_1 \in C_2, x_2 \in C_1)$. (b) An expansive equilibrium grasp satisfying the condition $(x_1 \in C_2^-, x_2 \in C_1^-)$.

a one-dimensional null space of internal grasp forces. The internal grasp forces, $\vec{f}=(f_1,f_2)$ such that $G\vec{f}=\vec{0}$, consist of opposing finger forces of variable magnitude, which may *not* lie within the friction cones. Since G has full row rank, it forms a *surjective mapping* onto $\mathcal{B}$'s wrench space at such grasps.

2 Planar Three-Finger Equilibrium Grasps

While two-finger robot hands form traditional industrial gripper designs, three-finger hands offer significant advantages as all-purpose minimalistic robot hands, designed for grasping planar objects of diverse shape and size. Based on line geometry, the three-finger equilibrium grasps involve three linearly dependent finger force lines that lie in a common two-dimensional subspace in Plücker coordinates. Since such a subspace forms a *flat pencil*, the finger force lines must intersect at a common point in $\mathbb{R}^2$. This insight leads to the following lemma that graphically characterizes the three-finger equilibrium grasps.

LEMMA 5.5 **(planar three-finger grasps)** *A 2-D object $\mathcal{B}$ is held at a feasible* **three-finger equilibrium grasp** *iff there exist feasible contact forces, $(f_1,f_2,f_3)\in C_1\times C_2\times C_3$, which positively span the origin of $\mathbb{R}^2$, such that the finger force lines intersect at a common point in $\mathbb{R}^2$.*

In the case of parallel finger forces,[2] *two forces point in the same direction while the third force must have an opposing direction and lie in the strip bounded by the unidirectional finger force lines* (Figure 5.5).

The special case of parallel finger forces is considered in the following example.

Example: Consider the three-finger grasp of a rectangular object along two parallel edges depicted in Figure 5.5(a). One finger contacts the bottom edge of $\mathcal{B}$ at x_1, the other two fingers contact the upper edge of $\mathcal{B}$ at x_2 and x_3. The torque component of the

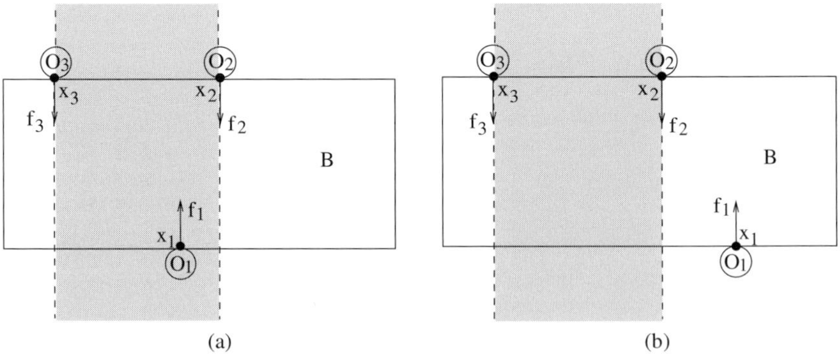

Figure 5.5 Parallel three-finger grasps demonstrating the graphical criterion for equilibrium feasibility: (a) A feasible equilibrium grasp, and (b) an infeasible equilibrium grasp.

[2] Three parallel finger force lines can be thought of as intersecting at infinity.

equilibrium grasp condition, Eq. (5.1), requires that the line underlying the finger force f_1 lie in the strip bounded by the lines underlying the unidirectional finger forces f_2 and f_3. Hence, the three-finger grasp depicted in Figure 5.5(a) forms a feasible equilibrium grasp. Once the opposing force at x_1 moves outside the strip bounded by x_2 and x_3, the fingers cannot generate zero net torque on $\mathcal{B}$ for any combination of finger force magnitudes (Figure 5.5(b)). ○

2.1 Frictionless Three-Finger Equilibrium Grasps

In the case of frictionless contacts, Lemma 5.5 applies to $\mathcal{B}$'s contact normals. A simple technique for constructing frictionless three-finger equilibrium grasps is illustrated in Figure 5.6. The *inscribed disc* of $\mathcal{B}$ is defined as the largest disc contained inside the object $\mathcal{B}$. We will see in Chapter 9 that such a disc generically touches the boundary of $\mathcal{B}$ at two or three isolated points. When the inscribed disc touches the boundary of $\mathcal{B}$ at three points, these points automatically form a three-contact equilibrium grasp (Figure 5.6(a)). When the inscribed disc touches the boundary of $\mathcal{B}$ at two points, any local splitting of one of the two contacts along opposite directions results in a three-contact equilibrium grasp (Figure 5.6(b)). More generally, the frictionless three-finger equilibrium grasps span a two-dimensional manifold in *contact space* (Appendix II). Hence, one can locally vary the position of two contacts, or locally split one contact of a two-finger equilibrium grasp, and then locally move the third contact as to maintain the three-finger equilibrium grasp.

C-space view of frictionless three-finger equilibrium grasps: Three finger bodies induce three c-obstacles in $\mathcal{B}$'s-space. When $\mathcal{B}$ is held by three finger bodies at a configuration q_0, the point q_0 lies at the intersection of the three c-obstacle surfaces (note that the three c-obstacle surfaces generically intersect at discrete points). Since the *wrench* generated by a normal finger force is collinear with the c-obstacle's outward normal at q_0, the c-obstacle outward normals are *linearly dependent* at a frictionless

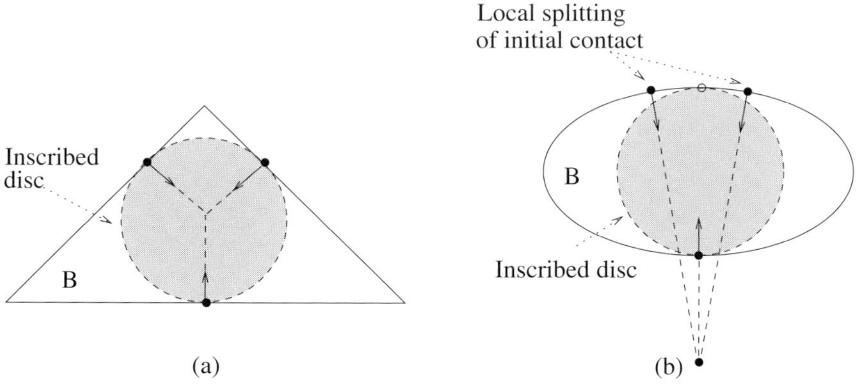

Figure 5.6 (a) The inscribed disc touches the boundary of $\mathcal{B}$ at three points. (b) When the inscribed disc touches the boundary of $\mathcal{B}$ at two points, any local splitting of one of the two contacts gives a three-contact equilibrium grasp.

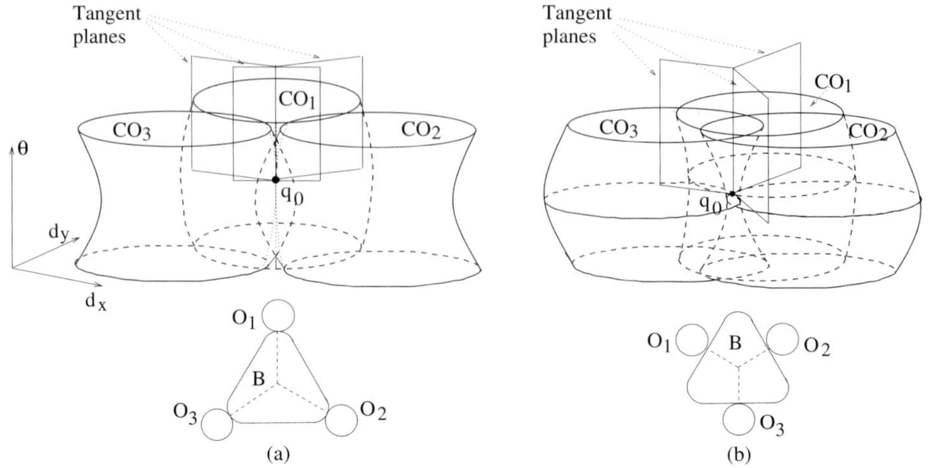

Figure 5.7 At a frictionless three-finger equilibrium grasp, the finger c-obstacle surfaces intersect at q_0 with linearly dependent c-obstacle normals at this point.

equilibrium grasp. As depicted in Figure 5.7, the c-obstacles' tangent planes at q_0 share a common line at the frictionless three-finger equilibrium grasps. The tangent vector $\dot{q} \in T_{q_0}\mathbb{R}^3$ collinear with this line represents pure instantaneous rotation of $\mathcal{B}$ about the intersection point of the three finger force lines. ○

2.2 Frictional Three-Finger Equilibrium Grasps

In the case of frictional contacts, the conditions of Lemma 5.5 attain the following form. At a frictional three-finger equilibrium grasp, the friction cones must contain three force lines that intersect at a common point in $\mathbb{R}^2$, such that their directions positively span the origin of $\mathbb{R}^2$. A graphical procedure for determining if a candidate contact arrangement forms a feasible equilibrium grasp is based on the following lemma (see Exercises).

LEMMA 5.6 *Let a 2-D object $\mathcal{B}$ be held at a frictional three-finger equilibrium grasp. If the finger force directions $\hat{f}_1, \hat{f}_2, \hat{f}_3$ are not all parallel, the contact force magnitudes are given up to a common scaling factor by*

$$\lambda_i = \hat{f}_{i+1} \times \hat{f}_{i+2} \qquad i = 1, 2, 3, \tag{5.2}$$

where index addition is taken modulo 3.

It follows from Lemma 5.6 that $\lambda_i = 0$ when $\hat{f}_{i+1}$ and $\hat{f}_{i+2}$ are parallel. Hence, when $\mathcal{B}$ is held at a feasible equilibrium grasp, $\lambda_i = 0$ implies that the grasp involves only two active contacts at x_{i+1} and x_{i+2}. At such a grasp, $\hat{f}_{i+1}$ and $\hat{f}_{i+2}$ are collinear with the line passing through x_{i+1} and x_{i+2}. Based on this geometric insight, the lines passing through the three contact pairs partition the plane into polygonal cells, such that $\lambda_1, \lambda_2, \lambda_3$ retain their sign within each cell. This *sign invariance* property is the basis for

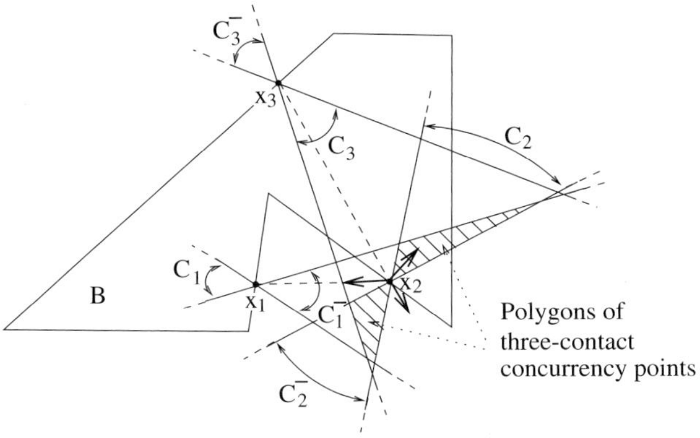

Figure 5.8 A feasible three-contact equilibrium grasp at x_1, x_2 and x_3 illustrating the graphical procedure.

a graphical procedure that identifies equilibrium feasibility of a candidate three-finger grasp. The procedure will use the following *double friction cone.*

DEFINITION 5.3 *Let C_i be the ith friction cone. The* **double friction cone**, $\mathbb{C}_i$, *is given by the union $\mathbb{C}_i = C_i \cup C_i^-$, where C_i^- is the negative reflection of C_i with respect to x_i* (Figure 5.8).

Graphical Procedure for Frictional Three-Finger Planar Equilibrium Grasps

1. Check if the three-contact arrangement contains a feasible two-finger equilibrium grasp. Continue if a two-finger equilibrium is infeasible.
2. Enumerate the nonempty polygons of the intersection $\mathbb{C}_1 \cap \mathbb{C}_2 \cap \mathbb{C}_3$, where $\mathbb{C}_i$ is the double friction cone at x_i $(i = 1, 2, 3)$.
3. Check if a force triplet in any of the nonempty polygons of $\mathbb{C}_1 \cap \mathbb{C}_2 \cap \mathbb{C}_3$ positively spans the origin of $\mathbb{R}^2$, using a single-point check in each polygon.
4. If the test succeeds in any of the nonempty polygons, the contact arrangement is a feasible three-finger equilibrium grasp. Otherwise, it is not a feasible equilibrium grasp.

Let us clarify some details of the graphical procedure. Consider the nonempty polygons of the intersection $\mathbb{C}_1 \cap \mathbb{C}_2 \cap \mathbb{C}_3$. The polygons containing force triplets that positively span the origin of $\mathbb{R}^2$ will be termed *positive span* polygons. When a three-contact arrangement does not contain a feasible two-finger equilibrium grasp, the lines passing through all contact pairs do not cross any nonempty positive span polygon of $\mathbb{C}_1 \cap \mathbb{C}_2 \cap \mathbb{C}_3$. Therefore, the signs of $\lambda_1, \lambda_2, \lambda_3$ remain *invariant* in each positive span polygon, and a single-point check of the signs of $\lambda_1, \lambda_2, \lambda_3$ suffices in these polygons. As the nonempty polygons of $\mathbb{C}_1 \cap \mathbb{C}_2 \cap \mathbb{C}_3$ contain all possible concurrency points of the finger force line triplets, the graphical procedure accounts for all possible three-finger equilibrium grasps and is thus a complete procedure.

Example: Consider the three-contact grasp of the polygon $\mathcal{B}$ depicted in Figure 5.8. A two-finger equilibrium is not feasible at this contact arrangement. Hence, one proceeds with Step 2 of the procedure. The intersection $\mathbb{C}_1 \cap \mathbb{C}_2 \cap \mathbb{C}_3$ consists of two nonempty polygons, $C_1^- \cap C_2^- \cap C_3$ and $C_1^- \cap C_2 \cap C_3$. These polygons share the point x_2 as a common vertex. Hence, one may apply Step 3 using only this vertex. The finger force lines passing through x_2 positively span the origin of $\mathbb{R}^2$. Based on the graphical technique, this contact arrangement forms a feasible three-finger equilibrium grasp of $\mathcal{B}$. ○

One last property is based on the grasp matrix, G, which is 3×6 in the case of three frictional contacts. Generically G has full row rank and a three-dimensional null space of internal grasp forces. The internal grasp forces (f_1, f_2, f_3) are spanned by three pairs of opposing finger forces (which may *not* lie in the respective friction cones). As discussed in Chapter 4, the internal grasp forces modulate the finger force magnitudes by a common scaling factor and rotate the finger force directions. Much like the frictional two-finger grasps, the grasp matrix G forms a *surjective mapping* onto $\mathcal{B}$'s wrench space in the case. The surjectivity of G holds for any higher number of frictional contacts.

3 Planar Four-Finger Equilibrium Grasps

The four-finger equilibrium grasps can immobilize almost any 2-D object based on pure rigid-body constraints. Moreover, they are able to immobilize objects in a manner that is *robust* with respect to small finger placement errors. These properties make the four-finger grasps an important choice when grasping objects under low-friction conditions using imprecise grasping systems.

Based on line geometry, the finger force lines should be linearly dependent at a four-finger equilibrium grasp. Since the space of Plücker coordinates of all planar lines is equivalent to $\mathbb{R}^3$, four force lines are always linearly dependent in their Plücker coordinates. However, this does not mean that every four-contact arrangement forms a feasible equilibrium grasp, since the coefficients $\lambda_1, \ldots, \lambda_4$ represent force magnitudes that should be positive semi-definite at a feasible equilibrium grasp. The following proposition characterizes the four-contact arrangements that form feasible equilibrium grasps (see Appendix I for a proof).

PROPOSITION 5.7 **(planar four-finger grasps)** *A 2-D object $\mathcal{B}$ is held at a feasible* **four-finger equilibrium grasp** *with all four fingers active iff there exist feasible contact forces, $(f_1, f_2, f_3, f_4) \in C_1 \times C_2 \times C_3 \times C_4$, satisfying the conditions:*

(i) The finger force directions positively span the origin of $\mathbb{R}^2$.
(ii) Each pair of finger forces generates opposite moments about the intersection point of the other pair of finger force lines.

To apply the proposition, one must verify that opposite moments are generated about all intersection points of the finger force lines. There can be up to *six* such intersection points. However, based on the proof of Proposition 5.7 in Appendix I, it suffices to

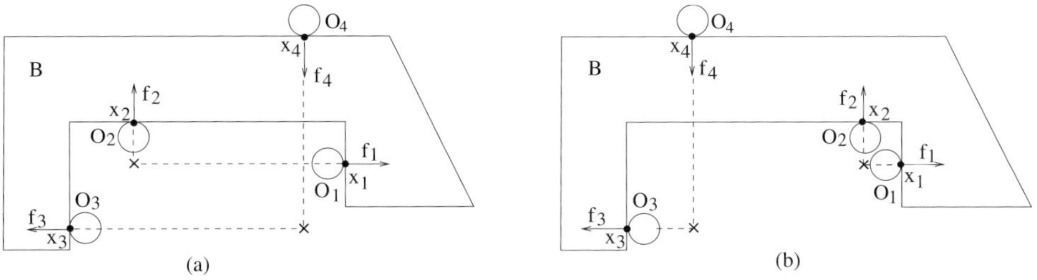

Figure 5.9 Top view of a four-contact arrangement that forms (a) an infeasible equilibrium grasp, and (b) a feasible equilibrium grasp.

check this condition only at *two* intersection points as follows. First, verify that the four finger force directions positively span the origin of $\mathbb{R}^2$. Then split the finger forces into two consecutive pairs according to counterclockwise ordering of their directions. A four-finger equilibrium is feasible iff each of the two pairs generates opposite moments about the intersection point of the other pair. This shorter test is illustrated in the following example.

Example: Figure 5.9(a) depicts a polygonal object $\mathcal{B}$, held by four disc fingers in a horizontal environment without gravity. The finger forces act along $\mathcal{B}$'s contact normals and their directions positively span the origin of $\mathbb{R}^2$. Hence, one proceeds to split the finger forces into two pairs according to a counterclockwise ordering of their directions. One pair is (f_1, f_2) while the other pair is (f_3, f_4). Since f_1 and f_2 generate the *same* clockwise moment about the intersection point of the lines underlying f_3 and f_4, the normal finger forces do *not* form a feasible equilibrium grasp of $\mathcal{B}$. Figure 5.9(b) depicts a different four-finger grasp of the same object $\mathcal{B}$. The positions of x_1 and x_3 are unchanged, but the contacts x_2 and x_4 moved to new locations. The finger force directions still positively span the origin of $\mathbb{R}^2$. Using the same splitting of the four forces into two pairs, (f_1, f_2) and (f_3, f_4), each pair now generates *opposite* moments about the intersection point of the lines underlying the other pair. This contact arrangement therefore forms a feasible equilibrium grasp of $\mathcal{B}$. ○

Some important properties of the four-finger equilibrium grasps associated with frictionless and frictional contacts are next described.

3.1 Frictionless Four-Finger Equilibrium Grasps

Any piecewise smooth object $\mathcal{B}$ can be held in a rich variety of frictionless four-finger equilibrium grasps. These grasps form a four-dimensional subset in *contact space* (Appendix II). Hence, starting from a nominal four-finger equilibrium grasp, one can independently vary all four contacts along the object's boundary while maintaining equilibrium grasp feasibility. The frictionless four-finger equilibrium grasps are thus *robust* with respect to small finger placement errors.

A simple technique for constructing frictionless four-finger equilibrium grasps is next described in the context of polygonal objects. When a polygon $\mathcal{B}$ is held by two fingers at an antipodal grasp, the fingers are located at two opposing vertices, at a

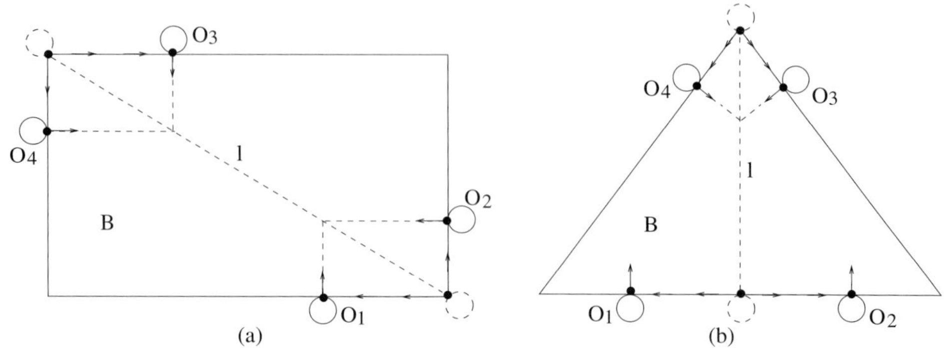

Figure 5.10 Two antipodal grasps and their splitting into four-finger equilibrium grasps.

vertex and an opposing edge or on two parallel edges of $\mathcal{B}$. In all of these cases, the *generalized contact normals* contain opposite vectors at the antipodal points. These vectors positively span the origin of $\mathbb{R}^2$, and this property is maintained by the normals to the edges of $\mathcal{B}$ adjacent to the antipodal points. Hence, suitable splitting of the antipodal points into two pairs lying on the adjacent edges would give four contact normals that positively span the origin of $\mathbb{R}^2$. Next consider the line, l, passing through the initial antipodal points. Split each antipodal point into a pair of points, such that $\mathcal{B}$'s inward contact normals at each pair intersect at a point on l. The two contact pairs generate opposing net wrenches on $\mathcal{B}$, thus resulting in a feasible four-finger equilibrium grasp of $\mathcal{B}$. The technique is illustrated in the following example.

Example: Consider the polygonal objects depicted in Figure 5.10. Two antipodal points are first identified along the boundary of these objects: the maximal grasp in the case of the rectangular object (Figure 5.10(a)) and the minimal grasp in the case of the triangular object (Figure 5.10(b)). Each contact is next split into a pair of contacts lying on opposite sides of the initial point, such that the inward contact normals intersect on the line l. The resulting contact arrangements form feasible four-finger equilibrium grasps. Note that the splitting need not be local in the case of polygonal objects. ○

An important property of the frictionless four-finger equilibrium grasps is their ability to restrain all piecewise smooth 2-D objects.[3] A rigid object $\mathcal{B}$ is said to be *immobilized* by stationary rigid finger bodies $\mathcal{O}_1, \ldots, \mathcal{O}_k$ when any local motion of $\mathcal{B}$ will cause its penetration into one of the stationary finger bodies. Since inter-body penetration is physically infeasible under the rigid body model, an immobilized object is fully secured by the fingers. We will see in Chapter 7 that frictionless equilibrium grasps are necessary for object immobilization. While object immobilization with two or three fingers requires suitable fingertip curvatures, four fingers can achieve object immobilization without any need to consider curvature effects.

[3] When a finger contacts a non-smooth convex vertex of $\mathcal{B}$, the fingertip must have a non-zero radius of curvature for object immobilization.

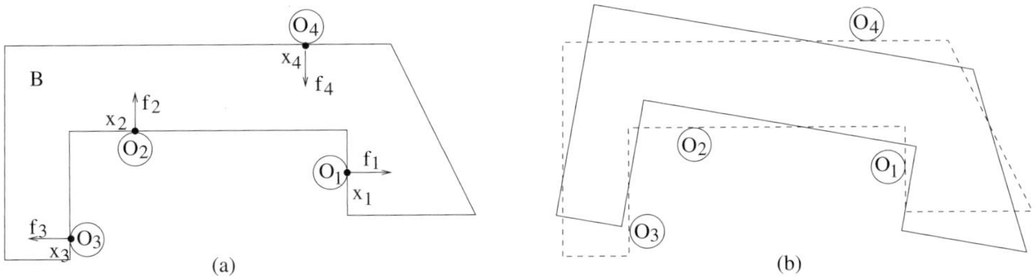

Figure 5.11 (a) Top view of an object held at an infeasible frictionless equilibrium grasp. (b) The object is not immobilized by the surrounding fingers.

Example: Consider the frictionless equilibrium grasp of the polygonal object $\mathcal{B}$ by four disc fingers, depicted in Figure 5.9(b). Based on tools to be developed in Chapter 7, the object is completely immobilized by the disc fingers at this grasp. Next consider the *infeasible* equilibrium grasp of the same polygonal object depicted in Figure 5.11(a). Since frictionless equilibrium grasp feasibility is necessary for object immobilization, the object is *not* immobilized at this grasp. Indeed, the object can escape the stationary fingers as illustrated in Figure 5.11(b). ○

3.2 Frictional Four-Finger Equilibrium Grasps

Given a candidate four-contact arrangement with frictional contacts, we wish to graphically determine if the contact arrangement forms an equilibrium grasp for some choice of feasible finger forces at the contacts. The graphical technique will be based on the following formula for the finger force magnitudes (see Exercises).

LEMMA 5.8 *Let a 2-D object $\mathcal{B}$ be held at a frictional four-finger equilibrium grasp. If the finger force lines do not intersect at a common point in $\mathbb{R}^2$ (in particular, the finger force lines may not be all parallel), up to a common scaling factor their magnitude is given by*

$$\lambda_i = (-1)^i \det \begin{bmatrix} \hat{f}_{i+1} & \hat{f}_{i+2} & \hat{f}_{i+3} \\ x_{i+1} \times \hat{f}_{i+1} & x_{i+2} \times \hat{f}_{i+2} & x_{i+3} \times \hat{f}_{i+3} \end{bmatrix} \qquad i = 1 \ldots 4, \tag{5.3}$$

where index addition is taken modulo 4.

The columns of the matrix in Eq. (5.3) are the Plücker coordinates of the finger force lines associated with the forces f_{i+1}, f_{i+2} and f_{i+3}. It follows that $\lambda_i = 0$ either when the three force lines intersect at a common point, or when two of the three force lines happen to be collinear. Hence, when a four-finger equilibrium grasp does *not* contain any feasible two- or three-finger equilibrium grasp, the signs of $\lambda_1, \ldots, \lambda_4$ remain invariant in each connected set of four force directions (this notion is clarified next). This *sign invariance* is the basis for the following graphical procedure for testing equilibrium feasibility of a candidate four-finger grasp.

Graphical Procedure for Frictional Four-Finger Planar Equilibrium Grasps

1. Check if the contact arrangement contains any two-finger or three-finger equilibrium grasp within the allowed friction cones. Continue if such equilibria are infeasible.
2. Check if the *contact normals* satisfy the four-finger equilibrium grasp conditions: (i) the finger forces positively span the origin of $\mathbb{R}^2$, and (ii) each pair of finger forces generates opposite moments about the intersection point of the other pair of finger force lines.
3. If the test succeeds on the *contact normals,* the contact arrangement forms a feasible equilibrium grasp. Otherwise, the contact arrangement is *not* a feasible equilibrium grasp.

To justify the graphical procedure, consider the case where the four friction cones, when placed at a common origin, span non-overlapping directions in $\mathbb{R}^2$. The friction cones can be unambiguously ordered in counterclockwise direction with increasing indices, and the equilibrium feasibility test of Proposition 5.7 can be applied by splitting each finger force quadruple into two consecutive pairs. Let us therefore split the friction cones into two consecutive pairs, (C_1, C_2) and (C_3, C_4). Next, consider the nonempty polygons of the intersections $\mathbb{C}_1 \cap \mathbb{C}_2$ and $\mathbb{C}_3 \cap \mathbb{C}_4$, where $\mathbb{C}_i = C_i \cup C_i^-$ is the double friction cone at x_i for $i = 1 \ldots 4$. The points of $\mathbb{C}_1 \cap \mathbb{C}_2$ represent all force directions $(\hat{f}_1, \hat{f}_2) \in C_1 \times C_2$, while the points of $\mathbb{C}_3 \cap \mathbb{C}_4$ represent all force directions $(\hat{f}_3, \hat{f}_4) \in C_3 \times C_4$. The set product, $(\mathbb{C}_1 \cap \mathbb{C}_2) \times (\mathbb{C}_3 \cap \mathbb{C}_4)$, represents all force directions $(\hat{f}_1, \hat{f}_2, \hat{f}_3, \hat{f}_4) \in C_1 \times C_2 \times C_3 \times C_4$.[4] Beyond Step 1 of the procedure, the signs of $\lambda_1, \ldots, \lambda_4$ remain positive over all choices of finger force directions within any connected component of the set $(\mathbb{C}_1 \cap \mathbb{C}_2) \times (\mathbb{C}_3 \cap \mathbb{C}_4)$ that contains a feasible equilibrium grasp. This geometric fact can be verified by noting that the entire friction cones C_1 and C_2 must generate *opposite* moments about all points of $\mathbb{C}_3 \cap \mathbb{C}_4$, while the entire friction cones C_3 and C_4 must generate *opposite* moments about all points of $\mathbb{C}_1 \cap \mathbb{C}_2$ (Figure 5.12(b)). Hence, any particular choice of four finger force directions from $(\mathbb{C}_1 \cap \mathbb{C}_2) \times (\mathbb{C}_3 \cap \mathbb{C}_4)$ suffices to determine equilibrium feasibility. Since the *contact normals* always pairwise intersect within $\mathbb{C}_1 \cap \mathbb{C}_2$ and $\mathbb{C}_3 \cap \mathbb{C}_4$, one can determine equilibrium feasibility using the four contact normals, as illustrated in the following example.

Example: Figure 5.12(a) depicts an elliptical object $\mathcal{B}$ that is held by four disc fingers via frictional contacts, with coefficients of friction $\mu_i = 0.3$ for $i = 1 \ldots 4$. In the absence of external influences, this contact arrangement does not support any two- or three-finger equilibrium grasp. Hence, one can verify equilibrium feasibility by checking only the *contact normals.* Since the contact normals do not positively span the origin of $\mathbb{R}^2$, this contact arrangement does *not* form a feasible equilibrium grasp for any choice of finger force directions within the allowed friction cones. Note that higher values of μ_i will eventually support two- and three-finger equilibrium grasps at the given contacts.

Next, consider a different four-contact grasp of $\mathcal{B}$ with the same amount of friction, shown in Figure 5.12(b). This contact arrangement does not support any two- or

[4] Its elements are pairs of points, $(p_1, p_2) \in \mathbb{R}^2 \times \mathbb{R}^2$.

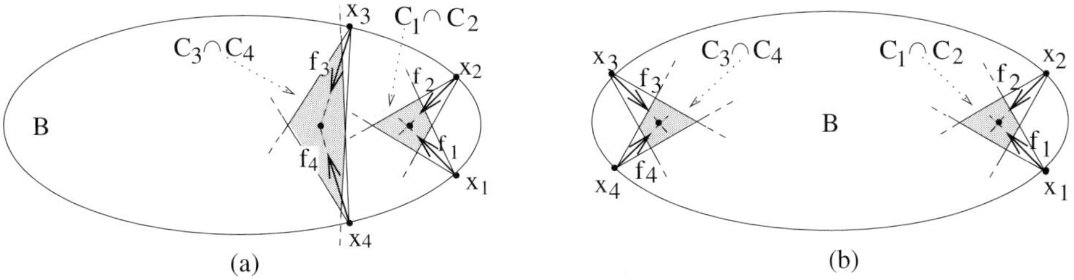

Figure 5.12 Top view of a frictional four-contact arrangement that (a) is an infeasible equilibrium grasp, and (b) forms a feasible equilibrium grasp.

three-finger equilibrium grasp. Hence, one proceeds to verify equilibrium feasibility using the *contact normals*. First, the contact normals positively span the origin of $\mathbb{R}^2$. Second, each pair of contact normals generates opposite moments about the intersection point of the other pair of contact normals. This contact arrangement therefore forms a feasible equilibrium grasp of $\mathcal{B}$. ○

The graphical procedure highlights the following property of the four-finger equilibrium grasps. See proof of a similar property that holds for 3-D equilibrium grasps (Theorem 5.16).

THEOREM 5.9 **(frictional four-finger grasps)** *Let a 2-D object $\mathcal{B}$ be held at a frictional four-finger grasp. If the friction cones at the contacts do not support any two- or three-finger equilibrium grasp but allow a four-finger equilibrium grasp, the equilibrium grasp can be realized with finger forces directed along the* **contact normals**.

Friction at the contacts affects the set of attainable two- and three-finger equilibrium grasps of any given object, as many such grasps can only be achieved when the finger forces are directed away from $\mathcal{B}$'s inward normals at the contacts. In the case of four-finger grasps, Theorem 5.9 implies that friction effects do *not* enlarge the set of attainable four-finger equilibrium grasps beyond the set afforded by frictionless contacts.

5.3 The Spatial Equilibrium Grasps

Let us first consider the number of fingers that are needed to establish *secure* equilibrium grasps of 3-D objects. Secure equilibrium grasps are achieved when the fingers are able to resist any external wrench disturbance that may act on the grasped object $\mathcal{B}$. This notion of *wrench resistance* will be discussed in Chapters 6 and 12. Under the soft-finger contact model, two fingers can already achieve wrench resistant grasps. In the case of hard-finger frictional contacts, three fingers can achieve wrench resistant grasps. A higher level of grasp security is achieved when the fingers are able to restrain all local motions of the grasped object based on rigid-body constraints. This notion of *immobilizing grasps* will be discussed in Chapters 6 and 7. In the case of frictionless contacts, seven fingers can achieve immobilizing grasps that also form wrench resistant grasps. However, we will see in Chapter 8 that when the fingertips' curvature is taken into account, four fingers can immobilize almost all 3-D objects, except objects that have parallel facets. The catalog will therefore include spatial equilibrium grasps having $k = 2, 3, 4$, and 7 contacts.

Line geometry provides the following graphical representation of the 3-D equilibrium grasps. The *Plücker coordinates* of a directed spatial line, l, are given by the pair $(\hat{l}, p \times \hat{l}) \in \mathbb{R}^6$, where $\hat{l}$ is the unit direction of l and p is a point on l. Based on Theorem 5.1, the finger force lines at a k-finger equilibrium grasp lie in a $(k-1)$-dimensional linear subspace in Plücker coordinates: a one-dimensional subspace at the two-finger equilibrium grasps, a two-dimensional subspace at the three-finger equilibrium grasps, a three-dimensional subspace at the four-finger equilibrium grasps and the full six-dimensional space at the seven-finger equilibrium grasps. The geometric description of these linear subspaces follows (see Bibliographical Notes).

Linear Subspaces of Spatial Lines

1. A single line in $\mathbb{R}^3$ spans a one-dimensional subspace in Plücker coordinates, consisting of the line itself.
2. Two lines in $\mathbb{R}^3$ span a two-dimensional subspace in Plücker coordinates:
   (a) When the two lines intersect, the subspace is the *flat pencil* of all spatial lines that lie in the plane of the two lines and pass through their intersection point.
   (b) When the two lines are skew relative to each other, the subspace consists only of these two lines (this case does not correspond to three-finger equilibrium grasps).
3. Three lines in $\mathbb{R}^3$ that do not belong to a common flat pencil span a three-dimensional subspace in Plücker coordinates:
   (a) When the three lines lie in a common plane, the subspace consists of all planar lines in this plane.
   (b) When the three lines intersect at a common point, p, the subspace forms a *solid pencil,* defined as all lines in $\mathbb{R}^3$ that pass through p.
   (c) When only two of the three lines intersect at a common point, the subspace consists of two flat pencils containing a common line passing through the flat pencils' base points.
   (d) When all three lines are skew relative to each other, the subspace spanned by the three lines consists of a quadratic surface called a *regulus* and discussed in the next paragraph.
4. Six lines in $\mathbb{R}^3$ that are skew relative to each other span the full six-dimensional space of Plücker coordinates of all spatial lines.

Let us discuss in some detail the *regulus,* listed in Case 3(d). The regulus is a quadratic surface made of straight lines in $\mathbb{R}^3$. Such a surface is said to form a *ruled surface.* The regulus has the following property. Recall from Section 3.3 that two spatial lines having reciprocal Plücker coordinates intersect in $\mathbb{R}^3$. Since the lines of a regulus span a three-dimensional subspace in Plücker coordinates, there exist three lines reciprocal to this subspace, called *generators*, such that the regulus consists of all spatial lines that intersect the three generators (see Lemma 5.12). Cases 3(a)–(c) correspond to different forms of degenerate reguli, since small perturbations of the three lines would destroy their special intersection arrangement and give the regulus. Note that four- and five-dimensional linear subspaces of spatial lines were omitted from the list. These subspaces correspond to five- and six-finger equilibrium grasps, which are not part of the catalog developed in this chapter.

We now apply the line geometry characterization of the spatial equilibrium grasps to the cases of $k = 2, 3, 4$, and 7 contacts.

1 Spatial Two-Finger Equilibrium Grasps

The line geometry condition for two-finger equilibrium grasps requires that the finger forces lie on a common line, which necessarily passes through the two contacts. Let us highlight some properties of the two-finger equilibrium grasps associated with frictionless and frictional contacts.

1.1 Frictionless Two-Finger Spatial Equilibrium Grasps

The frictionless two-finger equilibria form *antipodal grasps,* according to Definition 5.2. Piecewise smooth 3-D objects generically possess a *discrete* set of antipodal grasps. The non-generic cases are objects with parallel or concentric boundary facets (see Exercises). While 2-D objects can always be grasped along their minimum- and maximum-width segments, 3-D objects can be additionally grasped along their intermediate-width segment, as stated in the following lemma (see Appendix I for a proof).

LEMMA 5.10 *Every piecewise smooth 3-D object possesses at least* **three** *antipodal equilibrium grasps along its minimal-, intermediate- and maximal-width segments.*

The proof of Lemma 5.10 is based on the *mountain pass theorem.* Briefly, let $\sigma : \mathcal{M} \rightarrow \mathbb{R}$ be a continuous scalar-valued function defined on a compact manifold $\mathcal{M}$. If σ has two neighboring local minima in $\mathcal{M}$, the boundary between the two local minima contains a mountain pass, or a saddle point, of σ. The boundary between two neighboring local maxima similarly contains a saddle point. In the case of two-finger grasps, one uses the inter-finger distance function, $\sigma(x_1, x_2) = \|x_1 - x_2\|$, where x_1 and x_2 vary along the object surface. The antipodal grasp along the object's maximum-width segment corresponds to a pair of local maxima of σ. The mountain pass that lies between these two local maxima determines the antipodal grasp along the object's intermediate-width segment (see Appendix I).

Example: Consider the ellipsoid $\mathcal{B}$ shown in Figure 5.13. The principal axes of the ellipsoid form its minimum width, intermediate-width and maximum-width segments. A placement of two fingers at the endpoints of these segments gives three antipodal grasps. Note that the ellipsoid possesses exactly three antipodal grasps, which is a tight lower bound on the number of frictionless two-finger equilibria for any 3-D object. ○

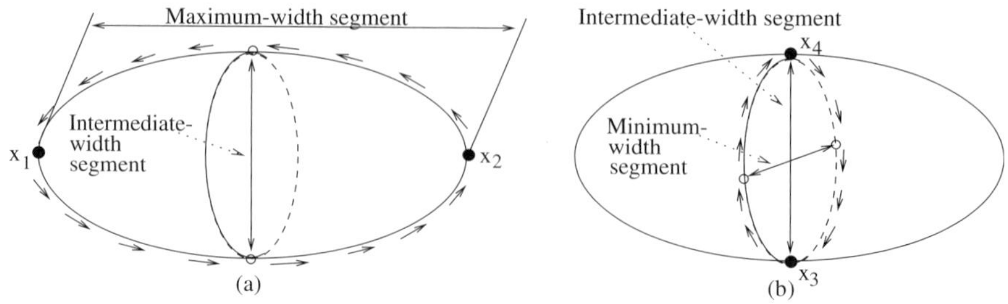

Figure 5.13 (a) The antipodal grasp along $\mathcal{B}$'s maximum-width segment. (b) The antipodal grasp along $\mathcal{B}$'s intermediate-width segment, as well as along $\mathcal{B}$'s minimum-width segment.

1.2 Frictional Two-Finger Spatial Equilibrium Grasps

The frictional two-finger equilibrium grasps require that the 3-D friction cones contain opposing forces collinear with the line passing through the two contacts. The graphical criterion for a two-finger equilibrium grasp is identical to the case of 2-D grasps: $x_2 \in C_1$ and $x_1 \in C_2$, or $x_2 \in C_1^-$ and $x_1 \in C_2^-$, where C_i and C_i^- are the friction cone and its reflection at x_i $(i = 1, 2)$.

Recall from Chapter 4 that $\mathcal{B}$'s *wrench space* at q_0 consists of all object wrenches, $\boldsymbol{w} \in T^*_{q_0}\mathbb{R}^6$. When $\mathcal{B}$ is held at a frictional two-finger equilibrium grasp, the collection of net wrenches that can be affected on $\mathcal{B}$ varies within a five-dimensional linear subspace in $\mathcal{B}$'s wrench space. This property can be observed using line geometry. Let l denote the line passing through the two contacts, x_1 and x_2. Every pair of feasible finger force lines passes through x_1 and x_2 and, hence, intersects the line l. As discussed in Section 3.3, intersecting spatial lines have reciprocal Plücker coordinates. The finger force lines are thus reciprocal to the line l. Let the tangent vector $\dot{q} = (v, \omega)$ represent instantaneous rotation of $\mathcal{B}$ about l. The reciprocality relation implies that the net object wrenches cannot impede the instantaneous rotation of $\mathcal{B}$ about l. That is, $\boldsymbol{w} \cdot \dot{q} = 0$ for all $\boldsymbol{w} \in \mathcal{W}_1 + \mathcal{W}_2$. The *net wrench cone* at a two-finger grasp, $\mathcal{W} = \mathcal{W}_1 + \mathcal{W}_2$, thus lies within a five-dimensional subspace of $\mathcal{B}$'s wrench space. The practical implication of this property is that fully secure grasps of 3-D objects require at least *three* frictional finger contacts.[5]

2 Spatial Three-Finger Equilibrium Grasps

The line geometry condition for three-finger equilibrium grasps requires that the finger force lines lie in a common *flat pencil.* A flat pencil forms a plane, Δ, which is embedded in $\mathbb{R}^3$ and necessarily passes through the three contact points. The characterization of the three-finger equilibrium grasps now reduces to the plane Δ. The finger force line directions must positively span the origin of Δ, and they must intersect at a common point within this plane. Let us highlight some properties of the spatial three-finger equilibrium grasps associated with frictionless and frictional contacts.

[5] Soft fingertips can additionally generate torques about the contact normals. As discussed in Chapter 4, these torques are typically small and provide little resistance against rotations of $\mathcal{B}$ about the line l.

2.1 Frictionless Three-Finger Spatial Equilibrium Grasps

Every piecewise smooth 3-D object $\mathcal{B}$ possesses two equilibrium grasps whose contacts form an equilateral triangle. The existence of such grasps can be established with the following c-space view of the finger c-obstacles.

Existence of minimal and maximal equilateral equilibrium grasps: Let three *point fingers* be located at the vertices of an equilateral triangle embedded in a fixed plane Δ, such that the triangle is much larger than $\mathcal{B}$. Now move the fingers linearly in Δ toward a common center point while maintaining an equilateral triangle formation. In $\mathcal{B}$'s c-space, the finger c-obstacles are initially disjoint but approach each other along fixed translation lines while retaining their shape and size. When the triangle's edge length equals the maximal width of $\mathcal{B}$, the finger c-obstacles pairwise touch for the first time. As the triangle keeps shrinking, $\mathcal{B}$ can still move through the triangular gap formed by the three point fingers. Once $\mathcal{B}$ can simultaneously touch the three fingers, the three finger c-obstacles touch at a common point, q_1, for the first time. Based on *Morse theory* (Appendix B), the finger c-obstacle normals must be linearly dependent at q_1. Since the finger c-obstacles approach q_1 from three directions, their outward normals positively span the origin at q_1. The point q_1 thus forms the maximal equilateral equilibrium grasp of $\mathcal{B}$. As the triangle keeps shrinking, the triangular gap formed by the three point fingers eventually becomes too small for $\mathcal{B}$ to pass. This event occurs when the three c-obstacles touch at a common point, q_2. Based on *Morse theory,* the c-obstacle normals must be linearly dependent at q_2, and they positively span the origin (see Figure 5.7(a) for an analogous event). The point q_2 thus forms the minimal equilateral equilibrium grasp of $\mathcal{B}$. ○

As discussed in Appendix II, the frictionless three-finger equilibrium grasps form a two-dimensional manifold in *contact space* (the equilateral grasps are two points on this manifold). Therefore, one can locally move any two of the three finger contacts along one-dimensional curves on the object surface and then adjust the location of the third finger as to regain the equilibrium grasp. This insight can serve to obtain three-finger equilibrium grasps by locally splitting one contact of a two-finger antipodal grasp, as illustrated in the following example.

Example of contact splitting: A 3-D object $\mathcal{B}$ is initially held at an antipodal grasp, at x_1 and x_2, as shown in Figure 5.14. Let the surface of $\mathcal{B}$ be approximated by a quadratic

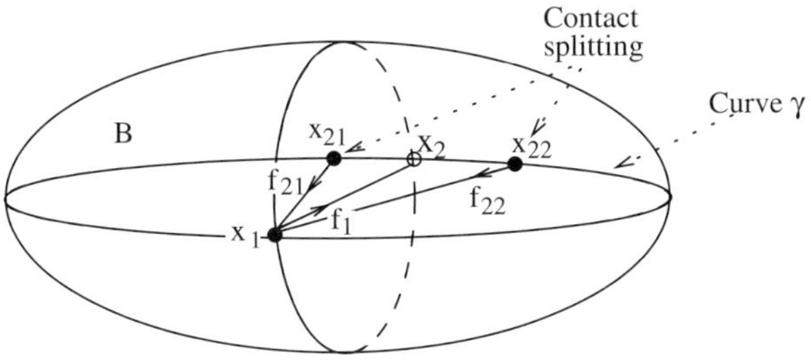

Figure 5.14 A three-finger equilibrium grasp generated by the contact splitting technique.

surface, denoted Q, at x_2. Let V be the plane orthogonal to the object surface at x_2, such that V is spanned by $\mathcal{B}$'s surface normal and one of $\mathcal{B}$'s principal directions of curvature at x_2. Let γ be the intersection curve of Q with V. Being a quadratic surface, the normal to Q along γ lies within the plane V. Hence, it is possible to split x_2 into two contacts along γ, x_{21} and x_{22}, such that the new contact normals intersect at a point on the original contact normal line. The opposing contact normal at x_1 lies on the original contact normal line. The points x_1, x_{21}, and x_{22} form a frictionless three-finger equilibrium grasp of $\mathcal{B}$, shown in Figure 5.14. ○

2.2 Frictional Three-Finger Spatial Equilibrium Grasps

The frictional three-finger equilibrium grasps can also be characterized within the plane Δ spanned by the three contacts. Define the *planar* double friction cones as $\mathbb{C}_i = (C_i \cup C_i^-) \cap \Delta$ for $i = 1, 2, 3$. For each nonempty polygon of $\mathbb{C}_1 \cap \mathbb{C}_2 \cap \mathbb{C}_3$, check if the force triplets in this polygon positively span the origin of Δ (using a single-point check). If the test succeeds, the contact arrangement forms a feasible equilibrium grasp. If the test fails over all nonempty polygons of $\mathbb{C}_1 \cap \mathbb{C}_2 \cap \mathbb{C}_3$, the contact arrangement is not a feasible equilibrium grasp.

Additional insight into the frictional three-finger grasps is provided by the grasp matrix, G, which is 6×9 in the case of three frictional contacts. Generically G has full row rank and a three-dimensional null space of internal grasp forces, $\vec{f} = (f_1, f_2, f_3)$ such that $G\vec{f} = \vec{0}$. The internal grasp forces are spanned by the pairs of opposing forces:

$$\text{span}\left\{ \begin{pmatrix} x_2 - x_1 \\ x_1 - x_2 \\ \vec{0} \end{pmatrix}, \begin{pmatrix} \vec{0} \\ x_3 - x_2 \\ x_2 - x_3 \end{pmatrix}, \begin{pmatrix} x_1 - x_3 \\ \vec{0} \\ x_3 - x_1 \end{pmatrix} \right\}.$$

The force lines of this basis lie within the plane Δ. When a three-finger contact arrangement forms a feasible equilibrium grasp, the internal grasp forces can modulate and rotate the equilibrium forces within the *planar* friction cones, $C_i \cap \Delta$ for $i = 1, 2, 3$. Since G has full row rank, it forms a *surjective mapping* onto $\mathcal{B}$'s wrench space. This surjectivity holds for any higher number of frictional contacts.

The generalized tripod rule: During quasistatic legged locomotion, a legged robot strives to maintain equilibrium against gravity while lifting a free leg to a new position. Consider the simplest case where the robot moves over a flat horizontal terrain. When the mechanism supports itself on two legs via frictional point contacts, the ground reaction forces generate only a five-dimensional net wrench cone on the mechanism. Hence, certain tip-over motions of the robot cannot be impeded by the contacts. When the mechanism supports itself on three well-placed legs via frictional contacts, the net wrench cone of the ground reaction forces is fully six-dimensional. In this case, the robot can vary its center of mass within a fully three-dimensional column in $\mathbb{R}^3$ while maintaining equilibrium against gravity. This property will become a *generalized tripod rule* in Chapter 11, where a robot hand supports 3-D objects against gravity via fingertip and palm contacts. ○

3 Spatial Four-Finger Equilibrium Grasps

The line geometry condition for the four-finger equilibrium grasps requires that the finger force lines lie in a three-dimensional linear subspace in Plücker coordinates. As described at the beginning of this section, such a subspace can take one of four possible forms. The first form involves four coplanar force lines. This case occurs only when the four contacts lie in a common plane in $\mathbb{R}^3$ and is basically a planar four-finger grasp embedded in $\mathbb{R}^3$. Hence, we will focus on the remaining three cases: the solid pencil, the intersecting flat pencils and the regulus. Combining these three cases with Theorem 5.1, we obtain the following characterization of the four-finger equilibrium grasps.

LEMMA 5.11 **(spatial four-finger grasps)** *A 3-D object $\mathcal{B}$ is held at a feasible* **four-finger equilibrium grasp** *iff there exist feasible contact forces,* $(f_1, f_2, f_3, f_4) \in C_1 \times C_2 \times C_3 \times C_4$, *which positively span the origin of* $\mathbb{R}^3$ *and form one of the line arrangements:*

1. *The finger force lines lie in a* **solid pencil**, *thus intersecting at a common point in* $\mathbb{R}^3$.
2. *The finger force lines lie in two* **intersecting flat pencils**, *thus forming two pairs of coplanar lines sharing a line passing through the two pencils' base points.*
3. *The finger force lines lie in a* **regulus**, *thus forming four skew lines. In this case, there exist three lines, or generators, such that the four finger force lines intersect the three generators.*

The possible four-finger equilibrium grasp arrangements are illustrated in the following example.

Example: Consider the successively more complex line arrangements shown in Figure 5.15 (the object $\mathcal{B}$ is not shown). The simplest case consists of four force lines that pass through a common point and positively span the origin of $\mathbb{R}^3$, shown in Figure 5.15(a). By splitting the finger force lines into two coplanar pairs then pulling the two pairs apart, we obtain the four-finger equilibrium grasp associated with two intersecting flat pencils, shown in Figure 5.15(b). By pulling each pair of force lines away from its common plane, such that the new lines lie in parallel planes at equal distances from the original plane, we obtain the four-finger equilibrium grasp associated

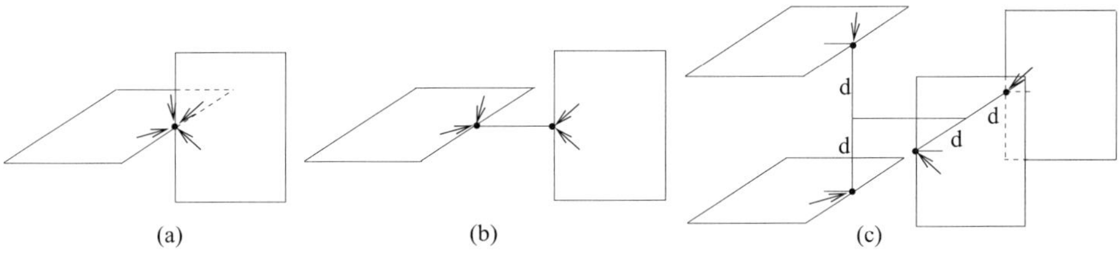

Figure 5.15 The finger force lines at a four-finger equilibrium grasp (the object $\mathcal{B}$ is not shown). (a) The four lines intersect at a common point. (b) The four lines lie in two flat pencils sharing a common line. (c) The four lines lie in a common regulus.

with a regulus, shown in Figure 5.15(c). One can verify that the force lines shown in Figure 5.15(c) lie in a common regulus (and hence form a four-finger equilibrium grasp), using a graphical test that is described next. ○

The graphical test that determines when four finger force lines form a feasible equilibrium grasp is based on the following lemma (see Appendix I for a proof).

LEMMA 5.12 **(regulus test)** *Four lines in* $\mathbb{R}^3$ *lie in a common* **regulus** *iff the projections of every line triplet on the plane orthogonal to the fourth line intersect at a common point.*

When the conditions of Lemma 5.12 are met, one can graphically construct three generators for the regulus containing the four lines as follows. Select a particular line triplet, and denote by P the plane orthogonal to the fourth line. The projections of the line triplet on P intersect at a common point, $p \in P$. Recall now that parallel or intersecting spatial lines have reciprocal Plücker coordinates. The line parallel to the fourth line and passing through p is reciprocal to the four lines and hence forms a generator for the regulus containing the four lines. Based on Lemma 5.12, equilibrium feasibility of four finger force lines can be graphically tested as follows (see Appendix I for a proof).

PROPOSITION 5.13 **(Four-Finger Grasp)** *Let a 3-D object* $\mathcal{B}$ *be held by four fingers along nonparallel force lines. The finger force lines form a feasible* **equilibrium grasp** *iff every triplet of finger force lines projects to a planar equilibrium grasp in the plane orthogonal to the fourth finger force line.*

The special case of parallel finger forces lines is discussed in the Exercises. The proposition is illustrated in the following example.

Example: Consider the finger force lines l_1, l_2, l_3, l_4 depicted in Figure 5.16(a), where l_1 and l_2 lie in parallel horizontal planes while l_3 and l_4 lie in parallel vertical planes. Each pair of planes is located $2d$ apart, such that the four lines span the same angle α within their planes (Figure 5.16(a)). Consider the line triplet l_1, l_2, l_3. Based on Proposition 5.13, at an equilibrium grasp these lines must project to a three-finger equilibrium grasp in the plane orthogonal to l_4. Let V denote the vertical plane containing the line l_3, depicted in Figure 5.16(b). Let l_0 be the line passing through the intersection points of l_1 and l_2 with the plane V, depicted in Figure 5.16(b). A simple

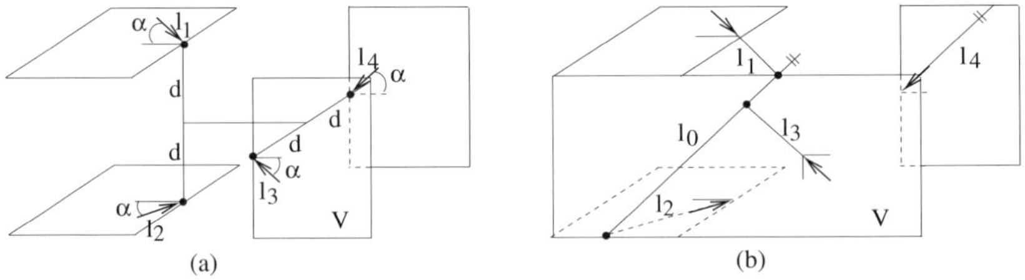

Figure 5.16 (a) Four force lines lying in a common regulus, and (b) the generator l_0 parallel to l_4 (the object $\mathcal{B}$ is not shown).

trigonometric computation confirms that l_0 is parallel to l_4 in this line arrangement. The projections of l_1, l_2, l_3 thus intersect at a common point in the plane orthogonal to l_4, which implies that l_0 is a generator for the regulus containing the lines l_1, l_2, l_3, l_4. Since the projections of l_1, l_2, l_3 positively span the origin of $\mathbb{R}^2$, they form a three-finger equilibrium grasp in the plane orthogonal to l_4. By symmetry of the grasp arrangement, the other line triplets similarly project to planar three-finger equilibrium grasps. By Proposition 5.13, the four force lines form a feasible four-finger equilibrium grasp. ○

3.1 Frictionless Four-Finger Spatial Equilibrium Grasps

Every piecewise smooth 3-D object $\mathcal{B}$ possesses a frictionless four-finger equilibrium grasp whose contacts form a regular tetrahedron. The existence of such an equilibrium grasp can be established with the following c-space view of the finger c-obstacles.

Existence of a tetrahedral equilibrium grasp: Let four *point fingers* be located at the vertices of a symmetric tetrahedron, such that the tetrahedron is much larger than $\mathcal{B}$. The finger c-obstacles corresponding to the initial finger placements are disjoint in $\mathcal{B}$'s c-space. Now move the fingers in a symmetric tetrahedral formation toward a common center point. The finger c-obstacles approach each other along pure translational motion while retaining their shape and size. The first critical event occurs when $\mathcal{B}$ can simultaneously touch two fingers for the first time. At this instant, the finger c-obstacles pairwise touch for the first time. As the tetrahedron keeps shrinking, a second critical event occurs when $\mathcal{B}$ can simultaneously touch three fingers for the first time. At this instant, every triplet of finger c-obstacles touches for the first time. The shrinking tetrahedron is eventually contained in $\mathcal{B}$. Hence, there exists a third critical event at which $\mathcal{B}$ simultaneously touches all four fingers for the first time. At this instant, the four finger c-obstacles touch at a common point, q_0, for the first time. Based on *Morse theory* (Appendix B), the c-obstacle normals must be linearly dependent at q_0. Since the c-obstacles approach q_0 from four directions, their outward normals positively span the origin at q_0. The point q_0 thus forms a tetrahedral equilibrium grasp $\mathcal{B}$. ○

The set of frictionless four-finger equilibrium grasps forms a five-dimensional manifold in *contact space* (Appendix II). Hence, starting from a nominal four-finger equilibrium grasp, one can locally vary the location of the contacts along the object surface using five independent position parameters while maintaining the equilibrium grasp. Such local motions of the contacts can also be used to obtain four-finger equilibrium grasp by splitting the contacts of two- or three-finger equilibrium grasps, as illustrated in the following example.

Example of contact splitting: The *inscribed ball* of a 3-D object $\mathcal{B}$ is defined as the largest ball contained inside the given object. As discussed in Chapter 9, the inscribed ball generically touches the object surface at two, three or four isolated points, as shown in Figure 5.17. When the inscribed ball touches the surface of $\mathcal{B}$ at two points, a local splitting of both contacts gives a four-finger equilibrium grasp (Figure 5.17(a)). Since every contact splitting involves two position parameters, say the direction of splitting and the distance between the two contacts, one can always generate

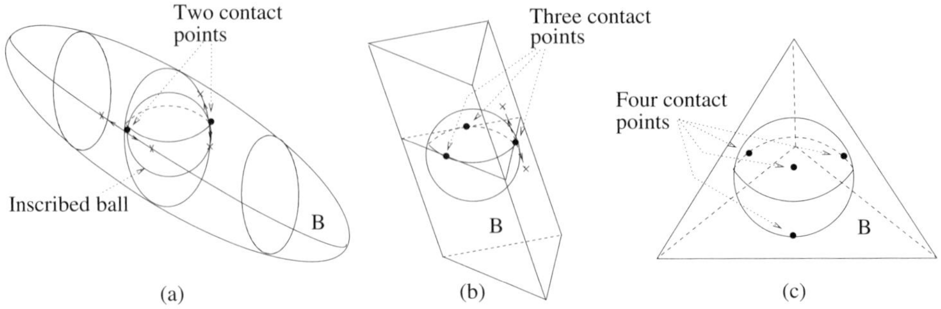

Figure 5.17 The inscribed ball of $\mathcal{B}$ touches its surface at (a) two, (b) three, and (c) four points. Cases (a) and (b) require contact splittings to become four-finger equilibrium grasps.

a frictionless four-finger equilibrium grasp by starting from the object's inscribed ball. When the inscribed ball touches the surface of $\mathcal{B}$ at three points, a local splitting of one of the three contacts gives a four-finger equilibrium grasp (Figure 5.17(b)). When the ball touches the surface of $\mathcal{B}$ at four points, the contact normals already form a concentric four-finger equilibrium grasp (Figure 5.17(c)). ○

3.2 Frictional Four-Finger Spatial Equilibrium Grasps

One would ideally like to graphically determine when the friction cones of a four-contact arrangement form a feasible equilibrium grasp. However, the frictional four-finger equilibria do not lend themselves to a simple graphical interpretation. Some graphical intuition can be gained from the following necessary condition for a four-finger equilibrium grasp. Recall that $\mathbb{C}_i = C_i \cup C_i^-$ denotes the double friction cone at x_i. When a spatial line l lies outside $\mathbb{C}_i$, all force lines of $\mathbb{C}_i$ generate the *same* moment sign about l. Based on this observation, let l_{ij} be the line passing through the contacts x_i and x_j. Consider the situation where l_{ij} lies outside the double friction cones at x_k and x_l, $\mathbb{C}_k$ and $\mathbb{C}_l$. The contact forces at x_i and x_j generate zero moment about l_{ij}. Hence, a necessary condition for a four-finger equilibrium grasp is that the contact forces at x_k and x_l generate *opposing* moments about l_{ij}. This necessary condition must hold with respect to all six lines l_{ij} such that $1 \leq i, j \leq 4$, as illustrated in the following example.

Example – Hand-supported stances: Figure 5.18 shows a rigid object $\mathcal{B}$ supported against gravity by three frictional contacts. The contacts are formed by a supporting robot hand, which is not shown. Let Δ be the plane spanned by the three contacts. The friction cones C_1, C_2, and C_3 lie strictly above Δ, as depicted in Figure 5.18. The *support problem* requires that we find all center-of-mass locations of the object $\mathcal{B}$, which keep the object in static equilibrium on the supporting contacts and thus prevent the object from escaping the robot hand. The situation can be interpreted as a four-finger equilibrium grasp, with gravity acting as a *virtual finger* at the object's center of mass. The necessary condition can now be applied to the support problem. Denote by l_{ij} the line passing through the contact points x_i and x_j. Since l_{ij} does not intersect the double friction cone $\mathbb{C}_k$, all force lines of $\mathbb{C}_k$ generate the same

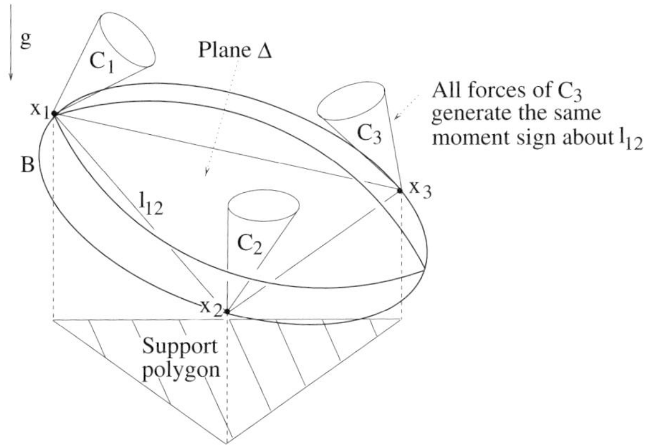

Figure 5.18 A rigid object $\mathcal{B}$ supported against gravity by three frictional contacts (the supporting robot hand is not shown). The object's statically stable center of mass locations occupy a subset of the vertical prism spanned by the support polygon.

moment sign about l_{ij}. It follows that $\mathcal{B}$'s center-of-mass must lie in a vertical half-space bounded by l_{ij}, such that the half-space contains the point x_k. By repeating this argument for the three contact pairs, the feasible center-of-mass locations must lie in the vertical prism spanned by the three supporting contacts. The cross section of this prism forms the *support polygon* of the hand-supported stance. We will see in Chapter 15 that the solution to the general support problem forms a *smaller subset* of the support polygon. ○

Let us end this section with a remark on the grasp matrix G, which is 6×12 in the case of four frictional contacts. Generically G has full row rank and hence forms a surjective mapping onto $\mathcal{B}$'s wrench space. However, G is already surjective in the case of three frictional contacts. One may therefore argue that *four fingers do not offer any qualitative advantage over three fingers* in frictional 3-D grasps, since both grasps can in principle generate any net wrench on the grasped object $\mathcal{B}$.[6] However, a three-finger hand can use its palm to *immobilize* 3-D objects in highly secure four-contact grasps. Moreover, manipulation tasks require at least one additional finger that can be repositioned while the object is held in a three-finger equilibrium grasp. Hence, both three and four-finger grasps have important roles in the design of all-purpose minimalistic robot hands.

4 Spatial Seven-Finger Equilibrium Grasps

The line geometry condition for seven-finger equilibrium grasps requires that the finger force lines be linearly dependent in Plücker coordinates. But the Plücker coordinates of seven spatial lines are vectors in $\mathbb{R}^6$, which are always linearly dependent in Plücker coordinates. The equilibrium condition thus reduces to the requirement that the linear

[6] We will see in Chapter 12 that secure *wrench resistant* grasps require a surjective grasp matrix, as well as feasible internal grasp forces at the given grasp.

dependency coefficients will be positive semi-definite. Let us describe a formula for these coefficients, which will be used to test if seven finger force lines form a feasible equilibrium grasp (see Exercises for a derivation of the formula).

LEMMA 5.14 *When seven lines, $l_i = (\hat{l}_i, x_i \times \hat{l}_i)$ for $i = 1 \ldots 7$, do not lie in a lower dimensional linear subspace in Plücker coordinates, their linear dependency coefficients in the equation $\lambda_1 l_1 + \cdots + \lambda_7 l_7 = \vec{0}$ are given up to a common scaling factor by*

$$\lambda_i = \det\begin{bmatrix} \hat{l}_{i+1} & \cdots & \hat{l}_{i+6} \\ x_{i+1} \times \hat{l}_{i+1} & & x_{i+6} \times \hat{l}_{i+6} \end{bmatrix} \qquad i = 1 \ldots 7, \tag{5.4}$$

where index addition is taken modulo 7.

Equilibrium grasp feasibility requires that the seven finger lines be linearly dependent with positive semi-definite coefficients. This leads to the following equilibrium feasibility test.

COROLLARY 5.15 (**seven-finger grasp**) *Let a 3-D object $\mathcal{B}$ be held along seven finger force lines. The finger force lines form a feasible* **seven-finger equilibrium grasp** *iff the coefficients $\lambda_1, \ldots, \lambda_7$ specified in Eq. (5.4) are all positive semi-definite or all negative semi-definite.*

Proof sketch: Consider the generic case where the coefficients $\lambda_1, \ldots, \lambda_7$ specified in Eq. (5.4) are all strictly positive or negative at a given grasp. Since $\lambda_i \neq 0$ for $i = 1 \ldots 7$, the 6×7 matrix $[l_1 \cdots l_7]$ has full row rank and a one-dimensional null space. The null space has the form $s \cdot (\sigma_1, \ldots, \sigma_7)$ for some fixed coefficients vector $(\sigma_1, \ldots, \sigma_7)$ and $s \in \mathbb{R}$. Since $\lambda_1, \ldots, \lambda_7$ must be positive semi-definite at a feasible equilibrium grasp, either $\sigma_1, \ldots, \sigma_7 > 0$ and then $\lambda_i = \sigma_i$ for $i = 1 \ldots 7$, or $\sigma_1, \ldots, \sigma_7 < 0$ and then $\lambda_i = -\sigma_i$ for $i = 1 \ldots 7$. □

Let us next highlight some properties of the seven-finger equilibrium grasps associated with frictionless and frictional contacts.

4.1 Frictionless Seven-Finger Spatial Equilibrium Grasps

The frictionless seven-finger equilibrium grasps are *robust* with respect to small finger placement errors, based on the following argument. Each finger contact varies with two position parameters along the object surface. The seven-finger contact arrangements are parametrized by $\mathbb{R}^{14}$, which forms the *contact space* of these grasps. The frictionless seven-finger equilibrium grasps span a fully 14-dimensional subset in contact space (Appendix II). Hence, starting from a nominal seven-finger equilibrium grasp, one can independently vary all seven contacts along the object's surface while maintaining equilibrium grasp feasibility.

Another important property of the frictionless seven-finger equilibrium grasps is their ability to *immobilize* 3-D objects. We will see in Chapters 7 and 8 that equilibrium grasps that involve six or fewer fingers require suitable contact curvature to achieve object immobilization. In contrast, when a 3-D object is held by seven fingers that are all essential for maintaining the equilibrium grasp, the object is fully immobilized by the fingers, without any need to consider curvature effects.

4.2 Frictional Seven-Finger Spatial Equilibrium Grasps

It is somewhat non-intuitive to observe that seven frictional contacts do *not* enlarge the collection of seven-finger equilibrium grasps beyond those already afforded by seven frictionless contacts. This property is the topic of the following theorem.

THEOREM 5.16 **(frictional seven-finger grasps)** *Let a 3-D object $\mathcal{B}$ be held via seven frictional contacts. If the contact arrangement forms a feasible equilibrium grasp such that all seven contacts are essential for maintaining the equilibrium grasp, the grasp can be established with the finger forces directed along $\mathcal{B}$'s surface normals at the contacts.*

Proof sketch: Starting from the feasible seven-finger equilibrium grasp, we will show that the finger force directions can be rotated while maintaining the equilibrium grasp, until they become aligned with $\mathcal{B}$'s inward normals at the contacts. Let $(\hat{f}_1, \ldots, \hat{f}_7) \in C_1 \times \cdots \times C_7$ be the finger force directions at the initial equilibrium grasp. Consider a continuous path of force directions, $\gamma(s) = (\hat{f}_1(s), \ldots, \hat{f}_7(s))$ for $s \in [0,1]$, which starts at $\gamma(0) = (\hat{f}_1, \ldots, \hat{f}_7)$ and ends at the inward normals, $\gamma(1) = (n_1, \ldots, n_7)$. Clearly, a path γ can be constructed such that the finger force directions rotate in a continuous manner within their respective friction cones. Each point along γ determines seven finger force lines, $\big(\hat{f}_i(s), x_i \times \hat{f}_i(s)\big)$ for $i = 1 \ldots 7$. Seen as vectors in $\mathbb{R}^6$, the seven finger force lines are linearly dependent for all $s \in [0,1]$. Now consider the finger force magnitudes $\lambda_1, \ldots, \lambda_7$ along γ. Initially $\lambda_i(\gamma(0)) > 0$ for $i = 1 \ldots 7$. Suppose that at some point along γ, one of these coefficients vanishes. Let s^* be the *first instant* this event happens, say to the coefficient λ_1. Based on Lemma 5.14:

$$\lambda_1(\gamma(s^*)) = \det \begin{bmatrix} \hat{f}_2(s^*) & \cdots & \hat{f}_7(s^*) \\ x_2 \times \hat{f}_2(s^*) & & x_7 \times \hat{f}_7(s^*) \end{bmatrix} = 0. \tag{5.5}$$

It follows from Eq. (5.5) that the six finger force lines, $\big(\hat{f}_i(s^*), x_i \times \hat{f}_i(s^*)\big)$ for $i = 2 \ldots 7$, are linearly dependent. Since $\lambda_i(\gamma(s^*)) > 0$ for $i = 2 \ldots 7$, the six fingers form a feasible equilibrium grasp of $\mathcal{B}$. But this contradicts our assumption that all seven contacts are *essential* for maintaining the equilibrium grasp. Hence, all seven coefficients $\lambda_1, \ldots, \lambda_7$ must remain positive along the entire path γ. Thus $\lambda_i(\gamma(1)) > 0$ for $i = 1 \ldots 7$, which implies that the contact normals $(n_1, \ldots, n_7)$ also form a feasible equilibrium grasp. □

Note that friction effects do enlarge the collection of 3-D equilibrium grasps attainable by six or fewer fingers beyond the set afforded by frictionless contacts. However, when an object is held by seven fingers, Theorem 5.16 implies that friction effects do *not* enlarge the collection of seven-finger equilibrium grasps of 3-D objects beyond the set afforded by seven frictionless contacts.

5.4 Equilibrium Grasps Involving Higher Numbers of Fingers

The catalog of equilibrium grasps ended with four fingers in the case of 2-D grasps and seven fingers in the case of 3-D grasps. When the number of fingers is higher, the equilibrium grasp can be maintained with a subset of the fingers whose number matches the catalog's upper limits. The reducibility of such grasps is based on *Carathéodory's theorem* (see Bibliographical Notes). The *convex hull* of a set of points $\{p_1, \ldots, p_k\}$ in $\mathbb{R}^m$ is defined as the smallest convex set containing these points, which forms a convex polyhedral set in $\mathbb{R}^m$.

Carathéodory's theorem: *If a point p_0 lies in the convex hull of a set of points $\{p_1, \ldots, p_k\}$ such that $k \geq m+1$ in $\mathbb{R}^m$, p_0 lies in the convex hull of at most* $\boldsymbol{m+1}$ **points** *from this set.*

To apply Carathéodory's theorem to k-finger grasps, recall that the object wrench space, $T^*_{q_0}\mathbb{R}^m$, is equivalent to $\mathbb{R}^m$. When an object $\mathcal{B}$ is held at an equilibrium grasp by $k > m+1$ fingers, at most $m+1$ of the finger wrenches are required to positively span the wrench space origin. We shall make this statement formal using the notion of essential fingers from Section 4.3. When an object $\mathcal{B}$ is held at a k-finger equilibrium grasp, a finger body $\mathcal{O}_i$ is *essential* for maintaining the equilibrium grasp when the net wrench cone spanned by feasible finger forces at the remaining $k-1$ contacts forms a pointed cone in $\mathcal{B}$'s wrench space. We now apply Carathéodory's theorem to equilibrium grasps having a high number of fingers.

PROPOSITION 5.17 (**equilibrium grasp reducibility**) *Let a rigid object $\mathcal{B}$ be held at an equilibrium grasp by $k > m+1$ fingers, where $m=3$ in 2-D grasps and $m=6$ in 3-D grasps. Then the equilibrium grasp can be maintained by at most* $\boldsymbol{m+1}$ **active fingers.**

Proof: Consider the equilibrium grasp equation (5.1):

$$\lambda_1 \begin{pmatrix} \hat{f}_1 \\ x_1 \times \hat{f}_1 \end{pmatrix} + \cdots + \lambda_k \begin{pmatrix} \hat{f}_k \\ x_k \times \hat{f}_k \end{pmatrix} = \vec{0} \qquad \lambda_i \geq 0, \;\; i = 1 \ldots k. \tag{5.6}$$

Since the finger force magnitudes are not all zero at the equilibrium, $\sum_{j=1}^{k} \lambda_i > 0$. Multiplying both sides of Eq. (5.6) by $1/\sum_{j=1}^{k} \lambda_j$ gives

$$\sigma_1 \begin{pmatrix} \hat{f}_1 \\ x_1 \times \hat{f}_1 \end{pmatrix} + \cdots + \sigma_k \begin{pmatrix} \hat{f}_k \\ x_k \times \hat{f}_k \end{pmatrix} = \vec{0} \qquad \sigma_i \geq 0, \; \textstyle\sum_{i=1}^{k} \sigma_i = 1,$$

where $\sigma_i = \lambda_i / \sum_{j=1}^{k} \lambda_j$ for $i = 1 \ldots k$. This shows that the zero net wrench lies in the convex hull of the finger wrenches, $(\hat{f}_i, x_i \times \hat{f}_i)$ for $i = 1 \ldots k$. By Carathéodory's theorem, the zero wrench lies in the convex hull of at most $m+1$ of these wrenches. The equilibrium grasp can thus be maintained by at most $m+1$ active fingers. □

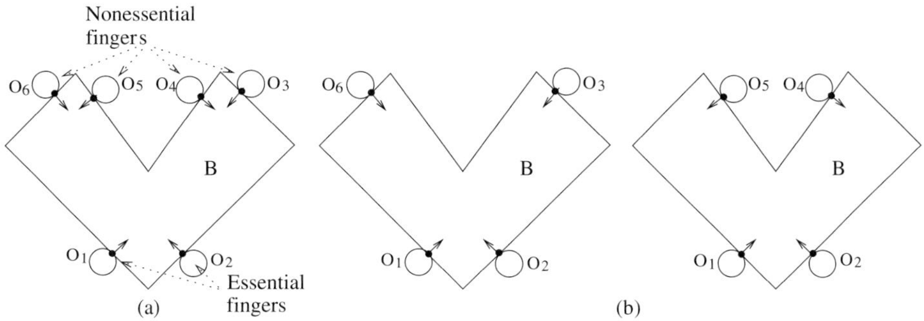

Figure 5.19 (a) A six-finger equilibrium grasp that contains nonessential fingers. (b) Two equilibrium grasps containing four essential fingers.

Proposition 5.17 implies that all 2-D equilibrium grasps can be maintained with at most *four* active fingers, while all 3-D equilibrium grasps require at most *seven* active fingers. This property is illustrated in the following example.

Example: Consider the six-finger grasp of the polygonal object $\mathcal{B}$ shown in Figure 5.19(a), where the finger forces act along the contact normals. One way to see that the contacts form a feasible equilibrium grasp is to note that the object is *immobilized* by the fingers at this grasp. As discussed in the next chapter, object immobilization can only be achieved with frictionless equilibrium grasps. Based on Proposition 5.17, the grasp contains at least two *nonessential* fingers that can be removed without affecting equilibrium feasibility. Since the finger force directions must positively span the origin of $\mathbb{R}^2$, the bottom fingers $\mathcal{O}_1$ and $\mathcal{O}_2$ are essential for maintaining the equilibrium grasp. Each of the remaining four fingers is nonessential. Figures 5.19(b) show two equilibrium grasps, obtained by removing the nonessential fingers $(\mathcal{O}_4, \mathcal{O}_5)$ or $(\mathcal{O}_3, \mathcal{O}_6)$ from the six-finger grasp. ○

Bibliographical Notes

The catalog of equilibrium grasps is based on notions of line geometry discussed in the Bibliographical Notes of Chapter 3. The graphical characterization of linear subspaces of lines, which correspond to equilibrium grasps that involve different number of fingers, appears in Dandurand [1] and Veblen and Young [2]. The interpretation of the equilibrium grasp condition as linear dependency of the finger contact force lines is based on Ponce, Sullivan, Sudsang, Boissonnat and Merlet [3].

The reducibility of grasps involving high number of fingers into at most four fingers in 2-D grasps and seven fingers in 3-D grasps, based on *Carathéodory's theorem,* appears in Mishra, Schwartz and Sharir [4]. Finally, Nirenberg [5] describes the *mountain pass theorem*, which is used in Appendix I and as an exercise to show that 2-D rigid objects possess minimal and maximal antipodal grasps, while 3-D rigid objects possess minimal, intermediate and maximal antipodal grasps.

Appendix I: Proof Details

This appendix contains proofs of key geometric properties underlying the catalog of equilibrium grasps. We begin with Corollary 5.2, which concerns the line geometric characterization of the equilibrium grasps.

Corollary 5.2 *The conditions of Theorem 5.1 are* **necessary and sufficient** *for equilibrium grasp feasibility of generic 2-D grasps having $2 \leq k \leq 3$ contacts, and generic 3-D grasps having $2 \leq k \leq 4$ contacts.*

Proof: First write the equilibrium grasp condition of Eq. (5.1) in matrix form:

$$\begin{bmatrix} \hat{f}_1 & \cdots & \hat{f}_k \\ x_1 \times \hat{f}_1 & & x_k \times \hat{f}_k \end{bmatrix} \begin{pmatrix} \lambda_1 \\ \vdots \\ \lambda_k \end{pmatrix} = H \begin{pmatrix} \lambda_1 \\ \vdots \\ \lambda_k \end{pmatrix} = \vec{0} \qquad \lambda_1, \ldots, \lambda_k \geq 0, \tag{5.7}$$

where H is $3 \times k$ in 2-D grasps and $6 \times k$ in 3-D grasps. Consider the upper $n \times k$ sub-matrix of H, $\bar{H} = \left[\hat{f}_1 \cdots \hat{f}_k\right]$, where $n = 2$ or 3. The columns of $\bar{H}$ are linearly dependent according to Condition (i) of Theorem 5.1. Since $k-1 \leq n$, $\text{rank}(\bar{H}) \leq k-1$. In 2-D grasps, $\text{rank}(\bar{H}) = k-1$ at any two-finger grasp and at any three-finger grasp with nonparallel finger force lines. In 3-D grasps, $\text{rank}(\bar{H}) = k-1$ at any two-finger grasp, at any three-finger grasp with nonparallel finger force lines and at any four-finger grasp whose finger force lines do not lie in parallel planes in $\mathbb{R}^3$. Note that these are all generic grasps.

Next consider the full matrix H. In 2-D grasps, H is $3 \times k$ with linearly dependent columns according to Condition (ii) of Theorem 5.1. Since $k \leq 3$, $\text{rank}(H) \leq k-1$ while $\text{rank}(\bar{H}) = k-1$. Hence, the lower row of H is linearly dependent on its upper two rows. Since $\hat{f}_1, \ldots, \hat{f}_k$ positively span the origin of $\mathbb{R}^2$, the vector $(\lambda_1, \ldots, \lambda_k)$ lies in the null space of H. Therefore, H satisfies the equilibrium equation (5.7). In 3-D grasps, H is $6 \times k$ with linearly dependent columns according to Condition (ii) of Theorem 5.1. Since $k \leq 4$, $\text{rank}(H) \leq k-1$ while $\text{rank}(\bar{H}) = k-1$. Hence, the lower three rows of H are linearly dependent on its upper three rows. Since $\hat{f}_1, \ldots, \hat{f}_k$ positively span the origin of $\mathbb{R}^3$, the vector $(\lambda_1, \ldots, \lambda_k)$ lies in the null space of H. Hence, H satisfies the equilibrium grasp condition of Eq. (5.7). □

The next proposition specifies the conditions for equilibrium feasibility of planar four-finger grasps.

Proposition 5.7 *A 2-D object $\mathcal{B}$ is held at a feasible* **four-finger equilibrium grasp** *with all four fingers active iff there exist feasible contact forces, $(f_1, f_2, f_3, f_4) \in C_1 \times C_2 \times C_3 \times C_4$, satisfying the conditions:*

(i) *The finger force directions positively span the origin of $\mathbb{R}^2$.*
(ii) *Each pair of finger forces generates opposite moments about the intersection point of the other pair of finger force lines.*

Proof: First consider the necessity of Conditions (i) and (ii). Let a four-contact arrangement form a feasible equilibrium grasp with four active finger forces. That is, $\lambda_1, \lambda_2, \lambda_3, \lambda_4 > 0$ in Eq. (5.1). The force component of Eq. (5.1) gives Condition (i). The torque component of Eq. (5.1) (a scalar quantity), $\lambda_1(x_1 \times \hat{f}_1) + \cdots + \lambda_4(x_4 \times \hat{f}_4) = 0$, has two non-zero summands when computed about the intersection point of any two finger force lines. The non-zero summands must have opposite signs, which gives Condition (ii).

Next assume that Conditions (i) and (ii) are met at a given grasp. Let the finger force indices be assigned according to a counterclockwise ordering of their directions. When these forces are placed at a common origin, the sector spanned by $(\hat{f}_1, \hat{f}_2)$ is disjoint from the sector spanned by $(\hat{f}_3, \hat{f}_4)$. Let l_i denote the ith finger force line. Let p_{12} denote the intersection point of l_1 and l_2, and let p_{34} denote the intersection point of l_3 and l_4. The set of net wrenches generated by varying the magnitudes of each pair of forces, $(\hat{f}_1, \hat{f}_2)$ and $(\hat{f}_3, \hat{f}_4)$, corresponds to two sectors of force lines. One sector is based at p_{12} and bounded by l_1 and l_2; the other sector is based at p_{34} and bounded by l_3 and l_4. Since f_1 and f_2 generate opposite moments about p_{34}, the sector based at p_{12} contains a force line that passes through p_{34}. Similarly, the sector based at p_{34} contains a force line that passes through p_{12}. Since the sectors spanned by $(\hat{f}_1, \hat{f}_2)$ and $(\hat{f}_3, \hat{f}_4)$ are disjoint, the two collinear forces must have *opposite* directions. Since we can freely modulate the magnitude of these opposing forces, there exists a combination of feasible finger forces that forms a four-finger equilibrium grasp. □

The remainder of the appendix concerns 3-D equilibrium grasps. The following lemma asserts that every 3-D object possesses three antipodal grasps along its minimal, intermediate and maximal segments.

Lemma 5.10. *Every piecewise smooth 3-D object possesses at least* **three** *antipodal equilibrium grasps along its minimal-, intermediate- and maximal-width segments.*

Proof sketch: Consider an object $\mathcal{B}$ that has a smooth surface. Using $\mathsf{bdy}(\mathcal{B})$ to denote the boundary of $\mathcal{B}$, the ordered pair $(x_1, x_2) \in \mathsf{bdy}(\mathcal{B}) \times \mathsf{bdy}(\mathcal{B})$ specifies the position of the two finger contacts. Let $\sigma(x_1, x_2) = \|x_1 - x_2\|$ be the inter-contact distance. Since $\sigma(x_1, x_2) = \sigma(x_2, x_1)$, an extremum of σ at (x_1, x_2) has a symmetric extremum at (x_2, x_1), associated with a switch of the finger positions. The gradient $\nabla\sigma$ is well defined in the set $x_1 \neq x_2$, and the extrema of σ in this set occur at points where $\nabla\sigma(x_1, x_2) = \vec{0}$. The extrema of σ correspond to antipodal grasps of $\mathcal{B}$ (see Exercises).[7] Hence, we will show that σ has at least *three* extrema in the set $x_1 \neq x_2$. Since $\sigma(x_1, x_2) = \sigma(x_2, x_1)$, this would imply that σ has three symmetric pairs of extrema.

The object surface, $\mathsf{bdy}(\mathcal{B})$, forms a compact two-dimensional manifold. Hence the product $\mathsf{bdy}(\mathcal{B}) \times \mathsf{bdy}(\mathcal{B})$ forms a compact four-dimensional manifold. Since σ is a continuous function, it attains a global minimum and maximum on $\mathsf{bdy}(\mathcal{B}) \times \mathsf{bdy}(\mathcal{B})$. The global minimum occurs on the two-dimensional subset $x_1 = x_2$. The global maximum occurs at the endpoints of $\mathcal{B}$'s maximum-width segment, at $p_1 = (x_1^0, x_2^0)$. Note that a

[7] This property holds for convex objects. When the object $\mathcal{B}$ is non-convex, only a subset of the extremum points of σ may correspond to antipodal grasps.

symmetric global maximum occurs when the fingers switch positions, at $p_2=(x_2^0,x_1^0)$. The maximum-width segment thus gives the first antipodal grasp of $\mathcal{B}$.

The *mountain pass theorem*, adapted to our purposes, is as follows. Let $\sigma(p):\mathcal{M}\to\mathcal{R}$ be a continuous function on a compact and connected manifold $\mathcal{M}$. Let σ have two isolated local minima at $p_1,p_2\in\mathcal{M}$. Let $\mathcal{D}_1,\mathcal{D}_2\subset\mathcal{M}$ be the basins of attraction of p_1 and p_2, each consisting of the points attracted to the local minimum by the flow of the gradient system $\dot{p}=-\nabla\sigma(p)$. Let $\bar{\mathcal{D}}_i$ denote the closure of $\mathcal{D}_i$ for $i=1,2$. If the set $\bar{\mathcal{D}}_1\cap\bar{\mathcal{D}}_2$ is nonempty, it represents a "mountain range" separating $\mathcal{D}_1$ from $\mathcal{D}_2$, on which σ attains higher values. The mountain pass theorem asserts that σ has a *saddle point* in $\bar{\mathcal{D}}_1\cap\bar{\mathcal{D}}_2$. Moreover, the saddle occurs at the point that minimizes the amount of climbing along all paths from p_1 to p_2. In our case, $\sigma(x_1,x_2)$ has two mountain tops at $p_1=(x_1^0,x_2^0)$ and $p_2=(x_2^0,x_1^0)$ separated by valleys. The saddle of σ therefore occurs at the point that minimizes the amount of descent among all paths connecting p_1 to p_2. The saddle and its symmetric counterpart occur at the endpoints of $\mathcal{B}$'s intermediate width segment, depicted in Figure 5.13(a). This segment gives the intermediate antipodal grasp of $\mathcal{B}$.

Let p_3 be the saddle point of σ associated with $\mathcal{B}$'s intermediate-width segment, and let p_4 be its symmetric counterpart. Based on *Morse theory* (Appendix B), the Hessian matrix $D^2\sigma$ has one positive and three negative eigenvalues at p_3 and p_4. It follows that σ is locally increasing along one direction (leading to p_1 and p_2) and is locally decreasing along the remaining three directions. The two saddles thus form global maxima in the sublevel set: $\mathcal{D}=\{(x_1,x_2)\in \mathsf{bdy}(\mathcal{B})\times\mathsf{bdy}(\mathcal{B}):\sigma(x_1,x_2)\le c\}$, where $c=\sigma(p_3)=\sigma(p_4)$. It can be verified that $\mathcal{D}$ forms a connected set, which therefore contains a path from p_3 to p_4. Since p_3 and p_4 form two mountain tops separated by common valleys within $\mathcal{D}$, σ must have a saddle in the valleys separating p_3 from p_4 within $\mathcal{D}$. The saddle and its symmetric counterpart occur at the endpoints of $\mathcal{B}$'s minimum-width segment, depicted in Figure 5.13(b). The latter segment gives the minimal antipodal grasp of $\mathcal{B}$. □

We proceed to consider the 3-D equilibrium grasps that involve four fingers. The following lemma describes a geometric condition under which four finger force lines lie in a common regulus.

Lemma 5.12 *Four lines in $\mathbb{R}^3$ lie in a common* **regulus** *iff the projections of every line triplet on the plane orthogonal to the fourth line intersect at a common point.*

Proof: First assume that the four lines – denoted l_1,l_2,l_3,l_4 – lie in a common regulus. We have to show that the projections of every line triplet – say l_1,l_2,l_3 on the plane orthogonal to l_4 – intersect at a common point. Let the z-axis of the world and object frames be aligned with $\hat{l}_4$. Under this choice the (x,y) plane is orthogonal to $\hat{l}_4$. Let $\boldsymbol{e}=(0,0,1)$ denote a unit vector aligned with the z-axis, so that $\hat{l}_4=\boldsymbol{e}$. Given a vector $v\in\mathbb{R}^3$, let $\tilde{v}\in\mathbb{R}^2$ denote the projection of v on the (x,y) plane. One can verify that $v\in\mathbb{R}^3$ satisfies the identity:

$$v\times\boldsymbol{e}=\begin{pmatrix} J\tilde{v}\\ 0\end{pmatrix} \qquad J=\begin{bmatrix}0 & 1\\ -1 & 0\end{bmatrix}.$$

By assumption the lines $l_i = (\hat{l}_i, x_i \times \hat{l}_i)$ for $i = 1 \ldots 4$ lie in a common three-dimensional subspace in Plücker coordinates. Hence, the following 6×4 matrix has linearly dependent columns:

$$A = \begin{bmatrix} \hat{l}_1 & \hat{l}_2 & \hat{l}_3 & \boldsymbol{e} \\ x_1 \times \hat{l}_1 & x_2 \times \hat{l}_2 & x_3 \times \hat{l}_3 & \begin{pmatrix} J\tilde{x}_4 \\ 0 \end{pmatrix} \end{bmatrix},$$

where we substituted $\hat{l}_4 = \boldsymbol{e}$ and $x_4 \times \boldsymbol{e} = \begin{pmatrix} J\tilde{x}_4 \\ 0 \end{pmatrix}$. Let us focus on the upper two rows and the bottom row of A. These three rows have a vanishing fourth component. Let $\bar{A}$ be the 3×3 submatrix consisting of the non-zero components of these three rows. Since $\bar{A}$ is a submatrix of A, the condition $A\begin{pmatrix} \lambda_1 \\ \vdots \\ \lambda_4 \end{pmatrix} = \vec{0}$ implies that $\bar{A}\begin{pmatrix} \lambda_1 \\ \lambda_2 \\ \lambda_3 \end{pmatrix} = \vec{0}$. Hence, $\bar{A}$ has linearly dependent columns:

$$\bar{A} = \begin{bmatrix} \tilde{l}_1 & \tilde{l}_2 & \tilde{l}_3 \\ \boldsymbol{e} \cdot (x_1 \times \hat{l}_1) & \boldsymbol{e} \cdot (x_2 \times \hat{l}_2) & \boldsymbol{e} \cdot (x_3 \times \hat{l}_3) \end{bmatrix} = \begin{bmatrix} \tilde{l}_1 & \tilde{l}_2 & \tilde{l}_3 \\ \tilde{x}_1 \times \tilde{l}_1 & \tilde{x}_2 \times \tilde{l}_2 & \tilde{x}_3 \times \tilde{l}_3 \end{bmatrix},$$

where $\tilde{l}_i$ is the (x, y) projection of $\hat{l}_i$, $\boldsymbol{e} \cdot (x_i \times \hat{l}_i) = \tilde{x}_i \times \tilde{l}_i$ based on the triple scalar product identity, and $\tilde{u} \times \tilde{v} = \tilde{u}^T J \tilde{v}$ for $\tilde{u}, \tilde{v} \in \mathbb{R}^2$. The columns of $\bar{A}$ are the Plücker coordinates of the planar lines obtained by projecting l_1, l_2, l_3 on the (x, y) plane. Since these columns are linearly dependent, the projected lines belong to a common flat pencil and, therefore, intersect at a common point in the (x, y) plane. In the special case where one of the three lines is parallel to l_4, the column of this line vanishes in $\bar{A}$. In this case, the columns of the remaining two lines are linearly dependent and hence project to a common line in the (x, y) plane.

Next, assume that the projections of every line triplet on the plane orthogonal to the fourth line intersect at a common point. We have to show that the lines l_1, l_2, l_3, l_4 lie in a common regulus. We will consider the generic case in which l_1, l_2, l_3, l_4 do not lie in parallel planes. Using the previous choice of world and object frames, let $\tilde{x}_0 \in \mathbb{R}^2$ be the intersection point of the projections of l_1, l_2, l_3 in the (x, y) plane, and let $x_0 = (\tilde{x}_0, 0)$ be an embedding of $\tilde{x}_0$ in $\mathbb{R}^3$. Let $l = (\boldsymbol{e}, x_0 \times \boldsymbol{e})$ be the Plücker coordinates of the line passing through x_0 along the direction $\boldsymbol{e}$. The reciprocal product of l with the columns of A is given by

$$\begin{aligned} (x_0 \times \boldsymbol{e}, \boldsymbol{e}) &\begin{bmatrix} \hat{l}_1 & \hat{l}_2 & \hat{l}_3 & \boldsymbol{e} \\ x_1 \times \hat{l}_1 & x_2 \times \hat{l}_2 & x_3 \times \hat{l}_3 & \begin{pmatrix} J\tilde{x}_4 \\ 0 \end{pmatrix} \end{bmatrix} \\ &= \big((x_0 \times \boldsymbol{e}) \cdot \hat{l}_1 + \boldsymbol{e} \cdot (x_1 \times \hat{l}_1), \ldots, (x_0 \times \boldsymbol{e}) \cdot \hat{l}_3 + \boldsymbol{e} \cdot (x_3 \times \hat{l}_3), 0\big) \\ &= \big(\tilde{l}_1 \times (\tilde{x}_1 - \tilde{x}_0), \ldots, \tilde{l}_3 \times (\tilde{x}_3 - \tilde{x}_0), 0\big) = (0, 0, 0, 0), \end{aligned}$$

since $\tilde{x}_i - \tilde{x}_0$ is collinear with $\tilde{l}_i$ for $i = 1, 2, 3$. The line l is thus *reciprocal* to l_1, l_2, l_3, l_4. By repeating this argument for the other line triplets, we obtain a total of four lines, which are all reciprocal to l_1, l_2, l_3, l_4. Since l_1, l_2, l_3, l_4 do not lie in parallel planes, they span at least a three-dimensional subspace in Plücker coordinates. Since the four reciprocal lines have the same directions as $l_1,\ldots,l_4$, they also span at least a three-dimensional subspace in Plücker coordinates. Since the two subspaces are mutually orthogonal, l_1,l_2,l_3,l_4 lie in a common three-dimensional subspace in Plücker coordinates, which forms a regulus. □

Based on Lemma 5.12, equilibrium feasibility of four finger force lines can be graphically determined as follows.

Proposition 5.13 *Let a 3-D object $\mathcal{B}$ be held by four fingers along nonparallel force lines. The finger force lines form a feasible* **equilibrium grasp** *iff every triplet of finger force lines projects to a planar equilibrium grasp in the plane orthogonal to the fourth finger force line.*

Proof: We will use the same notation introduced in the proof of Lemma 5.12, with $l_i = (\hat{l}_i, x_i \times \hat{l}_i)$ for $i = 1\ldots 4$. First assume that $\mathcal{B}$ is held in a 3-D equilibrium grasp by four active fingers. In this case, the columns of the 6×4 matrix A are linearly dependent with positive coefficients:

$$\begin{bmatrix} \hat{l}_1 & \hat{l}_2 & \hat{l}_3 & \boldsymbol{e} \\ x_1\times\hat{l}_1 & x_2\times\hat{l}_2 & x_3\times\hat{l}_3 & \begin{pmatrix} J\tilde{x}_4 \\ 0 \end{pmatrix} \end{bmatrix} \begin{pmatrix} \lambda_1 \\ \vdots \\ \lambda_4 \end{pmatrix} = \vec{0} \qquad \lambda_i > 0,\ \ i = 1\ldots 4.$$

Therefore, the columns of the 3×3 submatrix $\bar{A}$ are also linearly dependent with the same positive coefficients:

$$\begin{bmatrix} \tilde{f}_1 & \tilde{f}_2 & \tilde{f}_3 \\ \tilde{x}_1\times\tilde{f}_1 & \tilde{x}_2\times\tilde{f}_2 & \tilde{x}_3\times\tilde{f}_3 \end{bmatrix} \begin{pmatrix} \lambda_1 \\ \lambda_2 \\ \lambda_3 \end{pmatrix} = \vec{0} \qquad \lambda_i > 0,\ \ i = 1\ldots 3. \tag{5.8}$$

Since the columns of $\bar{A}$ are the projections of the force lines l_1,l_2,l_3 on the (x,y) plane, Eq. (5.8) implies that these force lines form a planar equilibrium grasp. In the special case where one of the force lines is parallel to l_4, its column vanishes in $\bar{A}$. In this case, the projections of the remaining two force lines form a two-finger equilibrium grasp in the (x,y) plane.

Next, assume that the projections of the force-line triplets form a three-finger equilibrium grasp in the plane orthogonal to the fourth force line. According to Lemma 5.12, this condition implies that the four spatial force lines have linearly dependent Plücker coordinates. Suppose, to the contrary, that the four finger forces do not form a spatial equilibrium grasp. Equivalently, the wrenches induced by the four finger forces do *not* positively span the wrench space origin. The net wrench cone spanned by the four finger wrenches is given by

$$\mathcal{W} = \left\{ \lambda_1 \begin{pmatrix} \hat{l}_1 \\ x_1\times\hat{l}_1 \end{pmatrix} + \cdots + \lambda_4 \begin{pmatrix} \hat{l}_4 \\ x_4\times\hat{l}_4 \end{pmatrix} : \lambda_1,\lambda_2,\lambda_3,\lambda_4 \geq 0 \right\}.$$

Since the four wrenches do not positively span the origin, $\mathcal{W}$ forms a pointed cone having the wrench space origin at its apex. By construction, every edge of $\mathcal{W}$ is aligned with one of its generating wrenches, $(\hat{l}_i, x_i\times\hat{l}_i)$ for $i = 1\ldots 4$. Since $\mathcal{W}$ is four-dimensional and convex, it possesses a separating three-dimensional hyperplane that passes through each of its edges. Let $\tilde{\mathcal{W}}$ be the projection of $\mathcal{W}$ on the three-dimensional subspace (f_x, f_y, τ_z), representing the wrenches generated by forces lying in the (x, y) plane. Since the (x, y) plane is orthogonal to $\hat{l}_4$, the projection maps the wrench $(\hat{l}_4, x_4 \times \hat{l}_4)$ to the origin. Since $(\hat{l}_4, x_4 \times \hat{l}_4)$ is an edge of $\mathcal{W}$, the projection maps the separating three-dimensional hyperplane of $\mathcal{W}$ passing through this edge to a separating two-dimensional plane of $\tilde{\mathcal{W}}$ in the (f_x, f_y, τ_z) subspace. It follows that the three projected wrenches, $(\tilde{f}_i, \tilde{x}_i \times \tilde{f}_i)$ for $i = 1 \ldots 3$, lie in a common half-space of the (f_x, f_y, τ_z) subspace and therefore cannot form a planar three-finger equilibrium grasp. As this contradicts our assumption, the original four wrenches must positively span the wrench space origin and therefore form a feasible equilibrium grasp of $\mathcal{B}$. □

Appendix II: The Dimension of the Set of Frictionless Equilibrium Grasps

This appendix characterizes the dimension of the set of frictionless equilibrium grasps of a rigid object $\mathcal{B}$, using the following *contact space.* Assume that $\mathcal{B}$ is bounded by a single curve in 2-D and a single surface in 3-D. Let the object boundary, $\mathsf{bdy}(\mathcal{B})$, be parametrized by a continuous function: $x(u) : \mathbb{R} \rightarrow \mathsf{bdy}(\mathcal{B})$ in 2-D grasps, and $x(u) : \mathbb{R}^2 \rightarrow \mathsf{bdy}(\mathcal{B})$ in 3-D grasps. Contact space is defined as follows.

DEFINITION 5.4 (**contact space**) *Let a rigid object $\mathcal{B}$ be held by k fingers whose contact positions are parametrized by $x_i = x(u_i)$ for $i = 1 \ldots k$.* **Contact space** *is the space of parameters $\mathcal{U} = (u_1, \ldots, u_k)$, where $\mathcal{U} = \mathbb{R}^k$ for 2-D grasps and $\mathcal{U} = \mathbb{R}^{2k}$ for 3-D grasps, with the suitable periodicity rules.*

Let $\mathcal{E}$ denote the set of parameters representing frictionless equilibrium grasps in contact space $\mathcal{U}$. The dimension of $\mathcal{E}$, denoted $\dim(\mathcal{E})$, is characterized in the following theorem.

THEOREM 5.18 (**dimension of frictionless equilibrium grasps**) *In 2-D grasps, the set $\mathcal{E}$ is generically discrete for two fingers,* $\dim(\mathcal{E}) = 2$ *for three fingers and* $\dim(\mathcal{E}) = k$ *for $k \geq 4$ fingers.*

In 3-D grasps, the set $\mathcal{E}$ is generically discrete for two fingers, $\dim(\mathcal{E}) = 3k - 7$ *for $3 \leq k \leq 6$ fingers and* $\dim(\mathcal{E}) = 2k$ *for $k \geq 7$ fingers.*

Note that the dimension of $\mathcal{E}$ matches the dimension of the ambient contact space for $k \geq 4$ fingers in 2-D grasps and $k \geq 7$ fingers in 3-D grasps. Some special cases are discussed later in this appendix.

Proof: Consider the generic case where a rigid object $\mathcal{B}$ is held along locally smooth pieces of its boundary. Recall that n_i is $\mathcal{B}$'s inward unit normal at x_i. When $\mathcal{B}$ is held at a fixed configuration q_0, $x_i = x(u_i)$ and $n_i = n(u_i)$ for $i = 1 \ldots k$. The wrench generated by application of a normal contact force, $f_i = \lambda_i n_i$, on $\mathcal{B}$ at x_i is given by $\boldsymbol{w}_i = \lambda_i(n_i, x_i \times n_i)$. Hence, we may write the equilibrium condition of Eq. (5.1) as

$$\lambda_1 \begin{pmatrix} n(u_1) \\ x(u_1) \times n(u_1) \end{pmatrix} + \ldots + \lambda_k \begin{pmatrix} n(u_k) \\ x(u_k) \times n(u_k) \end{pmatrix} = \vec{0} \qquad \lambda_1, \ldots, \lambda_k \geq 0. \quad (5.9)$$

Let $\tilde{\mathcal{U}}$ denote the composite space of parameters $(u_1, \ldots, u_k, \lambda_1, \ldots, \lambda_k) \in \mathbb{R}^{nk}$, where $n = 2$ or 3. The set of equilibrium grasps in $\tilde{\mathcal{U}}$, denoted $\tilde{\mathcal{E}}$, is characterized by the conditions:

$$\tilde{\mathcal{E}} = \left\{ \begin{pmatrix} \vec{u} \\ \vec{\lambda} \end{pmatrix} \in \tilde{\mathcal{U}} : \sum_{i=1}^{k} \lambda_i \begin{pmatrix} n(u_i) \\ x(u_i) \times n(u_i) \end{pmatrix} = \vec{0} \quad \text{and} \quad \lambda_1, \ldots, \lambda_k \geq 0 \right\},$$

where $\vec{u} = (u_1, \ldots, u_k)$ and $\vec{\lambda} = (\lambda_1, \ldots, \lambda_k)$. Let $\pi : \tilde{\mathcal{U}} \to \mathcal{U}$ be the *coordinate projection,* which maps points $(\vec{u}, \vec{\lambda}) \in \tilde{\mathcal{U}}$ to points $\vec{u} \in \mathcal{U}$. Then $\mathcal{E}$ is obtained by projecting $\tilde{\mathcal{E}}$ into $\mathcal{U}$, $\mathcal{E} = \pi(\tilde{\mathcal{E}})$.

The remainder of the proof focuses on 3-D grasps. First consider the case of $k \geq 3$ fingers. Eq. (5.9) imposes *six* scalar constraints in the $3k$-dimensional space $\tilde{\mathcal{U}}$. Let us verify that these constraints intersect transversally in the composite contact space, $\tilde{\mathcal{U}} = \mathbb{R}^{3k}$. The six constraints form a vector-valued mapping, h, which maps the equilibrium grasps in $\tilde{\mathcal{U}}$ to the origin in $\mathbb{R}^6$:

$$h(\vec{u}, \vec{\lambda}) = \sum_{i=1}^{k} \lambda_i \begin{pmatrix} n(u_i) \\ x(u_i) \times n(u_i) \end{pmatrix} : \mathbb{R}^{3k} \to \mathbb{R}^6. \tag{5.10}$$

For the six constraints to intersect transversally (and hence define a smooth $(3k - 6)$-dimensional manifold in $\tilde{\mathcal{U}}$), the $6 \times 3k$ Jacobian matrix, Dh, must have full rank at all points of $\tilde{\mathcal{E}}$. The Jacobian Dh has the form:

$$Dh = \begin{bmatrix} \lambda_1 \frac{d}{du_1} n_1 & \cdots & \lambda_k \frac{d}{du_k} n_k & n_1 & \cdots & n_k \\ \lambda_1 \frac{d}{du_1}(x_1 \times n_1) & \cdots & \lambda_k \frac{d}{du_k}(x_k \times n_k) & x_1 \times n_1 & \cdots & x_k \times n_k \end{bmatrix}. \tag{5.11}$$

Note that $\frac{d}{du_i} n_i$ and $\frac{d}{du_i} x_i \times n_i$ are 3×2 matrices in Eq. (5.11). The last k columns of Dh form the *grasp matrix,* G. The columns of G are linearly dependent at an equilibrium grasp. Hence, the rank of Dh at points of $\tilde{\mathcal{E}}$ is generically $\min\{6, 3k - 1\} = 6$ for $k \geq 3$ fingers, which is full rank. The six constraints thus intersect transversally in generic grasps that involve $k \geq 3$ fingers. The solution set $h(\vec{u}, \vec{\lambda}) = \vec{0}$ consequently forms a smooth $(3k - 6)$-dimensional manifold in $\tilde{\mathcal{U}}$. The set $\tilde{\mathcal{E}}$ is obtained by intersecting the latter manifold with the k-quadrant $\lambda_1, \ldots, \lambda_k \geq 0$. This intersection may introduce boundary components into the manifold, but the intersection does not change the dimension of the solution set. The set $\tilde{\mathcal{E}}$ thus forms a smooth $(3k - 6)$-dimensional manifold with boundary in $\tilde{\mathcal{U}}$.

In order to obtain the set of equilibrium grasps, $\mathcal{E}$, consider the projection $\pi : \tilde{\mathcal{U}} \to \mathcal{U}$. There are two cases to consider. When the grasp involves $3 \leq k \leq 6$ fingers, the function h specified in Eq. (5.10) is linearly *homogeneous* in $\lambda_1, \ldots, \lambda_k$. Hence, $\tilde{\mathcal{E}}$ consists of entire rays, each having the form $(u_1, \ldots, u_k, s \cdot \lambda_1, \ldots, s \cdot \lambda_k)$ for $s \geq 0$. The projection π maps each of these rays to a single point, $(u_1, \ldots, u_k) \in \mathcal{U}$. The dimension of $\mathcal{E} = \pi(\tilde{\mathcal{E}})$ is thus *one less* than the dimension of $\tilde{\mathcal{E}}$, which is $3k - 7$.[8] For $k \geq 7$ fingers, the function h is *linear* in $\lambda_1, \ldots, \lambda_k$. In this case, $\tilde{\mathcal{E}}$ consists of linear spaces of rays in $\lambda_1, \ldots, \lambda_k$. The projection π maps each of these subspaces to a single point, $(u_1, \ldots, u_k) \in \mathcal{U}$. The

[8] The set $\tilde{\mathcal{E}}$ forms a *cylindrical set* above the lower dimensional set $\mathcal{E}$.

dimension of the linear subspaces is $k-6$, and the dimension of $\mathcal{E}=\pi(\tilde{\mathcal{E}})$ is $(3k-6)-(k-6)=2k$. This establishes that the dimension of $\mathcal{E}$ is $3k-7$ for $3 \leq k \leq 6$ fingers and $2k$ for $k \geq 7$ fingers.

In the case of two-finger grasps, the six equations of Eq. (5.9) can be reduced into four independent equations directly in $\mathcal{U}$. Let $(t_{i1}(u_i), t_{i2}(u_i))$ be mutually orthogonal unit tangents to the surface of $\mathcal{B}$ at $x_i = x(u_i)$ $(i=1,2)$. A two-finger equilibrium grasp requires that the two contact normals be collinear with the line passing through x_1 and x_2. This requirement is captured by the four equations: $t_{i1} \cdot (x_1 - x_2) = 0$ and $t_{i2} \cdot (x_1 - x_2) = 0$ for $i = 1,2$. The four equations form a vector-valued mapping, $h(\vec{u})$, which maps the set of two-finger equilibrium grasps in $\mathcal{U} = \mathbb{R}^4$ to the origin in $\mathbb{R}^4$:

$$h(\vec{u}) = \begin{pmatrix} t_{i1}(u_1) \cdot (x_1(u_1) - x_2(u_2)) \\ t_{i2}(u_1) \cdot (x_1(u_1) - x_2(u_2)) \\ t_{i1}(u_2) \cdot (x_1(u_1) - x_2(u_2)) \\ t_{i2}(u_2) \cdot (x_1(u_1) - x_2(u_2)) \end{pmatrix} : \mathbb{R}^4 \to \mathbb{R}^4. \tag{5.12}$$

It can be verified that the 4×4 Jacobian of this mapping, Dh, has full rank when $x_1 \neq x_2$. In this case, the four constraints intersect transversally in $\mathcal{U} = \mathbb{R}^4$, and the solution set of Eq. (5.12) forms a zero-dimensional manifold in $\mathcal{U}$, which is a set of discrete points in $\mathcal{U}$. □

Special cases of parallel edges or parallel facets: The dimension of $\mathcal{E}$ specified in Theorem 5.18 holds for *generic* grasp arrangements. These grasps must possess a full rank Jacobian Dh in Eq. (5.11). A common special case occurs when a polygonal object is held along parallel edges or when a polyhedral object is held along parallel facets. In these cases Dh is rank deficient and the dimension of $\mathcal{E}$ increases, as illustrated in the following example.

Example: A rectangular object $\mathcal{B}$ is held along two parallel edges, e and e', as shown in Figure 5.20. When $\mathcal{B}$ is held by two fingers (Figure 5.20(a)), any antipodal positioning of the two fingers on e and e' satisfies the equilibrium grasp condition. The set $\mathcal{E}$ is thus one-dimensional (a line segment) rather than zero-dimensional in this grasp. When $\mathcal{B}$ is held by three fingers (Figure 5.20(b)), equilibrium requires that the force line of $\mathcal{O}_1$ lie within the strip bounded by the opposing force lines of $\mathcal{O}_2$ and

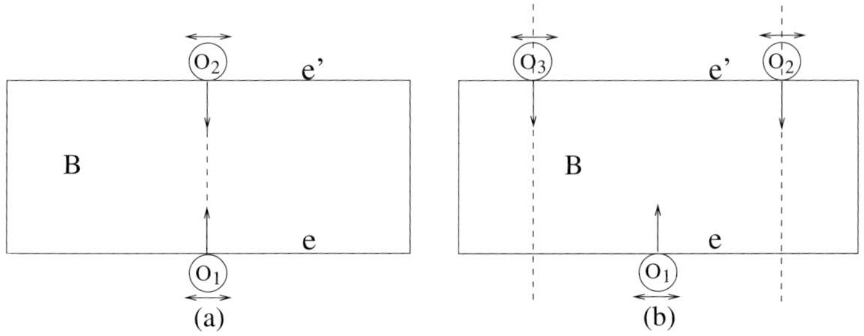

Figure 5.20 When $\mathcal{B}$ is held along two parallel edges, (a) $\dim(\mathcal{E}) = 1$ for two-finger grasps, (b) $\dim(\mathcal{E}) = 3$ for three-finger grasps along the parallel edges.

$\mathcal{O}_3$ (see Section 4.2.2). Since the fingers can independently vary within three intervals while maintaining the equilibrium grasp, the set $\mathcal{E}$ is three-dimensional rather than two-dimensional in this grasp. ○

Exercises

Section 5.1

Exercise 5.1: The vectors $v_1, \ldots, v_k \in \mathbb{R}^n$ are said to be *affinely independent* if, for any $v_i \in \{v_1, \ldots, v_k\}$, the $k-1$ vectors $v_1 - v_i, \ldots, v_{i-1} - v_i, v_{i+1} - v_i, \ldots, v_k - v_i$ are linearly independent. Prove that when affinely independent vectors $v_1, \ldots, v_k \in \mathbb{R}^n$ positively span the origin of $\mathbb{R}^n$ with positive coefficients, they positively span the *entire space* $\mathbb{R}^n$.

Solution: Since $v_1, \ldots, v_{n+1}$ are affinely independent, the $n \times (n+1)$ matrix $A = [v_1 \cdots v_{n+1}]$ has full row rank of n. The matrix A forms a linear mapping from $\mathbb{R}^{n+1}$ onto $\mathbb{R}^n$. Since A has full row rank, it maps a neighborhood about the origin in $\mathbb{R}^{n+1}$ onto a neighborhood about the origin in $\mathbb{R}^n$. Therefore, for any vector $u \in \mathbb{R}^n$ such that $\|u\| << 1$, there exists a coefficients vector $(s_1, \ldots, s_{n+1}) \in \mathbb{R}^{n+1}$ satisfying $u = s_1 v_1 + \ldots + s_{n+1} v_{n+1}$, such that $|s_i| << 1$ for $i = 1 \ldots n+1$. The coefficients vector $(\lambda_1, \ldots, \lambda_{n+1})$ lies in the null space of A. Hence we may add this vector to $(s_1, \ldots, s_{n+1})$ without affecting the value of A, $u = (\lambda_1 + s_1) v_1 + \ldots + (\lambda_{n+1} + s_{n+1}) v_{n+1}$. By assumption, the λ_i's are strictly positive. Hence, by shrinking the magnitude of u, we can ensure that $\lambda_i + s_i \geq 0$ for $i = 1 \ldots n+1$. Since u is an arbitrary vector in a neighborhood of the origin in $\mathbb{R}^n$, the vectors $v_1, \ldots, v_{n+1}$ positively span this neighborhood. Since A is homogenous with respect to scalar multiplication, the vectors $v_1, \ldots, v_{n+1}$ positively span the entire space $\mathbb{R}^n$. ○

Exercise 5.2: Use the moment labeling technique (Section 4.5) to verify that the four-finger grasp depicted in Figure 5.1 is *not* a feasible equilibrium grasp.

Section 5.2

Exercise 5.3: A polygonal object $\mathcal{B}$ has a non-smooth vertex at p. Verify that any vector from $\mathcal{B}$'s *generalized inward contact normal* at p can be realized as the contact normal of a properly placed finger at p.

Solution: First consider the case where p is a convex vertex of $\mathcal{B}$. Let v be any vector from $\mathcal{B}$'s generalized inward contact normal at p. It can be verified that v is orthogonal to a local separating line for $\mathcal{B}$, i.e., a line that passes through p and bounds a half plane containing $\mathcal{B}$. Since $\mathcal{B}$ is tangent to this line at p, a smooth finger tangent to this line from its other side at p would have v as its contact normal. Next, consider the case where p is a concave vertex of $\mathcal{B}$. In this case, the finger must have a sharp tip at p, and the finger edges incident to p must lie within the concavity formed by the edges of $\mathcal{B}$. Therefore, $\mathcal{B}$'s generalized inward contact normal at p is a *subset* of the finger's generalized outward contact normal at p (one must use $\mathcal{B}$'s inward edge normals and the finger's outward edge normals in this intersection). The finger's sharp tip can practically apply any contact force within the intersection of the object's and finger's generalized contact normals at p. Hence, $\mathcal{B}$'s generalized contact normal at p is fully realizable in this case too. ○

Exercise 5.4: Let a planar object $\mathcal{B}$ have a smooth boundary. Prove that every antipodal grasp of $\mathcal{B}$ occurs at an extremum of the function $\sigma(x_1,x_2)=\|x_1-x_2\|$, where $x_1,x_2\in$ bdy($\mathcal{B}$). (Using generalized contact normals, this result also holds for piecewise smooth objects.)

Solution: Let $\beta(s)$ parametrize the boundary of $\mathcal{B}$, such that $\beta'(s)$ is a unit vector tangent to the boundary of $\mathcal{B}$. Let (s_1,s_2) parametrize the position of the two contacts, $x_1=\beta(s_1)$ and $x_2=\beta(s_2)$. Then $\sigma(s_1,s_2)=\|\beta(s_1)-\beta(s_2)\|$, and the extrema of $\sigma(s_1,s_2)$ satisfy

$$\nabla\sigma(s_1,s_2)=\frac{1}{\|\beta(s_1)-\beta(s_2)\|}\begin{pmatrix}(\beta(s_1)-\beta(s_2))\cdot\beta'(s_1)\\ -(\beta(s_1)-\beta(s_2))\cdot\beta'(s_2)\end{pmatrix}=\begin{pmatrix}0\\0\end{pmatrix},$$

where we must assume that $s_1\neq s_2$ for $\nabla\sigma$ to be well defined. We see that $\nabla\sigma$ vanishes at points $x_1=\beta(s_1)$ and $x_2=\beta(s_2)$ where the segment $\overline{x_1x_2}$ is orthogonal to $\mathcal{B}$'s boundary. Equivalently, $\nabla\sigma$ vanishes at points where the unit normals are collinear with the segment $\overline{x_1x_2}$. In particular, $\nabla\sigma$ vanishes at the antipodal equilibrium grasps of $\mathcal{B}$. ○

Exercise 5.5*: Let S^{n-1} denote a set that is topologically equivalent to the $(n-1)$ dimensional sphere in $\mathbb{R}^n$. The *Bursuk–Ulum theorem* states that any continuous map $\sigma: S^{n-1}\to\mathbb{R}^n$ maps some pair of antipodal points to the *same* point in $\mathbb{R}^n$. That is, there exist two points $x_1,x_2\in S^{n-1}$ such that $x_2=-x_1$ and $\sigma(x_1)=\sigma(x_2)$. Let $\mathcal{B}$ be a planar object bounded by a simple closed loop. Does the Bursuk–Ulum theorem imply that $\mathcal{B}$ possesses at least one antipodal grasp?

Exercise 5.6: A planar object $\mathcal{B}$ has a smooth boundary, parametrized in counterclockwise direction by $\beta(s)$. Assume that $\beta(s)$ is an arc-length parametrization, so that $\beta'(s)$ has unit magnitude. The three-contact arrangements can be parametrized by the position of the three contacts, $x_i=\beta(s_i)$ for $i=1,2,3$. Prove that the set of contact triplets with co-intersecting contact normals satisfies the scalar constraint:

$$\det\begin{bmatrix}\beta'(s_1) & \beta'(s_2) & \beta'(s_3)\\ \beta(s_1)\cdot\beta'(s_1) & \beta(s_2)\cdot\beta'(s_2) & \beta(s_3)\cdot\beta'(s_3)\end{bmatrix}=0, \tag{5.13}$$

where $\beta'(s_i)$ is the unit tangent to the boundary of $\mathcal{B}$ at $\beta(s_i)$ for $i=1,2,3$.

Solution: Substituting $f_i=\lambda_i n_i$ for the finger forces in (5.1), the three-finger equilibria satisfy the condition:

$$\begin{bmatrix}n_1 & n_2 & n_3\\ x_1\times n_1 & x_2\times n_2 & x_3\times n_3\end{bmatrix}\begin{pmatrix}\lambda_1\\ \lambda_2\\ \lambda_2\end{pmatrix}=\vec{0}.$$

At a frictionless three-finger equilibrium, the contact normal lines intersect at a common point. Hence the vanishing of the determinant of the 3×3 matrix gives the condition for intersecting contact normals:

$$\det \begin{bmatrix} n_1 & n_2 & n_3 \\ x_1 \times n_1 & x_2 \times n_2 & x_3 \times n_3 \end{bmatrix} = 0.$$

The determinant can be expressed as a function of the contact parameters – s_1, s_2, s_3 – as follows. The lower-row terms satisfy $x_i \times n_i = x_i^T J^T n_i$ for $i = 1,2,3$. Since $n_i = J\beta'(s_i)$, substituting for x_i and n_i gives $x_i \times n_i = \beta(s_i)^T J^T J \beta'(s_i) = \beta(s_i) \cdot \beta'(s_i)$, where we used the identity $J^T J = I$. The upper-rows terms satisfy the identity $\det[n_i\ n_{i+1}] = n_i^T J^T n_{i+1} = \beta'(s_i)^T J^T \beta'(s_{i+1}) = \det[\beta'(s_i)\ \beta'(s_{i+1})]$. The condition that the contact normals intersect at a common point is thus captured by Eq. (5.13). ○

Exercise 5.7: Provide a proof sketch of Lemma 5.6.

Solution: The equilibrium condition can be written in matrix form as

$$\begin{bmatrix} \hat{f}_1 & \hat{f}_2 & \hat{f}_3 \\ x_1 \times \hat{f}_1 & x_2 \times \hat{f}_2 & x_3 \times \hat{f}_3 \end{bmatrix} \begin{pmatrix} \lambda_1 \\ \lambda_2 \\ \lambda_3 \end{pmatrix} = \vec{0}.$$

Since $(\hat{f}_1, \hat{f}_2, \hat{f}_3)$ are not all collinear, the upper two rows are linearly independent, and the 3×3 matrix thus has rank two at the equilibrium. Hence, the solution $(\lambda_1, \lambda_2, \lambda_3)$ spans a one-dimensional null space, which must be orthogonal to the three rows. Since the upper two rows are linearly independent, the bottom row is a linear combination of the upper two rows. Hence, up to a common scalar factor, the solution $(\lambda_1, \lambda_2, \lambda_3)$ is the cross product of the upper two rows, given by $(\lambda_1, \lambda_2, \lambda_3) = (\hat{f}_2 \times \hat{f}_3, \hat{f}_3 \times \hat{f}_1, \hat{f}_1 \times \hat{f}_2)$. ○

Exercise 5.8: Use the moment labeling technique of Chapter 4 to verify that the four-finger grasp of Figure 5.9(a) is *not* a feasible equilibrium grasp.

Solution: Consider the four-contact arrangement of Figure 5.9(a), depicted again in Figure 5.21(a). Using the notation of Section 4.5, $M^- = \emptyset$ while M^+ has a nonempty interior, as indicated in Figure 5.21(a). Since equilibrium requires that M^- and M^+ be either empty or subsets of a common line, this contact arrangement is not a feasible equilibrium grasp. The force lines corresponding to the net wrench cone of the four finger forces are indicated in the figure. The resulting collection of force lines does not contain any pair of opposing force lines, thus indicating that an equilibrium is not feasible for this grasp. An equilibrium is infeasible here since all variations of the finger force magnitudes generate a nonzero net torque on $\mathcal{B}$. ○

Exercise 5.9: Use the moment labeling technique to verify that the four-finger grasp of Figure 5.9(b) forms a feasible equilibrium grasp.

Solution: Consider the four-contact arrangement of Figure 5.9(b), depicted again in Figure 5.21(b). Now $M^- = M^+ = \emptyset$, and the net wrench cone consequently occupies the entire wrench space. In particular, there exists a positive linear combination of the

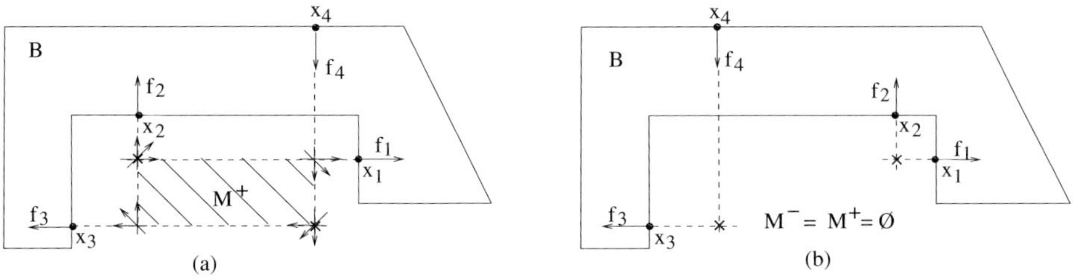

Figure 5.21 (a) A four-contact arrangement with $M^- = \emptyset$ and M^+ having a nonempty interior. (b) A four-contact arrangement with $M^- = M^+ = \emptyset$.

finger forces that spans the wrench space origin, so this contact arrangement is a feasible equilibrium grasp. ○

Exercise 5.10: Provide a proof sketch of Lemma 5.8.

Solution: Let $\boldsymbol{w}_i = (\hat{f}_i, x_i \times \hat{f}_i)$ be the wrench generated by the ith finger force on $\mathcal{B}$ $(i = 1 \ldots 4)$. The equilibrium condition can be written in matrix form as

$$\begin{bmatrix} \boldsymbol{w}_1 & \boldsymbol{w}_1 & \boldsymbol{w}_3 & \boldsymbol{w}_4 \end{bmatrix} \begin{pmatrix} \lambda_1 \\ \lambda_2 \\ \lambda_3 \\ \lambda_4 \end{pmatrix} = \vec{0}.$$

Since the four force lines do not intersect at a common point, some three of the four force lines do not intersect at a common point. Let the three nonintersecting lines be associated with the forces f_1, f_2, f_3. Since $\boldsymbol{w}_i = (\hat{f}_i, x_i \times \hat{f}_i)$ are the Plücker coordinates of the ith force line, the non-concurrency of the three force lines implies that $\det[\boldsymbol{w}_1\, \boldsymbol{w}_2\, \boldsymbol{w}_3] \neq 0$. The system of equations,

$$\begin{bmatrix} \boldsymbol{w}_1 & \boldsymbol{w}_1 & \boldsymbol{w}_3 \end{bmatrix} \begin{pmatrix} \lambda_1 \\ \lambda_2 \\ \lambda_3 \end{pmatrix} = -\lambda_4 \boldsymbol{w}_4,$$

can be solved using Cramer's rule:

$$\begin{pmatrix} \lambda_1 \\ \lambda_2 \\ \lambda_3 \end{pmatrix} = \frac{\lambda_4}{\det[\boldsymbol{w}_1\, \boldsymbol{w}_2\, \boldsymbol{w}_3]} \begin{pmatrix} -\det[\boldsymbol{w}_2\, \boldsymbol{w}_3\, \boldsymbol{w}_4] \\ \det[\boldsymbol{w}_1\, \boldsymbol{w}_3\, \boldsymbol{w}_4] \\ -\det[\boldsymbol{w}_1\, \boldsymbol{w}_2\, \boldsymbol{w}_4] \end{pmatrix}.$$

Expressing λ_4 as $\lambda_4 = s \cdot \det[\boldsymbol{w}_1\, \boldsymbol{w}_2\, \boldsymbol{w}_3]$ for $s \in \mathbb{R}$, we obtain Formula (5.3) with s as the common scaling factor. ○

Exercise 5.11: Consider the frictional four-finger grasp of an ellipse shown in Figure 5.22. The contact normals form a feasible four-finger equilibrium grasp, while other force directions do not. Does this contradict the graphical procedure for checking equilibrium feasibility of frictional four-finger grasps?

Solution: The graphical procedure requires that we first check if the four-contact arrangement supports any two- or three-finger equilibrium grasp. Recall that $\mathbb{C}_i = C_i \cup C_i^-$ is the double friction cone at x_i. In this example, $\mathbb{C}_1 \cap \mathbb{C}_3 \cap \mathbb{C}_4$ consists of two nonempty polygons. The force triplets in these polygons positively span the origin

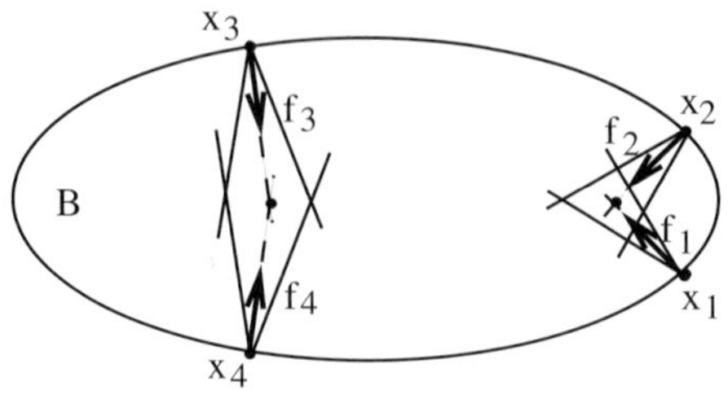

Figure 5.22 A frictional four-contact arrangement that forms a feasible equilibrium grasp along the contact normals but not along other force directions.

of $\mathbb{R}^2$. It follows that this contact arrangement supports three-finger equilibrium grasps, and the coefficients $\lambda_1, \ldots, \lambda_4$ are no longer sign invariant. ○

Exercise 5.12: According to Theorem 5.9, in the case of 2-D grasps, the frictional four-finger equilibrium grasps are identical to the frictionless four-finger equilibrium grasps. Explain why a similar equivalence relation does *not* hold for planar two- and three-finger equilibrium grasps.

Solution: Friction always *enlarges* the collection of planar two- and three-finger equilibrium grasps. First consider the two-finger grasps. A frictional equilibrium requires that each of the double friction cones will contain the opposing contact point. This condition is clearly less stringent than the requirement for antipodal contact normals: the frictionless two-finger grasps occur at discrete points, while the two-finger frictional grasps occur along contiguous boundary segments. Next consider the three-finger grasps. A frictional equilibrium requires that the three double friction cones have a nonempty intersection. This condition is again less stringent than the requirement that the three contact normals intersect at a common point. ○

Section 5.3

Exercise 5.13*: Based on the *mountain pass theorem* discussed in the proof of Lemma 5.10, prove that a smooth planar object always possesses minimal and maximal antipodal grasps.

Solution: Suppose $\mathcal{B}$ has no internal holes and is thus bounded by a single curve, $\mathsf{bdy}(\mathcal{B})$, which forms a loop. Let the ordered pair $(x_1, x_2) \in \mathsf{bdy}(\mathcal{B}) \times \mathsf{bdy}(\mathcal{B})$ denote the position of the finger contacts. Let $\sigma(x_1, x_2) = \|x_1 - x_2\|$ be the inter-finger distance. Since $\sigma(x_1, x_2) = \sigma(x_2, x_1)$, an extremum of σ at (x_1, x_2) has a symmetric extremum at (x_2, x_1). The product $T = \mathsf{bdy}(\mathcal{B}) \times \mathsf{bdy}(\mathcal{B})$ is a compact two-dimensional manifold topologically equivalent to a torus. Since σ is a continuous function, it attains a global minimum and maximum on T. The global minimum of zero occurs on the subset $x_1 = x_2$. The global maximum occurs at the endpoints of $\mathcal{B}$'s maximum-width segment, at $p_1 = (x_1^0, x_2^0)$ and its symmetric counterpart $p_2 = (x_2^0, x_1^0)$. The maximum-width segment thus gives the maximal antipodal grasp of $\mathcal{B}$.

According to the mountain pass theorem, the "valleys" separating the "mountain tops" p_1 and p_2 contain a saddle of σ, which occurs at the point that minimizes the amount of descent among all paths connecting p_1 to p_2. The saddle and its symmetric counterpart occur at the endpoints of $\mathcal{B}$'s minimum-width segment. This segment gives the minimal antipodal grasp of $\mathcal{B}$. ○

Exercise 5.14: Let a 3-D object $\mathcal{B}$ be grasped by two fingers via *frictionless* contacts. Describe the net wrench cone, $\mathcal{W} = \mathcal{W}_1 + \mathcal{W}_2$, associated with this grasp.

Solution: Let x_1 and x_2 be the finger contacts. The net wrench cone, $\mathcal{W} = \mathcal{W}_1 + \mathcal{W}_2$, is a two-dimensional sector based at $\mathcal{B}$'s wrench space origin. The edges of $\mathcal{W}$ are spanned by the wrenches generated by the normal forces at x_1 and x_2. ○

Exercise 5.15: Let a 3-D object $\mathcal{B}$ be grasped by two fingers via *frictional* contacts. Describe the net wrench cone, $\mathcal{W} = \mathcal{W}_1 + \mathcal{W}_2$, associated with this grasp.

Solution: Let x_1 and x_2 be the finger contacts. Set the world and object frame origins at the contacts' midpoint. Under this choice, $\dot{q} = (\vec{0}, x_1 - x_2)$ corresponds to instantaneous rotation of $\mathcal{B}$ about the line passing through x_1 and x_2. Since the wrenches of $\mathcal{W}$ are reciprocal to $\dot{q} = (\vec{0}, x_1 - x_2)$, all force lines representing the wrenches of $\mathcal{W}$ must intersect the line passing through x_1 and x_2. These force lines can be parametrized in terms of a plane $P(\phi)$ rotating with an angle ϕ about the line passing through x_1 and x_2. For each fixed value of ϕ, the sets $C_1 \cap P(\phi)$ and $C_2 \cap P(\phi)$ form planar friction cones. Assume these planar friction cones are nonempty over some $[\phi_{min}, \phi_{max}]$ interval. The wrenches generated by the forces lying in $C_i \cap P(\phi)$ span a two-dimensional sector based at $\mathcal{B}$'s wrench space origin. Let $\mathcal{W}_i(\phi)$ denote this wrench sector for $i = 1, 2$. Then $\mathcal{W}_1(\phi) + \mathcal{W}_2(\phi)$ is a four-dimensional tetrahedral cone based at the wrench space origin. The net wrench cone is thus a family of semi-infinite tetrahedral cones, $\mathcal{W}_1(\phi) + \mathcal{W}_2(\phi)$, parametrized by $\phi \in [\phi_{min}, \phi_{max}]$. Note that the force lines corresponding to each tetrahedral cone lie within the plane $P(\phi)$ and can be depicted with the moment labeling technique described in Chapter 4. ○

Exercise 5.16: Let $\mathcal{B}$ be a smooth 3-D convex object, and let $\sigma(x, y) = \|x - y\|$ for $x, y \in \text{bdy}(\mathcal{B})$ be the inter-finger distance function. Prove that the antipodal grasps of $\mathcal{B}$ occur at the extrema of σ at points $(x, y) \in \text{bdy}(\mathcal{B}) \times \text{bdy}(\mathcal{B})$ such that $x \neq y$.

Solution: Let the surface of $\mathcal{B}$ be parametrized by the smooth function $h(u) : \mathbb{R}^2 \to \mathbb{R}^3$. Let $(u_1, u_2) \in \mathbb{R}^2 \times \mathbb{R}^2$ parametrize the position of the two contacts, $x_1 = h(u_1)$ and $x_2 = h(u_2)$. Then $\sigma(u_1, u_2) = \|h(u_1) - h(u_2)\|$, and the extrema of $\sigma(u_1, u_2)$ satisfy the conditions:

$$\nabla g(u_1, u_2) = \frac{1}{\|h(u_1) - h(u_2)\|} \begin{pmatrix} Dh^T(u_1)(h(u_1) - h(u_2)) \\ -Dh^T(u_2)(h(u_1) - h(u_2)) \end{pmatrix} = \begin{pmatrix} \vec{0} \\ \vec{0} \end{pmatrix}. \tag{5.14}$$

The gradient $\nabla\sigma$ is well defined away from the set $u_1 = u_2$. The columns of the 3×2 Jacobian $Dh(u_i)$ span the tangent plane to $\mathcal{B}$'s surface at $x_i = h(u_i)$ $(i = 1, 2)$. Therefore, Eq. (5.14) implies that $\nabla\sigma$ vanishes at points x_1 and x_2 where the segment $\overline{x_1 x_2}$ – or, equivalently, the vector $h(u_1) - h(u_2)$ – is orthogonal to $\mathcal{B}$'s surface. Equivalently,

$\nabla\sigma$ vanishes at points where the unit normals lie on the line passing through the two contacts. This proves that the extrema of σ correspond to antipodal equilibrium grasps of $\mathcal{B}$. ○

Exercise 5.17: Let a 3-D object $\mathcal{B}$ be grasped by three fingers via *frictionless* contacts. Based on a c-space argument, the object possesses minimal and maximal equilateral equilibrium grasps. Can a similar c-space argument demonstrate that every *planar* object possesses minimal and maximal equilateral grasps?

Solution: Let us attempt to repeat the thought experiment of three-point fingers moving linearly toward a common center point in the planar environment while maintaining an equilateral triangle. In the spatial case, the plane spanned by the three fingers, Δ, partitions $\mathbb{R}^3$ into two half-spaces such that the fingers form a triangular gap embedded in Δ. In the planar case, Δ coincides with the environment. The triangular gap formed by three fingers consequently does not divide the planar environment into two half planes. For objects such as an ellipse, the shrinking three-finger gap involves only a single critical event – the first time that $\mathcal{B}$ can simultaneously touch the three fingers. The resulting equilibrium grasp is analogous to the minimal equilateral grasp of solid objects, since the c-space gap between the three c-obstacles shrinks to a single point at this instant. Every planar object possesses only a minimal equilateral frictionless equilibrium grasp. ○

Exercise 5.18: Let the surface of a 3-D object $\mathcal{B}$ be approximated a local quadratic surface, Q_i, which is tangent to $\mathcal{B}$ at x_i and has $\mathcal{B}$'s principal curvatures at this point. Let $\bar{n}_i$ be $\mathcal{B}$'s *outward* unit normal at x_i. Let V_i be the plane passing through x_i and spanned by $\bar{n}_i$ and one of Q_i's principal directions of curvatures at x_i. Prove that the unit normal to Q_i along its intersection curve with V_i lies within this plane.

Solution: Let t_{i1} and t_{i2} be the principal directions of curvatures of $\mathcal{B}$ at x_i, and let the scalars $\kappa_{\mathcal{B}_{i1}}$ and $\kappa_{\mathcal{B}_{i2}}$ be $\mathcal{B}$'s principal curvatures along these directions. The quadratic surface Q_i is the zero level-set of the scalar-valued function:

$$f(x) = \bar{n}_i \cdot (x - x_i) + \tfrac{1}{2}\left(\kappa_{\mathcal{B}_{i1}}(t_{i1} \cdot (x - x_i))^2 + \kappa_{\mathcal{B}_{i2}}(t_{i2} \cdot (x - x_i))^2\right) \quad x \in \mathbb{R}^3.$$

The gradient of f, evaluated at points on Q_i, gives the normal to Q_i:

$$\nabla f(x) = \bar{n}_i + \kappa_{\mathcal{B}_{i1}}(t_{i1} \cdot (x - x_i))t_{i1} + \kappa_{\mathcal{B}_{i2}}(t_{i2} \cdot (x - x_i))t_{i2} \quad x \in Q_i.$$

Now let V_i be the plane passing through x_i and spanned by t_{i1} and $\bar{n}_i$. Let $x(s)$ for $s \in \mathbb{R}$ be the intersection curve of Q_i with V_i. Since $x(s)$ lies within V_i, the vector $x(s) - x_i$ is orthogonal to the other tangent direction, t_{i2}. Therefore, $\nabla f(x(s)) = \bar{n}_i + \kappa_{\mathcal{B}_{i1}}(t_{i1} \cdot (x(s) - x_i))t_{i1}$, which implies that $\nabla f(x(s))$ lies in V_i for $s \in \mathbb{R}$. ○

Exercise 5.19: Let a 3-D object $\mathcal{B}$ be grasped via three *frictional* contacts. Under what condition does the 6×9 grasp matrix G have full rank?

Solution: Let the three contacts be located at the points x_1, x_2 and x_3. When the world and object frame origins are located at x_1, the grasp map formula (4.2) becomes

$$\mathcal{L}_G = \begin{bmatrix} I & I & I \\ O & [x_2 \times] & [x_3 \times] \end{bmatrix},$$

where O is a 3×3 zero matrix. If G is rank deficient, the columns of the grasp map must be conormal to some non-zero tangent vector $\dot{q} = (\vec{0}, \omega)$:

$$(\vec{0}, \omega) \begin{bmatrix} I & I & I \\ O & [x_2 \times] & [x_3 \times] \end{bmatrix} = -(\vec{0}, x_2 \times \omega, x_3 \times \omega) = (\vec{0}, \vec{0}, \vec{0}) \in \mathbb{R}^9.$$

It follows that G is rank deficient when x_2 and x_3 are collinear with ω. Since ω is an instantaneous rotation of $\mathcal{B}$ about an axis passing through $\mathcal{B}$'s origin at x_1, G is rank deficient only in the non-generic case where the three contacts lie on a common line. ◦

Exercise 5.20*: Figure 5.15(c) shows four spatial lines that lie in a common regulus. Can this graphical depiction be generalized to any quadruple of linearly dependent lines?

Exercise 5.21: A 3-D object $\mathcal{B}$ is held by four parallel finger force lines. Characterize the condition under which such a contact arrangement forms a feasible equilibrium grasp.

Solution: Equilibrium requires that the finger force directions positively span the origin of $\mathbb{R}^3$. Since the force directions are all parallel, they either consist of two opposing pairs or a unidirectional triplet and a fourth opposing direction. In the first case, the strips bounded by the unidirectional pairs must intersect along a common line. In the second case, the triangular prism bounded by the unidirectional triplet must contain the opposing force line. ◦

Exercise 5.22*: The set of frictionless four-finger equilibrium grasps of a 3-D object $\mathcal{B}$ forms a five-dimensional manifold in contact space. Obtain a graphical description of these grasps for a tetrahedral object (not necessarily a regular tetrahedron).

Solution sketch: Equilibrium requires that the projection of each triplet of finger forces on the plane orthogonal to the fourth finger force will form a three-finger equilibrium grasp in the projection plane. The collection of frictionless four-finger equilibrium grasps can thus be constructed by fixing one of the tetrahedral facets as the horizontal (x, y) plane, then sliding a horizontal plane along the orthogonal z-axis. At each z-axis location where the horizontal plane intersects the object $\mathcal{B}$, construct the 2-D region of the frictionless three-finger equilibrium grasps within this plane (see Section 5.2). The latter region determines three horizontal segments on the facets of $\mathcal{B}$ that support horizontal three-finger equilibrium grasps. By scanning the entire height of $\mathcal{B}$ in this manner, we obtain three strips on the non-horizontal facets of $\mathcal{B}$. The fingers contacting the non-horizontal facets can independently move up and down their respective strips with three independent parameters, and they can move horizontally with two coupled parameters in a way that maintains the horizontal three-finger equilibrium grasp condition. For each position of the three fingers, the position of the fourth finger on the horizontal facet of $\mathcal{B}$ is uniquely determined by the need to ensure zero net moment on $\mathcal{B}$ about the horizontal x and y axes (see Chapter 11). ○

Exercise 5.23: A 3-D object $\mathcal{B}$ is held in a nonparallel equilibrium grasp by four fingers via frictionless contacts, such that all four fingers are essential for maintaining the grasp. Verify that $\mathcal{B}$ is *immobilized* with respect to pure translational motions at the equilibrium grasp.

Solution: The finger forces associated with frictionless contacts act along the object's inward contact normal, n_i for $i=1\ldots4$. Since the force lines of a nonparallel grasp do not lie in parallel planes, the 3×4 matrix of force directions, $[n_1\, n_2\, n_3\, n_4]$, has full rank. Every linear velocity $v\in\mathbb{R}^3$ of $\mathcal{B}$ thus satisfies $v\cdot n_j\neq0$ for some $1\leq j\leq4$. At a four-finger equilibrium grasp, $\lambda_1 n_1+\ldots+\lambda_4 n_4=\vec{0}$. Hence, $\lambda_1 v\cdot n_1+\ldots+\lambda_4 v\cdot n_4=0$ for all $v\in\mathbb{R}^3$. The coefficients $\lambda_1,\ldots,\lambda_4$ represent finger force magnitudes. Since all fingers apply non-zero forces on $\mathcal{B}$, $\lambda_i>0$ for $i=1\ldots4$. It follows that $v\cdot n_j<0$ for some $1\leq j\leq4$. The jth c-obstacle outward normal is given by $\eta_j=(n_j, x_j\times n_j)$. Therefore, a pure translational motion of $\mathcal{B}$ along v would penetrate the jth c-obstacle, $(v,\vec{0})\cdot(n_j, x_j\times n_j)<0$. Since this condition holds for all $v\in\mathbb{R}^3$, $\mathcal{B}$ is immobilized with respect to all pure translational motions. ○

Exercise 5.24: Consider the four force lines shown in Figure 5.23. Two force lines lie in parallel horizontal planes, while the other two lie in parallel vertical planes. Show that the force lines lie in a common regulus by showing that a suitable linear combination of the finger wrenches gives a zero net wrench.

Solution: Let the world and object frame be located at the midpoint of the central line segment, with the x-axis aligned with this segment (Figure 5.23). Assuming that the central line segment has length of $2l$, the position of the four contacts is given by $x_1=(-l,0,d)$, $x_2=(-l,0,-d)$, $x_3=(l,d,0)$ and $x_4=(l,-d,0)$. Next, assume that the force directions span the same angle, α, with respect to the central line segment (the solution extends to asymmetric angles). For this symmetric choice, the force directions are given by $\hat{f}_1=(\cos\alpha,-\sin\alpha,0)$, $\hat{f}_2=(\cos\alpha,\sin\alpha,0)$, $\hat{f}_3=(-\cos\alpha,0,-\sin\alpha)$ and $\hat{f}_4=(-\cos\alpha,0,\sin\alpha)$. Assuming unit force magnitudes, $\lambda_i=1$ for $i=1\ldots4$, the

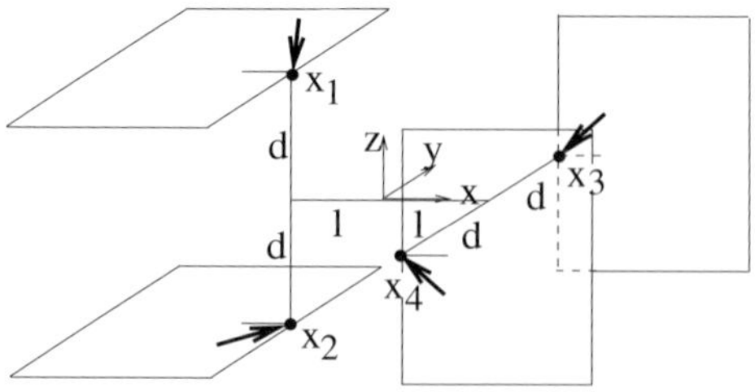

Figure 5.23 Four finger force lines that lie in a common regulus (the object $\mathcal{B}$ is not shown).

net force applied by the four fingers is zero. The net torque applied by the finger forces at x_1 and x_2 is

$$x_1 \times \hat{f}_1 + x_2 \times \hat{f}_2 = (x_2 - x_1) \times \begin{pmatrix} 0 \\ \sin\alpha \\ 0 \end{pmatrix} = 2d \sin\alpha \begin{pmatrix} 1 \\ 0 \\ 0 \end{pmatrix}.$$

The net torque applied by the finger forces at x_3 and x_4 is

$$x_3 \times \hat{f}_3 + x_4 \times \hat{f}_4 = (x_4 - x_3) \times \begin{pmatrix} 0 \\ 0 \\ \sin\alpha \end{pmatrix} = -2d \sin\alpha \begin{pmatrix} 1 \\ 0 \\ 0 \end{pmatrix}.$$

It follows that the net torque associated with unit force magnitudes is zero. The finger force lines are thus linearly dependent in Plücker coordinates and, hence, lie in a common regulus. ○

Exercise 5.25: Consider the four finger force lines shown in Figure 5.24(a). The finger force lines span a three-dimensional linear subspace in Plücker coordinates. Characterize the lines of the three-dimensional *orthogonal complement* subspace in Plücker coordinates.

Solution: Let W be the three-dimensional subspace spanned by the finger force lines in Plücker coordinates. The lines of this subspace represent the net wrenches that can act on $\mathcal{B}$. Let V be the three-dimensional orthogonal complement of W in $\mathbb{R}^6$. When the lines of V are interpreted as instantaneous twist axes, they are reciprocal to the lines of W. Recall now that reciprocal lines are either parallel or intersect in $\mathbb{R}^3$. The lines of V therefore consist of two complementary flat pencils sharing the same base points and the same common line as the flat pencils of W, as shown in Figure 5.24(b). Three linearly independent twist axes that form a basis for the subspace V are marked as e_1, e_2, e_3 in Figure 5.24(b). ○

Exercise 5.26: Provide a proof sketch of Lemma 5.14.

Solution: Let $l_i = (\hat{l}_i, x_i \times \hat{l}_i)$ for $i = 1 \ldots 7$. Since the seven lines do *not* lie in a lower-dimensional linear subspace of $\mathbb{R}^6$, six of the seven lines must be linearly independent in Plücker coordinates. Let us assume that these are the first six lines, $l_1, \ldots, l_6$ (the formula is the same for any other choice of six linearly independent lines). The first six

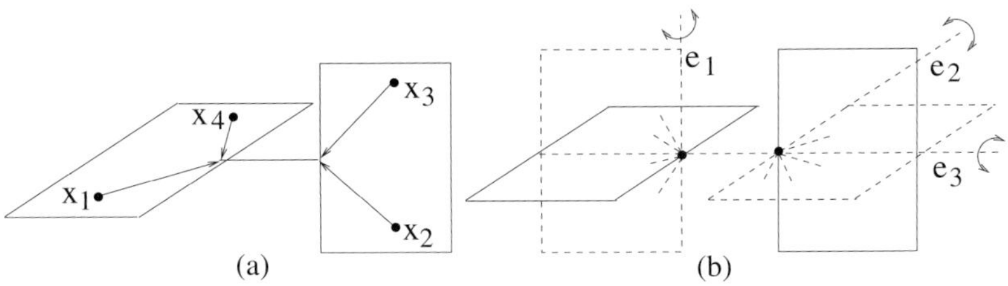

Figure 5.24 (a) Four finger force lines lying in two intersecting flat pencils (the object $\mathcal{B}$ is not shown). (b) The twists reciprocal to the finger wrenches also form two flat pencils.

lines thus satisfy $\det[l_1 \dots l_6] \neq 0$. We can now apply Cramer's rule to the equilibrium equation:

$$[l_1 \ \dots \ l_7] \begin{pmatrix} \lambda_1 \\ \vdots \\ \lambda_7 \end{pmatrix} = \vec{0} \implies [l_1 \ \dots \ l_6] \begin{pmatrix} \lambda_1 \\ \vdots \\ \lambda_6 \end{pmatrix} = -\lambda_7 l_7.$$

It can be verified that the solution for $(\lambda_1, \dots, \lambda_6)$ in terms of λ_7 is given by

$$\begin{pmatrix} \lambda_1 \\ \vdots \\ \lambda_6 \end{pmatrix} = \frac{\lambda_7}{\det[l_1 \dots l_6]} \begin{pmatrix} \det[l_2\, l_3\, l_4\, l_5\, l_6\, l_7] \\ \det[l_3\, l_4\, l_5\, l_6\, l_7\, l_1] \\ \det[l_4\, l_5\, l_6\, l_7\, l_1\, l_2] \\ \det[l_5\, l_6\, l_7\, l_1\, l_2\, l_3] \\ \det[l_6\, l_7\, l_1\, l_2\, l_3\, l_4] \\ \det[l_7\, l_1\, l_2\, l_3\, l_4\, l_5] \end{pmatrix}.$$

Expressing λ_7 as $\lambda_7 = s \cdot \det[l_1 \dots l_6]$ for $s \in \mathbb{R}$ gives Formula (5.4). ○

Exercise 5.27: Any seven-finger equilibrium grasp of a 3-D object can be realized with finger forces that act along the object surface normals (provided the contacts do not support an equilibrium grasp involving a smaller number of fingers). Explain why this property does *not* hold for 3-D grasps that involve a smaller number of fingers.

Solution: Seven finger force lines are always linearly dependent in Plücker coordinates. In contrast, for smaller number of fingers, the finger force lines associated with frictional contact may be linearly dependent in Plücker coordinates, but the contact normals need not be linearly dependent in Plücker coordinates. Since linear dependency is necessary for equilibrium grasp feasibility, equilibrium grasps that involve smaller numbers of fingers cannot be automatically realized along the contact normals. ○

Appendix II

Exercise 5.28: Consider the set $\mathcal{E}$ of frictionless k-finger equilibrium grasps in contact space. Explain why the dimension of $\mathcal{E}$ matches the dimension of the ambient contact space, $\mathcal{U}$, when k exceeds the dimension of the object's configuration space.

Solution: For 2-D grasps, $k \geq 4$ finger force lines are always linearly dependent as vectors in Plücker coordinates, which are equivalent to $\mathbb{R}^3$. For 3-D grasps, $k \geq 7$ finger force lines are always linearly dependent in Plücker coordinates, which are equivalent to $\mathbb{R}^6$. Hence, when the equilibrium grasp condition is satisfied with linear dependency coefficients $\lambda_1, \dots, \lambda_k \geq 0$, it is also satisfied in an open neighborhood of $\mathcal{U}$ surrounding this point. The set $\mathcal{E}$ thus has the same dimension as the ambient space $\mathcal{U}$ in these grasps. ○

Exercise 5.29: Extend the proof of Theorem 5.18 concerning the dimension of the set $\mathcal{E}$ of frictionless equilibrium grasps to the case of planar grasps.

Solution: In planar k-finger grasps, contact space is parametrized by $(u_1,\ldots,u_k)\in\mathcal{U}$, where $\mathcal{U}=\mathbb{R}^k$ with suitable periodicity rules. The composite space of parameters is parametrized by $(u_1,\ldots,u_k,\lambda_1,\ldots,\lambda)\in\tilde{\mathcal{U}}$, where $\tilde{\mathcal{U}}=\mathbb{R}^{2k}$. First consider $k>2$ fingers. The planar equilibrium grasps are determined by three scalar constraints. The three constraints generically intersect transversally in $\tilde{\mathcal{U}}$. Hence, they generically define a smooth $(2k-3)$-dimensional manifold, $\tilde{\mathcal{E}}$, in $\tilde{\mathcal{U}}$. The projection $\pi:\tilde{\mathcal{U}}\to\mathcal{U}$ maps $\tilde{\mathcal{E}}$ to $\mathcal{E}$. As argued in the proof of Theorem 5.18, the dimension of $\mathcal{E}$ is *one less* than the dimension of $\tilde{\mathcal{E}}$ for $k=3$ fingers, which gives $2\cdot 3-4=2$, and k for $k\geq 4$ fingers. As for the two-finger grasps, the set $\mathcal{E}$ is determined by the two scalar equations $n_1\times(x_1-x_2)=0$ and $n_2\times(x_1-x_2)=0$, where $u\times v=u^TJv$ and $J=\begin{bmatrix}0 & 1\\ -1 & 0\end{bmatrix}$. It can be verified that the two equations intersect transversally in $\mathcal{U}=\mathbb{R}^2$, hence $\mathcal{E}$ forms a discrete set in $\mathcal{U}$. ○

Exercise 5.30: Determine the dimension of the equilibrium set $\mathcal{E}$ when a polygonal object is held along two parallel edges by two or three fingers.

Solution: When a polygon $\mathcal{B}$ is held along two parallel edges, the contact normals can be written as $n_i=s_i\cdot n$ where $s_i=\pm 1$ according to the specific grasp arrangement. Substituting for the contact normals in the equilibrium grasp equation gives:

$$\sum_{i=1}^{k}\lambda_i s_i\cdot\begin{pmatrix}n\\ x(u_i)\times n\end{pmatrix}=\vec{0}\qquad \lambda_1,\ldots,\lambda_k\geq 0. \tag{5.15}$$

Since n and $s_1,\ldots,s_k$ are constants for a given grasp arrangement, the equilibrium grasps are specified by two scalar constraints in $\tilde{\mathcal{U}}=\mathbb{R}^{2k}$:

$$\sum_{i=1}^{k}\lambda_i s_i=0\quad\text{and}\quad\Big(\sum_{i=1}^{k}\lambda_i s_i x(u_i)\Big)\times n=0.$$

The two constraints form a vector-valued mapping, $h(\vec{u},\vec{\lambda})$, which maps the equilibrium grasps in $\tilde{\mathcal{U}}$ to the zero vector in $\mathbb{R}^2$:

$$h(\vec{u},\vec{\lambda})=\begin{pmatrix}\sum_{i=1}^{k}\lambda_i s_i\\ \sum_{i=1}^{k}\lambda_i s_i\cdot(x(u_i)\times n)\end{pmatrix}:\mathbb{R}^{2k}\to\mathbb{R}^2.$$

The two constraints intersect transversally and thus define a smooth $(2k-2)$-dimensional manifold in $\tilde{\mathcal{U}}$ when the $2\times 2k$ Jacobian matrix, Dh, possesses full rank at all points of $\tilde{\mathcal{E}}$. The Jacobian Dh has the form:

$$Dh=\begin{bmatrix}\vec{0} & \cdots & \vec{0} & s_1 & \cdots & s_k\\ \lambda_1(x_1'\times n) & \cdots & \lambda_k(x_k'\times n) & s_1x_1\times n & \cdots & s_kx_k\times n\end{bmatrix}.$$

Since $s_i=\pm 1$, the Jacobian Dh has full rank at all generic grasp arrangements. Hence, $\tilde{\mathcal{E}}$ generically forms a smooth $(2k-2)$-dimensional manifold in $\tilde{\mathcal{U}}$. The equilibrium set $\mathcal{E}$ is obtained by projecting the set $\tilde{\mathcal{E}}$ to contact space $\mathcal{U}$. Hence, its dimension is $(2\cdot 2-2)-1=1$ for two fingers, $(2\cdot 3-2)-1=3$ for three fingers and k for $k\geq 4$ fingers. ○

References

[1] A. Dandurand, "The rigidity of compound spatial grid," *Structural Topology*, vol. 10, pp. 41–56, 1984.

[2] O. V. Veblen and J. W. Young, *Projective Geometry*. Blaisdell Publishing, 1938.

[3] J. Ponce, S. Sullivan, A. Sudsang, J.-D. Boissonnat and J.-P. Merlet, "On computing four-finger equilibrium and force-closure grasps of polyhedral objects," *The International of Robotics Research*, vol. 16, no. 1, pp. 11–35, 1997.

[4] B. Mishra, J. T. Schwartz and M. Sharir, "On the existence and synthesis of multifinger positive grips," *Algorithmica*, vol. 2, pp. 541–558, 1987.

[5] L. Nirenberg, "Variational and topological methods in nonlinear problems," *Bulletin of the AMS*, vol. 4, no. 3, pp. 267–302, 1981.

Part II

Frictionless Rigid-Body Grasps and Stances

6 Introduction to Secure Grasps

Having developed in detail the theory of *equilibrium grasps,* our next step is to develop a theory of *secure grasps.* Secure grasps should be able to keep the grasped object safely contained within the robot hand under all possible disturbances that might arise while the object is transported and manipulated by the robot hand. Such disturbances routinely arise by movement of the robot arm on which the grasping mechanism is mounted and frequently occur when a grasping mechanism mounted on a robot arm interacts with the environment. This chapter introduces two complementary approaches to grasp security. The two approaches are known as *form and force closure* in the robotics literature, but we will use a clearer terminology in this chapter. The rigorous details of these concepts are developed in subsequent chapters.

The first approach for achieving secure grasps is the notion of *immobilizing grasps,* introduced in Section 6.1. Under this approach, the fingers are arranged so as to prevent any movement of the object with respect to the grasping fingers. The complementary grasp security approach is the notion of *wrench resistant grasps,* introduced in Section 6.2. Under this approach, the fingers are arranged so as to resist all possible forces and torques, or *wrenches,* that might disturb the object within the grasp. To complete the picture, we will see in Section 6.3 that for a high number of fingers, immobilizing grasps are perfectly *dual* to wrench resistant grasps (this duality breaks down for smaller numbers of fingers).

Immobilizing grasps can be analyzed using the geometric c-space techniques introduced in Part I of the book. In order to focus on the essential ideas, most of this chapter, as well as Part II of the book, considers frictionless contacts. Their study serves as a useful introduction to the key grasp safety issues, which are extended to frictional grasps in Part III of the book.

6.1 Immobilizing Grasps

The *mobility* of a mechanism composed of moving parts is the number of independent degrees of freedom needed to uniquely describe the internal motions of the mechanism's components. When a mechanism is immobile, it forms a rigid structure where no internal movement of its parts is possible. In grasp mechanics, we seek to analyze the mobility of a grasped object, $\mathcal{B}$, relative to stationary finger bodies, $\mathcal{O}_1, \ldots, \mathcal{O}_k$, under the frictionless rigid-body model. This type of mobility is most naturally analyzed in the object's c-space introduced in Chapter 2. The c-space view starts with the following set of free object configurations.

DEFINITION 6.1 (**free c-space**) *Let a rigid object $\mathcal{B}$ be held by rigid and stationary finger bodies $\mathcal{O}_1,\ldots,\mathcal{O}_k$. Let $\mathcal{CO}_1,\ldots,\mathcal{CO}_k$ be the finger c-obstacles in $\mathcal{B}$'s c-space. The* **free c-space** *is the set of configurations:*

$$\mathcal{F} = \mathbb{R}^m - \bigcup_{i=1}^{k} \mathsf{int}(\mathcal{CO}_i) \qquad m = 3 \text{ in 2-D grasps, } m = 6 \text{ in 3-D grasps}$$

where $\mathsf{int}(\mathcal{CO}_i)$ *denotes the interior of the finger c-obstacle* $\mathcal{CO}_i$.

The interior of $\mathcal{F}$ is an open subset of $\mathbb{R}^m$, which consists of contact-free configurations of the object $\mathcal{B}$. The boundary of $\mathcal{F}$, denoted $\mathsf{bdy}(\mathcal{F})$, consists of configurations at which the object $\mathcal{B}$ touches one or more of the finger bodies. It forms a subset of the finger c-obstacle boundaries: $\mathsf{bdy}(\mathcal{F}) \subseteq \cup_{i=1}^{k} \mathsf{bdy}(\mathcal{CO}_i)$. Any c-space path that lies in $\mathsf{bdy}(\mathcal{CO}_i)$ represents a motion of $\mathcal{B}$ that maintains surface contact with the stationary finger body $\mathcal{O}_i$. However, the object may penetrate neighboring finger bodies along such a path. Hence, the boundary of $\mathcal{F}$ forms a subset of the union of the c-obstacle boundaries. When $\mathcal{B}$ is contacted by k fingers at a configuration q_0, the point q_0 lies at the intersection of the finger c-obstacle boundaries, $q_0 \in \cap_{i=1}^{k} \mathsf{bdy}(\mathcal{CO}_i)$. The free c-space motions available to $\mathcal{B}$ are defined as follows.

DEFINITION 6.2 (**free motions**) *Let a rigid object $\mathcal{B}$ be held by rigid and stationary finger bodies $\mathcal{O}_1,\ldots,\mathcal{O}_k$ at a configuration point q_0. The* **free c-space motions** *of $\mathcal{B}$ at q_0 are all c-space paths that start at q_0 and lie in $\mathcal{F}$.*

The free c-space motions are paths along which the object $\mathcal{B}$ either breaks away or maintains surface contact with the stationary finger bodies (some examples are discussed later). When an object $\mathcal{B}$ has available free motions at a given grasp, there exist perturbing forces that can induce these motions and thus allow $\mathcal{B}$ to escape the grasp. Conversely, grasp safety is ensured when the grasping fingers allow no free motions of $\mathcal{B}$. This notion of *immobilizing grasps* is introduced in the following definition.

DEFINITION 6.3 (**immobilization**) *Let a rigid object $\mathcal{B}$ be held at a configuration q_0 by rigid and stationary finger bodies $\mathcal{O}_1,\ldots,\mathcal{O}_k$. The object is* **immobilized** *by the finger bodies when it has no free c-space motions at q_0.*

An object $\mathcal{B}$ is immobilized when the finger c-obstacles completely surround the object's configuration point q_0. The finger c-obstacles can isolate the point q_0 only when the physical fingers form a *frictionless equilibrium grasp* of the object $\mathcal{B}$. This important property is stated in the following theorem (see this chapter's appendix for a proof).

THEOREM 6.1 (**equilibrium grasp immobilization**) *A necessary condition for the immobilization of a rigid object $\mathcal{B}$ by rigid finger bodies $\mathcal{O}_1,\ldots,\mathcal{O}_k$ is that the fingers hold the object in a feasible* **frictionless equilibrium grasp.**

To understand the need for a frictionless equilibrium grasp, consider the finger c-obstacles in 2-D grasps. Let $\mathcal{S}_i$ denote the ith finger c-obstacle surface, let $\eta_i(q_0)$ denote the ith finger c-obstacle outward normal at q_0, and let $T_{q_0}\mathcal{S}_i$ be the tangent plane to S_i at q_0. The half-space that is bounded by the tangent plane $T_{q_0}\mathcal{S}_i$ and occupies the exterior of $\mathcal{CO}_i$ is given by

$$M_i(q_0) = \{\dot{q} \in T_{q_0}\mathbb{R}^m : \eta_i(q_0) \cdot \dot{q} \geq 0\}. \tag{6.1}$$

The half-space $M_i(q_0)$ represents the first-order approximation to the free c-space motions allowed by the stationary finger body $\mathcal{O}_i$ (Figure 6.1(a)). Let $M_{1...k}(q_0)$ denote the intersection of the free half-spaces associated with the k fingers:

$$M_{1...k}(q_0) = \cap_{i=1}^{k} M_i(q_0) = \{\dot{q} \in T_{q_0}\mathbb{R}^m \ : \ \eta_i(q_0) \cdot \dot{q} \geq 0, \ i = 1 \dots k\}. \tag{6.2}$$

The set $M_{1...k}(q_0)$ forms a convex cone of tangent vectors based at q_0. If $M_{1...k}(q_0)$ has a nonempty interior, tangent vectors $\dot{q}$ in the interior of $M_{1...k}(q_0)$ satisfy $\eta_i(q_0) \cdot \dot{q} > 0$ for $i = 1 \dots k$. It follows that the object $\mathcal{B}$ locally breaks away from all k fingers along such instantaneous motions. In order to prevent such escape motions, $M_{1...k}(q_0)$ must have an *empty interior* at the given grasp. For $M_{1...k}(q_0)$ to have an empty interior, two conditions must be met. The c-obstacle outward normals must be linearly dependent at q_0, and their directions must positively span the origin. The c-obstacle outward normals are collinear with the wrenches generated by finger forces that act along $\mathcal{B}$'s inward contact normals (Section 3.2). Immobilization, therefore, requires that the object $\mathcal{B}$ be held at a frictionless equilibrium grasp.

Frictionless equilibrium grasps are thus necessary for immobilization. However, some equilibrium grasps are *not* immobilizing grasps, as illustrated in the following example.

A non-immobilizing equilibrium grasp: Figure 6.1(b) shows an ellipse held by two disc fingers at a frictionless equilibrium grasp. Consider the first-order approximation to the finger c-obstacles at q_0, depicted in Figure 6.1(b). The world and object frames are set at the ellipse center, with the x-axis aligned with the ellipse's major axis. The contact

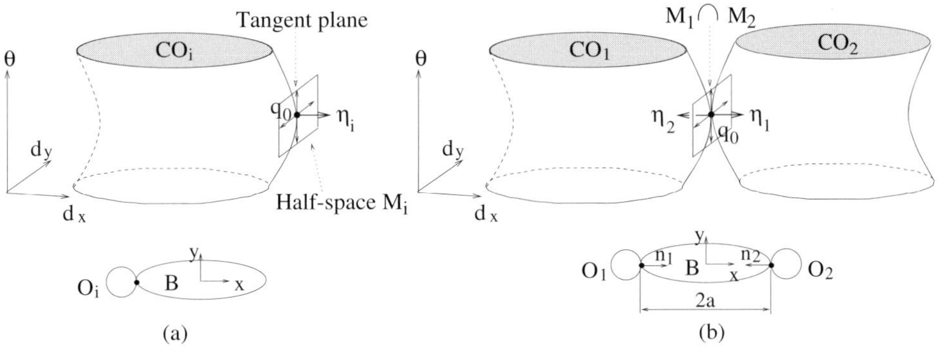

Figure 6.1 (a) The half-space $M_i(q_0)$ approximates the exterior of $\mathcal{CO}_i$ at q_0. (b) The first-order approximation of the object's free motions, $M_{1,2}(q_0) = M_1(q_0) \cap M_2(q_0)$, at a two-finger equilibrium grasp.

locations are $x_1 = (-a, 0)$ and $x_2 = (a, 0)$, where $2a$ is the ellipse's major axis length. The ellipse's inward contact normals are $n_1 = (1, 0)$ at x_1 and $n_2 = (-1, 0)$ at x_2. The finger c-obstacles outward normals at q_0 are given by

$$\eta_1(q_0) = \begin{pmatrix} n_1 \\ x_1 \times n_1 \end{pmatrix} = \begin{pmatrix} 1 \\ 0 \\ 0 \end{pmatrix} \quad \text{and} \quad \eta_2(q_0) = \begin{pmatrix} n_2 \\ x_2 \times n_2 \end{pmatrix} = \begin{pmatrix} -1 \\ 0 \\ 0 \end{pmatrix},$$

where $u \times v = u^T J v$ for $u, v \in \mathbb{R}^2$ and $J = \begin{bmatrix} 0 & 1 \\ -1 & 0 \end{bmatrix}$. Since $\eta_1 + \eta_2 = \vec{0}$ at the equilibrium grasp, the intersection $M_1 \cap M_2$ consists only of the tangent plane common to $\mathcal{CO}_1$ and $\mathcal{CO}_2$ at q_0, which has an empty interior:

$$\begin{aligned} M_{1,2}(q_0) &= M_1(q_0) \cap M_2(q_0) \\ &= \left\{ \dot{q} \in T_{q_0}\mathbb{R}^m : \begin{pmatrix} 0 \\ 1 \\ 0 \end{pmatrix} \cdot \begin{pmatrix} v \\ \omega \end{pmatrix} \geq 0 \ \text{ and } \ \begin{pmatrix} 0 \\ -1 \\ 0 \end{pmatrix} \cdot \begin{pmatrix} v \\ \omega \end{pmatrix} \right\} \qquad \dot{q} = (v, \omega). \end{aligned}$$

However, the two-finger equilibrium grasp of Figure 6.1(b) is *non-immobilizing,* since the ellipse can escape along any combination of vertical translation and rotation about its center. ○

The next example describes an immobilizing and hence secure equilibrium grasp.

An immobilizing equilibrium grasp: Figure 6.2(a) shows a triangular object $\mathcal{B}$ that is held in a frictionless equilibrium grasp by three disc fingers. First-order analysis implies that the object $\mathcal{B}$ is free to instantaneously rotate about the intersection point of the finger contact normals. But the object is clearly immobilized by the three fingers. As seen in the c-space geometry of Figure 6.2(b), the point q_0 is completely surrounded by the finger c-obstacles. How can this object be immobilized when the first-order analysis suggests that some instantaneous free motions are possible? The paradox can be resolved by realizing that the local free motions have *first-order* (velocities, or tangent vectors) and *second-order* (accelerations, or path curvature) properties. A proper analysis of mobility

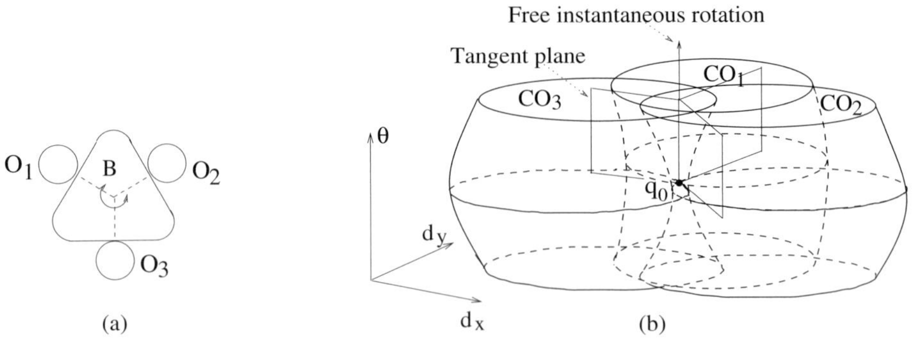

Figure 6.2 (a) A three-finger equilibrium grasp of a triangular object. (b) The finger c-obstacles completely surround the object's configuration point q_0.

must include *both* properties of the local free motions. When curvature effects are taken into account, the object $\mathcal{B}$ is fully immobilized by the three fingers. ○

Immobilization focuses on the rigid-body constraints imposed on the grasped object's free motions by the surrounding finger bodies. It is thus suited for ensuring grasp safety in *low-friction* situations or in situations where friction effects cannot be relied upon. When friction effects are included in the finger contact selection, a non-immobilizing grasp can still form a secure grasp. This complementary notion of grasp security is discussed next.

6.2 Wrench Resistant Grasps

Rather than focus on the grasped object's free motions, one can alternatively analyze the fingers' ability to resist perturbing forces and torques, or *wrenches*, that might pull the object away from the grasping fingers. When every possible disturbance wrench applied to $\mathcal{B}$ can be resisted by feasible finger forces, the grasped object can be safely kept within the grasping mechanism. The following definition of *wrench resistant grasps* applies to frictional, frictionless and soft point contact grasps.

DEFINITION 6.4 (**wrench resistant grasp**) *Let a rigid object $\mathcal{B}$ be held at a configuration q_0 by rigid finger bodies $\mathcal{O}_1, \ldots, \mathcal{O}_k$. Let $\mathcal{L}_G : C_1 \times \cdots \times C_k \to T^*_{q_0}\mathbb{R}^m$ ($m = 3$ or 6) be the grasp map at the given grasp. The grasp is* **wrench resistant** *when any external wrench, $\boldsymbol{w}_{ext} \in T^*_{q_0}\mathbb{R}^m$, applied to $\mathcal{B}$ can be resisted by feasible finger forces and possibly torques about the contact normals:*

$$\mathcal{L}_G(f_1, \ldots, f_k) + \boldsymbol{w}_{ext} = \vec{0} \qquad (f_1, \ldots, f_k) \in C_1 \times \ldots \times C_k, \tag{6.3}$$

where $C_1, \ldots, C_k$ are the generalized friction cones at the contacts.

A k-finger grasp is wrench resistant when the grasp map $\mathcal{L}_G$ forms a *surjective mapping* of the composite friction cone $C_1 \times \cdots \times C_k$ onto $\mathcal{B}$'s wrench space. In contrast with immobilizing grasps, wrench resistant grasps provide only a necessary condition for achieving secure grasps. Depending on the application, one must ensure that the grasping system actually generates the required reaction forces (this advanced topic is discussed in Part IV of the book). Like immobilizing grasps, wrench resistant grasps can only be achieved through *equilibrium grasps* – a frictionless equilibrium in the case of frictionless contacts, a frictional equilibrium in the case of frictional contacts and an equilibrium that additionally involves torques about the contact normals in the case of soft point contacts. This property is based on Proposition 4.1, which is repeated here.

Proposition 4.1 *A rigid object $\mathcal{B}$ is held by rigid finger bodies $\mathcal{O}_1, \ldots, \mathcal{O}_k$ at a feasible equilibrium grasp iff the net wrench cone $\mathcal{W}$ contains a nontrivial* **linear subspace** *at the given grasp.*

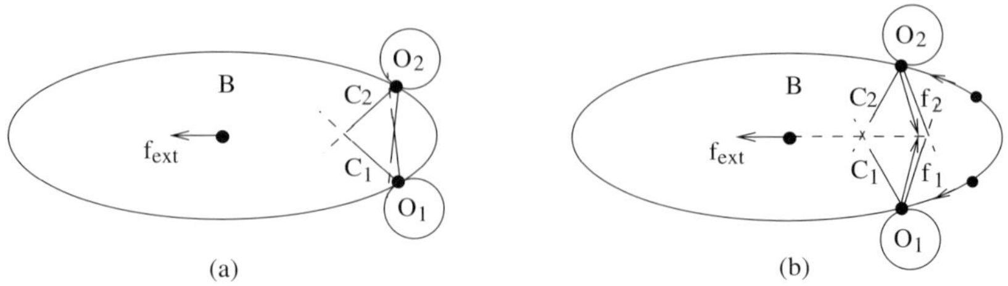

Figure 6.3 (a) A frictional two-finger arrangement that is not a feasible equilibrium grasp, and hence is *not* wrench resistant. (b) A frictional two-finger arrangement that is an equilibrium grasp, with internal grasp forces that ensure wrench resistance.

Since wrench resistant grasps span the object's entire wrench space, their net wrench cone satisfies the condition of Proposition 4.1, which leads to the following corollary.

COROLLARY 6.2 **(wrench resistant equilibrium grasps)** *A necessary condition for wrench resistance is that the fingers hold the object $\mathcal{B}$ at a* **feasible equilibrium grasp,** *according to the contact model that is assumed to hold at the contacts.*

The following example illustrates the necessity of establishing a frictional equilibrium grasp in order to achieve wrench resistance against external wrench disturbances.

Example: Figure 6.3 shows two grasps of an ellipse in the horizontal plane. First, consider the frictional two-finger grasp shown in Figure 6.3(a). The friction cones do *not* support an equilibrium grasp at the contacts, hence this is *not* a feasible equilibrium grasp. When an external force, f_{ext}, attempts to pull the ellipse away from the two fingers, there are no feasible finger forces, $f_1 \in C_1$ and $f_2 \in C_2$, that can possibly resist this force. Next, consider the frictional two-finger grasp of the same ellipse shown in Figure 6.3(b). The friction cones now support an equilibrium grasp at the contacts. As discussed in a Chapter 13, this grasp forms a secure, *wrench resistant* grasp. For instance, f_{ext} can now be resisted by finger forces f_1 and f_2 which lie in their friction cones. ○

Practical Implications of Wrench Resistant Grasps

The definition of wrench resistant grasps makes rather strong assumptions about the design and operation of the grasping system. The grasping system must be able to quickly sense the disturbing wrench, $\boldsymbol{w}_{ext}$, rapidly compute a new set of finger forces that will cancel the disturbance, and then react with the fingers so as to generate the required forces. Furthermore, the grasping system sensing and reaction cycle must keep pace with the time variations of the external disturbances. In light of these observations, one might consider the grasps described by Definition 6.4 to be *actively wrench resistant grasps.*

In some grasping mechanisms, and in most fixtures, some or all of the contact forces are not actively generated and controlled by the grasping system. Instead, the contact forces are generated by *passive* mechanical means. Such effects generally arise from two processes:

- **Preloaded grasps:** A *preloading* process, where an object $\mathcal{B}$ is initially pressed against stationary finger bodies by an external agent (e.g., a factory worker squeezes a part against fixturing elements), ending at an equilibrium grasp. The preloading process establishes non-zero contact forces that lie within the allowed friction cones. When the fingers are kept as stationary rigid bodies, any subsequent external wrench disturbance acting on $\mathcal{B}$ (e.g., a factory worker attempting a machining operation) induces a perturbation of the finger contact forces. As long as the perturbed contact forces remain within their respective friction cones, the equilibrium grasp can be maintained.
- **Compliant grasps:** When the fingertips are made of soft material (say rubber or Teflon), the grasped object $\mathcal{B}$ can locally move in response to perturbing wrenches, while the fingertip contacts are deformed. The finger contact forces arise from a *stiffness relationship* that describes how the contact deformations generate reaction forces. For small displacements of $\mathcal{B}$ with respect to an equilibrium configuration q_0, the reaction force at the ith contact can be modeled as

$$f_i = -K_i(q - q_0) + f_i^0, \tag{6.4}$$

where f_i^0 denotes the preload force at the ith contact, and K_i is the *stiffness matrix* at the ith contact. Eq. (6.4) can be seen as a model for a linear spring with spring constant K_i and initial spring compression f_i^0. When an object $\mathcal{B}$ is held in a compliant equilibrium grasp at a configuration q_0, external wrench disturbances are typically balanced at equilibrium points that lie in the vicinity of q_0. Compliant finger mechanisms, or feedback algorithms that induce a compliant relationship at the joints of the finger mechanisms, can also contribute to the grasping system compliance (see Chapter 18).

In order to account for secure grasps that are based on heterogeneous passive and active means, let us add the following, more lenient notion of *local* wrench resistant grasps.

DEFINITION 6.5 **(local wrench resistance)** *Let a rigid object $\mathcal{B}$ be held by rigid finger bodies $\mathcal{O}_1, \ldots, \mathcal{O}_k$ at an equilibrium grasp configuration q_0. The grasp is* **locally wrench resistant** *if the fingers can resist any external wrench in a bounded neighborhood centered at $\mathcal{B}$'s wrench space origin, possibly at a new equilibrium configuration that lies in a local neighborhood centered at q_0 in $\mathcal{B}$'s c-space.*

Local wrench resistant grasps are able to resist only a *bounded* set of external wrenches and may do so at nearby equilibrium points. This type of local grasp security is the best one can hope for in grasping systems whose contacts react according to passive force displacement laws. When such contact laws are implemented as fingertip feedback control laws, local wrench resistance is equivalent to the notion of local *disturbance rejection* from control theory.

6.3 Duality of Immobilizing and Wrench-Resistant Grasps under Frictionless Contact Conditions

Under frictionless contact conditions, a certain type of immobilizing grasps are perfectly dual to wrench resistant grasps, thus showing that the two approaches sometimes provide the same type of grasp security. The wrench resistance condition associated with frictionless contacts has a simple c-space interpretation. A frictionless finger force can be parametrized as $f_i = \lambda_i n_i$, where $\lambda_i \geq 0$ is the finger force magnitude and n_i is $\mathcal{B}$'s inward unit normal at x_i. The wrench induced by this force, $\boldsymbol{w}_i = \lambda_i(n_i, x_i \times n_i)$, is collinear with the finger c-obstacle outward normal at q_0, $\eta_i(q_0) = (n_i, x_i \times n_i)$. Using $\mathcal{W}$ for the net wrench cone that can be affected on $\mathcal{B}$ under frictionless contact conditions, wrench resistance takes the form:

$$\mathcal{W} = \left\{\lambda_1\eta_1(q_0) + \ldots + \lambda_k\eta_k(q_0) : \lambda_1, \ldots, \lambda_k \geq 0\right\} = T^*_{q_0}\mathbb{R}^m \qquad m = 3 \text{ or } 6.$$

The wrench resistance condition is thus equivalent to the requirement that the finger c-obstacle outward normals, seen as wrenches, positively span the entire object's wrench space.

Next, consider the complementary grasp security approach, which seeks to prevent the object's free motions by proper placement of rigid finger bodies. We will see in the next chapter that the simplest type of immobility, called *first-order immobilization,* is based on preventing all possible first-order instantaneous motions of the grasped object $\mathcal{B}$ as follows.

DEFINITION 6.6 (**first-order immobilization**) *Let a rigid object $\mathcal{B}$ be held at an equilibrium grasp configuration q_0 by rigid and stationary finger bodies $\mathcal{O}_1, \ldots, \mathcal{O}_k$. The object is* **first-order immobilized** *when it has no first-order free instantaneous motions at q_0:*

$$M_{1\ldots k}(q_0) = \left\{\dot{q} \in T_{q_0}\mathbb{R}^m : \eta_i(q_0) \cdot \dot{q} \geq 0,\ i = 1 \ldots k\right\} = \{0\}, \tag{6.5}$$

where $T_{q_0}\mathbb{R}^m$ is $\mathcal{B}$'s ambient tangent space at q_0, and $\{0\}$ is the origin of this space.

First-order immobilization requires at least four frictionless contacts in 2-D grasps and at least seven frictionless contacts in 3-D grasps. The following theorem asserts that the first-order immobilizing grasps are dual to the wrench resistant grasps (see this chapter's appendix for a proof).

THEOREM 6.3 (**equivalent secure grasps**) *Under frictionless contact conditions, a k-finger equilibrium grasp is* **wrench resistant** *iff it forms a* **first-order immobilizing grasp.**

The proof of Theorem 6.3 is based on the notion of *dual cones.* Let $\mathcal{W}$ be a convex cone of wrenches based at $\mathcal{B}$'s wrench space origin. Its *dual cone*, $\mathcal{W}^*$, is the cone of tangent vectors:[1]

$$\mathcal{W}^* = \left\{\dot{q} \in T_{q_0}\mathbb{R}^m : \boldsymbol{w} \cdot \dot{q} \leq 0 \ \text{ for all } \ \boldsymbol{w} \in \mathcal{W}\right\}. \tag{6.6}$$

[1] The *duality* refers to the action of a wrench or covector on tangent vectors, $\boldsymbol{w}(\dot{q}) = \boldsymbol{w} \cdot \dot{q}$ for $\dot{q} \in T_{q_0}\mathbb{R}^m$.

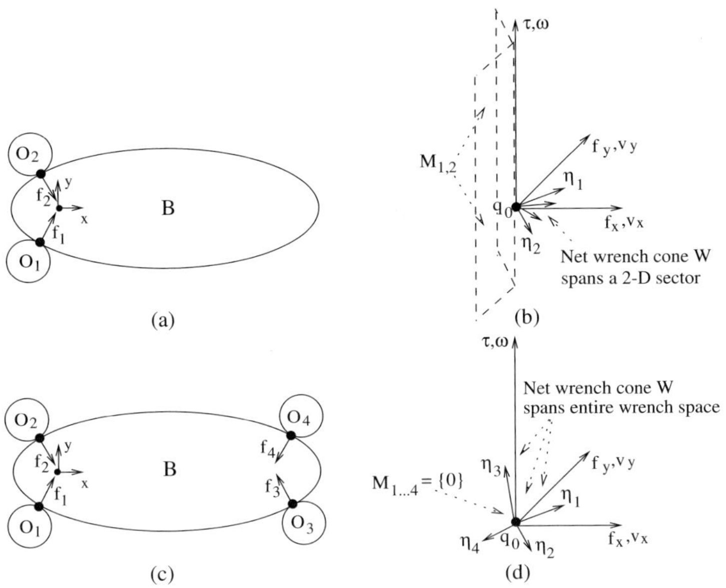

Figure 6.4 (a)–(b) The net wrench cone $\mathcal{W}$ and the relation $M_{1,2}(q_0) = -\mathcal{W}^*$ associated with two frictionless contacts. (c)–(d) A frictionless four-finger equilibrium grasp that is both wrench resistant and first-order immobilizing.

As shown in the appendix, the cone of first-order free motions, $M_{1...k}(q_0)$, is *dual* to the negated net wrench cone that can be affected on $\mathcal{B}$ under frictionless contact conditions, $M_{1...k}(q_0) = -\mathcal{W}^*$. Since $\mathcal{W} = T^*_{q_0}\mathbb{R}^m$ at a wrench resistant grasp, the duality implies that $M_{1...k}(q_0) = \{0\}$, which is first-order immobilization according to Eq. (6.5).

Example: Figure 6.4(a) depicts a frictionless two-finger grasp of an ellipse. The world and object frames are located at the intersection point of the finger force lines. The finger wrenches are given by $\boldsymbol{w}_1 = (f_1, 0)$ and $\boldsymbol{w}_2 = (f_2, 0)$, and their net wrench cone forms the 2-D horizontal sector depicted in Figure 6.4(b). Its dual cone, $M_{1,2}(q_0) = -\mathcal{W}^*$, forms the three-dimensional wedge of tangent vectors depicted in Figure 6.4(b). Next, consider the four-finger grasp of the same ellipse shown in Figure 6.4(c). The net wrench cone spans the object's entire wrench space, $\mathcal{W} = T^*_{q_0}\mathbb{R}^3$, which implies that the object $\mathcal{B}$ is held at a wrench resistant grasp (Figure 6.4(d)). Based on the duality property, $M_{1...4}(q_0) = -\mathcal{W}^* = \{0\}$, which implies that the object $\mathcal{B}$ is immobilized by first-order geometric effects at the four-finger grasp. ○

The next chapter will establish that first-order immobilization requires at least four frictionless contacts in 2-D grasps and at least seven frictionless contacts in 3-D grasps. For these high numbers of contacts, one can verify the safety of a candidate frictionless equilibrium grasp in two equivalent ways. Either one verifies the fingers' ability to resist all external wrenches that might act on $\mathcal{B}$, or, equivalently, one checks that all rigid-body velocities of $\mathcal{B}$ are prevented by the rigid finger bodies.

6.4 A Forward Look at the Chapters of Part II

This chapter provided an intuitive overview of the two fundamental approaches to grasp security. The remaining chapters of this part of the book will develop the analytical tools needed to analyze the mobility of an object held at a given equilibrium grasp and then show how these tools can be used to address other important issues in grasp mechanics associated with frictionless contacts. Chapter 7 will develop the theory of *first-order* immobilizing grasps, which analyzes the first-order properties, or velocities, of the free motions of an object held by multiple fingers. Chapter 8 will extend the framework of Chapter 7 to consider the second-order properties of the free motions and define *second-order* immobilizing grasps. Based on the methods introduced in Chapters 7 and 8, Chapter 9 will take up the following basic problem: *What is the minimum number of fingers required to immobilize arbitrary rigid objects*? A detailed analysis of this problem will be given for 2-D objects, where it will be seen that the minimal number of fingers depends on the fingertips geometry. A summary of the corresponding bounds for 3-D objects will also be given. Chapter 11 will extend the immobilization analysis to include gravitational effects. In this case, grasp safety amounts to ensuring that the fingers support the object at a *locally stable* equilibrium stance under the influence of gravity. The locally stable hand-supported stances are precisely *immobilizing grasps* under the interpretation of gravity as a virtual finger that acts at the object's center of mass. Therefore, the safety of a candidate hand-supported stance under the influence of gravity can be judged with the immobilization tools developed in Chapters 7 and 8. Finally, Chapter 10 will consider the notion of *cage formation* surrounding a desired grasp. Using cage formations, the process of closing the fingers about a desired immobilizing grasp can be performed in a robust manner, under huge uncertainty in the position of the fingers relative to the object. The geometric techniques introduced in this part of the book will form the foundation for the subsequent analysis of frictional grasps in Part III of the book.

Bibliographical Notes

The notion of immobilizing grasps is known as *form closure* in the robotics literature. This notion dates to the work of Reuleaux in the nineteenth century [1]. The notion of wrench resistant grasps is known as *force closure* in the robotics literature. While this notion is also deemed to originate with Reuleaux, the first formal discussion of force closure concepts can be found in the early twentieth century work of Somoff [2]. The study of form closure grasps in modern robotics started with the work of Lakshminarayana [3], while modern analysis of force closure grasps began with the work of Roth and the PhD studies of Salisbury [4] and Kerr [5].

Basic notions of *cone analysis* can be found in convex optimization texts, such as Boyd and Vandenberghe [6] and Ben-Tal and Nemirovski [7]. We are indebted to Nemirovski, who suggested the dual cones argument used in the chapter's appendix to conclude that frictionless equilibrium grasps are necessary for immobilizing grasps.

Appendix: Proof Details

This appendix contains proofs of two key properties concerning immobilizing and wrench resistant grasps. We begin with the property that every immobilizing grasp must form a frictionless equilibrium grasp.

Theorem 6.1 *A necessary condition for the immobilization of a rigid object $\mathcal{B}$ by rigid finger bodies $\mathcal{O}_1, \ldots, \mathcal{O}_k$ is that the fingers hold the object in a feasible* **frictionless equilibrium grasp.**

Proof: Recall that $\mathcal{B}$'s c-space is parametrized by $\mathbb{R}^m$ ($m = 3$ or 6). When an object $\mathcal{B}$ is held by finger bodies $\mathcal{O}_1, \ldots, \mathcal{O}_k$ at a configuration q_0, the point q_0 lies at the intersection of the k finger c-obstacle boundaries. Let $\eta_i(q_0)$ be the ith finger c-obstacle outward normal at q_0. The wrench induced by a normal force f_i that acts on $\mathcal{B}$ at x_i is given by $\boldsymbol{w}_i = \lambda_i \eta_i(q_0)$ for $\lambda_i \geq 0$. Let us show that when the object is *not* held at a frictionless equilibrium grasp, it can break away form all k fingers and, therefore, is *not* immobilized. The net wrench cone that can be affected on $\mathcal{B}$ by k frictionless finger contacts has the form:

$$\mathcal{W} = \left\{\boldsymbol{w} \in T^*_{q_0}\mathbb{R}^m \;:\; \boldsymbol{w} = \lambda_1 \eta_1(q_0) + \cdots + \lambda_k \eta_k(q_0),\;\; \lambda_1 \ldots \lambda_k \geq 0\right\}.$$

Based on Proposition 4.1, if a contact arrangement does *not* form a feasible equilibrium grasp, its net wrench cone $\mathcal{W}$ does *not* contain any nontrivial linear subspace. It follows that $\mathcal{W}$ must form a *pointed cone* – that is, a convex cone that is strictly convex at its vertex. The cone *dual* to $\mathcal{W}$, denoted $\mathcal{W}^*$, is the convex cone of tangent vectors:

$$\mathcal{W}^* = \left\{\boldsymbol{t} \in T_{q_0}\mathbb{R}^m \;:\; \boldsymbol{t} \cdot \boldsymbol{w} \leq 0 \;\text{ for all }\; \boldsymbol{w} \in \mathcal{W}\right\}.$$

The following is a key argument of the proof. If $\mathcal{W}$ forms a pointed cone, its dual cone $\mathcal{W}^*$ must possess a nonempty interior. Otherwise, $(\mathcal{W}^*)^* = \mathcal{W}$ must contain at least a one-dimensional linear subspace, which means that $\mathcal{W}$ *cannot* be a pointed cone. Hence, $\mathcal{W}^*$ must have a nonempty interior. Every tangent vector $\boldsymbol{t}$ from the interior of $\mathcal{W}^*$ satisfies

$$\boldsymbol{t} \cdot \boldsymbol{w} < 0 \qquad \text{for all } \boldsymbol{w} \in \mathcal{W}. \tag{6.7}$$

Let $\boldsymbol{t}$ be any tangent vector from the interior of $\mathcal{W}^*$, and let $\dot{q} = -\boldsymbol{t}$ represent a particular rigid body velocity of $\mathcal{B}$ that starts at q_0. Substituting $\dot{q} = -\boldsymbol{t}$ in Eq. (6.7), and noting that $\eta_i(q_0) \in \mathcal{W}$ for $i = 1 \ldots k$, gives the inequalities:

$$\eta_i(q_0) \cdot \dot{q} > 0 \qquad i = 1 \ldots k.$$

It follows that $\mathcal{B}$ moves away from all k finger bodies along the c-space trajectory $q(t)$ such that $q(0) = q_0$ and $\dot{q}(0) = \dot{q}$, implying that the object is *not* immobilized at the given grasp. Frictionless equilibrium grasps are thus necessary for object immobilization. □

The next theorem concerns the duality between wrench resistant and first-order immobilizing grasps (such grasps require a sufficiently high number of fingers).

Theorem 6.3 *Under frictionless contact conditions, a k-finger equilibrium grasp is* **wrench resistant** *iff it forms a* **first-order immobilizing** *grasp.*

Proof: The net wrench cone generated by the k fingers via frictionless contacts is given by $\mathcal{W} = \{\lambda_1\eta_1(q_0) + \ldots + \lambda_k\eta_k(q_0) : \lambda_1, \ldots, \lambda_k \geq 0\}$. Hence, the set of external wrenches that can be resisted by the fingers, $\mathcal{W}_{ext}$, is given by

$$\mathcal{W}_{ext} = \{-\boldsymbol{w} : \boldsymbol{w} \in \mathcal{W}\} = \big\{\lambda_1(-\eta_1(q_0)) + \ldots + \lambda_k(-\eta_k(q_0)) : \lambda_1, \ldots, \lambda_k \geq 0\big\}.$$

Based on Definition 6.6, the set of first-order free motions of the object $\mathcal{B}$ is obtained by intersecting the first-order free half-spaces:

$$M_{1\ldots k}(q_0) = \cap_{i=1}^{k} M_i(q_0) = \big\{\dot{q} \in T_{q_0}\mathbb{R}^m : \eta_i(q_0) \cdot \dot{q} \geq 0,\ i = 1 \ldots k\big\}.$$

Since $\boldsymbol{w}_{ext} = \lambda_1(-\eta_1(q_0)) + \ldots + \lambda_k(-\eta_k(q_0))$ for all $\boldsymbol{w}_{ext} \in \mathcal{W}_{ext}$, we obtain the following relation between $M_{1\ldots k}(q_0)$ and $\mathcal{W}_{ext}$:

$$\forall \boldsymbol{w}_{ext} \in \mathcal{W}_{ext} \quad \forall \dot{q} \in M_{1\ldots k}(q_0) \qquad \boldsymbol{w}_{ext} \cdot \dot{q} = \sum_{i=1}^{k} \lambda_i(-\eta_i(q_0)) \cdot \dot{q} \leq 0.$$

Based on the definition of *dual cones* in Eq. (6.6):

$$(\mathcal{W}_{ext})^* = \big\{\dot{q} \in T_{q_0}\mathbb{R}^m : \boldsymbol{w} \cdot \dot{q} \leq 0 \text{ for all } \boldsymbol{w} \in \mathcal{W}_{ext}\big\} = M_{1\ldots k}(q_0).$$

The cone of first-order free motions, $M_{1\ldots k}(q_0)$, is thus *dual* to the cone $\mathcal{W}_{ext}$ of external wrenches that can be resisted by k frictionless finger contacts. In particular, $\mathcal{W}_{ext} = T^*_{q_0}\mathbb{R}^m$ iff $M_{1\ldots k}(q_0) = \{0\}$. Hence, wrench resistance is dual to first-order immobilization. □

References

[1] F. Reuleaux, *The Kinematics of Machinery*. Macmillan, 1876, republished by Dover, 1963.

[2] P. Somoff, "Ueber Gebiete von Schraubengeschwindigkeiten eines starren Koerpers bei verschiedener Zahl von Stuetzflaechen," *Zeitschrift fuer Mathematik und Physik*, vol. 45, pp. 245–306, 1900.

[3] K. Lakshminarayana, "Mechanics of form closure," ASME, Tech. Rep. 78-DET-32, 1978.

[4] J. K. Salisbury, "Kinematic and force analysis of articulated hands," PhD thesis, Dept. of Mechanical Engineering, Stanford University, 1982.

[5] J. R. Kerr and B. Roth, "Analysis of multi-fingered hands," *International Journal of Robotics Research*, vol. 4, no. 4, pp. 3–17, 1986.

[6] S. Boyd and L. Vandenberghe, *Convex Optimization*. Cambridge University Press, 2004.

[7] A. Ben-Tal and A. S. Nemirovski, *Lectures on Modern Convex Optimization: Analysis, Algorithms, and Engineering Applications*. MOS-SIAM Series on Optimization, 2001.

7 First-Order Immobilizing Grasps

Immobilizing grasps maintain grasp security by preventing any movement of the grasped object with respect to the grasping robot hand. Immobilizing grasps are universally used in industrial fixturing, and they provide a simple and reliable approach for maintaining grasp security during robotic grasping and manipulation tasks. Immobilization theory begins with the observation made in the previous chapter that object immobilization can only be achieved with *frictionless equilibrium grasps*. Thus, consider a rigid object $\mathcal{B}$ held in a frictionless equilibrium grasp by stationary rigid finger bodies $\mathcal{O}_1, \ldots, \mathcal{O}_k$. While friction is allowed at the contacts, grasp immobilization seeks to secure the object based on rigid-body constraints imposed on the object's free motions by rigid finger bodies.

This chapter focuses on the *first-order geometry* of immobilizing grasps. Section 7.1 describes the *first-order free motions* available to a grasped object $\mathcal{B}$ with respect to the finger bodies. These free motions are c-space tangent vectors that represent the object's velocities, along which the object $\mathcal{B}$ breaks or maintains contact with the stationary finger bodies. Section 7.2 shows that the object's first-order free motions at a frictionless equilibrium grasp form a linear subspace of tangent vectors. The dimension of this subspace defines the grasp *first-order mobility index.* We will see in Section 7.3 that when the first-order mobility index is zero, the object is fully immobilized and, hence, secured by the grasping fingers. Moreover, the immobilization is *robust* with respect to small finger placement errors. Section 7.4 provides a graphical interpretation of the first-order mobility index of a given grasp arrangement.

The graphical interpretation of Section 7.4 will help us recognize important limitations of immobilization that is based on the bodies' first-order geometric properties. Grasp immobilization can also be achieved by a combination of first and second-order geometric effects. When the bodies' curvature is taken into account, a grasp that has a positive first-order mobility index (and thus is mobile according to the first-order theory) can be perfectly immobilizing. This part of grasp mobility theory will be considered in Chapter 8, while techniques for synthesizing minimal immobilizing grasps will be considered in Chapter 9.

7.1 The First-Order Free Motions

Grasp mobility analysis is based on the *free motions* available to the object $\mathcal{B}$ at a given grasp. When the object $\mathcal{B}$ is contacted by stationary finger bodies $\mathcal{O}_1, \ldots, \mathcal{O}_k$, the finger c-obstacles constrain the object's possible motions. The object is free to move within the *free c-space*, $\mathcal{F}$, which is given by

$$\mathcal{F} = \mathbb{R}^m - \cup_{i=1}^{k} \mathsf{int}(\mathcal{CO}_i) \qquad m = 3 \text{ in 2-D grasps, } m = 6 \text{ in 3-D grasps,}$$

where $\mathsf{int}(\mathcal{CO}_i)$ denotes the interior of the c-obstacle $\mathcal{CO}_i$. When $\mathcal{B}$ is held at a configuration q_0, its *free motions* are those c-space paths that start at q_0 and locally lie in $\mathcal{F}$. To analyze the object's free motions in the c-space framework, let us model the finger c-obstacles in terms of the following ith *c-obstacle signed distance function.*

DEFINITION 7.1 *Let $\mathcal{CO}_i$ be a finger c-obstacle and let $\mathcal{S}_i$ be its boundary. The* **c-obstacle distance function,** *$d_i(q): \mathbb{R}^m \to \mathbb{R}$, measures the signed distance of configuration points q from $\mathcal{S}_i$:*

$$d_i(q) = \begin{cases} -\mathtt{dst}(q, \mathcal{S}_i) & \text{if } q \in \mathsf{int}(\mathcal{CO}_i) \\ 0 & \text{if } q \in \mathcal{S}_i \\ \mathtt{dst}(q, \mathcal{S}_i) & \text{if } q \in \mathbb{R}^m - \mathcal{CO}_i \end{cases} \tag{7.1}$$

where $\mathtt{dst}(q, \mathcal{S}_i) = \min_{p \in \mathcal{S}_i}\{\|q - p\|\}$ is the minimal Euclidean distance of q from $\mathcal{S}_i$.

The finger c-obstacle boundary forms the zero level-set of d_i. We will assume that $\mathcal{S}_i$ is a smooth $(m - 1)$-dimensional manifold. This assumption is met when $\mathcal{B}$ and $\mathcal{O}_i$ have smooth boundaries and maintain point contact – for instance, when $\mathcal{B}$ is an ellipse and $\mathcal{O}_i$ is a disc. In more general cases – for instance, when $\mathcal{B}$ is a polygon – $\mathcal{S}_i$ is a piecewise smooth manifold. Such cases can be worked out with the non-smooth analysis tools described in Appendix A. When $\mathcal{S}_i$ forms a smooth manifold, ∇d_i is well defined in a local neighborhood that surrounds $\mathcal{S}_i$ in the ambient c-space $\mathbb{R}^m$. By construction, d_i is negative inside $\mathcal{CO}_i$, zero on $\mathcal{S}_i$ and positive outside $\mathcal{CO}_i$. Hence, $\nabla d_i(q)$ is collinear with the c-obstacle outward normal, $\eta_i(q)$, at all points $q \in \mathcal{S}_i$. Moreover, d_i is defined in terms of the Euclidean distance, and, consequently, $\|\nabla d_i(q)\| = 1$ at all points $q \in \mathcal{S}_i$ (see Bibliographical Notes). Thus, $\nabla d_i(q) = \hat{\eta}_i(q)$ at all points $q \in \mathcal{S}_i$, where $\hat{\eta}_i(q)$ is the c-obstacle unit outward normal at $q \in \mathcal{S}_i$.

We can now formulate the first-order properties of the free motion curves, which will lead to a *first-order mobility theory*. Let $\alpha(t)$ be a smooth c-space curve, chosen such that $\alpha(0) = q_0 \in \mathcal{S}_i$ and $\dot{\alpha}(0) = \dot{q}$. The *first-order Taylor expansion* of d_i along α has the form:

$$d_i(\alpha(t)) = d_i(q_0) + (\nabla d_i(q_0) \cdot \dot{q})t + \mathrm{o}(t^2) \qquad t \in (-\epsilon, \epsilon),$$

where $\epsilon > 0$ is a small parameter. When the Taylor expansion is evaluated at $q_0 \in \mathcal{S}_i$, the first-order approximation of d_i becomes

$$d_i(\alpha(t)) = (\hat{\eta}_i(q_0) \cdot \dot{q})t + \mathrm{o}(t^2) \qquad t \in (-\epsilon, \epsilon), \tag{7.2}$$

where we substituted $d_i(q_0) = 0$ and $\nabla d_i(q_0) = \hat{\eta}_i(q_0)$. The first-order free motions are characterized in terms of the first-order approximation of Eq. (7.2) as follows.

DEFINITION 7.2 **(single finger first-order free motions)** *Let a rigid object $\mathcal{B}$ be contacted by a stationary rigid finger body $\mathcal{O}_i$ at a configuration $q_0 \in \mathcal{S}_i$. The* **first-order free motions** *of $\mathcal{B}$ at q_0 are the half-space of tangent vectors:*

$$M_i^1(q_0) = \left\{\dot{q} \in T_{q_0}\mathbb{R}^m \ : \ \hat{\eta}_i(q_0) \cdot \dot{q} \geq 0\right\}.$$

The half-space's boundary, $T_{q_0}\mathcal{S}_i = \{\dot{q} \in T_{q_0}\mathbb{R}^m : \hat{\eta}_i(q_0) \cdot \dot{q} = 0\}$, *forms the* **first order roll–slide motions**. *Its interior,* $\{\dot{q} \in T_{q_0}\mathbb{R}^m : \hat{\eta}_i(q_0) \cdot \dot{q} > 0\}$, *is the set of* **first-order escape motions.**

Example: Figure 7.1(a) shows an ellipse $\mathcal{B}$ contacted by a disc finger $\mathcal{O}_i$. Figure 7.1(b) shows the c-space geometry of this grasp, with $q_0 \in \mathcal{S}_i$ being the ellipse's contact configuration. The half-space $M_i^1(q_0)$ forms the first-order approximation to the exterior of the finger c-obstacle $\mathcal{CO}_i$ at q_0. The half-space's boundary coincides with the c-obstacle tangent plane, $T_{q_0}\mathcal{S}_i$. Tangent vectors pointing into the interior of $M_i^1(q_0)$ represent first-order escape motions, while tangent vectors that lie in $T_{q_0}\mathcal{S}_i$ represent first order roll–slide motions. ○

The first-order free motions of $\mathcal{B}$ admit the following geometric interpretation. When $\dot{q} \in M_i^1(q_0)$ is a first-order escape motion, any path $\alpha(t)$ with $\alpha(0) = q_0$ and $\dot{\alpha}(0) = \dot{q}$ *locally lies in the free c-space* $\mathcal{F}$, for all $t \in [0, \epsilon]$. The motion of $\mathcal{B}$ along this path would cause it to separate from $\mathcal{O}_i$, no matter what the value of the higher derivatives of $d_i(q)$ along α. On the other hand, when $\dot{q} \in M_i^1(q_0)$ is a first order roll–slide motion, $\dot{q}$ is tangent to the finger c-obstacle boundary at q_0. In this case, it is *not possible* to determine based on first-order considerations whether α locally lies in $\mathcal{F}$ or locally penetrates the finger c-obstacle $\mathcal{CO}_i$ at q_0. This indeterminacy is illustrated in the following example.

Example: Consider the two c-space curves α and β depicted in Figure 7.1(b). Both curves start at $q_0 \in \mathcal{S}_i$ and have the same initial tangent vector, $\dot{\alpha}(0) = \dot{\beta}(0) \in T_{q_0}\mathcal{S}_i$. Both curves are thus equivalent to first order, and represent the same first order roll–slide motion of the object $\mathcal{B}$ with respect $\mathcal{O}_i$. However, α locally lies in the free c-space $\mathcal{F}$, while β locally penetrates the interior of the finger c-obstacle $\mathcal{CO}_i$. We will see later in this chapter that the free motions available to $\mathcal{B}$ at a *frictionless equilibrium grasp* are roll–slide to first order, much like the two curves depicted in Figure 7.1(b). ○

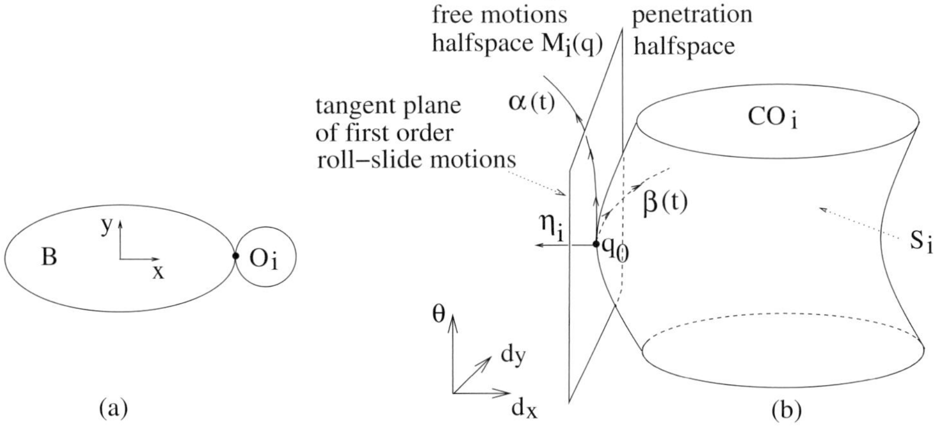

Figure 7.1 (a) An ellipse $\mathcal{B}$ contacted by a stationary disc finger $\mathcal{O}_i$. (b) The half-space $M_i^1(q_0)$ approximates the exterior of $\mathcal{CO}_i$ at q_0. While α and β are roll–slide motions to first order, α locally lies in $\mathcal{F}$ while β locally lies in the interior of $\mathcal{CO}_i$.

The definition of the first-order free motions is next extended to the case where an object $\mathcal{B}$ is contacted by multiple finger bodies. The object's first-order free motions must respect the half-space constraints imposed by all contacting fingers, as stated in the following definition.

DEFINITION 7.3 (**K-Finger first-order free motions**) *Let a rigid object $\mathcal{B}$ be contacted by rigid stationary finger bodies $\mathcal{O}_1, \ldots, \mathcal{O}_k$ at a configuration $q_0 \in \cap_{i=1}^{k} S_i$. The set of* **first-order free motions** *of $\mathcal{B}$ at q_0, denoted $M^1_{1\ldots k}(q_0)$, has the form:*

$$M^1_{1\ldots k}(q_0) = \cap_{i=1}^{k} M_i^1(q_0) = \left\{\dot{q} \in T_{q_0}\mathbb{R}^m \; : \; \hat{\eta}_i(q_0) \cdot \dot{q} \geq 0, \; i = 1 \ldots k\right\},$$

where $\hat{\eta}_i(q_0)$ is the ith finger c-obstacle outward unit normal at q_0 for $i = 1 \ldots k$.

Recall from Chapter 4 that a set C forms a *convex cone* if for any $v_1, v_2 \in C$, the linear combination $\lambda_1 v_1 + \lambda_2 v_2$ lies in C for all $\lambda_1, \lambda_2 \geq 0$. The set $M^1_{1\ldots k}(q_0)$ forms a *convex cone* of tangent vectors at q_0. This property follows from the observation that each half-space $M_i^1(q_0)$ is a convex cone based at q_0, and the intersection of convex cones based at the origin is a convex cone based at this point.

The set $M^1_{1\ldots k}(q_0)$ is defined in terms of the Euclidean inner product, which is usually *not* preserved by coordinate transformations. Hence, we must verify that $M^1_{1\ldots k}(q_0)$ is coordinate invariant under different choices of world and object reference frames. Any reasonable coordinate transformation forms a *diffeomorphism* – a smooth one-to-one and surjective mapping having a non-singular Jacobian at all points of its domain. The following proposition asserts that the set $M^1_{1\ldots k}(q_0)$ is coordinate invariant (see chapter's appendix for a proof).

PROPOSITION 7.1 (**coordinate invariance**) *Let q and $\bar{q}$ be two parametrizations of $\mathcal{B}$'s c-space, related by a coordinate transformation $q = h(\bar{q})$. If h is a diffeomorphism, the set of first-order free motions of $\mathcal{B}$ is* **coordinate invariant:**

$$\dot{q} \in M^1_{1\ldots k}(q_0) \quad \textit{iff} \quad \dot{\bar{q}} \in \bar{M}_{1\ldots k}(\bar{q}_0) \qquad\qquad q_0 = h(\bar{q}_0),$$

where $M^1_{1\ldots k}(q_0)$ and $\bar{M}_{1\ldots k}(\bar{q}_0)$ are the sets of first-order free motions in the q and $\bar{q}$ coordinates.

Different choices of the world and object frames give different parametrizations of $\mathcal{B}$'s c-space. As verified in the Exercises, the coordinate transformation induced by different reference frames are related by a standard diffeomorphism. The object's first-order free motions can thus be analyzed under any choice of world and object frames.

Proposition 7.1 can be justified as follows. The c-obstacle outward normal, $\eta_i(q_0)$, is collinear with the wrench generated by a normal finger force acting on $\mathcal{B}$ at x_i. The inner product $\eta_i(q_0) \cdot \dot{q}$ thus represents the *power* (or instantaneous work) applied to $\mathcal{B}$ by the wrench $\eta_i(q_0)$. From this perspective, the *sign* of the power applied to $\mathcal{B}$ by $\eta_i(q_0)$ should not depend on the choice of reference frames. To formally justify this intuition, recall from Section 3.2 that a cotangent vector (or wrench), $\boldsymbol{w} \in T^*_{q_0}\mathbb{R}^m$, acts as a linear function on tangent vectors: $\boldsymbol{w}(\dot{q}) = \boldsymbol{w} \cdot \dot{q}$ for $\dot{q} \in T_{q_0}\mathbb{R}^m$. When $\mathcal{B}$'s c-space is parametrized by different world and object frames, the value of $\boldsymbol{w}(\dot{q})$ must be preserved in the different c-space parametrizations. Under a diffeomorphism $q = h(\bar{q})$, tangent vectors transform by the rule, $\dot{q} = Dh\dot{\bar{q}}$, while cotangent vectors (or wrenches) transform by the rule, $\bar{w} = Dh^T w$. Therefore, $w \cdot \dot{q} = w^T(Dh\,\dot{\bar{q}}) = (w^T Dh)\dot{\bar{q}} = \bar{w} \cdot \dot{\bar{q}}$. Hence, $w \cdot \dot{q} \geq 0$ iff $\bar{w} \cdot \dot{\bar{q}} \geq 0$, and $w \cdot \dot{q} = 0$ iff $\bar{w} \cdot \dot{\bar{q}} = 0$.

7.2 The First-Order Mobility Index

A grasp *mobility index* is a coordinate invariant integer that measures the grasped object's mobility, or effective number of degrees of freedom, at a frictionless equilibrium grasp. The grasp *first-order mobility index* is based on the object's first-order free motions at the equilibrium grasp. The index is well defined when the fingers form an *essential equilibrium grasp.* Let us first recall the notion of essential finger contact from Chapter 4.

Essential finger contact: *Let an object $\mathcal{B}$ be held at a k-finger equilibrium grasp. A finger contact is* **essential** *for the grasp if the corresponding finger must generate a non-zero contact force in order to maintain the equilibrium grasp.*

Essential finger contacts can be identified as follows. At a k-finger equilibrium grasp, the fingers' net wrench cone contains a linear subspace passing through $\mathcal{B}$'s wrench space origin (Proposition 4.1). A particular finger is essential for the equilibrium grasp when the net wrench cone spanned by the remaining $k - 1$ fingers forms a pointed cone – that is, a cone that is strictly convex at its vertex located at $\mathcal{B}$'s wrench space origin. The essential k-finger equilibrium grasps are defined as follows.

DEFINITION 7.4 (**essential K-finger grasp**) *Let a rigid object $\mathcal{B}$ be held by rigid finger bodies $\mathcal{O}_1, \ldots, \mathcal{O}_k$ at a frictionless equilibrium grasp. The object is held at an* **essential equilibrium grasp** *under one of the conditions:*

(i) *There are $k \leq m + 1$ fingers, and all k fingers are essential for maintaining the equilibrium grasp.*

(ii) *There are $k > m + 1$ fingers, and $m + 1$ of the k fingers are essential for the equilibrium grasp, where $m = 3$ in 2-D grasps and $m = 6$ in 3-D grasps.*

The essential equilibrium grasps are generic in the following sense. Recall that *contact space* parametrizes the position of k frictionless contacts along the grasped object boundary (see Appendix II of Chapter 5). The k-finger equilibrium grasps form a subset, $\mathcal{E}$, in contact space. When an equilibrium grasp is essential, local perturbations of the contacts within the equilibrium set $\mathcal{E}$ also form essential equilibrium grasps. On the other hand, nonessential equilibrium grasps represent special finger arrangements that can be locally perturbed into essential grasps. These properties are illustrated in the following example.

Example: The rectangular object $\mathcal{B}$ shown in Figure 7.2(a) is held by four disc fingers via frictionless contacts. The grasp consists of two antipodal pairs, $(\mathcal{O}_1, \mathcal{O}_3)$

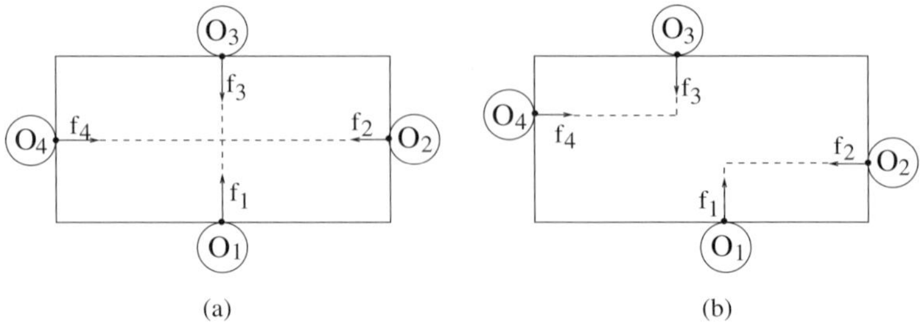

Figure 7.2 (a) Top view of a nonessential equilibrium grasp. (b) Local perturbation of the finger contacts form an essential equilibrium grasp, which remains essential under sufficiently small perturbations of the finger contacts.

and $(\mathcal{O}_2, \mathcal{O}_4)$, and hence forms a nonessential equilibrium grasp. However, any counterclockwise perturbation of $(\mathcal{O}_1, \mathcal{O}_3)$ accompanied by clockwise perturbation of $(\mathcal{O}_2, \mathcal{O}_4)$ forms an essential equilibrium grasp (Figure 7.2(b)). Moreover, the grasp depicted in Figure 7.2(b) remains essential under all sufficiently small perturbations of the finger contacts. ○

When an object $\mathcal{B}$ is held at an essential equilibrium grasp, its first-order free motions span a well-defined linear subspace which is tangent to the finger c-obstacles. This basic property of the first-order free motions is stated in the following proposition (see chapter's appendix for a proof).

PROPOSITION 7.2 (**Subspace property**) *Let a rigid object $\mathcal{B}$ be held in an essential k-finger equilibrium grasp at a configuration q_0. The object's first-order free motions, $M^1_{1...k}(q_0)$, span a* **linear subspace** *tangent to the finger c-obstacles, whose dimension is given by*

$$\dim\big(M^1_{1...k}(q_0)\big) = \max\big\{m - k + 1, 0\big\},$$

where $m = 3$ in 2-D grasps and $m = 6$ in 3-D grasps.

The proposition can be justified as follows. At an essential k-finger equilibrium grasp, the net wrench cone $\mathcal{W}$ forms a $(k-1)$-dimensional linear subspace in $\mathcal{B}$'s wrench space (see Appendix). We have seen in Chapter 6 that the net wrench cone $\mathcal{W}$ is *dual* to the cone formed by the first-order free motions, $M^1_{1...k}(q_0)$.[1] When $\mathcal{W}$ is embedded as a $(k-1)$-dimensional linear subspace in $\mathcal{B}$'s tangent space, $T_{q_0}\mathbb{R}^m$, its dual cone can be thought of as its orthogonal complement in $T_{q_0}\mathbb{R}^m$. Hence $M^1_{1...k}(q_0)$ forms a linear subspace, whose dimension is given by $\dim(M^1_{1...k}(q_0)) = m - (k-1) = m - k + 1$.

The set $M^1_{1...k}(q_0)$ forms a linear subspace tangent to the finger c-obstacles at q_0. Hence, the only first-order free motions available to $\mathcal{B}$ at a frictionless equilibrium grasp are *roll–slide motions* with respect to each of the finger bodies $\mathcal{O}_1, \ldots, \mathcal{O}_k$. The dimension of the subspace $M^1_{1...k}(q_0)$ defines the following *first-order mobility index.*

[1] Let $\mathcal{W}$ be a cone of wrenches based at $\mathcal{B}$'s wrench space origin. Its *dual cone*, $\mathcal{W}^*$, is the cone of tangent vectors given by $\mathcal{W}^* = \{\dot{q} \in T_{q_0}\mathbb{R}^m : \boldsymbol{w} \cdot \dot{q} \leq 0 \text{ for all } \boldsymbol{w} \in \mathcal{W}\}$.

DEFINITION 7.5 (**first-order mobility index**) *Let a rigid object $\mathcal{B}$ be held in an essential k-finger equilibrium grasp at a configuration q_0. The object's* **first-order mobility index, $m^1_{q_0}$**, *is the dimension of the linear subspace spanned by the object's first-order free motions:*

$$m^1_{q_0} = \dim\big(M^1_{1\ldots k}(q_0)\big) = \max\big\{m - k + 1, 0\big\}, \tag{7.3}$$

where $m = 3$ in 2-D grasps and $m = 6$ in 3-D grasps.

The first-order mobility index is *coordinate invariant,* since the set $M^1_{1\ldots k}(q_0)$ is coordinate invariant according to Proposition 7.1. Note that the index attains values only in the range $0 \leq m^1_{q_0} \leq m$ ($m = 3$ or 6), as illustrated in the following examples.

Example: Figure 7.3(a) depicts an ellipse $\mathcal{B}$ that is held in a frictionless equilibrium grasp by two disc fingers. The c-space geometry of this grasp is depicted in Figure 7.3(b), where $q_0 \in \mathcal{S}_1 \cap \mathcal{S}_2$ is the object's equilibrium grasp configuration. Substituting $m = 3$ and $k = 2$ in Eq. (7.3) gives

$$m^1_{q_0} = \max\{2, 0\} = 2.$$

The two-dimensional linear subspace $M^1_{1,2}(q_0)$ forms the tangent plane common to $\mathcal{S}_1$ and $\mathcal{S}_2$ at q_0 (Figure 7.3(b)). Based on a graphical technique developed in Section 7.4, every tangent vector $\dot{q} \in M^1_{1,2}(q_0)$ represents instantaneous rotation of $\mathcal{B}$ about the midpoint of the two contacts, coupled with instantaneous translation of $\mathcal{B}$ along the vertical y-axis. ○

Example: Figure 7.4(a) depicts a triangular object $\mathcal{B}$ that is held in a frictionless equilibrium grasp by three disc fingers. The c-space geometry of this grasp is depicted in Figure 7.4(b), where $q_0 \in \cap_{i=1}^3 \mathcal{S}_i$ is the object's equilibrium grasp configuration. Substituting $m = 3$ and $k = 3$ in Eq. (7.3) gives

$$m^1_{q_0} = \max\{1, 0\} = 1.$$

The linear subspace $M^1_{1,2,3}(q_0)$ forms the one-dimensional intersection of the tangent planes to $\mathcal{S}_1$, $\mathcal{S}_2$, and $\mathcal{S}_3$ at q_0. Based on a graphical technique discussed in Section 7.4, every tangent vector $\dot{q} \in M^1_{1,2,3}(q_0)$ represents instantaneous rotation of $\mathcal{B}$ about the intersection point of the finger contact normals. When the world and object

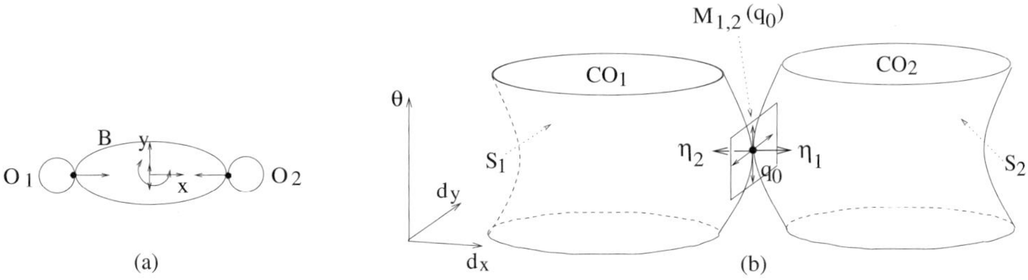

Figure 7.3 (a) Top view of an ellipse held in a frictionless two-finger equilibrium grasp. (b) The two-dimensional subspace $M^1_{1,2}(q_0)$ at the equilibrium grasp.

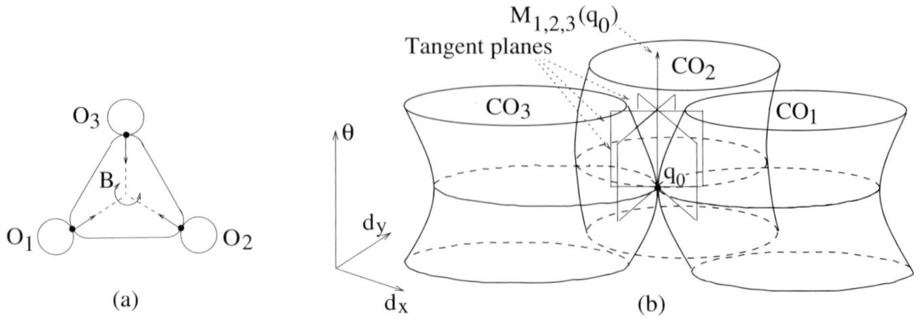

Figure 7.4 (a) Top view of a triangular object held in a frictionless three-finger equilibrium grasp. (b) The one-dimensional subspace $M^1_{1,2,3}(q_0)$ at the equilibrium grasp.

frame origins are located at this point, $M^1_{1,2,3}(q_0)$ forms the vertical line parallel to the c-space θ axis depicted Figure 7.4(b). ○

7.3 First-Order Immobilization

The dimension of $\mathcal{B}$'s c-space is $m=3$ in 2-D grasps and $m=6$ in 3-D grasps. Since m attains one of two fixed values, the first-order mobility index, $m^1_{q_0}=\max\{m-k+1,0\}$, depends only on the parameter k, which represents the number of grasping fingers. As the number of fingers increases, the value of $m^1_{q_0}$ (and hence the amount of first-order free motions available to $\mathcal{B}$) decreases. When $m^1_{q_0}=0$, the object has no first-order free motions and should, therefore, be fully immobilized by the finger bodies. This important means of achieving object immobilization is stated in the following theorem.

THEOREM 7.3 (**first-order immobilization**) *Let a rigid object $\mathcal{B}$ be held in an essential equilibrium grasp by rigid finger bodies $\mathcal{O}_1,\ldots,\mathcal{O}_k$ at a configuration q_0. If $m^1_{q_0}=0$ at the grasp, the object is* **fully immobilized** *by the grasping finger bodies.*

Proof: We have to show that the configuration point q_0 is completely surrounded by the finger c-obstacles. We will prove this fact using the *minimum distance function:*

$$d_{min}(q)=\min\left\{d_1(q),\ldots,d_k(q)\right\} \qquad q\in\mathbb{R}^m,$$

where $d_i(q)$ is the ith c-obstacle distance function specified in Eq. (7.1). While $d_{min}(q)$ is *not* differentiable at q_0, it forms a Lipschitz continuous function that can be analyzed with the tools described in Appendix A. In particular, d_{min} possesses a *generalized gradient* at q_0, denoted $\partial d_{min}(q_0)$, which is the convex combination of the gradients $\nabla d_i(q_0)$ for $i=1\ldots k$. The gradients are collinear with the c-obstacle outward unit normals, $\nabla d_i(q_0)=\hat{\eta}_i(q_0)$ for $i=1\ldots k$. Hence, the generalized gradient can be expressed as the convex combination:

$$\partial d_{min}(q_0)=\sum_{i=1}^{k}\sigma_i\hat{\eta}_i(q_0) \qquad 0\le\sigma_1,\ldots,\sigma_k\le 1,\ \textstyle\sum_{i=1}^{k}\sigma_i=1. \tag{7.4}$$

The generalized gradient $\partial d_{min}(q_0)$ forms a convex set in $\mathcal{B}$'s wrench space, $T^*_{q0}\mathbb{R}^m$. According to Theorem A.2 of Appendix A, $d_{min}(q)$ has a *strict local maximum* at q_0 when $\partial d_{min}(q_0)$ contains a small m-dimensional ball centered at the origin of $T^*_{q0}\mathbb{R}^m$. Since $M^1_{1\ldots k}(q_0) = \{0\}$ when $m^1_{q_0} = 0$, for any $\dot{q} \in T_{q0}\mathbb{R}^m$ there exists a c-obstacle outward normal $\hat{\eta}_i(q_0)$ such that $\hat{\eta}_i(q_0) \cdot \dot{q} < 0$. The latter inequality can be interpreted as the condition that $\hat{\eta}_i(q_0)$ lies in the interior of the half-space of wrenches:

$$H_{\dot{q}} = \left\{ \boldsymbol{w} \in T^*_{q0}\mathbb{R}^m \, : \, \boldsymbol{w} \cdot \dot{q} < 0 \right\},$$

which is bounded by an $(m - 1)$-dimensional plane passing through $\mathcal{B}$'s wrench space origin. Since $\hat{\eta}_i(q_0) \in \partial d_{min}(q_0)$ and $\dot{q}$ freely varies in $T_{q0}\mathbb{R}^m$, $\partial d_{min}(q_0)$ intersects the interior of *every half-space of wrenches* $H_{\dot{q}}$, whose boundary passes through $\mathcal{B}$'s wrench space origin. Since $\partial d_{min}(q_0)$ forms a convex set in $T^*_{q0}\mathbb{R}^m$, the latter condition can only occur when $\partial d_{min}(q_0)$ contains a small m-dimensional ball centered at $\mathcal{B}$'s wrench space origin (see Exercises). The function $d_{min}(q)$ thus has a strict local maximum at q_0.

We know that $d_{min}(q_0) = 0$, since $q_0 \in \cap_{i=1}^k S_i$. Hence, d_{min} must be *strictly negative* in a small m-dimensional neighborhood in $\mathcal{B}$'s c-space centered at q_0. Since $d_{min}(q) = \min\{d_1(q), \ldots, d_k(q)\}$, at each point q in this neighborhood, some $d_i(q)$ must attain negative value, which implies that q lies in the *interior* of the c-obstacle $\mathcal{CO}_i$. The point q_0 is thus completely surrounded by the finger c-obstacles, which proves that $\mathcal{B}$ is fully immobilized by the finger bodies $\mathcal{O}_1, \ldots, \mathcal{O}_k$. □

Theorem 7.3 offers the following object immobilization technique. Given an object $\mathcal{B}$, construct a frictionless equilibrium grasp that involves at least *four* fingers in 2-D grasps and at least *seven* fingers in 3-D grasps. When all fingers are essential for maintaining the equilibrium grasp, the object is fully immobilized by the finger bodies. Moreover, the object remains immobilized under all sufficiently small finger placement errors. The first-order immobilizing grasps are thus *robust*, as summarized in the following corollary (see Exercises).

COROLLARY 7.4 **(robust immobilization)** *Let a rigid object $\mathcal{B}$ be held at an essential equilibrium grasp by $k \geq 4$ fingers in $\mathbb{R}^2$, and $k \geq 7$ fingers in $\mathbb{R}^3$. The object is fully immobilized, and the immobilization is* **robust** *with respect to small contact placement errors.*

Example: An example of robust immobilization appears in Figure 7.2(b), which shows a rectangular object $\mathcal{B}$ held at an essential four-finger equilibrium grasp. The object is immobilized at this grasp and remains immobilized under small finger placement errors. ○

First order rigid-body constraints can thus be used to achieve secure and robust grasps. Such grasps can be established with no more than *four fingers* in 2-D grasps, and no more than *seven fingers* in 3-D grasps. These are *tight upper bounds* on the number of fingers, since, as we know from Chapter 6, at least four fingers are required for first-order immobilization in 2-D grasps, while at least seven fingers are required for first-order immobilization in 3-D grasps. However, we will see in subsequent chapters that these upper bounds are unnecessarily high for minimalistic robot hands, which can establish secure grasps using smaller numbers of fingers. First-order immobilization is common in industrial fixturing systems, as next described.

The 3-2-1 fixturing technique: Industrial fixturing systems typically lack sophisticated sensory feedback. Hence, the precise positioning of a workpiece $\mathcal{B}$ is part of the fixturing process. The classical 3-2-1 fixturing technique is based on the observation that the configuration of a 3-D workpiece $\mathcal{B}$ is uniquely determined when six bodies, called *locators,* are brought into point contact with $\mathcal{B}$. From $\mathcal{B}$'s c-space perspective, each locator induces a c-obstacle in $\mathcal{B}$'s c-space. When six locators are brought into contact with $\mathcal{B}$, its configuration is located at one of a discrete set of intersection points of the six c-obstacle boundaries. The 3-2-1 technique starts by placing the workpiece $\mathcal{B}$ on three upward-pointing locators which provide support against gravity. Next, two locators are brought into contact with $\mathcal{B}$ along a horizontal x-direction; then a single locator is brought into contact with $\mathcal{B}$ along a horizontal y-direction. Based on Theorem 7.3, first-order immobilization can be achieved at this stage with a *single* additional contact. However, traditional industrial fixtures employ three clamping devices:[2] the first forming an equilibrium grasp against the three upward-pointing locators, the second forming an equilibrium grasp against the two x-axis locators, and the third forming an equilibrium grasp against the single y-axis locator (Figure 7.5). The resulting fixture arrangement is first-order immobilizing as well as robust with respect to small contact placement errors. However, the technique requires *nine* fixturing elements, while *seven* such elements suffice to achieve robust immobilization in all generic cases. ○

A planar analog of the 3-2-1 technique is described in the following example.

Example: The planar version of the 3-2-1 technique is illustrated in Figure 7.5. The object $\mathcal{B}$ is first placed under the influence of gravity on two upward-pointing locators,

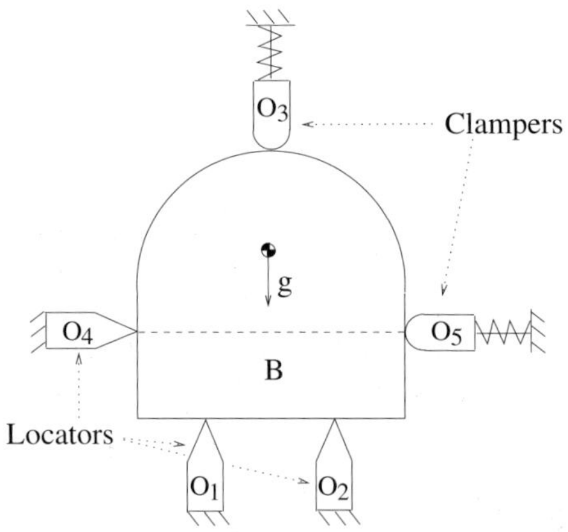

Figure 7.5 The planar version of the 3-2-1 fixturing technique involves three locators and two clampers.

[2] A *clamping device* consists of a rigid fingertip mounted on a high stiffness leadscrew. These devices approximate stationary rigid bodies preloaded against the workpiece $\mathcal{B}$.

$\mathcal{O}_1$ and $\mathcal{O}_2$, then a horizontal locator, $\mathcal{O}_4$, is brought into contact with $\mathcal{B}$. At this stage, $\mathcal{B}$ is located at a unique configuration in the plane, since three c-obstacle surfaces intersect at a discrete set of points. To secure the workpiece at this configuration, the clamper $\mathcal{O}_3$ is pressed downward against the upward-pointing locators; then the clamper $\mathcal{O}_5$ is pressed against the horizontal locator (Figure 7.5). The resulting fixture arrangement consists of two essential grasps: an essential three-finger grasp along the vertical axis and an essential two-finger grasp along the horizontal axis. Using the indices indicated in the figure, the subspace of first-order free motions associated with the three-finger grasp, $M^1_{1,2,3}(q_0)$, is one-dimensional. The subspace associated with the two-finger grasp, $M^1_{4,5}(q_0)$, is two-dimensional. The two subspaces intersect transversally in $\mathcal{B}$'s tangent space (see Exercises). Hence $M^1_{1\ldots5}(q_0) = M^1_{1,2,3}(q_0) \cap M^1_{4,5}(q_0) = \{0\}$, thus providing a first-order immobilizing grasp of $\mathcal{B}$. ◦

Theorem 7.3 and Corollary 7.4 achieve grasp security based on first-order geometric effects. However, they do not preclude the possibility that an object $\mathcal{B}$ be fully immobilized by a *smaller* number of fingers, due to second-order geometric effects. This possibility is captured by the second-order mobility index introduced in the next chapter.

7.4 Graphical Interpretation of the First-Order Mobility Index

This section describes the *instantaneous centers of rotation technique,* which can be used to graphically depict the first-order free motions of a 2-D rigid object $\mathcal{B}$ held by rigid finger bodies $\mathcal{O}_1, \ldots, \mathcal{O}_k$. We will use this technique to determine the first-order mobility index of objects held at different types of frictionless equilibrium grasps.

The technique is based on Mozzi-Chasles' theorem from Chapter 3. According to this theorem, every rigid-body velocity of $\mathcal{B}$, $\dot{q} \in T_{q_0}\mathbb{R}^3$, can be described as an instantaneous rotation of $\mathcal{B}$ about a point $p \in \mathbb{R}^2$. The point p defines the object's *instantaneous center of rotation* associated with $\dot{q}$. The position of p as a function of $\dot{q} = (v, \omega)$, where v and ω are $\mathcal{B}$'s linear and angular velocities, is given by the formula (see Exercises):

$$p = -\frac{1}{\omega} J v \qquad \dot{q} = (v, \omega), \; J = \begin{bmatrix} 0 & 1 \\ -1 & 0 \end{bmatrix}.$$

Using the identity $J^2 = -I$, the object's linear velocity can be expressed in terms of p and ω as $v = \omega J p$. The object's tangent space can thus be parametrized in terms of instantaneous centers of rotation as follows:

$$T_{q_0}\mathbb{R}^3 = \left\{ \dot{q} = \omega \cdot \begin{pmatrix} Jp \\ 1 \end{pmatrix} : p \in \mathbb{R}^2, \; \omega \in \mathbb{R} \right\}. \tag{7.5}$$

The parametrization specified in Eq. (7.5) contains three scalars, $p \in \mathbb{R}^2$ and $\omega \in \mathbb{R}$, which is consistent with the fact that $T_{q_0}\mathbb{R}^3$ is a three-dimensional vector space. In particular, instantaneous rotations about points p located at infinity represent instantaneous translations of the object $\mathcal{B}$.

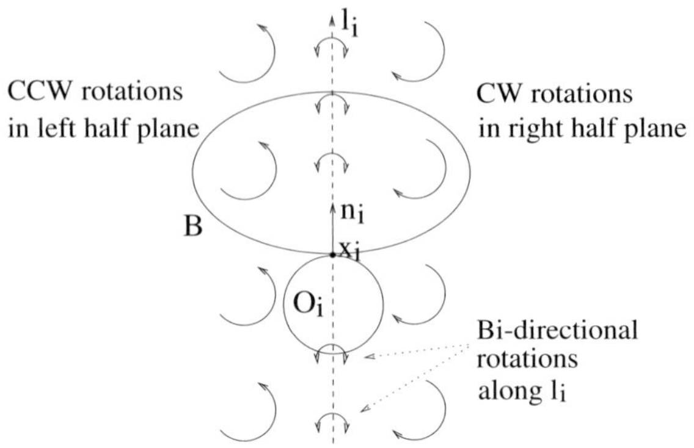

Figure 7.6 The half-space $M_i^1(q_0)$ consists of counterclockwise rotations in the left half plane, clockwise rotations in the right half plane, and bi-directional rotations along the contact normal line.

Let us next obtain a representation of the half-space of first-order free motions associated with a single finger body, $M_i^1(q_0) = \{\dot{q} \in T_{q_0}\mathbb{R}^3 : \eta_i(q_0)\cdot\dot{q} \geq 0\}$, in terms of instantaneous centers of rotation. The ith finger c-obstacle outward normal is given by $\eta_i(q_0) = (n_i, x_i \times n_i)$, where x_i is the contact point and n_i is $\mathcal{B}$'s inward unit normal at x_i (Chapter 2). Using Eq. (7.5), the half-space $M_i^1(q_0)$ is represented by the inequality:

$$\eta_i(q_0)\cdot\dot{q} = \omega\begin{pmatrix} n_i \\ x_i \times n_i \end{pmatrix}\cdot\begin{pmatrix} Jp \\ 1 \end{pmatrix} = \omega\big(n_i \times (p - x_i)\big) \geq 0 \qquad p \in \mathbb{R}^2,\ \omega \in \mathbb{R}, \quad (7.6)$$

where $u \times v = u^T J v$ such that $J = \begin{bmatrix} 0 & 1 \\ -1 & 0 \end{bmatrix}$. Based on Eq. (7.6), denote by l_i the directed line that passes through x_i along $\mathcal{B}$'s inward unit normal at this point. As shown in Figure 7.6, the half-space $M_i^1(q_0)$ consists of instantaneous counterclockwise rotations on the left side of l_i, instantaneous clockwise rotations on the right side of l_i, and instantaneous bi-directional rotations about all points of l_i (see also Chapter 3).

The graphical depiction of the first-order free motions of the object $\mathcal{B}$ at a k-finger grasp is summarized as follows (see Exercises).

Graphical depiction of first-order free motions in 2-D grasps: Consider the set $M_{1\ldots k}^1(q_0)$ associated with stationary finger bodies $\mathcal{O}_1, \ldots, \mathcal{O}_k$. The finger contact normal lines $l_1, \ldots, l_k$ partition the plane into polygons. When the k contacts agree on the direction of $\mathcal{B}$'s instantaneous rotation in a particular polygon, all points in the interior of this polygon represent *first-order escape motions* of $\mathcal{B}$ with respect to all k fingers. When the k contacts agree on the direction of $\mathcal{B}$'s instantaneous rotation along a lower dimensional set (a line or a single point), all points p in this set represent *first order roll–slide motions* of $\mathcal{B}$ at the given grasp. ◦

The graphical technique is next applied to frictionless equilibrium grasps (which are necessary for object immobilization). The set $M_{1\ldots k}^1(q_0)$ forms a linear subspace at such grasps (Proposition 7.2), which consists of first order roll–slide motion with respect to the grasping finger bodies. Graphically, these are instantaneous rotations of

$\mathcal{B}$ about points that lie at the *intersection* of the contact normal lines $l_1, \ldots, l_k$. Since these instantaneous rotations can be freely multiplied by $\omega \in \mathbb{R}$, each of these points represents a one-dimensional subspace of $M^1_{1\ldots k}(q_0)$. Since $M^1_{1\ldots k}(q_0)$ forms a linear subspace, one expects that the contact normal lines will intersect at a single point (possibly located at infinity) or lie on a common line in $\mathbb{R}^2$, as illustrated in the following examples.

Two-finger grasps: Figure 7.7 shows two-finger equilibrium grasps of several objects. In each of these grasps, the contact normal lines lie on common line l. According to the graphical technique, the object's first order roll–slide motions with respect to both fingers consist of instantaneous rotations about all points $p \in l$. Moreover, each of these instantaneous rotations is *first order roll–slide* for any angular velocity $\omega \in \mathbb{R}$. As the point p scans the line l from $-\infty$ to ∞ and ω varies in $\mathbb{R}$, we obtain a two-parameter family of instantaneous rotations. This gives the graphical representation of the two-dimensional linear subspace of first order roll–slide motions, $M^1_{1,2}(q_0)$, which is consistent with our earlier observation that $m^1_{q_0} = 2$ for these grasps. The depicted objects are all first-order *mobile,* since $m^1_{q_0} > 0$ at each of these grasps. However, intuition suggests that the grasp in Figure 7.7(a) is the most mobile, while the grasp in Figure 7.7(d) is the least mobile and is, in fact, a fully immobilizing grasp. This intuitive observation will be made precise in the next chapter. ○

Three-finger grasps: Figure 7.8 shows a triangular object $\mathcal{B}$ held by three disc fingers in two *essential* equilibrium grasps. The contact normal lines intersect at a single point p. Based on the graphical technique, every first order roll–slide motion available to $\mathcal{B}$

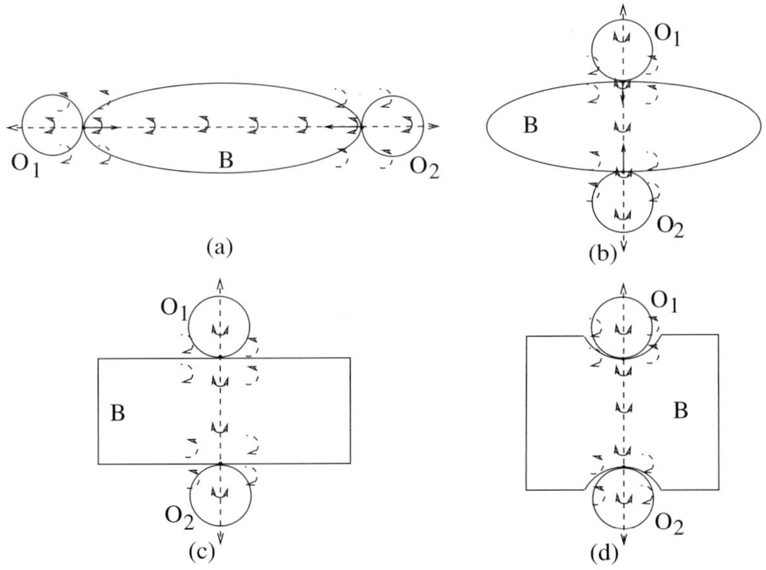

Figure 7.7 All two-finger equilibrium grasps have the same first-order mobility index of $m^1_{q_0} = 2$. (a)–(c) The object $\mathcal{B}$ is mobile in these grasps. (d) The object $\mathcal{B}$ is immobile at this grasp.

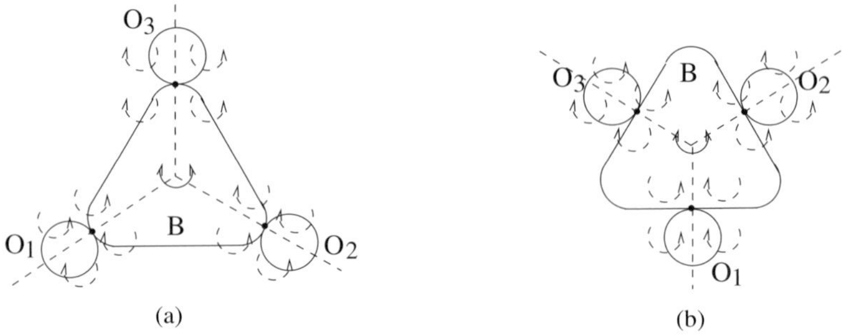

Figure 7.8 All three-finger equilibrium grasps have the same first-order mobility index of $m^1_{q_0} = 1$. (a) The object $\mathcal{B}$ is mobile at this grasp. (b) The object $\mathcal{B}$ is immobile at this grasp.

is an instantaneous rotation about p, with $\omega \in \mathbb{R}$. This one-parameter family graphically represents a one-dimensional linear subspace of first order roll–slide motions, $M^1_{1,2,3}(q_0)$, which is consistent with the fact that $m^1_{q_0} = 1$ for essential three-finger equilibrium grasps. Here, too, the object $\mathcal{B}$ is first-order *mobile,* since $m^1_{q_0} > 0$ in both grasps. However, intuition suggests that the triangular object is fully immobilized in the grasp of Figure 7.8(b), as will be justified in the next chapter. ○

The examples highlight two significant limitations of first-order immobilization. First, the number of fingersrequired to achieve full object immobilization based on first-order geometric effects seems excessively high for many grasping applications. Second, the first-order mobility index is rather crude in its ability to assess object immobility. In particular, it cannot differentiate between equilibrium grasps involving the same number of fingers. The second-order mobility theory described in the next chapter will help us differentiate between different equilibrium grasp choices that are first-order equivalent.

Bibliographical Notes

The c-obstacle distance function, $d_i(q)$, is defined in terms of the Euclidean distance of configuration points q from the ith c-obstacle boundary. The gradient of this function has unit magnitude, $\|\nabla d_i(q)\| = 1$, at any smooth point on the ith c-obstacle boundary. This fact is stated as Proposition 2.5.4 in Clarke's text on non-smooth analysis [1].

In classical line geometry, the first order roll–slide motions are represented by *reciprocal screws,* while the first-order escape motions are represented by *repelling screws* [2]. In modern robotics, this terminology is associated with the parametrization of $\mathcal{B}$'s c-space in terms of *exponential coordinates* [3, 4]. This book uses *hybrid coordinates*, $q = (d, \theta)$, to represent the object's c-space. However, the notion of first-order free motions is valid under any parametrization of $\mathcal{B}$'s c-space. As noted in Chapter 3, the graphical depiction of the first-order free motions of a rigid object $\mathcal{B}$ dates to Reuleaux [5] (1876). Here, we extended the single-body technique to the case where the object $\mathcal{B}$ is held by multiple finger bodies $\mathcal{O}_1, \ldots, \mathcal{O}_k$ at an equilibrium grasp.

The need to verify coordinate invariance of structures containing inner products of tangent vectors (representing instantaneous motions) and cotangent vectors (representing wrenches) is discussed with examples by Duffy [6]. Here, we made sure that the first-order mobility index of a rigid object held at a frictionless k-finger equilibrium grasp is invariant under different choices of world and object reference frames.

Finally, the theory of first-order immobilization extends to *kinematic chains* of interconnected rigid bodies. The extension of this theory to hinged polygons is discussed in references [7] and [8]. The extension of first-order immobilization to grasps of multiple rigid bodies is an active research area.

Appendix: Proof Details

This appendix contains proofs of two key properties of the first-order free motions. The following proposition establishes the coordinate invariance of the first-order free motions.

Proposition 7.1 *Let q and $\bar{q}$ be two parametrizations of $\mathcal{B}$'s c-space, related by a coordinate transformation $q = h(\bar{q})$. If h is a diffeomorphism, the set of first-order free motions of $\mathcal{B}$ is* **coordinate invariant:**

$$\dot{q} \in M^1_{1\ldots k}(q_0) \quad \textit{iff} \quad \dot{\bar{q}} \in \bar{M}_{1\ldots k}(\bar{q}_0) \qquad q_0 = h(\bar{q}_0),$$

where $M^1_{1\ldots k}(q_0)$ and $\bar{M}_{1\ldots k}(\bar{q}_0)$ are the sets of first-order free motions in the q and $\bar{q}$ coordinates.

Proof: First consider the coordinate invariance of the individual half-spaces, $M^1_i(q)$ for $i = 1\ldots k$. Since $\bar{q}$ and q parametrize the same c-space, $h(\bar{q})$ must map $\overline{\mathcal{CO}}_i$ to $\mathcal{CO}_i$ and $\bar{\mathcal{S}}_i$ to $\mathcal{S}_i$ for $i = 1\ldots k$. Let $\tilde{d}_i(\bar{q})$ be the composition of d_i with h, $\tilde{d}_i(\bar{q}) = d_i(h(\bar{q}))$ (note that $\tilde{d}_i$ is *not* the Euclidean distance in the $\bar{q}$ coordinates). Since $\tilde{d}_i$ is negative in the interior of $\overline{\mathcal{CO}}_i$, zero on $\bar{\mathcal{S}}_i$ and positive outside $\overline{\mathcal{CO}}_i$, the gradient $\nabla\tilde{d}_i(\bar{q})$ forms the outward normal to $\overline{\mathcal{CO}}_i$ at points $\bar{q} \in \bar{\mathcal{S}}_i$. Let $\bar{\alpha}(t)$ be a path in $\bar{q}$-space, chosen such that $\bar{\alpha}(0) = \bar{q}_0$ and $\dot{\bar{\alpha}}(0) = \dot{\bar{q}}$. This path is mapped by h to a path α in q-space, $\alpha(t) = h(\bar{\alpha}(t))$ such that $\alpha(0) = q_0$ and $\dot{\alpha}(0) = Dh(\bar{q}_0)\dot{\bar{q}} = \dot{q}$. Next consider the identity:

$$(d_i \circ \alpha)(t) = \Big(d_i \circ (h \circ \bar{\alpha})\Big)(t) = \Big((d_i \circ h) \circ \bar{\alpha}\Big)(t) = (\tilde{d}_i \circ \bar{\alpha})(t), \tag{7.7}$$

where '$\circ$' denotes function composition. Application of the chain rule to Eq. (7.7) gives

$$\nabla d_i(q) \cdot \dot{q} = \nabla\tilde{d}_i(\bar{q}) \cdot \dot{\bar{q}} \qquad \text{for all } q = h(\bar{q}) \text{ and } \dot{q} = Dh(\bar{q})\dot{\bar{q}}.$$

Hence, $\dot{q} \in M^1_i(q_0)$ iff $\dot{\bar{q}} \in \bar{M}_i(\bar{q}_0)$ for the individual half-spaces.

Next consider the coordinate invariance of the convex cone $M^1_{1\ldots k}(q_0) = \cap_{i=1}^k M^1_i(q_0)$. Each $M^1_i(q_0)$ is a half-space of $T_{q_0}\mathbb{R}^m$, and $M^1_i(q_0) = Dh(\bar{q}_0)\Big(\bar{M}_i(\bar{q}_0)\Big)$ for $i = 1\ldots k$. This implies that

$$M^1_{1\ldots k}(q_0) = \cap_{i=1}^k M^1_i(q_0) = \cap_{i=1}^k Dh(\bar{q}_0)(\bar{M}_i(\bar{q}_0)) = Dh(\bar{q}_0)\Big(\bar{M}_{1\ldots k}(\bar{q}_0)\Big),$$

since, in general, $g(\mathcal{U}_1 \bigcap \mathcal{U}_2) = g(\mathcal{U}_1) \bigcap g(\mathcal{U}_2)$ when g is an invertible function. In our case, $g = Dh(\bar{q}_0)$, and $Dh(\bar{q}_0)$ is invertible since h is a diffeomorphism. □

The next proposition establishes that at an essential equilibrium grasp, the set of first-order free motions forms a linear subspace tangent to the finger c-obstacles.

Proposition 7.2 *Let a rigid object $\mathcal{B}$ be held in an essential k-finger equilibrium grasp at a configuration q_0. The object's first-order free motions, $M^1_{1\ldots k}(q_0)$, span a* **linear subspace** *tangent to the finger c-obstacles, whose dimension is given by*

$$\dim\left(M^1_{1\ldots k}(q_0)\right) = \max\left\{m - k + 1, 0\right\},$$

where $m = 3$ in 2-D grasps and $m = 6$ in 3-D grasps.

Proof: The net wrench cone generated by k frictionless finger contacts is given by

$$\mathcal{W} = \left\{ \boldsymbol{w} \in T^*_{q_0}\mathbb{R}^m \ : \boldsymbol{w} = \lambda_1 \eta_1(q_0) + \ldots + \lambda_k \eta_k(q_0) \quad \lambda_1 \ldots \lambda_k \geq 0 \right\},$$

where $\eta_i(q_0)$ is the ith finger c-obstacle outward normal at $q_0 \in \mathcal{S}_i$. Let us first show that when $\mathcal{B}$ is held in an essential frictionless equilibrium grasp, the net wrench cone $\mathcal{W}$ forms a linear subspace in $T^*_{q_0}\mathbb{R}^m$. By definition of essential grasps, either $k \leq m+1$ and then all fingers are essential, or $k > m+1$ and then $m+1$ of the k fingers are essential. Let us therefore focus on k-finger grasp involving $k \leq m+1$ essential fingers, where $m = 3$ or 6. The finger wrenches are positively linearly dependent at the equilibrium grasp:

$$\lambda_1 \eta_1(q_0) + \ldots + \lambda_k \eta_k(q_0) = \vec{0} \qquad \lambda_1, \ldots, \lambda_k \geq 0,\ k \leq m+1. \tag{7.8}$$

Since all k fingers are essential, the coefficients $\lambda_1, \ldots, \lambda_k$ must be strictly positive in Eq. (7.8); otherwise the equilibrium can be maintained with a sub-collection of $k-1$ fingers. Since $\lambda_1, \ldots, \lambda_k > 0$, it follows from Eq. (7.8) that every negated finger wrench, $-\eta_i(q_0)$, can be expressed as a positive linear combination of the remaining $k-1$ finger wrenches. Hence, all linear combinations $\sigma_1 \eta_1(q_0) + \ldots + \sigma_k \eta_k(q_0)$ such that $\sigma_1, \ldots, \sigma_k \in \mathbb{R}$ can be written as positive semi-definite linear combinations of $\eta_1(q_0), \ldots, \eta_k(q_0)$. The net wrench cone $\mathcal{W}$ thus coincides with the linear subspace spanned by $\eta_1(q_0), \ldots, \eta_k(q_0)$.

Let us next show that the set $M^1_{1\ldots k}(q_0)$ forms a linear subspace of tangent vectors. The fingers' net wrench cone can be expressed in terms of the negated finger wrenches:

$$\mathcal{W} = \left\{ \boldsymbol{w} \in T^*_{q_0}\mathbb{R}^m \ : \boldsymbol{w} = \lambda_1(-\eta_1(q_0)) + \ldots + \lambda_k(-\eta_k(q_0)) \ : \lambda_1, \ldots, \lambda_k \geq 0 \right\}$$

(this expression is justified since $\mathcal{W}$ form a linear subspace of wrenches). The cone *dual* to $\mathcal{W}$, denoted $\mathcal{W}^*$, is the cone of tangent vectors:

$$\mathcal{W}^* = \left\{ \dot{q} \in T_{q_0}\mathbb{R}^m \ : \boldsymbol{w} \cdot \dot{q} \leq 0 \ \text{ for all } \ \boldsymbol{w} \in \mathcal{W} \right\}.$$

Since $\mathcal{W}$ is a positive linear combination of $-\eta_1, \ldots, -\eta_k$, its dual cone can be written as:

$$\mathcal{W}^* = \{\dot{q} \in T_{q_0}\mathbb{R}^m \ : \ \eta_i(q_0) \cdot \dot{q} \geq 0 \ \text{for} \ i = 1 \ldots k\} = M^1_{1\ldots k}(q_0).$$

In general, the cone dual to a linear subspace is the subspace's orthogonal complement. Hence, $M^1_{1\ldots k}(q_0)$ forms a linear subspace in $T_{q_0}\mathbb{R}^m$, which is the orthogonal complement to $\mathcal{W}$.[3] The subspace $M^1_{1\ldots k}(q_0)$ is tangent to the finger c-obstacles at q_0, since every $\dot{q} \in M^1_{1\ldots k}(q_0)$ satisfies $\eta_i(q_0) \cdot \dot{q} = 0$ for $i = 1 \ldots k$.

Finally, consider the dimension of $M^1_{1\ldots k}(q_0)$. Since $M^1_{1\ldots k}(q_0)$ is the orthogonal complement of $\mathcal{W}$, $\dim\left(M^1_{1\ldots k}(q_0)\right) = m - \dim(\mathcal{W})$. To determine the dimension of $\mathcal{W}$, consider the $m \times k$ matrix $W = [\eta_1(q_0) \ldots \eta_k(q_0)]$. Since $k \leq m+1$ and $\lambda_1\eta_1(q_0) + \ldots + \lambda_k\eta_k(q_0) = \vec{0}$ at the equilibrium grasp, the rank of W is *at most* $k-1$. Let us show that W has exactly rank $k-1$. If the rank of W is less than $k-1$, the null space of W is at least two-dimensional in the coordinates $(\lambda_1, \ldots, \lambda_k) \in \mathbb{R}^k$. This null space contains the strictly positive vector $(\lambda_1, \ldots, \lambda_k)$, as well as a positive semi-definite vector $(\lambda'_1, \ldots, \lambda'_k)$ such that $\lambda'_i = 0$ for some $i \in \{1, \ldots, k\}$. But this contradicts our assumption that all k fingers are essential for the equilibrium grasp. Therefore, the matrix W must have rank of $k-1$. The net wrench cone $\mathcal{W}$ thus forms a $(k-1)$-dimensional linear subspace, which implies that $\dim\left(M^1_{1\ldots k}(q_0)\right) = m - (k-1) = m - k + 1$. □

Exercises

Section 7.1

Exercise 7.1: Describe how the gradient of the c-obstacle distance function, ∇d_i, becomes a *generalized gradient,* ∂d_i, at the non-smooth points of the ith c-obstacle boundary (see tools provided in Appendix A).

Exercise 7.2*: Let $(\mathcal{F}_W, \mathcal{F}_B)$ and $(\bar{\mathcal{F}}_W, \bar{\mathcal{F}}_B)$ be two choices of world and object frames. Let q and $\bar{q}$ be the parametrization of $\mathcal{B}$'s c-space in terms of these two frame choices. Derive the formula for the coordinate transformation $q = h(\bar{q})$.

Exercise 7.3: Prove that the coordinate transformation of the previous exercise forms a *diffeomorphism.*

Exercise 7.4: Give an example where the inner product between two tangent vectors, $\dot{q}_1, \dot{q}_2 \in T_q\mathbb{R}^m$, is *not* preserved by the coordinate transformation associated with different choices of the world and object frames (see [6] for examples).

Section 7.2

Exercise 7.5: Do two-finger equilibrium grasps automatically form essential grasps?

[3] Orthogonality has to be interpreted as the *zero action* of covectors $\boldsymbol{w} \in \mathcal{W}$ on tangent vectors $\dot{q} \in T_{q_0}\mathbb{R}^m$.

Exercise 7.6*: Consider an essential equilibrium grasp involving $k \geq 3$ fingers. Prove that all local perturbations of the contacts within the equilibrium set $\mathcal{E}$ remain essential grasps.

Exercise 7.7: Consider the *non-essential* equilibrium grasp of the rectangular object $\mathcal{B}$ by four disc fingers, depicted in Figure 7.2(a). Does the set of first-order free motions, $M^1_{1...4}(q_0)$, spans a linear subspace of tangent vectors?

Exercise 7.8: Describe an example of a *nonessential* equilibrium grasp, whose set of first-order free motions spans a convex cone rather than a linear subspace of tangent vectors.

Exercise 7.9: Explain why a *single* nonessential finger cannot affect the first-order mobility of an object held at a k-finger grasp.

Section 7.3

Exercise 7.10: Let a convex set in $\mathbb{R}^m$ intersect the interior of every half-space whose boundary passes through the origin. Prove that this set contains a small m-dimensional ball centered at the origin (this fact is a key component in the proof of Theorem 7.3).

Exercise 7.11: Provide a proof sketch of Corollary 7.4, which asserts that first-order immobilization is *robust* with respect to small contact placement errors. (Hint: use the fact that essential equilibrium grasps remain essential under small contact perturbations.)

Exercise 7.12: Consider the planar version of the 3-2-1 fixturing technique depicted in Figure 7.5. Explain why the two subspaces of first order roll–slide motions, $M^1_{1,2,3}(q_0)$ and $M^1_{4,5}(q_0)$, intersect transversally in $\mathcal{B}$'s tangent space.

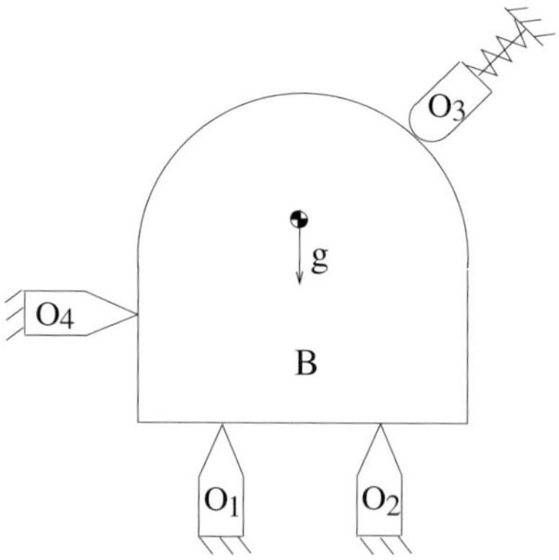

Figure 7.9 Four fixturing elements suffice to robustly immobilize any 2-D object $\mathcal{B}$.

Exercise 7.13: The planar version of the 3-2-1 technique depicted in Figure 7.5 uses a total of *six* fixturing elements. Explain why the object $\mathcal{B}$ can be robustly immobilized using only four fixturing elements, as shown in Figure 7.9.

Exercise 7.14: The 3-2-1 fixturing technique immobilizes 3-D objects using three essential equilibrium grasps arranged along mutually orthogonal directions. Determine the dimension of the subspaces of first-order free motions associated with the individual grasps. Prove that these subspaces intersect transversally in $\mathcal{B}$'s tangent space, so that $m^1_{q_0} = 0$ for the 3-2-1 fixture.

Exercise 7.15*: Prove that the 3-2-1 fixturing technique is robust with respect to small contact placement errors.

Section 7.4

Exercise 7.16: Sketch the set $M^1_{1\ldots k}(q)$ for planar grasps involving two and three fingers, in the case where the fingers do *not* form a frictionless equilibrium grasp.

Exercise 7.17: Verify that the instantaneous twists of a planar body $\mathcal{B}$ are pure rotations about points $p \in \mathbb{R}^2$ (hint: consider the *pitch* parameter in the case of planar body rotations).

Exercise 7.18: Let a planar body $\mathcal{B}$ be located at a configuration q. Prove that every instantaneous motion of $\mathcal{B}$, $\dot{q} = (v, \omega) \in T_q\mathbb{R}^3$, can be represented as an instantaneous rotation of $\mathcal{B}$ about a *center of rotation* located at the point $p = -\frac{1}{\omega} J v \in \mathbb{R}^2$.

Exercise 7.19*: Figure 7.10(a) shows a chain of two rigid polygons, $\mathcal{B}_1$ and $\mathcal{B}_2$, connected by a rotational joint. Prove that immobilization of the chain can only be achieved when the chain is held at a *frictionless equilibrium grasp* (this result extends to arbitrary chains).

Exercise 7.20: By extension of the theory described in this chapter, determine the minimum number of frictionless *point fingers* that allow first-order immobilization of the two-link chain depicted in Figure 7.10(a).

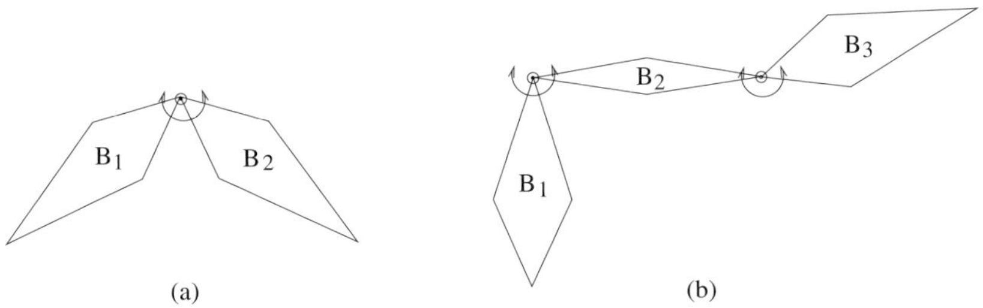

Figure 7.10 Serial chains of (a) two polygons connected by a rotational joint, and (b) three polygons connected by two rotational joints.

Exercise 7.21: Repeat the previous exercise for the three-link chain of Figure 7.10(b).

Exercise 7.22*: Determine the minimum number of frictionless *point fingers* that allow first-order immobilization of any 2-D serial chain of n polygons (see Bibliographical Notes).

References

[1] F. H. Clarke, *Optimization and Nonsmooth Analysis*. SIAM, 1990.

[2] M. S. Ohwovoriole and B. Roth, "An extension of screw theory," *Journal of Mechanical Design*, vol. 103, pp. 725–735, 1981.

[3] K. M. Lynch and F. C. Park, *Modern Robotics: Mechanics, Planning, and Control*. Cambridge University Press, 2017.

[4] R. M. Murray, Z. Li and S. S. Sastry, *A Mathematical Introduction to Robotic Manipulation*. CRC Press, 1994.

[5] F. Reuleaux, *The Kinematics of Machinery*. Macmillan, 1876, republished by Dover, 1963.

[6] J. Duffy, "The fallacy of modern hybrid control theory that is based on orthogonal complements of twist and wrench spaces," *Journal of Robotics Systems*, vol. 7, no. 2, pp. 139–144, 1990.

[7] J.-S. Cheong, K. Goldberg, M. H. Overmars, E. Rimon and A. F. van der Stappen, "Immobilizing hinged polygons," *International Journal of Computational Geometry and Applications*, vol. 17, no. 1, 2007.

[8] E. Rimon and A. F. van der Stappen, "Immobilizing 2D serial chains in form closure grasps," *IEEE Transactions on Robotics*, vol. 28, no. 1, pp. 32–43, 2012.

8 Second-Order Immobilizing Grasps

This chapter studies the role of second-order geometric effects in grasp mobility theory. The first-order immobilizing grasps rely on first-order geometric effects and require at least $m + 1$ fingers to achieve object immobilization, where $m = 3$ in 2-D grasps and $m = 6$ in 3-D grasps. When second-order geometric effects are taken into account, a robot hand can realize immobilizing grasps using a smaller number of fingers. Such grasps are first-order mobile, but second-order analysis of the object's free motions will indicate that the object is fully immobilized and hence secured by the grasping fingers. Moreover, the combined first and second-order geometric effects provide a complete grasp mobility theory, so there is no need to consider higher-order geometric effects (with the exception of two-finger grasps of 3-D objects). While friction is allowed at the contacts, immobilizing grasps seek to secure the object based on rigid-body constraints imposed on the object's free motions by the grasping finger bodies.

The second-order free motions of a rigid object $\mathcal{B}$ with respect to rigid finger bodies $\mathcal{O}_1, \ldots, \mathcal{O}_k$ are described in Section 8.1. These free motions represent pairs of object velocity and acceleration, $(\dot{q}, \ddot{q})$, along which the object breaks or maintains contact with the grasping fingers. The *relative c-space curvature form,* introduced in Section 8.2, will determine the second-order mobility of $\mathcal{B}$ at a given equilibrium grasp. The number of nonnegative eigenvalues of the relative c-space curvature form defines the grasp's *second-order mobility index.* We will see in Section 8.3 that when the second-order mobility index is zero, the object is fully immobilized and, hence, secured by the grasping fingers. Section 8.4 describes a graphical representation of the second-order mobility of an object held at various grasp arrangements.

8.1 The Second-Order Free Motions

This section characterizes the second-order free motions of a rigid object $\mathcal{B}$ held by rigid finger bodies $\mathcal{O}_1, \ldots, \mathcal{O}_k$ at a frictionless equilibrium grasp (recall that such grasps are necessary for object immobilization). When the object is held at a configuration q_0, its free motions are c-space curves that start at q_0 and locally lie in the free c-space $\mathcal{F}$. To study the second-order properties of the free motion curves, consider the *c-obstacle distance function,* $d_i(q)$, which measures the signed distance of configuration points q from the c-obstacle boundary $\mathcal{S}_i$ (see Chapter 7). Let $\alpha(t)$ be a smooth c-space curve, chosen such that $\alpha(0) = q_0 \in S_i$, $\dot{\alpha}(0) = \dot{q}$, and $\ddot{\alpha}(0) = \ddot{q}$ (Figure 8.1). The *second-order Taylor expansion* of d_i along α has the form:

$$d_i(\alpha(t)) = d_i(q_0) + \Big(\nabla d_i(q_0)\cdot\dot{q}\Big)t + \tfrac{1}{2}\Big(\dot{q}^T D^2 d_i(q_0)\dot{q} + \nabla d_i(q_0)\cdot\ddot{q}\Big)t^2 + O(t^3) \quad t \in (-\epsilon, \epsilon),$$

where $\epsilon > 0$ is a small parameter. Recall that $\nabla d_i(q_0)$ has unit magnitude and is collinear with the finger c-obstacle outward normal, so that $\nabla d_i(q_0) = \hat{\eta}_i(q_0)$, where $\hat{\eta}_i = \eta_i / \|\eta_i\|$. When the Taylor expansion is evaluated at $q_0 \in S_i$, the second-order approximation of d_i becomes

$$d_i(\alpha(t)) = \Big(\hat{\eta}_i(q_0)\cdot\dot{q}\Big)t + \tfrac{1}{2}\Big(\dot{q}^T D^2 d_i(q_0)\dot{q} + \hat{\eta}_i(q_0)\cdot\ddot{q}\Big)t^2 + O(t^3) \qquad t \in (-\epsilon, \epsilon), \quad (8.1)$$

where we substituted $d_i(q_0) = 0$ and $\nabla d_i(q_0) = \hat{\eta}_i(q_0)$.

Since $\hat{\eta}_i(q_0) \cdot \dot{q} > 0$ along *first-order escape motions,* the linear term in Eq. (8.1) determines the sign of $d_i(\alpha(t))$ in a small interval surrounding $t = 0$. The second-order term in Eq. (8.1) determines the sign of $d_i(\alpha(t))$ only along *first order roll–slide motions,* which satisfy the condition $\hat{\eta}_i(q_0)\cdot\dot{q} = 0$. The second-order term has a special geometric meaning along first order roll–slide motions. Recall that these motions are tangent to the ith finger c-obstacle, $\dot{q} \in T_{q_0}S_i$. Since $\nabla d_i(q) = \hat{\eta}_i(q)$ at all points $q \in S_i$, we have that $\dot{q}^T D^2 d_i(q_0)\dot{q} = \dot{q}^T [D\hat{\eta}_i(q_0)]\dot{q}$ for $\dot{q} \in T_{q_0}S_i$. The sign of $d_i(\alpha(t))$ along first order roll–slide motions is thus determined by the ith finger *c-obstacle curvature form:* $\kappa_i(q_0, \dot{q}) = \dot{q}^T [D\hat{\eta}_i(q_0)]\dot{q}$ (see Chapter 2).

The second-order properties of $\mathcal{B}$'s c-space curves are determined by their *velocity and acceleration* at q_0. The composite space of $\mathcal{B}$'s velocities and accelerations at a configuration point q is defined as follows. The ambient tangent space, $T_q\mathbb{R}^m$, consists of all tangents to paths $\alpha(t)$, evaluated at $\alpha(0) = q$. The *double tangent space,* denoted $T_{(q,\dot{q})}(T\mathbb{R}^m)$, consists of all tangents to paths $(\alpha(t), \dot{\alpha}(t))$, evaluated at $(\alpha(0), \dot{\alpha}(0)) = (q, \dot{q})$. The elements of $T_{(q,\dot{q})}(T\mathbb{R}^m)$ are vectors based at $(q, \dot{q})$, specified as velocity and acceleration pairs $(\dot{q}, \ddot{q}) \in \mathbb{R}^m \times \mathbb{R}^m$ (see additional discussion at the end of this section). The following simpler notation will be used.

Double tangent space: *The* **double tangent space** *of $\mathbb{R}^m$ at q, denoted $T_q^2\mathbb{R}^m$, forms the collection of all velocities and accelerations $(\dot{q}, \ddot{q}) \in \mathbb{R}^m \times \mathbb{R}^m$ based at q.*

Recall from Chapter 7 that object immobilization can only be achieved with frictionless equilibrium grasps. The free motions of $\mathcal{B}$ at a frictionless equilibrium grasp are roll–slide to first order. Hence, the following definition of the second-order free motions focuses on the first order roll–slide motions of $\mathcal{B}$ with respect to a single finger body $\mathcal{O}_i$.

DEFINITION 8.1 (**single finger second-order free motions**) *Let a rigid object $\mathcal{B}$ be contacted by a stationary rigid finger body $\mathcal{O}_i$ at a configuration q_0. The* **second-order free motions** *of $\mathcal{B}$ at q_0 are the subset of $T_{q_0}^2\mathbb{R}^m$:*

$$M_i^2(q_0) = \Big\{(\dot{q}, \ddot{q}) \in T_{q_0}^2\mathbb{R}^m \,:\, \hat{\eta}_i(q_0)\cdot\dot{q} = 0 \quad and \quad \dot{q}^T [D\hat{\eta}_i(q_0)]\dot{q} + \hat{\eta}_i(q_0)\cdot\ddot{q} \geq 0\Big\}.$$

Analogous to the first-order free motions, pairs $(\dot{q}, \ddot{q})$ satisfying $\hat{\eta}_i(q_0) \cdot \dot{q} = 0$ and $\dot{q}^T [D\hat{\eta}_i(q_0)]\dot{q} + \hat{\eta}_i(q_0)\cdot\ddot{q} = 0$ are **second order roll–slide motions**, *while pairs $(\dot{q}, \ddot{q})$*

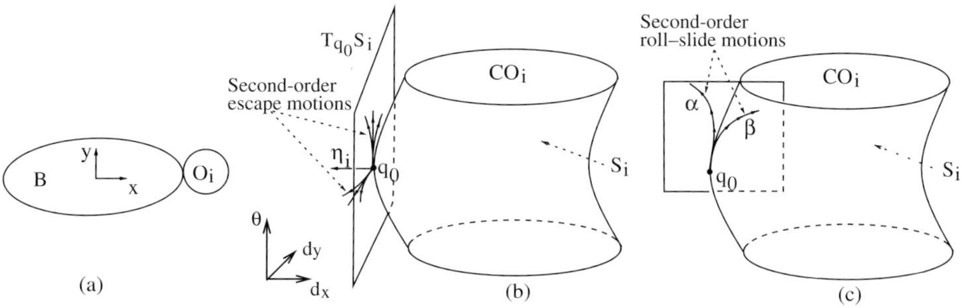

Figure 8.1 (a) Top view of an ellipse $\mathcal{B}$ contacted by a stationary disc finger $\mathcal{O}_i$. (b) Second-order escape motions. (c) While α and β are second-order roll–slide motions, α lies in $\mathcal{F}$ while β penetrates $\mathcal{CO}_i$.

satisfying $\hat{\eta}_i(q_0) \cdot \dot{q} = 0$ *and* $\dot{q}^T[D\hat{\eta}_i(q_0)]\dot{q} + \hat{\eta}_i(q_0) \cdot \ddot{q} > 0$ *are* **second-order escape motions.**

The second-order free motions partition the linear subspace of first order roll–slide motions, $M_i^1(q_0)$, into three disjoint subsets: second-order escape motions, second-order roll–slide motions and their complement set, which consists of second-order penetration motions. When $(\dot{q}, \ddot{q}) \in M_i^2(q_0)$ is a second-order escape motion, its corresponding path, $\alpha(t)$ such that $\alpha(0) = q_0$, $\dot{\alpha}(0) = \dot{q}$, $\ddot{\alpha}(0) = \ddot{q}$, *locally lies in the free c-space* $\mathcal{F}$ for all $t \in [0, \epsilon]$. Such curves are illustrated in the following example.

Example: Figure 8.1(a) shows an object $\mathcal{B}$ contacted by a stationary disc finger $\mathcal{O}_i$. Figure 8.1(b) schematically depicts the finger c-obstacle, $\mathcal{CO}_i$, at the object's contact configuration $q_0 \in \mathcal{S}_i$. The curves shown in Figure 8.1(b) are tangent to $\mathcal{S}_i$ at q_0 and thus represent first order roll–slide motions. These curves locally move into the free c-space $\mathcal{F}$ and, hence, represent *second-order escape motions.* The object $\mathcal{B}$ moves away from $\mathcal{O}_i$ along any first order roll–slide motion that forms a second-order escape motion. ○

When the pair $(\dot{q}, \ddot{q}) \in M_i^2(q_0)$ is a second order roll–slide motion, *it is not possible* to determine based on Eq. (8.1) whether $\alpha(t)$ locally lies in the free c-space $\mathcal{F}$ or locally penetrates the c-obstacle $\mathcal{CO}_i$. This indeterminacy is illustrated in the following example.

Indeterminacy of second order roll–slide motions: Figure 8.1(c) depicts two curves, α and β, which start at q_0 and have the same tangent to $\mathcal{S}_i$ at q_0: $\dot{\alpha}(0) = \dot{\beta}(0) \in T_{q_0}\mathcal{S}_i$. The two curves lie in a common vertical plane in Figure 8.1(c), such that both curves have the *same* curvature at q_0. The two curves satisfy the equation $\dot{q}^T[D\hat{\eta}_i(q_0)]\dot{q} + \hat{\eta}_i(q_0) \cdot \ddot{q} = 0$ and thus represent the *same* second order roll–slide motion. Yet, α locally lies in the free c-space $\mathcal{F}$, while β locally penetrates the c-obstacle $\mathcal{CO}_i$. ○

The definition of the second-order free motions is next extended to multi-finger grasps. The object's free motions must respect the velocity and acceleration constraints imposed by all contacting finger bodies, as stated in the following definition.

DEFINITION 8.2 (K**-finger second-order free motions**) *Let a rigid object $\mathcal{B}$ be contacted by stationary rigid finger bodies $\mathcal{O}_1,\ldots,\mathcal{O}_k$ at a configuration $q_0 \in \cap_{i=1}^{k} \mathcal{S}_i$. The set of* **second-order free motions** *of $\mathcal{B}$ at q_0, denoted $M^2_{1\ldots k}(q_0)$, has the form:*

$$M^2_{1\ldots k}(q_0) = \{(\dot{q},\ddot{q}) \in T^2_{q_0}\mathbb{R}^m \, : \, \eta_i(q_0)\cdot\dot{q} = 0 \\ \text{and } \dot{q}^T[D\hat{\eta}_i(q_0)]\dot{q} + \hat{\eta}_i(q_0)\cdot\ddot{q} \geq 0 \text{ for } i = 1\ldots k\},$$

where $\hat{\eta}_i(q_0)$ is the ith finger c-obstacle outward unit normal and $\dot{q}^T[D\hat{\eta}_i(q_0)]\dot{q}$ the ith finger c-obstacle curvature form at q_0.

Much like first-order free motions, the set $M^2_{1\ldots k}(q_0)$ is defined in terms of the Euclidean inner product, which is usually *not* preserved by coordinate transformations. One must therefore verify that the set $M^2_{1\ldots k}(q_0)$ does not depend on the choice of the world and object frames. The proof of the following proposition appears in the chapter's appendix.

PROPOSITION 8.1 (**coordinate invariance**) *Let q and $\bar{q}$ be two parametrizations of $\mathcal{B}$'s c-space, related by a coordinate transformation $q = h(\bar{q})$. If h is a diffeomorphism, the set of second-order free motions of $\mathcal{B}$ is* **coordinate invariant:**

$$(\dot{q},\ddot{q}) \in M^2_{1\ldots k}(q_0) \quad \textit{iff} \quad (\dot{\bar{q}},\ddot{\bar{q}}) \in \bar{M}^2_{1\ldots k}(\bar{q}_0) \qquad q_0 = h(\bar{q}_0),$$

where $M^2_{1\ldots k}(q_0)$ and $\bar{M}^2_{1\ldots k}(\bar{q}_0)$ are the second-order free motion sets in the q and $\bar{q}$ coordinates.

The second-order free motions can thus be analyzed under any choice of world and object frames. Proposition 8.1 can be explained as follows. The ith c-obstacle outward normal, $\eta_i(q_0)$, can be interpreted as a wrench, $\boldsymbol{w}_i(q_0)$, generated by a force that acts along $\mathcal{B}$'s inward contact normal at x_i. The *power* (or instantaneous work) applied to $\mathcal{B}$ by this wrench is given by the inner product: $\boldsymbol{w}_i(q_0)\cdot\dot{q}$. Consider a path $\alpha(t)$ that lies in $\mathcal{S}_i$, such that $\alpha(0) = q_0$ and $\dot{\alpha}(0) = \dot{q} \in T_{q_0}\mathcal{S}_i$. The *change of power* transmitted to the object $\mathcal{B}$ along α is given by the derivative: $\frac{d}{dt}(\boldsymbol{w}_i(q)\cdot\dot{q}) = \dot{\boldsymbol{w}}_i\cdot\dot{q} + \boldsymbol{w}_i\cdot\ddot{q}$. The *wrench derivative*, $\dot{\boldsymbol{w}}_i$, represents the instantaneous change in $\boldsymbol{w}_i$ along α, evaluated at $\alpha(0) = q_0$. It can be written as the sum of two components: $\dot{\boldsymbol{w}}_i = (\frac{d}{dt}\|\eta_i\|)\hat{\eta}_i + \|\eta_i\|[D\hat{\eta}_i]\dot{q}$, where $\boldsymbol{w}_i = \|\eta_i\|\hat{\eta}_i$. When evaluated along α, the change of power transmitted to the object $\mathcal{B}$ has the form:

$$\dot{\boldsymbol{w}}_i\cdot\dot{q} + \boldsymbol{w}_i\cdot\ddot{q} = \|\eta_i\|\left(\dot{q}^T[D\hat{\eta}_i(q_0)]\dot{q} + \hat{\eta}_i(q_0)\cdot\ddot{q}\right). \tag{8.2}$$

When $\boldsymbol{w}_i\cdot\dot{q} = 0$, Eq. (8.2) determines the *sign* of the power transmitted to the object $\mathcal{B}$ in a small time interval surrounding $t = 0$. The latter physical quantity should not depend on the choice of world and object frames. Since the right side of Eq. (8.2) defines the set $M^2_i(q_0)$, the intersection $M^2_{1\ldots k}(q_0) = \cap_{i=1}^{k} M^2_i(q_0)$ must be independent on the choice of world and object frames.

The tangent bundle*: The *tangent bundle* of $\mathbb{R}^m$, denoted $T\mathbb{R}^m$, is defined as the disjoint union of the tangent spaces $T_q\mathbb{R}^m$, as the base point q varies in $\mathbb{R}^m$ ($T\mathbb{R}^m$ is equivalent to $\mathbb{R}^m \times \mathbb{R}^m$). The *double tangent space*, $T_{(q,\dot{q})}T\mathbb{R}^m$, is the tangent space of the manifold $T\mathbb{R}^m$ at $(q,\dot{q})$. Based on this viewpoint, the *double cotangent space,* denoted $T^*_{(q,\dot{q})}T\mathbb{R}^m$, is defined as the vector space of linear functions that act on pairs $(\dot{q},\ddot{q}) \in T_{(q,\dot{q})}T\mathbb{R}^m$. Each element of $T^*_{(q,\dot{q})}T\mathbb{R}^m$ forms a row vector: $(\dot{q}^T Dw(q), w(q))$, which acts on $(\dot{q},\ddot{q})$ as specified by Eq. (8.2): $\dot{q}^T Dw(q)\dot{q} + w(q) \cdot \ddot{q}$. The sign of this action is coordinate invariant, thus providing a formal justification for the coordinate invariance of the set of second-order free motions. ○

8.2 The Second-Order Mobility Index

The *second-order mobility index* is a coordinate invariant integer that measures the amount of second-order free motions available to an object $\mathcal{B}$ held at an equilibrium grasp. We may assume that the object is grasped by $k \le m+1$ fingers ($m=3$ or 6), since equilibrium grasps that involve higher numbers of fingers are generically immobilizing based on first-order geometric effects (see Chapter 7). When an object $\mathcal{B}$ is held by finger bodies $\mathcal{O}_1,\ldots,\mathcal{O}_k$ under frictionless contact conditions, the finger wrenches have the form:

$$w_i = \lambda_i \hat{\eta}_i(q_0) \qquad i=1\ldots k,$$

where $\hat{\eta}_i(q_0)$ is the ith finger c-obstacle outward unit normal at q_0, and $\lambda_i \ge 0$ is proportional to the finger force magnitude applied to $\mathcal{B}$ at x_i. At an equilibrium grasp, the finger wrenches satisfy the equilibrium condition:

$$\lambda_1\hat{\eta}_1(q_0) + \cdots + \lambda_k\hat{\eta}_k(q_0) = \vec{0} \qquad \lambda_1,\ldots,\lambda_k \ge 0. \tag{8.3}$$

The coefficients $\lambda_1,\ldots,\lambda_k$ play a key role in the definition of the second-order mobility index. To ensure that these coefficients are well defined at a given grasp, assume that all fingers are *essential* for maintaining the equilibrium grasp (see Chapter 7). When an object $\mathcal{B}$ is held at an essential equilibrium grasp, the coefficients $\lambda_1,\ldots,\lambda_k$ are strictly positive. Otherwise the equilibrium can be maintained with a sub-collection of $k-1$ fingers. Moreover, their value is uniquely determined up to a common scaling factor, as stated in the following lemma (see Exercises).

LEMMA 8.2 *Let a rigid object $\mathcal{B}$ be held at a frictionless k-finger equilibrium grasp. The coefficients $\lambda_1,\ldots,\lambda_k$ in Eq.* (8.3) *are strictly positive and unique up to a common scaling factor iff the finger bodies $\mathcal{O}_1,\ldots,\mathcal{O}_k$ are all essential for maintaining the equilibrium grasp.*

When an object $\mathcal{B}$ is held at an essential equilibrium grasp, its first-order free motions, $M^1_{1\ldots k}(q_0)$, span a linear subspace tangent to the finger c-obstacles at q_0 (Proposition 7.2 of Chapter 7). A c-space path $\alpha(t)$ such that $\alpha(0)=q_0$ and $\dot{\alpha}(0) \in M^1_{1\ldots k}(q_0)$ can either locally lie in the free c-space $\mathcal{F}$ or locally penetrate one of the finger c-obstacles $\mathcal{CO}_i$ for some $i \in \{1,\ldots,k\}$. The finger c-obstacles curvature will determine which paths are locally free and which paths are physically infeasible. The second-order geometry of the c-obstacle boundary, $\mathcal{S}_i$, is captured by its *curvature form:*

$$\kappa_i(q_0,\dot{q}) = \dot{q}^T D\hat{\eta}_i(q_0)\dot{q} \qquad \dot{q} \in T_{q_0}\mathcal{S}_i.$$

The equilibrium grasp coefficients $\lambda_1, \ldots, \lambda_k$ specified by Eq. (8.3) determine the following weighted sum of the finger c-obstacle curvature forms.

DEFINITION 8.3 (**C-space relative curvature form**) *Let a rigid object $\mathcal{B}$ be held at an essential k-finger equilibrium grasp configuration q_0, with equilibrium grasp coefficients $\lambda_1, \ldots, \lambda_k$. The* **c-space relative curvature form** *at q_0, denoted $\kappa_{\text{rel}}(q_0,\dot{q})$, is the quadratic form of tangent vectors:*

$$\kappa_{\text{rel}}(q_0,\dot{q}) = \textstyle\sum_{i=1}^{k} \lambda_i \kappa_i(q_0,\dot{q}) = \dot{q}^T \left[\sum_{i=1}^{k} \lambda_i D\hat{\eta}_i(q_0)\right]\dot{q} \qquad \dot{q} \in M^1_{1\ldots k}(q_0), \qquad (8.4)$$

where $\kappa_i(q_0,\dot{q})$ is the ith finger c-obstacle curvature form at q_0.

Based on the uniqueness of the coefficients $\lambda_1, \ldots, \lambda_k$, the c-space relative curvature form, $\kappa_{\text{rel}}(q_0,\dot{q})$, is well defined at an essential equilibrium grasp. The number of positive semi-definite eigenvalues of $\kappa_{\text{rel}}(q_0,\dot{q})$ determines the grasp's *second-order mobility index,* which is defined as follows.

DEFINITION 8.4 (**second-order mobility index**) *Let a rigid object $\mathcal{B}$ be held at an essential k-finger equilibrium grasp configuration q_0. The object's* **second-order mobility index, $m^2_{q_0}$**, *is the number of positive semi-definite eigenvalues of the matrix associated with the c-space relative curvature form $\kappa_{\text{rel}}(q_0,\dot{q})$ at the given grasp.*

Let us highlight an important property of the second-order mobility index. The c-space relative curvature form, $\kappa_{\text{rel}}(q_0,\dot{q})$, is defined on the linear subspace of tangent vectors $M^1_{1\ldots k}(q_0)$. The dimension of $M^1_{1\ldots k}(q_0)$ is captured by the first-order mobility index, $m^1_{q_0}$. When $m^1_{q_0} = 0$, the object $\mathcal{B}$ is *completely immobilized* by first-order geometric effects, and the second-order mobility index carries no immediately useful information. When $m^1_{q_0} > 0$, the second-order mobility index attains values in the range $0, \ldots, m^1_{q_0}$. In these grasps the object $\mathcal{B}$ is *not* immobilized by first-order geometric effects but may become fully immobilized by second-order geometric effects. The second-order mobility index is thus useful for 2-D grasps that involve two or three fingers and 3-D grasps having two, three, four, five or six fingers. This possibility of using a small number of fingers to form minimalistic grasps will be considered in the next chapter.

Coordinate invariance: The ith finger c-obstacle curvature is *not* invariant under different choices of world and object frames (see Figure 2.3). However, the c-space relative curvature form, $\kappa_{\text{rel}}(q_0,\dot{q})$, measures the *relative* curvature of the finger c-obstacles at q_0. Consequently, the second-order mobility index is coordinate invariant, as stated in Proposition 8.7 which appears in the chapter's appendix. The coordinate invariance of $m^2_{q_0}$ can be justified as follows. A smooth function $\varphi(q): \mathbb{R}^m \to \mathbb{R}$ has a *critical point* at q_0 when $\nabla\varphi(q_0) = \vec{0}$. The *Morse index* of φ at q_0 is the number of negative eigenvalues of its Hessian matrix, $D^2\varphi(q_0)$, at the critical point q_0. The

Morse index is known to be coordinate invariant. We will see later in this chapter that the function $d_{min}(q) = \min\{d_1(q), \ldots, d_k(q)\}$ has a non-smooth extremum point at the equilibrium grasp configuration q_0. Moreover, the function $d_{min}(q)$ possesses a *generalized Hessian* at q_0 (see Appendix A). The second-order mobility index, $m^2_{q_0}$, measures the number of positive semi-definite eigenvalues of the generalized Hessian of $d_{min}(q)$ at q_0. Since the Morse index is coordinate invariant, so must be $m^2_{q_0}$, which measures the complementary number of eigenvalues of the generalized Hessian at q_0. ○

8.3 Second-Order Immobilization

A rigid object $\mathcal{B}$ is fully immobilized by the grasping fingers based on first-order geometric effects when $m^1_{q_0} = 0$. This section will establish two results. When $m^1_{q_0} > 0$ and $m^2_{q_0} = 0$, the object is *fully immobilized* by the grasping fingers. Conversely, when $m^1_{q_0} > 0$ and $m^2_{q_0} > 0$, the object can escape the grasping fingers. Based on this insight, we will be able to conclude that the two mobility indices, $m^1_{q_0}$ and $m^2_{q_0}$, completely characterize the mobility of all generic equilibrium grasp arrangements. We begin with the second-order immobilization theorem.

THEOREM 8.3 (**second-order immobilization**) *Let a rigid object $\mathcal{B}$ be held in an essential equilibrium grasp by rigid finger bodies $\mathcal{O}_1, \ldots, \mathcal{O}_k$ at a configuration q_0. If the first-order mobility index is positive, $m^1_{q_0} > 0$, while the second-order mobility index is zero, $m^2_{q_0} = 0$, the object is* **fully immobilized** *by the grasping finger bodies.*

Proof: We have to show that the object's configuration point, q_0, is completely surrounded by the finger c-obstacles. Consider the minimum distance function introduced in Chapter 7:

$$d_{min}(q) = \min\left\{d_1(q), \ldots, d_k(q)\right\} \qquad q \in \mathbb{R}^m.$$

Each $d_i(q)$ is negative inside the finger c-obstacle $\mathcal{CO}_i$, zero on its boundary $\mathcal{S}_i$, and positive outside $\mathcal{CO}_i$. When the object $\mathcal{B}$ is held by k finger bodies, $d_i(q_0) = 0$ for $i = 1 \ldots k$, and consequently $d_{min}(q_0) = 0$. Hence, q_0 is surrounded by the finger c-obstacles when $d_{min}(q)$ has a *strict local maximum* at q_0 – that is, when $d_{min}(q) < 0$ in a small m-dimensional ball centered at q_0.

While $d_{min}(q)$ is non-differentiable at q_0, it possesses a *generalized gradient* at q_0 (see Appendix A). The generalized gradient, denoted $\partial d_{min}(q_0)$, is the convex combination of the usual gradients $\nabla d_i(q_0)$ for $i = 1 \ldots k$. Since each $\nabla d_i(q_0)$ is collinear with the c-obstacle outward unit normal at q_0:

$$\partial d_{min}(q_0) = \sum_{i=1}^{k} \sigma_i \hat{\eta}_i(q_0) \qquad 0 \le \sigma_1, \ldots, \sigma_k \le 1, \ \textstyle\sum_{i=1}^{k} \sigma_i = 1. \tag{8.5}$$

The generalized gradient $\partial d_{min}(q_0)$ forms a convex set in $\mathcal{B}$'s wrench space, $T^*_{q_0}\mathbb{R}^m$. When the object $\mathcal{B}$ is held at an equilibrium grasp, the generalized gradient $\partial d_{min}(q_0)$

contains the origin of $\mathcal{B}$'s wrench space. Let $\lambda_1, \ldots, \lambda_k$ be the coefficients expressing the origin at this convex combination:

$$\lambda_1 \hat{\eta}_1(q_0) + \cdots + \lambda_k \hat{\eta}_1(q_0) = \vec{0} \qquad 0 \leq \lambda_1, \ldots, \lambda_k \leq 1, \; \textstyle\sum_{i=1}^{k} \lambda_i = 1. \tag{8.6}$$

As discussed in Appendix A, the function $d_{min}(q)$ has a *generalized Hessian* at q_0. The generalized Hessian, denoted $\partial^2 d_{min}(q_0, \boldsymbol{w})$, forms a collection of $m \times m$ matrices, each based at a particular wrench $\boldsymbol{w} \in \partial d_{min}(q_0)$:

$$\partial^2 d_{min}(q_0, \boldsymbol{w}) = \sum_{i=1}^{k} \sigma_i D^2 d_i(q_0) \qquad \boldsymbol{w} \in \partial d_{min}(q_0),$$

where $D^2 d_i(q_0)$ is the Hessian matrix of the c-obstacle distance function $d_i(q)$, while the coefficients $\sigma_1, \ldots, \sigma_k$ are determined by the convex combination of Eq. (8.5) associated with the wrench $\boldsymbol{w}$. When the generalized Hessian is evaluated at $\boldsymbol{w} = \vec{0}$, we obtain the $m \times m$ matrix:

$$\partial^2 d_{min}(q_0, \vec{0}) = \sum_{i=1}^{k} \lambda_i D^2 d_i(q_0), \tag{8.7}$$

where the coefficients $\lambda_1, \ldots, \lambda_k$ are specified by Eq. (8.6). Recall from Section 8.1 that $\dot{q}^T D^2 d_i(q_0)\dot{q} = \dot{q}^T [D\hat{\eta}_i(q_0)]\dot{q}$ for $\dot{q} \in T_{q_0}\mathcal{S}_i$. Hence, Eq. (8.7) implies that the quadratic form $\dot{q}^T[\partial^2 d_{min}(q_0, \vec{0})]\dot{q}$ is a *positive multiple* of $\kappa_{\text{rel}}(q_0, \dot{q}) = \dot{q}^T[\sum_{i=1}^{k} \lambda_i D\hat{\eta}_i(q_0)]\dot{q}$.

According to Theorem A.2, $d_{min}(q)$ has a *strict local maximum* at q_0 when $\partial d_{min}(q_0)$ is embedded in a linear subspace, V, such that $\partial^2 d_{min}(q_0, \vec{0})$ is *negative definite* along the orthogonal complement subspace $V^{\perp}$. That is, $\dot{q}^T[\partial^2 d_{min}(q_0, \vec{0})]\dot{q} < 0$ for all $\dot{q} \in V^{\perp}$. The generalized gradient, $\partial d_{min}(q_0)$, spans the fingers' *net wrench cone* at the equilibrium grasp (see Chapter 4). The net wrench cone forms a $(k-1)$-dimensional linear subspace at an essential k-finger equilibrium grasp (see Chapter 7). The subspace V thus corresponds to the fingers' net wrench cone. Its orthogonal complement[1] is the subspace of first-order free motions at q_0, $V^{\perp} = M^1_{1\ldots k}(q_0)$. If $m^2_{q_0} = 0$ at the given grasp, we have that $\kappa_{\text{rel}}(q_0, \dot{q}) < 0$ for all $\dot{q} \in M^1_{1\ldots k}(q_0)$. Since $\kappa_{\text{rel}}(q_0, \dot{q})$ is a positive multiple of $\dot{q}^T[\partial^2 d_{min}(q_0, \vec{0})]\dot{q}$, the generalized Hessian of $d_{min}(q)$ is negative definite on $V^{\perp}$. The function $d_{min}(q)$ thus has a strict local maximum at q_0, which proves that q_0 is completely surrounded by the finger c-obstacles. □

Theorem 8.3 offers the following insight. When $m^2_{q_0} = 0$ at a k-finger grasp, all c-space paths $\alpha(t)$ that start at q_0 with $\dot{\alpha}(0) = \dot{q} \in M^1_{1\ldots k}(q_0)$ will locally penetrate one or more of the finger c-obstacles that meet at q_0, irrespective of the path's acceleration at q_0:

$$\alpha(t) \in \text{int}(\mathcal{CO}_i) \qquad i \in \{1, \ldots, k\} \text{ and } t \in (0, \epsilon],$$

where $\text{int}(\mathcal{CO}_i)$ is the interior of the ith finger c-obstacle. Since any such penetration motion is prevented by the rigidity of the object and finger bodies, the object $\mathcal{B}$ is fully

[1] Orthogonality refers the zero action of a wrench $\mathbf{w}$ on tangent vectors, $\boldsymbol{w}(\dot{q}) = \boldsymbol{w} \cdot \dot{q} = 0$ for all $\dot{q} \in M^1_{1\ldots k}(q_0)$.

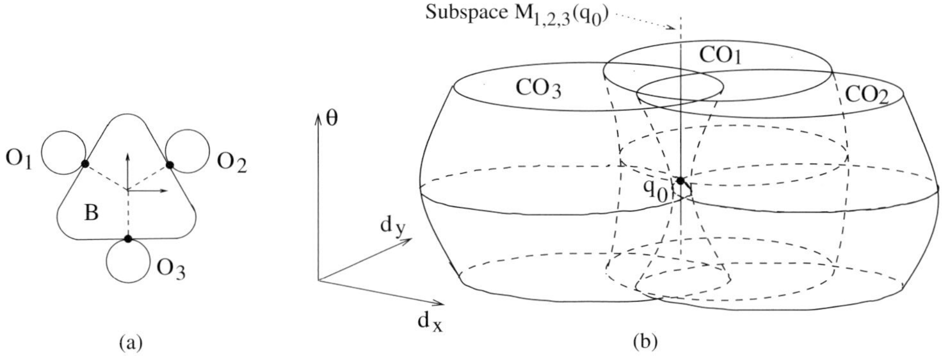

Figure 8.2 An immobilizing three-finger equilibrium grasp with $m^1_{q_0}=1$ and $m^2_{q_0}=0$.

immobilized by the grasping fingers, even though it is mobile to first order. This insight is illustrated in the following example.

Example: Figure 8.2(a) shows a triangular object $\mathcal{B}$ held at an equilibrium grasp by three disc fingers. The c-space geometry of this grasp is sketched in Figure 8.2(b), where q_0 is $\mathcal{B}$'s equilibrium grasp configuration. The set of first-order free motions, $M^1_{1,2,3}(q_0)$, spans a one-dimensional linear subspace, which is tangent to the three finger c-obstacles at q_0. Hence, $m^1_{q_0}=1$, and the object is mobile to first-order at this grasp. Based on a graphical technique described later in this chapter, $m^2_{q_0}=0$ at this grasp. The finger c-obstacles thus completely surround the point q_0 in Figure 8.2(b), and the object is fully immobilized at this grasp. ○

To complete the grasp mobility theory, consider the situation where $m^1_{q_0}$ and $m^2_{q_0}$ are both positive at a given grasp. The c-space relative curvature form, $\kappa_{\text{rel}}(q_0,\dot{q})$, is said to be *non-degenerate* when it possesses only non-zero eigenvalues (the generic case). The following mobility theorem complements the first and second immobilization theorems.

THEOREM 8.4 **(grasp mobility)** *Let a rigid object $\mathcal{B}$ be held at an essential equilibrium grasp by rigid finger bodies $\mathcal{O}_1,\ldots,\mathcal{O}_k$ at q_0. If $\kappa_{\text{rel}}(q_0,\dot{q})$ is non-degenerate and the two mobility indices are positive, $m^1_{q_0}>0$ and $m^2_{q_0}>0$, the object $\mathcal{B}$ can* **escape** *the grasping fingers.*

Proof: Consider the minimum distance function, $d_{min}(q)=\min\{d_1(q),\ldots,d_k(q)\}$. Each d_i is negative inside the finger c-obstacle $\mathcal{CO}_i$, zero on its boundary $\mathcal{S}_i$, and positive outside $\mathcal{CO}_i$. Since $\mathcal{B}$ is held at an equilibrium grasp, the origin can be expressed as a convex combination of the finger c-obstacle outward unit normals:

$$\lambda_1\hat{\eta}_1(q_0)+\cdots+\lambda_k\hat{\eta}_k(q_0)=\vec{0} \qquad 0\le\lambda_1,\ldots,\lambda_k\le 1,\ \textstyle\sum_{i=1}^k\lambda_i=1.$$

The function d_{min} has a non-smooth *saddle point* at q_0 under the following two conditions (see Appendix A). First, the coefficients $\lambda_1,\ldots,\lambda_k$ must be uniquely specified and strictly positive. This is indeed the case for essential equilibrium grasps. Second, the c-space relative curvature form must be strictly positive along some first order roll–slide motion:

$$\kappa_{\text{rel}}(q_0, \dot{q}) > 0 \qquad \text{for some } \dot{q} \in M^1_{1\ldots k}(q_0). \tag{8.8}$$

When $m^1_{q_0} > 0$, the subspace $M^1_{1\ldots k}(q_0)$ is nontrivial. When $m^2_{q_0} > 0$, the c-space relative curvature form, $\kappa_{\text{rel}}(q_0, \dot{q})$, has at least one positive semi-definite eigenvalue on this subspace. Since $\kappa_{\text{rel}}(q_0, \dot{q})$ is assumed to be non-degenerate, it has at least one positive eigenvalue, whose eigenvector satisfies Eq. (8.8). Hence, the function $d_{min}(q)$ possesses a saddle point at $q = q_0$.

Consider a c-space path $\alpha(t)$ that starts at q_0, such that $\dot{\alpha}(0) = \dot{q} \in M^1_{1\ldots k}(q_0)$ and $\kappa_{\text{rel}}(q_0, \dot{q}) > 0$. Since d_{min} has a saddle point at q_0, along this particular path $d_{min}(\alpha(t))$ has a *strict local minimum* at $t = 0$, irrespective of the path's acceleration at $t = 0$. Since $d_{min}(q) = \min\{d_1(q), \ldots, d_k(q)\}$ such that $d_i(q_0) = 0$ for $i = 1 \ldots k$, each function $d_i(\alpha(t))$ has a strict local minimum at $t = 0$. Thus, $d_i(\alpha(t)) > 0$ for $t \in (0, \epsilon]$ $(i = 1 \ldots k)$, implying that $\alpha(t)$ is an *escape path* that lies in the free c-space $\mathcal{F}$ for $t \in [0, \epsilon]$. □

Theorem 8.4 offers the following insight. When $m^1_{q_0} > 0$ and $m^2_{q_0} > 0$ at a k-finger grasp, there exist c-space paths $\alpha(t)$ that start at q_0 and form first order roll–slide motions, $\dot{\alpha}(0) = \dot{q} \in M^1_{1\ldots k}(q_0)$, such that $\kappa_{\text{rel}}(q_0, \dot{q}) > 0$. For any such path, there exists an acceleration, $\ddot{\alpha}(0) = \ddot{q}$, such that the path $\alpha(t)$ having this acceleration locally lies in the free c-pace:

$$\alpha(t) \in \mathcal{F} \qquad t \in [0, \epsilon].$$

The existence of such escape paths is illustrated in the following example.

Example: Figure 8.3(a) shows the triangular object $\mathcal{B}$ of the previous example, which is now held at a different three-finger equilibrium grasp. The c-space geometry of this grasp is sketched in Figure 8.3(b), where q_0 is $\mathcal{B}$'s equilibrium grasp configuration. The set of first-order free motions, $M^1_{1,2,3}(q_0)$, spans a one-dimensional linear subspace tangent to the finger c-obstacles at q_0. Hence, $m^1_{q_0} = 1$ for this grasp. Based on a graphical

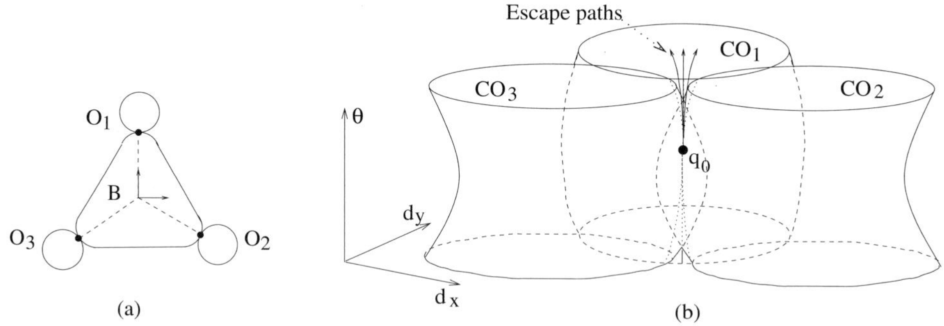

Figure 8.3 (a) A non-immobilizing three-finger equilibrium grasp with $m^1_{q_0} = 1$ and $m^2_{q_0} = 1$. (b) Some escape paths that start at q_0 and move into the upper solid cone which forms part of the free c-space $\mathcal{F}$.

technique described in the next section, $m^2_{q_0} = 1$ for this grasp. In contrast with the previous example, the finger-obstacles are convex along the θ-axis at q_0 (Figure 8.3(b)). The two solid cones that meet at q_0 form part of the free c-space $\mathcal{F}$. Any escape path that starts at q_0 must be tangent to the θ-axis at q_0 (since the subspace of first-order free motions is aligned with this axis), and then move into one of the two solid cones that meet at q_0, as depicted in Figure 8.3(b). ○

To summarize, when the number of fingers exceeds the object's c-space dimension, $k \geq 4$ fingers in 2-D grasps and $k \geq 7$ fingers in 3-D grasps, the first-order mobility index is generically zero and the object is immobilized by first-order geometric effects. For smaller numbers of fingers, $k = 2, 3$ fingers in 2-D grasps and $k = 2, \ldots, 6$ fingers in 3-D grasps, the first-order mobility index is always positive. In these cases, the second-order mobility index determines the grasp's mobility. When the second-order mobility index is zero, the object is immobilized by second-order geometric effects. Conversely, when the two mobility indices are positive, the object is generically mobile and can escape the grasping fingers. An important observation is that the two mobility indices cover all possible numbers of fingers. The two mobility indices, $m^1_{q_0}$ and $m^2_{q_0}$, thus *completely characterize* the mobility of all generic equilibrium grasp arrangements.

8.4 Graphical Depiction of Second-Order Mobility

This section describes a graphical technique that determines the second-order mobility of objects held at various 2-D grasps. The graphical technique is based on the one-to-one correspondence between tangent vectors in the ith finger c-obstacle tangent plane, $T_{q_0}\mathcal{S}_i$, and instantaneous rotations of the object $\mathcal{B}$ about all points on the finger contact normal line, denoted l_i (see Chapter 7). Based on this correspondence, the ith finger c-obstacle curvature form, $\kappa_i(q_0, \dot{q})$ such that $\dot{q}$ varies in $T_{q_0}\mathcal{S}_i$, can be parametrized by the position $\mathcal{B}$'s frame origin on the line l_i. Let $\mathcal{B}$'s frame origin be located at a distance ρ_i from the contact point x_i along the line l_i, such that $\rho_i > 0$ when $\mathcal{B}$'s origin is located on the object's side of x_i, and $\rho_i < 0$ when $\mathcal{B}$'s origin is located on the finger's side of the contact (Figure 8.4(a)). The following lemma provides a formula for the ith c-obstacle curvature form in terms of ρ_i.

LEMMA 8.5 *Let $\mathcal{B}$'s frame origin be located at a distance ρ_i from x_i along the line l_i. Let $\mathcal{S}_i$ be the ith finger c-obstacle boundary, expressed in the c-space coordinates associated with this object frame assignment. The curvature of $\mathcal{S}_i$ along instantaneous rotation about $\mathcal{B}$'s frame origin, $\dot{q} = (0, \omega)$, is given by*

$$\kappa_i(q_0, (0, \omega)) = \frac{(\kappa_{\mathcal{B}_i}\rho_i - 1)(\kappa_{\mathcal{O}_i}\rho_i + 1)}{\kappa_{\mathcal{B}_i} + \kappa_{\mathcal{O}_i}}\omega^2 = \frac{(\rho_i - r_{\mathcal{B}_i})(\rho_i + r_{\mathcal{O}_i})}{r_{\mathcal{B}_i} + r_{\mathcal{O}_i}}\omega^2, \tag{8.9}$$

where $\kappa_{\mathcal{B}_i}$, $\kappa_{\mathcal{O}_i}$, $r_{\mathcal{B}_i}$ and $r_{\mathcal{O}_i}$ are the curvatures and radii-of-curvature of $\mathcal{B}$ and $\mathcal{O}_i$ at x_i.

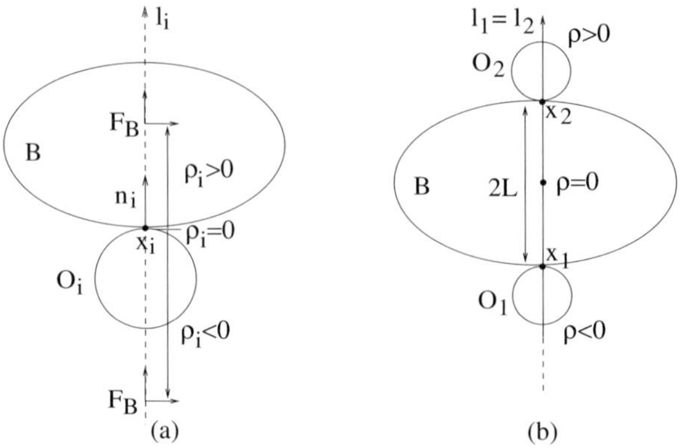

Figure 8.4 (a) The parametrization of $T_{q_0}\mathcal{S}_i$ in terms of instantaneous rotations of $\mathcal{B}$ about points located on the line l_i, at a distance ρ_i from x_i. (b) The parametrization of $T_{q_0}\mathcal{S}_1 = T_{q_0}\mathcal{S}_2$ in terms of a common parameter ρ.

Proof: According to Theorem 2.7, the *curvature form* of the ith finger c-obstacle surface is given by

$$\kappa_i(q_0,\dot{q}) = \frac{1}{\|\eta_i(q_0)\|} \cdot \frac{1}{\kappa_{\mathcal{B}_i}+\kappa_{\mathcal{O}_i}}$$
$$\dot{q}^T \begin{bmatrix} \kappa_{\mathcal{B}_i}\kappa_{\mathcal{O}_i} I & -\kappa_{\mathcal{B}_i}\kappa_{\mathcal{O}_i} JRb_{c_i} \\ -\kappa_{\mathcal{B}_i}\kappa_{\mathcal{O}_i}(JRb_{c_i})^T & (\kappa_{\mathcal{O}_i}Rb_i - n_i)^T(\kappa_{\mathcal{B}_i}Rb_i + n_i) \end{bmatrix} \dot{q} \quad \dot{q} \in T_{q_0}\mathcal{S}_i, \tag{8.10}$$

where $\eta_i(q_0)$ is the ith c-obstacle outward normal at q_0, $\kappa_{\mathcal{B}_i}$ and $\kappa_{\mathcal{O}_i}$ are the curvatures of $\mathcal{B}$ and $\mathcal{O}_i$ at x_i, b_i is the contact point position expressed in $\mathcal{B}$'s frame, b_{c_i} is $\mathcal{B}$'s center of curvature expressed in $\mathcal{B}$'s frame, and n_i is $\mathcal{B}$'s inward unit normal at x_i. The 2×2 matrices in Eq. (8.10) are $\mathcal{B}$'s orientation matrix R, the identity matrix I and $J = \begin{bmatrix} 0 & 1 \\ -1 & 0 \end{bmatrix}$.

The term Rb_i which appears in Eq. (8.10) is the vector from $\mathcal{B}$'s frame origin to x_i, expressed in the fixed world frame. When $\mathcal{B}$'s frame origin is located at a distance ρ_i from x_i on the line l_i, one obtains that $Rb_i = -\rho_i n_i$. Since $\eta_i = (n_i, Rb_i \times n_i)$ (Theorem 2.5), we have that $\|\eta_i(q_0)\| = 1$ when $Rb_i = -\rho_i n_i$. Substituting $Rb_i = -\rho_i n_i$ and $\|\eta_i(q_0)\| = 1$ in Eq. (8.10) gives the simpler formula for the ith finger c-obstacle curvature form:

$$\kappa_i(q_0,\dot{q}) = \frac{1}{\kappa_{\mathcal{B}_i}+\kappa_{\mathcal{O}_i}} \dot{q}^T \begin{bmatrix} \kappa_{\mathcal{B}_i}\kappa_{\mathcal{O}_i} I & -\kappa_{\mathcal{B}_i}\kappa_{\mathcal{O}_i} JRb_{c_i} \\ -\kappa_{\mathcal{B}_i}\kappa_{\mathcal{O}_i}(JRb_{c_i})^T & (\kappa_{\mathcal{O}_i}\rho_i + 1)(\kappa_{\mathcal{B}_i}\rho_i - 1) \end{bmatrix} \dot{q} \quad \dot{q} \in T_{q_i}\mathcal{S}_i. \tag{8.11}$$

Evaluating $\kappa_i(q_0,\dot{q})$ along the tangent vector $\dot{q} = (0,\omega)$ gives Formula (8.9). □

In order to proceed with the graphical technique, let $\kappa_i(q_0,\rho_i)$ denote the ith finger c-obstacle curvature form specified by Lemma 8.5, evaluated along unit magnitude instantaneous rotation, $\dot{q} = (0,\omega)$ such that $|\omega| = 1$. The possible signs attained by $\kappa_i(q_0,\dot{q})$ as $\dot{q}$ varies in $T_{q_0}\mathcal{S}_i$ are *equivalent* to the signs attained by $\kappa_i(q_0,\rho_i)$, as ρ_i

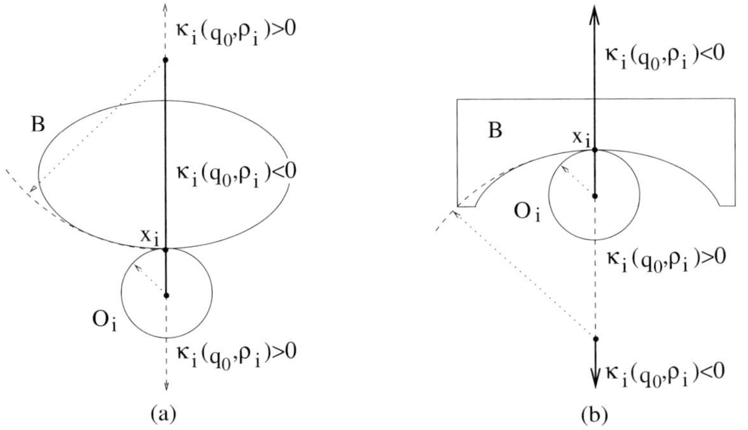

Figure 8.5 (a) When $\mathcal{B}$ and $\mathcal{O}_i$ are convex at x_i, $\kappa_i(q_0, \rho_i) < 0$ in the interval between the bodies' centers of curvature at x_i. (b) When $\mathcal{B}$ is concave and $\mathcal{O}_i$ is convex at x_i, $\kappa_i(q_0, \rho_i) > 0$ in the interval between the bodies' centers of curvature at x_i.

varies in the extended real line, $-\infty \le \rho_i \le +\infty$. There are now two cases to consider. In the first case, the object and finger bodies are convex at the contacts. In this case $r_{\mathcal{B}_i} \ge 0$ and $r_{\mathcal{O}_i} \ge 0$. Based on Formula (8.9), $\kappa_i(q_0, \rho_i)$ is *negative* in the interval between the centers of curvature of $\mathcal{B}$ and $\mathcal{O}_i$ at x_i, and *positive* in the semi-infinite intervals that start at the centers of curvature and extend to infinity on both sides of the ith contact (Figure 8.5(a)). In the second case, one of the contacting bodies, say $\mathcal{B}$, is concave at the contact point x_i. In this case, $r_{\mathcal{B}_i} < 0$ while $r_{\mathcal{O}_i} \ge 0$, such that $r_{\mathcal{O}_i} \le |r_{\mathcal{B}_i}|$; otherwise, $\mathcal{B}$ and $\mathcal{O}_i$ interpenetrate at x_i. Since the bodies are assumed to maintain point contact at x_i, we may assume that $r_{\mathcal{O}_i} < |r_{\mathcal{B}_i}|$. The denominator, $r_{\mathcal{B}_i} + r_{\mathcal{O}_i}$, is thus *negative* in Formula (8.9). Hence, $\kappa_i(q_0, \rho_i)$ is *positive* in the interval bounded by the bodies' centers of curvature at x_i and *negative* in the complementary semi-infinite intervals (Figure 8.5(b)). This graphical insight is next applied to various two- and three-finger grasps.

Two-finger grasps: The subspace of first-order free motions, $M_{1,2}(q_0)$, is two-dimensional and consists of instantaneous rotations of $\mathcal{B}$ about all points of the line $l_1 = l_2$. Hence $m^1_{q_0} = 2$ and $0 \le m^2_{q_0} \le 2$ at the two-finger grasps. Let ρ parametrize the points of this line by their distance from the midpoint of the finger contacts, such that $\rho < 0$ on $\mathcal{O}_1$'s side of the midpoint and $\rho > 0$ on $\mathcal{O}_2$'s side of the midpoint (Figure 8.4(b)). If the distance between the contact points is $2L$, then $\rho_1 = L + \rho$ and $\rho_2 = L - \rho$. Substituting for $\kappa_1(q_0, \rho_1)$ and $\kappa_2(q_0, \rho_2)$ according to Lemma 8.5, we obtain the c-space relative curvature form:

$$\kappa_{\text{rel}}(q_0, \rho) = \frac{\Big(\rho_1(\rho)\kappa_{\mathcal{B}_1} - 1\Big)\Big(\rho_1(\rho)\kappa_{\mathcal{O}_1} + 1\Big)}{\kappa_{\mathcal{B}_1} + \kappa_{\mathcal{O}_1}} + \frac{\Big(\rho_2(\rho)\kappa_{\mathcal{B}_2} - 1\Big)\Big(\rho_2(\rho)\kappa_{\mathcal{O}_2} + 1\Big)}{\kappa_{\mathcal{B}_2} + \kappa_{\mathcal{O}_2}}, \tag{8.12}$$

where the equilibrium coefficients $\lambda_1 = \lambda_2 = \frac{1}{2}$ have been omitted. Each summand in Eq. (8.12) is quadratic in ρ and, hence, has two real roots. Each summand therefore partitions the line $l_1 = l_2$ into at most *three* intervals according to its sign. Let I_- and I_+ be the negative and positive intervals of $\kappa_1(q_0, \rho)$. Let J_- and J_+ be the respective intervals of $\kappa_2(q_0, \rho)$. The possible overlaps of these intervals determine the grasp's second-order mobility index as follows (see Exercises).

Second-order mobility of two-finger grasps: *Let a rigid object $\mathcal{B}$ be held by identical finger bodies, $\mathcal{O}_1$ and $\mathcal{O}_2$, at an equilibrium grasp configuration q_0.*

(i) When $\mathcal{B}$, $\mathcal{O}_1$ and $\mathcal{O}_2$ are **convex** at the contacts, $m^2_{q_0} \geq 1$. If I_- and J_- overlap, $m^2_{q_0} = 1$. If I_- and J_- do not overlap and $\mathcal{B}$ has identical curvature at the contacts, $m^2_{q_0} = 2$.

(ii) When $\mathcal{B}$ is **concave** at the contacts, $0 \leq m^2_{q_0} \leq 1$. If I_+ and J_+ do not overlap and $\mathcal{B}$ has identical curvature at the contacts, $m^2_{q_0} = 0$ and the object is immobilized.

(iii) When $\mathcal{O}_1$ and $\mathcal{O}_2$ are **concave** at the contacts, $0 \leq m^2_{q_0} \leq 1$. If I_+ and J_+ do not overlap and $\mathcal{B}$ has identical curvature at the contacts, $m^2_{q_0} = 0$ and the object is immobilized.

The following examples illustrate the graphical technique for cases (i) and (ii).

Example: Figure 8.6 illustrates case (i), where $\mathcal{B}$, $\mathcal{O}_1$ and $\mathcal{O}_2$ are convex at the contact points. There are two possible sub-cases. In Figure 8.6(a), the intervals I_- and J_- overlap and consequently $m^2_{q_0} = 1$. The object $\mathcal{B}$ can escape the fingers along pure horizontal translation but is prevented from rotating about points in the interval where I_- and J_- overlap. In Figure 8.6(b), the intervals I_- and J_- do not overlap. The object $\mathcal{B}$ has identical curvature at the contacts, and, therefore, $m^2_{q_0} = 2$ at this grasp. The object can escape the fingers along any combination of horizontal translation and rotation about any point on the line $l_1 = l_2$. ○

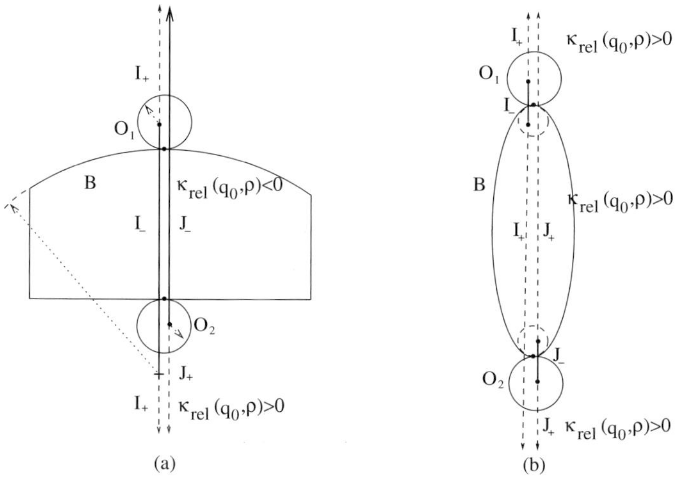

Figure 8.6 When $\mathcal{B}$, $\mathcal{O}_1$, and $\mathcal{O}_2$ are convex at the contacts: (a) $m^2_{q_0} = 1$ when I_- and J_- overlap. (b) $m^2_{q_0} = 2$ when I_- and J_- do not overlap and $\mathcal{B}$ has identical curvature at the contacts.

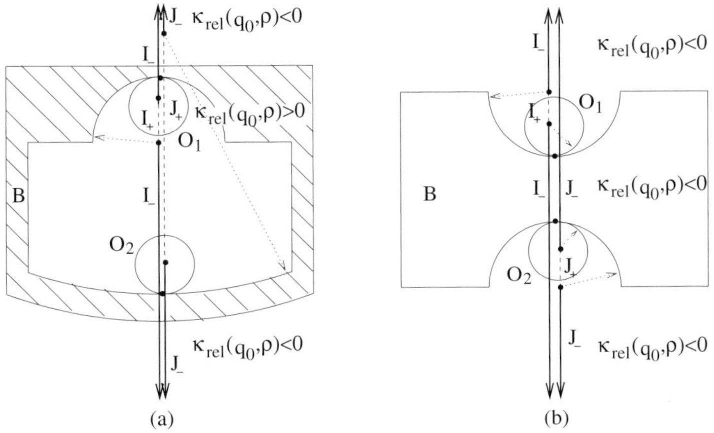

Figure 8.7 When $\mathcal{B}$ is concave at the contacts: (a) $m^2_{q0} = 1$ when I_+ and J_+ overlap, and (b) $m^2_{q0} = 0$ when I_+ and J_+ do not overlap and $\mathcal{B}$ has identical curvature at the contacts.

Example: Figure 8.7 illustrates case (ii), where convex finger bodies $\mathcal{O}_1$ and $\mathcal{O}_2$ are placed at opposing concavities of the object $\mathcal{B}$. In Figure 8.7(a), the two fingers push outward against the object's boundary, creating a segment where the intervals I_+ and J_+ overlap. As a result, $m^2_{q0} = 1$ at this grasp. The object can locally escape the grasping fingers along paths that start with instantaneous rotation about any point in the interval where $I_+ \cap J_+$. In the example of Figure 8.7(b), the two fingers apply compressive forces on $\mathcal{B}$, such that the intervals I_+ and J_+ do not overlap. The object $\mathcal{B}$ has identical curvature at the contacts, and, therefore, $m^2_{q0} = 0$ at this grasp. The object is thus fully immobilized at this grasp. ○

Three-finger grasps: When the three fingers form an essential equilibrium grasp (the generic case), the subspace of first-order free motions, $M^1_{1,2,3}(q_0)$, is one-dimensional. Hence $m^1_{q0} = 1$, and $0 \le m^2_{q0} \le 1$ at such grasps. The subspace $M^1_{1,2,3}(q_0)$ consists of instantaneous rotations of $\mathcal{B}$ about the common intersection point of the lines underlying the finger contact normals, denoted p. Let $\rho_i(p)$ denote the signed distance of p from the contact point x_i for $i = 1,2,3$. Based on Lemma 8.5, the c-space relative curvature form is given by

$$\begin{aligned}\kappa_{\text{rel}}(q_0, \rho_1(p), \rho_2(p), \rho_3(p)) &= \sum_{i=1}^{3} \lambda_i \kappa_i(q_0, \rho_i(p)) \\ &= \sum_{i=1}^{3} \lambda_i \frac{\Big(\rho_i(p)\kappa_{\mathcal{B}_i} - 1\Big)\Big(\rho_i(p)\kappa_{\mathcal{O}_i} + 1\Big)}{\kappa_{\mathcal{B}_i} + \kappa_{\mathcal{O}_1}}. \end{aligned} \tag{8.13}$$

The sign of $\kappa_{\text{rel}}(q_0, \rho_1, \rho_2, \rho_3)$ at p determines the grasp's second-order mobility index. When the sign is positive, $m^2_{q0} = 1$. When the sign is negative, $m^2_{q0} = 0$ and the object is fully immobilized. The graphically determinable cases occur when the three summands in Eq. (8.13) have the same sign, as illustrated in the following example.

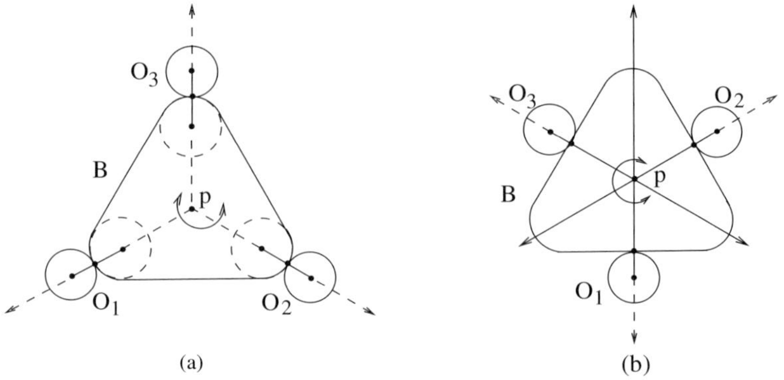

Figure 8.8 A three-finger equilibrium grasp that is (a) non-immobilizing with $m_{q0}^2 = 1$, and (b) immobilizing with $m_{q0}^2 = 0$. Solid segments represent negative sign of the respective finger c-obstacle curvature form.

Example: Figure 8.8(a) depicts a triangular object $\mathcal{B}$ that is held by three disc fingers along its tips (see Figure 8.3(b) for the c-space geometry of this grasp). The negative interval of each finger c-obstacle curvature form lies between the bodies' centers of curvatures, along the respective contact normal line l_i. The concurrency point p lies outside the negative intervals. The three summands in Eq. (8.13) are *positive,* and $m_{q0}^2 = 1$ at this grasp. The object can escape the fingers along a local motion that starts with instantaneous rotation about p.

Figure 8.8(b) depicts a different grasp of the object $\mathcal{B}$, which is now held by three disc fingers along its flat edges (see Figure 8.2(b) for the c-space geometry of this grasp). When a convex finger body $\mathcal{O}_i$ contacts a flat edge of $\mathcal{B}$ at x_i, the negative interval of each finger c-obstacle curvature form extends from x_i to infinity on the object's side of the contact, along the respective contact normal line l_i. The concurrency point p thus lies within the negative interval of each finger contact normal line l_i. The three summands in Eq. (8.13) are *negative,* and $m_{q0}^2 = 0$ at this grasp. The object is thus fully immobilized at this grasp. ○

The next chapter will generalize the second-order immobilization examples. We will see that almost all 2-D objects can be immobilized by two fingers when the object has opposing concavities, and by three fingers having sufficiently flat fingertips when the object has no such concavities. In a similar manner, we will see that curvature effects can be used to immobilize almost all 3-D objects using four fingers having sufficiently flat fingertips. Minimalistic robot hand designs can thus apply second-order immobilization theory in order to achieve the widest grasping capabilities with the smallest number of fingers.

Bibliographical Notes

Second-order or curvature based immobilization was developed by Rimon and Burdick [1, 2], in order to explain a series of paradoxes that occur in frictionless grasp

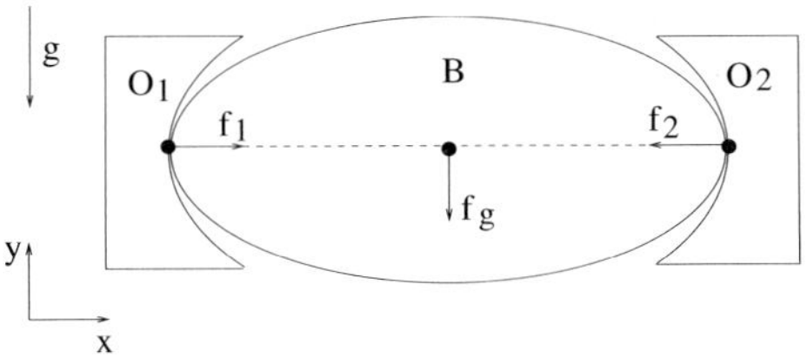

Figure 8.9 Illustration of the paradox where second-order immobilization under frictionless contact conditions is not able to resist the gravitational force that acts on $\mathcal{B}$.

arrangements. These paradoxes were described by Featherstone [3] and Mason [4]. Consider, for example, the rigid ellipse $\mathcal{B}$ that is held at an equilibrium grasp by two rigid finger bodies, $\mathcal{O}_1$ and $\mathcal{O}_2$, as depicted in Figure 8.9. Under frictionless contact conditions, the fingers can apply only horizontal forces, f_1 and f_2, at the contacts. If the object is held under the influence of gravity, the finger forces cannot oppose the vertical gravitational force f_g that acts on $\mathcal{B}$. However, the object $\mathcal{B}$ is fully immobilized when the two finger bodies are sufficiently concave at the contacts. The fingers should therefore be able to statically resist any external influence that acts on $\mathcal{B}$. This paradox can be explained in terms of local material deformations at the contacts [5].

However, the need to include local material deformations highlights a missing gap in second-order immobilization theory. There must be a way to explain the forces of restraint that arise in second-order immobilizing grasps under the ideal rigid-body model. We offer the following conjecture. Recall from Section 6.3 that *first-order immobilizing grasps* are dual to *wrench resistant grasps* under frictionless contact conditions. Similarly, there must exist a notion of statically generated *wrench derivative* that can be statically realized by rigid finger bodies at a frictionless equilibrium grasp. The *second-order immobilizing grasps* must be dual to *second-order wrench resistant grasps,* which are able to statically resist external disturbances that act on $\mathcal{B}$ with combinations of finger forces and finger force derivatives at the contacts.

Appendix: Proof Details

This appendix contains proofs of key properties concerning the second-order free motions and the second-order mobility index. The first lemma is the coordinate invariance of the second-order free motions, which requires the following transformation rule.

LEMMA 8.6 *Let two parametrizations of B's c-space be related by a diffeomorphism, $q = h(\bar{q})$. The velocity-and-acceleration mapping induced by h, mapping $(\dot{\bar{q}}, \ddot{\bar{q}}) \in T^2_{\bar{q}}\mathbb{R}^m$ to $(\dot{q}, \ddot{q}) \in T^2_q\mathbb{R}^m$, is given by*

$$\begin{pmatrix} \dot{q} \\ \ddot{q} \end{pmatrix} = \begin{pmatrix} [Dh(\bar{q})]\dot{\bar{q}} \\ \left[D^2h(\bar{q})\right](\dot{\bar{q}},\dot{\bar{q}}) + \left[Dh(\bar{q})\right]\ddot{\bar{q}} \end{pmatrix}, \tag{8.14}$$

where $Dh(\bar{q})$ is the Jacobian of h, and $D^2h(\bar{q})$ is a vector-valued bilinear map that acts on $(\dot{\bar{q}},\dot{\bar{q}})$.

Proof: Let $T\mathbb{R}^m$ and $\overline{T\mathbb{R}^m}$ denote the *tangent bundle* of $\mathbb{R}^m$ in the q and $\bar{q}$ coordinates, which are equivalent to $\mathbb{R}^m \times \mathbb{R}^m$. The *tangent bundle mapping* induced by h, $H: T\mathbb{R}^m \to \overline{T\mathbb{R}^m}$, has the form:

$$H(\bar{q},\dot{\bar{q}}) = \begin{pmatrix} h(\bar{q}) \\ Dh(\bar{q})\dot{\bar{q}} \end{pmatrix}.$$

The Jacobian of $H(\bar{q},\dot{\bar{q}})$ forms a $2m \times 2m$ matrix, $DH(\bar{q},\dot{\bar{q}})$, which provides the transformation rule:

$$[DH(\bar{q},\dot{\bar{q}})]\begin{pmatrix} \dot{\bar{q}} \\ \ddot{\bar{q}} \end{pmatrix} = \begin{bmatrix} Dh(\bar{q}) & O \\ D^2h(\bar{q})(\dot{\bar{q}},\cdot) & Dh(\bar{q}) \end{bmatrix}\begin{pmatrix} \dot{\bar{q}} \\ \ddot{\bar{q}} \end{pmatrix} = \begin{pmatrix} [Dh(\bar{q})]\dot{\bar{q}} \\ \left[D^2h(\bar{q})\right](\dot{\bar{q}},\dot{\bar{q}}) + \left[Dh(\bar{q})\right]\ddot{\bar{q}} \end{pmatrix},$$

where O is a zero $m \times m$ matrix. □

The following proposition establishes the coordinate invariance of the second-order free motions.

Proposition 8.1 *Let q and $\bar{q}$ be two parametrizations of $\mathcal{B}$'s c-space, related by a coordinate transformation $q = h(\bar{q})$. If h is a diffeomorphism, the set of second-order free motions of $\mathcal{B}$ is* **coordinate invariant:**

$$\dot{q} \in M^2_{1\ldots k}(q_0) \quad \textit{iff} \quad \dot{\bar{q}} \in \bar{M}^2_{1\ldots k}(\bar{q}_0) \qquad q_0 = h(\bar{q}_0),$$

where $M^2_{1\ldots k}(q_0)$ and $\bar{M}^2_{1\ldots k}(\bar{q}_0)$ are the second-order free motion sets in the q and $\bar{q}$ coordinates.

Proof: First consider the case where the object $\mathcal{B}$ is contacted by a single finger body $\mathcal{O}_i$. We have to show that $(\dot{q},\ddot{q}) \in M^2_i(q_0)$ iff $(\dot{\bar{q}},\ddot{\bar{q}}) \in \bar{M}^2_i(\bar{q}_0)$, where $q_0 = h(\bar{q}_0)$. Let $d_i(q)$ and $\bar{d}_i(\bar{q})$ be the signed distance functions from the ith finger c-obstacle boundary, $\mathcal{S}_i$ and $\overline{\mathcal{S}}_i$, in the two c-space coordinates. Let $(\dot{q},\ddot{q})$ lie in $M^2_i(q_0)$:

$$\nabla d_i(q_0)\cdot\dot{q} = 0 \quad \text{and} \quad \dot{q}^T[D^2 d_i(q_0)]\dot{q} + \nabla d_i(q_0)\cdot\ddot{q} \geq 0. \tag{8.15}$$

Substituting for $(\dot{q},\ddot{q})$ in Eq. (8.15) according to the transformation rule (8.14) gives two equations:

$$\nabla d_i(q_0)^T Dh(\bar{q}_0)\dot{\bar{q}} = 0 \tag{8.16}$$

and

$$\dot{\bar{q}}^T Dh(\bar{q}_0)^T[D^2 d_i(q_0)]Dh(\bar{q}_0)\dot{\bar{q}} + \nabla d_i(q_0)^T\left([D^2h(\bar{q}_0)](\dot{\bar{q}},\dot{\bar{q}}) + Dh(\bar{q}_0)\ddot{\bar{q}}\right) \geq 0. \tag{8.17}$$

We now show that Equations (8.16) and (8.17) imply that the pair $(\dot{\bar{q}},\ddot{\bar{q}})$ lies in $\bar{M}_i^2(\bar{q}_0)$. That is, $\nabla\bar{d}_i(\bar{q}_0)\cdot\dot{\bar{q}}=0$ and $\dot{\bar{q}}^T\,[D^2\bar{d}_i(\bar{q}_0)]\dot{\bar{q}}+\nabla\bar{d}_i(\bar{q}_0)\cdot\ddot{\bar{q}}\geq 0$. All the ensuing arguments are bi-directional, so this proves the single-finger case.

Consider the induced distance function in $\bar{q}$-space: $\tilde{d}_i(\bar{q})=d_i(h(\bar{q}))$ (note that $\tilde{d}_i$ is *not* the Euclidean distance in the $\bar{q}$ coordinates). Since q and $\bar{q}$ parametrize the same c-space, the diffeomorphism h maps the interior of $\overline{\mathcal{CO}}_i$ to the interior of $\mathcal{CO}_i$, $\bar{\mathcal{S}}_i$ to $\mathcal{S}_i$, and the exterior of $\overline{\mathcal{CO}}_i$ to the exterior of $\mathcal{CO}_i$. Hence $\tilde{d}_i$ is negative in the interior of $\overline{\mathcal{CO}}_i$, zero on its boundary $\overline{\mathcal{S}}_i$, and positive outside $\overline{\mathcal{CO}}_i$. In particular, $\nabla\tilde{d}_i(\bar{q})$ is collinear with the finger c-obstacle outward normal at $\bar{q}\in\overline{\mathcal{S}}_i$. Using the chain rule, the formulas for $\nabla\tilde{d}_i$ and $D^2\tilde{d}_i$ are

$$\nabla\tilde{d}_i(\bar{q})=[Dh(\bar{q})]^T\,\nabla d_i(h(\bar{q}))$$

and

$$D^2\tilde{d}_i(\bar{q})=Dh(\bar{q})^T\,D^2d_i(q)Dh(\bar{q})+\nabla d_i(q)^T\,D^2h(\bar{q}).$$

Substituting the terms $\nabla\tilde{d}_i(\bar{q})$ and $D^2\tilde{d}_i(\bar{q})$ in Formulas (8.16) and (8.17), the pairs $(\dot{\bar{q}},\ddot{\bar{q}})$, which are mapped under h to the set $M_i^2(q_0)$, are given by

$$\nabla\tilde{d}_i(\bar{q}_0)\cdot\dot{\bar{q}}=0 \quad\text{and}\quad \dot{\bar{q}}^T[D^2\tilde{d}_i(\bar{q}_0)]\dot{\bar{q}}+\nabla\tilde{d}_i(\bar{q}_0)\cdot\ddot{\bar{q}}\geq 0\,. \tag{8.18}$$

The last step is to show that $\nabla\tilde{d}_i(\bar{q}_0)$ and $D^2\tilde{d}_i(\bar{q}_0)$ can be replaced by $\nabla\bar{d}_i(\bar{q}_0)$ and $D^2\bar{d}_i(\bar{q}_0)$, which define the set $\bar{M}_i^2(\bar{q}_0)$. Since $\bar{d}_i(\bar{q})$ is defined in terms of the Euclidean distance, its gradient has unit magnitude. Since $\nabla\bar{d}_i$ and $\nabla\tilde{d}_i$ satisfy the relation:

$$\nabla\bar{d}_i(\bar{q})=\frac{1}{\|\nabla\tilde{d}_i(\bar{q})\|}\nabla\tilde{d}_i(\bar{q}) \qquad \bar{q}\in\overline{\mathcal{S}}_i, \tag{8.19}$$

we obtain from Eq. (8.19) that $\nabla\tilde{d}_i(\bar{q})\cdot\dot{\bar{q}}=0$ iff $\nabla\bar{d}_i(\bar{q})\cdot\dot{\bar{q}}=0$ at points $\bar{q}\in\overline{\mathcal{S}}_i$.

The formula for $D^2\bar{d}_i(\bar{q}_0)$ is as follows. Close to $\overline{\mathcal{S}}_i$ and along paths $\bar{q}(t)$ which are *normal* to it, $\nabla\bar{d}_i(\bar{q}(t))$ is a unit magnitude vector aligned with the normal direction. Hence, $[D^2\bar{d}_i(\bar{q}_0)]\dot{\bar{q}}=0$ for any $\dot{\bar{q}}\perp T_{\bar{q}_0}\overline{\mathcal{S}}_i$. Using the chain rule on Eq. (8.19) gives

$$D^2\bar{d}_i(\bar{q})=D\Big(\frac{1}{\|\nabla\tilde{d}(\bar{q})\|}\nabla\tilde{d}(\bar{q})\Big)=\frac{1}{\|\nabla\tilde{d}(\bar{q})\|}[I\,-\,\nabla\bar{d}_i\nabla\bar{d}_i^T\,]D^2\tilde{d}_i(\bar{q}).$$

Given a unit-magnitude vector $\hat{v}$, $[I\,-\,\hat{v}\hat{v}^T]\dot{\bar{q}}$ is the projection of $\dot{\bar{q}}$ on the subspace orthogonal to $\hat{v}$. The tangent vectors $\dot{\bar{q}}\in T_{\bar{q}_0}\overline{\mathcal{S}}_i$ satisfy the identity: $\dot{\bar{q}}=[I-\nabla\bar{d}_i\nabla\bar{d}_i^T\,]\dot{\bar{q}}+(\nabla\bar{d}_i\,\cdot\,\dot{\bar{q}})\nabla\bar{d}_i$. Since $D^2\bar{d}_i(\bar{q}_0)$ annihilates tangent vectors collinear with $\nabla\bar{d}_i(\bar{q}_0)$, the formula for $D^2\bar{d}_i(\bar{q}_0)$ is

$$D^2\bar{d}_i(\bar{q}_0)=\frac{1}{\|\nabla\tilde{d}(\bar{q}_0)\|}[I\,-\,\nabla\bar{d}_i\nabla\bar{d}_i^T\,]D^2\tilde{d}_i(\bar{q}_0)[I\,-\,\nabla\bar{d}_i\nabla\bar{d}_i^T\,]. \tag{8.20}$$

Based on Eq. (8.20), we conclude that for $\dot{\bar{q}}\in T_{\bar{q}_0}\overline{\mathcal{S}}_i$:

$$\dot{\bar{q}}^T\,D^2\bar{d}_i(\bar{q}_0)\dot{\bar{q}}+\nabla\bar{d}_i(\bar{q}_0)\cdot\ddot{\bar{q}}=\frac{1}{\|\nabla\tilde{d}(\bar{q})\|}\{\dot{\bar{q}}^T\,D^2\tilde{d}_i(\bar{q}_0)\dot{\bar{q}}+\nabla\tilde{d}_i(\bar{q}_0)\cdot\ddot{\bar{q}}\}.$$

Replacing $\nabla\tilde{d}_i(\bar{q}_0)$ and $D^2\tilde{d}_i(\bar{q}_0)$ by $\nabla\bar{d}_i(\bar{q}_0)$ and $D^2\bar{d}_i(\bar{q}_0)$ in Eq. (8.18), we obtain that the set $\bar{M}_i^2(\bar{q}_0)$ in the $\bar{q}$ coordinates is mapped under h to the set $M_i^2(q_0)$ in the q coordinates.

When the object $\mathcal{B}$ is held by finger bodies $\mathcal{O}_1,\dots,\mathcal{O}_k$ at a frictionless equilibrium grasp, the subspaces of first order roll–slide motions in the two c-space coordinates are given by

$$M^2_{1\dots k}(q_0)=\cap_{i=1}^k M_i^2(q_0) \quad \text{and} \quad \bar{M}^2_{1\dots k}(\bar{q}_0)=\cap_{i=1}^k \bar{M}_i^2(\bar{q}_0).$$

Let $F_{\bar{q}}(\dot{\bar{q}},\ddot{\bar{q}})$ denote the velocity-and-acceleration mapping specified in Lemma 8.6:

$$F_{\bar{q}}(\dot{\bar{q}},\ddot{\bar{q}})=[D^2h(\bar{q})](\dot{\bar{q}},\dot{\bar{q}})+[Dh(\bar{q})]\ddot{\bar{q}}.$$

We have shown that $\bar{M}_i^2(\bar{q}_0)$ is mapped under $F_{\bar{q}}$ to $M_i^2(q_0)$ for each finger body $\mathcal{O}_i$. In general, given a mapping F and two sets $\mathcal{U}_1,\mathcal{U}_2$, the relation

$$F(\mathcal{U}_1\cap\mathcal{U}_2)=F(\mathcal{U}_1)\cap F(\mathcal{U}_2) \tag{8.21}$$

holds when the mapping F is invertible. In our case, the Jacobian $Dh(\bar{q}_0)$ is non-singular since h is a diffeomorphism. Hence, $F_{\bar{q}}$ has a well-defined inverse of the form:

$$\begin{pmatrix}\dot{\bar{q}}\\ \ddot{\bar{q}}\end{pmatrix}=F_{\bar{q}}^{-1}(\dot{q},\ddot{q})=\begin{pmatrix}[Dh(\bar{q})]^{-1}\dot{q}\\ [Dh(\bar{q})]^{-1}\left(\ddot{q}-[D^2h(\bar{q})]\left([Dh(\bar{q})]^{-1}\dot{q},\,[Dh(\bar{q})]^{-1}\dot{q}\right)\right)\end{pmatrix}.$$

Since $F_{\bar{q}}$ is invertible, the relation (8.21) implies that $M^2_{1\dots k}(q_0)=F_{\bar{q}_0}\left(\bar{M}^2_{1\dots k}(\bar{q}_0)\right)$ at $q_0=h(\bar{q}_0)$. This proves the coordinate invariance of the second-order free motions. □

The next proposition establishes the coordinate invariance of the second-order mobility index.

PROPOSITION 8.7 *Let q and $\bar{q}$ be two parametrizations of $\mathcal{B}$'s c-space, and let $m^2_{q_0}$ and $m^2_{\bar{q}_0}$ be the second-order mobility indices of $\mathcal{B}$ at a k-finger grasp in these coordinates. Then $m^2_{q_0}=m^2_{\bar{q}_0}$.*

Proof: Let $d_i(q)$ and $\bar{d}_i(\bar{q})$ be the distance functions from the ith finger c-obstacle boundary, $\mathcal{S}_i$ and $\overline{\mathcal{S}}_i$, in the respective c-space coordinates. We may assume that q and $\bar{q}$ are related by a diffeomorphism, $q=h(\bar{q})$, where $q_0=h(\bar{q}_0)$ is the equilibrium grasp configuration. The equilibrium grasp condition in the two c-space coordinates have the forms:

$$\lambda_1\nabla d_1(q_0)+\cdots+\lambda_k\nabla d_k(q_0)=\vec{0} \qquad \lambda_1,\dots,\lambda_k\geq 0 \tag{8.22}$$

and

$$\bar{\lambda}_1\nabla\bar{d}_1(\bar{q}_0)+\cdots+\bar{\lambda}_k\nabla\bar{d}_k(\bar{q}_0)=\vec{0} \qquad \bar{\lambda}_1,\dots,\bar{\lambda}_k\geq 0. \tag{8.23}$$

Using the equilibrium grasp coefficients of Equations (8.22) and (8.23), the c-space relative curvature forms in the two c-space coordinates are

$$\kappa_{\text{rel}}(q_0,\dot{q}) = \sum_{i=1}^{k} \lambda_i \dot{q}^T D^2 d_i(q_0)\dot{q} \qquad \dot{q} \in M^1_{1\ldots k}(q_0)$$

and

$$\kappa_{\text{rel}}(\bar{q}_0,\dot{\bar{q}}) = \sum_{i=1}^{k} \bar{\lambda}_i \dot{\bar{q}}^T D^2 \bar{d}_i(\bar{q}_0)\dot{\bar{q}} \qquad \dot{\bar{q}} \in \bar{M}^1_{1\ldots k}(\bar{q}_0).$$

Our first step is to express $\bar{\lambda}_i$ in terms of λ_i for $i = 1\ldots k$. Consider the induced distance function: $\tilde{d}_i(\bar{q}) = d_i(h(\bar{q}))$. As noted in the proof of Proposition 8.1, $\nabla\tilde{d}_i(\bar{q})$ is collinear with the finger c-obstacle outward normal at, so that $\nabla\bar{d}_i(\bar{q}) = \nabla\tilde{d}_i(\bar{q})/\|\nabla\tilde{d}_i(\bar{q})\|$ at points $\bar{q} \in \overline{\mathcal{S}}_i$. Using the chain rule, the first and second derivative of $\tilde{d}_i$ are

$$\nabla\tilde{d}_i(\bar{q}) = [Dh(\bar{q})]^T \nabla d_i(h(\bar{q})), \tag{8.24}$$

where Dh is the $m \times m$ Jacobian of h, and

$$D^2\tilde{d}_i(\bar{q}) = [Dh(\bar{q})]^T D^2 d_i(h(\bar{q}))[Dh(\bar{q})] + \nabla d_i(h(\bar{q}))^T D^2 h(\bar{q}), \tag{8.25}$$

where D^2h is a vector-valued bilinear map that acts on $(\dot{\bar{q}},\dot{\bar{q}})$. Multiplying both sides of the equilibrium grasp equation (8.22) by $[Dh(\bar{q}_0)]^T$, then applying Formula (8.24), gives

$$\lambda_1\nabla\tilde{d}_1(\bar{q}_0) + \ldots + \lambda_k\nabla\tilde{d}_k(\bar{q}_0) = \vec{0} \qquad \lambda_1,\ldots,\lambda_k \geq 0. \tag{8.26}$$

Since $\nabla\bar{d}_i(\bar{q}_0) = \nabla\tilde{d}_i(\bar{q}_0)/\|\nabla\tilde{d}_i(\bar{q}_0)\|$ for $i = 1\ldots k$, it follows from Eq. (8.26) that $\bar{\lambda}_i = \|\nabla\tilde{d}_i\| \cdot \lambda_i$ for $i = 1\ldots k$.

Our next step is to express the quadratic form $\dot{\bar{q}}^T D^2\bar{d}_i(\bar{q}_0)\dot{\bar{q}}$ in terms of $\tilde{d}_i(\bar{q})$. Based on the proof of Proposition 8.1:

$$D^2\bar{d}_i(\bar{q}) = \frac{1}{\|\nabla\tilde{d}_i(\bar{q})\|}[I - \nabla\bar{d}_i\nabla\bar{d}_i^T]D^2\tilde{d}_i(\bar{q})[I - \nabla\bar{d}_i\nabla\bar{d}_i^T] \qquad \bar{q} \in \overline{\mathcal{S}}_i.$$

Since $[I - \nabla\bar{d}_i\nabla\bar{d}_i^T]\dot{\bar{q}} = \dot{\bar{q}}$ for $\dot{\bar{q}} \in T_{\bar{q}}\overline{\mathcal{S}}_i$, we obtain the quadratic form at $\bar{q}_0$:

$$\dot{\bar{q}}^T D^2\bar{d}_i(\bar{q}_0)\dot{\bar{q}} = \frac{1}{\|\nabla\tilde{d}_i(\bar{q}_0)\|}\dot{\bar{q}}^T[D^2\tilde{d}_i(\bar{q}_0)]\dot{\bar{q}} \qquad \dot{\bar{q}} \in T_{\bar{q}_0}\overline{\mathcal{S}}_i.$$

The $\bar{q}$-space relative curvature form can thus be written as

$$\begin{aligned}\kappa_{\text{rel}}(\bar{q}_0,\dot{\bar{q}}) &= \textstyle\sum_{i=1}^{k} \bar{\lambda}_i\dot{\bar{q}}^T D^2\bar{d}_i(\bar{q}_0)\dot{\bar{q}} \\ &= \textstyle\sum_{i=1}^{k} \frac{\bar{\lambda}_i}{\|\nabla\tilde{d}(\bar{q}_0)\|}\dot{\bar{q}}^T[D^2\tilde{d}_i(\bar{q}_0)]\dot{\bar{q}} = \dot{\bar{q}}^T\left[\sum_{i=1}^{k}\lambda_i D^2\tilde{d}_i(\bar{q}_0)\right]\dot{\bar{q}}\end{aligned} \qquad \dot{\bar{q}} \in T_{\bar{q}_0}\overline{\mathcal{S}}_i, \tag{8.27}$$

where we substituted $\bar{\lambda}_i = \|\nabla \tilde{d}_i\| \cdot \lambda_i$. The final step is to show that the quadratic form on the right side of Eq. (8.27) is equivalent to $\kappa_{\text{rel}}(q_0, \dot{q}) = \dot{q}^T [\sum_{i=1}^k \lambda_i D^2 d_i(q_0)] \dot{q}$. Using the transformation rule of Eq. (8.25):

$$\begin{aligned}\sum_{i=1}^k \lambda_i D^2 \tilde{d}_i(\bar{q}_0) &= \sum_{i=1}^k \lambda_i \big([Dh(\bar{q}_0)]^T D^2 d_i(h(\bar{q}_0))[Dh(\bar{q}_0)] \\ &\quad + \nabla d_i(h(\bar{q}_0))^T D^2 h(\bar{q}_0)\big) \\ &= [Dh(\bar{q}_0)]^T \left(\sum_{i=1}^k \lambda_i D^2 d_i(h(\bar{q}_0))\right) [Dh(\bar{q}_0)] \\ &\quad + \left(\sum_{i=1}^k \lambda_i \nabla d_i(q_0)\right)^T D^2 h(\bar{q}_0) \\ &= [Dh(\bar{q}_0)]^T \left(\sum_{i=1}^k \lambda_i D^2 d_i(q_0)\right) [Dh(\bar{q}_0)]\end{aligned}$$

since $q_0 = h(\bar{q}_0)$ and $\sum_{i=1}^k \lambda_i \nabla d_i(q_0) = \vec{0}$ at the equilibrium grasp. Substituting for $\sum_{i=1}^k \lambda_i D^2 \tilde{d}_i(\bar{q}_0)$ in Eq. (8.27), then using the transformation rule $\dot{q} = Dh(\bar{q}_0)\dot{\bar{q}}$, gives

$$\begin{aligned}\kappa_{\text{rel}}(\bar{q}_0, \dot{\bar{q}}) &= \dot{\bar{q}}^T [Dh(\bar{q}_0)]^T \left(\sum_{i=1}^k \lambda_i D^2 d_i(q_0)\right) [Dh(\bar{q}_0)] \dot{\bar{q}} \\ &= \sum_{i=1}^k \lambda_i \dot{q}^T D^2 d_i(q_0) \dot{q} \qquad q_0 = h(\bar{q}_0).\end{aligned}$$

Since $Dh(\bar{q}_0)$ maps the subspace $\bar{M}^1_{1\ldots k}(\bar{q}_0)$ to the subspace $M^1_{1\ldots k}(q_0)$ according to Proposition 7.1, it follows from the last equation that the two relative curvature forms are equivalent: $\kappa_{\text{rel}}(q_0, \dot{q}) = \kappa_{\text{rel}}(\bar{q}_0, \dot{\bar{q}})$ for $q_0 = h(\bar{q}_0)$ and $\dot{q} = Dh(\bar{q}_0)\dot{\bar{q}}$. Therefore, $m^2_{q_0} = m^2_{\bar{q}_0}$. □

Exercises

Section 8.1

Exercise 8.1: A rigid object $\mathcal{B}$ is held by $k \le m + 1$ fingers at a frictionless equilibrium grasp ($m = 3$ or 6), such that all k fingers are *essential* for maintaining the grasp. Prove Lemma 8.2, which states that the equilibrium grasp coefficients $\lambda_1, \ldots, \lambda_k$ are positive and uniquely determined up to a common scaling factor at such grasps.

Solution: As noted in the discussion preceding Lemma 8.2, the coefficients $\lambda_1, \ldots, \lambda_k$ are strictly positive at an essential equilibrium grasp. These coefficients are unique up to a common scaling factor iff the finger c-obstacle unit normals, $\hat{\eta}_1, \ldots, \hat{\eta}_k$, are *affinely independent.* That is, $\{\hat{\eta}_i - \hat{\eta}_j\}_{1 \le i \le k, i \ne j}$ are linearly independent for any fixed $j \in \{1, \ldots, k\}$. Hence, it suffices to show that the vectors $\{\hat{\eta}_i - \hat{\eta}_j\}_{i=1,\ldots,k, i \ne j}$ are linearly independent, say for $j=1$, when the fingers are all essential for maintaining the equilibrium grasp. Suppose to the contrary, that the vectors $\{\hat{\eta}_2 - \hat{\eta}_1, \ldots, \hat{\eta}_k - \hat{\eta}_1\}$ are linearly dependent. Let Δ be the set in $\mathbb{R}^{k-1}$, defined as the *convex hull* of the origin and points located at a unit distance along the coordinate axes in this space:

$$\Delta = \left\{ \begin{pmatrix} \sigma_2 \\ \vdots \\ \sigma_k \end{pmatrix} \in \mathbb{R}^{k-1} : \sigma_i \geq 0 \text{ for } i = 2 \ldots k,\ \sigma_2 + \cdots + \sigma_k \leq 1 \right\}.$$

Since $\lambda_1, \ldots, \lambda_k > 0$, the equilibrium grasp coefficients can be normalized such that $0 \leq \lambda_i \leq 1$ for $i = 1 \ldots k$ and $\sum_{i=1}^{k} \lambda_i = 1$. The vector of normalized coefficients $\vec{\lambda} = (\lambda_2, \ldots, \lambda_k)$ thus lies in Δ. The equilibrium grasp equation (8.3) implies that $\vec{\lambda}$ forms a particular solution of the linear system of equations in the variables $(\sigma_2, \ldots, \sigma_k)$:

$$\left[(\hat{n}_2 - \hat{n}_1) \cdots (\hat{n}_k - \hat{n}_1)\right] \vec{\sigma} = -\hat{n}_1 \qquad \vec{\sigma} = (\sigma_2, \ldots, \sigma_k). \tag{8.28}$$

The particular solution $\vec{\lambda} = (\lambda_2, \ldots, \lambda_k)$ is part of an affine subspace of solutions, spanned by $\vec{\lambda} + \vec{\mu}$, where $\left[(\hat{n}_2 - \hat{n}_1) \ldots (\hat{\eta}_k - \hat{\eta}_1)\right] \vec{\mu} = \vec{0}$. The general solution, $\vec{\lambda} + \vec{\mu}$, forms an unbounded affine subspace, while Δ is a bounded set. Hence, Eq. (8.28) has a solution, denoted $\vec{\sigma}^*$, which lies on a boundary facet of Δ. Every facet of Δ is a convex combination of $k - 1$ vertices. If $\vec{\sigma}^*$ lies on a facet of Δ that contains the origin, the jth entry of $\vec{\sigma}^*$ that corresponds to the vertex of Δ *opposite* the facet is zero. In this case, the finger body $\mathcal{O}_j$ is nonessential. If $\vec{\sigma}^*$ lies on the facet of Δ *opposite* the origin, we have that $\sum_{i=2}^{k} \sigma_i^* = 1$, and the finger body $\mathcal{O}_1$ is nonessential. Hence, at least one finger is nonessential, which contradicts our assumption that all fingers are essential. The vectors $\{\hat{\eta}_2 - \hat{\eta}_1, \ldots, \hat{\eta}_k - \hat{\eta}_1\}$ are thus linearly independent, and consequently $\lambda_1, \ldots, \lambda_k$ are unique up to a common scaling factor. ○

Exercise 8.2: A rectangular object $\mathcal{B}$ is held at an equilibrium grasp by four disc fingers, such that the finger contact normals intersect at the object's center (Figure 8.10). While the grasp is nonessential, its set of first-order free motions, $M^1_{1,2,3,4}(q_0)$, spans a linear subspace. Determine the subspace dimension. Can second-order geometric effects provide full object immobilization at this grasp?

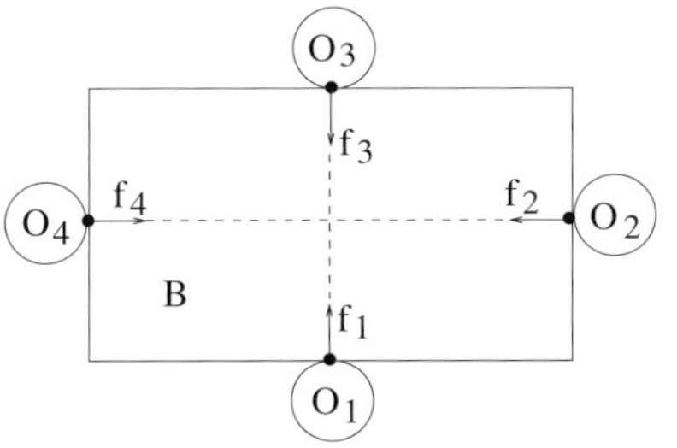

Figure 8.10 A rectangular object held at a concentric four-finger equilibrium grasp.

Exercise 8.3*: The instantaneous change of power, $\dot{w} \cdot \dot{q} + w \cdot \ddot{q}$, can be interpreted as the action of a *double cotangent vector,* $(\dot{q}^T Dw(q), w(q))$ based at $(q, \dot{q})$, on the vector space of double tangent vectors, $(\dot{q}, \ddot{q}) \in \mathbb{R}^m \times \mathbb{R}^m$, based at the same point. Verify that this action does *not* depend on the specific c-space parametrization. (Hint: under a diffeomorphism $q = h(\bar{q})$, tangent vectors transform by the rule: $\dot{q} = Dh\dot{\bar{q}}$, acceleration vectors transform by the rule: $\ddot{q} = [D^2h(\bar{q})](\dot{\bar{q}}, \dot{\bar{q}}) + [Dh(\bar{q})]\ddot{\bar{q}}$ (Lemma 8.6), cotangent vectors transform by the rule: $\bar{w} = Dh^T w$, while *wrench derivatives* transform by the rule: $D\bar{w} = Dh^T Dw Dh + D^2h(\cdot, \cdot)w)$.

Solution: Based on the transformation rules, we must verify that $\dot{q}^T Dw(q)\dot{q} + w(q) \cdot \ddot{q} = \dot{\bar{q}}^T D\bar{w}(\bar{q})\dot{\bar{q}} + \bar{w}(\bar{q}) \cdot \ddot{\bar{q}}$, where $q = h(\bar{q})$:

$$\begin{aligned} \dot{q}^T Dw(q)\dot{q} + w(q) \cdot \ddot{q} &= \dot{\bar{q}}^T Dh^T Dw Dh\dot{\bar{q}} + w^T \left([D^2h(\bar{q})](\dot{\bar{q}}, \dot{\bar{q}}) + [Dh(\bar{q})]\ddot{\bar{q}}\right) \\ &= \dot{\bar{q}}^T \left(Dh^T Dw Dh + D^2h(\cdot, \cdot)w\right)\dot{\bar{q}} + \bar{w}(\bar{q}) \cdot \ddot{\bar{q}} \\ &= \dot{\bar{q}}^T D\bar{w}(\bar{q})\dot{\bar{q}} + \bar{w}(\bar{q}) \cdot \ddot{\bar{q}}. \end{aligned}$$

○

Section 8.2

Exercise 8.4: When $m^1_{q_0} > 0$ and $m^2_{q_0} = 0$, any path $\alpha(t)$ with $\alpha(0) = q_0$, $\dot{\alpha}(0) = \dot{q} \in M^1_{1\ldots k}(q_0)$ and $\ddot{\alpha}(0) = \ddot{q} \in \mathbb{R}^m$ locally penetrates one of the finger c-obstacles that meet at q_0. Verify this fact using the second-order Taylor expansion of the finger c-obstacle distance functions.

Exercise 8.5: Determine the possible values of the second-order mobility index, $m^2_{q_0}$, at a two-finger equilibrium grasp of a 3-D object $\mathcal{B}$.

Exercise 8.6: Can the second-order mobility index, $m^2_{q_0}$, be zero at a two-finger equilibrium grasp of a 3-D object $\mathcal{B}$? This is the only case where *third-order* geometric effects determine the immobilizing grasps. (As discussed in Section 2.4, the ith finger c-obstacle has zero curvature along instantaneous rotation of $\mathcal{B}$ about the contact normal.)

Section 8.4

Exercise 8.7: Consider the finger c-obstacle curvature form specified in Lemma 8.5:

$$\kappa_i(q_0, (0, \omega)) = \frac{(\kappa_{\mathcal{B}_i}\rho_i - 1)(\kappa_{\mathcal{O}_i}\rho_i + 1)}{\kappa_{\mathcal{B}_i} + \kappa_{\mathcal{O}_i}}\omega^2 = \frac{(\rho_i - r_{\mathcal{B}_i})(\rho_i + r_{\mathcal{O}_i})}{r_{\mathcal{B}_i} + r_{\mathcal{O}_i}}\omega^2.$$

As ρ_i approaches $\pm\infty$ one obtains

$$\lim_{\rho_i \to \pm\infty} \kappa_i(q_0, (0, \omega)) = \lim_{\rho_i \to \pm\infty} \frac{\kappa_{\mathcal{B}_i}\kappa_{\mathcal{O}_i}}{\kappa_{\mathcal{B}_i} + \kappa_{\mathcal{O}_i}}\rho_i^2\omega^2 = \lim_{\rho_i \to \pm\infty} \frac{1}{r_{\mathcal{B}_i} + r_{\mathcal{O}_i}}\rho_i^2\omega^2 = +\infty.$$

Does this means that $\kappa_i(q_0, (0, \omega))$ attains arbitrarily large values as $\rho_i \to \pm\infty$? (Hint: instantaneous rotation of $\mathcal{B}$ about points at infinity represents instantaneous translation of $\mathcal{B}$. The curvature along instantaneous translation at q_0 is $\kappa_i(q_0,(v,0)) = \kappa_{\mathcal{B}_i}\kappa_{\mathcal{O}_i}/(\kappa_{\mathcal{B}_i} + \kappa_{\mathcal{O}_i})$.)

Exercise 8.8: Characterize the sign of the finger c-obstacle curvature form, $\kappa_i(q_0,\rho_i)$, as a function of ρ_i when a flat edge of $\mathcal{B}$ touches a stationary finger body $\mathcal{O}_i$.

Solution: The finger c-obstacle curvature form is negative for all values of ρ_i, except at $\rho_i = \pm\infty$ where it is zero. The matrix associated with $\kappa_i(q_0,\rho_i)$ has an eigenvector with zero eigenvalue along instantaneous translation of $\mathcal{B}$ parallel to the object's flat edge. ○

Exercise 8.9: A flat edge of $\mathcal{B}$ touches a stationary finger body $\mathcal{O}_i$. Can $\mathcal{B}$ break contact along a local motion that starts with instantaneous rotation about points located in the interval where $\kappa_i(q_0,\rho_i)$ is negative? (Hint: consider the role of $\ddot{q}$ in the second-order Taylor expansion of the c-obstacle distance function $d_i(q)$.)

Exercise 8.10: Repeat the previous exercise for the case where a a vertex of a polygonal object $\mathcal{B}$ touches a stationary finger body $\mathcal{O}_i$.

Exercise 8.11: Prove case (i) of the graphical technique concerning the second-order mobility of two-finger grasps.

Solution: case (i) assumes that $\mathcal{B}$, $\mathcal{O}_1$ and $\mathcal{O}_2$ are convex at the contacts. Substituting $\rho_1 = \pm\infty$ and $\rho_2 = \pm\infty$ in Eq. (8.12) gives:

$$\kappa_{\text{rel}}(q_0, \pm\infty, \pm\infty) = \frac{\kappa_{\mathcal{B}_1}\kappa_{\mathcal{O}_1}}{\kappa_{\mathcal{O}_1} + \kappa_{\mathcal{B}_1}}(\rho_1|_{\pm\infty})^2 + \frac{\kappa_{\mathcal{B}_2}\kappa_{\mathcal{O}_2}}{\kappa_{\mathcal{O}_2} + \kappa_{\mathcal{B}_2}}(\rho_2|_{\pm\infty})^2 \geq 0, \quad (8.29)$$

since $\kappa_{\mathcal{B}_i} \geq 0$ and and $\kappa_{\mathcal{O}_i} \geq 0$ when $\mathcal{B}$ and $\mathcal{O}_i$ are convex at the contact ($i = 1,2$). It follows that $m^2_{q_0} \geq 1$, since we have found a test direction $\dot{q}$ for which $\kappa_{\text{rel}}(q_0,\rho_1,\rho_2)$ is nonnegative. If I_- and J_- overlap, both summands in Eq. (8.12) are negative for any pure rotation whose axis lies in the overlap region. It follows that $\kappa_{\text{rel}}(q_0,\rho_1,\rho_2)$ attains both negative and positive signs and, thus, $m^2_{q_0} = 1$. Figure 8.6(a) illustrates this case.

Next assume that $\mathcal{B}$ has identical curvature at the grasp points, and $I_- \bigcap J_- = \emptyset$. Suppose, to the contrary, that $\kappa_{\text{rel}}(q_0,\rho_1,\rho_2)$ is negative for some choice of instantaneous rotation point on l. The rotation point must lie in the region where I_- and J_+ overlap or where I_+ and J_- overlap. Symmetry implies that such a rotation must occur at *two* symmetric locations. Substituting $\rho_i = 0$ in $\kappa_i(q_0,\rho_i)$ gives that $\kappa_i(q_0,0) < 0$ for $i = 1,2$. Each contact point, therefore, lies in its respective negative interval. Since $I_- \bigcap J_- = \emptyset$, there must be an interval between the contact points where the positive intervals, I_+ and J_+, overlap. Since both summands are positive in the overlap interval, $\kappa_{\text{rel}}(q_0,\rho_1,\rho_2)$ is positive in this interval. Thus, $\kappa_{\text{rel}}(q_0,\rho_1(\rho),\rho_2(\rho))$, a *quadratic* polynomial in ρ, crosses zero at *four* distinct points – an impossibility. We conclude that $\kappa_{\text{rel}}(\rho_1,\rho_2)$ is never negative, and $m^2_{q_0} = 2$. Figure 8.6(b) illustrates this case. ○

Exercise 8.12: Prove case (ii) of the graphical technique concerning the second-order mobility of two-finger equilibrium grasps.

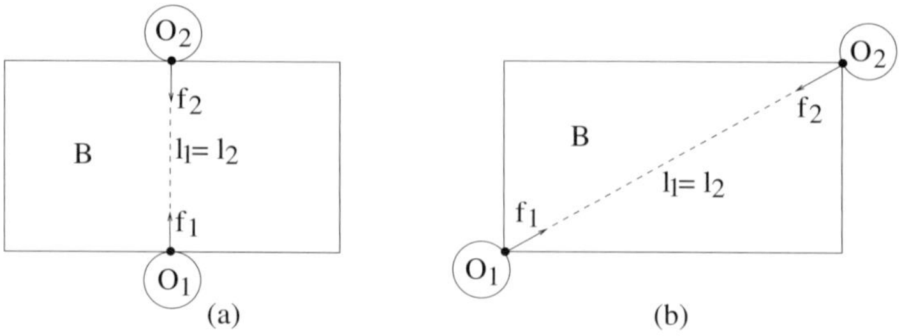

Figure 8.11 A rectangular object held at a two-finger equilibrium grasp along (a) parallel edges, and (b) opposing vertices.

Solution: Case (ii) assumes that one of the contacting bodies, say $\mathcal{B}$, is concave at the two contacts. In this case, $\kappa_{\mathcal{B}_i} < 0$ for $i = 1,2$ as illustrated in Figure 8.7. It must be true that $\left|\kappa_{\mathcal{B}_i}\right| \leq \kappa_{\mathcal{O}_i}$ for $i = 1,2$; otherwise, $\mathcal{B}$ and $\mathcal{O}_i$ would interpenetrate at the respective contact. It follows that both summands in Eq. (8.29) are negative at $\rho_i = \pm\infty$. Hence, $\kappa_{\text{rel}}(q_0, \rho_1, \rho_2)$ is negative along instantaneous translation, and, consequently, $0 \leq m^2_{q_0} \leq 1$. It is exactly unity when I_+ and J_+ overlap, since in this case there also exist instantaneous rotations for which $\kappa_{\text{rel}}(q_0, \rho_1, \rho_2)$ is positive. This case is illustrated in Figure 8.7(a). When $\mathcal{B}$ has identical curvature at the contacts such that $I_+ \cap J_+ = \emptyset$, a symmetry argument indicates that $m^2_{q_0} = 0$. This case is illustrated in Figure 8.7(b). ○

Exercise 8.13: Prove the first part of case (iii) of the graphical technique concerning the second-order mobility of two-finger grasps, that $0 \leq m^2_{q_0} \leq 1$ when the finger bodies $\mathcal{O}_1$ and $\mathcal{O}_2$ are concave at the contacts.

Exercise 8.14: Prove the second part of case (iii) of the graphical technique concerning the second-order mobility of two-finger grasps, that $m^2_{q_0} = 0$ if additionally I_+ and J_+ do not overlap and $\mathcal{B}$ has identical curvature at the contacts.

Exercise 8.15: A rectangular object $\mathcal{B}$ is held at a two-finger equilibrium grasp along parallel edges, as shown in Figure 8.11(a). Determine the second-order mobility index, $m^2_{q_0}$, of this grasp.

Exercise 8.16: Can the object $\mathcal{B}$ escape the fingers in the grasp of Figure 8.11(a), along a motion that starts with some combination of instantaneous translation parallel to the object edges, and *non-zero* instantaneous rotation about a point on the line $l_1 = l_2$?

Exercise 8.17: A rectangular object $\mathcal{B}$ is held by two disc fingers placed at opposing vertices, as shown in Figure 8.11(b). Determine the second-order mobility index, $m^2_{q_0}$, of this grasp.

Exercise 8.18: Sketch an argument showing that every polygonal object $\mathcal{B}$ can be immobilized by two suitably concave fingers placed at two opposing vertices of the object.

Exercise 8.19: A polygonal object $\mathcal{B}$ is held in a *compressive* three-finger equilibrium grasp along nonparallel edges (the finger forces are inward pointing at the contacts). Use the graphical technique to prove that $\mathcal{B}$ is immobilized at such grasps.

Exercise 8.20: A polygonal object $\mathcal{B}$ is held in an *expansive* three-finger equilibrium grasp along nonparallel edges (the finger forces are outward pointing at the contacts). Use the graphical technique to prove that $\mathcal{B}$ is *not* immobilized at such grasps.

References

[1] E. Rimon and J. W. Burdick, "A configuration space analysis of bodies in contact – (I): 1st order mobility," *Mechanisms and Machine Theory*, vol. 30, no. 6, pp. 897–912, 1995.

[2] E. Rimon and J. W. Burdick, "A configuration space analysis of bodies in contact – (II): 2nd order mobility," *Mechanisms and Machine Theory*, vol. 30, no. 6, pp. 913–928, 1995.

[3] E. Rimon and J. W. Burdick, "Mobility of bodies in contact – I: A 2nd order mobility index for multiple-finger grasps," *IEEE Transactions on Robotics and Automation*, vol. 14, no. 5, pp. 696–708, 1998.

[4] R. Featherstone, "Dynamics of rigid body systems with multiple concurrent contacts," in *Robotics Research: The Third International Symposium*, O. D. Faugeras and G. Giralt, Eds., MIT Press, pp. 189–196, 1986.

[5] M. T. Mason, *Mechanics of Robotic Manipulation*. MIT Press, 2001.

9 Minimal Immobilizing Grasps

What is the minimal number of fingers needed to ensure that a robot hand can securely grasp a wide range of objects? The answer to this question has practical implications for robot hand design. The answer to this question may depend upon several factors:

- the desired type of grasp security: immobilization, wrench resistance, or caging;
- the type of object surface: smooth, piecewise smooth, or polyhedral surface;
- the type of contact between the fingers and object: frictionless or frictional contacts;
- the type of fingertip geometry: sharp tipped, convex, or concave fingertips.

This chapter analyzes the minimum number of fingers required to achieve *immobilizing grasps* while assuming frictionless contacts between the fingers and object. The chapter makes the reasonable assumption that the robot hand can realize any desired set of finger contacts on the object's surface. The finger bodies can thus be placed at any desired set of contacts on the object's surface. The following *minimal grasps* will be explored in this chapter.

DEFINITION 9.1 *A* **minimal immobilizing grasp** *deploys the minimum number of fingers that are necessary to hold all rigid objects in an immobilizing and hence secure grasp.*

One should be aware that objects with rotational or translational symmetry cannot be fully immobilized via frictionless contacts (see Bibliographical Notes). However, the minimal immobilizing grasps are required to restrain such objects up to their symmetry.

In analyzing the minimal immobilizing grasps, one can choose the type of immobilization which is to be achieved by the grasp. Section 9.1 analyzes the minimum number of fingers needed to realize first-order immobilizing grasps of 2-D objects. We will see that four fingers form the minimal first-order immobilizing grasps of 2-D objects. However, we have seen that curvature effects can be used to immobilize objects with fewer numbers of fingers. Section 9.2 introduces the *inscribed disc,* which will be used in Sections 9.3 and 9.4 to construct minimal second-order immobilizing grasps. We will see that three convex fingers, or two suitably concave fingers, form the minimal second-order immobilizing grasps of 2-D objects. The basic principles behind immobilization of 2-D objects can be extended to 3-D objects. Section 9.5 discusses the minimal immobilizing grasps of polyhedral objects. Using the *inscribed ball* of these objects, we will see that fours fingers with sufficiently flat fingertips suffice to immobilize all polyhedral objects. The chapter also contains two appendices, which provide proof details.

9.1 Minimal First-Order Immobilizing Grasps

This section considers the minimal first-order immobilizing grasps of 2-D objects. First-order immobilization requires that each finger c-obstacle, $\mathcal{CO}_i$, will possess a well-defined tangent plane at the grasp configuration, q_0. To ensure this property, we will assume that at least one of the contacting bodies, the object $\mathcal{B}$ or the finger body $\mathcal{O}_i$, is locally smooth at the contact point x_i. The first-order mobility index of an essential k-finger equilibrium grasp is given by $m^1_{q_0} = \max\{m - k + 1, 0\}$, where $m = 3$ in 2-D grasps (see Chapter 7). Hence, $m^1_{q_0} = 0$, and $\mathcal{B}$ is immobilized by first-order effects at any essential equilibrium grasp involving at least four fingers. Moreover, first-order immobilization makes no special requirements on the shape of the fingertips, hence we might as well assume *sharp-tipped* fingers. Since $m^1_{q_0} > 0$ for two- and three-finger grasps, four fingers form the minimal first-order immobilizing grasps, if we can establish that every 2-D object can be held at an essential four-finger equilibrium grasp.

The existence of essential four-finger equilibrium grasps can be established using the *contact splitting technique* of Chapter 5. Recall that every 2-D object $\mathcal{B}$ has at least two pairs of antipodal points on its boundary, each forming a two-finger equilibrium grasp. The technique works for objects without circular symmetry at the antipodal points, as specified in the following definition.

DEFINITION 9.2 *A 2-D object $\mathcal{B}$ has* **no circular symmetry** *when at least one pair of its antipodal points has nonconcentric circles of curvature.*

To illustrate the contact splitting technique (and refresh the notion of circle of curvature), consider the antipodal points x_1 and x_2 of the ellipse depicted in Figure 9.1. To construct an essential four-finger equilibrium grasp and, hence, an immobilizing grasp of the ellipse, place two finger contacts on both sides of x_1, at x_{11} and x_{12}. Similarly, place two finger contacts on both sides of x_2, at x_{21} and x_{22}. Each of these contact pairs is generated by *local splitting* of an antipodal point into two contacts. For the ellipse depicted in Figure 9.1, any local splitting of the antipodal points generates an essential

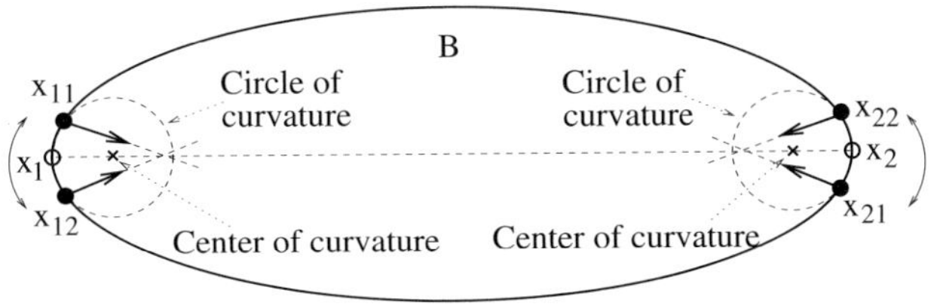

Figure 9.1 An essential four-finger equilibrium grasp constructed by local splitting of two antipodal points, x_1 and x_2, into two pairs of finger contacts, (x_{11}, x_{12}) and (x_{21}, x_{22}).

four-finger equilibrium grasp. For more general objects, the technique works well for objects that are locally smooth at the antipodal contacts, as stated in the following proposition.

PROPOSITION 9.1 (**existence of four-finger grasp**) *Every 2-D object $\mathcal{B}$ that has no circular symmetry can be held by an* **essential four-finger equilibrium grasp.**

Proof sketch: Let x_1 and x_2 be antipodal points on the object boundary, denoted $\text{bdy}(\mathcal{B})$. First consider the local splitting of x_1. Let $\alpha(s)$ parametrize the object boundary by arclength, such that $\alpha(0)=x_1$. Thus, $\alpha'(s)$ is the unit tangent to $\text{bdy}(\mathcal{B})$ at $\alpha(s)$. Denote by $t(\alpha(s))$ and $n(\alpha(s))$ the unit tangent and outward unit normal to $\text{bdy}(\mathcal{B})$ at $\alpha(s)$.[1] The object *curvature* at $\alpha(s)$ is given by the scalar $\kappa_{\mathcal{B}}(\alpha(s))=\alpha''(s)\cdot n(\alpha(s))$, while the object *radius of curvature* is given by $r_{\mathcal{B}}(\alpha(s))=1/\kappa_{\mathcal{B}}(\alpha(s))$. Let $t_1=t(\alpha(0))$ and $n_1=n(\alpha(0))$ denote the object's tangent and normal at x_1, and let $\kappa_{\mathcal{B}_1}$ denote the object curvature at x_1. The Taylor expansions of $\alpha(s)$ and $n(\alpha(s))$ are given by (see Exercises):

$$\alpha(s)=x_1+st_1+\tfrac{1}{2}s^2\kappa_{\mathcal{B}_1}n_1+o(s^3)$$

and

$$n(\alpha(s))=n_1+s\kappa_{\mathcal{B}_1}t_1+\tfrac{1}{2}s^2\left(\kappa'_{\mathcal{B}_1}t_1+(\kappa_{\mathcal{B}_1})^2n_1\right)+o(s^3),$$

where $\kappa'_{\mathcal{B}_1}=\frac{d}{ds}\big|_{s=0}\kappa_{\mathcal{B}}(\alpha(s))$ is the object's *curvature derivative* at x_1. Note that $\kappa'_{\mathcal{B}_1}$ attains finite values along a smooth object boundary.

According to Lemma 5.2.4, the object $\mathcal{B}$ is held in an *essential* four-finger equilibrium grasp when (i) the finger forces positively span the origin of $\mathbb{R}^2$; and (ii) each pair of finger forces generates opposite moments about the intersection point of the lines underlying the contact forces of the remaining two finger contacts.

Condition (i): Any local splitting of x_1 into two points located on both sides of x_1 corresponds to $s=-\epsilon_1$ and $s=\epsilon_2$, where ϵ_1 and ϵ_2 are small positive parameters. Let $x_{11}=\alpha(-\epsilon_1)$, $x_{12}=\alpha(\epsilon_2)$, $n_{11}=n(\alpha(-\epsilon_1))$ and $n_{12}=n(\alpha(\epsilon_2))$. The Taylor expansion of $n(\alpha(s))$ satisfies the relation $n(\alpha(s))\cdot t_1=s\kappa_{\mathcal{B}_1}+o(s^2)$. Hence, $n_{11}\cdot t_1$ and $n_{12}\cdot t_1$ possess *opposite signs* for sufficiently small ϵ_1 and ϵ_2. It follows that n_{11} points into the quadrant spanned by $(-\epsilon_1\kappa_{\mathcal{B}_1}t_1,n_1)$, while n_{12} points into the quadrant spanned by $(\epsilon_2\kappa_{\mathcal{B}_1}t_1,n_1)$. Similarly, let t_2 and n_2 denote the object's unit tangent and outward unit normal at the antipodal point x_2, and let $\kappa_{\mathcal{B}_2}$ denote the object's curvature at x_2. For any sufficiently local splitting of x_2, using ϵ_3 and ϵ_4, the contact normal n_{21} points into the quadrant spanned by $(-\epsilon_3\kappa_{\mathcal{B}_2}t_2,n_2)$, while the contact normal n_{22} points into the quadrant $(\epsilon_4\kappa_{\mathcal{B}_2}t_2,n_2)$. Since n_1 and n_2 have opposite directions, the four contact normals positively span the origin of $\mathbb{R}^2$, as required by Condition (i).

Condition (ii): Let l denote the line that passes through x_1 and x_2. Let l_{11} and l_{12} denote the lines underlying the contact normals n_{11} and n_{12} (Figure 9.2). Let z_1 denote the object's *center of curvature* at x_1 (located at a distance of $r_{\mathcal{B}_1}$ from x_1 on l), and

[1] The finger forces act along the object's *inward* contact normals. However, for consistency with the standard definition of curvature, let us use the object's *outward* contact normals.

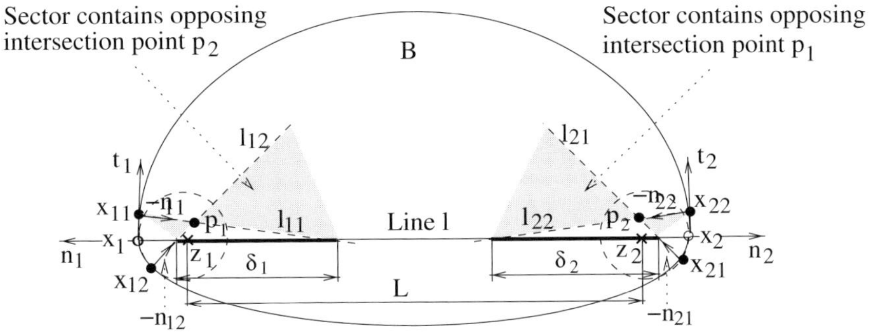

Figure 9.2 Local splitting of antipodal points x_1 and x_2 at which $\mathcal{B}$ has non-zero *curvature derivative*. The contact normal lines, (l_{11}, l_{12}) and (l_{21}, l_{22}), intersect at points p_1 and p_2, which lie in local neighborhoods of the object centers of curvature z_1 and z_2. The points x_{11}, x_{12}, x_{21}, and x_{22} support an essential four-finger equilibrium grasp of $\mathcal{B}$.

let p_1 denote the intersection point of l_{11} and l_{12} (Figure 9.2). In general, p_1 lies in an arbitrarily small neighborhood of z_1 for a sufficiently local splitting of x_1. Using similar notation at the antipodal point x_2, the lines l_{21} and l_{22} intersect at a point p_2 that lies in an arbitrarily small neighborhood of z_2. Since the object $\mathcal{B}$ does not possess circular symmetry at x_1 and x_2, the centers of curvature z_1 and z_2 are separated by some positive distance L (Figure 9.2).

Consider the sector of directions based at p_1 and spanned by the contact normals n_{11} and n_{12}. This sector contains the normal direction n_1 for any sufficiently local splitting of x_1. Any point in the interior of this sector has the property that the contacts normals n_{11} and n_{12} generate *opposite* moments about this point. This property also holds for any point in the negated sector based at p_1 and spanned by $-n_{11}$ and $-n_{12}$. If the union of the two sectors contains the opposite center of curvature z_2 in its interior, n_{11} and n_{12} generate *opposite* moments about any point in a sufficiently small neighborhood of z_2. Since l_{21} and l_{22} intersect at p_2, which lies arbitrarily close to z_2, the same property holds at p_2 for sufficiently local splitting of x_2. Similarly, n_{21} and n_{22} generate *opposite* moments about p_1, for sufficiently local splitting of x_1 and x_2 (Figure 9.2).

To complete the proof of Condition (ii), we have to show that the object's centers of curvature, z_1 and z_2, lie within the respective double sectors based at p_1 and p_2. A simple computation can determine which portion of the line l lies *outside* the double sector based at p_i for $i = 1, 2$ (see Exercises). As depicted in Figure 9.2, this portion of l forms a line segment of length δ_i. As the point p_i approaches the object's center of curvature, z_i, the length δ_i approaches zero. All sufficiently local splittings of x_1 and x_2 thus ensure that $\delta_i < L$ for $i = 1, 2$, which completes the proof of Condition (ii). □

Extension to piecewise smooth objects: The contact splitting technique can also be used to construct essential four-finger equilibrium grasps of piecewise smooth objects, such as polygons. Every polygon has at least one pair of antipodal points that do not lie on parallel edges. For example, the polygon $\mathcal{B}$ depicted in Figure 9.3

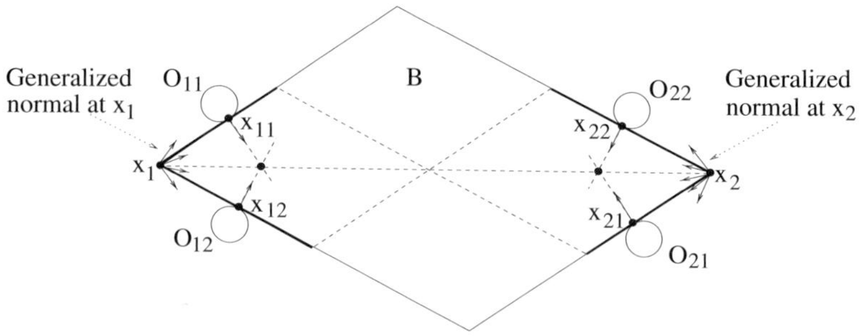

Figure 9.3 Any placement of four fingers along the thick edge segments adjacent to the antipodal vertices x_1 and x_2 forms an essential four-finger equilibrium grasp of the polygon.

can be held at the vertices x_1 and x_2, whose *generalized inward normals* contain opposing directions.[2] Using contact splitting, one can determine entire edge segments that support an essential four-finger equilibrium grasp of the polygon, as shown in Figure 9.3. Consequently, we have the following result concerning the minimal first-order immobilizing grasps.

THEOREM 9.2 **(minimal four-finger grasp)** *The minimal first-order immobilizing grasp of all piecewise smooth planar objects that have no circular symmetry requires* **four fingers.**

Let us end this section with a remark on the robustness of the first-order immobilizing grasps. Theorem 5.18 of Chapter 5 describes the dimension of the set of frictionless equilibrium grasps, $\mathcal{E}$, in *contact space.* In the case of four-finger grasps, contact space can be thought of as $\mathbb{R}^4$. The set $\mathcal{E}$ forms a four-dimensional set with nonempty interior in this space. Hence, starting at any four-finger grasp in the interior of $\mathcal{E}$, each of the four finger contacts can independently move by a small amount along the object boundary while maintaining an essential four-finger equilibrium grasp of the object $\mathcal{B}$. The minimal first-order immobilizing grasps are thus *robust* with respect to small finger placement errors.

9.2 The Maximal Inscribed Disc

This section defines the inscribed disc and discusses its relevant properties, which will help us construct minimal second-order immobilizing grasps. Any rigid object $\mathcal{B}$ forms a compact set with nonempty interior and piecewise smooth boundary. Given a point y inside the object $\mathcal{B}$, consider the minimum distance of y from the object boundary:

$$\text{dst}(y, \text{bdy}(\mathcal{B})) = \min_{x \in \text{bdy}(\mathcal{B})} \{\|x - y\|\} \qquad y \in \mathcal{B}.$$

[2] The *generalized inward normal* at a non-smooth vertex is the convex hull of the inward unit normals to the edges of $\mathcal{B}$ that meet at the vertex.

The function $\mathsf{dst}(y, \mathsf{bdy}(\mathcal{B}))$ is Lipschitz continuous (see Appendix A). It forms a continuous function on a compact set and, hence, attains at least one maximum inside the object $\mathcal{B}$. Since $\mathsf{dst}(y, \mathsf{bdy}(\mathcal{B}))$ is identically zero on $\mathsf{bdy}(\mathcal{B})$ and strictly positive in the interior of $\mathcal{B}$, its local maxima must occur in the interior of $\mathcal{B}$. Intuitively, every local maximum of $\mathsf{dst}(y, \mathsf{bdy}(\mathcal{B}))$ determines a *locally maximal inscribed disc,* which is defined as follows.

DEFINITION 9.3 **(inscribed disc)** *A* **locally maximal inscribed disc** *of a 2-D object $\mathcal{B}$, denoted $\mathcal{D}(y)$, is a disc centered at y that lies inside $\mathcal{B}$, whose radius is maximal among all discs contained in $\mathcal{B}$ with center in a neighborhood of y.*

The *inscribed disc* has several properties that are useful for constructing equilibrium grasps. The first is that it touches the object's boundary at a small number of points. See Appendix I for a proof of the following lemma.

LEMMA 9.3 *The set of points where an inscribed disc $\mathcal{D}(y)$ touches the object boundary, denoted $\mathcal{I}(y)$, generically contains* **two or three** *isolated points.*

The set $\mathcal{I}(y)$ contains higher number of points when the object $\mathcal{B}$ possesses circular symmetry. A regular polygon has a discrete circular symmetry (Figure 9.5), while an object whose boundary contains concentric circular segments has a continuous circular symmetry. In all other cases, $\mathcal{I}(y)$ contains two or three points, as illustrated in the following example.

Example: The ellipse $\mathcal{B}$ depicted in Figure 9.4(a) has an inscribed disc that touches its boundary at two antipodal points. More generally, any convex object has a unique inscribed disc, except when the object's boundary contains opposing parallel edges (see Exercises). Figure 9.4(b) shows that a non-convex object can have several inscribed discs. ○

The inscribed disc has the useful property that it automatically determines an equilibrium grasp of the object $\mathcal{B}$. To explain this property, recall that the *convex hull* of a set $\mathcal{X}$, denoted $\mathrm{co}(\mathcal{X})$, is the smallest convex set containing this set. See Appendix I for the proof of the following lemma.

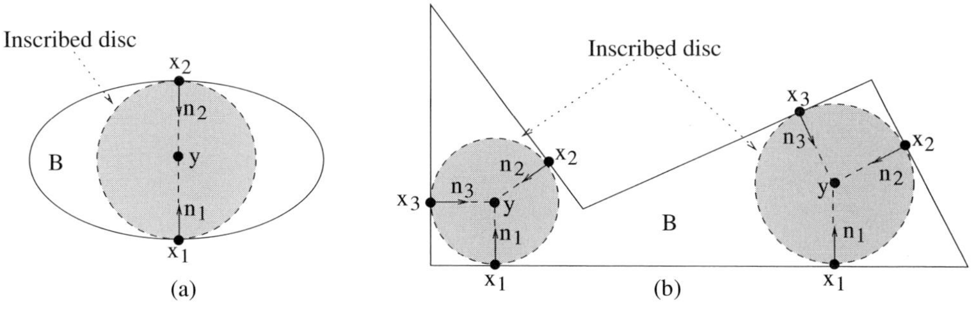

Figure 9.4 The inscribed disc generically touches the object's boundary at two or three points: (a) Any convex object without parallel edges has a unique inscribed disc. (b) Non-convex objects can have several inscribed discs.

LEMMA 9.4 **(convex hull property)** *Let $\mathcal{D}(y)$ be an inscribed disc of an object $\mathcal{B}$. The inward unit normals at the points $x \in \mathcal{I}(y)$ point toward the point y, and when placed at a common origin, the convex hull of these normals contains the origin: $\vec{0} \in \text{co}\{n(x) : x \in \mathcal{I}(y)\}$.*

If $\mathcal{B}$ has a piecewise smooth boundary and the inscribed disc touches its boundary at a non-smooth point, $x \in \mathcal{I}(y)$, the normal $n(x)$ should be taken from the *generalized inward normal* at this point. To show that the inscribed disc determines a feasible equilibrium grasp of $\mathcal{B}$, we will need *Carathéodory's theorem* (see Section 5.4).

Carathéodory's theorem: *If a point p_0 lies in the convex hull of a set of points $\mathcal{X}$ in $\mathbb{R}^n$, p_0 lies in the convex hull of at most* $\boldsymbol{n+1}$ **points** *from this set.*

Based on Carathéodory's theorem, the inscribed disc can be used to generate two- or three-finger equilibrium grasps of any 2-D object, as stated in the following proposition.

PROPOSITION 9.5 **(inscribed disc grasps)** *Every inscribed disc, $\mathcal{D}(y)$, determines a feasible* **two- or three-finger equilibrium grasp** *of the object $\mathcal{B}$.*

Proof: Let an inscribed disc $\mathcal{D}(y)$ touch bdy($\mathcal{B}$) at the points of $\mathcal{I}(x)$. Based on Lemma 9.4, the origin of $\mathbb{R}^2$ lies in the convex hull of $\mathcal{B}$'s inward unit normals, $n(x)$ for $x \in \mathcal{I}(y)$. If $\mathcal{I}(y)$ contains infinite number of points, the origin lies in the convex hull of two or three unit normals, according to Carathéodory's theorem. Hence, there exist two or three points in $\mathcal{I}(y)$, such that $\mathcal{B}$'s inward unit normals at these points satisfy one of the two equations: $\lambda_1 n(x_1) + \lambda_2 n(x_2) = \vec{0}$ or $\lambda_1 n(x_1) + \lambda_2 n(x_2) + \lambda_3 n(x_3) = \vec{0}$, where $0 \le \lambda_1, \ldots, \lambda_k \le 1$ and $\sum_{i=1}^{k} \lambda_i = 1$. Application of normal finger forces of magnitudes $\lambda_1, \ldots, \lambda_k$ at these points will generate zero net wrench on $\mathcal{B}$, as required for an equilibrium grasp. □

Example: The inscribed disc of the regular five-sided polygon depicted in Figure 9.5 touches its boundary at five points. Any three of the five points whose inward

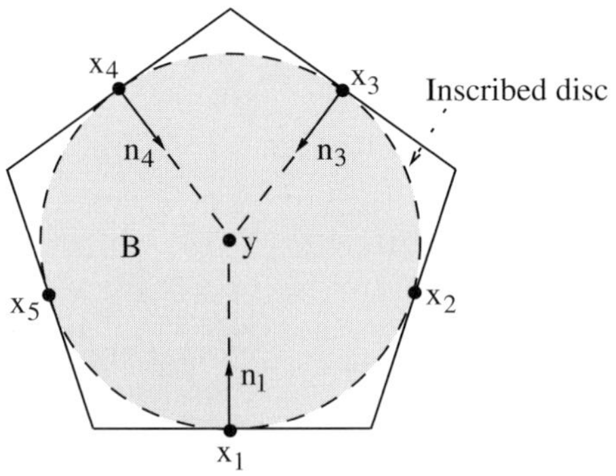

Figure 9.5 The regular five-sided polygon can be held by several three-finger equilibrium grasps determined by its inscribed disc.

unit normals positively span the origin of $\mathbb{R}^2$ form an equilibrium grasp of the polygon. ○

The construction of minimal second-order immobilizing grasps will use the following relation between the inscribed disc radius and $\mathcal{B}$'s radius of curvature at the points $x \in \mathcal{I}(y)$.

LEMMA 9.6 **(inscribed disc radius)** *Let an inscribed disc of $\mathcal{B}$, $\mathcal{D}(y)$, touch the object boundary at a locally convex point, $x \in \mathcal{I}(y)$. The inscribed disc radius,* $\mathsf{dst}(y, \mathsf{bdy}(\mathcal{B}))$*, forms a* **lower bound** *on the object's radius of curvature, $r_{\mathcal{B}}(x)$, at this point:*

$$r_{\mathcal{B}}(x) \geq \mathsf{dst}(y, \mathsf{bdy}(\mathcal{B})) \qquad x \in \mathcal{I}(y).$$

The lower bound is strict when $\mathcal{B}$'s center of curvature at x does not lie at the inscribed disc center y (the generic case).

Lemma 9.6 is based on the fact that the inscribed disc lies entirely inside the object $\mathcal{B}$, and its proof is left as an exercise.

9.3 Minimal Second-Order Immobilizing Grasps

The next two sections study the minimal second-order immobilizing grasps of 2-D objects. This section considers smooth objects. The next section will consider polygons that form an important type of piecewise smooth objects. The first part of this section shows that three convex finger bodies can immobilize all smooth 2-D objects. The second part shows that two sufficiently concave fingers can immobilize all smooth 2-D objects.

Recall the graphical representation of the ith finger c-obstacle curvature form, $\kappa_i(q_0, \dot{q})$, from Chapter 8. This technique depicts the object's *first order roll–slide motions* with respect to a stationary finger body $\mathcal{O}_i$ as instantaneous rotations of $\mathcal{B}$ about all points on the extended contact normal line, l_i. The points of l_i are parametrized by their distance ρ_i from the contact point x_i, such that $\rho_i > 0$ on $\mathcal{B}$'s side of the contact and $\rho_i < 0$ on $\mathcal{O}_i$'s side of the contact. Using this parametrization, Lemma 8.5 specifies the following formula for $\kappa_i(q_0, \dot{q})$.

Finger c-obstacle curvature form: *Let $\mathcal{B}$'s frame origin be located at a distance ρ_i from x_i along the line l_i. Let $\mathcal{S}_i$ be the ith finger c-obstacle boundary, expressed in the c-space coordinates associated with this object frame assignment. The curvature of $\mathcal{S}_i$ along instantaneous rotation about $\mathcal{B}$'s frame origin, $\dot{q} = (0, \omega)$, is given by*

$$\kappa_i(q_0, (0, \omega)) = \frac{(\rho_i - r_{\mathcal{B}_i})(\rho_i + r_{\mathcal{O}_i})}{r_{\mathcal{B}_i} + r_{\mathcal{O}_i}} \omega^2 \qquad -\infty \leq \rho_i \leq \infty, \tag{9.1}$$

where $r_{\mathcal{B}_i}$ and $r_{\mathcal{O}_i}$ are the radii of curvature of $\mathcal{B}$ and $\mathcal{O}_i$ at x_i.

When a rigid object $\mathcal{B}$ is held by rigid finger bodies $\mathcal{O}_1, \ldots, \mathcal{O}_k$ at a frictionless equilibrium grasp (which is necessary for immobilization), the object is *immobilized* by curvature effects when the second-order mobility index of the grasp, $m^2_{q_0}$, is zero. For $m^2_{q_0}$ to be zero, the relative c-space curvature, $\kappa_{\text{rel}}(q_0,\dot{q}) = \lambda_1\kappa_1(q_0,\dot{q}) + \cdots + \lambda_k\kappa_k(q_0,\dot{q})$, must satisfy the condition:

$$\kappa_{\text{rel}}(q_0,\dot{q}) < 0 \qquad \dot{q} \in M^1_{1\ldots k}(q_0),$$

where $M^1_{1\ldots k}(q_0)$ is the linear subspace of first-order free motions at the k-finger grasp. Let us use the relative c-space curvature to determine the minimal second-order immobilizing grasps of smooth 2-D objects.

9.3.1 Minimal Immobilization Using Three Convex Fingers

The inscribed disc of a 2-D object $\mathcal{B}$ generically touches its boundary at two or three points (Lemma 9.3). When the inscribed disc touches the object boundary at three points, these points automatically determine an immobilizing grasp, as stated in the following proposition.

PROPOSITION 9.7 (**three-finger immobilization, case I**) *Let $\mathcal{D}(y)$ be an inscribed disc of a rigid object $\mathcal{B}$. If $\mathcal{D}(y)$ touches the object boundary at three points such that $\mathcal{B}$'s centers of curvature at these points do not all lie at y (the generic case), the object $\mathcal{B}$ is* **immobilized** *by placing three convex finger bodies of any shape at these points.*

Proof: Let $x_1, x_2, x_3 \in \mathcal{I}(y)$ be the points where the inscribed disc $\mathcal{D}(y)$ touches the object boundary. Placing three fingers at these points forms a feasible equilibrium grasp according to Proposition 9.5. We may assume that all three fingers are essential for maintaining the equilibrium grasp; otherwise, two of the three fingers must form an antipodal grasp, and this case is considered below. The subspace of first-order free motions, $M^1_{1,2,3}(q_0)$, is one-dimensional and consists of instantaneous rotations of $\mathcal{B}$ about the inscribed disc center y. For $m^2_{q_0}$ to be zero and thus ensure immobilization of $\mathcal{B}$, we must verify that

$$\kappa_{\text{rel}}(q_0,\dot{q}) = \lambda_1\kappa_1(q_0,\dot{q}) + \lambda_2\kappa_2(q_0,\dot{q}) + \lambda_3\kappa_3(q_0,\dot{q}) < 0 \qquad \dot{q} \in M^1_{1,2,3}(q_0), \quad (9.2)$$

where $\kappa_i(q_0,\dot{q})$ is the ith finger c-obstacle curvature form at q_0, and the coefficients $\lambda_1, \lambda_2, \lambda_3$ are determined by the equilibrium grasp condition.

First consider the case where the object boundary is locally convex at the three contacts (Figure 9.6(a)). In this case, $r_{\mathcal{B}_i} \geq 0$ for $i = 1,2,3$. According to Lemma 9.6, $r_{\mathcal{B}_i} \geq \mathsf{dst}(y, \mathsf{bdy}(\mathcal{B}))$ at the three points. Since $\mathcal{B}$'s centers of curvature do not all lie at the point y, at least one of the contacts, say x_1, satisfies the strict inequality $r_{\mathcal{B}_1} > \mathsf{dst}(y, \mathsf{bdy}(\mathcal{B}))$. The object's center of curvature at this point, z_1, thus lies *beyond* the inscribed disc center on the line l_1 (Figure 9.6(a)). Since $\rho_1 = \mathsf{dst}(y, \mathsf{bdy}(\mathcal{B}))$ in Formula (9.1) for $\kappa_1(q_0,(0,\omega))$, the term $\rho_1 - r_{\mathcal{B}_1}$ is strictly negative, while the other terms in Formula (9.1) are positive. Hence, $\kappa_1(q_0,(0,\omega)) < 0$. Since $\kappa_i(q_0,(0,\omega)) \leq 0$ for $i = 2,3$ based on Lemma 9.6, $\kappa_{\text{rel}}(q_0,\dot{q}) < 0$ for $\dot{q} \in M^1_{1,2,3}(q_0)$, which implies that $m^2_{q_0} = 0$ at this grasp.

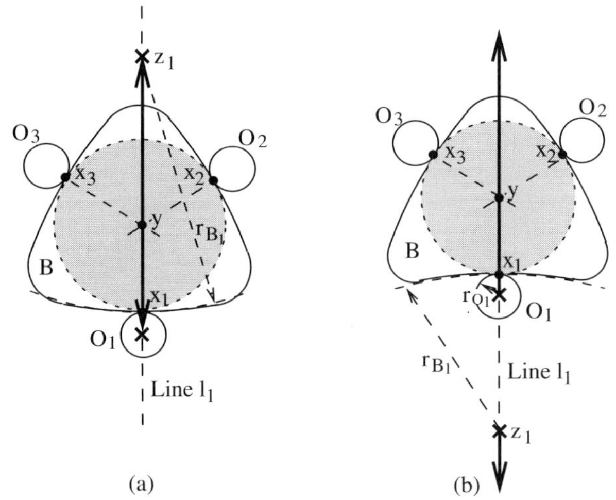

Figure 9.6 Second-order immobilization using three convex fingers. (a) The inscribed disc touches three locally convex points of $\mathcal{B}$. (b) The inscribed disc touches two locally convex points and one locally concave point of $\mathcal{B}$. The thick lines are second-order penetration axes of $\mathcal{B}$ with respect to the finger body $\mathcal{O}_1$.

Next consider the case where the object boundary is concave at some of the three contacts, say at x_1 (Figure 9.6(b)). In this case, $r_{\mathcal{B}_1} < 0$, while $r_{\mathcal{B}_i} \geq 0$ for $i = 2, 3$. The radius of curvature of a convex finger body $\mathcal{O}_1$ placed at x_1, $r_{\mathcal{O}_1} \geq 0$, must satisfy the inequality $r_{\mathcal{O}_1} \leq |r_{\mathcal{B}_1}|$; otherwise, the two bodies would interpenetrate at the contact. Generically $r_{\mathcal{O}_1} < |r_{\mathcal{B}_1}|$ and, hence, $r_{\mathcal{O}_1} + r_{\mathcal{B}_1} < 0$ in Formula (9.1) for $\kappa_1(q_0, (0, \omega))$. Since $\rho_1 = \mathsf{dst}(y, \mathsf{bdy}(\mathcal{B}))$, the terms $\rho_1 - r_{\mathcal{B}_1}$ and $\rho_1 + r_{\mathcal{O}_1}$ are positive in Formula (9.1), and consequently $\kappa_1(q_0, (0, \omega)) < 0$. Since $\kappa_i(q_0, (0, \omega)) \leq 0$ for $i = 2, 3$ based on Lemma 9.6, $\kappa_{\text{rel}}(q_0, \dot{q}) < 0$ for $\dot{q} \in M^1_{1,2,3}(q_0)$, which implies that $m^2_{q_0} = 0$ at this grasp. □

Proposition 9.7 asserts that three convex fingers of any shape can be used to immobilize the object $\mathcal{B}$. The finger curvatures play a key role in the other generic case, where an inscribed disc touches the object boundary at two points, which form antipodal grasp points of $\mathcal{B}$. The following lemma asserts that these two points can be locally converted into an essential three-finger equilibrium grasp of $\mathcal{B}$ (see Appendix II for a proof).

LEMMA 9.8 **(local splitting)** *If an inscribed disc of $\mathcal{B}$ touches its boundary at two points, x_1 and x_2, any local splitting of x_1 into two points, followed by suitable local shift of x_2 to a new point along* $\mathsf{bdy}(\mathcal{B})$*, forms an* **essential three-finger equilibrium grasp** *of $\mathcal{B}$.*

The lemma is illustrated in the following example.

Example: Consider the inscribed disc of the object $\mathcal{B}$ depicted in Figure 9.7, which touches the object boundary at two antipodal points, x_1 and x_2. To obtain an essential

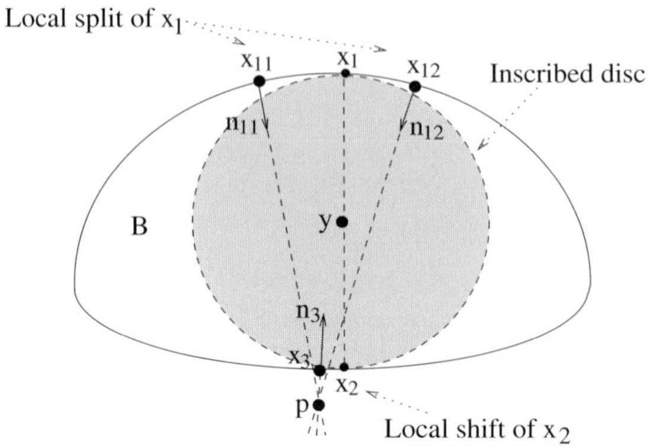

Figure 9.7 The inscribed disc touches the boundary of $\mathcal{B}$ at two antipodal points, x_1 and x_2. An essential three-finger equilibrium grasp of $\mathcal{B}$ is obtained by locally splitting x_1 into x_{11} and x_{12} (which can be asymmetric), followed by local shift of x_2 to x_3.

three-finger equilibrium grasp, start with any local splitting of x_1 into two points, x_{11} and x_{12}. The lines underlying the contact normals n_{11} and n_{12} intersect at a point p. When $\mathcal{B}$'s center of curvature at the antipodal point, x_2, does not lie at p, one can locally shift the point x_2 to a new point, x_3, whose contact normal line also passes through p. The contact normals n_{11}, n_{12}, n_3 positively span the origin of $\mathbb{R}^2$, such that no two of them are collinear. Hence, they form an essential three-finger equilibrium grasp of $\mathcal{B}$. ○

The next proposition asserts that the grasp obtained by the technique of Lemma 9.8 is guaranteed to be immobilizing when sufficiently flat finger bodies are placed at the contacts.

PROPOSITION 9.9 (**three-finger immobilization, case II**) *Let $\mathcal{D}(y)$ be an inscribed disc of a rigid object $\mathcal{B}$. If $\mathcal{D}(y)$ touches the object boundary at two points such that $\mathcal{B}$'s circles of curvature are nonconcentric at these points (the generic case), the object $\mathcal{B}$ is* **immobilized** *by placing three sufficiently flat convex finger bodies in the vicinity of the antipodal points.*

Proof sketch: Let x_1 and x_2 be the antipodal points where the inscribed disc touches the boundary of $\mathcal{B}$. According to Lemma 9.8, one can obtain an essential three-finger equilibrium grasp of $\mathcal{B}$ by local splitting of x_1 into x_{11} and x_{12}, followed by suitable local shift of x_2 to x_3. Let us therefore place convex finger bodies $\mathcal{O}_1$, $\mathcal{O}_2$, and $\mathcal{O}_3$ at these points. The subspace of first-order free motions, $M^1_{1,2,3}(q_0)$, is one-dimensional and consists of instantaneous rotations of $\mathcal{B}$ about the concurrency point, p, of the contact normal lines (Figure 9.8). For $m^2_{q_0}$ to be zero and thus ensure immobilization of $\mathcal{B}$, we must verify that

$$\begin{aligned}\kappa_{\text{rel}}(q_0,(0,\omega)) = \lambda_1\kappa_{11}(q_0,(0,\omega)) + \lambda_2\kappa_{12}(q_0,(0,\omega))\\ + \lambda_3\kappa_3(q_0,(0,\omega)) < 0 \qquad \dot{q} \in M^1_{1,2,3}(q_0),\end{aligned}$$

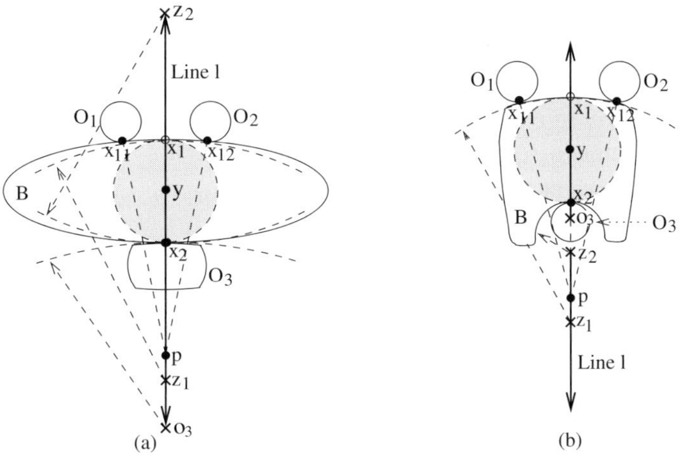

Figure 9.8 Second-order immobilization using three convex fingers. In both cases, two finger bodies $\mathcal{O}_1$ and $\mathcal{O}_2$ are placed at x_{11} and x_{12}; then a sufficiently flat finger body $\mathcal{O}_3$ is placed at the opposite point x_2. The object is immobilized when p lies in the interval of second-order penetration axes of $\mathcal{B}$ with respect to $\mathcal{O}_3$, marked as thick lines.

where $\kappa_{11}(q_0,(0,\omega))$, $\kappa_{12}(q_0,(0,\omega))$ and $\kappa_3(q_0,(0,\omega))$ are the finger c-obstacle curvature forms at q_0, and $\lambda_1, \lambda_2, \lambda_3$ are determined by the equilibrium grasp condition.

We will show that $\kappa_{1j}(q_0,(0,\omega))$ can be made arbitrarily small for $j = 1,2$, while $\kappa_3(q_0,(0,\omega))$ can be made strictly negative and hence dominate the sign of $\kappa_{\text{rel}}(q_0,(0,\omega))$. The point p approaches $\mathcal{B}$'s center of curvature, z_1, for a sufficiently local splitting of x_1. Hence, the distance of p from x_{1j}, which is the parameter ρ_{1j} in Formula (9.1), approaches the object's radius of curvature $r_{\mathcal{B}_1}$ for a sufficiently local splitting of x_1. Since $\text{bdy}(\mathcal{B})$ is locally smooth at x_1, its radius of curvature at x_{1j}, denoted $r_{\mathcal{B}_{1j}}$, also approaches $r_{\mathcal{B}_1}$ for a sufficiently local splitting of x_1. Hence, the term $\rho_{1j} - r_{\mathcal{B}_{1j}}$ in Formula (9.1), and consequently $\kappa_{1j}(q_0,(0,\omega))$ for $j = 1,2$, can be made arbitrarily small for a sufficiently local splitting of x_1.

The remainder of the proof verifies that $\kappa_3(q_0,(0,\omega)) < 0$. We may assume that $\mathcal{B}$ is *strictly convex* at one of the antipodal points, say at x_1. To simplify the proof, assume that the concurrency point p lies along the line l passing through x_1 and x_2 (a continuity argument extends the proof to the general case). Let us therefore place a convex finger body $\mathcal{O}_3$ at the opposite point x_2 and consider two possible cases.

First consider the case where $\mathcal{B}$ is *convex* at x_2 (Figure 9.8(a)). The second-order *penetration* axes of $\mathcal{B}$ form the interval between the object's center of curvature, z_2, and the finger's center of curvature, o_3, on the line l (Figure 9.8(a)). Based on Formula (9.1), $\kappa_3(q_0,(0,\omega)) < 0$ when p lies in the interior of this interval. By choosing the finger body $\mathcal{O}_3$ to be *sufficiently flat* at x_2, its center of curvature, o_3, lies sufficiently far from the contact point x_2, such that the second-order penetration interval contains the point z_1 in its interior (Figure 9.8(a)). Since p approaches z_1 for a sufficiently local splitting of x_1, the point p also lies in the interior of the second-order penetration interval.

Based on Formula (9.1), this insight can be translated into an algebraic statement that $\kappa_3(q_0,(0,\omega)) < 0$ for instantaneous rotation of $\mathcal{B}$ about p.

Next consider the case where $\mathcal{B}$ is *concave* at x_2 (Figure 9.8(b)). The second-order *free axes* of $\mathcal{B}$ lie between the object's center of curvature, z_2, and the finger's center of curvature, o_3 (Figure 9.8(b)). However, it is always true that $r_{\mathcal{O}_3} \leq |r_{\mathcal{B}_2}|$, where $r_{\mathcal{O}_3}$ is the radius of curvature of $\mathcal{O}_3$ at x_2. The finger's center of curvature, o_3, therefore lies between x_2 and z_2 on the line l (Figure 9.8(b)). We may now choose a *sufficiently flat* finger body $\mathcal{O}_3$, such that o_3 lies arbitrarily close to z_2. That is, we can make the interval of second-order *free axes* arbitrarily short, making sure that z_1 lies *outside* of this interval (Figure 9.8(b)). Hence, both z_1 and p lie in the interior of the second-order penetration interval for a sufficiently local splitting of x_1. Based on Formula (9.1), this insight translates to an algebraic statement that $\kappa_3(q_0,(0,\omega)) < 0$ for instantaneous rotation of $\mathcal{B}$ about p. □

Example: Both objects depicted in Figure 9.8 have an inscribed disc that touches their boundary at two antipodal points x_1 and x_2. The immobilization of both objects starts by a splitting x_1 into two points, x_{11} and x_{12}. For simplicity, the splitting is symmetric so that the contact normals, n_{11} and n_{12}, intersect at a point p located on the line l. Next, two convex finger bodies, $\mathcal{O}_{11}$ and $\mathcal{O}_{12}$, are placed at x_{11} and x_{12}. In the case of Figure 9.8(a), the object $\mathcal{B}$ is convex at x_2. Hence, the placement of a sufficiently flat finger body $\mathcal{O}_3$ at x_2 ensures an immobilizing grasp of $\mathcal{B}$. In the case of Figure 9.8(b), the object $\mathcal{B}$ is concave at x_2. In this case, even a *point finger* placed at x_2 ensures full immobilization of $\mathcal{B}$ at this grasp. ○

To summarize, when the inscribed disc touches the object boundary at three points, the object is immobilized by placing three convex fingers of any shape at these points (Proposition 9.7). When the inscribed disc touches the object boundary at two points, these points can be locally converted into a three-finger equilibrium grasp, and the object is immobilized by placing three sufficiently flat convex fingers at these points (Proposition 9.9). This leads to the following minimal second-order immobilizing grasp of smooth 2-D objects.

THEOREM 9.10 **(minimal three-finger grasp)** *The minimal second-order immobilizing grasp of all smooth 2-D objects that do not have circular symmetry requires* **three convex finger bodies**, *provided that the finger bodies are sufficiently flat at the contacts.*

The smaller number of fingers afforded by second-order effects requires *precise placement* of the finger contacts in order to form frictionless equilibrium grasps, which are necessary for object immobilization. When the fingers are placed inaccurately but sufficiently close to the immobilizing contacts, the object is automatically *caged* by the fingers (see next chapter). Moreover, all contacts possess some amount of friction. We will see in Part III of the book that sufficiently small finger placement errors result in a frictional *wrench resistant* grasp. Robot hands that use three independent fingers, or two independent fingers and a palm, thus offer *minimalistic hand designs* that can safely grasp almost all smooth 2-D objects.

9.3.2 Minimal Immobilization Using Two Concave Fingers

Minimal immobilization with three convex fingers merely requires that the fingers be sufficiently flat at the contacts. However, in many grasping applications it is conceivable that the geometry of the fingers or fixtures can be chosen in advance, according to the class of objects to be handled by the grasping system. The following theorem asserts that when finger pre-shaping is allowed, the minimal immobilization of all 2-D objects can be achieved using only two possibly concave fingers.

THEOREM 9.11 (**minimal two-finger grasp**) *When the fingers are allowed to have any curvature at the contacts, the minimal second-order immobilizing grasp of all smooth 2-D objects that do not have circular symmetry requires* **two possibly concave finger bodies**.

The proof, which appears in Appendix II, begins with an antipodal equilibrium grasp of $\mathcal{B}$, at x_1 and x_2. The first-order mobility index of the grasp is $m^1_{q_0} = m - k + 1 = 2$, since $m = 3$ and there are $k = 2$ contacts. The subspace of first-order free motions, $M^1_{1,2}(q_0)$, is two-dimensional and consists of instantaneous rotations of $\mathcal{B}$ about all points of the extended line, l, which passes through x_1 and x_2. The object is immobilized when the second-order mobility index of the grasp, $m^2_{q_0}$, is zero. For $m^2_{q_0}$ to be zero, the relative c-space curvature must satisfy the condition:

$$\kappa_{\text{rel}}(q_0, \dot{q}) = \lambda_1 \kappa_1(q_0, \dot{q}) + \lambda_2 \kappa_2(q_0, \dot{q}) < 0 \qquad \dot{q} \in M^1_{1,2}(q_0),$$

where $\lambda_1 = \lambda_2$ at the two-finger equilibrium grasp.

To achieve second-order immobilization, the proof sets the finger curvatures as follows. When the object $\mathcal{B}$ is *concave* at x_i, one places a convex finger body $\mathcal{O}_i$ at x_i, with positive radius of curvature $r_{\mathcal{O}_i} = |r_{\mathcal{B}_i}| - \epsilon$, where $r_{\mathcal{B}_i}$ is $\mathcal{B}$'s radius of curvature at x_i, and ϵ is a small positive parameter. When the object $\mathcal{B}$ is *convex* at x_i, one places a concave finger body $\mathcal{O}_i$ at x_i, with negative radius of curvature $r_{\mathcal{O}_i} = -(r_{\mathcal{B}_i} + \epsilon)$. Based on Formula (9.1) for $\kappa_i(q_0, (0, \omega))$, the second-order *free axes* of $\mathcal{B}$ with respect to $\mathcal{O}_i$ lie between the object and finger centers of curvature at x_i (Figure 9.9). The ϵ-matching of the object and finger curvatures at the contacts makes the *free axes* intervals arbitrarily short. Since $\mathcal{B}$'s circles of curvature are nonconcentric at the grasp points, sufficiently short free-axes intervals do not overlap on the line l (Figure 9.9). This property can be translated into an algebraic statement, which, based on Formula (9.1), gives that $\kappa_{\text{rel}}(q_0, \dot{q}) < 0$ for all $\dot{q} \in M^1_{1,2}(q_0)$.

Example: The ellipse depicted in Figure 9.9(a) is held along its major axis by two concave finger bodies, $\mathcal{O}_1$ and $\mathcal{O}_2$. When the curvature of these fingers closely matches the object's curvature at the contacts, the second-order *free* axes of the ellipse with respect to each finger do not overlap on the line l, which implies full immobilization of the ellipse at this grasp. The object depicted in Figure 9.9(b) is convex at x_1 and concave at x_2. To immobilize such an object, a closely matching concave finger body $\mathcal{O}_1$ is placed at x_1, while a closely matching convex finger body $\mathcal{O}_2$ is placed at x_2. The second-order *free axes* of $\mathcal{B}$ with respect to each finger lie between the object and finger

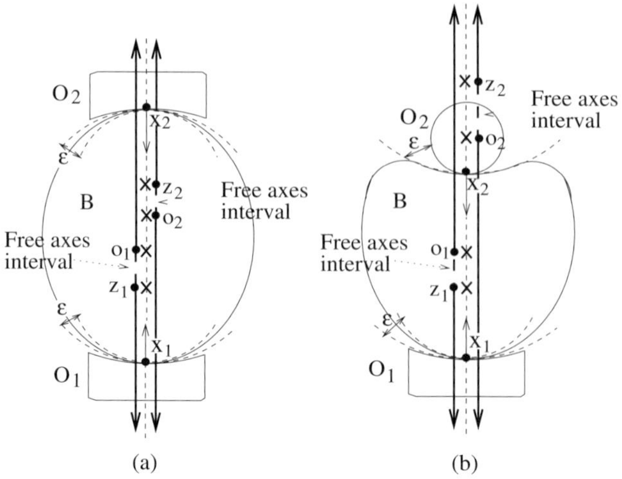

Figure 9.9 (a) Two closely matching concave finger bodies $\mathcal{O}_1$ and $\mathcal{O}_2$ are placed at the convex object points x_1 and x_2. (b) A closely matching concave finger body $\mathcal{O}_1$ is placed at the convex object point x_1, while a closely matching convex finger body $\mathcal{O}_2$ is placed at the concave object point x_2. In both cases, the object is immobilized when its second-order free axes intervals with respect to each finger, marked by dashed lines, do not overlap on the line l.

centers of curvature, z_i and o_i, at each contact. Since these intervals are non-overlapping, the object $\mathcal{B}$ is fully immobilized at this grasp. ○

9.4 Minimal Second-Order Immobilization of Polygons

This section considers the minimal second-order immobilization of polygonal objects, with the understanding that similar principles apply to all piecewise smooth objects. The inscribed disc of a polygon $\mathcal{B}$ generically touches its boundary at two or three isolated points (two points occur only when the polygon has parallel edges). If any of these points is located at a concave vertex of the polygon, a *pointed fingertip* must be placed at this point, as shown in Figure 9.10(a).[3] Let us therefore characterize the c-obstacle induced by a pointed fingertip, $\mathcal{O}_i$, which is placed at a concave vertex of $\mathcal{B}$.

The polygonal object $\mathcal{B}$ can be locally represented as the union of two sub-shapes, each bounded by one of the two edges of $\mathcal{B}$ that meet at the concave vertex x_i (Figure 9.10(b)). Let $\mathcal{CO}_{i1}$ and $\mathcal{CO}_{i2}$ denote the c-obstacles corresponding to the two sub-shapes of $\mathcal{B}$. The finger c-obstacle forms the union: $\mathcal{CO}_i = \mathcal{CO}_{i1} \cup \mathcal{CO}_{i2}$ (see Chapter 2). The boundary of each c-obstacle, $\mathcal{CO}_{ij}$, is locally smooth at the object's configuration, q_0, as long as the pointed fingertip maintains point contact with $\mathcal{B}$. The boundary of the full finger c-obstacle, $\mathcal{CO}_i$, thus consists of two smooth surfaces,

[3] A pointed fingertip can realize contact forces along any vector in the object's *generalized inward normal* at the concave vertex; see Figure 9.10(a).

bdy($\mathcal{CO}_{i1}$) and bdy($\mathcal{CO}_{i2}$), that meet along a curve formed by the intersection of these surfaces. The latter curve contains the point q_0, and motion along this curve represents local rotation of $\mathcal{B}$ about the sharp fingertip.

The set of first-order free motions of $\mathcal{B}$ with respect to a pointed fingertip, $M_i^1(q_0)$, is the intersection of the free motion half-spaces associated with $\mathcal{CO}_{i1}$ and $\mathcal{CO}_{i2}$. Thus, $M_i^1(q_0)$ forms a *convex wedge* in the ambient tangent space $T_{q_0}\mathbb{R}^3$. This insight leads to the following proposition, which is analogous to an earlier proposition concerning smooth 2-D objects.

PROPOSITION 9.12 (**nonparallel edges**) *Let an inscribed disc of a polygon $\mathcal{B}$ touch its boundary at three points, such that no two points are located on parallel edges (the generic case). The polygon is* **immobilized** *by placing three convex finger bodies of any shape at these points, provided that a pointed fingertip is placed at every concave vertex of $\mathcal{B}$.*

Proof sketch: First, consider the case where the inscribed disc touches three edges of $\mathcal{B}$. Since no two points are located on parallel edges of $\mathcal{B}$, three convex finger bodies placed at these points form an essential three-finger equilibrium grasp of $\mathcal{B}$ (Proposition 9.5). The subspace of first-order free motions, $M_{1,2,3}^1(q_0)$, is one-dimensional and consists of instantaneous rotations of $\mathcal{B}$ about the inscribed disc center. Since $\mathcal{B}$ is locally smooth at the contacts, the proof of Proposition 9.7 applies in this case. The polygonal object $\mathcal{B}$ is thus immobilized by *second-order* effects at this grasp, even when point fingers are placed at the contacts.

Next, consider the case where the inscribed disc touches at least one concave vertex of $\mathcal{B}$, at x_1. For simplicity, let the points x_2 and x_3 be interior edge points of $\mathcal{B}$ (Figure 9.10(a)). Let us place a *pointed fingertip*, $\mathcal{O}_1$, at x_1 and smooth convex finger bodies at x_2 and x_3. The first-order free motions of $\mathcal{B}$ with respect to $\mathcal{O}_1$ form a *convex wedge* in the ambient tangent space $T_{q_0}\mathbb{R}^3$. This wedge can be thought of as the intersection of two half-spaces whose bounding planes pass through the origin of $T_{q_0}\mathbb{R}^3$, associated with the c-obstacles $\mathcal{CO}_{11}$ and $\mathcal{CO}_{12}$ described earlier. The first-order

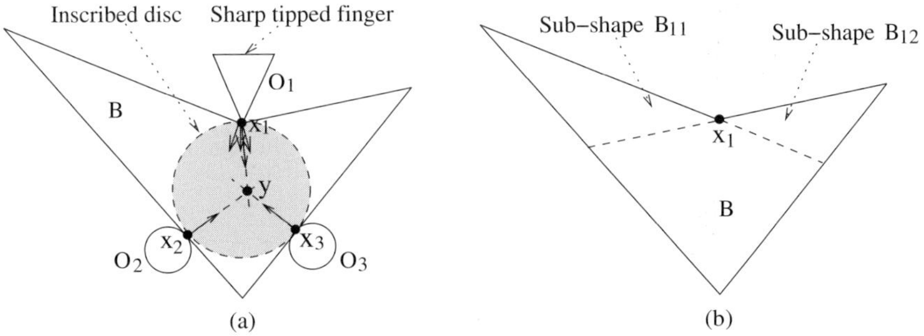

Figure 9.10 (a) The inscribed disc touches the polygon's boundary at three points, such that x_1 is a concave vertex of $\mathcal{B}$. (b) The polygonal object can be locally represented at x_1 as the union of two sub-shapes, $\mathcal{B} = \mathcal{B}_{11} \cup \mathcal{B}_{12}$, which gives a finger c-obstacle of the form $\mathcal{CO}_i = \mathcal{CO}_{i1} \cup \mathcal{CO}_{i2}$.

free motions of $\mathcal{B}$ at the equilibrium grasp, denoted $M^1_{11,12,2,3}(q_0)$, are given by the intersection of *four* half-spaces:

$$M^1_{11,12,2,3}(q_0) = M^1_{11}(q_0) \cap M^1_{12}(q_0) \cap M^1_2(q_0) \cap M^1_3(q_0).$$

To evaluate this intersection arrangement, recall that each c-obstacle normal is collinear with the wrench generated by a force that acts along $\mathcal{B}$'s inward contact normal (Chapter 2). In the case of the c-obstacles $\mathcal{CO}_{11}$ and $\mathcal{CO}_{12}$, the contact forces act along the edges' inward normals at x_1. Since all four contact forces are *essential* for the equilibrium grasp, $M^1_{11,12,2,3}(q_0) = \{0\}$. The object $\mathcal{B}$ is thus immobilized by *first-order* effects. □

When a pointed fingertip is placed at a concave vertex of a polygonal object, three fingers are able to immobilize the polygon based on *first-order* geometric effects (Figure 9.10(a)). This conclusion seems to contradict the principle that at least four fingers are required for first-order immobilization of any 2-D object. However, this situation happens only when a *pointed fingertip* contacts a concave vertex of the polygon. Such a contact can be interpreted as two edge contacts that coincide at a common vertex, which explains why first-order effects appear in these grasps. Note that such grasps are *robust* with respect to small finger placement errors along the polygon edges and small orientation placement errors of the pointed fingertip at the concave vertex of the polygon.

Next, consider the minimal second-order immobilization of polygons with parallel edges. We will consider for simplicity only convex polygons. When the inscribed disc touches two parallel edges of $\mathcal{B}$, three-finger immobilization can be achieved by a slight variation on the *contact splitting technique,* described in Lemma 9.8. The technique is illustrated in Figure 9.11, which shows the inscribed disc touching two parallel edges of a rectangular object $\mathcal{B}$, at x_1 and x_2. To immobilize this object, first split the point x_1 into two contacts, x_{11} and x_{12}; then move these points to the vertices at the ends of the edge containing this point. Next, place *smooth* convex finger bodies, $\mathcal{O}_1$ and $\mathcal{O}_2$, at these vertices (Figure 9.11). The normal to the finger boundary at each vertex uniquely determines the finger force direction at the vertex. Moreover, one can verify that each finger c-obstacle is locally smooth at q_0, as long as the finger contact normal does not coincide with the polygon's edge normals at the vertex. The finger c-obstacle curvature forms, $\kappa_{1j}(q_0, \dot{q})$ for $j = 1, 2$, can be obtained by substituting $r_{\mathcal{B}_{1j}} = 0$ for $\mathcal{B}$'s radius of curvature at x_{1j} in Formula (9.1). This insight leads to the following lemma, whose proof is left as an exercise.

LEMMA 9.13 **(parallel edges)** *Every convex polygon with parallel edges can be* **immobilized** *by three sufficiently flat convex finger bodies, provided that locally smooth fingers are placed at the vertices of the polygon* (Figure 9.11).

When a polygon with parallel edges is *non-convex,* one can slide the inscribed disc between the parallel edges until a third contact is established. At this inscribed disc position, one of the parallel-edge points can be replaced by a new contact point, which always exists along the opposite direction on the polygon boundary (see Exercises).

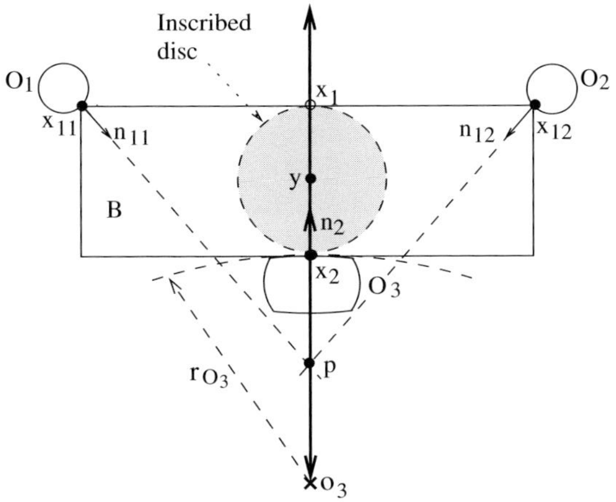

Figure 9.11 A polygon with parallel edges is immobilized by splitting the point x_1 in a nonlocal manner, placing finger bodies $\mathcal{O}_1$ and $\mathcal{O}_2$ at x_{11} and x_{12}. This is followed by placement of a sufficiently flat finger body $\mathcal{O}_3$ at x_2, such that the point p lies in the interval of second-order penetration axes of $\mathcal{B}$ with respect to $\mathcal{O}_3$ (marked as a thick line).

Based on Proposition 9.12 and Lemma 9.13 (and its extension to non-convex polygons), the minimal second-order immobilizing grasps of polygonal objects requires the same number of fingers required for smooth 2-D objects, as summarized in the following theorem.

THEOREM 9.14 **(minimal three-finger grasp of polygons)** *The minimal second-order immobilizing grasp of all polygonal objects, even with parallel edges, requires* **three sufficiently flat convex finger bodies**.

Finally, when the fingers can be pre-shaped according to the class of polygonal objects at hand, the minimal second-order immobilizing grasps require only two possibly concave fingers, as stated in the following theorem.

THEOREM 9.15 **(minimal two-finger grasp of polygons)** *The minimal second-order immobilizing grasp of all polygonal objects, even with parallel edges, requires only* **two possibly concave finger bodies**.

The proof of Theorem 9.15 is a slight variation on the proof of Theorem 9.11. Rather than giving a formal proof of the theorem, consider the following example.

Example: The polygon $\mathcal{B}$ depicted in Figure 9.12 is held at a two-finger equilibrium grasp by a sharp-tipped finger at the concave vertex, x_1, and a disc finger at the opposing interior edge point, x_2. The finger contact at x_1 can be interpreted as two edge contacts that coincide at x_1. Hence, there are two c-obstacles, $\mathcal{CO}_{11}$ and $\mathcal{CO}_{12}$, associated with the sharp-tipped finger at x_1, and a third c-obstacle, $\mathcal{CO}_2$, associated with the disc finger at x_2. The first-order mobility index of the grasp is $m^1_{q_0} = m - k + 1 = 1$, since $m = 3$, and there are effectively $k = 3$ finger contacts at this grasp. The subspace of first-order

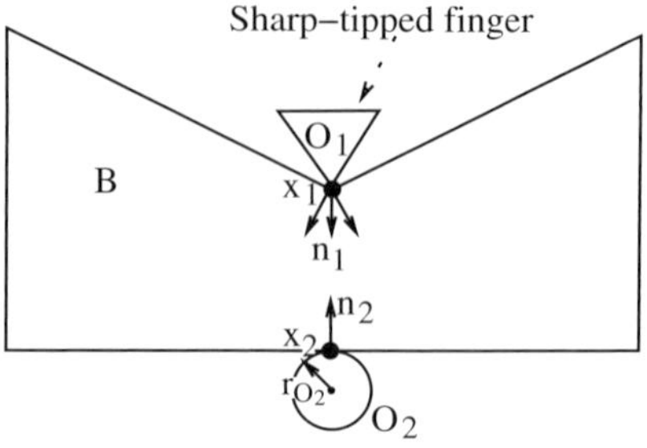

Figure 9.12 Immobilization of a polygonal object using a sharp-tipped finger $\mathcal{O}_1$ at x_1 and a disc finger $\mathcal{O}_2$ at the opposing edge point x_2. The object is immobilized by second-order effects even when point fingers are placed at the contacts.

free motions, $M^1_{11,12,2}(q_0)$, is one-dimensional and consists of instantaneous rotations of $\mathcal{B}$ about the concave vertex x_1. The polygon is immobilized when the relative c-space curvature form satisfies the condition:

$$\kappa_{\text{rel}}(q_0,\dot{q}) = \lambda_{11}\kappa_{11}(q_0,\dot{q}) + \lambda_{12}\kappa_{12}(q_0,\dot{q}) + \lambda_2\kappa_2(q_0,\dot{q}) < 0 \qquad \dot{q} \in M^1_{11,12,2}(q_0), \tag{9.3}$$

where $\kappa_{1j}(q_0,\dot{q})$ for $j=1,2$ and $\kappa_2(q_0,\dot{q})$ are the finger c-obstacle curvature forms at q_0, while $\lambda_{11},\lambda_{12},\lambda_2$ are the equilibrium grasp coefficients. The sharp-tipped finger, $\mathcal{O}_1$, can be modeled as having zero radius of curvature at its tip so that $r_{\mathcal{O}_{1j}}=0$ for $j=1,2$. Substituting $r_{\mathcal{O}_{1j}}=0$ and $r_{\mathcal{B}_{1j}}=\infty$ for the flat edges of $\mathcal{B}$ at x_1 in Formula (9.1) gives $\kappa_{1j}(q_0,(0,\omega))=0$ for $j=1,2$. The sign of $\kappa_{\text{rel}}(q_0,\dot{q})$ in Eq. (9.3) is thus determined by the sign of $\kappa_2(q_0,\dot{q})$. Substituting $r_{\mathcal{B}_2}=\infty$ for the flat edge of $\mathcal{B}$ at x_2 in Formula (9.1) gives $\kappa_2(q_0,(0,\omega)) = -(\rho_2 + r_{\mathcal{O}_2})$. The parameter ρ_2 measures the distance of $\mathcal{B}$'s instantaneous rotation axis at x_1 from x_2. The parameter ρ_2 is positive since x_1 is located on $\mathcal{B}$'s side of x_2, and the disc finger radius, $r_{\mathcal{O}_2}$, is also positive. Hence, we obtain that $\kappa_2(q_0,(0,\omega)) < 0$ and, consequently, $\kappa_{\text{rel}}(q_0,\dot{q}) < 0$ for all $\dot{q} \in M^1_{11,12,2}(q_0)$. The polygonal object $\mathcal{B}$ is thus immobilized by second-order effects, even when *point fingers* are placed at the two contacts. ○

9.5 Minimal Second-Order Immobilization of Polyhedral Objects

We will see in Part III of the book that in the case of 3-D grasps, three fingers can realize an alternative grasp security criterion of *wrench resistance* via frictional contacts. However, the amount of friction at the contacts depends on multiple factors, which can often be only crudely approximated in practical grasping systems. To increase the level of grasp security, imagine that the palm of a three-finger robot hand acts as a fourth contact. Alternatively, consider a four-finger robot hand. Can four independent finger bodies be used to secure all 3-D objects at an *immobilizing grasp?*

The immobilization of 3-D objects using only four fingers is clearly based on second-order effects, since first-order immobilization requires at least seven fingers. The construction of such immobilizing grasps will use the *inscribed ball* of $\mathcal{B}$. It is defined as a ball centered at y, which lies inside $\mathcal{B}$ and whose radius is maximal among all balls contained in $\mathcal{B}$ with their center in a neighborhood of y. The inscribed ball of $\mathcal{B}$ shares many properties with the inscribed disc. In particular, its center y forms a local maximum of the Lipschitz continuous function $\mathsf{dst}(y, \mathsf{bdy}(\mathcal{B}))$. Hence, in the same manner shown for 2-D objects, the points where an inscribed ball touches the polyhedral object surface, $x \in \mathcal{I}(y)$, can be used to realize a frictionless equilibrium grasp of $\mathcal{B}$. The inscribed ball generically touches the object surface at two, three or four isolated points (see Appendix I). The following example illustrates the special cases of two and three points, as well as the generic case of four points.

Example: The inscribed ball of the rectangular box depicted in Figure 9.13(a) touches its surface at two points. This non-generic case occurs only when the polyhedron has parallel facets. The inscribed ball of the triangular prism depicted in Figure 9.13(b) touches its surface at three points. This is also a non-generic case, since the polyhedron must have three coaxial facets. The generic case where the inscribed ball touches the polyhedron's boundary at four points is illustrated in Figure 9.13(c). Note that the inscribed ball of any convex polyhedron can touch only interior facet points. ○

The inscribed ball of a *non-convex* polyhedron, $\mathcal{B}$, can touch concave edges and vertices. At a concave edge of $\mathcal{B}$, the polyhedron locally forms the union of two half-spaces, each bounded by one of the two facets that meet at this edge. At a concave vertex of $\mathcal{B}$, the polyhedron locally forms the union of half-spaces, each bounded by one of the facets that meet at the vertex. The mobility analysis proceeds by regarding each half-space as giving rise to its own finger c-obstacle. A concave edge of $\mathcal{B}$ thus has the effect of a *double* facet contact, while a generic concave vertex of $\mathcal{B}$ has the effect of a *triple* facet contact. To show that four convex fingers suffice to immobilize all polyhedral objects, let us summarize the finger c-obstacle curvature formula from Chapter 2.

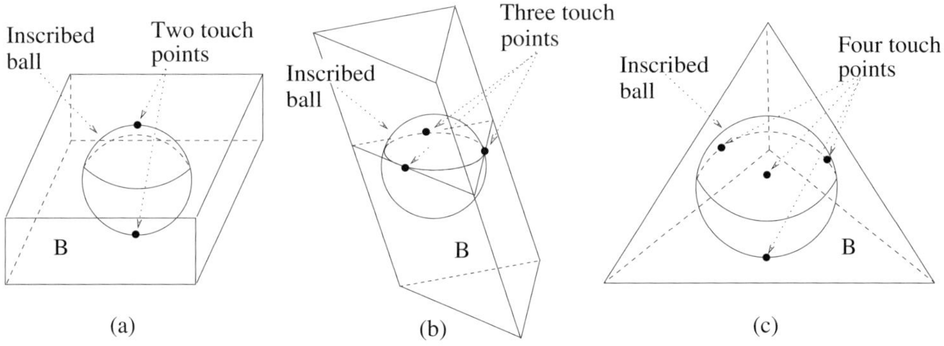

Figure 9.13 (a) The inscribed ball of a rectangular box touches its boundary at two points, located on parallel facets of $\mathcal{B}$. (b) The inscribed ball of a triangular prism touches its boundary at three points, located on three coaxial facets of $\mathcal{B}$. (c) The generic case is shown, where the inscribed ball touches the boundary of a polyhedral object at four points.

Finger c-obstacle curvature: Recall that the c-space of a 3-D object is parametrized by $\mathbb{R}^6$. Each finger c-obstacle, $\mathcal{CO}_i$, forms a six-dimensional set in this space. Hence, each finger c-obstacle boundary, $\mathcal{S}_i$, forms a five-dimensional manifold in $\mathbb{R}^6$. Let $q_0 \in \mathcal{S}_i$ be the object's contact configuration with a stationary finger body $\mathcal{O}$, and let x_i be the object contact point. Assuming that $\mathcal{S}_i$ is locally smooth at q_0, let $T_{q_0}\mathcal{S}_i$ denote the five-dimensional tangent space of $\mathcal{S}_i$ at q_0. The *curvature form* of $\mathcal{S}_i$ at q_0, denoted $\kappa_i(q,\dot{q})$, measures the change of the finger c-obstacle outward unit normal along tangent directions $\dot{q} \in T_{q_0}\mathcal{S}_i$. The curvature of $\mathcal{S}_i$ depends on the surface curvatures of the contacting bodies, $\mathcal{B}$ and $\mathcal{O}_i$, at x_i. These curvatures are determined by the linear maps, $L_{\mathcal{B}_i}$ and $L_{\mathcal{O}_i}$, which act on tangents to the respective body surface at x_i and give the change in the surface normal along a given tangent direction. The curvature form of $\mathcal{S}_i$ is given by

$$
\begin{aligned}
\kappa_i(q_0,\dot{q}) = \frac{1}{\|\eta_i\|}\dot{q}^T &\left(\begin{bmatrix} I & -[\rho_i\times] \\ O & [n_i\times] \end{bmatrix}^T \right.\\
&\times \begin{bmatrix} L_{\mathcal{B}}\left[L_{\mathcal{B}_i}+L_{\mathcal{O}_i}\right]^{-1}L_{\mathcal{O}_i} & -L_{\mathcal{O}_i}\left[L_{\mathcal{B}_i}+L_{\mathcal{O}_i}\right]^{-1} \\ -\left[L_{\mathcal{B}_i}+L_{\mathcal{O}_i}\right]^{-1}L_{\mathcal{O}_i} & -\left[L_{\mathcal{B}_i}+L_{\mathcal{O}_i}\right]^{-1} \end{bmatrix}\begin{bmatrix} I & -[\rho_i\times] \\ O & [n_i\times] \end{bmatrix}\\
&\left. + \begin{bmatrix} O & O \\ O & -\left([\rho_i\times]^T[n_i\times]\right)_s \end{bmatrix}\right)\dot{q} \qquad \dot{q}\in T_{q_0}\mathcal{S}_i, \qquad (9.4)
\end{aligned}
$$

where ρ_i is the vector from $\mathcal{B}$'s frame origin to x_i expressed in the world frame, n_i is $\mathcal{B}$'s inward unit normal at x_i, $\eta_i = (n_i, \rho_i \times n_i)$ is the finger c-obstacle outward normal at q_0, I is the 3×3 identity matrix, O is the 3×3 zero matrix and $([\rho_i\times]^T[n_i\times])_s = \frac{1}{2}([\rho_i\times]^T[n_i\times] + [n_i\times]^T[\rho_i\times])$.

Example: Let a stationary finger body $\mathcal{O}_i$ touch a facet of a polyhedron $\mathcal{B}$ at x_i. The finger c-obstacle curvature form is obtained by substituting $L_{\mathcal{B}_i} = O$ in Formula (9.4), where O is a 3×3 zero matrix. Writing the c-space tangent vectors as $\dot{q} = (v,\omega) \in T_{q_0}\mathcal{S}_i$, where v and ω are $\mathcal{B}$'s linear and angular velocities, the curvature form of $\mathcal{S}_i$ associated with a flat facet of $\mathcal{B}$ is

$$
\begin{aligned}
\kappa_i(q_0,(v,\omega)) = &\left((v-\rho_i\times\omega)^T,(n_i\times\omega)^T\right)\begin{bmatrix} O & -I \\ -I & -L_{\mathcal{O}_i}^{-1} \end{bmatrix}\begin{pmatrix} v-\rho_i\times\omega \\ n_i\times\omega \end{pmatrix}\\
&-\omega^T([\rho_i\times]^T[n_i\times])_s\,\omega,
\end{aligned}
$$

where we omitted the multiplicative factor $1/\|\eta_i\|$. Some insight can be gained by considering instantaneous translations of $\mathcal{B}$, $\dot{q} = (v,\vec{0})$ for $\dot{q} \in T_{q_0}\mathcal{S}_i$. These translations are spanned by linear velocities of $\mathcal{B}$ that are orthogonal to the contact normal n_i. The finger c-obstacle curvature is *zero* along such instantaneous translations of $\mathcal{B}$. The finger c-obstacle boundary, $\mathcal{S}_i$, is therefore *flat* along pure translations of $\mathcal{B}$ that maintain contact with the stationary finger body $\mathcal{O}_i$. The geometric shape of $\mathcal{S}_i$ is thus analogous to a *ruled surface*. That is, each fixed orientation slice of $\mathcal{S}_i$ forms a *planar patch* in $\mathcal{B}$'s c-space, and $\mathcal{S}_i$ is the union of these planar patches. ○

The next three propositions consider the minimal second-order immobilization of all polyhedral objects, according to three different cases. The first case concerns polyhedral objects without parallel or coaxial facets.

PROPOSITION 9.16 **(four-finger immobilization, case I)** *Let an inscribed ball of a polyhedron $\mathcal{B}$ touch its surface at four points. If the four points are not located on parallel or coaxial facets of $\mathcal{B}$ (the generic case), the polyhedron is* **immobilized** *by placing four convex finger bodies of any shape at these points.*

Note the following caveats. When a finger body is placed at a convex vertex or edge of $\mathcal{B}$, the finger body must be locally smooth at the contact point, to ensure a locally smooth finger c-obstacle boundary at q_0. When a finger body is placed at a concave vertex or edge of $\mathcal{B}$, the finger body must be pointed or sharp bladed at the contact point.

Proof: First consider the case where the inscribed ball touches four facet points of $\mathcal{B}$. The four points are not located on parallel or coaxial facets of $\mathcal{B}$. Hence, four convex finger bodies placed at these points form an essential four-finger equilibrium grasp of $\mathcal{B}$ (Proposition 9.5). The first-order mobility index of the grasp is given by $m^1_{q_0} = m - k + 1 = 3$, where $m = 6$ and $k = 4$ is the number of finger contacts. The subspace of first-order free motions, $M^1_{1,2,3,4}(q_0)$, is thus three-dimensional at this grasp. To identify this subspace, let $\mathcal{B}$'s frame origin be located at the inscribed ball center. The ith finger c-obstacle outward normal at q_0 is given by $\eta_i = (n_i, \rho_i \times n_i)$, where n_i is $\mathcal{B}$'s inward unit normal at x_i, and ρ_i is the vector from $\mathcal{B}$'s frame origin to x_i. Since ρ_i is collinear with n_i, we obtain that $\eta_i = (n_i, \vec{0})$ for $i = 1,2,3,4$. The subspace $M^1_{1,2,3,4}(q_0)$ is orthogonal to the c-obstacle normals $\eta_1, \eta_2, \eta_3, \eta_4$ and consists of instantaneous rotations of $\mathcal{B}$ about the inscribed ball center, $\dot{q} = (\vec{0}, \omega)$ for $\omega \in \mathbb{R}^3$.

The object $\mathcal{B}$ is immobilized when the second-order mobility index of the grasp, $m^2_{q_0}$, is zero. Equivalently, we must show that

$$\kappa_{\text{rel}}(q_0, \dot{q}) = \lambda_1 \kappa_1(q_0, \dot{q}) + \cdots + \lambda_4 \kappa_4(q_0, \dot{q}) < 0 \qquad \dot{q} = (\vec{0}, \omega),\ \omega \in \mathbb{R}^3,$$

where $\kappa_i(q_0, \dot{q})$ is the ith finger c-obstacle curvature form at q_0, and $\lambda_1, \lambda_2, \lambda_3, \lambda_4$ are the equilibrium grasp coefficients. For a choice of $\mathcal{B}$'s frame origin at the inscribed ball center, $\rho_i = -\rho_i n_i$, where ρ_i measures the distance of the inscribed ball center from the contact point x_i, such that $\rho_i > 0$ when $\mathcal{B}$'s frame origin lies on the object side of x_i. Substituting $L_{\mathcal{B}_i} = O$, $\rho_i = -\rho_i n_i$, and $\|\eta_i\| = 1$ in formula (9.4) gives

$$\kappa_{\text{rel}}\big(q_0, (\vec{0}, \omega)\big) = -\sum_{i=1}^{4} \lambda_i (n_i \times \omega)^T [\rho_i I + L^{-1}_{\mathcal{O}_i}](n_i \times \omega) \qquad \omega \in \mathbb{R}^3. \qquad (9.5)$$

Since the contacts are facets points, we might as well use *spherical fingertips* of radius $r_{O_i} \geq 0$ for $i = 1,2,3,4$. For such fingers $L^{-1}_{\mathcal{O}_i} = r_{O_i} I$. The parameter ρ_i is equal to the inscribed ball radius, denoted R, and is positive since the inscribed ball center lies on $\mathcal{B}$'s side of each contact point x_i. Formula (9.5) thus becomes

$$\kappa_{\text{rel}}(q_0,(0,\omega)) = -\sum_{i=1}^{4} \lambda_i(R + r_{\mathcal{O}_i})\|n_i \times \omega\|^2 \qquad \omega \in \mathbb{R}^3. \tag{9.6}$$

The matrix formed by the contact normals, $[n_1, n_2, n_3, n_4]$, has full row rank at an essential four-finger equilibrium grasp. Hence, for each $\omega \in \mathbb{R}^3$, at least one of the summands in Eq. (9.6) must be strictly negative. This implies that $\kappa_{\text{rel}}(q_0,(0,\omega)) < 0$ for all $\omega \in \mathbb{R}^3$, as required for immobilization.

Next consider the case where the inscribed ball touches a concave vertex or edge point of $\mathcal{B}$ (assume one such point), while the other three points touch facets of $\mathcal{B}$. A concave vertex is generically formed by three facets of $\mathcal{B}$. Hence, the placement of a *sharp-tipped finger* at the vertex can be interpreted as a triple facet contact. When four fingers are placed at the contacts, the first-order mobility index becomes $m^1_{q_0} = m - k + 1 = 1$, since $m = 6$ and there are effectively $k = 6$ finger contacts. The subspace $M^1_{1,2,3,4}(q_0)$ consists of instantaneous rotations of $\mathcal{B}$ about a single axis in $\mathbb{R}^3$. Similarly, the placement of a pointed or sharp-bladed finger at a concave edge of $\mathcal{B}$ can be interpreted as a double facet contact. When four fingers are placed at the contacts, the first-order mobility index becomes $m^1_{q_0} = m - k + 1 = 2$, since now there are effectively $k = 5$ finger contacts. The subspace $M^1_{1,2,3,4}(q_0)$ thus consists of instantaneous rotations of $\mathcal{B}$ about axes that span a plane in $\mathbb{R}^3$. The mobility analysis proceeds by showing that $\kappa_{\text{rel}}(q_0,\dot{q}) < 0$ for all instantaneous rotations of $\mathcal{B}$ in the respective subspace $M^1_{1,2,3,4}(q_0)$ (see Exercises). □

Example: The inscribed ball of the tetrahedral object $\mathcal{B}$ depicted in Figure 9.13(c) touches its surface at four interior facet points. The placement of four convex fingers of any shape, even *point fingers,* fully immobilizes the tetrahedral object at this grasp. ○

The generic polyhedra considered in Proposition 9.16 are immobilized by four convex finger bodies of any shape, provided that pointed or sharp-bladed fingers are placed at the object concavities. The finger curvatures become important when the inscribed ball touches the polyhedron's surface at two or three points. The immobilization of polyhedra with coaxial facets is considered in the second proposition.

PROPOSITION 9.17 **(four-finger immobilization, case II)** *Let an inscribed ball touch three coaxial facets of a polyhedron $\mathcal{B}$. By sliding the inscribed ball along the three facets, the ball determines four points, such that $\mathcal{B}$ is* **immobilized** *by placing four sufficiently flat convex finger bodies at these points.*

Proof sketch: Consider the triangular prism, $\mathcal{B}$, depicted in Figure 9.14. The inscribed ball touches three side facets of $\mathcal{B}$. To immobilize this object, first slide the inscribed ball upward until it touches the upper facet of $\mathcal{B}$. Let y denote the inscribed ball center at its upward position. Let x_0, x_1, x_2 denote the points where the ball touches the side facets at its upward position, and let x_3 denote the point where the inscribed ball touches the upper facet (Figure 9.14). We wish to locate a new point on the object surface, which will allow us to *eliminate* one of side-facet points from the grasp.

Let Δ be the plane spanned by the points x_0, x_1, x_2. Let $d_y(x)$ be the minimal distance of points x that lie in the portion of bdy($\mathcal{B}$) below Δ, measured from the inscribed ball center y (Figure 9.14). The *generalized inward normal* of $\mathcal{B}$ at a non-smooth point, $x \in$ bdy($\mathcal{B}$), is defined as the convex hull of the inward unit normals to the facets of $\mathcal{B}$ that meet at x. Using tools of non-smooth analysis (see Appendix A), it can be verified that $d_y(x)$ has at least one extremum point that lies below Δ. For instance, the point x_4 on the lower-right edge of $\mathcal{B}$ is such an extremum point (Figure 9.14). It can also be verified that at an extremum point of $d_y(x)$, the generalized inward normal of $\mathcal{B}$ contains a vector pointing toward y. Hence, we can eliminate the point x_0 and establish an essential four-finger equilibrium grasp of $\mathcal{B}$ at x_1, x_2, x_3, and x_4 (Figure 9.14).

The subspace of first-order free motions, $M^1_{1,2,3,4}(q_0)$, consists of instantaneous rotations of $\mathcal{B}$ about the inscribed ball center (see proof of Proposition 9.16). Second-order immobilization requires that $\kappa_{\text{rel}}(q_0,(0,\omega)) = \sum_{i=1}^4 \lambda_i \kappa_i(q_0,(0,\omega)) < 0$ for all $\omega \in \mathbb{R}^3$. Let us place identical *spherical fingers* of radius $r_{\mathcal{O}} \geq 0$ at the facet points x_1, x_2, and x_3. Setting $\mathcal{B}$'s frame origin at the inscribed ball center, $\boldsymbol{\rho}_i = -\rho_i n_i$ for $i = 1,2,3$. Moreover, each ρ_i equals the inscribed ball radius, $\rho_i = R$ for $i = 1,2,3$. Substituting for $\boldsymbol{\rho}_i$ as well as $L^{-1}_{\mathcal{O}_i} = r_{\mathcal{O}} I$, $L_{\mathcal{B}_i} = O$ (a 3×3 zero matrix) and $\|\eta_i\| = 1$ in Formula (9.4) gives $\kappa_i(q_0,(0,\omega)) = -(R + r_{\mathcal{O}})\|n_i \times \omega\|^2$ for $i = 1,2,3$.

Next consider the curvature form $\kappa_4(q_0,(0,\omega))$, associated with a locally smooth finger body, $\mathcal{O}_4$, placed at x_4. When x_4 is a vertex of $\mathcal{B}$, one substitutes $L_{\mathcal{B}_4} = \infty I$ in

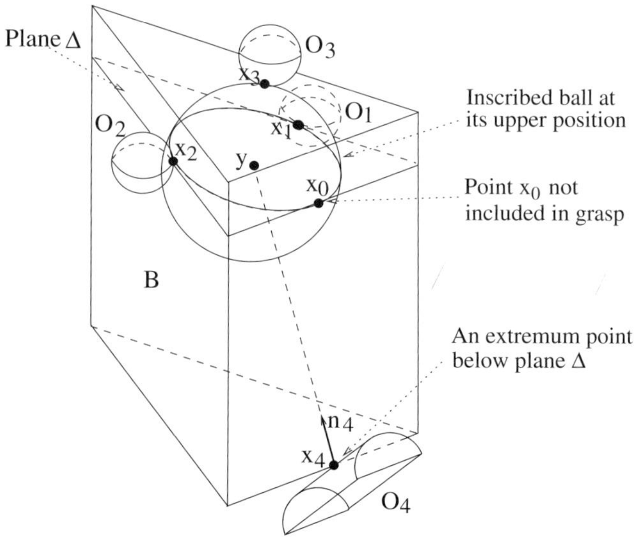

Figure 9.14 Minimal immobilization of a triangular prism using four convex fingers. The point x_4 is an extremum of $d_y(x)$ in the portion of bdy($\mathcal{B}$) that lies below the plane Δ. Immobilization requires sufficiently flat spherical fingers at the facet points x_1, x_2, and x_3, and a locally smooth finger body at x_4.

Formula (9.4). When x_4 is an interior edge point of $\mathcal{B}$, one substitutes $(n_4 \times \omega)^T L_{\mathcal{B}_4}(n_4 \times \omega) = 0$ when $n_4 \times \omega$ is aligned with the edge, and $(n_4 \times \omega)^T L_{\mathcal{B}_4}(n_4 \times \omega) = \infty$ when $n_4 \times \omega$ is transversal to the edge. We proceed with the case where x_4 is a vertex of $\mathcal{B}$ (an edge can be treated in a similar manner). Since $\mathcal{B}$'s frame origin is located at the inscribed ball center, $\rho_4 = -\rho_4 n_4$. Substituting $L_{\mathcal{B}_4} = \infty I$, $\rho_4 = -\rho_4 n_4$ and $\|\eta_i\| = 1$ in Formula (9.4) gives $\kappa_4(q_0, (0, \omega)) = \rho_4 (n_4 \times \omega)^T [\rho_4 L_{\mathcal{O}_4} + I](n_4 \times \omega)$. The relative c-space curvature form is thus given by

$$\kappa_{\text{rel}}(q_0, (0, \omega)) = -(R + r_{\mathcal{O}}) \cdot \sum_{i=1}^{3} \lambda_i \|n_i \times \omega\|^2 + \lambda_4 \rho_4 (n_4 \times \omega)^T [\rho_4 L_{\mathcal{O}_4} + I](n_4 \times \omega) \quad \omega \in \mathbb{R}^3.$$

The matrix $\sum_{i=1}^{3} \lambda_i [n_i \times]^T [n_i \times]$ is positive definite, since $\lambda_1, \lambda_2, \lambda_3 > 0$ and the contact normals n_1, n_2, n_3 are not coplanar (the generic case). Let $\sigma_{min} > 0$ denote the minimal eigenvalue of the latter matrix. Since x_4 is a vertex of $\mathcal{B}$, we may place a *flat finger* at this point (a cylindrical finger can be placed at an edge point as shown in Figure 9.14). Substituting $L_{\mathcal{O}_4} = O$ in the relative c-space curvature form gives $\kappa_{\text{rel}}(q_0, (0, \omega)) \leq (-(R + r_{\mathcal{O}})\sigma_{min} + \lambda_4 \rho_4)\|\omega\|^2$. The parameter ρ_4 is positive, since the inscribed ball center lies on $\mathcal{B}$'s side of x_4. However, ρ_4 is bounded by the diameter of $\mathcal{B}$,[4] while $r_{\mathcal{O}}$ can be made arbitrarily large. Choosing *sufficiently flat* spherical fingers at x_1, x_2, and x_3, with radius satisfying the inequality, $R + r_{\mathcal{O}} > \lambda_4 \rho_4 / \sigma_{min}$, gives $\kappa_{\text{rel}}(q_0, (0, \omega)) < 0$ for all $\omega \in \mathbb{R}^3$, as required for second-order immobilization. □

The immobilization of polyhedra with parallel facets is considered in the third proposition.

PROPOSITION 9.18 **(four-finger immobilization, case III)** *Let an inscribed ball touch two parallel facets of a polyhedron $\mathcal{B}$. By sliding the inscribed ball between the two facets, the ball determines four points, such that $\mathcal{B}$ is* **immobilized** *by placing sufficiently flat convex finger bodies at these points.*

Rather than giving a formal proof of Proposition 9.18, let us illustrate the immobilization technique with an example.

Example: Figure 9.15 shows a rectangular box, $\mathcal{B}$, with the inscribed ball touching its front and rear facets. The immobilization starts by sliding the inscribed ball to the left until it touches the left facet of $\mathcal{B}$. The inscribed ball is next moved upward while maintaining triple contact with the facets of $\mathcal{B}$ until it touches the upper facet of $\mathcal{B}$. At this stage, the inscribed ball touches the four points x_0, x_1, x_2, and x_3 (Figure 9.15). Now we have the situation discussed in the proof of Proposition 9.17. Namely, there exists a new point on the lower surface of $\mathcal{B}$, at the vertex x_4, which replaces the facet point x_0. The generalized inward normal of $\mathcal{B}$ at x_4 contains a vector, n_4, which points toward the inscribed ball center at its upper-left position (Figure 9.15). The placement of four convex finger bodies at x_1, x_2, x_3, x_4 forms an essential four-finger equilibrium grasp of $\mathcal{B}$. The subspace of first-order free motions of this grasp is three-dimensional and consists of all instantaneous rotations of $\mathcal{B}$ about the inscribed ball center. Second-order

[4] The *diameter* of $\mathcal{B}$ is the maximal distance between any two points on $\text{bdy}(\mathcal{B})$.

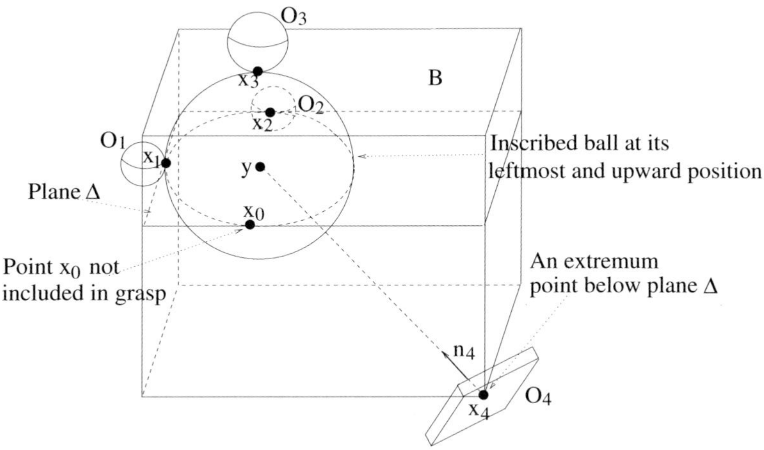

Figure 9.15 Minimal immobilization of a rectangular box using four convex finger bodies. The vertex x_4 is an extremum point of $d_y(x)$ in the portion of bdy($\mathcal{B}$) below the plane Δ. Immobilization requires sufficiently flat finger bodies at the facet points x_1, x_2, and x_3 and a locally smooth finger body at x_4.

immobilization requires sufficiently flat convex finger bodies at x_1, x_2, x_3 and a locally smooth convex finger body at the vertex x_4. The algebraic details of this statement are the same as discussed in the proof of Proposition 9.17. ○

To summarize, polyhedral objects without parallel or coaxial facets can be immobilized by four convex finger bodies of any shape (Proposition 9.16). Polyhedral objects with parallel or coaxial facets can be immobilized by four sufficiently flat convex finger bodies (Propositions 9.17 and 9.18). Since a smaller number of convex finger bodies *cannot* immobilize arbitrary polyhedral objects (see Exercises), we arrive at the following conclusion concerning the minimal second-order immobilization of all polyhedral objects.

THEOREM 9.19 **(minimal four-finger grasp of polyhedra)** *The minimal second-order immobilizing grasp of all polyhedral objects requires* **four convex finger bodies,** *provided that the finger bodies are sufficiently flat when the contacts are located on parallel or coaxial facets.*

The smaller number of fingers afforded by second-order effects (first-order immobilization requires seven fingers) comes at the price of *precise finger placement,* in order to ensure frictionless equilibrium grasps that are necessary for object immobilization. When small finger placement errors occur in the process of realizing an immobilizing grasp, the fingers will automatically form a *cage*, which limits the object's free motions to a small neighborhood centered at the grasp configuration q_0 (see next chapter). Moreover, we will see in Part III of the book that when friction is present at the contacts, small finger placement errors automatically result in a frictional *wrench resistant* grasp, which provides a complementary measure of grasp security. Robot hands that use four independent fingers, or three fingers and a palm, thus offer *minimalistic hand designs* that can safely grasp almost all 3-D polyhedral objects.

Bibliographical Notes

The question of how many fingers are needed to securely immobilize all rigid objects was first considered by Mishra, Schwartz and Sharir [1]. They noted that certain *exceptional objects,* such as bodies of revolution, cannot be fully immobilized by frictionless finger contacts. The exceptional objects were subsequently classified by Selig and Rooney [2]. Based on first-order effects, Markenscoff, Ni and Papadimitriou showed that all 3-D objects without rotational symmetry can be immobilized by seven frictionless point fingers [3]. Using curvature effects, Czyzowicz, Stojmenovic and Urrutia showed that three frictionless point fingers can immobilize all polygons without parallel edges, while four frictionless point fingers can immobilize all polyhedra without parallel or coaxial facets [4]. The geometric setting of configuration space used to describe the minimal immobilizing grasps, as well as their extension to minimal immobilization of general piecewise smooth 2-D and 3-D objects, appear in the authors' papers [5, 6].

Further reduction in the number of fingers is possible when the fingers can be preshaped according the class of objects at hand. In fact, a combination of a *flat finger* a *spherical concave finger* and a *cylindrical concave finger* can fully immobilize all polyhedral objects, provided that the concave fingers are suitably curved at the contacts. This result has been obtained by Seo, Yim and Kumar [7]. Thus, when finger preshaping is permissible, the minimal second-order immobilizing grasp of all polyhedral objects involves only three suitably concave fingers.

Finally, many robot hand designs strive to be minimal in terms of their number of fingers. See, for instance, Dollar, Howe and coworkers [8, 9]. A classical example is the *Salisbury Hand*, which consists of three independent finger mechanisms attached to a common palm [10, 11]. When such a three-finger hand grasps 3-D objects, friction effects ensure *wrench resistance,* which we will meet in Part III of the book. This chapter established that robot hands can use three independent fingers and a palm to immobilize almost all 3-D objects. Minimalistic robot hand designs can thus exploit the synergy that exists between friction effects, which ensure wrench resistant grasps, and rigid-body constraints that ensure immobilizing grasps.

Appendix I: Details Concerning the Inscribed Disc

This appendix contains proofs of two properties of the inscribed disc or ball of a rigid object $\mathcal{B}$. While these properties hold for general piecewise smooth objects, we assume for simplicity that $\mathcal{B}$ has a smooth boundary, represented as the zero-level set of a smooth real-valued function, $\textsf{bdy}(\mathcal{B}) = \{x \in \mathbb{R}^n : f(x) = 0\}$, such that $\nabla f(x)$ is non-zero at all points $x \in \textsf{bdy}(\mathcal{B})$. The inscribed disc or ball is determined by local maxima of the Lipschitz continuous function $\textsf{dst}(y, \textsf{bdy}(\mathcal{B}))$ (which is differentiable almost everywhere). The *generalized gradient* of this function, denoted $\partial\textsf{dst}(y, \textsf{bdy}(\mathcal{B}))$, forms the following set of vectors (see Appendix A). If $\textsf{dst}(y, \textsf{bdy}(\mathcal{B}))$ is differentiable at $y \in \mathcal{B}$, there exists a unique closest point, $x \in \textsf{bdy}(\mathcal{B})$, and $\partial\textsf{dst}(y, \textsf{bdy}(\mathcal{B}))$ reduces to the unit vector pointing from x to y. If $\mathsf{dst}(y, \mathsf{bdy}(\mathcal{B}))$ is not differentiable at $y \in \mathcal{B}$, its generalized gradient is the *convex hull* of the unit vectors associated with the points $x \in \mathsf{bdy}(\mathcal{B})$ that are closest to y:

$$\partial \mathsf{dst}(y, \mathsf{bdy}(\mathcal{B})) = \mathsf{co}\left\{ \frac{1}{\|y - x\|}(y - x) \, : \, \text{for } x \in \mathcal{I}(y) \right\}, \tag{9.7}$$

where co denotes convex hull and $\mathcal{I}(y)$ is the set of points $x \in \mathsf{bdy}(\mathcal{B})$ closest to y.

Let us start with the proof of Lemma 9.3, which characterizes the number of points where an inscribed disc or ball touches the boundary of a rigid body $\mathcal{B}$.

Lemma 9.3 *Let $\mathcal{B}$ form a compact n-dimensional body with nonempty interior, $n = 2$ or 3. The inscribed disc or ball of $\mathcal{B}$ generically touches its boundary at $2 \leq k \leq n + 1$ isolated points.*

Proof: The inscribed disc or ball is centered at an interior point, $y \in \mathcal{B}$, which forms a local maximum of $\mathsf{dst}(y, \mathsf{bdy}(\mathcal{B}))$. A necessary condition for a non-smooth extremum point of $\mathsf{dst}(y, \mathsf{bdy}(\mathcal{B}))$ is $\vec{0} \in \partial \mathsf{dst}(y, \mathsf{bdy}(\mathcal{B}))$ (see Appendix A). Since $\partial \mathsf{dst}(y, \mathsf{bdy}(\mathcal{B}))$ is the convex hull of vectors associated with the points $x \in \mathcal{I}(y)$, any local maximum point y must have $k \geq 2$ closest points on $\mathsf{bdy}(\mathcal{B})$ in order to satisfy the condition $\vec{0} \in \partial \mathsf{dst}(y, \mathsf{bdy}(\mathcal{B}))$. This gives the lower bound $k \geq 2$.

To establish the upper bound, let $\mathcal{I}(y) = \{x_1, \ldots, x_k\}$ be the closest points to y on $\mathsf{bdy}(\mathcal{B})$. The requirement that $x_1, \ldots, x_k \in \mathsf{bdy}(\mathcal{B})$ is captured by k scalar constraints: $f(x_i) = 0$ for $i = 1 \ldots k$. Assume that $\nabla f(x)$ is pointing into the interior of $\mathcal{B}$ at all points $x \in \mathsf{bdy}(\mathcal{B})$. Since each point $x_i \in \mathcal{I}(y)$ is locally closest to y on $\mathsf{bdy}(\mathcal{B})$, the *Lagrange multiplier rule* implies that $\nabla f(x_i)$ must be collinear with the vector pointing from x_i to y, which gives the equations:

$$\frac{1}{\|\nabla f(x_i)\|}\nabla f(x_i) = \frac{1}{\|y - x_i\|}(y - x_i) \qquad i = 1 \ldots k. \tag{9.8}$$

Note that Eq. (9.8) specifies a total of $k \cdot (n-1)$ scalar constraints. Last, the requirement that the inscribed disc or ball center point, y, be equidistant from the closest points $x_1, \ldots, x_k$ is captured by the $k-1$ scalar equations: $\|y - x_1\| = \cdots = \|y - x_k\|$.

Thus we have a total of $k + k \cdot (n-1) + (k-1)$ scalar constraints in the variables $(x_1, \ldots, x_k, y) \in \mathbb{R}^{(k+1)n}$. The manifolds determined by these constraints generically intersect in a transversal manner. The generic dimension of the intersection manifold is $(k+1) \cdot n - (kn + k - 1) = n - k + 1$. The solution manifold is nonempty, since $\mathsf{dst}(y, \mathsf{bdy}(\mathcal{B}))$ is a continuous function and hence attains at least one maximum that necessarily lies in the interior of $\mathcal{B}$. Hence, the dimension of the solution manifold is either zero, which corresponds to a set of isolated points, or strictly positive which gives the upper bound $k \leq n + 1$, where $n = 2$ or 3. □

The next lemma describes the feasibility of holding a rigid object $\mathcal{B}$ at a frictionless equilibrium grasp determined by its inscribed disc or ball.

Lemma 9.4 *Let $\mathcal{D}(y)$ be an inscribed disc or ball of an object $\mathcal{B}$. The inward unit normals at the points $x \in \mathcal{I}(y)$ point toward the point y, and when placed at a common origin, the convex hull of these normals contains the origin.*

Proof: Let $y \in \mathcal{B}$ be a local maximum of $\mathsf{dst}(y, \mathsf{bdy}(\mathcal{B}))$ (note that y necessarily lies in the interior of $\mathcal{B}$). We will show that the unit vectors, $(y - x_i)/\|y - x_i\|$ for $x_i \in \mathcal{I}(y)$, are collinear with $\mathcal{B}$'s inward unit normals, $n(x_i)$ for $x_i \in \mathcal{I}(y)$. Every point $x_i \in \mathcal{I}(y)$ locally minimizes the function $\varphi_y(x) = \|x - y\|$ over the scalar constraint $f(x) = 0$, which describes the object boundary. Using the *Lagrange multiplier rule*, a local minimum of $\varphi_y(x)$ satisfies the condition:

$$\nabla\varphi(x_i) = \frac{1}{\|y - x_i\|}(y - x_i) = \sigma\nabla f(x_i) \qquad \text{for some } \sigma \in \mathbb{R}.$$

Since $\nabla f(x_i)$ is orthogonal to the object boundary at x_i, so must be the unit vector $(y - x_i)/\|y - x_i\|$. By construction, the line segment joining x_i and y lies inside $\mathcal{B}$, otherwise it would cross the boundary of $\mathcal{B}$ at a point closer to y than x_i. Hence, the vector $y - x_i$ points into the interior of $\mathcal{B}$ at x_i and, consequently,

$$\frac{1}{\|y - x_i\|}(y - x_i) = n(x_i) \qquad x_i \in \mathcal{I}(y).$$

Substituting for these unit vectors in Eq. (9.7) gives $\partial\mathsf{dst}(y, \mathsf{bdy}(\mathcal{B})) = \mathrm{co}\{n(x_1), \ldots, n(x_k)\}$ for $x_i \in \mathcal{I}(y)$. As already stated in the proof of Lemma 9.3, any non-smooth extremum point of $\mathsf{dst}(y, \mathsf{bdy}(\mathcal{B}))$ satisfies the condition $\vec{0} \in \partial\mathsf{dst}(y, \mathsf{bdy}(\mathcal{B}))$. Hence, $\vec{0} \in \mathrm{co}\{n(x) : x \in \mathcal{I}(y)\}$, as required for an equilibrium grasp. □

Appendix II: Details Concerning Minimal Second-Order Immobilization of 2-D Objects

This appendix contains proofs of Lemma 9.8 and Theorem 9.11 from Section 9.3. The following lemma considers the conversion of an antipodal grasp into an essential three-finger equilibrium grasp.

Lemma 9.8 *If an inscribed disc of $\mathcal{B}$ touches its boundary at two points, x_1 and x_2, any local splitting of x_1 into two points, followed by suitable local shift of x_2 to a new point along* $\mathsf{bdy}(\mathcal{B})$, *forms an* **essential three-finger equilibrium grasp** *of $\mathcal{B}$.*

Proof: An essential three-finger equilibrium grasp satisfies two conditions. The finger force directions positively span the origin of $\mathbb{R}^2$, and the finger force lines intersect at a common point, such that no two of these lines are collinear. Let us prove the lemma for the generic case where $\mathcal{B}$ has nonconcentric circles of curvature at x_1 and x_2. In this case, $\mathcal{B}$ must have non-zero curvature at one of the antipodal points, say at x_1; otherwise $\mathcal{B}$'s circles of curvature are concentric at infinity. Consider a local splitting of x_1 into two points, x_{11} and x_{12}, located on both sides of x_1 (Figure 9.7). Let p denote the intersection point of the lines underlying the normals $n(x_{11})$ and $n(x_{12})$. We will first show that the antipodal point, x_2, can be locally moved along $\mathsf{bdy}(\mathcal{B})$ to a new point, x_3, such that the line underlying the normal, $n(x_3)$, passes through p.

Consider the mapping $\psi : \mathsf{bdy}(\mathcal{B}) \times \mathbb{R} \to \mathbb{R}^2$, given by $\psi(x, \sigma) = x + \sigma n(x)$, where $\sigma \in \mathbb{R}$ and $n(x)$ is $\mathcal{B}$'s inward unit normal at $x \in \mathsf{bdy}(\mathcal{B})$. Since x_1 and x_2 are antipodal points, $\mathcal{B}$'s center of curvature at x_1, denoted z_1, can be written as $z_1 = x_2 + \sigma_1 n(x_2)$ for some $\sigma_1 \in \mathbb{R}$. Thus, $z_1 = \psi(x_2, \sigma_1)$. In general, the Jacobian of ψ, $D\psi(x,\sigma)$, is singular only at points (x,σ) whose image, $\psi(x,\sigma)$, coincides with $\mathcal{B}$'s center of curvature at $x \in \text{bdy}(\mathcal{B})$. In particular, $D\psi(x_2,\sigma)$ is singular only at (x_2,σ_2) such that $z_2 = x_2 + \sigma_2 n(x_2)$ is $\mathcal{B}$'s center of curvature at x_2. Since, by assumption, $z_1 \neq z_2$, $D\psi(x,\sigma)$ is non-singular at (x_2,σ_1). By the *inverse function theorem,* ψ is a local diffeomorphism (i.e., a local one-to-one and onto mapping) of a small open neighborhood centered at (x_2,σ_1) onto an open neighborhood, $\mathcal{D}_{z_1}$, centered at z_1 in $\mathbb{R}^2$. For a sufficiently local splitting of x_1, the intersection point p lies in $\mathcal{D}_{z_1}$. By definition of ψ, each point $p \in \mathcal{D}_{z_1}$ can be written as $p = x_3 + \sigma_3 n(x_3)$, for some $x_3 \in \text{bdy}(\mathcal{B})$ that is close to x_2 and some σ_3 that is close to σ_1. That is, one can locally move x_2 to a new point, $x_3 \in \text{bdy}(\mathcal{B})$, at which the line underlying the normal $n(x_3)$ intersects the point p.

Next, we show that the normals $n(x_{11}), n(x_{12}), n(x_3)$ positively span the origin of $\mathbb{R}^2$. Let $\alpha(s)$ parametrize the object boundary by arclength, such that $\alpha(0) = x_1$. Let t_1 and n_1 be $\mathcal{B}$'s unit tangent and inward unit normal at x_1. The first-order approximation of $\mathcal{B}$'s inward unit normal in the vicinity of x_1 is given by $n(\alpha(s)) = n_1 + s\kappa_{\mathcal{B}_1} t_1$, where $\kappa_{\mathcal{B}_1}$ is $\mathcal{B}$'s curvature at x_1. Any local splitting of x_1 into x_{11} and x_{12} corresponds to $s = -\epsilon_{11}$ and $s = \epsilon_{12}$ for small parameters $\epsilon_{11}, \epsilon_{12} > 0$. Since $\kappa_{\mathcal{B}_1} \neq 0$, the normals $n(x_{11}) = n_1 - \epsilon_{11}\kappa_{\mathcal{B}_1} t_1$, $n(x_{12}) = n_1 + \epsilon_{12}\kappa_{\mathcal{B}_1} t_1$ and $n(x_2) = -n_1$ positively span the origin of $\mathbb{R}^2$, such that no two of these normals are collinear. This property is next extended to the normals $n(x_{11})$, $n(x_{12})$, and $n(x_3)$.

When $\mathcal{B}$ is held in a three-finger equilibrium grasp, the contact force magnitudes are given up to a common scaling factor by $\lambda_i = n(x_{i+1}) \times n(x_{i+2})$ for $i = 1,2,3$, where index addition is taken modulo 3 (Lemma 5.6). Using this formula, consider the mapping $\vec{\lambda}(x) : \text{bdy}(\mathcal{B}) \to \mathbb{R}^3$ given by $\vec{\lambda}(x) = (n(x_{12}) \times n(x), n(x) \times n(x_{11}), n(x_{11}) \times n(x_{12}))$. It maps points $x \in \text{bdy}(\mathcal{B})$ to the equilibrium grasp coefficients $\vec{\lambda} = (\lambda_1, \lambda_2, \lambda_3)$ associated with fixed contacts x_{11} and x_{12}, and a *variable contact* $x \in \text{bdy}(\mathcal{B})$. Since $n(x_{11}), n(x_{12}), n(x_2)$ positively span the origin of $\mathbb{R}^2$, $\vec{\lambda}(x)$ maps the point x_2 to a vector of positive coefficients, $\vec{\lambda}^*$. Since $\vec{\lambda}(x)$ is a continuous function, the inverse image of an open neighborhood centered at $\vec{\lambda}^*$ under $\vec{\lambda}(x)$ forms an open interval of $\text{bdy}(\mathcal{B})$, centered at x_2. Any sufficiently local splitting of x_1 such that x_3 lies in this interval ensures that the normals $n(x_{11}), n(x_{12}), n(x_3)$ positively span the origin of $\mathbb{R}^2$, such that no two of these normals are collinear. □

The next theorem considers the minimal immobilization of 2-D objects using two possibly concave fingers.

Theorem 9.11 *When the fingers are allowed to have any curvature at the contacts, the minimal second-order immobilizing grasp of all smooth 2-D objects that do not have circular symmetry requires* **two possibly concave finger bodies**.

Proof: The immobilization of $\mathcal{B}$ starts with the placement of two finger bodies, $\mathcal{O}_1$ and $\mathcal{O}_2$, at antipodal points $x_1, x_2 \in \text{bdy}(\mathcal{B})$, such that $\mathcal{B}$ has nonconcentric circles of curvature at these points. The first-order mobility index of the grasp is $m^1_{q_0} = m - k + 1 = 2$, since $m = 3$ and there are $k = 2$ finger contacts. The subspace of first-order free motions, $M^1_{1,2}(q_0)$, is two-dimensional and consists of instantaneous rotations of $\mathcal{B}$

about all points of the extended line, l, which passes through x_1 and x_2. The object $\mathcal{B}$ is immobilized when the relative c-space curvature satisfies the condition:

$$\kappa_{\text{rel}}(q_0,(0,\omega)) = \lambda_1\kappa_1(q_0,(0,\omega)) + \lambda_2\kappa_2(q_0,(0,\omega)) < 0 \qquad \dot{q} = (0,\omega) \in M^1_{1,2}(q_0),$$

where $\lambda_1 = \lambda_2$ at the two-finger equilibrium grasp. In particular, we will use $\lambda_1 = \lambda_2 = 1$.

Consider the generic situation, where the object $\mathcal{B}$ has non-zero curvature at x_1 and x_2. When $\mathcal{B}$ is *concave* at x_i ($r_{\mathcal{B}_i} < 0$), select a convex finger body $\mathcal{O}_i$ with radius of curvature $r_{\mathcal{O}_i} = (-r_{\mathcal{B}_i}) - \epsilon > 0$, such that $\epsilon < |r_{\mathcal{B}_i}|$. When $\mathcal{B}$ is *convex* at x_i ($r_{\mathcal{B}_i} > 0$), select a concave finger body $\mathcal{O}_i$ with radius of curvature $r_{\mathcal{O}_i} = (-r_{\mathcal{B}_i}) - \epsilon < 0$ (Figure 9.16). In both cases, ϵ is a small positive parameter. Substituting $r_{\mathcal{O}_i} = (-r_{\mathcal{B}_i}) - \epsilon$ in Formula (9.1) for $\kappa_i(q_0,(0,\omega))$ gives

$$\kappa_i(q_0,(0,\omega)) = -\frac{1}{\epsilon}(\rho_i - r_{\mathcal{B}_i})(\rho_i - r_{\mathcal{B}_i} - \epsilon)\cdot\omega^2 \qquad i = 1,2. \tag{9.9}$$

The parameter ρ_i measures the distance of $\mathcal{B}$'s instantaneous rotation axis from the contact point, x_i, such that $\rho_i > 0$ on $\mathcal{B}$'s side of x_i and $\rho_i < 0$ on $\mathcal{O}_i$'s side of x_i (Figure 9.16). The parameters ρ_1 and ρ_2 can be expressed in terms of a common variable as follows. Let ρ measure the distance of $\mathcal{B}$'s instantaneous rotation axis from the midpoint of x_1 and x_2 along l, such that ρ is negative toward x_1 and positive toward x_2 (Figure 9.16). Let $2L$ be the distance between x_1 and x_2. Then $\rho_1 = L + \rho$ and $\rho_2 = L - \rho$. Substituting for ρ_1 and ρ_2 in Eq. (9.9) gives the second-order immobilization condition

$$\begin{aligned}\kappa_{\text{rel}}(q_0,(0,\omega)) = -\frac{1}{\epsilon}\big((\rho + L - r_{\mathcal{B}_1})(\rho + L - r_{\mathcal{B}_1} - \epsilon)\\ + (\rho - L + r_{\mathcal{B}_2})(\rho - L + r_{\mathcal{B}_2} + \epsilon)\big)\cdot\omega^2 < 0 \qquad \rho \in \mathbb{R}.\end{aligned} \tag{9.10}$$

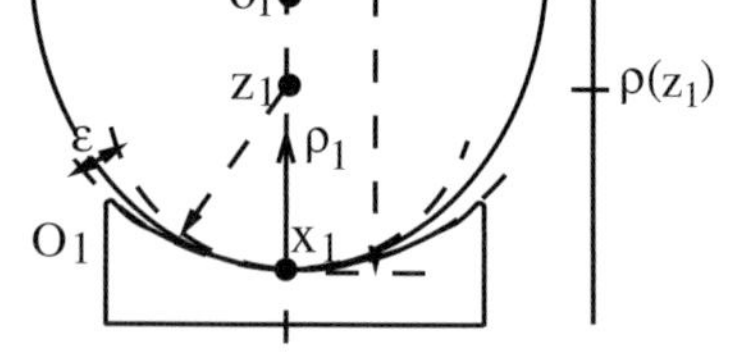

Figure 9.16 At a two-finger equilibrium grasp, the relative c-space curvature form becomes a quadratic polynomial in ρ. Two closely matching concave finger bodies, $\mathcal{O}_1$ and $\mathcal{O}_2$, make the polynomial strictly negative for all $\rho \in \mathbb{R}$, which ensures second-order immobilization of $\mathcal{B}$.

Let z_i denote the position of $\mathcal{B}$'s center of curvature at x_i for $i=1,2$. The ρ_i coordinate of z_i is $\rho_i(z_i)=r_{\mathcal{B}_i}$ for $i=1,2$. Hence, the ρ coordinates of z_1 and z_2 are $\rho(z_1)= r_{\mathcal{B}_1}-L$ and $\rho(z_2)=L-r_{\mathcal{B}_2}$. Writing Eq. (9.10) in terms of $\rho(z_1)$ and $\rho(z_2)$ gives the immobilization condition:

$$\kappa_{\text{rel}}(q_0,(0,\omega))=-\big((\rho-\rho(z_1))^2+(\rho-\rho(z_2))^2+\epsilon(\rho(z_1)-\rho(z_2))\big)<0 \qquad \rho\in\mathbb{R}, \tag{9.11}$$

where we omitted the positive factors $1/\epsilon$ and ω^2. Eq. (9.11) is quadratic in ρ. The maximum value attained by this polynomial is $-\big(\frac{1}{2}(\rho(z_1)-\rho(z_2))^2+\epsilon(\rho(z_1)-\rho(z_2))\big)$. Since $z_1\neq z_2$ and, hence, $\rho(z_1)\neq\rho(z_2)$, the condition of Eq. (9.11) holds true when ϵ is chosen according to the inequality:

$$\tfrac{1}{2}\left|\rho(z_1)-\rho(z_2)\right|^2-\epsilon\left|\rho(z_1)-\rho(z_2)\right|>0.$$

Or, equivalently, $\epsilon<\frac{1}{2}\left|\rho(z_1)-\rho(z_2)\right|=\frac{1}{2}\|z_1-z_2\|$. Thus, if the finger curvatures are chosen according to the rule $r_{\mathcal{O}_i}=(-r_{\mathcal{B}_i})-\epsilon$, such that $\epsilon<\left|r_{\mathcal{B}_i}\right|$ and $\epsilon<\frac{1}{2}\|z_1-z_2\|$, the relative c-space curvature is strictly negative for all $\rho\in\mathbb{R}$, as required for second-order immobilization. ○

Exercises

Exercise 9.1: A 3-D object bounded by a *surface of revolution* can be immobilized up to its rotational symmetry via frictionless finger contacts. Are there any other exceptional objects that *cannot* be fully immobilized via frictionless contacts?

Section 9.1

Exercise 9.2: Let $\alpha(s)$ parametrize the boundary of a smooth 2-D object $\mathcal{B}$ by arclength. Let $n(\alpha(s))$ be the object's outward unit normal at $\alpha(s)$. Develop the Taylor expansions $\alpha(s)=\alpha(0)+st_1+\frac{1}{2}s^2\kappa_{\mathcal{B}_1}n_1+o(s^3)$ and $n(\alpha(s))=n_1+s\kappa_{\mathcal{B}_1}t_1+\frac{1}{2}s^2\big(\kappa'_{\mathcal{B}_1}t_1+(\kappa_{\mathcal{B}_1})^2n_1\big)+o(s^3)$, where $\kappa_{\mathcal{B}_1}$ and $\kappa'_{\mathcal{B}_1}$ are the object's curvature and curvature derivative at $\alpha(0)$.

Exercise 9.3: Consider the line l that passes through two antipodal points, x_1 and x_2, of a smooth 2-D object $\mathcal{B}$. Based on the Taylor expansions of $\alpha(s)$ and $n(\alpha(s))$, show that the line underlying the contact normal $n(\alpha(s))$ intersects the line l at a distance given by $\sigma(s)=1/(\kappa_{\mathcal{B}}(\alpha(s))+\frac{1}{2}s\kappa'_{\mathcal{B}}(\alpha(s)))$. (Note that $\sigma(s)$ approaches the object's radius of curvature, $r_{\mathcal{B}}(\alpha(s))=1/\kappa_{\mathcal{B}}(\alpha(s))$, as the parameter s approaches zero.)

Exercise 9.4*: A k-finger robot hand grasps a polygonal object, $\mathcal{B}$, using one *flat* fingertip that can maintain *line contact* with $\mathcal{B}$, while all other fingertips maintain *point contact* with $\mathcal{B}$. Do the minimal first-order immobilizing grasps still require four fingers? (Consider the c-obstacle boundary when a flat fingertip maintains line contact with $\mathcal{B}$.)

Exercise 9.5: Every polygonal object $\mathcal{B}$ has a pair of antipodal vertices, whose generalized inward normals contain opposing directions (Figure 9.3). Starting at such a pair of vertices, determine which edge segments of $\mathcal{B}$ support an essential four-finger equilibrium grasp of $\mathcal{B}$.

Section 9.2

Exercise 9.6: The inscribed disc of a *regular k-sided polygon* $\mathcal{B}$ touches its boundary at k isolated points (Figure 9.5). Explain why two or three of these points must have inward contact normals that positively span the origin of $\mathbb{R}^2$.

Exercise 9.7: Can the inscribed disc of a polygon $\mathcal{B}$ touch any of its vertices? (Consider generic cases where the inscribed disc touches the object's boundary at two or three points.)

Exercise 9.8: Prove that any convex 2-D object $\mathcal{B}$ that has no opposing parallel edge contains a *unique* inscribed disc. (Hint: when the object $\mathcal{B}$ is convex, linear translation between any two inscribed discs can be performed inside $\mathcal{B}$.)

Exercise 9.9*: In continuation of the previous exercise, prove that the inscribed disc of any convex polygon $\mathcal{B}$ can be computed as a *convex optimization* problem. (The inscribed disc can therefore be computed by extremely efficient optimization algorithms.)

Exercise 9.10: Can a *non-convex* polygon $\mathcal{B}$ have a unique inscribed disc?

Exercise 9.11: An inscribed disc of a 2-D object touches a point at which the object boundary is locally smooth and convex. Prove that the inscribed disc's radius forms a *lower bound* on the object's radius of curvature at this point. (This exercise proves Lemma 9.6.)

Exercise 9.12: An inscribed disc of a 2-D object $\mathcal{B}$ touches its boundary at two points x_1 and x_2. Explain why the object boundary must be strictly convex at least at one of these points. For instance, $\mathcal{B}$ cannot be concave at both x_1 and x_2.

Solution sketch: Consider the inscribed disc as a rigid body, $\mathcal{D}$, that can freely move in the interior of $\mathcal{B}$. Since $\mathcal{D}$ is locally maximal, it must be fully immobilized by $\mathsf{bdy}(\mathcal{B})$; otherwise, $\mathcal{D}$ can locally break away from $\mathsf{bdy}(\mathcal{B})$ and move into the interior of $\mathcal{B}$. If $\mathcal{B}$ is non-convex at both points x_1 and x_2, $\mathcal{D}$ is "held"' by two "fingers," one strictly convex (the concave portion of $\mathsf{bdy}(\mathcal{B})$ surrounding x_1), while the other non-concave (the non-convex portion of $\mathsf{bdy}(\mathcal{B})$ surrounding x_2). Hence, $\mathcal{D}$ has a second-order escape motion that is basically local sliding in the direction orthogonal to the segment connecting x_1 and x_2. Hence, $\mathcal{D}$ cannot be a locally maximal inscribed disc of $\mathcal{B}$ in this case. ○

Section 9.3

Exercise 9.13*: Let an inscribed disc of a convex 2-D object $\mathcal{B}$ touch its boundary at two antipodal points that lie on parallel edges. Using nonlocal splitting of one of these

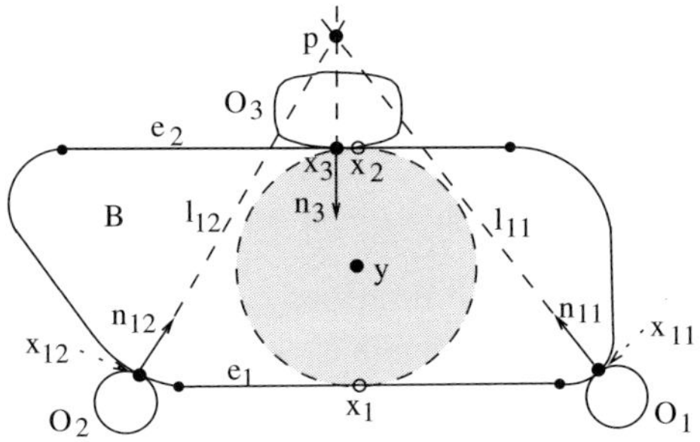

Figure 9.17 Immobilization of a smooth object with parallel edges, e_1 and e_2. The point x_1 is split into two points, x_{11} and x_{12}, which lie beyond the ends of e_1. The point x_2 is subsequently shifted along e_2 to a new point x_3, where the contact normal n_3 passes through the point p where the lines underlying the contact normals n_{11} and n_{12} intersect.

points, prove that $\mathcal{B}$ can be fully immobilized by three sufficiently flat convex finger bodies.

Solution sketch: Let the inscribed disc of $\mathcal{B}$ touch two parallel edges, e_1 and e_2, at the antipodal points x_1 and x_2 (Figure 9.17). Let the endpoints of e_1 be denoted x'_{11} and x'_{12}, such that x'_{12} is the point whose projection onto e_2 falls in the interior of e_2. The immobilization of $\mathcal{B}$ starts with the placement of a convex finger body $\mathcal{O}_1$ at x'_{11}, followed by local shift of this finger along bdy($\mathcal{B}$) outward with respect to e_1, to a nearby point x_{11} (Figure 9.17). Let l_{11} denote the line underlying the contact normal n_{11}, and let l denote the line orthogonal to e_1, which passes through x'_{12}. Note that l crosses the interior of e_2. The line l_{11} intersects the line l at some point p'. Now, place a second convex finger body, $\mathcal{O}_2$, at x'_{12}, then slide this finger along bdy($\mathcal{B}$) outward with respect to e_1, to a nearby point x_{12} (Figure 9.17). For a *fixed* line l_{11}, the line underlying the contact normal n_{12}, denoted l_{12}, intersects l_{11} at a point p close to p'. Since the projection of p' on e_2 falls in the interior of e_2, the projection of all points p sufficiently close to p' also fall in the interior of e_2. Hence, e_2 contains a point, x_3, whose edge normal yields a line that is concurrent with l_{11} and l_{12} at p. Placing a third convex finger body, $\mathcal{O}_3$, at x_3 gives an essential three-finger equilibrium grasp of $\mathcal{B}$.

The subspace of first-order free motions at the grasp, $M^1_{1,2,3}(q_0)$, consists of instantaneous rotations of $\mathcal{B}$ about the concurrency point p. The object is immobilized if $m^2_{q_0} = 0$ or, equivalently, if $\kappa_{\text{rel}}(q_0, \dot{q}) < 0$ for all $\dot{q} \in M^1_{1,2,3}(q_0)$. The relative c-space curvature along $\dot{q} = (0, \omega)$ is given by $\kappa_{\text{rel}}(q_0, (0, \omega)) = \lambda_1 \kappa_{11}(q_0, (0, \omega)) + \lambda_2 \kappa_{12}(q_0, (0, \omega)) + \lambda_3 \kappa_3(q_0, (0, \omega))$, where $\lambda_1, \lambda_2, \lambda_3 > 0$ are determined by the equilibrium grasp condition. Based on Formula (9.1), $\kappa_{11}(q_0, (0, \omega))$ and $\kappa_{12}(q_0, (0, \omega))$ have finite values, determined by the parameters ρ_1 and ρ_2 that measure the distance of p from the contact points x_{11} and x_{12}. Substituting $r_{\mathcal{B}_3} = \infty$ for $\mathcal{B}$'s radius of curvature at x_3 in Formula (9.1) gives: $\kappa_3(q_0, (0, \omega)) = -(\rho_3 + r_{\mathcal{O}_3})$. One can now derive a lower bound on $r_{\mathcal{O}_3}$ (i.e., a flatness requirement on $\mathcal{O}_3$), so that $\kappa_3(q_0, (0, \omega))$ becomes a *large negative* number, which ensures that $\kappa_{\text{rel}}(q_0, (0, \omega)) < 0$ for all $\dot{q} \in M^1_{1,2,3}(q_0)$. ○

Section 9.4

Exercise 9.14: Provide a proof sketch for Lemma 9.13, that every convex polygon with parallel edges can be immobilized by three sufficiently flat convex finger bodies, provided that locally smooth finger bodies are placed at the vertices of the polygon.

Solution sketch: Let the inscribed disc touch two parallel edges of the polygon, e_1 and e_2, at the antipodal points x_1 and x_2. The immobilization starts by placing two smooth convex finger bodies, $\mathcal{O}_1$ and $\mathcal{O}_2$, at the vertices x_{11} and x_{12} at the ends of the edge containing the point x_1 (Figure 9.11). By rotating the two finger bodies about their vertices, the lines underlying the finger contact normals intersect at a point p, whose horizontal projection can lie anywhere in the horizontal overlap of e_1 and e_2. Hence, placing a third finger body, $\mathcal{O}_3$, at the projection of p on the edge containing the antipodal point x_2 gives a three-finger equilibrium grasp of $\mathcal{B}$ (Figure 9.11). The three finger forces are essential for maintaining the equilibrium grasp. The first-order mobility index of the grasp is therefore $m^1_{q_0} = 1$. The one-dimensional subspace of first-order free motions, $M^1_{1,2,3}(q_0)$, consists of instantaneous rotations of $\mathcal{B}$ about the concurrency point p (Figure 9.11).

The polygon is immobilized when $\kappa_{\text{rel}}(q_0, \dot{q}) < 0$ for all $\dot{q} \in M^1_{1,2,3}(q_0)$. Recall that $\kappa_{\text{rel}}(q_0, (0, \omega)) = \lambda_1 \kappa_1(q_0, (0, \omega)) + \lambda_2 \kappa_2(q_0, (0, \omega)) + \lambda_3 \kappa_3(q_0, (0, \omega))$, where $\lambda_1, \lambda_2, \lambda_3 > 0$ are determined by the equilibrium grasp condition. Using Formula (9.1), $\kappa_1(q_0, (0, \omega))$ and $\kappa_2(q_0, (0, \omega))$ have finite values determined by the parameters ρ_1 and ρ_2, which measure the distance of p from the contact points x_{11} and x_{12}. The finger body $\mathcal{O}_3$ contacts a flat edge of $\mathcal{B}$. Substituting $r_{\mathcal{B}_3} = \infty$ in Formula (9.1) gives $\kappa_3(q_0, (0, \omega)) = -(\rho_3 + r_{\mathcal{O}_3})$. The parameter ρ_3 measures the distance of p from x_3 and is negative, since p is located on the finger's side of x_3 (Figure 9.11). However, ρ_3 is finite at the given grasp. On the other hand, the finger's radius of curvature at the contact, $r_{\mathcal{O}_3} > 0$, can be made arbitrarily large by choosing a sufficiently flat finger body $\mathcal{O}_3$. Hence, $\kappa_3(q_0, (0, \omega))$ can be made arbitrarily large and *negative*, thus ensuring that $\kappa_{\text{rel}}(q_0, \dot{q}) < 0$ for all $\dot{q} \in M^1_{1,2,3}(q_0)$. ○

Exercise 9.15: Provide a proof sketch that every *non-convex* polygon with parallel edges can be immobilized by three sufficiently flat convex finger bodies, provided that locally smooth finger bodies are placed at the convex vertices of the polygon.

Solution sketch: Consider the non-convex polygon depicted in Figure 9.18. First, slide the inscribed disc to the left, between the parallel edges, until a new contact is established. At this stage, the inscribed disc touches the polygon's boundary at the points x_0, x_1, and x_2. Next consider the minimal distance function, $d_y(x)$, which measures the minimal distance from the inscribed disc center y of points x in the portion of $\mathsf{bdy}(\mathcal{B})$ between x_1 and x_2 on the right side of these points. Let x_3 be the vertex where $d_y(x)$ attains a global maximum on this portion of $\mathsf{bdy}(\mathcal{B})$. Using tools of non-smooth analysis (see Appendix A), it can be verified that the generalized normal to $\mathsf{bdy}(\mathcal{B})$ at x_3 contains a direction that passes through y. The point x_0 can thus be replaced by the opposing vertex point x_3. The object is immobilized at the resulting three-finger

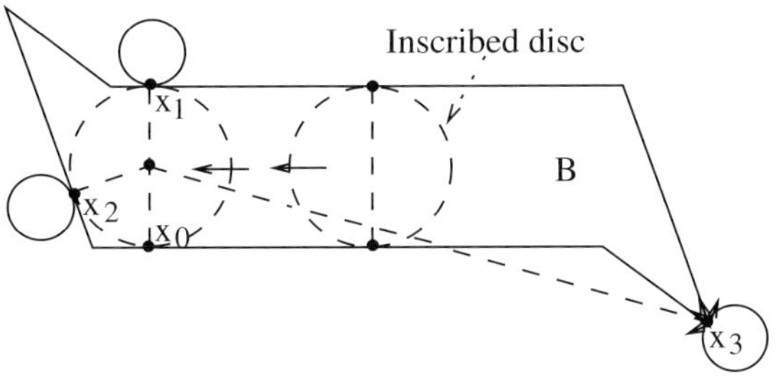

Figure 9.18 The inscribed disc of a non-convex polygon with parallel edges. The inscribed disc can slide to the left until a new contact is established at x_2. Then the point x_0 can be replaced by the opposing vertex point x_3. The object is immobilized at the resulting three-finger equilibrium grasps, provided that sufficiently flat convex finger bodies are placed at the interior edge contact points x_1 and x_2.

equilibrium grasps, provided that sufficiently flat convex finger bodies are placed at the edge contact points x_1 and x_2. ○

Exercise 9.16: Can the polygon depicted in Figure 9.12 be immobilized by two sufficiently concave fingers placed at antipodal vertices of the polygon? (Justify your answer in terms of the fingers' radii of curvature at the vertices.)

Section 9.5

Exercise 9.17: The inscribed ball of 3-D objects generically touches their boundary at two, three or four isolated points. Justify the statement that the inscribed ball of generic polyhedral objects touches their boundary only at four isolated points. ○

Exercise 9.18: Let an inscribed ball of a polyhedron touch its surface at three interior facet points and at a concave vertex. Characterize the subspace of first order roll-free motions, $M^1_{1,2,3,4}(q_0)$, when four finger bodies are placed at these points.

Solution: The subspace $M^1_{1,2,3,4}(q_0)$ is one-dimensional and consists of instantaneous rotations of $\mathcal{B}$ about an axis that passes through the inscribed ball center and the concave vertex of the polyhedron. ○

Exercise 9.19: In continuation of the previous exercise, show that the placement of three spherical fingertips at the facet points and a sharp-tipped finger at the concave vertex immobilizes the polyhedron based on second-order effects.

Exercise 9.20: Let an inscribed ball of a polyhedron $\mathcal{B}$ touch its surface at three interior facet points and at a concave edge point. Characterize the subspace of first-order free motions, $M^1_{1,2,3,4}(q_0)$, when four finger bodies are placed at these points.

Exercise 9.21: In continuation of the previous exercise, show that the placement of three spherical fingertips at the facet points and a curved-blade finger at the concave edge point immobilizes the polyhedron based on second-order effects.

Solution sketch: First, place spherical finger bodies of radius $r_{\mathcal{O}_i} \geq 0$ at the facet points x_1, x_2, and x_3. These fingers satisfy $L_{\mathcal{O}_i}^{-1} = r_{\mathcal{O}_i} I$ for $i = 1, 2, 3$. Next, place a sharp-bladed finger body, $\mathcal{O}_4$, at the concave edge point x_4. The parameter ρ_i in Formula (9.5) is equal to the inscribed ball radius, R, and is positive at the four contacts. Hence, Formula (9.5) becomes

$$\kappa_{\text{rel}}(q_0, (0, \omega)) = -\left(\sum_{i=1}^{3} \lambda_i (R + r_{\mathcal{O}_i}) \|n_i \times \omega\|^2 + \lambda_4 (n_4 \times \omega)^T [RI + L_{\mathcal{O}_4}^{-1}](n_4 \times \omega)\right).$$

Let the finger body $\mathcal{O}_4$ have a *circular blade* of radius $r_{\mathcal{O}_4} > 0$. Any planar cross-section of such a finger along a direction transversal to the blade forms a sharp-tipped planar finger body, which has zero radius of curvature at x_4. Hence, $(n_4 \times \omega)^T L_{\mathcal{O}_4}^{-1}(n_4 \times \omega) = r_{\mathcal{O}_4} \|n_4 \times \omega\|^2$, such that $r_{\mathcal{O}_4} > 0$ or $r_{\mathcal{O}_4} = 0$ according to the direction of the vector $n_4 \times \omega$ relative to the finger's blade direction. It follows that each summand in $\kappa_{\text{rel}}(q_0, (0, \omega))$ is negative semi-definite. Since the 3×4 matrix formed by the contact normals, $[n_1\, n_2\, n_3\, n_4]$, has full row rank, $\kappa_{\text{rel}}(q_0, (0, \omega)) < 0$ for all $\dot{q} = (0, \omega) \in M^1_{1,2,3,4}(q_0)$, which implies second-order immobilization. ○

Exercise 9.22: Consider the four-finger grasp of the triangular prism, $\mathcal{B}$, depicted in Figure 9.14. Show that $\mathcal{B}$ can be immobilized by three *point fingers* placed at the interior facet points, x_1, x_2, x_3, and a suitable finger body placed at x_4. (A concave finger body at x_4 can thus replace the need for sufficiently flat convex fingers at the facet points.)

Exercise 9.23: Explain why three convex finger bodies *cannot* immobilize arbitrary polyhedral objects. (Consider instantaneous translational motions of a convex polyhedron, held by three convex finger bodies at a frictionless equilibrium grasp.)

Exercise 9.24*: When the finger bodies can be pre-shaped according the polyhedral object at hand, can *two* suitably concave finger bodies immobilize all polyhedral objects, based on second-order effects? (Consider fingers with non-spherical contacting surfaces.)

References

[1] B. Mishra, J. T. Schwartz and M. Sharir, "On the existence and synthesis of multifinger positive grips," *Algorithmica*, vol. 2, pp. 541–558, 1987.

[2] J. Selig and J. Rooney, "Reuleaux pairs and surfaces that cannot be gripped," *International Journal of Robotics Research*, vol. 8, no. 5, pp. 79–86, 1989.

[3] X. Markenscoff, L. Ni and C. H. Papadimitriou, "The geometry of grasping," *International Journal of Robotics Research*, vol. 9, no. 1, pp. 61–74, 1990.

[4] J. Czyzowicz, I. Stojmenovic and J. Urrutia, "Immobilizing a shape," *International Journal of Computational Geometry and Applications*, vol. 9, no. 2, pp. 181–206, 1999.

[5] E. Rimon and J. W. Burdick, “New bounds on the number of frictionless fingers required to immobilize planar objects,” *Journal of Robotic Systems*, vol. 12, no. 6, pp. 433–451, 1995.

[6] E. Rimon, “A curvature based bound on the number of frictionless fingers required to immobilize three-dimensional objects,” *IEEE Transactions on Robotics and Automation*, vol. 14, no. 5, pp. 709–717, 2001.

[7] J. Seo, M. Yim and V. Kumar, “A theory on grasping objects using effectors with curved contact surfaces and its applications to whole-arm grasping,” *International Journal of Robotics Research*, vol. 35, pp. 1080–1102, 2016.

[8] A. M. Dollar and R. D. Howe, “The highly adaptive SDM hand: Design and performance evaluation,” *International Journal of Robotics Research*, vol. 29, pp. 585–597, 2010.

[9] J. Borrs-Sol and A. M. Dollar, “Dimensional synthesis of a three-fingered dexterous hand for maximal manipulation workspace,” *International Journal of Robotics Research*, vol. 34, no. 14, pp. 1731–1746, 2016.

[10] J. K. Salisbury and B. Roth, “Kinematic and force analysis of articulated mechanical hands,” *Journal of Mechanisms, Transmissions and Automation in Design*, vol. 105, no. 1, pp. 35–41, 1983.

[11] J. K. Salisbury, “Kinematic and force analysis of articulated hands,” PhD thesis, Dept. of Mechanical Engineering, Stanford University, 1982.

10 Multi-Finger Caging Grasps

Earlier chapters described the theory of *equilibrium grasps,* which was then followed by the theory of *immobilizing grasps.* This chapter continues with *caging theory,* which extends the notion of immobilizing grasps to allow an object $\mathcal{B}$ some amount of mobility relative to the surrounding finger bodies. Caging theory is particularly useful during the grasping phase, where a robot hand has to establish a secure equilibrium grasp of an object $\mathcal{B}$. The robot first places the fingers in a *cage formation* about the object and then closes the fingers while maintaining a cage formation until the object is securely grasped by the robot hand. Pre-opening the fingers in a cage formation allows initial placement of the fingers in the largest possible area surrounding the object, while subsequent closing of the cage allows the robot hand to grasp the object without any need for accurate finger positioning.

Let us begin with an intuitive definition of *cage formations.*

DEFINITION 10.1 **(cage formation)** *A* **cage formation** *is a placement of stationary rigid finger bodies $\mathcal{O}_1, \ldots, \mathcal{O}_k$ around a rigid object $\mathcal{B}$, such that $\mathcal{B}$ cannot move arbitrarily far from its original position without penetrating one of the finger bodies.*

In grasp mechanics, one would like to synthesize cage formations that can be closed until the object is securely grasped by the fingers. The robot hand is modeled as a collection of rigid finger bodies, $\mathcal{O}_1, \ldots, \mathcal{O}_k$, connected via linkages to a common base, called the hand's *palm.* The palm is modeled as a rigid body that is free to move in $\mathbb{R}^n$, where $n = 2$ in 2-D grasps and $n = 3$ in 3-D grasps. The rigid object $\mathcal{B}$ has known geometry and lies stationary at a given location. The *caging problem* is then:

> Given a desired k-finger immobilizing grasp of a rigid object $\mathcal{B}$, determine all cage formations that can be continuously closed to the desired immobilizing grasp, while the object $\mathcal{B}$ is kept caged by the finger bodies.

For the specification of the desired immobilizing grasp, we rely on the immobilization theory of Chapters 6–8. Note that any immobilizing grasp must form a frictionless equilibrium grasp of $\mathcal{B}$ (Theorem 6.1). Some cage formations are illustrated in the following example.

Example: Figure 10.1(a) shows an object $\mathcal{B}$ with two concavities. Two disc fingers placed within the concavities initially form a cage around this object, since the object cannot escape to infinity when the finger bodies are held fixed. Subsequent

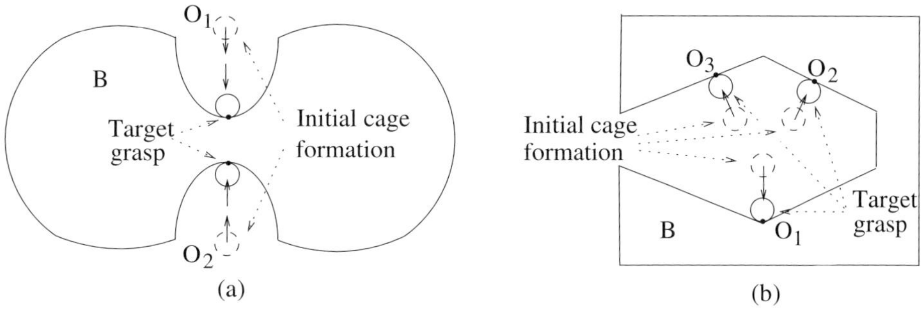

Figure 10.1 Two cage formations leading to immobilizing grasps: (a) A two-finger squeezing cage. (b) A three-finger expanding cage. (The hands' palm is not shown.)

closing of the fingers reaches an immobilizing grasp while the object remains caged by the two fingers. Figure 10.1(b) shows an object $\mathcal{B}$ with an inner cavity. Three disc fingers placed within this cavity initially form a cage, since the object cannot escape to infinity when the finger bodies are held fixed. Subsequent opening of the fingers reaches an immobilizing grasp while the object remains caged by the three fingers. Note that there are two types of cage formations: *squeezing* cages, whose fingers close inward in order to immobilize the object (Figure 10.1(a)), and *expanding* cages, whose fingers open outward in order to immobilize the object (Figure 10.1(b)). ○

This chapter introduces the notions of caging theory in the simplest context of planar k-finger robot hands whose fingers are governed by a scalar parameter. Sections 10.1 and 10.2 introduce this class of one-parameter robot hands and their four-dimensional configuration space. Section 10.3 describes the collection of cage formations that surround a given immobilizing grasp in terms of a *caging set* in the hand's configuration space. Section 10.4 establishes the core result of caging theory, that any break in a multi-finger cage corresponds to a *frictionless equilibrium grasp* of $\mathcal{B}$. In order to provide intuitive insight as to how one identifies cage formations, Section 10.5 describes a graphical technique that renders the two-finger caging set of polygonal objects as two *caging regions* surrounding the target immobilizing grasp.

10.1 Robot Hands Governed by a Scalar Shape Parameter

A planar k-finger robot hand is modeled as a collection of 2-D rigid finger bodies, $\mathcal{O}_1, \ldots, \mathcal{O}_k$, connected via linkages to a common 2-D rigid body that forms the hand's *palm*. The hand's palm moves freely in $\mathbb{R}^2$, while the finger bodies move along a single degree of freedom relative to the palm. In other words, while the finger linkages may have several degrees of freedom, we will study the cage formations in terms of the hand's palm configuration, and a scalar parameter that uniquely determines the hand's shape relative to its palm. Such one-parameter robot hands are defined as follows.

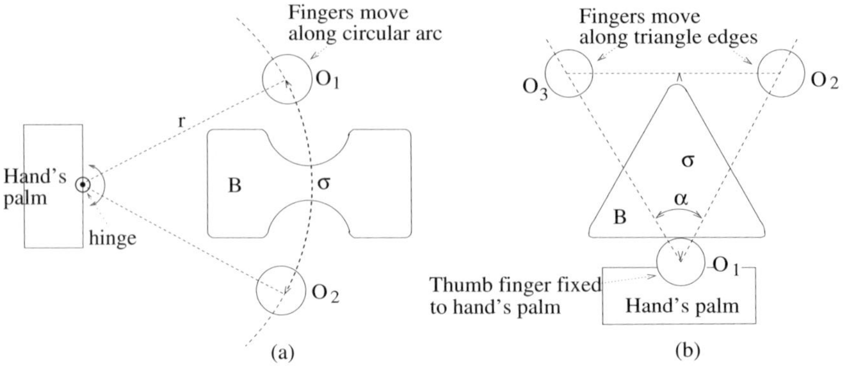

Figure 10.2 (a) A conceptual one-parameter hinged jaw gripper, where r is a fixed parameter. (b) A conceptual one-parameter thumb-and-two-fingers hand, where α is a fixed parameter.

DEFINITION 10.2 (**one-parameter robot hand**) *A planar k-finger robot hand whose palm moves freely in $\mathbb{R}^2$ while its finger bodies move along a single degree of freedom, $\sigma \in \mathbb{R}$, relative to the hand's palm forms a* **one-parameter robot hand.**

Some classical one-parameter robot hands are illustrated in the following example.

Example: The industrial *parallel jaw gripper* can be modeled as two fingers bodies, $\mathcal{O}_1$ and $\mathcal{O}_2$, that move along a common straight line passing through the finger centers (Figure 10.1(a)). The hand's opening parameter is defined as the inter-finger distance. A *hinged jaw gripper* can be modeled as depicted in Figure 10.2(a), where the finger bodies $\mathcal{O}_1$ and $\mathcal{O}_2$ move along a circular arc centered at the hinge. The hand's opening parameter σ is defined as the circular inter-finger distance. Another example is the *thumb-and-two-fingers hand* depicted in Figure 10.2(b). The thumb finger, $\mathcal{O}_1$, remains stationary relative to the hand's palm, while the finger bodies $\mathcal{O}_2$ and $\mathcal{O}_3$ move along a triangle edges according to a common height parameter σ. Note that four degrees of freedom allow arbitrary placement of two disc fingers in $\mathbb{R}^2$. Hence, two freely moving disc fingers can also be interpreted as a one-parameter robot hand. ○

While this chapter considers one-parameter robot hands, one should be aware that caging theory concerns *general* robot hands (see Bibliographical Notes). For now, note that the choice of the hand's opening parameter, σ, is itself a *free parameter* that can be made anew for each grasp operation. For instance, the radius r in Figure 10.2(a) and the angle α in Figure 10.2(b) may be set according to the particular desired grasp.

10.2 Configuration Space of One-Parameter Robot Hands

The configuration space of a one-parameter robot hand consists of the position and orientation of the hand's palm, $q \in \mathbb{R}^3$, and the fingers opening parameter, $\sigma \in \mathbb{R}$. The hand's configuration space is thus parametrized by $(q, \sigma) \in \mathbb{R}^4$. From the robot hand's point of view, the object $\mathcal{B}$ can be interpreted as a *stationary obstacle.* That is, while the object $\mathcal{B}$ can be disturbed by the closing fingers, the object movements will not break the cage nor affect the eventual arrival of the robot hand to the desired immobilizing grasp of the object $\mathcal{B}$ (see Figure 10.4).

The *c-space obstacle* corresponding to a stationary object $\mathcal{B}$, denoted $\mathcal{CB}$, is defined as the set of hand configurations at which one or more of the finger bodies $\mathcal{O}_1, \ldots, \mathcal{O}_k$ intersect the stationary object $\mathcal{B}$. This c-space obstacle consists of the union, $\mathcal{CB} = \cup_{i=1}^{k} \mathcal{CB}_i$, where each $\mathcal{CB}_i$ is the set of hand configurations at which the finger body $\mathcal{O}_i$ intersects the stationary object $\mathcal{B}$. Based on this interpretation, each $\mathcal{CB}_i$ will be called a *finger c-obstacle.* The notion of free c-space from Chapter 6 now becomes the hand's *free c-space,* $\mathcal{F}$, defined as the complement of the finger c-obstacle interiors: $\mathcal{F} = \mathbb{R}^4 - \cup_{i=1}^{k} \mathsf{int}(\mathcal{CB}_i)$, where int denotes set interior.

The hand's free c-space forms a *stratified set* in $\mathbb{R}^4$, consisting of a finite union of disjoint manifolds called *strata* (see Appendix B). The interior of $\mathcal{F}$ consists of open sets in $\mathbb{R}^4$, which form the 4-D strata of $\mathcal{F}$. In these open sets, none of the fingers touch the object. The boundary of $\mathcal{F}$ – or, equivalently, the boundary of $\mathcal{CB} = \cup_{i=1}^{k} \mathcal{CB}_i$ – consists of the following lower-dimensional strata. The 3-D strata of $\mathcal{F}$ form the boundaries of the individual finger c-obstacles, and they correspond to single-finger contacts with $\mathcal{B}$. The 2-D strata of $\mathcal{F}$ form the intersection of pairs of three-dimensional strata, and they correspond to two-finger contacts with $\mathcal{B}$. When the robot hand has at least three fingers, the 1-D strata of $\mathcal{F}$ correspond to three-finger contacts with $\mathcal{B}$. When the robot hand has at least four fingers, the zero-dimensional strata of $\mathcal{F}$ are isolated points that correspond to four-finger contacts with $\mathcal{B}$.

Example: The lower part of Figure 10.3 shows a two-finger hand that interacts with a stationary elliptical object $\mathcal{B}$. The hand's opening parameter, σ, is defined as the inter-finger distance (the hand's palm is not shown). The hand's free c-space, $\mathcal{F}$, forms a four-dimensional stratified set that can be depicted in terms of its fixed-σ slices. Three fixed-σ slices of $\mathcal{F}$ are illustrated in Figure 10.3. Each of the fixed-σ slices contains three types of strata: the open sets that form three-dimensional strata, the

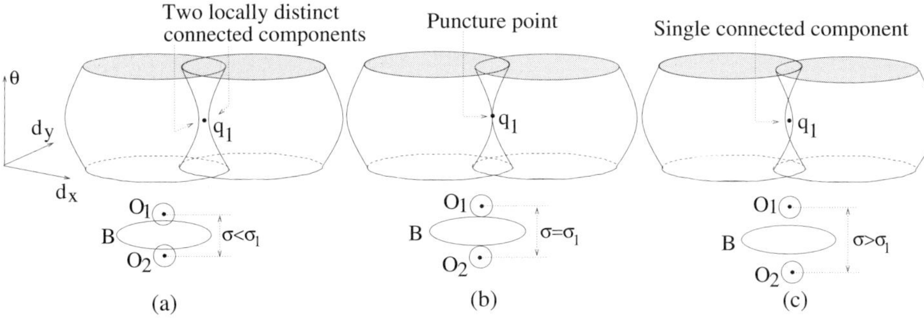

Figure 10.3 Three σ-slices of a two-finger hand's c-space, showing two locally distinct components joining at a puncture point: (a) $\mathcal{F}|_{\sigma_1 - \epsilon}$, (b) $\mathcal{F}|_{\sigma_1}$, and (c) $\mathcal{F}|_{\sigma_1 + \epsilon}$ for some $\epsilon > 0$.

boundaries of the individual finger c-obstacle slices that form two-dimensional strata, and the intersection curves of finger c-obstacle pairs that form one-dimensional strata. The actual strata of the hand's free c-space, $\mathcal{F}$, are formed by sweeping each stratum along the σ-axis. Hence, their dimension is larger by one than the dimension of the strata depicted in Figure 10.3. ○

10.3 C-Space Representation of Cage Formations

This section characterizes the cage formations that surround an immobilizing grasp of $\mathcal{B}$ as a *caging set* in the hand's configuration space. Let $\mathcal{F}|_\sigma$ denote the fixed-σ slice of the free c-space $\mathcal{F}$, and let $\mathcal{CB}_i|_\sigma$ denote the fixed-σ slice of the finger c-obstacle $\mathcal{CB}_i$. The k-finger cage formations are defined in the hand's c-space as follows.

DEFINITION 10.3 (**C-space cage**) *A one-parameter robot hand located at (q,σ) forms a* **cage** *about $\mathcal{B}$ when the hand's palm configuration, q, lies in a bounded path-connected component of $\mathcal{F}|_\sigma$, which is completely surrounded by the finger c-obstacle slices $\mathcal{CB}_1|_\sigma, \ldots, \mathcal{CB}_k|_\sigma$.*

When a path-connected component of $\mathcal{F}|_\sigma$ is completely surrounded by the finger c-obstacle slices $\mathcal{CB}_1|_\sigma, \ldots, \mathcal{CB}_k|_\sigma$, no motion of the *fixed-shape* hand can take it arbitrarily far from its initial position without some of its fingers colliding with the object $\mathcal{B}$. Equivalently, when the finger bodies are held fixed, a freely moving object $\mathcal{B}$ has *bounded mobility* when surrounded by the stationary finger bodies. To better understand the notion of c-space cages, consider the following relation between immobilizing grasps and cage formations.

THEOREM 10.1 (**local C-space cages**) *Let a rigid object $\mathcal{B}$ be immobilized by a one-parameter robot hand, located at (q_0,σ_0). All hand configurations (q,σ) sufficiently close to (q_0,σ_0) in the hand's free c-space $\mathcal{F}$ form* **cages** *about $\mathcal{B}$.*

In practical terms, when a one-parameter robot hand immobilizes an object $\mathcal{B}$, any slight finger opening in the case of squeezing grasps, or any slight finger closing in the case of expanding grasps, will automatically form a cage around this object.

Proof sketch: Immobilization of $\mathcal{B}$ at (q_0,σ_0) means that the point q_0 is an isolated point of $\mathcal{F}|_{\sigma_0}$, surrounded by the finger c-obstacle slices $\mathcal{CB}_1|_{\sigma_0}, \ldots, \mathcal{CB}_k|_{\sigma_0}$. The point q_0 can therefore be surrounded by a small sphere of radius δ centered at q_0, such that the sphere lies in the interior of the union $\cup_{i=1}^k \mathcal{CB}_i|_{\sigma_0}$. The sphere a fortiori lies in the interior of the full c-space obstacle, $\mathcal{CB} = \cup_{i=1}^k \mathcal{CB}_i$. The sphere and the boundary of $\mathcal{CB}$ are disjoint closed sets in $\mathbb{R}^4$, separated from each other by some finite distance. The sphere can therefore be locally swept along the σ axis in $\mathbb{R}^4$, forming a small 3-D cylindrical set that lies in the interior of $\mathcal{CB}$. The cylindrical set forms the outer boundary of a 4-D neighborhood of $\mathcal{F}$, centered at (q_0,σ_0). Every hand configuration (q,σ) in this neighborhood lies inside a sphere of radius δ centered at q_0 (each sphere is the fixed-σ slice of the cylindrical set), such that the sphere lies in the interior of the union $\cup_{i=1}^{k} \mathcal{CB}_i|_\sigma$. Each of these hand configurations thus forms a *cage* about $\mathcal{B}$. □

What would happen when the fingers of a one-parameter robot hand pull away from an immobilizing grasp of $\mathcal{B}$? If the hand's configuration at the immobilizing grasp is (q_0, σ_0), the point q_0 is an isolated point of $\mathcal{F}|_{\sigma_0}$, which is completely surrounded by the finger c-obstacle slices $\mathcal{CB}_1|_{\sigma_0}, \ldots, \mathcal{CB}_k|_{\sigma_0}$. For a *squeezing grasp,* the c-obstacle regions $\mathcal{CB}_1|_\sigma, \ldots, \mathcal{CB}_k|_\sigma$ move apart in the σ-slices just above σ_0, since the fingers spread apart in physical space. The free c-space $\mathcal{F}$ consequently expands into a three-dimensional cavity in these neighboring slices. Eventually the spreading fingers will reach a critical finger opening, σ_1, which allows the object to barely escape the cage formed by the fingers. This event is marked by the appearance of a *puncture point* on the boundary of the closed cavity in $\mathcal{F}|_{\sigma_1}$ (Figure 10.3).

Stratified Morse theory describes the appearance of such a puncture point as follows (see Appendix B for terminology). The changes in the topology of the fixed-σ slices of $\mathcal{F}$ occur locally, at the *critical points* of the scalar valued *projection function,* $\pi : \mathcal{F} \rightarrow \mathbb{R}$, given by

$$\pi(q, \sigma) = \sigma \qquad\qquad (q, \sigma) \in \mathbb{R}^4.$$

When σ varies in the open interval between adjacent *critical values* of π, the level sets $\mathcal{F}|_c = \{(q, \sigma) \in \mathcal{F} : \pi(q, \sigma) = c\}$ are topologically equivalent (homeomorphic) to each other. In particular, the path connectivity of these level sets is preserved between critical values of π. Any path connectivity change in the fixed-σ slices of $\mathcal{F}$ must occur locally, at a critical point of π in $\mathcal{F}$. The *puncture point* on the boundary of the cavity in $\mathcal{F}|_{\sigma_1}$ is such a critical point of π, since two locally distinct connected components of the slice $\mathcal{F}|_{\sigma_1}$ meet at this point. A critical point of this type is said to form a *join point,* as illustrated in the following example.

Example: Consider the two-finger hand grasping an ellipse depicted at the bottom of Figure 10.3. Let (q_1, σ_1) mark the hand's configuration shown in Figure 10.3(b). The slice $\mathcal{F}|_{\sigma_1 - \epsilon}$ has two locally distinct components in the vicinity of q_1 (Figure 10.3(a)). These two connected components meet at q_1 in the slice $\mathcal{F}|_{\sigma_1}$ (Figure 10.3(b)), and become a single connected component in the slice $\mathcal{F}|_{\sigma_1 + \epsilon}$ (Figure 10.3(c)). The point (q_1, σ_1) is thus a *join point* of the level sets of π in $\mathcal{F}$. ○

Let us use the insight provided by stratified Morse theory to characterize the collection of cage formations that surround a given immobilizing grasp of $\mathcal{B}$, which is assumed to be a *squeezing grasp.* Consider the subset of $\mathcal{F}$ that starts at the isolated point q_0 in the slice $\mathcal{F}|_{\sigma_0}$, becomes an isolated three-dimensional cavity in the slices $\mathcal{F}|_\sigma$ for $\sigma_0 < \sigma < \sigma_1$ and ends at the slice $\mathcal{F}|_{\sigma_1}$, which contains the puncture point. As one stacks these fixed-σ slices of $\mathcal{F}$, the union of the isolated regions forms the following *caging set* in the hand's configuration space.

DEFINITION 10.4 **(caging set)** *Let (q_0, σ_0) be an immobilizing hand configuration of a rigid object $\mathcal{B}$. Among all join points with $\sigma > \sigma_0$, the* **puncture point**, *(q_1, σ_1), is the one with the smallest σ from which it is possible to reach (q_0, σ_0) along a σ-decreasing* *path in $\mathcal{F}$. The* **c-space caging set**, *denoted $\mathcal{K}$, is the path-connected component of the union $\bigcup_{\sigma_0 \le \sigma \le \sigma_1} \mathcal{F}|_\sigma$ that contains the point (q_0, σ_0). The caging set is similarly defined for expanding grasps.*

The puncture point, (q_1, σ_1), is thus the first join point that can be connected to (q_0, σ_0) by a σ-decreasing path. The set $\mathcal{K}$ is surrounded by the finger c-obstacles and is bounded by the σ-slice that passes through the puncture point. The set $\mathcal{K}$ forms the basis for the following methodology for establishing secure equilibrium grasps.

> **Caging to grasping:** Place the fingers at any initial hand configuration in $\mathcal{K}$. Then continuously decrease the fingers' opening parameter, σ, while respecting the object's rigidity under frictionless contact conditions. The hand is guaranteed to reach the target immobilizing grasp, while the object $\mathcal{B}$ is guaranteed to remain caged throughout the finger closing process.

The caging-to-grasping methodology is *robust* under practical finger placement errors. The robot hand needs only place the fingers within the caging regions surrounding the target immobilizing grasp, with the finger opening smaller than the maximal finger opening allowed by the caging set puncture point. The entire finger-closing process is then guaranteed to end at a secure immobilizing grasp of the object $\mathcal{B}$, as illustrated in the following example.

Example: Figure 10.4(a) depicts an object $\mathcal{B}$ having two concavities. Based on graphical tools developed later in this chapter, the two disc fingers are initially located in the caging regions of the target immobilizing grasp shown in Figure 10.4(a). During the initial finger-closing process, the finger $\mathcal{O}_1$ happens to establish contact with $\mathcal{B}$ (Figure 10.4(a)). While this finger subsequently shifts the object rendered with dashed lines, the object cannot escape the cage formed by the two fingers. Eventually the other finger, $\mathcal{O}_2$, establishes contact with $\mathcal{B}$ (Figure 10.4(b)). Under frictionless contact conditions, the two fingers continue to slide along the object boundary while pushing the object to the right, until the immobilizing grasp is established (Figure 10.4(c)). Arrival to

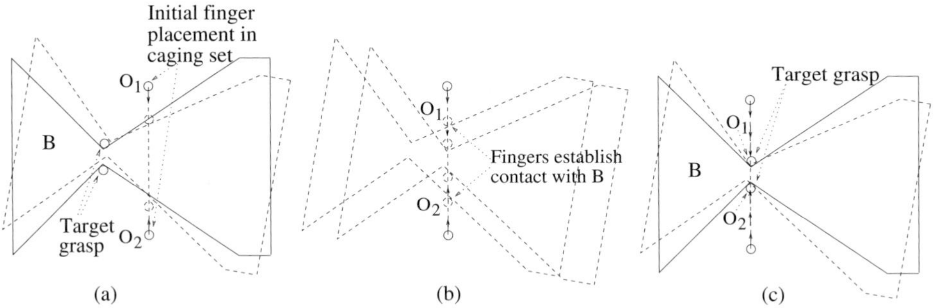

Figure 10.4 (a) The initial placement of disc fingers $\mathcal{O}_1$ and $\mathcal{O}_2$ is in the caging regions surrounding a target grasp. (b) The closing fingers establish contact with $\mathcal{B}$ and then (c) slide along the object edges until the target grasp is reached (intermediate object positions are rendered with dashed lines).

the immobilizing grasp is guaranteed from all initial hand configurations in the c-space caging set, and is not affected by any shifting of $\mathcal{B}$ during the finger closing process. ○

Composite caging sets: The immobilizing grasps of a one-parameter robot hand occur at discrete hand configurations (see Exercises). Hence, every c-space caging set $\mathcal{K}$ is associated with a specific immobilizing grasp of the object $\mathcal{B}$. However, it may happen that the puncture point of $\mathcal{K}$ connects the current caging set with a neighboring region of $\mathcal{F}$ that forms the caging set of a neighboring immobilizing grasp of $\mathcal{B}$. Continuing in this manner, it is possible to synthesize *composite cage formations,* which retract to one of several immobilizing grasps. Note that every composite cage is upper bounded by some maximal finger opening, beyond which the hand (or the object in the equivalent interpretation) can escape to infinity. ○

Let us end this section by mentioning two caveats. Certain 2-D objects possess circular symmetry or parallel edge (Figure 10.10). Such objects can be immobilized only up to their circular or parallel-edge symmetry, and the cage-closing process may reach any of these conditionally immobilizing grasps. Another issue concerns the possibility of non-negligible friction at the contacts. Friction effects may cause finger jamming during the cage-closing process, thus halting the fingers at a frictional equilibrium grasp. Certain measures must therefore be taken in order to avoid finger jamming during the cage-closing process (see Bibliographical Notes).

10.4 The Caging Set Puncture Point

This section establishes that the puncture point of a c-space caging set, $\mathcal{K}$, forms a *frictionless equilibrium grasp* of the object $\mathcal{B}$, involving some or all of the k fingers. For simplicity, assume the case of *squeezing grasps,* but the same property holds for expanding grasps. Recall that the puncture point of $\mathcal{K}$ is a join point of the level sets of the projection function $\pi : \mathcal{F} \to \mathbb{R}$, at which two locally distinct connected component of the fixed-σ slices of $\mathcal{F}$ meet and become a single connected component. Let us establish that every join point (and, hence, the puncture point of $\mathcal{K}$) corresponds to a frictionless equilibrium grasp of the object $\mathcal{B}$.

Let p denote any critical point of π in $\mathcal{F}$, and let $c = \pi(p)$ be the value of π at this point. Stratified Morse theory characterizes the *type* of a given critical point, in terms of the behavior of π on two complementary subsets of $\mathcal{F}$ (see Appendix B). The first set is the stratum of $\mathcal{F}$ that contains the point p, denoted $\mathcal{S}$. The other set, the *normal slice* of $\mathcal{F}$ at p, is constructed as follows. Let $D(p)$ be a small disc centered at p, which intersects the stratum $\mathcal{S}$ only at p and is transversal to $\mathcal{S}$.[1] The *normal slice,* denoted $E(p)$, is the set $E(p) = D(p) \cap \mathcal{F}$. The behavior of π on $\mathcal{S}$ is characterized by the *Morse index,* denoted ν. It is defined as the number of negative eigenvalues of the Hessian matrix $D^2\pi(p)$, evaluated on the stratum $\mathcal{S}$. The behavior of π on $E(p)$ is determined by the *lower half link* set, denoted l^-. It is defined as the intersection of $E(p)$ with

[1] The dimension of $D(p)$ is equal to the dimension of the ambient c-space minus the dimension of $\mathcal{S}$.

the level set just below the critical value, $\mathcal{F}|_{c-\epsilon} = \{(q,\sigma) \in \mathcal{F} : \pi(q,\sigma) = c - \epsilon\}$. Thus, $l^- = E(p) \cap \mathcal{F}|_{c-\epsilon}$, where $\epsilon > 0$ is a small parameter. According to Corollary B.3 in Appendix B, a critical point p forms a *join point* of π under one of two conditions:

(i) The Morse index is unity, $\nu = 1$, and the lower half link l^- is empty.
(ii) The Morse index is zero, $\nu = 0$, and the lower half link l^- is disconnected.

When an object $\mathcal{B}$ is caged and eventually grasped by a k-finger robot hand, the stratum $\mathcal{S}$ represents hand configurations at which a sub-collection of $r \leq k$ finger bodies maintain contact with the object $\mathcal{B}$. The stratum $\mathcal{S}$ thus forms the intersection of several finger c-obstacle boundaries: $\mathcal{S} = \cap_{i=1}^{r} \text{bdy}(\mathcal{CB}_i)$. This property can be used to establish that when l^- is nonempty at a critical point, it must form a *connected set* (see Lemma 10.5 in this chapter's Appendix). Hence, only Condition (i) $\nu = 1$ and $l^- = \emptyset$, can possibly hold at the join points of the function π in $\mathcal{F}$.

The next lemma specifies under what condition $l^- = \emptyset$ at a critical point of π in $\mathcal{F}$. Each finger c-obstacle, $\mathcal{CB}_i$, occupies a four-dimensional set in the hand's configuration space. Let $\eta_i(p) \in T_p\mathbb{R}^4$ denote the finger c-obstacle outward unit normal at $p \in \text{bdy}(\mathcal{CB}_i)$.

LEMMA 10.2 (**lower half link test**) *Let $f : \mathcal{F} \to \mathbb{R}$ be a Morse function,[2] such that f is the restriction to $\mathcal{F}$ of a smooth function $\tilde{f} : \mathbb{R}^4 \to \mathbb{R}$. Let p be a critical point of f in a stratum $\mathcal{S}$ of $\mathcal{F}$, such that $\mathcal{S} = \cap_{i=1}^{r} \text{bdy}(\mathcal{CB}_i)$ for $1 \leq r \leq k$. A necessary condition for an* **empty lower half link** *at p, $l^- = \emptyset$, is that:*

$$\nabla\tilde{f}(p) = \lambda_1 \eta_1(p) + \ldots + \lambda_r \eta_r(p) \qquad \lambda_1, \ldots, \lambda_r \geq 0 \tag{10.1}$$

such that $\lambda_1, \ldots, \lambda_r$ are not all zero. A sufficient condition for $l^- = \emptyset$ is that $\lambda_1, \ldots, \lambda_r$ are strictly positive in Eq. (10.1).

The proof of Lemma 10.2 appears in the chapter's Appendix and is based on the following argument. When l^- is empty at p, the function f must be non-decreasing along any c-space path that starts at p and lies in the normal slice, $E(p)$. Let $\alpha(t)$ be such a path, with $\alpha(0) = p$ and $\dot{\alpha}(0) = \dot{p} \in E(p)$. By the chain rule, $\frac{d}{dt}\big|_{t=0} f(\alpha(t)) = \nabla\tilde{f}(p) \cdot \dot{p} \geq 0$ for all tangent vectors $\dot{p} \in E(p)$. Since $E(p)$ is a subset of $\mathcal{F}$, the tangent vectors $\dot{p} \in E(p)$ lie in the *tangent cone* at p, defined as $T(p) = \{\dot{p} : \eta_i(p) \cdot \dot{p} \geq 0$ for $i = 1 \ldots r\}$. Taking the weighted sum of the products $\eta_i(p) \cdot \dot{p}$ with weights $\lambda_i \geq 0$ gives $\sum_{i=1}^{r} \lambda_i \eta_i(p) \cdot \dot{p} \geq 0$ for all $\dot{p} \in T(p)$. Next, consider the complementary cone in $T_p\mathbb{R}^4$, spanned by the finger c-obstacle outward normals at p: $N(p) = \{\sum_{i=1}^{r} \lambda_i \eta_i(p) : \lambda_i \geq 0$ for $i = 1 \ldots r\}$. Every vector $\eta \in N(p)$ satisfies $\eta \cdot \dot{p} \geq 0$ for all $\dot{p} \in T(p)$. As shown in the appendix, $N(p)$ contains all vectors having this property at p, and consequently $\nabla\tilde{f}(p) \in N(p)$ as specified by Eq. (10.1).

A direct consequence of Lemma 10.2 is the following characterization of all join points, and hence the caging set *puncture point,* as frictionless equilibrium grasps of the object $\mathcal{B}$.

[2] Morse functions are generic smooth functions that possess non-degenerate critical points (see Appendix B).

THEOREM 10.3 (**fundamental caging theorem**) *Let $\mathcal{F}$ be the free c-space of a k-finger hand governed by an opening parameter σ. Let $\pi : \mathcal{F} \to \mathbb{R}$ be the projection function $\pi(q,\sigma) = \sigma$. Then π generically forms a Morse function on $\mathcal{F}$, and in this case all* **join points** *of π in $\mathcal{F}$ correspond to r-finger* **frictionless equilibrium grasps** *of $\mathcal{B}$, where $2 \le r \le k$.*

Proof: A join point, $p_1 = (q_1, \sigma_1) \in \mathcal{F}$, must satisfy the conditions $\nu = 1$ and $I^- = \emptyset$. Since $I^- = \emptyset$ at p_1, the necessary condition of Lemma 10.2 holds at this point,

$$\nabla\pi(p_1) = \lambda_1 \boldsymbol{\eta}_1(p_1) + \ldots + \lambda_r \boldsymbol{\eta}_r(p_1) \qquad \lambda_1, \ldots, \lambda_r \ge 0 \tag{10.2}$$

such that $\lambda_1, \ldots, \lambda_r$ are not all zero. The gradient of the projection function, $\pi(q,\sigma) = \sigma$, is given by $\nabla\pi = (0,0,0,1)$. Hence, we will focus on showing that the first three components of Eq. (10.2) correspond to an r-finger frictionless equilibrium grasp of the object $\mathcal{B}$.

The k-finger hand moves as a fixed-shape rigid body in each σ-slice of $\mathcal{F}$. One can therefore interpret the ith finger c-obstacle slice, $\mathcal{CB}_i|_{\sigma_1}$, as the c-obstacle of a freely moving object $\mathcal{B}$, interacting with an obstacle formed by a stationary finger body $\mathcal{O}_i$. Thus, $\mathcal{CB}_i|_{\sigma_1} = \mathcal{CO}_i$, where $\mathcal{CO}_i$ is the c-obstacle induced in $\mathcal{B}$'s c-space by the stationary finger body $\mathcal{O}_i$. Under this interpretation, the hand's palm configuration, $q \in \mathbb{R}^3$, can be used to denote the configuration of the object $\mathcal{B}$, while the finger bodies $\mathcal{O}_1, \ldots, \mathcal{O}_r$ are kept stationary.

A frictionless equilibrium grasp of a rigid object $\mathcal{B}$ by finger bodies $\mathcal{O}_1, \ldots, \mathcal{O}_r$ can be formulated in $\mathcal{B}$'s c-space as follows (see Chapter 4). When $\mathcal{B}$ is located at a configuration q, the wrench affected on $\mathcal{B}$ by a finger force that acts through a frictionless contact is given by $\boldsymbol{w}_i = \lambda_i \hat{\eta}_i(q)$ for some $\lambda_i \ge 0$, where $\hat{\eta}_i(q)$ is the ith c-obstacle outward unit normal at q. When an object $\mathcal{B}$ is held at a frictionless equilibrium grasp, the net wrench applied on $\mathcal{B}$ by the finger bodies must be zero:

$$\lambda_1 \hat{\eta}_1(q) + \ldots + \lambda_r \hat{\eta}_r(q) = \vec{0} \qquad \lambda_1, \ldots, \lambda_r \ge 0 \tag{10.3}$$

such that $\lambda_1, \ldots, \lambda_r$ are not all zero.

In Eq. (10.2), the first three components of each finger c-obstacle outward normal, $\boldsymbol{\eta}_i(p_1) \in T_{p_1}\mathbb{R}^4$, are a positive multiple of the c-obstacle outward unit normal, $\hat{\eta}_i(q_1) \in T_{q_1}\mathbb{R}^3$, where $p_1 = (q_1, \sigma_1)$. One can therefore substitute in Eq. (10.2) the vector $\hat{\eta}_i(q_1)$ for the first three components of $\boldsymbol{\eta}_i(p_1)$ $(i = 1 \ldots r)$, and a zero vector for the first three components of $\nabla\pi$. Under the identification of $\mathcal{CB}_i|_{\sigma_1}$ with $\mathcal{CO}_i$, we obtain the equilibrium grasp condition of Eq. (10.3). □

The puncture point of the c-space caging set thus forms a frictionless equilibrium grasp of $\mathcal{B}$. One can therefore identify the object's frictionless equilibrium grasps, classify which of these equilibrum grasps forms a *join point,* and then test for the existence of a σ-decreasing path from each candidate join point to the target immobilizing grasp. This approach is taken in the next section for the case of two-finger hands.

10.5 Graphical Depiction of Two-Finger Cage Formations

This section provides an intuitive feel for the caging-to-grasping process, in the form of a graphical technique that determines the two-finger caging regions surrounding a target immobilizing grasp of a polygonal object $\mathcal{B}$. The technique determines the puncture point grasp, which is then used to render the c-space caging set as two *caging regions* surrounding the immobilizing grasp. The technique assumes two *point or disc fingers* that move freely in $\mathbb{R}^2$ with a total of four degrees of freedom. Two freely moving point or disc fingers can be interpreted as a one-parameter robot hand: the line segment connecting the two fingers forms the hand's *palm* with configuration $q \in \mathbb{R}^3$, while the inter-finger distance defines the hand's opening parameter σ.

The graphical technique searches for the puncture point of a target immobilizing grasp in *contact space*. This space parametrizes the two finger contacts along the object boundary in terms of two parameters, u_1 and u_2, such that $x(u_i) : [0,1] \to \mathsf{bdy}(\mathcal{B})$ describes the ith finger contact position along the object boundary. The endpoints of each unit interval parametrize the same point on the object boundary. *Contact space* is defined as the unit square, $\mathcal{U} = \{(u_1, u_2) \in [0,1] \times [0,1]\}$, with the periodicity rules $0 \times [0,1] = 1 \times [0,1]$ and $[0,1] \times 0 = [0,1] \times 1$.

The search for the puncture point of a target immobilizing grasp inspects the local minima and saddle points of the inter-finger distance function, $d(u_1, u_2)$, which is defined in contact space as follows.

DEFINITION 10.5 *The* **inter-finger distance function,** $d(u_1,u_2) : \mathcal{U} \to \mathbb{R}$, *measures the fingers' opening over all two-finger contacts in* $\mathcal{U}$, $d(u_1,u_2) = \|x(u_1) - x(u_2)\|$ *for* $(u_1,u_2) \in \mathcal{U}$.

Let us pause to clarify the relation of the function $d(u_1,u_2)$ to the hand's opening parameter σ. Recall that two freely moving point or disc fingers can be modeled as a one-parameter robot hand, with the hand's configuration space parametrized by $(q,\sigma) \in \mathbb{R}^4$. Consider the following submanifold of the hand's configuration space.

DEFINITION 10.6 *The* **double-contact submanifold** *in the two-finger hand's c-space consists of all hand configurations,* $(q,\sigma) \in \mathbb{R}^4$, *at which both fingers touch the object* $\mathcal{B}$: $\mathcal{S} = \mathsf{bdy}(\mathcal{CB}_1) \cap \mathsf{bdy}(\mathcal{CB}_2)$, *where* $\mathsf{bdy}(\mathcal{CB}_1)$ *and* $\mathsf{bdy}(\mathcal{CB}_2)$ *are the finger c-obstacle boundaries.*

The submanifold $\mathcal{S}$ is two-dimensional and is fully parametrized by contact space $\mathcal{U}$. The inter-finger distance function, $d(u_1,u_2)$, is simply the restriction of the projection function, $\pi(q,\sigma) = \sigma$, to the submanifold $\mathcal{S}$, which is parametrized in turn by $(u_1,u_2) \in \mathcal{U}$.

Let us return to the graphical technique. The function $d(u_1,u_2)$ has the property that the frictionless two-finger equilibrium grasps of $\mathcal{B}$ form *extremum points* of this function in contact space. These grasps can be graphically identified using the *generalized inward normal* at the vertices of $\mathcal{B}$, with the understanding that when a point finger touches a convex vertex of $\mathcal{B}$, one should conceptually model the finger as a small disc touching the vertex (Figure 10.5). Under this interpretation, the two-finger

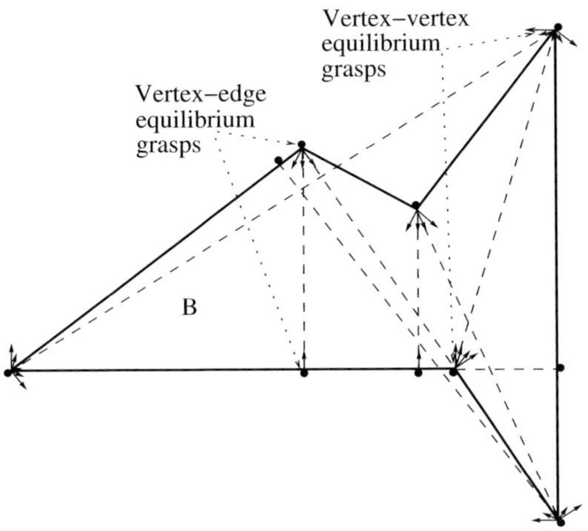

Figure 10.5 A polygonal object having four vertex–edge and six vertex–vertex equilibrium grasps (two vertex–vertex grasps are not shown).

frictionless equilibrium grasps of a polygon $\mathcal{B}$ can be of three types. *Vertex–vertex* grasps where the two fingers touch opposing vertices of $\mathcal{B}$, *vertex–edge* grasps where one finger touches a vertex while the other finger touches an opposing edge interior point, and *edge–edge* grasps where the two fingers touch parallel edges of $\mathcal{B}$. The first two types are illustrated in the following example.

Example: The polygon $\mathcal{B}$ depicted in Figure 10.5 has six vertex–vertex equilibrium grasps, having opposing generalized inward normals at the vertices. The polygon also has four vertex–edge equilibrium grasps, having a generalized inward normal at the vertex and an opposing edge inward contact normal. All of these grasps are non-smooth extremum points of $d(u_1, u_2)$, which can form local minima, saddles and local maxima in contact space $\mathcal{U}$. ○

The following lemma gives a contact-space criterion that identifies candidate puncture point grasps of a given object $\mathcal{B}$. The lemma is stated for *squeezing grasps,* but an analogous lemma exists for expanding grasps.

LEMMA 10.4 (**minima and saddles in $\mathcal{U}$**) *Let a polygonal object $\mathcal{B}$ be held by two point or disc fingers. Every immobilizing grasp of $\mathcal{B}$ is a* **local minimum** *of $d(u_1, u_2)$ in $\mathcal{U}$, while every join point grasp is a* **saddle point** *of $d(u_1, u_2)$ in $\mathcal{U}$.*

A proof sketch of Lemma 10.4 appears in the chapter's appendix. The immobilizing grasps of $\mathcal{B}$ can be of two types. Vertex–edge grasps, where one finger touches a concave vertex of $\mathcal{B}$ while the other finger touches an opposing edge interior point of $\mathcal{B}$ (Figure 10.6(a)). And vertex–vertex grasps, where the two fingers touch opposing concave vertices of $\mathcal{B}$ (Figure 10.6(b)). The join point grasps can also be of two possible types: vertex–edge grasps, where one finger touches a convex vertex of $\mathcal{B}$ while the other finger touches an opposing edge interior point of $\mathcal{B}$ (Figure 10.6(c)), and vertex–vertex

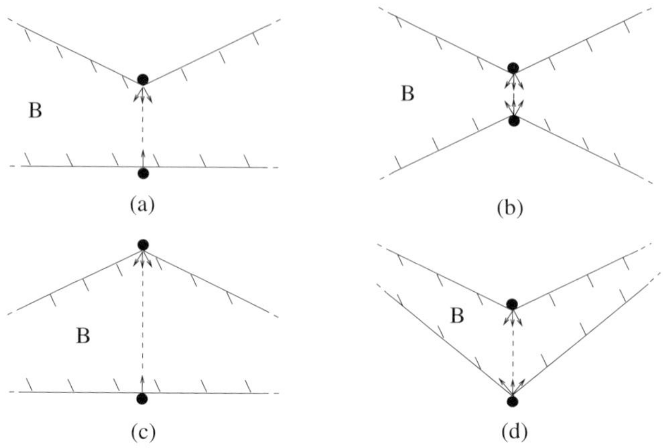

Figure 10.6 The two types of immobilizing grasps involve (a) a concave vertex and an edge, or (b) two concave vertices of B. The two types of join point grasps involve (c) a convex vertex and an edge, or (d) a convex vertex and a concave vertex of B.

grasps, where one finger touches a convex vertex of $\mathcal{B}$ while the other finger touches an opposing concave vertex of $\mathcal{B}$ (Figure 10.6(d)).

The graphical technique accepts as input a target immobilizing grasp of a polygon $\mathcal{B}$ with finger opening $\sigma = \sigma_0$, and determines the caging set puncture point. The technique is described for *squeezing grasps* but can be easily adapted to expanding grasps.

Graphical Detection of the Caging Set Puncture Point

1. Identify all two-finger equilibrium grasps of $\mathcal{B}$ with finger opening $\sigma > \sigma_0$.
2. Retain those equilibrium grasps that are *saddle points* of $d(u_1, u_2)$ in $\mathcal{U}$, sorted by increasing σ.
3. Starting with the saddle point having the lowest $\sigma > \sigma_0$ value, test each saddle point for the existence of a σ-decreasing path leading to the target immobilizing grasp:
   - 3.1 Identify the object boundary segments that lead each finger from its saddle-point contact to the immobilizing grasp contact.
   - 3.2 Slide both fingers along the object boundary segments, while minimizing the function $d(u_1, u_2)$ (i.e., squeeze the fingers) until reaching a local minimum.
   - 3.3 If the local minimum is the target immobilizing grasp, the saddle point is the caging set *puncture point.*
   - 3.4 If the local minimum forms some other immobilizing grasp, resume Step 3 with the next saddle point.
   - 3.5 If the local minimum is not a feasible equilibrium grasp, retract one finger away from $\mathcal{B}$ at the local minimum, while holding the other finger fixed, until the finger hits a new edge of $\mathcal{B}$ (Figure 10.7).

   Resume Step 3.1, using the two edges currently contacted by the fingers.

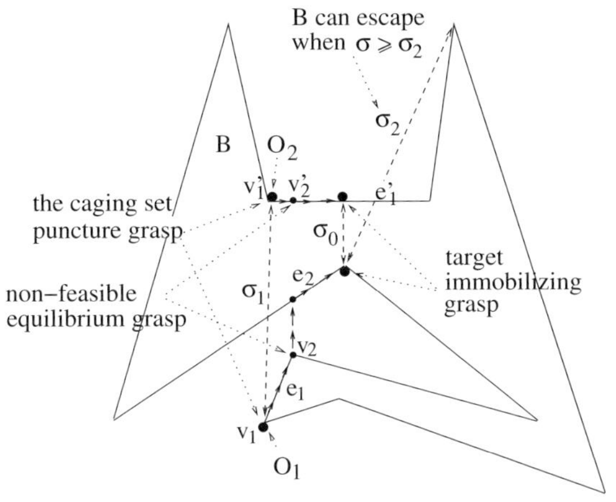

Figure 10.7 Example of the graphical technique, starting from the saddle point grasp at the vertices v_1 and v_1' with finger opening σ_1, and ending at the target immobilizing grasp with finger opening σ_0.

Example: The graphical technique is demonstrated on the polygon $\mathcal{B}$ depicted in Figure 10.7. The polygon has to be immobilized at its center with finger opening σ_0. The polygon has several saddle point grasps, and the search starts with the saddle point having the *lowest* finger opening, $\sigma_1 > \sigma_0$, located at v_1 and v_1'. The fingers initially slide on the edges, e_1 and e_1', while minimizing the function, $d(u_1, u_2)$, until reaching the vertex v_2 and the edge interior point v_2'. It is a local minimum of $d(u_1, u_2)$ in $\mathcal{U}$ but *not* a feasible equilibrium grasp. Hence the search proceeds with Step 3.5 which retracts the lower finger, $\mathcal{O}_1$, upward until it hits the new edge, e_2. The fingers resume their simultaneous sliding along e_2 and e_1' while minimizing the function $d(u_1, u_2)$, eventually reaching the target immobilizing grasp. The initial saddle point with finger opening σ_1 thus forms the caging set *puncture point.* At finger openings just above σ_1, the two fingers form a *composite cage* that retracts to one of *two* immobilizing grasps of $\mathcal{B}$. Eventually, when the finger opening reaches σ_2, the object can escape to infinity. ○

Having computed the puncture point of a target immobilizing grasp, the c-space caging set, $\mathcal{K}$, can be rendered as two *caging regions* surrounding the immobilizing grasp.[3] Let (q_0, σ_0) be the target immobilizing grasp, and let (q_1, σ_1) be its puncture point grasp. First, determine the object boundary segments that lie within the caging regions as follows (Figure 10.8). Place the two fingers at the immobilizing grasp; then slide both fingers along the object boundary while increasing the finger opening from σ_0 to σ_1. Mark the object's boundary touched by the fingers during this process, as γ_1 and γ_2 (Figure 10.8). Next, slide one finger along γ_1 while marking the curve located at a perpendicular distance of σ_1 on the object's opposite side. Denote by γ_1' the portion of this curve bounded by the endpoints of γ_2. Repeat this process with the curve γ_2, this time generating a curve γ_2' with endpoints at those of γ_1. One caging region is bounded

[3] The caging regions require a technical condition that holds for all reasonable objects. See Exercise 11.17.

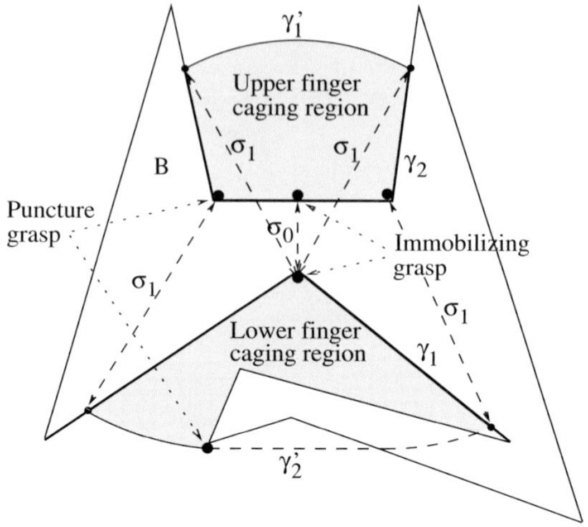

Figure 10.8 Graphical rendering of the caging regions surrounding a target immobilizing grasp. First the object boundary curves γ_1 and γ_2 are identified; then the opposing curves γ'_1 and γ'_2 provide the outer boundary of the two caging regions.

by γ_1 and γ'_2, while the other caging region is bounded by γ_2 and γ'_1, as illustrated in the following example.

Example: The polygonal object of the previous example is depicted again in Figure 10.8. Starting at the target immobilizing grasp, the boundary segments touched by the fingers while increasing the finger opening from σ_0 to σ_1, γ_1 and γ_2, are depicted as thick curves in Figure 10.8. The tracing of γ_1 at a perpendicular distance σ_1 yields the curve γ'_1, whose endpoints coincide with those of γ_2. A similar tracing of γ_2 gives the curve γ'_2, whose endpoints coincide with those of γ_1. Note that both curve tracings involve rotation of the perpendicular segment at the object vertices, with the rotation angle determined by the edge normals at the vertex. The lower caging region is bounded by γ_1 and γ'_2, as well as by a portion of the object's boundary that protrudes into this region. The upper caging region is bounded by γ_2 and γ'_1. Every placement of the two point fingers in the caging regions with finger opening $\sigma \leq \sigma_1$ will retract to the target immobilizing grasp. The two caging regions, coupled with the constraint $\sigma \leq \sigma_1$ thus represent the c-space caging set $\mathcal{K}$. ○

Bibliographical Notes

The earliest formulation of the caging problem was posed by Kuperberg in 1990 [1]:

> Let $\mathcal{B}$ be a polygonal object in $\mathbb{R}^2$ and $\mathcal{P}$ a set of fixed points which lies in the complement of the interior of $\mathcal{B}$. The points of $\mathcal{P}$ *capture* $\mathcal{B}$ if the object cannot be moved arbitrarily far from its original position without at least one point of $\mathcal{P}$ penetrating the interior of $\mathcal{B}$. Design an algorithm for finding a set of capturing points for $\mathcal{B}$.

Kuperberg's formulation suggests that the finger placements and object motions are two separate processes. When the caging problem is formulated in the hand's configuration space, these two processes can be described as a single process in the hand's configuration space.

Caging theory was first described in the context of one-parameter robot hands by Rimon and Blake [2]. Techniques that compute all two-finger cage formations of polygonal objects were developed by Sudsang et al. [3], Vahedi and van der Stappen [4] and Allen et al. [5]. Given a polygon $\mathcal{B}$ with n edges, these techniques compute all two-finger cage formations in $O(n^2 \log n)$ time. Note that certain objects might have $n/2$ "teeth" arranged along opposing edges. If the distance between the two edges increases away from the target immobilizing grasp, there can be $O(n^2)$ intermediate puncture points until the object can escape the hand to infinity. These techniques also construct a data structure that requires $O(n^2)$ space, which is capable of answering in $O(\log n)$ time whether a given two-finger placement forms a cage about the polygonal object. The development of caging algorithms for 3-D objects is an active research area in robotics; see, for instance, Pipattanasomporn et al. [6, 7].

Extension of caging theory to multiparameter robot hands is another active research area in robotics. The work of Rodriguez [8] offers insight into such potential generalizations of caging theory. For instance, given a target immobilizing grasp of an object $\mathcal{B}$, one may seek to determine all initial cage formations from which a *caging path* leads the fingers to the target immobilizing grasp. Caging theory can also serve as a robust object manipulation paradigm. For instance, Pereira, Campos and Kumar [9] discuss how teams of autonomous mobile robots (which replace the finger bodies) can transfer objects among obstacles using multi-agent cage formations.

Finally, the successful closing of a multi-finger cage to a target immobilizing grasp requires negligible friction at the contacts. Friction may cause *finger jamming* during the cage closing process, resulting in a frictional equilibrium grasp. Notable approaches that can prevent finger jamming include active control that keeps the finger loading levels sufficiently light when they slide along the object boundary, and fingertips that can change their friction properties according to the stage of the caging-to-grasping process.

Appendix: Proof Details

This appendix contains proofs of three lemmas. The first lemma concerns the lower half link, l^-, at the critical points of the projection function, $\pi(q,\sigma) = \sigma$, in the hand's free c-space $\mathcal{F}$.

LEMMA 10.5 *Let $\pi : \mathcal{F} \to \mathbb{R}$ be the projection function $\pi(q,\sigma) = \sigma$. Let p be a critical point of π on a stratum $\mathcal{S} = \cap_{i=1}^{r} \mathsf{bdy}(\mathcal{CB}_i)$ in $\mathcal{F}$. If the lower half link of π at p is nonempty, $l^- \neq \emptyset$, it must be a* **connected set**.

Proof: Recall that $\eta_i(p) \in T_p\mathbb{R}^4$ denotes the outward unit normal to the finger c-obstacle $\mathcal{CB}_i$, at a point $p = (q,\sigma) \in \mathsf{bdy}(\mathcal{CB}_i)$ in $\mathbb{R}^4$. The first-order approximation of the hand's free c-space $\mathcal{F}$ at p consists of the collection of tangent vectors:

$$T(p) = \left\{ \dot{p} \in T_p\mathbb{R}^4 : \ \eta_i(p) \cdot \dot{p} \geq 0 \text{ for } i = 1, \ldots, r \right\}. \tag{10.4}$$

Since $T(p)$ is the intersection of half-spaces, it forms a convex cone based at the origin of $T_p\mathbb{R}^4$, termed the *tangent cone* of $\mathcal{F}$ at p.

The normal slice of $\mathcal{F}$ at p, $E(p)$, is obtained by intersecting $\mathcal{F}$ with a disc $D(p)$ of suitable dimension. To a first-order approximation, $E(p) = T(p) \cap D(p)$. Since $T(p)$ and $D(p)$ are convex sets, $E(p)$ is also a convex set. The lower half link, l^-, is obtained by intersecting $E(p)$ with the level set $\pi^{-1}(c_0 - \epsilon) = \{(q, \sigma) : \pi(q, \sigma) = c_0 - \epsilon\}$, where $c_0 = \pi(p)$. Since π is a linear function, l^- is the intersection of the convex level set $\pi^{-1}(c_0 - \epsilon)$ with the normal slice $E(p)$. The intersection of two convex sets is either empty or a convex set. Hence, when l^- is nonempty, it must be a convex set that always forms a connected set. □

The next lemma characterizes the condition under which the lower half link, l^-, is empty at a critical point of a function f. We will need the following notion of *polar cones.* Let C_1 and C_2 be two convex cones based at the origin of $\mathbb{R}^m$. Then C_1 is said to be *polar* to C_2 if every vector $v \in C_1$ satisfies $w \cdot v \leq 0$ for all vectors $w \in C_2$.

Lemma 10.2 *Let $f : \mathcal{F} \to \mathbb{R}$ be a Morse function, such that f is the restriction to $\mathcal{F}$ of a smooth function $\tilde{f} : \mathbb{R}^4 \to \mathbb{R}$. Let p be a critical point of f in a stratum $\mathcal{S}$ of $\mathcal{F}$, such that $\mathcal{S} = \cap_{i=1}^{r} \mathrm{bdy}(\mathcal{CB}_i)$ for $1 \leq r \leq k$. A necessary condition for an* **empty lower half link** *at p, $l^- = \emptyset$, is that*

$$\nabla\tilde{f}(p) = \lambda_1 \eta_1(p) + \ldots + \lambda_r \eta_r(p) \qquad \lambda_1, \ldots, \lambda_r \geq 0, \tag{10.5}$$

such that $\lambda_1, \ldots, \lambda_r$ are not all zero. A sufficient condition for $l^- = \emptyset$ is that $\lambda_1, \ldots, \lambda_r$ are strictly positive in Eq. (10.1).

Proof: Assume that l^- is empty at p. Let $T(p)$ be the *tangent cone,* defined earlier in Eq. (10.4). The lower half link of f at p is given by $l^- = E(p) \cap f^{-1}(c_0 - \epsilon)$, where $E(p)$ is the normal slice of f at p, and $c_0 = f(p)$. Let $T_p^{\perp}\mathcal{S}$ denote the linear subspace of $T_p\mathbb{R}^4$ spanned by the finger c-obstacle outward normals $\eta_1(p), \ldots, \eta_r(p)$. It is the orthogonal complement of the tangent space $T_p\mathcal{S}$ in $T_p\mathbb{R}^4$. We may assume that $E(p)$ is the intersection of a small disc contained in $T_p^{\perp}\mathcal{S}$ with $\mathcal{F}$. Since l^- is empty at p, the function f must be non-decreasing along any c-space path that starts at p and lies in $E(p)$.

The collection of tangent vectors based at p and pointing into $E(p)$, given by $T(p) \cap T_p^{\perp}\mathcal{S}$, forms a *sub-cone* of $T(p)$. For any c-space path $\alpha(t)$ that starts at p and lies in $E(p)$, the tangent $\dot{p} = \dot{\alpha}(0)$ lies in the sub-cone $T(p) \cap T_p^{\perp}\mathcal{S}$. Since l^- is empty, it must be that $\frac{d}{dt}\big|_{t=0} f(\alpha(t)) = \nabla\tilde{f}(p) \cdot \dot{p} \geq 0$ for all $\dot{p} \in T(p) \cap T_p^{\perp}\mathcal{S}$. This condition is equivalent to the requirement that $\nabla\tilde{f}(p)$ would lie in the cone *polar* to the negated cone $-T(p)$ within the subspace $T_p^{\perp}\mathcal{S}$. The cone polar to $-T(p)$ can be characterized as follows. Consider the cone that is positively spanned by the finger c-obstacle outward normals $\eta_1(p), \ldots, \eta_r(p)$: $N(p) = \{\sum_{i=1}^{r} \lambda_i \eta_i(p) : \lambda_i \geq 0 \text{ for } i = 1, \ldots, r\}$. Let $-N(p)$ be the negated cone consisting of all vectors $-\eta$ such that $\eta \in N(p)$. A key property is that $T(p)$ is *polar* to $-N(p)$ in the ambient tangent space $T_p\mathbb{R}^4$. Hence, the sub-cone $T(p) \cap T_p^{\perp}\mathcal{S}$ is polar to $-N(p)$ within the subspace $T_p^{\perp}\mathcal{S}$. The cone polar to $-T(p)$ within $T_p^{\perp}\mathcal{S}$ is therefore $N(p)$. Since $\nabla\tilde{f}(p)$ lies in the cone polar to $-T(p)$ within $T_p^{\perp}\mathcal{S}$, $\nabla\tilde{f}(p)$ must belong to $N(p)$, which gives the condition specified in Eq. (10.5).

Next consider the case of strictly positive coefficients $\lambda_1, \ldots, \lambda_r$ in Eq. (10.5). Since $T_p^{\perp}\mathcal{S}$ is spanned by $\eta_1(p), \ldots, \eta_r(p)$, any non-zero tangent vector $\dot{p} \in T_p^{\perp}\mathcal{S}$ satisfies $\eta_i \cdot \dot{p} \neq 0$ for some $i \in \{1, \ldots, r\}$. This property also holds for all tangent vectors $\dot{p}$ that lie in the sub-cone $T(p) \cap T_p^{\perp}\mathcal{S}$. Since $T(p) \cap T_p^{\perp}\mathcal{S}$ is a sub-cone of $T(p)$, every $\dot{p}$ in the sub-cone satisfies $\eta_i(p) \cdot \dot{p} \geq 0$ for $i = 1 \ldots r$. The latter two facts imply that every vector $\dot{p} \in T(p) \cap T_p^{\perp}\mathcal{S}$ satisfies $\eta_i(p) \cdot \dot{p} > 0$ for some $i \in \{1, \ldots, r\}$. Since $\frac{d}{dt}\big|_{t=0} f(\alpha(t)) = \nabla\tilde{f}(p) \cdot \dot{p} = \sum_{i=1}^{r} \lambda_i(\eta_i(p) \cdot \dot{p})$ according to Eq. (10.5), it must be true that $\frac{d}{dt}\big|_{t=0} f(\alpha(t)) > 0$. Thus, f is strictly increasing along any c-space path α that starts at p and lies in $E(p)$, which implies that $I^- = \emptyset$. □

The last lemma characterizes the immobilizing and join point grasps as local minima and saddle points of the inter-finger distance function $d(u_1, u_2)$ in contact space $\mathcal{U}$.

Lemma 10.4 *Let a polygonal object $\mathcal{B}$ be held by two point or disc fingers. Every immobilizing grasp of $\mathcal{B}$ is a* **local minimum** *of $d(u_1, u_2)$ in $\mathcal{U}$, while every join point grasp is a* **saddle point** *of $d(u_1, u_2)$ in $\mathcal{U}$.*

Proof sketch: Let $(q_0, \sigma_0) \in \mathbb{R}^4$ be an immobilizing grasp of $\mathcal{B}$, with (u_1^0, u_2^0) the corresponding point in $\mathcal{U}$. Immobilization means that q_0 is completely surrounded by the finger c-obstacle slices $\mathcal{CB}_1|_{\sigma_0}$ and $\mathcal{CB}_2|_{\sigma_0}$ in the σ_0-slice of $\mathcal{F}$. As σ increases in the interval $[\sigma_0, \sigma_0 + \epsilon]$, the c-obstacle slices $\mathcal{CB}_1|_{\sigma}$ and $\mathcal{CB}_2|_{\sigma}$ move away from each other and form a three-dimensional cavity in each σ-slice of $\mathcal{F}$. The boundary of each cavity consists of the finger c-obstacles surfaces, as well as the intersection curves of these surfaces. The intersection curves form a closed loop for a sufficiently small ϵ. The closed loop maintains a fixed σ-value and, hence, represents one closed contour of $d(u_1, u_2)$ in $\mathcal{U}$. As σ increases in $[\sigma_0, \sigma_0 + \epsilon]$, the corresponding closed contours of $d(u_1, u_2)$ surround the point (u_1^0, u_2^0) with increasing σ value. The function $d(u_1, u_2)$ thus has a local minimum at (u_1^0, u_2^0).

Let $(q_1, \sigma_1) \in \mathbb{R}^4$ be a join point grasp of $\mathcal{B}$, with (u_1^1, u_2^1) the corresponding point in $\mathcal{U}$. First consider the interval $[\sigma_1 - \epsilon, \sigma_1]$. When σ increases in this interval, each σ-slice of $\mathcal{F}$ contains a three-dimensional cavity that is about to join another connected component of $\mathcal{F}$ at $\sigma = \sigma_1$. For ϵ sufficiently small, the finger c-obstacle surfaces intersect along two distinct curves in a local neighborhood of q_1 in $\mathcal{F}|_{\sigma}$, one from the inside of the cavity and one from the outside of the cavity. The two curves maintain a fixed σ-value and hence represent locally distinct contour segments of $d(u_1, u_2)$ in $\mathcal{U}$. As σ increases in the interval $[\sigma_1 - \epsilon, \sigma_1]$, the corresponding contour segments of $d(u_1, u_2)$ approach the point (u_1^1, u_2^1). Next, consider the interval $[\sigma_1, \sigma_1 + \epsilon]$. As σ increases in this interval, each σ-slice of $\mathcal{F}$ contains an expanding puncture in a local neighborhood of q_1 in $\mathcal{F}|_{\sigma}$. The c-obstacle surfaces in each slice $\mathcal{F}|_{\sigma}$ intersect along two distinct curves in this local neighborhood. The two curves maintain a fixed σ-value and hence represent another pair of locally distinct contour segments of $d(u_1,u_2)$ in $\mathcal{U}$. As σ increases in the interval $[\sigma_1,\sigma_1+\epsilon]$ the corresponding contour segments of $d(u_1,u_2)$ move away from (u_1^1,u_2^1). The point (u_1^1,u_2^1) is thus surrounded by two families of contour segment pairs. One family approaches the point (u_1^1,u_2^1) with σ increasing in the interval $[\sigma_1-\epsilon,\sigma_1]$; the other family moves away from (u_1^1,u_2^1) with σ increasing in the interval $[\sigma_1,\sigma_1+\epsilon]$. The function $d(u_1,u_2)$ thus has a saddle point at (u_1^1,u_2^1). □

Exercises

Section 10.1

Exercise 10.1: Explain how target immobilizing grasp of a 2-D object $\mathcal{B}$ specified in the world frame can be transformed to a target hand configuration, (q_0,σ_0), of a one-parameter robot hand.

Exercise 10.2: Can a system of three freely moving disc fingers in $\mathbb{R}^2$ be interpreted as a multi-finger hand? How many parameters describe the fingers' shape relative to the hand's palm in such a system?

Exercise 10.3: Identify the two-finger squeezing and expanding cages of the polygonal object $\mathcal{B}$ depicted in Figure 10.9, assuming a system of two freely moving point or fingers in $\mathbb{R}^2$ (provide a qualitative sketch of the caging sets).

Exercise 10.4*: Assuming a one-parameter k-finger robot hand, can the *squeezing cages* and *expanding cages* be formally distinguished in the hand's free c-space $\mathcal{F}$?

Solution Sketch: Recall that $\eta_i(p)\in T_p\mathbb{R}^4$ denotes the outward unit normal to the finger c-obstacle $\mathcal{CB}_i$ at $p=(q,\sigma)\in \text{bdy}(\mathcal{CB}_i)$. The boundary of $\mathcal{CB}_i$ is the level set of a signed distance function, denoted $d(q,\sigma)$, which measures the distance between the finger body $\mathcal{O}_i$ and the stationary object $\mathcal{B}$. Up to a positive multiplicative factor, $\eta_i(p)$ can be written as

$$\eta_i(p)=\begin{pmatrix}\frac{\partial}{\partial q}d(q,\sigma)\\ \frac{\partial}{\partial \sigma}d(q,\sigma)\end{pmatrix}\qquad p=(q,\sigma).$$

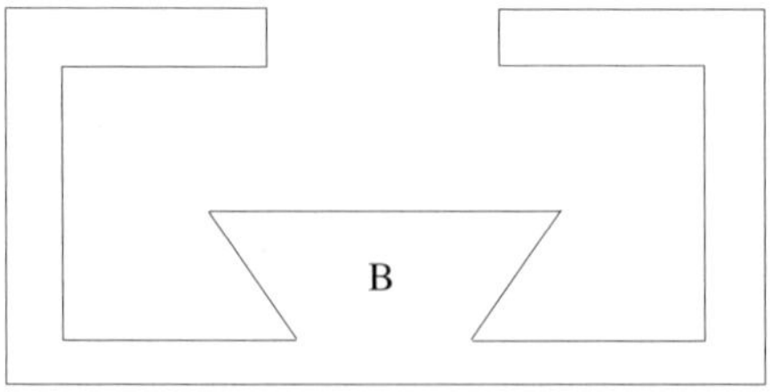

Figure 10.9 A polygonal object possessing several squeezing and expanding cages.

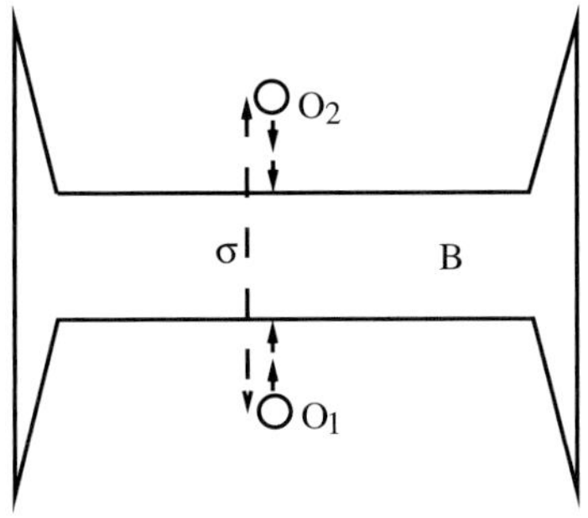

Figure 10.10 A cage formation that does *not* retract to a unique immobilizing grasp.

At a *squeezing grasp,* $\frac{\partial}{\partial \sigma} d(q_0, \sigma_0) > 0$ for $i = 1 \ldots k$. At an *expanding grasp,* $\frac{\partial}{\partial \sigma} d(q_0, \sigma_0) < 0$ for $i = 1 \ldots k$. ○

Section 10.3

Exercise 10.5: Figure 10.10 shows an initial two-finger cage formation that does *not* retract to a unique immobilizing grasp of $\mathcal{B}$. Is it true that in such special cases the cage retracts to a *bounded mobility grasp* of the object $\mathcal{B}$?

Exercise 10.6*: Every immobilizing grasp of $\mathcal{B}$ is locally surrounded by cage formations when the robot hand operates as a one-parameter system (Theorem 10.1). Can this result be extended to multi-parameter robot hands that initially immobilize the object $\mathcal{B}$?

Exercise 10.7: Let $\pi : \mathcal{F} \to \mathbb{R}$ be the projection function, $\pi(q, \sigma) = \sigma$. Prove that every k-finger equilibrium grasp of $\mathcal{B}$ is a *critical point* of π in $\mathcal{F}$. (Hint: let $\mathcal{S} = \cap_{i=1}^{k} \text{bdy}(\mathcal{CB}_i)$ be the stratum of $\mathcal{F}$ containing the equilibrium grasp configuration, $p = (q, \sigma)$. Then p is a critical point of π if $\nabla \pi \cdot \dot{p} = 0$ for all tangent vectors $\dot{p} \in T_p\mathcal{S}$.)

Exercise 10.8: A one-parameter robot hand immobilizes an object $\mathcal{B}$ at (q_0, σ_0), such that all fingers are essential for maintaining the grasp. Prove that the projection function, $\pi : \mathcal{F} \to \mathbb{R}$ such that $\pi(q, \sigma) = \sigma$, has a local minimum at (q_0, σ_0) at a *squeezing* immobilizing grasp and a local maximum at an *expanding* immobilizing grasp.

Exercise 10.9*: Consider a system of two freely moving disc fingers in $\mathbb{R}^2$. Prove that every two-finger cage about an object $\mathcal{B}$ either retracts or expands into an immobilizing grasp of $\mathcal{B}$, without any break in the cage during the finger closing or opening process.

Exercise 10.10: Figure 10.11(a) depicts a one-parameter robot hand whose three point fingers open while maintaining an equilateral triangle formation. Does the hand form a cage that prevents the object $\mathcal{B}$ from escaping to infinity at the finger opening indicated in the figure?

Exercise 10.11: Repeat the previous exercise for the same three-finger hand, this time locking the fingers at the larger finger opening indicated in Figure 10.11(b).

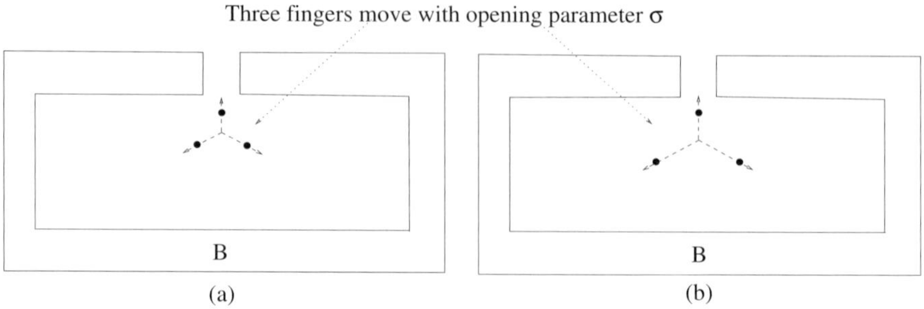

Figure 10.11 Three-finger caging of a polygonal object $\mathcal{B}$ along a bounded σ interval, which is not associated with any immobilizing grasp of $\mathcal{B}$.

Exercise 10.12: Identify the interval of finger openings that forms a three-finger cage of $\mathcal{B}$. (This exercise demonstrates that one-parameter robot hands with three fingers do *not* possess the two-finger caging to grasping monotonicity property.)

Section 10.4

Exercise 10.13: A one-parameter robot hand has three point or disc fingers. Characterize the immobilizing grasps of a polygonal object $\mathcal{B}$.

Solution: These are all the two-finger immobilizing grasps, together with two types of generic three-finger immobilizing grasps. Assuming *squeezing grasps,* the first type are three-finger equilibrium grasps along three distinct edges of $\mathcal{B}$. The second type are three-finger equilibrium grasps at which one finger is located at a *concave* vertex of $\mathcal{B}$, while the other two fingers are located on two (not necessarily distinct) edges of $\mathcal{B}$. ○

Exercise 10.14: A one-parameter robot hand has three point or disc fingers. Characterize the puncture-point grasps of a polygonal object $\mathcal{B}$.

Solution: These are all the two-finger puncture point grasps, together with one generic type of a three-finger equilibrium grasps. At a puncture point, one finger touches a *convex* vertex of $\mathcal{B}$, while the other two fingers are located on two (not necessarily distinct) edges of $\mathcal{B}$. Note that at a two-finger puncture grasp, the remaining finger may happen to touch the object without actively participating in the equilibrium grasp. ○

Exercise 10.15: The triangular polygon $\mathcal{B}$ depicted in Figure 10.12 is initially immobilized by three disc fingers. The three fingers are governed by a single opening parameter σ, as shown in the figure. Determine the puncture point of the c-space caging set associated with the immobilizing grasp. (Hint: it is a three-finger equilibrium grasp of $\mathcal{B}$).

Exercise 10.16: The puncture point of the c-space caging set forms a frictionless equilibrium grasp of $\mathcal{B}$ (Theorem 10.3). Explain why for 2-D grasps, a puncture point generically involves *two or three fingers,* even when the immobilizing grasp may involve higher number of fingers.

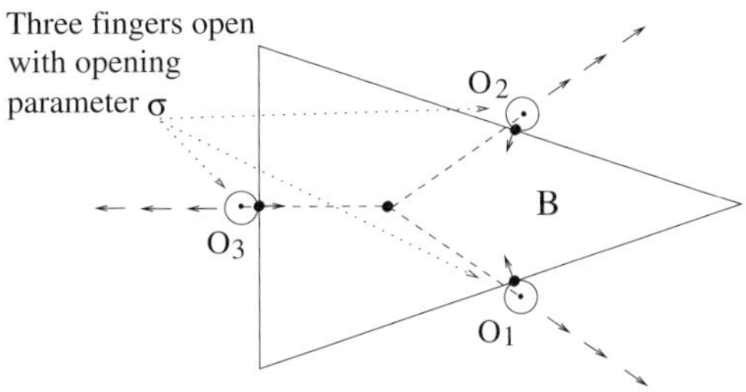

Figure 10.12 A three-finger robot hand whose fingers open while maintaining a similar triangle formation. A three-finger immobilizing grasp whose puncture point involves all three fingers.

Section 10.5

Exercise 10.17: Execute the graphical technique on the polygonal object $\mathcal{B}$ depicted in Figure 10.7, starting from the saddle point grasp with finger opening σ_2.

Exercise 10.18: The previous exercise established the existence of a σ-decreasing path from the σ_2 saddle point grasp to the immobilizing grasp with finger opening σ_0. Does the σ_2 grasp form the puncture point of the c-space caging set of the σ_0 immobilizing grasp?

Exercise 10.19: Explain how the graphical technique can be used to compute the caging regions of two disc fingers rather than two point fingers.

Solution sketch: The polygonal object $\mathcal{B}$ can be interpreted as a stationary obstacle. By uniformly expanding the object $\mathcal{B}$ by the fingers' radius and then approximating the expanded object by a polygon, we obtain two point fingers that interact with a polygon. ○

Exercise 10.20*: The graphical technique first computes all two-finger equilibrium grasps of the polygonal object $\mathcal{B}$. Can this step be executed in quasi-linear time?

Exercise 10.21: The graphical technique asserts in Step 3.5 that when a local minimum of $d(u_1, u_2)$ in $\mathcal{U}$ is *not* a feasible equilibrium grasp of $\mathcal{B}$, one finger can be retracted away from the stationary object $\mathcal{B}$. Justify this assertion.

Solution sketch: A local minimum of $d(u_1, u_2)$ in $\mathcal{U}$ is also a local minimum of the projection function, $\pi(q, \sigma) = \sigma$, in the double-contact submanifold, $\mathcal{S} = \text{bdy}(\mathcal{CB}_1) \cap \text{bdy}(\mathcal{CB}_2)$. If a local minimum of $d(u_1, u_2)$ in $\mathcal{U}$ is *not* a feasible equilibrium grasp of $\mathcal{B}$, the object configuration point, q_0, is *not* a local minimum of $\pi(q, \sigma)$ in the ambient free c-space $\mathcal{F}$. Hence, there exists a σ-decreasing path in $\mathcal{F}$ that starts at $(q_0, \sigma_0) \in \mathcal{S}$ and moves into the interior of $\mathcal{F}$. Since $\nabla\pi(q, \sigma) = (0, 0, 0, 1)$, such a path can maintain fixed $q = q_0$ coordinates while moving parallel to the σ-axis. In physical space, the hand's palm frame remains fixed while one finger moves away from the stationary object $\mathcal{B}$ along the respective object contact normal. ○

Exercise 10.22: Explain why the local minima of $d(u_1, u_2)$ in contact space $\mathcal{U}$ always form feasible equilibrium grasps when the polygon $\mathcal{B}$ is convex.

Exercise 10.23: Consider the rendering of the c-space caging set $\mathcal{K}$ as two *caging regions* in $\mathbb{R}^2$. Under what condition every placement of the two fingers on the boundary curves γ_1 and γ_2 with finger opening $\sigma \leq \sigma_1$ is guaranteed to retract to the immobilizing grasp at (q_0, σ_0)?

Solution: Let I_1 and I_2 be the contact space intervals parametrizing the boundary curves γ_1 and γ_2. That is, γ_i is parametrized by $x(u_i): I_i \to \mathbb{R}^2$ for $u_i \in I_i$ $(i = 1, 2)$. Then $I_1 \times I_2$ is a rectangular region in contact space $\mathcal{U}$. Let $(u_1^0, u_2^0) \in I_1 \times I_2$ correspond to the immobilizing grasp (q_0, σ_0). Denote by $\text{Basin}(u_1^0, u_2^0)$ the basin of attraction of the local minimum of σ at (u_1^0, u_2^0). That is, $\text{Basin}(u_1^0, u_2^0)$ is the set of points (u_1, u_2) attracted by the negated gradient flow $(\dot{u}_1, \dot{u}_2) = -\nabla d(u_1, u_2)$ to the local minimum (u_1^0, u_2^0). The rectangle $I_1 \times I_2$ is *admissible* with respect to (u_1^0, u_2^0) when all points (u_1, u_2) in $I_1 \times I_2$ with finger opening $d(u_1, u_2) \leq \sigma_1$ belong to $\text{Basin}(u_1^0, u_2^0)$. When $I_1 \times I_2$ is admissible with respect to (u_1^0, u_2^0), every placement of the two fingers on the boundary segments γ_1 and γ_2 with finger opening $\sigma \leq \sigma_1$ will retract to the immobilizing grasp at (q_0, σ_0). ○

Exercise 10.24*: Extend the graphical technique to one-parameter robot hands having *three* point or disc fingers that maintain equilateral or similar triangle formations.

Hint: Consider the contact spaces $\mathcal{U}_{i,i+1}$ for $i = 1, 2, 3$, where index addition is taken modulo 3. Each contact space $\mathcal{U}_{i,i+1}$ parametrizes the contacts of fingers i and $i + 1$ along the object boundary. Each contact space $\mathcal{U}_{i,i+1}$ contains *obstacles,* where finger $i + 2$ penetrates the object. The full contact space of the three-finger hand, $\mathcal{U}$, consists of the pairwise spaces $\mathcal{U}_{1,2}$, $\mathcal{U}_{2,3}$, and $\mathcal{U}_{3,1}$, glued together along the contact-space obstacle boundaries.

Bibliographical Notes

Exercise 10.25: Given a team of k autonomous mobile robots and a warehouse object $\mathcal{B}$, design a robust object transfer strategy that requires only local interagent distance measurements.

References

[1] W. Kuperberg, "Problems on polytopes and convex sets," in *DIMACS Workshop on Polytopes*, Rutgers University, pp. 584–589, 1990.

[2] E. Rimon and A. Blake, "Caging planar bodies by 1-parameter two-fingered gripping systems," *International Journal of Robotics Research*, vol. 18, no. 3, pp. 299–318, 1999.

[3] P. Pipattanasomporn and A. Sudsang, "Two-finger caging of convex polygons," in *IEEE Int. Conf. on Robotics and Automation*, pp. 2137–2142, 2006.

[4] M. Vahedi and A. F. van der Stappen, "Caging polygons with two and three fingers," *International Journal of Robotics Research*, vol. 27, no. 11–12, pp. 1308–1324, 2008.

[5] T. Allen, J. Burdick and E. Rimon, "Two-fingered caging of polygonal objects using contact space search," *IEEE Transactions on Robotics*, vol. 31, no. 5, pp. 1164–1179, 2015.

[6] P. Pipattanasomporn and A. Sudsang, "Two-finger caging of nonconvex polytopes," *IEEE Transactions on Robotics*, vol. 27, no. 2, 324–333, 2011.

[7] P. Pipattanasomporn, T. Makapunyo and A. Sudsang, "Multifinger caging using dispersion constraints," *IEEE Transactions on Robotics*, vol. 32, no. 4, pp. 1033–1041, 2016.

[8] A. Rodriguez, M. T. Mason and S. Ferry, "From caging to grasping," in *Robotics: Science and Systems (RSS)*, 2011.

[9] G. A. S. Pereira, M. F. M. Campos and V. Kumar, "Decentralized algorithms for multi-robot manipulation via caging," *International Journal of Robotics Research*, vol. 23, no. 7–8, pp. 783–795, 2004.

11 Frictionless Hand-Supported Stances under Gravity

This chapter extends grasp mobility theory to *frictionless equilibrium stances,* where a rigid object is supported against gravity by the palm and fingers of a robot hand. Based on the c-space geometry of such stances, a measure of stance security takes the form of *local stability,* which persists under small disturbances that may act on the supported object. For example, the three-finger robot hand depicted in Figure 11.1 supports a large rigid object against gravity via frictionless palm and fingertip contacts. The object is neither immobilized nor caged by the robot hand, but the hand supports the object at a local minimum of its gravitational potential energy, with a rather large margin of stability at this particular stance.

The chapter studies the stability of hand-supported stances by treating gravity as a *virtual finger* that forms part of the grasping system. When gravity is incorporated in this manner, the supported object free motions are determined by rigid-body constraints as well as conservation of energy. This insight will lead to analytic tests that assess the stability of hand-supported stances, analogous to the first and second-order immobilization tests developed in previous chapters. A notion of *gravity cages* analogous to the notion of caging grasps is discussed in the Bibliographical Notes.

Sections 11.1 and 11.2 describe basic facts concerning equilibrium stances under the influence of gravity. When a supporting robot hand forms a fixed terrain, the supported

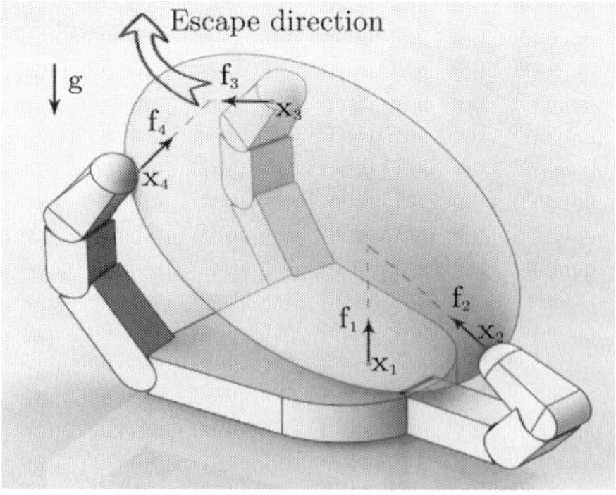

Figure 11.1 A three-finger robot hand supports a large rigid object via frictionless palm and fingertip contacts. The object is neither immobilized nor caged by the robot hand, but supported at a locally stable stance that persists under small disturbances.

object free c-space, $\mathcal{F}$, forms a *stratified set.* The object's equilibrium stances are *critical points* of its gravitational potential energy, $U(q)$, in the object's free c-space $\mathcal{F}$. Equilibrium stances at the *local minima* of $U(q)$ are locally stable as well as robust against small wrench disturbances. Based on these facts, Sections 11.3 and 11.4 develop analytic tests for the local minima of $U(q)$ in the free c-space $\mathcal{F}$, analogous to the first and second-order immobilization conditions discussed in previous chapters. Sections 11.5 and 11.6 apply the stance stability tests to the following *support problem.* A rigid object $\mathcal{B}$ with a variable center of mass is supported against gravity by a set of frictionless contacts. Determine the object center-or-mass positions that ensure stance stability on the supporting contacts.

11.1 C-Space Representation of Equilibrium Stances

This section characterizes the equilibrium stances of a rigid object, $\mathcal{B}$, supported against gravity by stationary rigid bodies $\mathcal{O}_1, \ldots, \mathcal{O}_k$, which represent the robot hand. When the robot hand is held stationary, the object's *free c-space* has the form:

$$\mathcal{F} = \mathbb{R}^m - \cup_{i=1}^{k} \text{int}(\mathcal{CO}_i) \qquad m = 3 \text{ in 2-D stances, } m = 6 \text{ in 3-D stances,}$$

where $\text{int}(\mathcal{CO}_i)$ denotes the interior of the c-obstacle induced by the supporting body $\mathcal{O}_i$. The free c-space $\mathcal{F}$ forms a *stratified set,* which consists of smooth manifolds of varying dimension. Each manifold represents one possible contact combination of $\mathcal{B}$ with the supporting bodies. Consider, for instance, the strata of $\mathcal{F}$ associated with a 2-D object $\mathcal{B}$ supported by 2-D bodies $\mathcal{O}_1, \ldots, \mathcal{O}_k$. The three-dimensional strata of $\mathcal{F}$ are open subsets of the ambient c-space parametrized by $\mathbb{R}^3$. The two-dimensional strata of $\mathcal{F}$ consist of portions of the c-obstacle boundaries, and they correspond to single-body contacts with $\mathcal{B}$. The one-dimensional strata of $\mathcal{F}$ are the intersection curves of the c-obstacle surfaces, and they correspond to two-body contacts with $\mathcal{B}$. The zero-dimensional strata are isolated points that correspond to three-body contacts with $\mathcal{B}$. The strata of $\mathcal{F}$ are illustrated in the following example.

Example: Figure 11.2(a) depicts a rigid object $\mathcal{B}$ supported against gravity by two stationary rigid bodies $\mathcal{O}_1$ and $\mathcal{O}_2$ (the full robot hand is not shown). The strata formed by the supporting bodies in $\mathcal{B}$'s c-space are sketched in Figure 11.2(b). The three-dimensional stratum is the complement of the c-obstacles in $\mathbb{R}^3$. The two-dimensional strata are the nonintersecting portions of the c-obstacle surfaces. There are also two one-dimensional strata, which form the intersection curves of the c-obstacle surfaces. ○

In order to characterize the object equilibrium stances we need to study the *critical points* of the object's gravitational potential energy, $U(q)$, in the stratified set $\mathcal{F}$. Let b_{cm} denote the position of $\mathcal{B}$'s center of mass in the object reference frame. When $\mathcal{B}$ is located at a configuration q, the position of its center of mass in the fixed world frame is given by the rigid body transformation, $x_{cm}(q) = R(\theta)b_{cm} + d$, where $q = (d, \theta) \in \mathbb{R}^m$ is the object configuration. Using this notation, the object's gravitational potential energy has the following form.

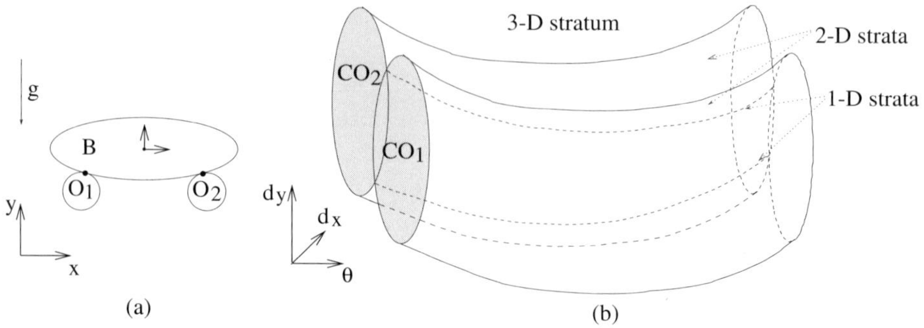

Figure 11.2 (a) A rigid object $\mathcal{B}$ supported against gravity by stationary rigid bodies $\mathcal{O}_1$ and $\mathcal{O}_2$ (the full robot hand is not shown). (b) A schematic view of the c-obstacles and the strata of the free c-space $\mathcal{F}$ (the θ-axis is periodic in 2π).

Gravitational potential energy: *The gravitational potential energy of a rigid object $\mathcal{B}$ is the function $U(q): \mathbb{R}^m \to \mathbb{R}$ given by*

$$U(q) = \mathsf{m}g\left(\boldsymbol{e} \cdot x_{\mathsf{cm}}(q)\right) \qquad x_{cm}(q) = R(\theta)b_{cm} + d,\ q = (d, \theta),$$

where m *is $\mathcal{B}$'s mass, g is the gravitational constant and $\boldsymbol{e}$ is the vertical upward direction: $\boldsymbol{e} = (0,1)$ in 2-D stances, and $\boldsymbol{e} = (0,0,1)$ in 3-D stances.*

The gravitational potential measures the height of $\mathcal{B}$'s center of mass in the fixed world frame. It is a smooth function of q that can be analyzed with the tools of stratified Morse theory (see Appendix B). Under stratified Morse theory, the critical points of $U(q)$ in $\mathcal{F}$ are the *union* of the critical points on the individual strata of $\mathcal{F}$. Let $\mathcal{S}$ denote a particular stratum of $\mathcal{F}$, say the surface of a particular c-obstacle. Then $U(q)$ has a *critical point* at $q \in \mathcal{S}$ when its gradient, $\nabla U(q)$, is orthogonal to the stratum $\mathcal{S}$ in the ambient c-space $\mathbb{R}^m$:

$$\nabla U(q) \cdot \dot{q} = 0 \qquad \text{for all } \dot{q} \in T_q\mathcal{S}. \tag{11.1}$$

Note that every zero-dimensional stratum of $\mathcal{F}$ automatically forms a critical point of $U(q)$.

When the object $\mathcal{B}$ is supported by the bodies $\mathcal{O}_1, \ldots, \mathcal{O}_k$ at an equilibrium stance, the net wrench on $\mathcal{B}$ due to gravity and the supporting contact forces must be zero. The gravitational wrench affecting $\mathcal{B}$ is given by $-\nabla U(q)$. Under frictionless contact conditions, the supporting contacts act along $\mathcal{B}$'s inward contact normals. The wrenches induced by such forces are collinear with the c-obstacle outward normals (see Chapter 3). Based on these facts, the feasible equilibrium stances are characterized as follows.

Equilibrium stance condition: *Let a rigid object $\mathcal{B}$ be contacted by stationary rigid bodies $\mathcal{O}_1, \ldots, \mathcal{O}_k$ at a configuration q_0. The object is supported at a feasible* **equilibrium stance** *under the influence of gravity when*

$$\lambda_1\hat{\eta}_1(q_0) + \cdots + \lambda_k\hat{\eta}_k(q_0) - \nabla U(q_0) = \vec{0} \qquad \lambda_1, \ldots, \lambda_k \geq 0, \tag{11.2}$$

where $\hat{\eta}_1(q_0), \ldots, \hat{\eta}_k(q_0)$ are the supporting bodies c-obstacle outward unit normals at q_0, and $U(q)$ is $\mathcal{B}$'s gravitational potential energy.

Since gravity applies a non-zero force on $\mathcal{B}$, at least one of the supporting contacts must be active at a feasible equilibrium stance. The equilibrium stance condition therefore holds true only at points on the boundary of the free c-space $\mathcal{F}$. The condition specified by Eq. (11.2) can be equivalently stated as follows. The *net wrench cone* of the supporting contacts is defined as the collection of net wrenches that can be affected on $\mathcal{B}$ by the supporting contacts:

$$\mathcal{W}(q_0)=\left\{w \in T^*_{q_0}\mathbb{R}^m\ :\ w=\lambda_1\hat{\eta}_1(q_0)+\ldots+\lambda_k\hat{\eta}_k(q_0)\ \text{ for }\ \lambda_1,\ldots,\lambda_k\geq 0\right\}.$$

Based on Eq. (11.2), a feasible equilibrium stance is characterized by the condition that $\nabla U(q_0)$ must lie in $\mathcal{W}(q_0)$ (Figure 11.3(c)). This condition implies that feasible equilibrium stances are *critical points* of $U(q)$, as stated in the following proposition.

PROPOSITION 11.1 **(equilibrium stances)** *Under the influence of gravity, the feasible equilibrium stances of a rigid object $\mathcal{B}$ on supporting rigid bodies $\mathcal{O}_1, \ldots, \mathcal{O}_k$ are* **critical points** *of the object gravitational potential energy $U(q)$ in the free c-space $\mathcal{F}$.*

Proof sketch: The object $\mathcal{B}$ is supported by the bodies $\mathcal{O}_1, \ldots, \mathcal{O}_k$ at a configuration q_0, which lies in the stratum $\mathcal{S}=\cap_{i=1}^{k}\text{bdy}(\mathcal{CO}_i)$. The tangent space to $\mathcal{S}$ at q_0 is given by

$$T_{q_0}\mathcal{S}=\{\dot{q}\in T_{q_0}\mathbb{R}^m\ :\ \hat{\eta}_i(q_0)\cdot\dot{q}=0\ \text{ for } i=1\ldots k\}.$$

It follows that the linear combination $\sum_{i=1}^{k}\lambda_i\hat{\eta}_i(q_0)$ satisfies

$$\Big(\sum_{i=1}^{k}\lambda_i\hat{\eta}_i(q_0)\Big)\cdot\dot{q}=0\quad \text{for all } \dot{q}\in T_{q_0}\mathcal{S}. \tag{11.3}$$

The equilibrium stance condition takes the form: $\sum_{i=1}^{k}\lambda_i\hat{\eta}_i(q_0)-\nabla U(q_0)=\vec{0}$. Substituting for $\sum_{i=1}^{k}\lambda_i\hat{\eta}_i(q_0)$ in Eq. (11.3) gives $\nabla U(q_0)\cdot\dot{q}=0$ for all $\dot{q}\in T_{q_0}\mathcal{S}$, which implies that $U(q)$ has a critical point at q_0 according to Eq. (11.1). □

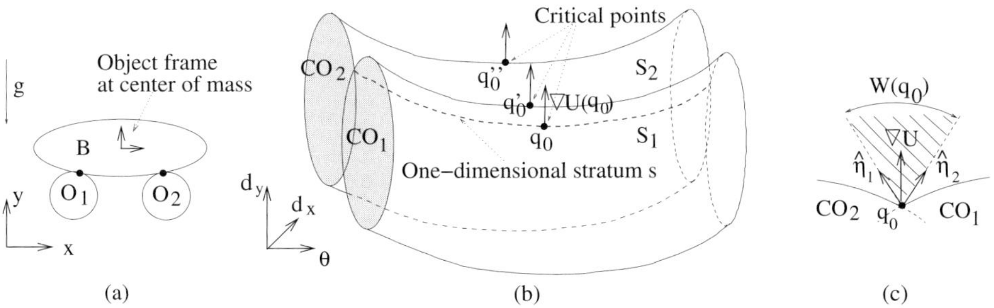

Figure 11.3 (a) A rigid object $\mathcal{B}$ supported against gravity by two stationary rigid bodies $\mathcal{O}_1$ and $\mathcal{O}_2$ (the full robot hand is not shown). (b) Schematic view of the critical points of $U(q)$ in $\mathcal{F}$. (c) The condition $\nabla U(q_0)\in\mathcal{W}(q_0)$ identifies the feasible equilibrium stances as critical points of $U(q)$ in $\mathcal{F}$.

Example: Figure 11.3(a) shows a rigid object $\mathcal{B}$ supported against gravity by stationary rigid bodies $\mathcal{O}_1$ and $\mathcal{O}_2$ (the full robot hand is not shown). The supporting bodies c-obstacles, $\mathcal{CO}_1$ and $\mathcal{CO}_2$, are sketched in Figure 11.3(b). The object configuration point, q_0, lies in the one-dimensional stratum $\mathcal{S} = \mathsf{bdy}(\mathcal{CO}_1) \cap \mathsf{bdy}(\mathcal{CO}_2)$. The supporting contacts net wrench cone, $\mathcal{W}(q_0)$, is spanned by the c-obstacle outward unit normals, $\hat{\eta}_1(q_0)$ and $\hat{\eta}_2(q_0)$. Since $\nabla U(q_0) \in \mathcal{W}(q_0)$ as depicted in Figure 11.3(c), the point q_0 is a critical point of $U(q)$. Note that $U(q)$ has two additional critical points, $q_0' \in \mathsf{bdy}(\mathcal{CO}_1)$ and $q_0'' \in \mathsf{bdy}(\mathcal{CO}_2)$, which correspond to single-support equilibrium stances on $\mathcal{O}_1$ and $\mathcal{O}_2$. ○

The feasible equilibrium stances of $\mathcal{B}$ form a *subset* of the critical points of $U(q)$ in $\mathcal{F}$. The remaining critical points correspond to infeasible equilibrium stances, where some of the contacts apply *adhesive forces* along the contact normals. While such forces are not feasible under the contact models used in grasp mechanics (Chapter 4), they can be incorporated into the c-space framework in other grasping systems that support such forces.

In order to identify which critical points of $U(q)$ correspond to locally stable stances, we have to verify under what conditions $U(q)$ forms a *Morse function,* which must possess non-degenerate critical points in $\mathcal{F}$ (Appendix B). A critical point of $U(q)$ is *non-degenerate* when it satisfies two conditions. Let $\mathcal{S}$ be the stratum of $\mathcal{F}$ containing a critical point q_0. The first condition requires that $\nabla U(q_0)$ must *not* be normal to any of the neighboring strata at q_0. When q_0 lies in a stratum $\mathcal{S}$ that corresponds to supporting bodies $\mathcal{O}_1, \ldots, \mathcal{O}_k$, the neighboring strata represent stances where $\mathcal{B}$ is supported by sub-collections of these bodies. The non-degeneracy condition requires that none of the supporting bodies be idle at the equilibrium stance, as illustrated in Figure 11.4. The second condition requires that $U(q)$ possess a non-degenerate second derivative along the stratum $\mathcal{S}$ at q_0. That is, the Hessian matrix $D^2U(q_0)$ must have non-zero eigenvalues when evaluated on the tangent space $T_{q_0}\mathcal{S}$.

Example: Figure 11.4(a) shows an object $\mathcal{B}$ supported by segments $\mathcal{O}_1$ and $\mathcal{O}_2$ of a robot hand via frictionless contacts x_1 and x_2. The object's center of mass lies exactly above x_2, which provides a supporting force f_2, while the contact point x_1 is idle at this stance. The equilibrium stance configuration, q_0, lies in the stratum $\mathcal{S} = \mathsf{bdy}(\mathcal{CO}_1) \cap \mathsf{bdy}(\mathcal{CO}_2)$, which is the intersection curve of the c-obstacles surfaces associated with

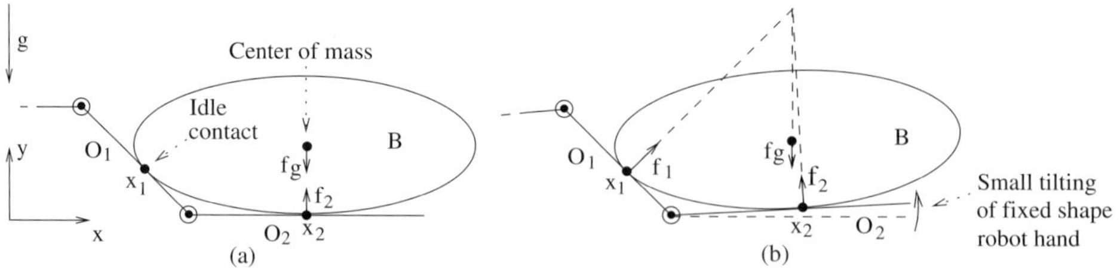

Figure 11.4 (a) A two-contact stance where x_2 is active while x_1 is idle. (b) Any small upward tilting of the robot hand results in a two-contact stance where both contacts are active.

$\mathcal{O}_1$ and $\mathcal{O}_2$. While $\nabla U(q_0)$ is orthogonal to the stratum $\mathcal{S}$ at q_0, it is also orthogonal to the neighboring stratum, $\mathcal{S}_2 = \text{bdy}(\mathcal{CO}_2)$, at this point. Hence, $U(q)$ *fails* to be Morse at q_0. However, the equilibrium stances generically form non-degenerate critical points of $U(q)$. Figure 11.4(b) shows that any small upward tilting of the fixed-shape robot hand forms an equilibrium stance that involves both contacts. ○

The chapter proceeds under the assumption that the object's gravitational potential energy forms a *Morse function* on the free c-space $\mathcal{F}$, which is the generic case.

11.2 The Stable Equilibrium Stances

This section establishes that among the critical points of the object's gravitational potential $U(q)$ in $\mathcal{F}$ (which correspond to equilibrium stances), the local minima of $U(q)$ are *locally stable* with respect to position and velocity perturbations, as well as *locally robust* against disturbances caused by small tiltings of the fixed-shape robot hand. The local stability property is based on the following conservation of energy principle.

Total mechanical energy: *The total mechanical energy of a rigid object $\mathcal{B}$ at the state $(q, \dot{q})$ has the form*

$$E(q,\dot{q}) = U(q) + \mathcal{K}(q,\dot{q}) \qquad (q,\dot{q}) \in \mathbb{R}^m \times \mathbb{R}^m,$$

where $U(q)$ is the gravitational potential energy of $\mathcal{B}$, while $\mathcal{K}(q,\dot{q})$ is the kinetic energy of $\mathcal{B}$, given by the quadratic form

$$\mathcal{K}(q,\dot{q}) = \tfrac{1}{2}\dot{q}^T M(q)\dot{q},$$

where $M(q)$ is $\mathcal{B}$'s $m \times m$ rigid-body mass matrix, $m = 3$ in 2-D stances and $m = 6$ in 3-D stances.

The object's kinetic energy, $\mathcal{K}(q,\dot{q})$, is positive definite since the rigid-body mass matrix, $M(q)$, is positive definite. Hence, when $U(q)$ has a strict local minimum at q_0, $E(q,\dot{q})$ has a strict local minimum at the zero-velocity state $(q_0, \vec{0})$. When the object and supporting bodies are perfectly rigid and interact through frictionless contacts with purely elastic collisions,[1] we have the following conservation of energy property.

Conservation of energy: *Let a rigid object $\mathcal{B}$ move under the influence of gravity while maintaining continuous or intermittent contact with stationary rigid bodies $\mathcal{O}_1, \ldots, \mathcal{O}_k$ via frictionless contacts and perfectly elastic collisions. Then the object's total mechanical energy, $E(q,\dot{q})$, remains* **constant** *during its motion.*

In practice, some amount of friction is always present at the contacts, while inelastic collisions always take place when $\mathcal{B}$ bounces against the supporting bodies. The object's total mechanical energy is therefore non-increasing rather than constant dur-

[1] An elastic collision between nonrigid bodies spans a finite time interval, which shrinks to a discrete time instant at the limit where the bodies are perfectly rigid.

ing its motions. The equilibrium stances are zero-velocity *equilibrium states* of the object's rigid-body dynamics. The local stability of these equilibrium states is defined as follows.

DEFINITION 11.1 **(local stability)** *A zero-velocity equilibrium state of $\mathcal{B}$, $(q_0, \vec{0})$, is* **locally stable** *if for every local neighborhood $\mathcal{V}$ centered at $(q_0, \vec{0})$ in $\mathcal{B}$'s state space, $(q, \dot{q}) \in \mathbb{R}^m \times \mathbb{R}^m$, there exists a smaller neighborhood centered at $(q_0, \vec{0})$, $\mathcal{V}' \subseteq \mathcal{V}$, such that all trajectories $(q(t), \dot{q}(t))$ that start in $\mathcal{V}'$ at $t=0$ remain in $\mathcal{V}$ for all $t \geq 0$.*

The local minima of the object gravitational potential energy form locally stable equilibrium stances, as stated in the following theorem.

THEOREM 11.2 **(stance stability)** *Let a rigid object $\mathcal{B}$ be supported at an equilibrium stance configuration q_0 by stationary rigid bodies $\mathcal{O}_1, \ldots, \mathcal{O}_k$. If q_0 is a non-degenerate local minimum of $U(q)$ (the generic case), the zero-velocity equilibrium state $(q_0, \vec{0})$ is* **locally stable.**

Proof sketch: When $U(q)$ has non-degenerate local minimum at q_0, the object's total mechanical energy, $E(q, \dot{q})$, has a strict local minimum at $(q_0, \vec{0})$. Let $\mathcal{D}$ be the basin of attraction of $(q_0, \vec{0})$ under the negative gradient flow:

$$\frac{d}{dt}\begin{pmatrix} q(t) \\ \dot{q}(t) \end{pmatrix} = -\nabla E(q(t), \dot{q}(t)).$$

That is, $\mathcal{D}$ is the set of initial states whose negative gradient trajectories converge to $(q_0, \vec{0})$. Consider the bounded energy sets,

$$\mathcal{U}_c = \{(q, \dot{q}) \in \mathcal{F} \times \mathbb{R}^m : E(q, \dot{q}) \leq c\} \qquad c \in [U(q_0), U(q_0) + \epsilon],$$

where ϵ is a small positive parameter. For a sufficiently small ϵ, each of these sets forms a single connected component that contains the point $(q_0, \vec{0})$ and lies within the basin of attraction $\mathcal{D}$. The sets $\mathcal{U}_c$ for $c \in [U(q_0), U(q_0) + \epsilon]$ form open neighborhoods about $(q_0, \vec{0})$, which shrink about $(q_0, \vec{0})$ as ϵ approaches zero. Hence, given a local neighborhood $\mathcal{V}$ centered at $(q_0, \vec{0})$, there exists a sufficiently small ϵ such that $\mathcal{U}_c \subseteq \mathcal{V}$ for all $c \in [U(q_0), U(q_0) + \epsilon]$. By conservation of energy, any trajectory that starts in $\mathcal{V}' = \mathcal{U}_c$ stays in this set (and hence in $\mathcal{V}$) for all $t \geq 0$. □

Note that local stability does *not* require that the perturbed object state will converge back to a zero-velocity equilibrium stance at q_0. In practice, some amount of friction will be present at the contacts. Friction effects will slow down the object's trajectory, forcing convergence to some zero-velocity equilibrium stance in the vicinity of q_0.

Example: Figure 11.5 depicts a rigid object $\mathcal{B}$ supported by segments $\mathcal{O}_1$ and $\mathcal{O}_2$ of a robot hand. The instantaneous free motions of $\mathcal{B}$ with respect to the supporting bodies $\mathcal{O}_1$ and $\mathcal{O}_2$ are graphically depicted in Figure 11.5 (see Chapter 7). Observe that the height of $\mathcal{B}$'s center of mass either increases or remains constant along its instantaneous free motions. Based on this observation as well as tools developed later in this chapter, the stance forms a local minimum of $U(q)$. The object is thus locally stable under small

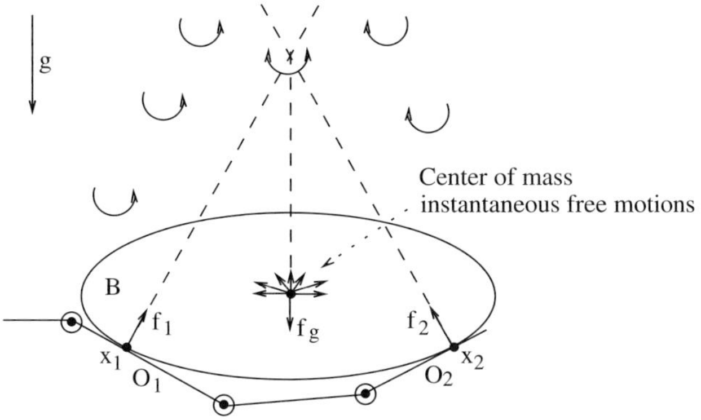

Figure 11.5 A hand-supported equilibrium stance of an object $\mathcal{B}$, located at a local minimum of $U(q)$ as indicated by the object's center of mass instantaneous free motions.

perturbations of its equilibrium state, which may include local contact breakage with respect to the supporting robot hand. ○

The local minima of the object gravitational potential energy possess another type of security measure, associated with the object's response to external wrench disturbances.

DEFINITION 11.2 **(local robustness)** *A locally stable equilibrium stance of a rigid object $\mathcal{B}$ at q_0 is* **locally robust** *if there exists a local neighborhood $\mathcal{U}$ centered at q_0, such that all sufficiently small wrench disturbances that act on $\mathcal{B}$ induce locally stable equilibrium stances located inside $\mathcal{U}$.*

Local robustness means that the supported object $\mathcal{B}$ may reach nearby equilibrium stances in response to sufficiently small wrench disturbances, such that the new equilibrium stances inherit the local stability property of the undisturbed equilibrium stance. The local minima of $U(q)$ are locally robust, as stated in the following proposition.

PROPOSITION 11.3 **(stance robustness)** *Let a rigid object $\mathcal{B}$ be supported at q_0 by stationary rigid bodies $\mathcal{O}_1, \ldots, \mathcal{O}_k$. If q_0 forms a non-degenerate local minimum of $U(q)$ (the generic case), the zero-velocity equilibrium state $(q_0, \vec{0})$ is locally stable and* **robust** *with respect to sufficiently small external wrench disturbances.*

A proof sketch of Proposition 11.3 appears in the chapter's appendix. The proof establishes that small wrench disturbances induce new equilibrium stances that lie in the *same stratum* of $\mathcal{F}$ associated with the unperturbed equilibrium stance. This property leads to the following insight. A stratum of $\mathcal{F}$ is generically zero-dimensional for $k \geq 3$ supporting contacts in 2-D stances and $k \geq 6$ supporting contacts in 3-D stances. When an equilibrium stance is located at a zero-dimensional stratum of $\mathcal{F}$, the object will remain stationary in response to sufficiently small disturbances, as illustrated in the following example.

Example: Figure 11.6 shows a rigid object $\mathcal{B}$ supported against gravity by three segments of a robot hand, $\mathcal{O}_1$, $\mathcal{O}_2$, and $\mathcal{O}_3$. The equilibrium stance configuration, q_0, is

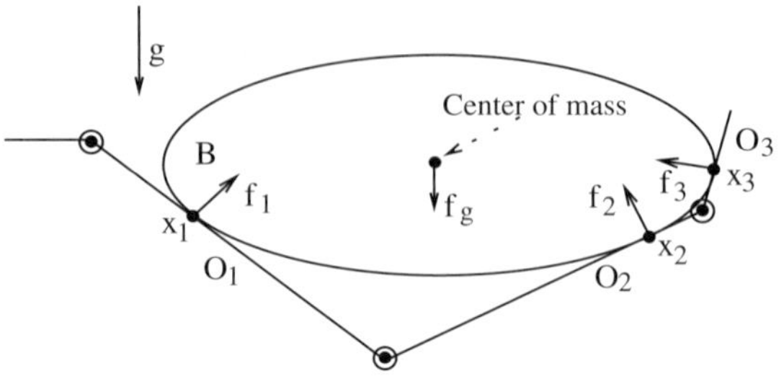

Figure 11.6 A hand-supported equilibrium stance of an object $\mathcal{B}$, via three frictionless contacts x_1, x_2 and x_3. The contacts are able to resist sufficiently small wrench disturbances without incurring any object motion relative to the robot hand.

located at a zero-dimensional stratum of $\mathcal{F}$. Based on tools discussed in Section 11.3, the point q_0 forms a non-degenerate local minimum of $U(q)$. The stance is therefore robust with respect to sufficiently small wrench disturbances. For instance, the object will remain stationary at the depicted stance under all sufficiently small tiltings of the robot hand. ○

The local minima of the object gravitational potential thus form equilibrium stances that possess two measures of security: local stability with respect to position and velocity perturbations, and robustness with respect to small external wrench disturbances. The geometric conditions identifying the local minima of $U(q)$ in $\mathcal{F}$ are discussed next.

11.3 The Stance Stability Test

The secure equilibrium stances of a rigid object $\mathcal{B}$ are located at the local minima of $U(q)$ in the object's free c-space $\mathcal{F}$. In order to translate this insight into useful synthesis tools, we need to determine when a candidate equilibrium stance forms a local minimum of $U(q)$ in $\mathcal{F}$. This section develops such a local minimum test, which will form the stance stability test.

Stratified Morse theory characterizes a local minimum of $U(q)$ in terms of its behavior on two complementary subsets of $\mathcal{F}$ (see Appendix B). Let $q_0 \in \mathcal{S}$ be the object equilibrium stance configuration, where the stratum $\mathcal{S}$ is associated with supporting bodies $\mathcal{O}_1, \ldots, \mathcal{O}_k$. The first set is the stratum $\mathcal{S}$. The other set, the *normal slice* of $\mathcal{F}$ at q_0, is constructed as follows. Let $D(q_0)$ be a small disc that intersects the stratum $\mathcal{S}$ only at q_0 and is transversal to $\mathcal{S}$.[2] The *normal slice,* denoted $E(q_0)$, is the set: $E(q_0) = D(q_0) \cap \mathcal{F}$. The behavior of $U(q)$ on $\mathcal{S}$ is characterized by its *Morse index* at q_0, denoted ν. It is defined as the number of negative eigenvalues of the restriction of the Hessian matrix $D^2U(q_0)$ to the tangent space $T_{q_0}\mathcal{S}$. The behavior of $U(q)$ on $E(q_0)$ is determined by its *lower half link* set, denoted l^-. It is defined as the intersection of $E(q_0)$ with the level set just below q_0: $U^{-1}(c_0 - \epsilon) = \{q \in \mathbb{R}^m : U(q) = c_0 - \epsilon\}$.

[2] The dimension of $D(q_0)$ equals the dimension of the ambient c-space minus the dimension of $\mathcal{S}$.

Thus, $l^- = E(q_0) \cap U^{-1}(c_0 - \epsilon)$, where $c_0 = U(q_0)$ and ϵ is a small positive parameter. According to Corollary B.2 of Appendix B, a non-degenerate critical point of $U(q)$ at q_0 forms a *local minimum* when the following two conditions are met:

(i) The Morse index at q_0 is zero, $\nu = 0$.
(ii) The lower half link set at q_0 is empty, $l^- = \emptyset$.

The condition $\nu = 0$ is trivially satisfied when $\mathcal{S}$ is a zero-dimensional stratum of $\mathcal{F}$. When $\mathcal{S}$ is at least one-dimensional, the condition $\nu = 0$ forms the classical *second-derivative test* for a local minimum of $U(q)$ along the stratum $\mathcal{S}$. The condition $l^- = \emptyset$ is a *first derivative test,* which verifies that $U(q)$ has a local minimum with respect to the neighboring strata at q_0. Let us next develop geometric tests for the two local minimum conditions.

11.3.1 Testing for $l^- = \emptyset$ at an Equilibrium Stance

The geometric test for $l^- = \emptyset$ is based on the following notion of *essential equilibrium stance,* which is analogous to the notion of essential equilibrium grasp from Chapter 7.

DEFINITION 11.3 **(essential equilibrium stance)** *Let a rigid object $\mathcal{B}$ be supported by stationary rigid bodies $\mathcal{O}_1, \ldots, \mathcal{O}_k$ at an equilibrium stance. The object is held at an* **essential equilibrium stance** *under one of the conditions –*

(i) There are $k \leq m$ supporting bodies, and all k bodies are essential for maintaining the equilibrium stance;
(ii) There are $k > m$ supporting bodies, and m of these bodies are essential for maintaining the equilibrium stance,

where $m = 3$ in 2-D stances and $m = 6$ in 3-D stances.

The essential equilibrium stances are *generic.* For instance, the stances depicted in Figures 11.3(a), 11.5 and 11.6 are all essential equilibrium stances. In contrast, the stance depicted in Figure 11.4(a) contains an idle contact and is therefore nonessential. The geometric test for $l^- = \emptyset$ is stated in the following proposition.

PROPOSITION 11.4 **(lower half link test)** *Let $q_0 \in \mathcal{S}$ be a critical point of the object gravitational potential, $U(q)$, where $\mathcal{S}$ is the stratum of $\mathcal{F}$ associated with supporting bodies $\mathcal{O}_1, \ldots, \mathcal{O}_k$. A necessary condition for $l^- = \emptyset$ is that q_0 forms a* **feasible equilibrium stance**:

$$\nabla U(q_0) = \sum_{i=1}^{k} \lambda_i \hat{\eta}_i(q_0) \qquad \lambda_1, \ldots, \lambda_k \geq 0, \tag{11.4}$$

where $\hat{\eta}_1(q_0), \ldots, \hat{\eta}_k(q_0)$ are the supporting bodies c-obstacle outward unit normals at q_0. A sufficient condition for $l^- = \emptyset$ is that q_0 forms an **essential equilibrium stance.**

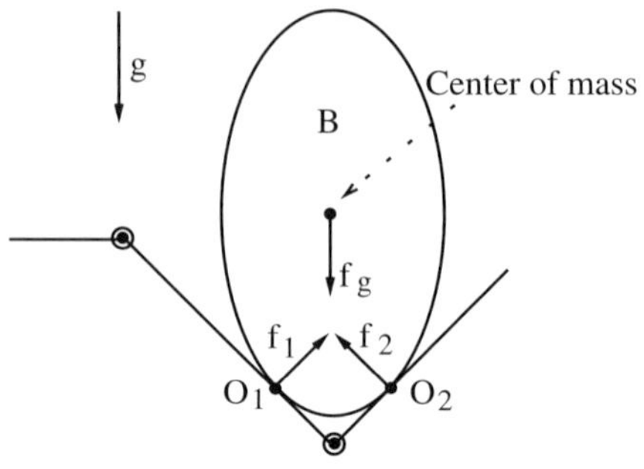

Figure 11.7 A generic two-contact stance whose two contacts are essential for maintaining the equilibrium stance. The lower half link set, l^-, is empty at this stance.

The proof of Proposition 11.4 appears in the chapter's appendix and is based on the following argument. By construction, $l^- = E(q_0) \cap U^{-1}(c_0 - \epsilon)$, where $E(q_0)$ is the normal slice of $\mathcal{F}$ at q_0 and $c_0 = U(q_0)$. If l^- is empty, $U(q)$ must be non-decreasing along any c-space path $q(t)$ that starts at q_0 and lies in $E(q_0)$. Let $C(q_0)$ denote the convex cone of tangent vectors that are based at q_0 and point into $E(q_0)$. Then $l^- = \emptyset$ implies that $\frac{d}{dt}\big|_{t=0} U(q(t)) = \nabla U(q_0) \cdot \dot{q} \geq 0$ for all $\dot{q} \in C(q_0)$. Next, consider the contacts' net wrench cone, $\mathcal{W}(q_0) = \{\sum_{i=1}^{k} \lambda_i \hat{\eta}_i(q_0) : \lambda_1, \ldots, \lambda_k \geq 0\}$. Every wrench $\boldsymbol{w} \in -\mathcal{W}(q_0)$ satisfies $\boldsymbol{w} \cdot \dot{q} \leq 0$ for all $\dot{q} \in C(q_0)$. Hence, the negated wrench cone $-\mathcal{W}(q_0)$ is *dual* to $C(q_0)$. Since $-\nabla U(q_0)$ satisfies the same duality condition, it must lie in $-\mathcal{W}(q_0)$. Equivalently, $\nabla U(q_0)$ must lie in $\mathcal{W}(q_0)$, which is the condition specified by Eq. (11.4).

When the object $\mathcal{B}$ is supported by $k \leq m$ contacts, an equilibrium stance is essential and $l^- = \emptyset$ when $\nabla U(q_0)$ lies in the *relative interior* of the net wrench cone $\mathcal{W}(q_0)$, as illustrated in the next example.

Example: Consider the two-contact equilibrium stance of an object $\mathcal{B}$ depicted in Figure 11.7. The object's configuration, q_0, lies in a one-dimensional stratum $\mathcal{S}$ of $\mathcal{F}$. The net wrench cone, $\mathcal{W}(q_0)$, forms a two-dimensional sector in $\mathcal{B}$'s wrench space, spanned by the c-obstacle outward normals at q_0. Since $\lambda_1 = \lambda_2 = \frac{1}{2}$ by the stance's symmetry, $\nabla U(q_0)$ lies in the relative interior of $\mathcal{W}(q_0)$. It is therefore an essential equilibrium stance with $l^- = \emptyset$. Intuitively, $l^- = \emptyset$ implies that $U(q)$ is strictly increasing along any c-space path that starts at q_0 and lies in $\mathcal{F}$, such that the path is transversal to $\mathcal{S}$ at q_0. The object $\mathcal{B}$ breaks at least one contact along any such path. ○

11.3.2 Testing for $\nu = 0$ at an Equilibrium Stance

The condition $\nu = 0$ requires that $U(q)$ possess a local minimum along motions of $\mathcal{B}$ that maintain contact with all supporting bodies $\mathcal{O}_1, \ldots, \mathcal{O}_k$. Such motions lie in the stratum $\mathcal{S} = \cap_{i=1}^{k} \text{bdy}(\mathcal{CO}_i)$. Hence, ν is possibly positive only when the dimension of $\mathcal{S}$ is at least unity. When the c-obstacles associated with the supporting bodies intersect transversally in $\mathcal{B}$'s c-space (the generic case), the test for $\nu = 0$ is required for $1 \le k < 3$ supporting contacts in 2-D stances, and $1 \le k < 6$ supporting contacts in 3-D stances.

When the stratum $\mathcal{S}$ is at least one-dimensional, the condition $\nu = 0$ is equivalent to the second-derivative test:

$$\left.\frac{d^2}{dt^2}\right|_{t=0} U(q(t)) > 0 \qquad q(0) = q_0,\ q(t) \in \mathcal{S} \text{ for } t \in (-\epsilon, \epsilon).$$

This condition involves the velocities and accelerations of $\mathcal{B}$, since by the chain rule,

$$\left.\frac{d^2}{dt^2}\right|_{t=0} U(q(t)) = \left.\frac{d}{dt}\right|_{t=0} (\nabla U(q(t)) \cdot \dot{q}(t)) = \dot{q}^T D^2 U(q_0)\dot{q} + \nabla U(q_0) \cdot \ddot{q}, \qquad (11.5)$$

where $q(0) = q_0$, $\dot{q} = \dot{q}(0)$ and $\ddot{q} = \ddot{q}(0)$. Since $\mathcal{S} = \cap_{i=1}^{k} \mathsf{bdy}(\mathcal{CO}_i)$, a c-space path that lies in $\mathcal{S}$ also lies in the individual c-obstacle boundaries, $q(t) \in \mathsf{bdy}(\mathcal{CO}_i)$ for $i = 1 \dots k$. The object's acceleration along $q(t)$ therefore depends on the c-obstacle *curvature forms*. From Chapter 2, the ith c-obstacle *curvature form* at q_0 is given by

$$\kappa_i(q_0, \dot{q}) = \dot{q}^T D\hat{\eta}_i(q_0)\dot{q} \qquad \dot{q} \in T_{q_0}\mathsf{bdy}(\mathcal{CO}_i),$$

where $\hat{\eta}_i(q_0)$ is the ith c-obstacle outward unit normal at q_0. Much like the relative c-space curvature form associated with second-order immobilizing grasps, the following *gravity relative curvature form* will provide the geometric test for $\nu = 0$.

DEFINITION 11.4 *Let a rigid object $\mathcal{B}$ be supported by stationary rigid bodies $\mathcal{O}_1, \dots, \mathcal{O}_k$ at an essential equilibrium stance configuration, $q_0 \in \mathcal{S}$, such that $1 \le k < m$. The* **gravity relative curvature form** *associated with $\mathcal{B}$'s gravitational potential $U(q)$ is the weighted sum,*

$$\kappa_U(q_0, \dot{q}) = \sum_{i=1}^{k} \lambda_i \kappa_i(q_0, \dot{q}) - \dot{q}^T D^2 U(q_0)\dot{q} \qquad \dot{q} \in T_{q_0}\mathcal{S}, \qquad (11.6)$$

where $\lambda_1, \dots \lambda_k \ge 0$ are determined by the equilibrium stance condition of Eq. (11.2), $\kappa_i(q_0, \dot{q})$ is the ith c-obstacle curvature form at q_0, and $D^2U(q_0)$ is the Hessian of $U(q)$ at q_0.

Let us verify that $\kappa_U(q_0, \dot{q})$ is well defined. The coefficients $\lambda_1, \dots, \lambda_k$ are proportional to the contact force magnitudes at the equilibrium stance. These coefficients are uniquely determined by Eq. (11.2) at an essential equilibrium stance that involve $1 \le k < m$ contacts. Each c-obstacle curvature form, $\kappa_i(q_0, \dot{q})$, is defined on the ith c-obstacle tangent space. The tangent space $T_{q_0}\mathcal{S}$ is defined as the intersection of the supporting c-obstacle tangent spaces at q_0. Hence, each $\kappa_i(q_0, \dot{q})$ is well defined for $\dot{q} \in T_{q_0}\mathcal{S}$, and, consequently, $\kappa_U(q_0, \dot{q})$ is well defined on the tangent space $T_{q_0}\mathcal{S}$.

The following proposition asserts that the condition $\nu = 0$ is determined by the *sign* attained by $\kappa_U(q_0, \dot{q})$, as $\dot{q}$ varies in $T_{q_0}\mathcal{S}$.

PROPOSITION 11.5 (**Morse index test**) *Let a rigid object $\mathcal{B}$ be supported at an essential k-body equilibrium stance, $q_0 \in \mathcal{S}$, such that $1 \leq k < m$. Then $\nu = 0$ iff the gravity relative curvature form is* **negative definite** *on $T_{q_0}\mathcal{S}$: $\kappa_U(q_0, \dot{q}) < 0$ for all $\dot{q} \in T_{q_0}\mathcal{S}$.*

Proof: Consider a c-space path $q(t)$ that lies in $\mathcal{S}$ for $t \in (-\epsilon, \epsilon)$, such that $q(0) = q_0$. Since $q(t)$ lies in $\mathcal{S}$, its tangent vector $\dot{q}(t)$ satisfies $\hat{\eta}_i(q(t)) \cdot \dot{q}(t) = 0$ for $t \in (-\epsilon, \epsilon)$, where $\hat{\eta}_i(q)$ is the ith c-obstacle outward unit normal at q. Taking the derivative of this expression:

$$\hat{\eta}_i(q(t)) \cdot \ddot{q}(t) + \dot{q}^T(t) D\hat{\eta}_i(q(t))\dot{q}(t) = 0 \qquad t \in (-\epsilon, \epsilon). \tag{11.7}$$

Next, consider the second derivative of $U(q(t))$ at $t = 0$, specified in Eq. (11.5). In this equation, $\nabla U(q_0) = \sum_{i=1}^{k} \lambda_i \hat{\eta}_i(q_0)$ at the equilibrium stance configuration q_0. Hence,

$$\left.\frac{d^2}{dt^2}\right|_{t=0} U(q(t)) = \dot{q}^T D^2 U(q_0)\dot{q} + \sum_{i=1}^{k} \lambda_i \hat{\eta}_i(q_0) \cdot \ddot{q}.$$

Substituting for $\hat{\eta}_i(q_0) \cdot \ddot{q}$ according to Eq. (11.7) gives

$$\left.\frac{d^2}{dt^2}\right|_{t=0} U(q(t)) = \dot{q}^T \left(D^2 U(q_0) - \sum_{i=1}^{d} \lambda_i D\hat{\eta}_i(q_0) \right) \dot{q} = -\kappa_U(q_0, \dot{q}).$$

Thus, $U(q(t))$ has a local minimum at $t = 0$ iff $\kappa_U(q_0, \dot{q}) < 0$, where $q(0) = q_0$ and $\dot{q} = \dot{q}(0)$. This condition must hold for all paths $q(t)$ that start at q_0 and lie in $\mathcal{S}$. Since $T_{q_0}\mathcal{S}$ is the collection of all tangents to these paths at q_0, the local-minimum condition is given by $\kappa_U(q_0, \dot{q}) < 0$ for all $\dot{q} \in T_{q_0}\mathcal{S}$. □

Gravity as a virtual finger: The gravity relative curvature form, $\kappa_U(q_0, \dot{q})$, is strikingly similar to the c-space relative curvature form, defined in Chapter 8 as

$$\kappa_{\text{rel}}(q_0, \dot{q}) = \sum_{i=1}^{k} \lambda_i \kappa_i(q_0, \dot{q}) \qquad \dot{q} \in M^1_{1\ldots k}(q_0),$$

where the coefficients $\lambda_1, \ldots, \lambda_k \geq 0$ are determined by the equilibrium grasp condition, and $M^1_{1\ldots k}(q_0)$ is the linear subspace of first-order free motions at the given grasp. Consider the level set of $U(q)$ that passes through q_0: $\mathcal{S}_U = \{q \in \mathbb{R}^m : U(q) = U(q_0)\}$. One can interpret the level set $\mathcal{S}_U$ as the boundary of a *virtual c-obstacle,* which arises from a *virtual finger* applying a gravitational force at $\mathcal{B}$'s center of mass. Based on conservation of energy, all c-space trajectories of $\mathcal{B}$ that start at the zero-velocity equilibrium stance, $(q_0, \vec{0})$, cannot cross $\mathcal{S}_U$ into higher value of $U(q)$. The term $\dot{q}^T D^2 U(q_0)\dot{q}$ in $\kappa_U(q_0, \dot{q})$ represents the *curvature form* of the level set $\mathcal{S}_U$. Hence, $\kappa_U(q_0, \dot{q})$ can be thought of as measuring the relative c-space curvature of the supporting c-obstacles and the virtual c-obstacle associated with gravity.[3] ○

[3] Technically, the curvature form of $\mathcal{S}_U$ is given by $\frac{-1}{\|\nabla U(q_0)\|}\dot{q}^T D^2 U(q_0)\dot{q}$ for $\dot{q} \in T_{q_0}\mathcal{S}_U$. The interpretation of $\kappa_U(q_0, \dot{q})$ as a c-space relative curvature requires that $\lambda_1, \ldots, \lambda_k$ be modulated by the common factor $\|\nabla U(q_0)\|$.

11.3.3 Summary of Stance Stability Test

Let us summarize the stance stability test in terms of the geometric conditions that must be satisfied by the object $\mathcal{B}$ and the supporting bodies $\mathcal{O}_1, \ldots, \mathcal{O}_k$. Let $q_0 \in \mathcal{S}$ be an equilibrium stance configuration of $\mathcal{B}$, where $\mathcal{S}$ is the c-space stratum associated with the supporting bodies $\mathcal{O}_1, \ldots, \mathcal{O}_k$. The test requires that q_0 be a non-degenerate critical point of $U(q)$. This condition is met when the object is supported at an essential equilibrium stance (the generic case), such that the matrix $\sum_{i=1}^{k} \lambda_i D\hat{\eta}_i(q_0) - D^2U(q_0)$ of $\kappa_U(q_0, \dot{q})$ has non-zero eigenvalues on the tangent space $T_{q_0}\mathcal{S}$ (the generic case). We can now summarize the stance stability test.

THEOREM 11.6 **(stance stability test)** *Let a rigid object $\mathcal{B}$ be supported at an equilibrium stance configuration, $q_0 \in \mathcal{S}$, by stationary rigid bodies $\mathcal{O}_1, \ldots, \mathcal{O}_k$, For $1 \leq k < m$ supporting contacts, the equilibrium stance is* **locally stable** *if the object is supported at an essential equilibrium stance, such that $\kappa_U(q_0, \dot{q}) < 0$ for all $\dot{q} \in T_{q_0}\mathcal{S}$, where $\kappa_U(q_0, \dot{q})$ is the gravity relative curvature form associated with $U(q)$.*

For $k \geq m$ supporting contacts, the equilibrium stance is **locally stable** *if the object is supported at an essential equilibrium stance.*

Theorem 11.6 contains the two components of the local minimum test, $I^- = \emptyset$ and $\nu = 0$, and its proof is omitted. We can now more fully explain the two parts of the stance stability test. The stratum $\mathcal{S}$ corresponds to motions of $\mathcal{B}$ that maintain contact with all supporting bodies. The gravity relative curvature form, $\kappa_U(q_0, \dot{q})$, verifies that $U(q)$ has a local minimum along the stratum $\mathcal{S}$ at q_0. It is a classical *second-derivative test.* However, one must also consider the possibility that $\mathcal{B}$ may break contact with some or all of the supporting bodies. At an essential equilibrium stance, $U(q)$ has a local minimum with respect to such contact breaking motions. This *first-derivative test* complements the second-derivative test for the local minima of $U(q)$ on the stratified set $\mathcal{F}$.

The remainder of this chapter illustrates how one applies the stance stability test by considering the following *support problem.* An object $\mathcal{B}$ with a variable center of mass is supported against gravity by a given set of frictionless contacts. Characterize the region of $\mathcal{B}$'s center-of-mass locations where the object is stably supported by the contacts. To this end, we first summarize the formulas for the stance stability test.

11.4 Formulas for the Stance Stability Test

We need formulas for the gradient and Hessian matrix of $U(q)$, expressions for the c-obstacle outward unit normals and formulas for the c-obstacle curvature forms. Recall that b_{cm} denotes the position of $\mathcal{B}$'s center of mass in its body frame. The world coordinates of $\mathcal{B}$'s center of mass, denoted x_{cm}, are given by $x_{\text{cm}}(q) = R(\theta)b_{cm} + d$, where $q = (d, \theta)$ is $\mathcal{B}$'s configuration. The term $\rho_{\text{cm}}(\theta) = R(\theta)b_{cm}$ is the vector from $\mathcal{B}$'s frame origin to its center of mass expressed in the fixed world frame. The gradient and Hessian of $U(q)$ are specified in the following lemma whose proof appears in the chapter's appendix.

LEMMA 11.7 (∇U **and** D^2U) *Let* $U(q) = \mathsf{mg}(\boldsymbol{e} \cdot x_{\mathsf{cm}}(q))$ *be the gravitational potential of a rigid object* $\mathcal{B}$*. The gradient of* $U(q)$ *in the case of 3-D stances has the form*

$$\nabla U(q) = \mathsf{mg} \begin{pmatrix} \boldsymbol{e} \\ \rho_{\mathsf{cm}}(\theta) \times \boldsymbol{e} \end{pmatrix} = \mathsf{mg} \left(\begin{array}{c} \boldsymbol{e} \\ \begin{pmatrix} y_{\mathsf{cm}} \\ -x_{\mathsf{cm}} \\ 0 \end{pmatrix} \end{array} \right) \qquad q = (d, \theta) \in \mathbb{R}^6, \tag{11.8}$$

where $\boldsymbol{e} = (0, 0, 1)$ *is the vertical upward direction, and* $\rho_{\mathsf{cm}} = (x_{\mathsf{cm}}, y_{\mathsf{cm}}, z_{\mathsf{cm}})$*. The* 6×6 *Hessian matrix of* $U(q)$ *has the form*

$$D^2U(q) = \mathsf{mg} \begin{bmatrix} O & O \\ O & \left([\rho_{\mathsf{cm}}(\theta)\times][\boldsymbol{e}\times]\right)_s \end{bmatrix} \qquad q = (d, \theta) \in \mathbb{R}^6, \tag{11.9}$$

where O *is a* 3×3 *matrix of zeroes, and* $([\rho_{\mathsf{cm}}\times][\boldsymbol{e}\times])_s$ *is the* 3×3 *symmetric matrix*

$$\left([\rho_{\mathsf{cm}}(\theta)\times][\boldsymbol{e}\times]\right)_s = \begin{bmatrix} -z_{\mathsf{cm}} & 0 & \frac{1}{2}x_{\mathsf{cm}} \\ 0 & -z_{\mathsf{cm}} & \frac{1}{2}y_{\mathsf{cm}} \\ \frac{1}{2}x_{\mathsf{cm}} & \frac{1}{2}y_{\mathsf{cm}} & 0 \end{bmatrix} \qquad \rho_{\mathsf{cm}} = (x_{\mathsf{cm}}, y_{\mathsf{cm}}, z_{\mathsf{cm}}),$$

where $\left([\rho_{\mathsf{cm}}(\theta)\times][\boldsymbol{e}\times]\right)_s = \frac{1}{2}([\rho_{\mathsf{cm}}(\theta)\times]^T[\boldsymbol{e}\times] + [\boldsymbol{e}\times]^T[\rho_{\mathsf{cm}}(\theta)\times])$.[4]

The gradient of $U(q)$ depends only on the horizontal coordinates of $\mathcal{B}$'s center of mass, (x_{cm}, y_{cm}). Hence, when $\mathcal{B}$ is supported against gravity at a feasible equilibrium stance, the object's center of mass can vary in a vertical line in $\mathbb{R}^3$ without affecting equilibrium feasibility. The Hessian matrix, $D^2U(q)$, additionally depends on the *height* of $\mathcal{B}$'s center of mass, z_{cm}.

In the case of 2-D stances, the gradient of $U(q)$ has the form

$$\nabla U(q) = \mathsf{mg} \begin{pmatrix} \boldsymbol{e} \\ \rho_{\mathsf{cm}}(\theta) \times \boldsymbol{e} \end{pmatrix} = \mathsf{mg} \begin{pmatrix} \boldsymbol{e} \\ x_{\mathsf{cm}} \end{pmatrix} \qquad q = (d, \theta) \in \mathbb{R}^3, \tag{11.10}$$

where $\rho_{\mathsf{cm}} = (x_{\mathsf{cm}}, y_{\mathsf{cm}})$, $\boldsymbol{e} = (0, 1)$ is the vertical upward direction, and $\rho_{\mathsf{cm}} \times \boldsymbol{e} = \rho_{\mathsf{cm}}{}^T J \boldsymbol{e}$, where $J = \left[\begin{smallmatrix} 0 & 1 \\ -1 & 0 \end{smallmatrix}\right]$. The Hessian of $U(q)$ forms the 3×3 symmetric matrix,

$$D^2U(q) = -\mathsf{mg} \begin{bmatrix} O & \vec{0} \\ \vec{0}^T & \rho_{\mathsf{cm}}(\theta) \cdot \boldsymbol{e} \end{bmatrix} = -\mathsf{mg} \begin{bmatrix} O & \vec{0} \\ \vec{0}^T & y_{\mathsf{cm}} \end{bmatrix} \qquad q = (d, \theta) \in \mathbb{R}^3, \tag{11.11}$$

where O is a 2×2 matrix of zeroes and $\rho_{\mathsf{cm}} = (x_{\mathsf{cm}}, y_{\mathsf{cm}})$. Here, too, $\nabla U(q)$ depends only on the horizontal coordinate of $\mathcal{B}$'s center-of-mass position, x_{cm}, while $D^2U(q)$ depends on the *height* of $\mathcal{B}$'s center of mass, y_{cm}.

The critical points of $U(q)$ satisfy the condition $\nabla U(q_0) \in \mathcal{W}(q_0)$, where the net wrench cone, $\mathcal{W}(q_0)$, is spanned by the c-obstacle outward unit normals $\hat{\eta}_1(q_0), \ldots,$

4 $A_s = \frac{1}{2}(A + A^T)$, and $[u\times]$ is the skew-symmetric matrix satisfying $[u\times]v = u \times v$ for $u, v \in \mathbb{R}^3$.

$\hat{\eta}_k(q_0)$. The ith c-obstacle outward unit normal at a point $q \in \text{bdy}(\mathcal{CO}_i)$ has the form (Theorem 2.5):

$$\hat{\eta}_i(q) = \frac{1}{c_i}\begin{pmatrix} n_i \\ \rho_i(\theta) \times n_i \end{pmatrix} \qquad q = (d, \theta) \in \mathbb{R}^m,\ m = 3 \text{ or } 6, \tag{11.12}$$

where n_i is $\mathcal{B}$'s inward unit normal at the contact point x_i, $\rho_i(\theta) = R(\theta)b_i$ is the vector from $\mathcal{B}$'s origin to x_i, and $c_i = \sqrt{1 + \|\rho_i \times n_i\|^2}$ is a normalization factor. In 2-D stances, $\rho_i \times n_i = \rho_i^T J n_i$, where $J = \begin{bmatrix} 0 & 1 \\ -1 & 0 \end{bmatrix}$.

The gravity relative curvature form, $\kappa_U(q_0, \dot{q})$, contains the weighted sum of the c-obstacle curvature forms $\kappa_i(q_0, \dot{q})$ for $i = 1 \dots k$. In the case of 2-D stances, we have the following graphical formula for $\kappa_i(q_0, \dot{q})$ (see Chapter 8). Let $\mathcal{S}_i = \text{bdy}(\mathcal{CO}_i)$. The tangent vectors $\dot{q} \in T_{q_0}\mathcal{S}_i$ correspond to instantaneous rotations of $\mathcal{B}$ about points on the contact normal line, denoted l_i. Let $\mathcal{B}$'s frame origin be located at a distance ρ_i from the contact point x_i on l_i, such that $\rho_i > 0$ on $\mathcal{B}$'s side of x_i and $\rho_i < 0$ on $\mathcal{O}_i$'s side of this point. When $\mathcal{B}$'s frame origin is located at a distance ρ_i from x_i on l_i, the ith c-obstacle curvature along instantaneous rotation $\dot{q} = (0, \omega)$ is given by

$$\kappa_i(q_0, (0, \omega)) = \frac{(\rho_i - r_{\mathcal{B}_i})(\rho_i + r_{\mathcal{O}_i})}{r_{\mathcal{B}_i} + r_{\mathcal{O}_i}}\omega^2, \tag{11.13}$$

where $r_{\mathcal{B}_i}$ and $r_{\mathcal{O}_i}$ are the radii of curvature of $\mathcal{B}$ and $\mathcal{O}_i$ at the contact point x_i. The sign attained by $\kappa_i(q_0, \dot{q})$ as $\dot{q}$ varies in the ith c-obstacle tangent space can thus be determined by evaluating Eq. (11.13) for instantaneous rotations parametrized by $-\infty \le \rho_i \le \infty$.

In the case of 3-D stances, the formula for $\kappa_i(q_0, \dot{q})$ depends on the surface curvature of the object $\mathcal{B}$ and the supporting body $\mathcal{O}_i$ at the contact point x_i. These curvatures are represented by linear maps, $L_{\mathcal{B}_i}$ and $L_{\mathcal{O}_i}$, which act on the tangent plane of the respective surface at x_i, and yield the change in the surface normal along a given tangent direction. The ith c-obstacle curvature form is given by the formula (see Section 2.4):

$$\kappa_i(q_0, \dot{q}) = \frac{1}{c_i}\dot{q}^T \left(\begin{bmatrix} I & -[\rho_i \times] \\ O & [n_i \times] \end{bmatrix}^T \begin{bmatrix} L_{\mathcal{B}_i}\left[L_{\mathcal{O}_i} + L_{\mathcal{B}_i}\right]^{-1} L_{\mathcal{O}_i} & -L_{\mathcal{O}_i}\left[L_{\mathcal{O}_i} + L_{\mathcal{B}_i}\right]^{-1} \\ -\left[L_{\mathcal{O}_i} + L_{\mathcal{B}_i}\right]^{-1} L_{\mathcal{O}_i} & -\left[L_{\mathcal{O}_i} + L_{\mathcal{B}_i}\right]^{-1} \end{bmatrix} \begin{bmatrix} I & -[\rho_i \times] \\ O & [n_i \times] \end{bmatrix} + \begin{bmatrix} O & O \\ O & -([\rho_i \times]^T [n_i \times])_s \end{bmatrix} \right) \dot{q} \qquad \dot{q} \in T_{q_0}\mathcal{S}_i,\ \mathcal{S}_i = \text{bdy}(\mathcal{CO}_i), \tag{11.14}$$

where c_i, ρ_i and n_i were previously specified, I is a 3×3 identity matrix, O is a 3×3 matrix of zeroes and $([\rho_i \times]^T [n_i \times])_s = \frac{1}{2}([\rho_i \times]^T [n_i \times] + [n_i \times]^T [\rho_i \times])$.

11.5 The Stable Equilibrium Region of 2-D Stances

This section applies the stance stability test to the 2-D *support problem.* A 2-D rigid object $\mathcal{B}$ with a variable center of mass is supported against gravity via frictionless contact by stationary rigid bodies $\mathcal{O}_1, \ldots, \mathcal{O}_k$ in a vertical plane. For a given set of supporting contacts, determine the object center-of-mass positions that ensure stance stability on the supporting contacts. We begin with a scheme for computing the region of feasible equilibrium stances, denoted $\mathcal{E}(q_0)$; then determine the subset of $\mathcal{E}(q_0)$ that ensures stance stability on the supporting contacts.

11.5.1 A Scheme for Computing the Stance Equilibrium Region

Using Formula (11.10), the gravitational wrench that acts on $\mathcal{B}$ at q_0 can be written in the form

$$\nabla U(q_0) = \mathsf{m}g \begin{pmatrix} \boldsymbol{e} \\ \boldsymbol{\rho}_{\mathsf{cm}} \times \boldsymbol{e} \end{pmatrix} = \mathsf{m}g \left\{ \begin{pmatrix} 0 \\ 1 \\ 0 \end{pmatrix} + x_{\mathsf{cm}} \begin{pmatrix} 0 \\ 0 \\ 1 \end{pmatrix} \right\}, \tag{11.15}$$

where $\boldsymbol{e} = (0,1)$ and $x_{cm} = \boldsymbol{\rho}_{\mathsf{cm}} \times \boldsymbol{e}$ is the horizontal coordinate of $\mathcal{B}$'s center of mass expressed in $\mathcal{B}$'s reference frame. Since x_{cm} is a free parameter in Eq. (11.15), the possible gravitational wrenches span an *affine line* in $\mathcal{B}$'s wrench space, denoted L (Figure 11.8). Let (f_x, f_y, τ) denote the object's wrench space coordinates, where (f_x, f_y) and τ are force and torque components. Then L is perpendicular to the (f_x, f_y)-plane and passes through the point $(f_x, f_y, \tau) = (\boldsymbol{e}, 0)$ in $\mathcal{B}$'s wrench space (Figure 11.8).

Equilibrium stance feasibility requires that the gravitational wrench, $\nabla U(q_0)$, lie in the contacts' net wrench cone, $\mathcal{W}(q_0)$. Every intersection point of L with $\mathcal{W}(q_0)$ thus determines one value of x_{cm}, at which the contacts support the object $\mathcal{B}$ at a feasible equilibrium stance. Each value of x_{cm} determines a *vertical line* in $\mathbb{R}^2$ that belongs to the equilibrium region $\mathcal{E}(q_0)$. To derive a geometric test for the intersection of L with $\mathcal{W}(q_0)$, let us scale the gravitational wrench so that $\mathsf{m}g = 1$ in Eq. (11.15) (this scaling amounts to a choice of energy units). The intersection of L with $\mathcal{W}(q_0)$ can be tested as follows.

LEMMA 11.8 **(2-D equilibrium stance feasibility)** *A rigid object $\mathcal{B}$ is supported by stationary rigid bodies $\mathcal{O}_1, \ldots, \mathcal{O}_k$ at a* **feasible equilibrium stance** *iff the vertical upward direction, $\boldsymbol{e} = (0,1)$, lies in the positive span of $\mathcal{B}$'s inward contact normals $n_1, \ldots, n_k$.*

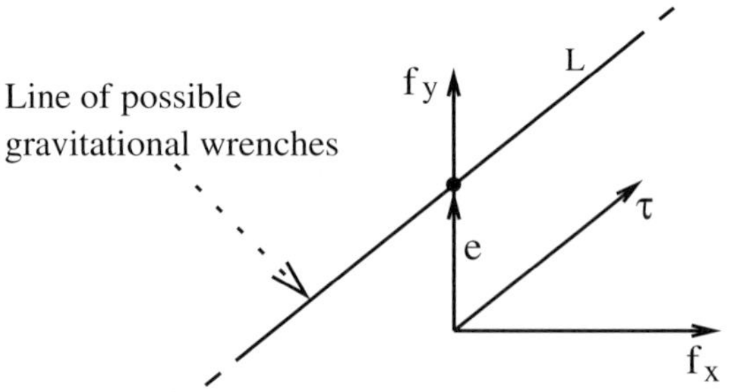

Figure 11.8 The affine line L of gravitational wrenches depicted in $\mathcal{B}$'s wrench space. The line L is perpendicular to the (f_x, f_y)-plane and passes through the point $(\boldsymbol{e}, 0)$.

Proof: The affine line L is orthogonal to the (f_x, f_y) plane in $\mathcal{B}$'s wrench space. Hence, L intersects $\mathcal{W}(q_0)$ iff the projections of L and $\mathcal{W}(q_0)$ intersect within the (f_x, f_y) plane. The projection of L on the (f_x, f_y) plane is the single point $\boldsymbol{e} = (0, 1)$. Since $\mathcal{W}(q_0)$ is spanned by the c-obstacle outward unit normals $\hat{\eta}_1(q_0), \ldots, \hat{\eta}_k(q_0)$, its projection on the (f_x, f_y) plane is the positive span of the projection of the c-obstacle outward unit normals. Eq (11.12) specifies the formula $\hat{\eta}_i(q_0) = (n_i, \boldsymbol{\rho}_i \times n_i)/c_i$, where n_i is $\mathcal{B}$'s inward unit normal at the ith contact point. The projection of $\hat{\eta}_i(q_0)$ on the (f_x, f_y) plane is thus a positive multiple of n_i. Hence, L intersects $\mathcal{W}(q_0)$ iff $\boldsymbol{e}$ lies in the positive span of $\mathcal{B}$'s inward contact normals $n_1, \ldots, n_k$. □

When the geometric test specified in Lemma 11.8 holds true at a given stance, the *stance equilibrium region,* denoted $\mathcal{E}(q_0)$, forms one of the following sets in the vertical plane.

PROPOSITION 11.9 **(the region $\mathcal{E}(q_0)$)** *The* **stance equilibrium region,** *$\mathcal{E}(q_0)$, forms a vertical line at a single contact stance; a vertical line at a two-contact stance with nonparallel contact normals; a vertical strip or half plane at a three-contact stance; and a vertical strip, half plane, or the entire vertical plane at stances supported by $k \geq 4$ contacts.*

Proof: Since $\mathcal{W}(q_0)$ forms a convex cone based at $\mathcal{B}$'s wrench space origin, the intersection $L \cap \mathcal{W}(q_0)$ is a single point or a single interval. At a one-contact stance, $\mathcal{W}(q_0)$ spans a ray based at $\mathcal{B}$'s wrench space origin. If $L \cap \mathcal{W}(q_0)$ is nonempty, it must be a single point, as illustrated in Figure 11.9(a). In this case, $\mathcal{E}(q_0)$ forms a vertical line. At a two-contact stance, $\mathcal{W}(q_0)$ generically spans a 2-D sector based at $\mathcal{B}$'s wrench space origin. If $L \cap \mathcal{W}(q_0)$ is nonempty, it is generically a single point, as illustrated in Figure 11.9(b). In this case, too, $\mathcal{E}(q_0)$ forms a vertical line. At a three-contact stance, $\mathcal{W}(q_0)$ generically spans a 3-D convex cone based at $\mathcal{B}$'s wrench space origin. If $L \cap \mathcal{W}(q_0)$ is nonempty, it is generically a finite or semi-infinite interval, as illustrated in Figure 11.9(c). In this case, $\mathcal{E}(q_0)$ forms a vertical strip or a vertical half plane. At stances supported by $k \geq 4$ contacts, if $\mathcal{W}(q_0)$ spans a subset of $\mathcal{B}$'s wrench space, $\mathcal{E}(q_0)$ is the same as in the three-contact stances. If $\mathcal{W}(q_0)$ spans the entire wrench space, the line L lies in $\mathcal{W}(q_0)$, and $\mathcal{E}(q_0)$ is the entire vertical plane. □

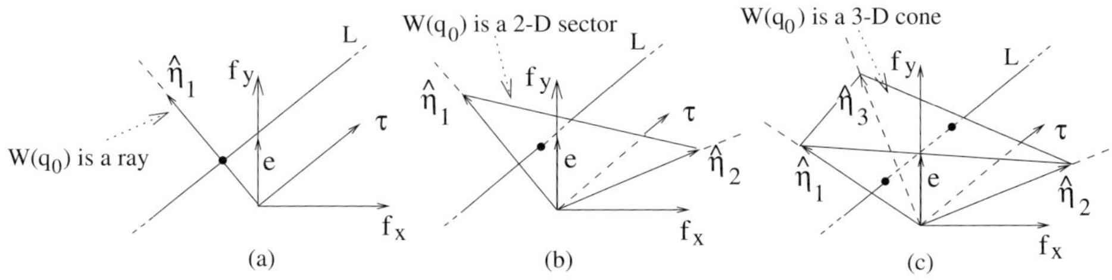

Figure 11.9 The line L of gravitational wrenches and the net wrench cone $\mathcal{W}(q_0)$, shown in $\mathcal{B}$'s wrench space. (a) For a single contact, $L \cap \mathcal{W}(q_0)$ is at most a single point. (b) For two contacts, $L \cap \mathcal{W}(q_0)$ is generically a single point. (c) For three contacts, $L \cap \mathcal{W}(q_0)$ is generically a finite or semi-infinite interval.

Proposition 11.9 asserts that the stance equilibrium region, $\mathcal{E}(q_0)$, always forms a connected set of vertical lines. The stability region forms a subset of $\mathcal{E}(q_0)$ as next characterized for single-support stances, and then for stances supported by three or more contacts.

11.5.2 The Stability Region of Single-Support 2-D Stances

The stability of such stances has applications beyond robot hands, which usually deploy multiple finger and palm contacts when supporting objects against gravity (Figure 11.1). Equilibrium stance feasibility requires that $\mathcal{B}$'s inward unit normal at the contact be collinear with the vertical upward direction: $n_1 = \boldsymbol{e}$. The stance equilibrium region is simply the entire vertical line that passes through the contact point, denoted l_1.

Next consider the stability region of one-contact stances. The first part of the stance stability test, $l^- = \emptyset$, is automatically satisfied at any center-of-mass position in $\mathcal{E}(q_0)$ (see Exercises). Let us therefore consider the second part of the stance stability test, $\kappa_U(q_0, \dot{q}) < 0$ for all $\dot{q} \in T_{q_0}\mathcal{S}$, where $\mathcal{S} = \mathsf{bdy}(\mathcal{CO}_1)$. Using Formulas (11.11) and (11.13), and noting that $\lambda_1 = 1$ at a single-support stance, we obtain the stance stability test,

$$\kappa_U(q_0, (0, \omega)) = \frac{(\rho_1 - r_{\mathcal{B}_1})(\rho_1 + r_{\mathcal{O}_1})}{r_{\mathcal{B}_1} + r_{\mathcal{O}_1}} + y_{cm} < 0 \qquad -\infty \leq \rho_1 \leq \infty,\ \omega = 1,$$

where ρ_1 is the signed distance of $\mathcal{B}$'s frame origin from the contact point x_1 and y_{cm} is the height of $\mathcal{B}$'s center of mass in the object reference frame. Let h_{cm} denote the height of $\mathcal{B}$'s center of mass above the contact point x_1. The parameter y_{cm} can be written as $y_{cm} = h_{\mathsf{cm}} - \rho_1$, which gives the equivalent stance stability test:

$$\kappa_U(q_0, (0, \omega)) = \frac{(\rho_1 - r_{\mathcal{B}_1})(\rho_1 + r_{\mathcal{O}_1})}{r_{\mathcal{B}_1} + r_{\mathcal{O}_1}} + (h_{\mathsf{cm}} - \rho_1) < 0 \qquad -\infty \leq \rho_1 \leq \infty,\ \omega = 1. \tag{11.16}$$

The height of $\mathcal{B}$'s center of mass, h_{cm}, is a free parameter in this test. The denominator, $r_{\mathcal{B}_1} + r_{\mathcal{O}_1}$, attains a negative sign when the object $\mathcal{B}$ or the supporting body $\mathcal{O}_1$ are concave at the contact point (see Chapter 8).

Since $\kappa_U(q_0, \rho_1)$ is linear in h_{cm}, the *stance stability region,* denoted $\mathcal{E}_S(q_0)$, is either empty or a *lower half line* of l_1. It is now a matter of elementary algebra to determine which values of h_{cm} satisfy the condition specified in Eq. (11.16). The possible cases are summarized in Table 11.1. The first row asserts that a convex object $\mathcal{B}$ is unstable on a single convex support. The second and third rows correspond to a convex object $\mathcal{B}$ supported by a concave body $\mathcal{O}_1$ and a concave object $\mathcal{B}$ supported by a convex body $\mathcal{O}_1$ (Figure 11.10). The last row asserts a classical result: which when $\mathcal{B}$ rests on a flat horizontal support, its center of mass must lie below its center of curvature for stability (see Bibliographical Notes).

Example: Figure 11.10(a) illustrates an equilibrium stance associated with the second row of Table 11.1, where a convex object $\mathcal{B}$ lies on a concave support. Figure 11.10(b) illustrates an equilibrium stance associated with the third row of Table 11.1, where

Table 11.1 The stability region of the one-contact stances, where $\kappa_{\mathcal{O}_1} = 1/r_{\mathcal{O}_1}$ and $\kappa_{\mathcal{B}_1} = 1/r_{\mathcal{B}_1}$ are the curvatures of $\mathcal{O}_1$ and $\mathcal{B}$ at the contact point.

Support	Object	Curvatures	Stability
Convex	Convex	$\kappa_{O_1} > 0,\ \kappa_{B_1} > 0$	Always unstable
Concave	Convex	$\kappa_{O_1} < 0,\ \kappa_{B_1} > 0$	Stable iff $h_{cm} < r_{\mathcal{B}_1}$
Convex	Concave	$\kappa_{O_1} > 0,\ \kappa_{B_1} < 0$	Stable iff $h_{cm} < r_{\mathcal{B}_1}$
Convex	Flat	$\kappa_{O_1} > 0,\ \kappa_{B_1} = 0$	Always unstable
Flat	Convex	$\kappa_{O_1} = 0,\ \kappa_{B_1} > 0$	Neutrally stable under pure translation Otherwise stable iff $h_{cm} < r_{\mathcal{B}_1}$

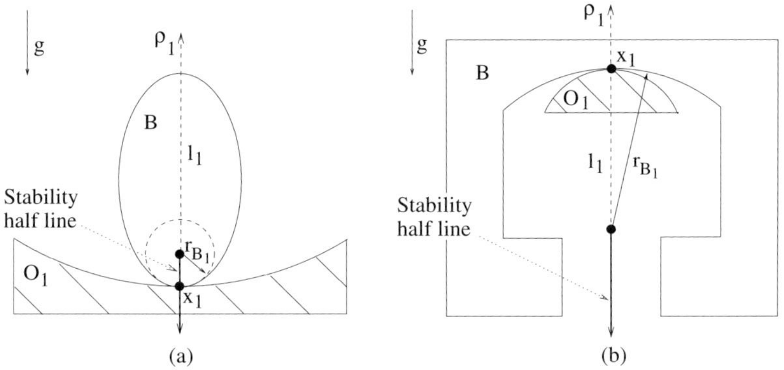

Figure 11.10 The stability half lines of single support equilibrium stances: (a) A convex object $\mathcal{B}$ rests on a concave support. (b) A locally concave object $\mathcal{B}$ rests on a convex support.

a locally concave object $\mathcal{B}$ rests on a convex support. This case forms a 2-D model for a cylindrical robot finger attempting to lift an object $\mathcal{B}$ through a handle. Both stances satisfy the stability rule $h_{cm} < r_{\mathcal{B}_1}$, which must also hold when $\mathcal{B}$ lies on a flat horizontal support. ○

11.5.3 Stable 2-D Stances Involving Three or More Contacts

When a 2-D rigid object $\mathcal{B}$ is supported by $k \geq 3$ frictionless contacts, its configuration q_0 generically lies in a zero-dimensional stratum of $\mathcal{F}$. In this case stance stability is determined only by the first part of the stance stability test, $l^- = \emptyset$, since the condition $v = 0$ is trivially satisfied at such stances. Let us focus on three-contact stances and then generalize to stances supported by higher numbers of contacts.

The affine line L at a three-contact stance can intersect the net wrench cone $\mathcal{W}(q_0)$ along a finite or semi-infinite interval (Figure 11.9(c)). In the first case $\mathcal{E}(q_0)$ forms a *vertical strip,* having the following boundary lines. Let l_i denote the line underlying the contact normal n_i at x_i; let p_{ij} denote the intersection point of the lines l_i and l_j; and let l_{ij} denote the vertical line passing through p_{ij} (Figure 11.11). When $\boldsymbol{e}$ lies in the positive span of n_i and n_j, the vertical line l_{ij} forms one of the two boundary lines of $\mathcal{E}(q_0)$. This condition holds for two pairs of contact normals as shown in Figure 11.11(a). When L

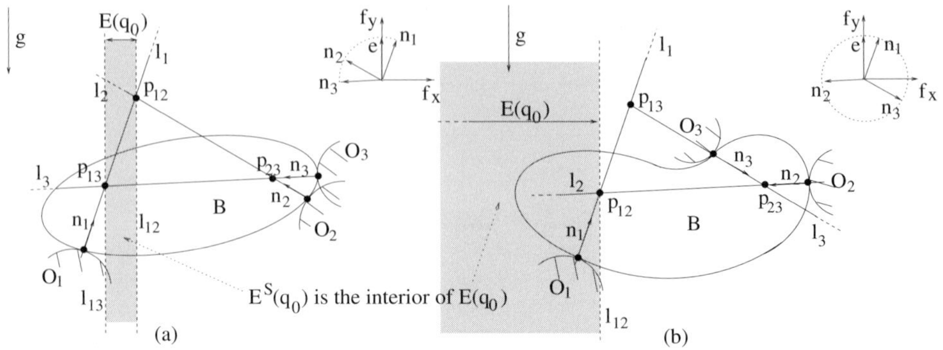

Figure 11.11 The equilibrium region of three-contact stances: (a) n_1, n_2, n_3 positively span a portion of the vertical plane. (b) n_1, n_2, n_3 span the entire vertical plane.

intersects $\mathcal{W}(q_0)$ along a semi-infinite interval, $\boldsymbol{e}$ lies in the positive span of exactly one pair of contact normals, n_i and n_j, as shown in Figure 11.11(b). In this case $\mathcal{E}(q_0)$ forms the *vertical half plane* bounded by l_{ij}, lying on the side of l_{ij} which does *not* contain the points p_{ik} and p_{jk}.

Example: Figure 11.11(a) depicts a three-contact stance whose contact normals positively span a subset of the vertical plane. The equilibrium region of this stance, $\mathcal{E}(q_0)$, forms the vertical strip depicted in Figure 11.11(a). Since $\boldsymbol{e}$ lies in the positive span of the pairs (n_1, n_2) and (n_1, n_3), the vertical strip is bounded by the lines l_{12} and l_{13}. Figure 11.11(b) depicts a three-contact stance whose contact normals positively span the entire vertical plane. The equilibrium region of this stance, $\mathcal{E}(q_0)$, forms the vertical half plane depicted in Figure 11.11(b). Since $\boldsymbol{e}$ is positively spanned by n_1 and n_2, the half plane is bounded by the vertical line l_{12}, such that the half plane lies on the side of l_{12} that does *not* contain the points p_{13} and p_{23}. ○

Next consider the stability region of three-contact stances. The stance stability test requires that the object $\mathcal{B}$ be supported at an essential equilibrium stance. Equivalently, the gravitational wrench $\nabla U(q_0)$ must lie in the interior of the net wrench cone $\mathcal{W}(q_0)$. Since the line L represents the possible gravitational wrenches, all points of L in the interior of $\mathcal{W}(q_0)$ correspond to stable stances. The stability region of the three-contact stances thus forms the *interior* of the respective equilibrium region, as illustrated in the following example.

Example: The stance stability region, $\mathcal{E}_S(q_0)$, occupies the interior of the vertical strip in Figure 11.11(a) and the interior of the vertical half plane in Figure 11.11(b). Note that both stability regions consist of *full vertical lines*. However, the contacts' ability to resist wrench disturbances diminishes with the height of $\mathcal{B}$'s center of mass. We will observe a similar phenomenon in the case of frictional hand-supported stances in Chapter 14. ○

The equilibrium region of stances supported by $k \geq 4$ contacts can be computed by taking the *convex hull* of the equilibrium regions associated with all contact triplets (see Exercises). The convex hull always forms a connected set of vertical lines: a strip, a half plane, or the entire vertical plane. Much like the three-contact stances, the stability region forms the *interior* of the respective equilibrium region at an essential equilibrium stance.

11.6 The Stable Equilibrium Region of 3-D Stances

This section applies the stance stability test to the 3-D *support problem.* A 3-D rigid object $\mathcal{B}$ with a variable center of mass is supported against gravity by stationary rigid bodies $\mathcal{O}_1, \ldots, \mathcal{O}_k$, which interact with the object $\mathcal{B}$ through frictionless contacts. Determine the object's center-of-mass locations guaranteeing stance stability while the object is supported by the given set of contacts. We begin with a scheme for computing the *stance equilibrium region,* $\mathcal{E}(q_0)$, and then describe the *stance stability region,* $\mathcal{E}_S(q_0) \subseteq \mathcal{E}(q_0)$, for the following three cases: (1) single-support stances, which model a robot hand attempting to carry an object by inserting a finger into a handle; (2) three contact stances, which will be used to demonstrate that the classical *tripod rule* no longer holds on uneven terrains formed by a supporting robot hand; and (3) six contact stances, which is the smallest number of supporting contacts that keep the object stationary while resisting external wrench disturbances.

11.6.1 A Scheme for Computing the Stance Equilibrium Region

Using Formula (11.8), the gravitational wrench that acts on $\mathcal{B}$ can be written in the form

$$\nabla U(q_0) = \mathsf{mg}\begin{pmatrix} \boldsymbol{e} \\ \rho_{\mathsf{cm}} \times \boldsymbol{e} \end{pmatrix}$$

$$= \mathsf{mg}\left\{\begin{pmatrix} \boldsymbol{e} \\ \vec{0} \end{pmatrix} - x_{\mathsf{cm}}\begin{pmatrix} \vec{0} \\ \begin{pmatrix} 0 \\ 1 \\ 0 \end{pmatrix} \end{pmatrix} + y_{\mathsf{cm}}\begin{pmatrix} \vec{0} \\ \begin{pmatrix} 1 \\ 0 \\ 0 \end{pmatrix} \end{pmatrix}\right\} \qquad (x_{\mathsf{cm}}, y_{cm}) \in \mathbb{R}^2, \tag{11.17}$$

where $\boldsymbol{e} = (0, 0, 1)$ and $\rho_{\mathsf{cm}} = (x_{\mathsf{cm}}, y_{\mathsf{cm}}, z_{\mathsf{cm}})$. The horizontal coordinates of $\mathcal{B}$'s center of mass, $(x_{\mathsf{cm}}, y_{\mathsf{cm}})$, are free parameters in Eq. (11.17). Hence, the possible gravitational wrenches span an *affine plane* in $\mathcal{B}$'s wrench space, denoted L. Let $(f, \tau) \in \mathbb{R}^6$ denote the object's wrench space coordinates, where $f = (f_x, f_y, f_z)$ and $\tau = (\tau_x, \tau_y, \tau_z)$ are force and torque components. Then L passes through the point $(\boldsymbol{e}, \vec{0})$ and is orthogonal to the four-dimensional subspace spanned by (f_x, f_y, f_z, τ_z) in $\mathcal{B}$'s wrench space.

Equilibrium stance feasibility requires that the affine plane L intersect the net wrench cone $\mathcal{W}(q_0)$. As a preliminary step, the gravitational wrench is scaled so that $\mathsf{mg} = 1$ in Eq. (11.17). The intersection of L with $\mathcal{W}(q_0)$ is characterized as follows.

LEMMA 11.10 (**3-D equilibrium stance feasibility**) *A rigid object $\mathcal{B}$ is supported by stationary rigid bodies $\mathcal{O}_1, \ldots, \mathcal{O}_k$ at a* **feasible equilibrium stance** *iff the vertical upward direction $\boldsymbol{e} = (0,0,1)$ lies in the positive span of $\mathcal{B}$'s inward contact normals:*

$$\boldsymbol{e} = \lambda_1 n_1 + \ldots + \lambda_k n_k \qquad \lambda_1, \ldots, \lambda_k \geq 0$$

such that the supporting contact forces apply net torque with zero z-component:

$$\tau_z = \boldsymbol{e} \cdot \sum_{i=1}^{k} \lambda_i (\rho_i(\theta) \times n_i) = 0$$

where $\rho_i(\theta) = R(\theta) b_i$ is the vector from $\mathcal{B}$'s frame origin to the contact point x_i for $i = 1 \ldots k$.

Proof: The gravitational wrench that acts on $\mathcal{B}$ always has a zero τ_z-component. Equilibrium feasibility thus requires that the net torque applied by the contact forces on $\mathcal{B}$ will have a zero τ_z-component. Next consider the five-dimensional subspace of wrenches having zero τ_z-component: $Z = \{(\vec{f}, \vec{\tau}) \in \mathbb{R}^6 : \tau_z = 0\}$. The intersection of Z with the net wrench cone $\mathcal{W}(q_0)$ forms a convex cone based at $\mathcal{B}$'s wrench space origin, denoted $\mathcal{W}_Z(q_0)$. Let V be the three-dimensional subspace of Z, consisting of pure forces: $V = \{(f, \tau) \in Z : f = (f_x, f_y, f_z) \in \mathbb{R}^3$ and $\tau = \vec{0}\}$. The affine plane L specified by Eq. (11.17) is orthogonal to V within the subspace Z. Hence L and $\mathcal{W}_Z(q_0)$ intersect in $\mathcal{B}$'s wrench space iff their projections on V intersect each other. The projection of L on V is the vector $\boldsymbol{e} = (0,0,1)$. Using an argument analogous to the one given for 2-D stances, the projection of $\mathcal{W}_Z(q_0)$ on V is the positive span of $\mathcal{B}$'s inward unit normals at the contacts. Hence, L intersects $\mathcal{W}_Z(q_0)$ iff $\boldsymbol{e}$ lies in the positive span of $\mathcal{B}$'s inward contact normals $n_1, \ldots, n_k$. □

When equilibrium stances are feasible on a given set of supporting contacts, the stance equilibrium region forms one of the following sets of vertical lines in $\mathbb{R}^3$.

PROPOSITION 11.11 (**3-D stance equilibrium region**) *Let a rigid object $\mathcal{B}$ be supported by stationary rigid bodies $\mathcal{O}_1, \ldots, \mathcal{O}_k$ at q_0. The* **stance equilibrium region,** *$\mathcal{E}(q_0)$, generically forms a vertical line for $1 \leq k \leq 4$ contacts; a vertical strip or half plane for $k = 5$ contacts; a vertical prism with a convex polygonal cross section for $k = 6$ contacts; and a vertical prism with a convex polygonal cross section or the entire $\mathbb{R}^3$ for $k \geq 7$ contacts.*

Proof sketch: Consider the possible intersections of the affine plane L with the net wrench cone $\mathcal{W}(q_0)$. At a single-contact stance, $\mathcal{W}(q_0)$ spans a ray in $\mathcal{B}$'s wrench space. If $L \cap \mathcal{W}(q_0) \neq \emptyset$, the intersection occurs at a single point. At a two-contact stance, $\mathcal{W}(q_0)$ spans a 2-D sector. If $L \cap \mathcal{W}(q_0) \neq \emptyset$, the intersection generically occurs at a single point. Similarly, for three and four contacts, the intersection $L \cap \mathcal{W}(q_0)$ generically occurs at a single point. In all of these cases, $\mathcal{E}(q_0)$ forms a single vertical line. At a five-contact stance, $\mathcal{W}(q_0)$ spans a 5-D convex cone based at $\mathcal{B}$'s wrench space origin. If $L \cap \mathcal{W}(q_0) \neq \emptyset$, the intersection occurs along a bounded or semi-infinite interval. Consequently, $\mathcal{E}(q_0)$ forms a vertical strip or half plane in $\mathbb{R}^3$. At a six-contact stance,

$\mathcal{W}(q_0)$ spans a full 6-D convex cone based at $\mathcal{B}$'s wrench space origin. If $L \cap \mathcal{W}(q_0) \neq \emptyset$, the intersection occurs along a convex two-dimensional polygon. Consequently, $\mathcal{E}(q_0)$ forms a vertical prism in $\mathbb{R}^3$, having a convex polygonal cross section. When $\mathcal{B}$ is supported by $k \geq 7$ contacts, $\mathcal{W}(q_0)$ still spans a 6-D convex cone. When $\mathcal{W}(q_0)$ spans a subset of $\mathcal{B}$'s wrench space, $\mathcal{E}(q_0)$ is the same as in the six-contact stances.[5] □

Relation to the support polygon: Proposition 11.11 asserts that the stance equilibrium region, $\mathcal{E}(q_0)$, always forms a connected set of vertical lines having a convex cross section (a point, an interval or a convex polygon). In robotic legged locomotion, the classical *support polygon* is defined as the convex region spanned by the horizontal projection of the supporting contacts. Under frictionless contact conditions, the support polygon determines the stance equilibrium region on flat horizontal terrains or, more generally, on terrains having vertical upward-pointing contact normals. However, a supporting robot hand usually forms an *uneven terrain,* and the stance equilibrium region is *not* related to the support polygon. ○

The stability region forms a subset of $\mathcal{E}(q_0)$ as next characterized for 3-D stances supported by one, three and six contacts.

11.6.2 The Stability Region of Single-Support 3-D Stances

A one-contact stance forms a feasible equilibrium when the object's inward contact normal is collinear with the vertical upward direction: $n_1 = \boldsymbol{e}$. In this case the stance equilibrium region, $\mathcal{E}(q_0)$, spans the vertical line that passes through the contact point, denoted l_1. Next consider the stance stability region. The first part of the stance stability test, $l^- = \emptyset$, is automatically satisfied at any center-of-mass position in $\mathcal{E}(q_0)$. Thus, consider the second part of the stance stability test, $\kappa_U(q_0, \dot{q}) < 0$ for all $\dot{q} \in T_{q_0}\mathcal{S}$, where $\mathcal{S} = \text{bdy}(\mathcal{CO}_1)$. Let us characterize the tangent space $T_{q_0}\mathcal{S}$ and then evaluate the sign of $\kappa_U(q_0, \dot{q})$ on this subspace.

Let $\mathcal{B}$'s frame origin be located on the vertical line l_1. Under this choice of $\mathcal{B}$'s frame origin, the c-obstacle outward unit normal at q_0 is the vector $\hat{\eta}_1(q_0) = (n_1, \vec{0}) \in \mathbb{R}^6$, and the tangent space $T_{q_0}\mathcal{S}$ has the form:

$$T_{q_0}\mathcal{S} = \left\{\dot{q} = (v, \omega) \in T_{q_0}\mathbb{R}^6 \, : \, (n_1, \vec{0}) \cdot (v, \omega) = n_1 \cdot v = 0\right\}.$$

Consider the sign attained by $\kappa_U(q_0, \dot{q})$ on the tangent space $T_{q_0}\mathcal{S}$. When $\mathcal{B}$'s frame origin is located on the line l_1 at a signed distance ρ_1 from the contact point x_1, $\boldsymbol{\rho}_{\mathsf{cm}} = (0, 0, h_{\mathsf{cm}} - \rho_1)$ where h_{cm} is the height of $\mathcal{B}$'s center of mass above the contact point. Substituting for $\boldsymbol{\rho}_{\mathsf{cm}}$ in Formula (11.9) for $D^2U(q_0)$ gives

$$D^2U(q_0) = \begin{bmatrix} O & O \\ O & A \end{bmatrix} \qquad A = \begin{bmatrix} \rho_1 - h_{\mathsf{cm}} & 0 & 0 \\ 0 & \rho_1 - h_{\mathsf{cm}} & 0 \\ 0 & 0 & 0 \end{bmatrix}, \tag{11.18}$$

[5] When $\mathcal{W}(q_0)$ spans the entire wrench space, the supporting contacts form a *first-order immobilizing grasp* of the object $\mathcal{B}$. Technically, these are no longer hand-supported stances.

Table 11.2 Stability conditions for a 3-D object $\mathcal{B}$ supported by a single body.

Support	Object	Stability
Convex	Convex	Always unstable
Concave	Convex	Stable iff $h_{\text{cm}} < \min\{r^1_{\mathcal{B}}, r^2_{\mathcal{B}}\}$
Saddle	Convex	Always unstable
Convex	Concave	Stable iff $h_{\text{cm}} < \min\{r^1_{\mathcal{B}}, r^2_{\mathcal{B}}\}$
Concave	Concave	Infeasible surface contact
Saddle	Concave	Infeasible surface contact
Convex	Saddle	Always unstable
Concave	Saddle	Case not possible
Saddle	Saddle	Stable iff $h_{\text{cm}} < \min\{r^1_{\mathcal{B}}, r^2_{\mathcal{B}}\}$
Convex	Flat	Always unstable
Flat	Convex	Neutrally stable under pure translation Otherwise stable iff $h_{\text{cm}} < \min\{r^1_{\mathcal{B}}, r^2_{\mathcal{B}}\}$

where O is a 3×3 zero matrix. Recall from Chapter 2 that $\kappa_1(q_0, \dot{q}) = 0$ along the tangent vector $\dot{q} = (\vec{0}, n_1) \in T_{q_0}\mathcal{S}$, which represents instantaneous rotation of $\mathcal{B}$ about the contact normal. Since $\dot{q}^T D^2U(q_0)\dot{q} = 0$ along this particular tangent vector, we obtain that $\kappa_U(q_0, \dot{q}) = \kappa_1(q_0, \dot{q}) - \dot{q}^T D^2U(q_0)\dot{q} = 0$ along $\dot{q} = (\vec{0}, n_1)$. This property implies that $\kappa_U(q_0, \dot{q})$ is at most negative semi-definite at any single-support stance.

The possible single contact stability regions are summarized in Table 11.2. A surface is *convex* when its principal curvatures at the contact are positive, *concave* when its principal curvatures are negative, and *saddle-like* when its principal curvatures have mixed signs at the contact point. As seen in Table 11.2, the stability region, $\mathcal{E}_S(q_0)$, is nonempty when one of the contacting surfaces is convex while the other is concave, when both surfaces are saddle-like, and when a convex object $\mathcal{B}$ rests on a flat horizontal support. The height of $\mathcal{B}$'s center of mass, h_{cm}, appears linearly in Eq. (11.18). Hence, the stability region $\mathcal{E}_S(q_0)$ always forms a *lower half line* of $\mathcal{E}(q_0)$, given by $h_{\text{cm}} < \min\{r^1_{\mathcal{B}}, r^2_{\mathcal{B}}\}$, where $r^1_{\mathcal{B}}$ and $r^2_{\mathcal{B}}$ are the object's principal radii of curvature at the contact point. The case of a convex object lying on a flat horizontal support is a classical result (see Bibliographical Notes). The following example illustrates the case of a saddle-like object resting on a saddle-like support.

Example: Figure 11.12 shows a U-shaped object $\mathcal{B}$ balanced on an inverted U-shaped supporting body $\mathcal{O}_1$ under the influence of gravity. In this case the object $\mathcal{B}$ and the supporting body $\mathcal{O}_1$ have saddle-shaped surfaces at the contact. According to Row 9 of Table 11.2, the object's stability is ensured when its center of mass lies below its minimum radius of curvature at the contact. That is, $r^1_{\mathcal{B}} > 0$ and $r^2_{\mathcal{B}} < 0$ at the contact, and $h_{\text{cm}} < r^2_{\mathcal{B}}$ for stability. This example can model a robot hand attempting to lift a rigid object $\mathcal{B}$ by inserting a curved finger, $\mathcal{O}_1$, into a handle attached to the object $\mathcal{B}$. This task can be carried out in a locally stable manner, as long as the object's center of mass lies within the stability lower half line of the single-support stance. ○

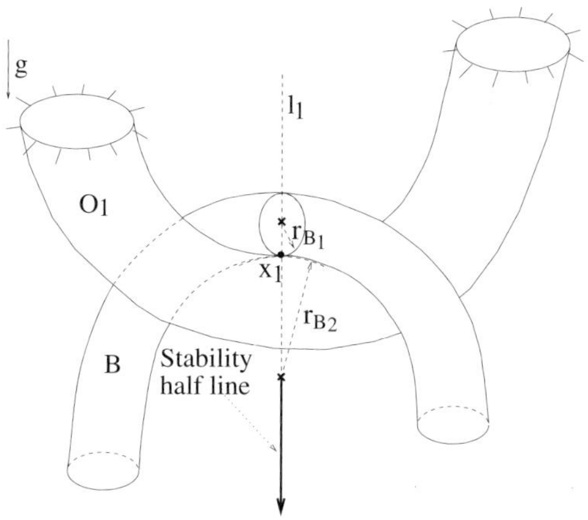

Figure 11.12 The stability half line of a saddle-like object $\mathcal{B}$ resting on a saddle-like support $\mathcal{O}_1$. Note that $r^1_{\mathcal{B}} > 0$ while $r^2_{\mathcal{B}} < 0$ due to the saddle-like shape of $\mathcal{B}$ at x_1.

11.6.3 The Stability Region of Three-Contact 3-D Stances

At a three-contact stance, equilibrium feasibility requires that the contact normals positively span the vertical upward direction:

$$\boldsymbol{e} = \lambda_1 n_1 + \lambda_2 n_2 + \lambda_3 n_3 \qquad \lambda_1, \lambda_2, \lambda_3 \geq 0, \tag{11.19}$$

such that the contact forces, $f_i = \lambda_i n_i$ for $i = 1, 2, 3$, apply zero net torque about the vertical axis in $\mathbb{R}^3$. The zero torque requirement can be graphically characterized as follows. Consider a generic stance whose contact normal lines l_1, l_2, and l_3 do not lie in three parallel planes. Such stances are said to have *skew contact normals.* The projections of l_1 and l_2 on the horizontal (x, y) plane intersect at a point p. Let l denote the vertical line erected at p (Figure 11.13). Since l_1 and l_2 pass through l, the contact forces f_1 and f_2 generate zero torque about this line. Equilibrium feasibility requires that the contact force f_3 also generates zero torque about l. This is possible only if the (x, y) projection of l_3 passes through p in the (x, y) plane. Equilibrium stance feasibility thus requires that the three contact normals positively span the vertical direction $\boldsymbol{e}$, such that the lines l_1, l_2, and l_3 intersect a common vertical line l (Figure 11.13).

Remark: Consider the gravitational line, l_g, which passes through $\mathcal{B}$'s center of mass along the vertical direction $\boldsymbol{e}$. At a three-contact equilibrium stance, the contact force lines l_1, l_2, and l_3, and the line l_g are linearly dependent in Plücker coordinates. As discussed in Chapter 5, the four lines must lie on a common ruled surface called a *regulus*. One way to define the regulus is as follows. Given three skew lines in $\mathbb{R}^3$ called *generators,* the regulus spanned by these generators is the collection of all lines that intersect the three generators. The vertical line l is thus one generator of the regulus containing the lines l_1, l_2, l_3, and l_g. ○

The equilibrium region of three-contact stances having skew contact normals generically forms a single vertical line according to Proposition 11.11. Let $\mathcal{B}$'s frame origin be

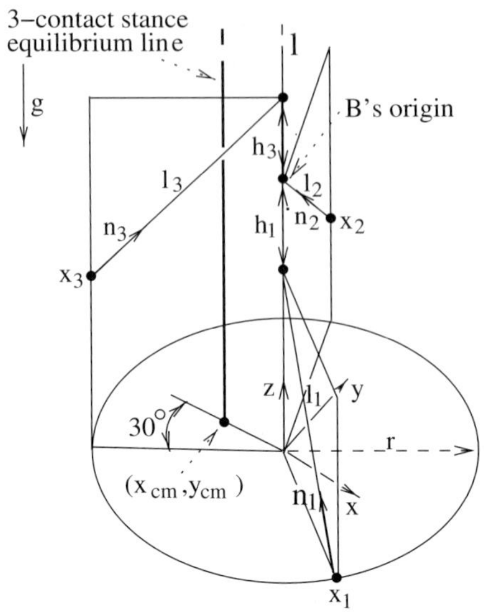

Figure 11.13 The equilibrium line of this three-contact stance passes through a point located $r/2\sqrt{3}$ away from the vertical line l along the direction $(-\cos 30, \sin 30)$ in the horizontal (x, y) plane (the supported object is not shown).

located on the vertical line l. Let h_i denote the height of the intersection point of the contact normal line l_i with l (Figure 11.13). The horizontal coordinates of the stance equilibrium line, $(x_{\text{cm}}, y_{\text{cm}})$, are determined by the τ_x and τ_y components of the equilibrium stance equation (see Exercises):

$$\begin{pmatrix} x_{\text{cm}} \\ y_{\text{cm}} \end{pmatrix} = -\sum_{i=1}^{3} h_i \lambda_i \cdot (n_i)_{xy}, \tag{11.20}$$

where $(n_i)_{xy}$ is the horizontal projection of the contact normal n_i for $i = 1, 2, 3$, and the coefficients $\lambda_1, \lambda_2, \lambda_3 \geq 0$ are determined by the force balance specified in Eq. (11.19).

Example: Figure 11.13 shows a three-contact stance whose contacts are arranged at 120° angles on a vertical cylinder of radius r (the supported object is not shown). The contact points x_1 and x_3 are located at heights of $r/2$ below and above the middle contact point x_2. The contact normals n_1, n_2, and n_3 point toward the cylinder's central axis, l, at an upward 45° angle with respect to the horizontal (x, y) plane. The three contact normals positively span the vertical direction $\boldsymbol{e}$, such that the contact normal lines l_1, l_2, and l_3 intersect the vertical line l. The stance therefore possesses a nonempty equilibrium line. Under a choice of $\mathcal{B}$'s frame origin at the intersection point of l_2 with l (Figure 11.13), $h_1 = -r/2$, $h_2 = 0$ and $h_3 = r/2$, while $\lambda_i = \sqrt{2}/3$ for $i = 1, 2, 3$. Using Formula (11.20), the stance equilibrium line passes through the point $(x_{\text{cm}}, y_{\text{cm}}) = r\big((n_1)_{xy} - (n_3)_{xy}\big)/3\sqrt{2}$, located $r/2\sqrt{3}$ away from the vertical line l along the direction $(-\cos 30, \sin 30)$ in the (x, y) plane (Figure 11.13). ○

The stability region of three-contact stances with skew contact normals, $\mathcal{E}_S(q_0) \subseteq \mathcal{E}(q_0)$, depends on the *curvatures* of the object and supporting bodies. When the curvatures allow stable center-of-mass locations (for instance, when the supporting bodies are concave at the contacts), the stability region $\mathcal{E}_S(q_0)$ forms a *lower half line* of the equilibrium line $\mathcal{E}(q_0)$. Much like the one-contact stances, this property is based on the linear dependency of the Hessian matrix $D^2U(q_0)$ on the height of $\mathcal{B}$'s center of mass, measured along the stance equilibrium line. Computation of the stability half line requires evaluation of the gravity relative curvature form, $\kappa_U(q_0, \dot{q})$, on the tangent space $T_{q_0}\mathcal{S}$, where $\mathcal{S} = \cap_{i=1}^{3} \text{bdy}(\mathcal{CO}_i)$ is the stratum associated with the supporting bodies c-obstacles.

Relation to the support polygon: Consider a robot hand that supports a rigid object $\mathcal{B}$ via three *frictional* contacts, such that the friction cones contain the vertical upward direction $\boldsymbol{e}$. The stance equilibrium region, $\mathcal{E}(q_0)$, is the vertical prism spanned the stance's support polygon. This property is known as the *tripod rule.* Next consider the gradual decrease of the coefficient of friction at the contacts. As long as all friction cones contain the vertical upward direction $\boldsymbol{e}$, the stance equilibrium region remains fixed. Once one or more of the friction cones no longer contain the vertical direction $\boldsymbol{e}$, the stance equilibrium region starts to shrink within the vertical prism spanned by the support polygon, always forming a connected set of vertical lines having a convex cross section. At the limit of *frictionless contacts,* the region $\mathcal{E}(q_0)$ shrinks to a single vertical line when a frictionless equilibrium stance is feasible. The tripod rule therefore does *not* hold under frictionless contact conditions, except when the supports happen to posses vertical contact normals. ○

11.6.4 The Stability Region of Six-Contact 3-D Stances

When a rigid object $\mathcal{B}$ is supported against gravity by six contacts, its configuration q_0 generically lies in a zero-dimensional stratum of $\mathcal{F}$. The equilibrium region, $\mathcal{E}(q_0)$, forms a 3-D *vertical prism* having a convex polygonal cross section according to Proposition 11.11. Since q_0 is located at a zero-dimensional stratum of $\mathcal{F}$, the stance stability region, $\mathcal{E}_S(q_0) \subseteq \mathcal{E}(q_0)$, is determined only by the first part of the stance stability test, $I^- = \emptyset$. According to Theorem 11.6, $I^- = \emptyset$ when the object $\mathcal{B}$ is supported at an essential equilibrium stance. Equivalently, $\nabla U(q_0)$ must lie in the interior of the contacts' net wrench cone $\mathcal{W}(q_0)$. The latter condition is satisfied at all center-of-mass locations in the interior of $\mathcal{E}(q_0)$, hence $\mathcal{E}_S(q_0)$ forms the *interior* of the stance equilibrium region.

The stability region of six-contact stances has two notable properties. The first property concerns its robustness. The object $\mathcal{B}$ can locally shift its center of mass about any point in $\mathcal{E}_S(q_0)$ without affecting the stance's equilibrium feasibility or its local stability. The object $\mathcal{B}$ will therefore remain stationary and reasonably secure when placed under small contact placement errors on a supporting robot hand. The second property concerns the structure of the region $\mathcal{E}_S(q_0)$ that consists of full vertical lines. While the height of $\mathcal{B}$'s center of mass does *not* affect the local stability of the six-contact stances, their ability to resist wrench disturbances diminishes with the height of $\mathcal{B}$'s center of mass. We will observe a similar property in the context of frictional hand-supported stances in Chapter 14.

Let us describe how one computes the horizontal cross section of the equilibrium region of a six-contact stance, termed the *stance equilibrium polygon.* The ith c-obstacle outward normal at q_0 is given by $\eta_i(q_0) = (n_i, \rho_i \times n_i) \in \mathbb{R}^6$, while $\nabla U(q_0) = (\boldsymbol{e}, \boldsymbol{\rho}_{\text{cm}} \times \boldsymbol{e}) \in \mathbb{R}^6$. The equilibrium equation of a six-contact stance has the form:

$$\begin{bmatrix} n_1 & \cdots & n_6 \\ \rho_1 \times n_1 & & \rho_6 \times n_6 \end{bmatrix} \begin{pmatrix} \lambda_1 \\ \vdots \\ \lambda_6 \end{pmatrix} = \begin{pmatrix} \boldsymbol{e} \\ \boldsymbol{\rho}_{\text{cm}} \times \boldsymbol{e} \end{pmatrix} \qquad \lambda_1, \ldots, \lambda_6 \geq 0, \tag{11.21}$$

where the coefficients $\lambda_1, \ldots, \lambda_6$ represent the contact force magnitudes. The object center-of-mass position, $\boldsymbol{\rho}_{\text{cm}}$, is a free parameter in Eq. (11.21). Generic six-contact stances possess linearly independent c-obstacle normals at q_0. At such stances, the solution $(\lambda_1, \ldots, \lambda_6)$ is obtained by *Cramer's rule:*

$$\lambda_i = \frac{1}{D} \det \begin{bmatrix} n_1 & \cdots & \boldsymbol{e} & \cdots & n_6 \\ \rho_1 \times n_1 & & \boldsymbol{\rho}_{\text{cm}} \times \boldsymbol{e} & & \rho_6 \times n_6 \end{bmatrix} \qquad i = 1 \ldots 6,$$

such that $\nabla U(q_0) = (\boldsymbol{e}, \boldsymbol{\rho}_{\text{cm}} \times \boldsymbol{e})$ occupies the ith column and $D = \det[\eta_1(q_0) \cdots \eta_6(q_0)]$. For a given set of supporting contacts, the stance equilibrium polygon is determined by the inequalities: $\lambda_1(\boldsymbol{\rho}_{\text{cm}}), \ldots, \lambda_6(\boldsymbol{\rho}_{\text{cm}}) \geq 0$. Since $\boldsymbol{\rho}_{\text{cm}} \times \boldsymbol{e} = (y_{\text{cm}}, -x_{\text{cm}}, 0)$, each of the six constraints is linear in $\mathcal{B}$'s horizontal center-of-mass coordinates, $(x_{\text{cm}}, y_{\text{cm}})$. The intersection of the six inequalities defines a convex polygon in the horizontal (x, y) plane, which is bounded by up to six edges. Each edge consists of five-contact equilibrium stances where one of the six contacts becomes idle, while each vertex corresponds to a four-contact equilibrium stance where two of the six contacts are idle.

One can use this insight to determine the vertices of the stance equilibrium polygon as follows. For each force-line quadruple, check if the four contact normals positively span the vertical upward direction $\boldsymbol{e}$, such that the horizontal projections of the four contact normals form a *feasible equilibrium* in the (x, y) plane. The latter test can be performed graphically as described in Chapter 5. If a force-line quadruple forms a feasible equilibrium stance, its equilibrium line (a vertex of the stance equilibrium polygon) can be computed with the formula

$$\begin{pmatrix} x_{\text{cm}} \\ y_{\text{cm}} \end{pmatrix} = J \cdot \sum_{i=1}^{4} \lambda_i (\rho_i \times n_i)_{xy} \qquad J = \begin{bmatrix} 0 & 1 \\ -1 & 0 \end{bmatrix}, \tag{11.22}$$

where $(\cdot)_{xy}$ denotes horizontal projection on the horizontal (x, y) plane. The technique is illustrated in the following example.

Example: Consider the six-contact stance with skew contact normals depicted in Figure 11.14 (the supported object is not shown). The six contacts are arranged at 120° angles on two circles embedded in the horizontal (x, y) plane, having unit radius and centers located 1.25 units apart. The contact normals are tangent to the vertical

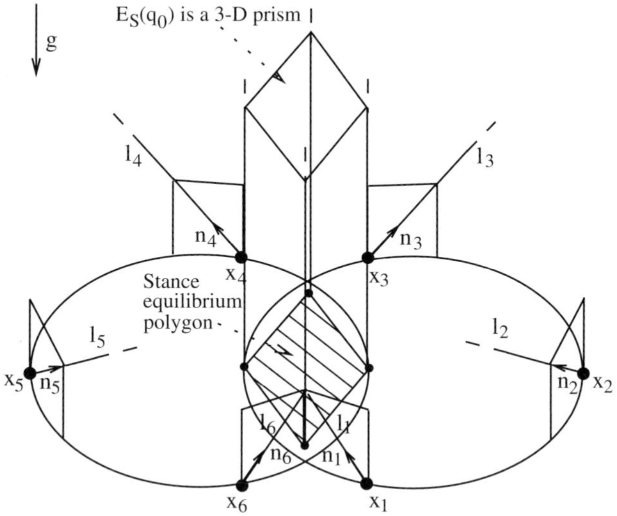

Figure 11.14 A six-contact stance with skew contact normals (the supported object is not shown). The stance equilibrium polygon has four vertices determined by the line quadruples (l_1, l_2, l_3, l_5), (l_1, l_2, l_5, l_6), (l_2, l_3, l_4, l_5), and (l_2, l_4, l_5, l_6). The stance stability region is the interior of the 3-D prism spanned by the equilibrium polygon.

cylinders passing through the two circles and point upward at a 45° angle above the (x, y) plane. Three contact normals point in counterclockwise direction on the left circle, while the other three contact normals point in clockwise direction on the right circle. To determine equilibrium feasibility of this stance, we must check equilibrium feasibility of all fifteen force-line quadruples. In this example, eleven quadruples fail to form a feasible equilibrium within the (x, y) plane. The remaining four quadruples positively span the vertical upward direction $\boldsymbol{e}$ and also form a feasible equilibrium within the (x, y) plane. These are (l_1, l_2, l_3, l_5), (l_1, l_2, l_5, l_6), (l_2, l_3, l_4, l_5), and (l_2, l_4, l_5, l_6). The four quadruples determine four vertices of the stance equilibrium polygon. Their location has been computed with Formula (11.22), and the stance equilibrium region forms the four-sided prism plotted in Figure 11.14. ○

Relation to the support polygon: Under mild conditions on the terrain formed by the supporting robot hand, called *tame terrains*, the stance equilibrium polygon lies inside the stance's support polygon. However, the stance equilibrium polygon occupies only a subset of the stance's support polygon (Figure 11.14). The support polygon is thus no longer suitable for assessing the security of hand-supported stances. ○

Bibliographical Notes

The chapter is based on Rimon, Mason and Burdick [1]. This paper provides the analysis principles of equilibrium stances supported by multiple frictionless bodies under the influence of gravity. It focuses on legged robot postures over uneven terrains, but the same principles apply to robot hands which form uneven terrains when supporting rigid objects against gravity. The first and second-order conditions of the stance stability test were obtained without Stratified Morse Theory, as first and second-order *stability cells* by Trinkle et al. [2].

The stability of single-support stances is closely related to the stability of floating bodies such as ships, submersibles, dirigibles and hot air balloons. A floating body is supported against gravity by its net buoyancy force and lies at a stable equilibrium when its center of mass is located below the body's center-of-buoyancy curve [3, chap. 10]. Stable stances are also useful in computer vision to identify likely object poses before pickup by a robot hand [4]. The stability of *multiple-contact* stances lies strictly in the realm of robotics applications, such as carrying and manipulating large nonprehensile objects by robot hands. Hand-supported stances of several interacting rigid bodies carried by a robot hand under the influence of gravity is a formidable challenge; see Lynch et al. [5]. Another challenge concerns *gravity cages* formed by the supporting bodies and a virtual finger associated with gravity. Can the *caging grasps* from Chapter 10 be adapted into gravity cages whose depth measures the security of a candidate hand-supported stance? See Mahler et al. [6].

Finally, this chapter focused on stances having a fixed set of contacts and a variable center of mass. What would happen to the stance stability region when other key parameters are varied? For instance, an understanding of the stability region of an object having a fixed center of mass and continuously varying supporting contacts is highly useful for in-hand object manipulation under the influence of gravity. Such problems fall under the general framework of *bifurcation theory,* where a system's stability regions are mapped as a function of key parameters of the system [3, 7].

Appendix: Proof Details

This appendix contains proofs of two key properties of the gravitational potential $U(q)$ on the stratified set $\mathcal{F}$. The first property concerns the robustness of the local minima of $U(q)$ with respect to external wrench disturbances.

Proposition 11.3 *Let a rigid object $\mathcal{B}$ be supported at an equilibrium stance configuration q_0 by stationary rigid bodies $\mathcal{O}_1,\dots,\mathcal{O}_k$. If q_0 forms a non-degenerate local minimum of $U(q)$, the zero-velocity equilibrium state $(q_0,\vec{0})$ is* **robust** *with respect to sufficiently small external wrench disturbances.*

Proof sketch: Consider the case where the external wrench disturbances arise from local shifts of the object's center of mass. Let b^0_{cm} denote the initial position of $\mathcal{B}$'s center of mass, and let Δb_{cm} denote a small change in $\mathcal{B}$'s center-of-mass position, both expressed in $\mathcal{B}$'s reference frame. The object center-of-mass position, b_{cm}, appears as a parameter in $\mathcal{B}$'s gravitational potential $U(q)$:

$$U(q,b_{cm}) = \mathsf{m}g\boldsymbol{e}\cdot x_{\mathsf{cm}}(q) = \mathsf{m}g\boldsymbol{e}\cdot(R(\theta)b_{cm}+d) \qquad q=(d,\theta)\in\mathbb{R}^m,\ m=3 \text{ or } 6.$$

We have to show that small Δb_{cm} induce small changes in the equilibrium stance position, such that the new equilibrium stance remains stable.

The object's equilibrium stance configuration, q_0, lies in the stratum $\mathcal{S} = \cap_{i=1}^{k}$ bdy$(\mathcal{CO}_i)$, associated with the supporting bodies $\mathcal{O}_1, \ldots, \mathcal{O}_k$. Since $U(q, b_{cm}^0)$ has a non-degenerate local minimum at q_0, $\nabla U(q_0, b_{cm}^0)$ is not orthogonal to any of the neighboring strata at q_0. Since $U(q, b_{cm})$ varies smoothly in b_{cm}, a small change Δb_{cm} incurs a small change in ∇U. Hence, the perturbed gradient, $\nabla U(q, b_{cm}^0 + \Delta b_{cm})$, is still *not* normal to any of the neighboring strata at q_0. Any critical point of $U(q, b_{cm}^0 + \Delta b_{cm})$ must therefore lie in the *same* stratum $\mathcal{S}$ for all sufficiently small Δb_{cm}. When $\mathcal{S}$ is zero-dimensional, $\mathcal{B}$ remains at the same configuration q_0 in response to Δb_{cm}. When $\mathcal{S}$ is at least one-dimensional, $\mathcal{B}$ shifts to a new equilibrium configuration located within $\mathcal{S}$ in response to Δb_{cm}.

Thus, we need to consider the case where $\mathcal{S}$ is at least one-dimensional. We have to show that the new equilibrium induced by Δb_{cm} lies in a local neighborhood of $\mathcal{S}$ centered at q_0. The *unperturbed* equilibrium stance satisfies the condition

$$\lambda_1^0 \hat{\eta}_1(q_0) + \ldots + \lambda_k^0 \hat{\eta}_k(q_0) - \nabla U(q_0, b_{cm}^0) = \vec{0} \qquad \lambda_1^0, \ldots, \lambda_k^0 \geq 0, \tag{11.23}$$

where $\hat{\eta}_1(q_0), \ldots, \hat{\eta}_k(q_0)$ are the c-obstacle outward unit normals at q_0. The *perturbed* equilibrium stance induced by Δb_{cm} satisfies the equilibrium condition:

$$\lambda_1 \hat{\eta}_1(q) + \cdots + \lambda_k \hat{\eta}_k(q) - \nabla U(q, b_{cm}^0 + \Delta b_{cm}) = \vec{0} \qquad q \in \mathcal{S}. \tag{11.24}$$

Denote by Δq_0 the change in the equilibrium stance configuration in response to Δb_{cm}. Denote by $\vec{\lambda}_0 = (\lambda_1^0, \ldots, \lambda_k^0)$ the unperturbed equilibrium stance coefficients, and by $\Delta \vec{\lambda}_0$ the change in these coefficients at the new equilibrium stance induced by Δb_{cm}. Implicit differentiation of Eq. (11.24) with respect to the variables $(q, \vec{\lambda}, b_{cm})$, followed by substitution of Eq. (11.23), gives the first-order approximation:

$$\begin{aligned} &\left[\textstyle\sum_{i=1}^{k} \lambda_i^0 D\hat{\eta}_i(q_0) - D^2 U(q_0, b_{cm}^0)\right] \Delta q_0 \\ &\quad + \left[\hat{\eta}_1(q_0) \cdots \hat{\eta}_k(q_0)\right] \Delta \vec{\lambda}_0 = \left[\frac{\partial}{\partial b_{cm}} \nabla U(q_0, b_{cm}^0)\right] \Delta b_{cm} \end{aligned}$$

where Δq_0 varies in $T_{q_0}\mathcal{S}$. Define the matrix multiplying Δq_0 as $K_U(q_0, b_{cm}^0)$:

$$\left[K_U(q_0, b_{cm}^0)\right] \Delta q_0 + \left[\hat{\eta}_1(q_0) \cdots \hat{\eta}_k(q_0)\right] \Delta \vec{\lambda}_0 = \left[\frac{\partial}{\partial b_{cm}} \nabla U(q_0, b_{cm}^0)\right] \Delta b_{cm}. \tag{11.25}$$

The matrix $K_U(q_0, b_{cm}^0)$ acts on $\Delta q_0 \in T_{q_0}\mathcal{S}$ and represents the restriction of the Hessian matrix, $-D^2 U(q_0, b_{cm}^0)$, to the tangent space $T_{q_0}\mathcal{S}$. Let us therefore establish the dimension of this matrix. Since $\mathcal{B}$ is supported by k bodies at q_0, the dimension of $\mathcal{S}$ is generically $m - k$, where m is the ambient c-space dimension. The matrix $K_U(q_0, b_{cm}^0)$ is thus $m \times (m - k)$. Since $U(q, b_{cm}^0)$ has a non-degenerate local minimum at q_0, the restriction of $D^2 U(q_0, b_{cm}^0)$ to $T_{q_0}\mathcal{S}$ has full rank of $m - k$. Hence, $K_U(q_0, b_{cm}^0)$ has full rank of $m - k$.

Next consider the composite $m \times m$ matrix: $[K_U(q_0, b_{cm}^0)\ \hat{\eta}_1(q_0) \ldots \hat{\eta}_k(q_0)]$, which acts on vectors $(\Delta q_0, \Delta \vec{\lambda}_0)$. We may assume that the c-obstacle outward normals

$\hat{\eta}_1(q_0),\ldots,\hat{\eta}_k(q_0)$ are linearly independent at q_0. Moreover, it can be verified that $K_U(q_0,b^0_{cm})$ maps tangent vectors $\Delta q_0 \in T_{q_0}\mathcal{S}$ into the tangent space $T_{q_0}\mathcal{S}$. Since the subspace spanned by $\hat{\eta}_1(q_0),\ldots,\hat{\eta}_k(q_0)$ and the tangent space $T_{q_0}\mathcal{S}$ are orthogonal complements, the matrix $[K_U(q_0,b^0_{cm})\ \hat{\eta}_1(q_0)\cdots\hat{\eta}_k(q_0)]$ has full rank of m. That is, $K_U\Delta q_0 + [\hat{\eta}_1(q_0)\ldots\hat{\eta}_k(q_0)]\Delta\vec{\lambda}_0 = \vec{0}$ only when $(\Delta q_0,\Delta\vec{\lambda}_0) = (\vec{0},\vec{0})$. The composite change, $(\Delta q_0,\Delta\vec{\lambda}_0)\in\mathbb{R}^m$, can therefore be expressed as a continuous function of Δb_{cm}:

$$\begin{pmatrix}\Delta q_0\\ \Delta\vec{\lambda}_0\end{pmatrix} = \left[K_U(q_0,b^0_{cm})\ \hat{\eta}_1(q_0)\cdots\hat{\eta}_k(q_0)\right]^{-1}\left[\frac{\partial}{\partial b_{cm}}\nabla U(q_0,b^0_{cm})\right]\Delta b_{cm}.$$

The new equilibrium stance configuration, $q_0 + \Delta q_0$, as well as the new equilibrium coefficients, $\vec{\lambda}_0 + \Delta\vec{\lambda}_0$, thus vary continuously with b_{cm}. Based on this continuity, all sufficiently small Δb_{cm} induce equilibrium stances in a local neighborhood of $\mathcal{S}$ centered at q_0.

Let us next show that the new equilibrium at $q_0 + \Delta q_0$ forms a local minimum of $U(q,b^0_{cm} + \Delta b_{cm})$, and hence a locally stable stance, for all sufficiently small Δb_{cm}. As discussed in Section 11.3, $U(q,b^0_{cm} + \Delta b_{cm})$ has a local minimum at $q_0 + \Delta q_0$ when it satisfies the conditions $\nu = 0$ and $I^- = \emptyset$ at this point. The condition $\nu = 0$ requires that the restriction of $D^2U(q_0 + \Delta q_0, b^0_{cm})$ to the tangent space of $\mathcal{S}$ at $q_0 + \Delta q_0$ will have positive eigenvalues. This condition holds at (q_0,b^0_{cm}) and, by continuity of the eigenvalues of $D^2U(q,b^0_{cm})$ in q, still holds in a local neighborhood of (q_0,b^0_{cm}). The condition $I^- = \emptyset$ holds when $q_0 + \Delta q_0$ forms an essential equilibrium stance (Proposition 11.4). Equivalently, $I^- = \emptyset$ when $\hat{\eta}_1(q_0 + \Delta q_0),\ldots,\hat{\eta}_k(q_0 + \Delta q_0)$ are linearly independent, such that $\nabla U(q_0 + \Delta q_0)$ lies in the interior of the net wrench cone $\mathcal{W}(q_0 + \Delta q_0)$ (Proposition 11.4). The net wrench cone $\mathcal{W}(q)$ is spanned by the c-obstacle outward unit normals, $\hat{\eta}_1(q),\ldots,\hat{\eta}_k(q)$, which vary continuously in $q\in\mathcal{S}$. The gradient $\nabla U(q,b_{cm})$ varies continuously in $q\in\mathcal{S}$ and b_{cm}. Since $U(q,b^0_{cm})$ has a non-degenerate local minimum at q_0, $\nabla U(q_0,b^0_{cm})$ lies in the interior of $\mathcal{W}(q_0)$. By continuity, $\nabla U(q_0+\Delta q_0, b^0_{cm} + \Delta b_{cm})$ lies in the interior of $\mathcal{W}(q_0 + \Delta q_0)$ for all sufficiently small $\Delta q_0 \in T_{q_0}\mathcal{S}$ and sufficiently small Δb_{cm}. Thus, $\nu = 0$ and $I^- = \emptyset$ at the new equilibrium, and $q_0 + \Delta q_0$ forms a local minimum of $U(q,b^0_{cm} + \Delta b_{cm})$. □

The next proposition provides a geometric test for $I^- = \emptyset$ on the stratified set that forms the object's free c-space $\mathcal{F}$. The proposition uses the following notion of *dual cones.* Let $\mathcal{W}$ be a convex cone of wrenches based at $\mathcal{B}$'s wrench space origin. Let C be a convex cone of tangent vectors based at $\mathcal{B}$'s tangent space origin. Then $\mathcal{W}$ is said to be *dual* to C if every $\boldsymbol{w}\in\mathcal{W}$ satisfies $\boldsymbol{w}\cdot\dot{q}\le 0$ for all $\dot{q}\in C$.

Proposition 11.4 *Let $q_0\in\mathcal{S}$ be a critical point of the object's gravitational potential, $U(q)$, where $\mathcal{S}$ is the stratum of $\mathcal{F}$ associated with supporting bodies $\mathcal{O}_1,\ldots,\mathcal{O}_k$. A necessary condition for $I^- = \emptyset$ is that q_0 forms a* **feasible equilibrium stance***:*

$$\nabla U(q_0) = \sum_{i=1}^{k}\lambda_i\hat{\eta}_i(q_0) \qquad \lambda_1,\ldots,\lambda_k\ge 0,$$

where $\hat{\eta}_1(q_0),\ldots,\hat{\eta}_k(q_0)$ are the supporting bodies c-obstacle outward unit normals at q_0. A sufficient condition for $I^- = \emptyset$ is that q_0 forms an **essential equilibrium stance.**

Proof: Let us first prove the necessary condition for $l^- = \emptyset$. The lower half link is defined as $l^- = E(q_0) \cap U^{-1}(c_0 - \epsilon)$, where $E(q_0)$ is the normal slice of $\mathcal{F}$ at q_0 and $c_0 = U(q_0)$. We may assume that $E(q_0)$ is the intersection of a small k-dimensional disc normal to the stratum $\mathcal{S}$ at q_0 with $\mathcal{F}$. If l^- is empty, $U(q)$ must be non-decreasing along any c-space path that starts at q_0 and lies in $E(q_0)$. Let $C(q_0)$ denote the collection of tangent vectors that are based at q_0 and point into $E(q_0)$. This collection can be characterized as follows. The object's *first-order free motions* at q_0 form the convex cone of tangent vectors:

$$M^1_{1\ldots k}(q_0) = \{\dot{q} \in T_{q_0}\mathbb{R}^m : \hat{\eta}_i(q_0) \cdot \dot{q} \geq 0 \text{ for } i = 1 \ldots k\}. \tag{11.26}$$

Hence, $C(q_0) = M^1_{1\ldots k}(q_0) \cap \text{span}\{\hat{\eta}_1(q_0), \ldots, \hat{\eta}_k(q_0)\}$, which forms a *sub-cone* of $M^1_{1\ldots k}(q_0)$.

Let $q(t)$ be a c-space path that starts at q_0 and lies in $E(q_0)$. Its tangent $\dot{q}(0) = \dot{q}$ lies in $C(q_0)$. Since l^- is empty, by the chain rule $\frac{d}{dt}\big|_{t=0} U(q(t)) = \nabla U(q_0) \cdot \dot{q} \geq 0$ for *all* $\dot{q} \in C(q_0)$. It follows that $\nabla U(q_0)$ lies in the wrench cone *dual* to the negated cone $-C(q_0)$. Let us therefore show that the cone *dual* to $-C(q_0)$ is the net wrench cone $\mathcal{W}(q_0) = \{\sum_{i=1}^k \lambda_i \hat{\eta}_i(q_0) : \lambda_1, \ldots, \lambda_k \geq 0\}$, associated with the supporting contacts.

Each wrench $\boldsymbol{w} \in \mathcal{W}(q_0)$ acts as a covector on tangent vectors $\dot{q} \in T_{q_0}\mathbb{R}^m$. Moreover, all wrenches $\boldsymbol{w} \in \mathcal{W}(q_0)$ satisfy $\boldsymbol{w} \cdot \dot{q} = 0$ for all $\dot{q} \in T_{q_0}\mathcal{S}$. The wrenches of $\mathcal{W}(q_0)$ can thus be interpreted as acting only on the orthogonal complement of $T_{q_0}\mathcal{S}$ in $T_{q_0}\mathbb{R}^m$, a subspace denoted $T^{\perp}_{q_0}\mathcal{S}$. The cone of tangent vector $C(q_0)$ is a subset of $T^{\perp}_{q_0}\mathcal{S}$. Based on Eq. (11.26), the cone $C(q_0)$ is *dual* to the negated wrench cone $-\mathcal{W}(q_0)$:

$$\forall \boldsymbol{w} \in -\mathcal{W}(q_0) \quad \boldsymbol{w} \cdot \dot{q} \leq 0 \quad \text{for all } \dot{q} \in C(q_0).$$

Since $C(q_0)$ is dual to $-\mathcal{W}(q_0)$, $-C(q_0)$ is dual to $\mathcal{W}(q_0)$ when the wrenches of $\mathcal{W}(q_0)$ are interpreted as acting on the subspace $T^{\perp}_{q_0}\mathcal{S}$. Since $\nabla U(q_0)$ is known to lie in the cone dual to $-C(q_0)$, $\nabla U(q_0)$ lies in $\mathcal{W}(q_0)$, which proves the necessity of a feasible equilibrium stance.

Let us next prove the sufficient condition for $l^- = \emptyset$. Consider, for simplicity, the case of $k \leq m$ contacts, where $m = 3$ or 6. In this case all coefficients $\lambda_1, \ldots, \lambda_k$ are strictly positive at an essential equilibrium stance. Since $T^{\perp}_{q_0}\mathcal{S}$ is the orthogonal complement of $T_{q_0}\mathcal{S}$ in the ambient tangent space $T_{q_0}\mathbb{R}^m$, any non-zero tangent vector $\dot{q} \in T^{\perp}_{q_0}\mathcal{S}$ satisfies $\hat{\eta}_i \cdot \dot{q} \neq 0$ for some $i \in \{1, \ldots, k\}$. Since $C(q_0)$ is a subset of $T^{\perp}_{q_0}\mathcal{S}$, every $\dot{q} \in C(q_0)$ satisfies $\hat{\eta}_i(q_0) \cdot \dot{q} \geq 0$ for all $i = 1 \ldots k$. Based on these two facts, every $\dot{q} \in C(q_0)$ satisfies $\hat{\eta}_i(q_0) \cdot \dot{q} > 0$ for some $i \in \{1, \ldots, k\}$. Since $\frac{d}{dt}\big|_{t=0} U(q(t)) = \nabla U(q_0) \cdot \dot{q} = \sum_{i=1}^k \lambda_i(\hat{\eta}_i(q_0) \cdot \dot{q})$ by the chain rule, it must be true that $\frac{d}{dt}\big|_{t=0} U(q(t)) > 0$. Thus, $U(q)$ is strictly increasing along any c-space path that starts at q_0 and lies in $E(q_0)$, which implies that $l^- = \emptyset$. □

Exercises

Section 11.1

Exercise 11.1: Verify that the gravitational potential $U(q)$ cannot have critical points in the interior of the free c-space $\mathcal{F}$.

Exercise 11.2: Explain why every zero-dimensional stratum of $\mathcal{F}$ automatically forms a critical point of U (consider the tangent space to a zero-dimensional stratum of $\mathcal{F}$).

Exercise 11.3: Explain why the *full set* of critical points of $U(q)$ in $\mathcal{F}$ correspond to equilibrium stances whose supporting contacts apply any combination of bi-directional forces along the contact normals.

Exercise 11.4: Consider the perturbed two-contact stance depicted in Figure 11.4(b). The contact forces are collinear with the vectors $n_1 = (0,1)$ and $n_2 = (\epsilon, 1)$, where ϵ is a small positive parameter. Verify that ∇U is normal to the stratum $\mathcal{S}_1 \cap \mathcal{S}_2$ at this stance but is *not* normal to the neighboring strata $\mathcal{S}_1$ and $\mathcal{S}_2$ for all sufficiently small $\epsilon > 0$.

Section 11.2

Exercise 11.5: A rigid object $\mathcal{B}$ is supported by stationary bodies $\mathcal{O}_1, \ldots, \mathcal{O}_k$ at a feasible equilibrium stance configuration, q_0. Prove that if q_0 is a *local minimum* of $U(q)$ in $\mathcal{F}$, the object must remain at a zero-velocity equilibrium state as long as there are no external disturbances.

Solution: When q_0 is a local minimum of $U(q)$, $(q_0, \vec{0})$ is a local minimum of the object's total mechanical energy, $E(q,\dot{q}) = \mathcal{K}(q,\dot{q}) + U(q)$. It follows that any feasible state space trajectory $(q(t),\dot{q}(t))$ that starts at $(q_0,\vec{0})$ satisfies $\frac{d}{dt}E(q(t),\dot{q}(t)) \geq 0$ for $t \in [0,\epsilon)$. However, by conservation of energy $\frac{d}{dt}E(q(t),\dot{q}(t)) \leq 0$ during this time interval. Hence, $E(q(t),\dot{q}(t)) = E(q_0,\vec{0}) = U(q_0)$ for $t \in [0,\epsilon)$, which implies that $(q(t),\dot{q}(t)) = (q_0,\vec{0})$ for all $t \geq 0$. ○

Exercise 11.6: Figure 11.5 depicts a two-contact equilibrium stance of an object $\mathcal{B}$, located at a local minimum of $U(q)$ that is therefore locally stable. Show that the same instantaneous free motions can also appear when $\mathcal{B}$ is located at an unstable saddle point of $U(q)$.

Section 11.3

Exercise 11.7: Do the essential equilibrium stances form non-degenerate critical points of the gravitational potential $U(q)$ in the object's free c-space $\mathcal{F}$?

Solution: The non-degenerate critical points of $U(q)$ in $\mathcal{F}$ must satisfy two conditions. First, $\nabla U(q_0)$ must *not* be normal to any of the neighboring strata of $\mathcal{F}$ meeting at q_0. This condition is met by an essential equilibrium stance. Second, the Hessian matrix $D^2U(q_0)$ must be non-degenerate along the stratum of $\mathcal{F}$ containing q_0. This condition is *not* automatically met by the essential equilibrium stances. ○

Exercise 11.8*: Verify that an equilibrium stance supported by $k \leq m$ contacts ($m = 3$ or 6) is essential iff $\hat{\eta}_1(q_0), \ldots, \hat{\eta}_k(q_0)$ are linearly independent, such that the coefficients $\lambda_1, \ldots, \lambda_k$ in the equilibrium stance condition $\sum_{i=1}^{k} \lambda_i \hat{\eta}_i(q_0) - \nabla U(q_0) = \vec{0}$ are strictly positive.

Exercise 11.9: Based on the previous exercise, explain why the essential equilibrium stances are generic.

Exercise 11.10: Explain why $I^{-}=\emptyset$ at any single-support stance.

Exercise 11.11: Consider the three-contact equilibrium stance depicted in Figure 11.6. Explain why $I^{-}=\emptyset$ for this stance.

Section 11.4

Exercise 11.12: Let $U(q)=\mathrm{mg}\Big(\boldsymbol{e}\cdot x_{\mathrm{cm}}(q)\Big)$ be the gravitational potential of a 3-*D* object $\mathcal{B}$, where $\boldsymbol{e}=(0,0,1)$. Obtain the formulas for $\nabla U(q)$ and $D^2U(q)$ specified in Lemma 11.7.

Solution: Assume the scaling $\mathrm{m}g=1$. Let $q(t)=(d(t),\theta(t))$ be a c-space trajectory, and let $\dot q(t)=(v,\omega)$ be its tangent vector. Since $\boldsymbol{x}_{\mathrm{cm}}(q)=R(\theta)b_{cm}+d$, $\dot{\boldsymbol{x}}_{\mathrm{cm}}=\dot R b_{cm}+v$, and subsequently $\dot U=(\boldsymbol{e}\cdot(\dot R b_{cm}+v))$. But $\dot R b_{cm}=\dot R R^T R b_{cm}=\omega\times R b_{cm}=\omega\times\boldsymbol{\rho}_{\mathrm{cm}}$, where we used the skew-symmetry of the matrix $\dot R R^T$ and the definition of ω in terms of this matrix. Thus, $\dot U=(\boldsymbol{e}\cdot(\omega\times\boldsymbol{\rho}_{\mathrm{cm}}+v))$. Using the triple scalar product identity $\boldsymbol{e}\cdot(\omega\times\boldsymbol{\rho}_{\mathrm{cm}})=(\boldsymbol{\rho}_{\mathrm{cm}}\times\boldsymbol{e})\cdot\omega$:

$$\frac{d}{dt}U(q(t))=\begin{pmatrix}\boldsymbol{e}^T & (\boldsymbol{\rho}_{\mathrm{cm}}\times\boldsymbol{e})^T\end{pmatrix}\begin{pmatrix}v\\ \omega\end{pmatrix}. \tag{11.27}$$

Since $\frac{d}{dt}U(q(t))=\nabla U(q)\cdot\dot q$ by the chain rule, it follows from Eq. (11.27) that $\nabla U(q)=(\boldsymbol{e},\boldsymbol{\rho}_{\mathrm{cm}}\times\boldsymbol{e})$. In order to compute the Hessian matrix, $D^2U(q)$, consider the time derivative of $\nabla U(q(t))$:

$$\frac{d}{dt}\nabla U(q(t))=\begin{pmatrix}0\\ \dot R b_{cm}\times\boldsymbol{e}\end{pmatrix}=\begin{pmatrix}0\\ (\omega\times\boldsymbol{\rho}_{\mathrm{cm}})\times\boldsymbol{e}\end{pmatrix}. \tag{11.28}$$

Since $\frac{d}{dt}\nabla U(q(t))=D^2U(q)\dot q$ by the chain rule, $\dot q^T D^2U(q)\dot q=\dot q^T\frac{d}{dt}\nabla U(q(t))=w\cdot((\omega\times\boldsymbol{\rho}_{\mathrm{cm}})\times\boldsymbol{e})$, according to Eq. (11.28). Using the triple scalar product identity, $\dot q^T D^2U(q)\dot q=(\omega\times\boldsymbol{\rho}_{\mathrm{cm}})\cdot(\boldsymbol{e}\times\omega)=\omega^T[\boldsymbol{\rho}_{\mathrm{cm}}\times][\boldsymbol{e}\times]\omega=\omega^T([\boldsymbol{\rho}_{\mathrm{cm}}\times][\boldsymbol{e}\times])_s\omega$, which gives the lower-right term in Formula (11.9) for $D^2U(q)$. ○

Exercise 11.13: Obtain Formulas (11.10)–(11.11) for $\nabla U(q)$ and $D^2U(q)$ for a 2-D rigid object $\mathcal{B}$.

Solution: Substituting $\boldsymbol{\rho}_{\mathrm{cm}}=(x_{cm},y_{cm},0)$ and $\boldsymbol{e}=(0,1,0)$ in Eq. (11.8) gives Formula (11.10) for $\nabla U(q)$. Denote by $\vec\omega$ the angular velocity vector in $\mathbb{R}^3$, whose axis is perpendicular to the plane and whose magnitude is the scalar ω. Let $\boldsymbol{e}=(0,1,0)$ denote the embedding of the vertical direction in 2-D, $\boldsymbol{e}=(0,1)$, in $\mathbb{R}^3$. Substituting for $\vec\omega$ and $\boldsymbol{e}=(0,1,0)$ in (11.9) gives $\dot q^T D^2U(q)\dot q=(\vec\omega\times\boldsymbol{\rho}_{\mathrm{cm}})\cdot(\boldsymbol{e}\times\vec\omega)=\boldsymbol{\rho}_{\mathrm{cm}}\cdot(\vec\omega\times(\vec\omega\times\boldsymbol{e}))$. Use of the identity $[a\times]^2=aa^T-\|a\|^2I$, gives $\vec\omega\times(\vec\omega\times\boldsymbol{e})=[\vec\omega\times]^2\boldsymbol{e}=[\vec\omega\vec\omega^T-\omega^2I]\boldsymbol{e}$. Since $\vec\omega\perp\boldsymbol{e}$, $\vec\omega\times(\vec\omega\times\boldsymbol{e})=-\omega^2\boldsymbol{e}$. Hence, $\dot q^T D^2U(q)\dot q=-\omega^2\boldsymbol{\rho}_{\mathrm{cm}}\cdot\boldsymbol{e}$, which gives the lower-right term in Formula (11.11) for $D^2U(q)$. ○

Section 11.5 (2-D stances)

Exercise 11.14: The stance equilibrium region, $\mathcal{E}(q_0)$, is generically a vertical line for $k=1,2$ supporting contacts and a vertical strip or half plane for $k=3$ supporting contacts. Characterize the region $\mathcal{E}(q_0)$ relative to the contacts for each of these cases.

Exercise 11.15: The stance equilibrium region, $\mathcal{E}(q_0)$, can span the entire vertical plane for $k \geq 4$ supporting contacts. Do the supporting contacts form a *first-order immobilizing grasp* at such stances?

Exercise 11.16: Verify the stability rules for the single-support equilibrium stances listed in the first, second and third rows of Table 11.2.

Exercise 11.17: A 2-D rigid object $\mathcal{B}$ is supported by three contacts having vertical contact normals, $n_1 = n_2 = n_3 = \boldsymbol{e}$. Determine the stance equilibrium region, based on the intersection of the affine line L with the net wrench cone $\mathcal{W}(q_0)$.

Exercise 11.18: A 2-D rigid object $\mathcal{B}$ is supported at a feasible three-contact equilibrium stance, such that two of the three contact normals are antiparallel, say $n_1 = -n_2$. Determine the equilibrium region of this stance.

Solution: Since the stance forms a feasible equilibrium, the vertical direction $\boldsymbol{e}$ must lie in the positive span of two contact normals, say n_1 and n_3. Denote by l'_{13} the vertical line passing through the intersection point of the lines l_1 and l_3. Then $\mathcal{E}(q_0)$ is the *vertical half plane* bounded by l'_{13} that does not contain the intersection point of l_2 and l_3. ○

Exercise 11.19*: Under what condition the feasible equilibrium region of a three-contact stance forms a vertical half plane?

Solution: The vertical lines that form the stance's equilibrium region, $\mathcal{E}(q_0)$, correspond to intersection points of the affine line L with the net wrench cone $\mathcal{W}(q_0)$. The horizontal cross section of $\mathcal{E}(q_0)$ is unbounded iff L intersects $\mathcal{W}(q_0)$ along a semi-infinite interval. The affine line L is perpendicular to the (f_x, f_y) plane in $\mathcal{B}$'s wrench space. Hence, L intersects $\mathcal{W}(q_0)$ along a semi-infinite interval iff the projections of the contact wrenches on the (f_x, f_y) plane positively span the entire (f_x, f_y) plane. The projections of the contact wrenches on the (f_x, f_y) plane are the contact normals n_1, n_2, and n_3. The feasible equilibrium region thus forms a vertical half plane iff the three contact normals positively span the (f_x, f_y) plane. ○

Section 11.6 (3-D stances)

Exercise 11.20: Verify that a 3-D stance forms a feasible equilibrium iff $\boldsymbol{e}$ lies in the positive span of the object's inward contact normals, such that the horizontal projections of the contact forces form a feasible *equilibrium* within the horizontal (x, y) plane.

Exercise 11.21: Explain why the horizontal cross section of the stance stability region always forms a convex polygon in the (x,y) plane.

Exercise 11.22: A 3-D object $\mathcal{B}$ lies on horizontal supports, not necessarily located at the same heights. Does the *support polygon* form the horizontal cross section of the stance equilibrium region?

Hint: Consider the line that passes through two contact points. If the remaining contacts generate the *same* moment sign about this line, the object's center of mass must generate an opposing moment sign for an equilibrium to exist.

Exercise 11.23: The horizontal cross section polygon of the stance stability region always lies within the convex hull of the (x,y) projection of the contacts – true or false?

Exercise 11.24*: A 3-D object $\mathcal{B}$ is supported by two contacts having skew contact normals. If the stance has a nonempty equilibrium line, $\mathcal{E}(q_0) \neq \emptyset$, under what condition does it have a nonempty stability half line $\mathcal{E}_S(q_0)$?

Solution: If $\mathcal{E}(q_0) \neq \emptyset$, the contact normal lines, l_1 and l_2, lie in a common vertical plane, $\mathcal{P}$. Since l_1 and l_2 have skew contact normals, they intersect at a point $p \in \mathcal{P}$, and $\mathcal{E}(q_0)$ is the vertical line passing through p. To determine the stability region, $\mathcal{E}_S(q_0) \subseteq \mathcal{E}(q_0)$, we have to characterize the tangent space $T_{q_0}\mathcal{S}$ and then evaluate $\kappa_U(q_0,\dot{q})$ on this subspace. At a two-contact stance, q_0 lies in the stratum of $\mathcal{F}$ formed by the intersection of two c-obstacle boundaries, $\mathcal{S}_1$ and $\mathcal{S}_2$. Each c-obstacle boundary is a five-dimensional manifold in $\mathbb{R}^6$. Hence, $\mathcal{S}$ forms a four-dimensional manifold in $\mathbb{R}^6$. The tangent space $T_{q_0}\mathcal{S} = T_{q_0}\mathcal{S}_1 \cap T_{q_0}\mathcal{S}_2$ forms a four-dimensional linear subspace of $T_{q_0}\mathbb{R}^6$. When $\mathcal{B}$'s frame origin is located at p, the tangent space $T_{q_0}\mathcal{S}$ has the following basis vectors. Let v be a horizontal unit vector perpendicular to the vertical plane $\mathcal{P}$ at p. Consider the four vectors in the ambient tangent space $T_{q_0}\mathbb{R}^6$:

$$u_1 = (v,\vec{0}) \quad u_2 = (\vec{0}, v \times \boldsymbol{e}) \quad u_3 = (\vec{0}, v) \quad u_4 = (\vec{0}, \boldsymbol{e}).$$

The vector u_1 represents instantaneous translation of $\mathcal{B}$ in the horizontal direction v. The other three vectors represent instantaneous rotation of $\mathcal{B}$ about three orthogonal axes passing through p. All four vectors are orthogonal to the c-obstacle normals, $\eta_1(q_0) = (n_1,\vec{0})$ and $\eta_2(q_0) = (n_2,\vec{0})$, and they are linearly independent. Hence, they form a basis for the tangent space $T_{q_0}\mathcal{S}$. In the following, $B = [u_1\ u_2\ u_3\ u_4]$ is the 6×4 matrix whose columns are the basis vectors for $T_{q_0}\mathcal{S}$.

Our goal is to derive an expression for the 4×4 matrix representing $\kappa_U(q_0,\dot{q})$ on $T_{q_0}\mathcal{S}$. The gravity relative curvature form is given by

$$\kappa_U(q_0,\dot{q}) = \lambda_1 \kappa_1(q_0,\dot{q}) + \lambda_2 \kappa_2(q_0,\dot{q}) - \dot{q}^T D^2 U(q_0) \dot{q} \qquad \dot{q} \in T_{q_0}\mathcal{S},$$

where λ_1 and λ_2 are determined by the equilibrium condition of Eq. (11.2). First consider the Hessian $D^2U(q_0)$. The object's center of mass lies on the vertical line l passing through p, while $\mathcal{B}$'s frame origin is located at p. Hence, $\rho_{\text{cm}} = (0,0,z_{\text{cm}})$, such that $z_{\text{cm}} < 0$ at points of l below p. Substituting for ρ_{cm} in the Hessian formula (11.9), the restriction of $D^2U(q_0)$ to $T_{q_0}\mathcal{S}$ is given by

$$B^T D^2U(q_0)B = \begin{bmatrix} 0 & 0 & 0 & 0 \\ 0 & -z_{\text{cm}} & 0 & 0 \\ 0 & 0 & -z_{\text{cm}} & 0 \\ 0 & 0 & 0 & 0 \end{bmatrix}. \tag{11.29}$$

Denote by L_i the 6×6 matrix of the c-obstacle curvature form $\kappa_i(q_0, \dot{q})$, specified in Eq. (11.14). The 4×4 matrix representing $\kappa_U(q_0, \dot{q})$ on $T_{q_0}\mathcal{S}$, denoted C, is given by

$$C = \lambda_1 B^T L_1 B + \lambda_2 B^T L_2 B + z_{\text{cm}} \begin{bmatrix} 0 & 0 & 0 & 0 \\ 0 & 1 & 0 & 0 \\ 0 & 0 & 1 & 0 \\ 0 & 0 & 0 & 0 \end{bmatrix} \qquad \lambda_1, \lambda_2 \geq 0. \tag{11.30}$$

Stance stability requires that the matrix C be negative definite: $\dot{q}^T C \dot{q} < 0$ for all $\dot{q} \in T_{q_0}\mathcal{S}$. The behavior of C on the subspace spanned by u_1 and u_4 depends only on the curvature of the supporting bodies at the contacts. If C is negative definite on the latter subspace, $\mathcal{E}_S(q_0)$ forms a *lower half line* of the vertical line l passing through p, since z_{cm} appears linearly in Eq. (11.30) such that $z_{\text{cm}} < 0$ at points of l below p. ○

Exercise 11.25: Derive Formula (11.20) for the horizontal position of a three-contact stance equilibrium line, assuming skew contact normals.

Solution: When the object $\mathcal{B}$ is supported by contact forces $f_i = \lambda_i n_i$ for $i = 1, 2, 3$, the torque part of the equilibrium stance condition of Eq. (11.2) is given by

$$\sum_{i=1}^{3} \lambda_i (\rho_i \times n_i) = \rho_{\text{cm}} \times \boldsymbol{e}. \tag{11.31}$$

Let $\mathcal{B}$'s frame origin be located on the vertical line l passing through l_1, l_2, and l_3. Then h_i is the height of the point where the contact line l_i intersects the line l relative to $\mathcal{B}$'s origin. The vector ρ_i from $\mathcal{B}$'s origin to x_i can be written as $\rho_i = h_i \boldsymbol{e} + h_i' n_i$, where h_i' is the signed distance of the intersection point of l_i with l from x_i along the contact line l_i. Substituting $\rho_i = h_i \boldsymbol{e} + h_i' n_i$ in Eq. (11.31) gives $\boldsymbol{e} \times (\sum_{i=1}^{3} h_i \lambda_i n_i + \rho_{\text{cm}}) = \vec{0}$. The net torque vector $\sum_{i=1}^{3} h_i \lambda_i n_i + \rho_{\text{cm}}$ is therefore vertical, and its horizontal coordinates must be zero as specified in Formula (11.20). ○

Exercise 11.26: When a 3-D object $\mathcal{B}$ is supported by three contacts at a feasible equilibrium stance, the contact normal lines l_1, l_2, l_3, and the gravity line l_g lie on a common regulus. Determine three generators for this regulus.

Solution: Consider the vertical line l that intersects the contact normal lines l_1, l_2, and l_3. Since l is parallel to l_g, it intersects the gravity line at infinity. Hence, l is one generator for the regulus. Two more generators can be found as follows. Assume skew contact normal lines l_1, l_2, and l_3. Consider the projection of l_2 and l_3 on the plane orthogonal to l_1. Since the contact lines are skew, the projections of l_1 and l_2 must intersect at a point.

Hence, there exists a line parallel to l_1, denoted l'_1, which intersects l_2 and l_3. Since l'_1 is parallel to l_1, it intersects l_1 at infinity. The net torque affecting $\mathcal{B}$ about l'_1 must be zero at the equilibrium stance. Hence, the gravity line, l_g, must also intersect the line l'_1. The line l'_1 is therefore another generator for the regulus. One can similarly construct a line l'_2 (or l'_3) that is parallel to l_2 and intersects the other three lines. The lines l', l'_1, and l'_2 are three generators for the regulus. ○

Exercise 11.27*: A 3-D object $\mathcal{B}$ is supported by three contacts having vertical contact normals, $n_1 = n_2 = n_3 = \boldsymbol{e}$. The stance equilibrium region is the vertical triangular prism spanned by the three contacts. Show that the stance stability region, $\mathcal{E}_S(q_0)$, forms a vertical sub-prism of $\mathcal{E}(q_0)$.

Solution: The object's configuration point, q_0, lies on a three-dimensional stratum $\mathcal{S}$ of $\mathcal{F}$. The tangent space $T_{q_0}\mathcal{S}$ is three-dimensional and can be parametrized as follows. Let v be a horizontal unit vector. The linearly independent tangent vectors $u_1 = (v, \vec{0})$, $u_2 = (v \times \boldsymbol{e}, \vec{0})$ and $u_3 = (\vec{0}, \boldsymbol{e})$ represent instantaneous horizontal translations and rotation of $\mathcal{B}$ about the vertical direction $\boldsymbol{e}$. Each tangent vector u_i is orthogonal to the three c-obstacle normals, $\eta_i(q_0) = (\boldsymbol{e}, \boldsymbol{\rho}_i \times \boldsymbol{e})$ for $i = 1,2,3$. Hence u_1, u_2, and u_3 form a basis for $T_{q_0}\mathcal{S}$. Using the Hessian formula (11.9), we find that

$$u_i^T D^2U(q_0)u_j = 0 \qquad i, j \in \{1,2,3\}.$$

Thus $D^2U(q_0)$ vanishes on the tangent space $T_{q_0}\mathcal{S}$. This observation can be interpreted as follows. At an equilibrium stance, the gradient $\nabla U(q_0)$ is orthogonal to the tangent space $T_{q_0}\mathcal{S}$, $\nabla U(q_0) \cdot \dot{q} = 0$ for all $\dot{q} \in T_{q_0}\mathcal{S}$. This means that any combination of instantaneous horizontal translation and instantaneous rotation about $\boldsymbol{e}$ does not change the height of $\mathcal{B}$'s center of mass. The vanishing of $D^2U(q_0)$ on $T_{q_0}\mathcal{S}$ implies that any velocity-and-acceleration pair, $(\dot{q}, \ddot{q}) \in T_{q_0}\mathcal{S} \times \mathbb{R}^6$, also does not change the height of $\mathcal{B}$'s center of mass at q_0. Since $D^2U(q_0)$ vanishes on $T_{q_0}\mathcal{S}$, the gravity relative curvature form is given by

$$\kappa_U(q_0, \dot{q}) = \sum_{i=1}^{3} \lambda_i \kappa_i(q_0, \dot{q}) \qquad \dot{q} \in T_{q_0}\mathcal{S}.$$

Based on Formula (11.14), the ith c-obstacle curvature form can be written as $\kappa_i(q_0, \dot{q}) = \frac{1}{c_i}\tilde{\kappa}_i(q_0, \dot{q})$, where $c_i = \sqrt{1 + \|\rho_i \times n_i\|^2}$. The gravity relative curvature form can thus be written as

$$\kappa_U(q_0, \dot{q}) = \sum_{i=1}^{3} \sigma_i \tilde{\kappa}_i(q_0, \dot{q}) \qquad \dot{q} \in T_{q_0}\mathcal{S},$$

where $\sigma_i = \lambda_i / c_i$ for $i = 1,2,3$. The coefficients σ_i for $i = 1,2,3$ depend on the horizontal location of $\mathcal{B}$'s center of mass, $(x_{\text{cm}}, y_{\text{cm}})$, but *not* on the height of $\mathcal{B}$'s center of mass, z_{cm}. The stance stability condition can therefore be written as

$$\kappa_U(q_0, \dot{q}) = \sum_{i=1}^{3} \sigma_i(x_{\text{cm}}, y_{\text{cm}}) \tilde{\kappa}_i(q_0, \dot{q}) < 0 \qquad \dot{q} \in T_{q_0}\mathcal{S}. \qquad (11.32)$$

The height of $\mathcal{B}$'s center of mass, z_{cm}, does not appear in Eq. (11.32). The stance stability region associated with three horizontal supports therefore generically forms a *vertical sub-prism* of $\mathcal{E}(q_0)$. ○

Exercise 11.28: A 3-D object $\mathcal{B}$ lies on three horizontal and locally flat supports. Show that the gravity relative curvature form satisfies $\kappa_U(q_0,\dot{q})=0$ for all $\dot{q}\in T_{q_0}\mathcal{S}$, at all center-of-mass locations in $\mathcal{E}(q_0)$ (thus implying *neutral stability* with respect to local object motions on horizontal and locally flat supports).

Solution: The Hessian $D^2U(q_0)$ vanishes identically on $T_{q_0}\mathcal{S}$. Hence, $\kappa_U(q_0,\dot{q})=\sum_{i=1}^{3}\lambda_i\kappa_i(q_0,\dot{q})$ for $\dot{q}\in T_{q_0}\mathcal{S}$. Consider the individual c-obstacle curvature forms, $\kappa_i(q_0,\dot{q})$ for $i=1,2,3$. Let L_i be the 3×3 matrix representing the restriction of $\kappa_i(q_0,\dot{q})$ to $T_{q_0}\mathcal{S}$. Substituting $L_{\mathcal{O}_i}=O$ for the flat supports in Formula (11.14) for $\kappa_i(q_0,\dot{q})$ gives

$$L_i=\frac{1}{c_i}\,[u_1\;u_2\;u_3]^T\left(\begin{bmatrix} O & O\\ O & -[\rho_i\times]^T\left[L_{\mathcal{B}_i}\right]^{-1}[\rho_i\times]\end{bmatrix}\right.$$
$$\left.+\begin{bmatrix} O & O\\ O & -\left([\rho_i\times]^T[e\times]\right)_s\end{bmatrix}\right)[u_1\;u_2\;u_3]=O,$$

where $u_1=(v,\vec{0})$, $u_2=(v\times e,\vec{0})$ and $u_3=(\vec{0},e)$ represent instantaneous horizontal translations and rotation of $\mathcal{B}$ about e. To explain the vanishing of $\kappa_i(q_0,\dot{q})$, assume small planar patches at the supporting contacts. Any local motion of $\mathcal{B}$ consisting of horizontal translation and rotation about the vertical axis maintains contact with the flat supports. It follows that at every point $q\in\mathcal{S}$ in the vicinity of q_0, the tangent space $T_q\mathcal{S}$ is embedded in the stratum $\mathcal{S}$. The stratum $\mathcal{S}$ is thus locally flat at q_0, which explains the vanishing of the c-obstacle curvatures at q_0. All the terms comprising $\kappa_U(q_0,\dot{q})$ thus vanish on $T_{q_0}\mathcal{S}$. ○

Exercise 11.29: A 3-D rigid object $\mathcal{B}$ is supported by six contacts located on a common horizontal plane, such that the contact normals are *upward pointing,* $e\cdot n_i\geq 0$ for $i=1\ldots 6$. Prove that the stance equilibrium polygon must lie inside the stance's support polygon.

Exercise 11.30*: A 3-D rigid object $\mathcal{B}$ is supported by six contacts. Under what condition the stance equilibrium polygon is *unbounded?*

Solution: The vertical lines that form of the stance equilibrium region, $\mathcal{E}(q_0)$, correspond to intersection points of the affine plane L with the net wrench cone $\mathcal{W}(q_0)$. The stance equilibrium polygon is unbounded iff L intersects $\mathcal{W}(q_0)$ along at least one semi-infinite line. Let V be the subspace spanned by (f_x,f_y,f_z) in $\mathcal{B}$'s wrench space. The affine plane L lies in the five-dimensional linear subspace spanned by $(f_x,f_y,f_z,\tau_x,\tau_y)$ and is perpendicular to V within this subspace. Hence, L intersects $\mathcal{W}(q_0)$ along a semi-infinite line iff the projections of the contact wrenches on V positively span the subspace V which is equivalent to $\mathbb{R}^3$. The projections of the wrenches applied by the supporting contacts on $\mathcal{B}$ are positive multiples of the contact normals $n_1,\ldots,n_6$. The stance equilibrium polygon is thus unbounded iff $\mathcal{B}$'s inward contact normals, placed at a common origin, positively span $\mathbb{R}^3$. ○

References

[1] E. Rimon, R. Mason, J. W. Burdick and Y. Or, "A general stance stability test based on stratified morse theory with application to quasistatic locomotion planning," *IEEE Transactions on Robotics and Automation*, vol. 24, pp. 626–641, 2008.

[2] J. C. Trinkle, A. O. Farahat and P. F. Stiller, "Second-order stability cells of a frictionless rigid body grasped by rigid fingers," in *IEEE International Conference on Robotics and Automation*, pp. 2815–2821, 1994.

[3] T. Poston and I. Stewart, *Catastrophe Theory and Its Applications*. Pitman, 1978.

[4] D. J. Kriegman, "Computing stable poses of piecewise smooth objects," *CVGIP: Image Understanding*, vol. 55, no. 2, pp. 109–118, 1992.

[5] J. Bernheisel and K. M. Lynch, "Stable transport of assemblies by pushing," *IEEE Transactions on Robotics*, vol. 22, no. 4, 740–750, 2006.

[6] J. Mahler, F. T. Pokorny, Z. McCarthy, A. F. van der Stappen and K. Goldberg, Energy-bounded caging: Formal definition and 2D energy lower bound algorithm based on weighted alpha shapes, *IEEE Robotics and Automation Letters*, pp. 508–515, 2016.

[7] J. Guckenheimer and P. Holmes, *Nonlinear Oscillations, Dynamical Systems, and Bifurcations of Vector Fields*. Springer-Verlag, 1983.

Part III

Frictional Rigid-Body Grasps and Stances

12 Wrench-Resistant Grasps

Chapter 7 introduced two basic types of secure grasps: immobilizing grasps associated with frictionless contacts and wrench resistant grasps associated with frictional contacts. The immobilizing grasps are based on pure rigid-body constraints. The wrench resistant grasps depend on friction effects and involve a much larger set of contact forces afforded by the Coulomb friction model. This chapter highlights the basic properties of the wrench resistant grasps, whose definition is repeated here.

DEFINITION 12.1 (**wrench resistance**) *Let a rigid object $\mathcal{B}$ be held at a configuration q_0 by rigid finger bodies $\mathcal{O}_1, \ldots, \mathcal{O}_k$ via frictional contacts with generalized friction cones $C_1, \ldots, C_k$. The grasp is* **wrench resistant** *if the grasp map, $\mathcal{L}_G : C_1 \times \ldots \times C_k \to T^*_{q_0}\mathbb{R}^m$, forms a surjective mapping onto $\mathcal{B}$'s wrench space, where $m = 3$ in 2-D grasps and $m = 6$ in 3-D grasps.*

A multi-finger grasp is *wrench resistant* when any disturbance wrench that may act on the grasped object $\mathcal{B}$ can be balanced by feasible finger forces and possibly moments at the contacts. There are practical limits to the magnitude of forces and moments that can be applied at the contacts before the finger-joint actuators reach their limits. While such limits are ignored in this chapter, this chapter describes a grasp force optimization technique that can be used to minimize the finger force magnitudes.

Section 12.1 characterizes the wrench resistant grasps in terms of properties of the *grasp matrix,* which represents the grasp map in terms of contact force components. Section 12.2 formulates a wrench resistance test for arbitrary grasps in terms of *linear matrix inequalities* (LMI), for which there exist efficient computational algorithms. Section 12.3 demonstrates the practical utility of the LMI technique on the problem of *grasp force optimization,* which can be formulated as a convex optimization problem. Section 12.4 takes a control system point of view of wrench resistant grasps. A multi-finger grasp can be viewed as a control system whose feasible finger forces provide the control inputs. A grasp is *locally controllable* when feasible finger forces can steer the grasped object to arbitrary desired configurations in a local neighborhood of the initial grasp. We will see that wrench resistant grasps are locally controllable and possess a variety of stabilizing feedback control laws.

12.1 Wrench Resistance and Internal Grasp Forces

We know from Chapter 7 that all wrench resistant grasps must form feasible equilibrium grasps of the object $\mathcal{B}$. This necessary condition is now extended into a full test for assessing the wrench resistance ability of a given grasp. Let the object $\mathcal{B}$ be held by finger bodies $\mathcal{O}_1, \ldots, \mathcal{O}_k$ at a configuration q_0. The grasp matrix is based on selection of contact frames, attached to the grasped object $\mathcal{B}$ at the contacts (see Chapter 4). Using such contact frames, one parametrizes each finger force (and possibly the finger moment about the contact normal) in terms of p independent components. The composite vector of all k finger force components is given by $\vec{f} = (f_1, \ldots, f_k) \in \mathbb{R}^{kp}$. The *grasp matrix* is the $m \times kp$ matrix $G(q_0)$ that maps the finger contact force components to a net object wrench: $\boldsymbol{w} = G(q_0)\vec{f}$, where $m = 3$ in 2-D grasps and $m = 6$ in 3-D grasps. The null space of $G(q_0)$ consists of *internal grasp forces* that generate zero net wrench on the object $\mathcal{B}$.

The following theorem describes necessary and sufficient conditions for wrench resistant grasps. A *non-marginal equilibrium grasp* is defined as an equilibrium grasp that can be maintained with finger forces that lie in the interior of the respective friction cones.

THEOREM 12.1 **(wrench resistance condition)** *Let a rigid object $\mathcal{B}$ be held at a frictional k-finger equilibrium grasp. The grasp is* **wrench resistant** *iff (i) the grasp matrix $G(q_0)$ is surjective on $\mathcal{B}$'s wrench space, and (ii) the grasp matrix $G(q_0)$ possesses internal grasp forces that lie in the interior of the composite friction cone $C_1 \times \cdots \times C_k$.*[1]

Proof: Let us first verify that Conditions (i) and (ii) imply wrench resistance. Condition (i) specifies that $G(q_0)$ maps the finger contact forces surjectively onto $\mathcal{B}$'s wrench space. Hence, for any external wrench that acts on $\mathcal{B}$, $\boldsymbol{w}_{ext}$, there exists a particular vector of finger contact forces, $\vec{f}_p \in \mathbb{R}^{kp}$, which satisfies the equilibrium grasp condition:

$$G(q_0)\vec{f}_p + \boldsymbol{w}_{ext} = \vec{0}. \tag{12.1}$$

The vector $\vec{f}_p$ that solves Eq. (12.1) need not be feasible. Condition (ii) specifies that the grasp has a vector of internal forces, $\vec{f}_{int} \in \mathbb{R}^{kp}$ such that $G(q_0)\vec{f}_{int} = \vec{0}$, that lies in the interior of the composite friction cone $C_1 \times \cdots \times C_k$. Let us show that $\vec{f}_{int}$ can be used to realize contact forces of the form: $\vec{f} = \vec{f}_p + s \cdot \vec{f}_{int}$, such that $\vec{f}$ lies in $C_1 \times \cdots \times C_k$ for sufficiently large positive s. First note that

$$\lim_{s \to \infty} \frac{\vec{f}_p + s \cdot \vec{f}_{int}}{s} = \vec{f}_{int} \in \mathsf{int}(C_1 \times \cdots \times C_k)$$

where $\mathsf{int}(\cdot)$ denotes set interior. The mapping $\vec{f}(s) = (\vec{f}_p + s \cdot \vec{f}_{int})/s : \mathbb{R}^+ \to \mathbb{R}^{kp}$ is a continuous function of s. Since the interior of $C_1 \times \cdots \times C_k$ is an open set in $\mathbb{R}^{kp}$, its inverse image under $\vec{f}(s)$ is an open set in $\mathbb{R}^+$. The set of feasible contact forces, $C_1 \times \cdots \times C_k$, forms a *convex cone* in $\mathbb{R}^{kp}$. Hence, the inverse image of $C_1 \times \cdots \times C_k$

[1] The *interior* of $C_1 \times \cdots \times C_k$ is the open set $\mathsf{int}(C_1) \times \cdots \times \mathsf{int}(C_k)$, where $\mathsf{int}(\cdot)$ denotes set interior. It consists of all finger forces that lie in the interior of the respective friction cones.

under $\vec{f}(s)$ forms a semi-infinite interval in $\mathbb{R}^+$, parametrized by $s \geq s_{min}$ for some fixed s_{min}. The finger contact forces, $\vec{f}(s) = \vec{f}_p + s \cdot \vec{f}_{int}$, thus form a feasible solution of Eq. (12.1) for all $s \geq s_{min}$, which implies wrench resistance.

To show the necessity of Conditions (i) and (ii), assume that the object $\mathcal{B}$ is held at a wrench resistant grasp. The grasp matrix $G(q_0)$ must be surjective in order that a solution to Eq. (12.1) will exist for arbitrary external wrenches that act on $\mathcal{B}$. To show the necessity of Condition (ii), choose finger contact forces that lie in the *interior* of the friction cones, $\vec{f}_1 \in \text{int}(C_1 \times \cdots \times C_k)$. Let $\boldsymbol{w}$ be the net wrench generated on $\mathcal{B}$ by $\vec{f}_1$: $\boldsymbol{w} = G(q_0)\vec{f}_1$. Since the grasp is assumed to be wrench resistant, there exist feasible contact forces, $\vec{f}_2 \in C_1 \times \cdots \times C_k$, such that $G(q_0)\vec{f}_2 = -\boldsymbol{w}$. Hence, $G(q_0)(\vec{f}_1 + \vec{f}_2) = \boldsymbol{w} - \boldsymbol{w} = \vec{0}$, which implies that $\vec{f}_1 + \vec{f}_2$ is a vector of internal grasp forces. Let us show that $\vec{f}_1 + \vec{f}_2$ lies in the interior of $C_1 \times \cdots \times C_k$. Since $\vec{f}_1 \in \text{int}(C_1 \times \cdots \times C_k)$, all vectors $\vec{f}_1 + \Delta\vec{f}$ such that $\|\Delta\vec{f}\|$ is sufficiently small also lie in $\text{int}(C_1 \times \cdots \times C_k)$. Since $\vec{f}_1 + \Delta\vec{f}$ and $\vec{f}_2$ lie in the convex cone $C_1 \times \cdots \times C_k$, their sum $(\vec{f}_1 + \Delta\vec{f}) + \vec{f}_2 = (\vec{f}_1 + \vec{f}_2) + \Delta\vec{f}$ also lies in this cone for all $\Delta\vec{f}$. Since $\Delta\vec{f} \in \mathbb{R}^{kp}$ is arbitrary with a sufficiently small magnitude, $\vec{f}_1 + \vec{f}_2$ lies in the interior of $C_1 \times \cdots \times C_k$ and Condition (ii) is thus necessary for wrench resistance. □

A surjective grasp matrix is clearly required at any wrench resistant grasp. While it is always possible to find a solution $\vec{f}_p$ for Eq. (12.1) when $G(q_0)$ is surjective, the contact forces must lie in their respective friction cones. Let us graphically describe how Condition (ii) ensures the existence of feasible contact forces. Assume there exists a particular combination of contact forces, $\vec{f}_p$, that solves Eq. (12.1) for a given external wrench, such that $\vec{f}_p$ does not necessarily lie in the composite friction cone $C_1 \times \cdots \times C_k$. If the grasp possesses internal forces that lie in the interior of the composite friction cone, $G(q_0)\vec{f}_{int} = \vec{0}$ such that $\vec{f}_{int} \in \text{int}(C_1 \times \cdots \times C_k)$, then as the fingers squeeze harder along the internal force subspace, there is a corresponding increase in the magnitude of the normal contact force components, which allows the grasp to support greater tangential force components at the contacts (Figure 12.1). For sufficiently strong squeezing forces, the net contact force, $\vec{f}_{net} = \vec{f}_p + \vec{f}_{int}$, is guaranteed to lie in $C_1 \times \cdots \times C_k$, as illustrated in the following example.

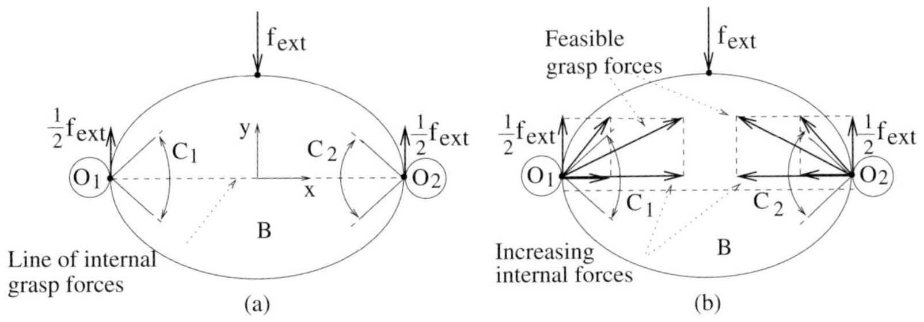

Figure 12.1 (a) A frictional two-finger grasp with physically unrealizable tangential forces applied by $\mathcal{O}_1$ and $\mathcal{O}_2$. (b) As the internal grasp forces increase, the net finger forces eventually lie within the allowed friction cones and are therefore feasible.

Example: Figure 12.1 depicts a two-finger equilibrium grasp of an object $\mathcal{B}$ via antipodal frictional contacts. A perturbing force, f_{ext}, acts along the negative y-axis that is orthogonal to the contact normals. A particular solution to the equilibrium grasp equation would be the application of tangential finger forces of magnitudes $\|f_{ext}\|/2$ along the positive y-axis. However, pure tangential forces are *not* feasible under the Coulomb friction constraint at the contacts. But this grasp possesses a one-dimensional subspace of internal grasp forces that physically correspond to equal magnitude squeezes along the contact normals. As the fingers increase their squeezing forces, more and more tangential forces, which are needed to resist the external force, can be supported by the frictional contacts. At a sufficiently large squeeze force magnitude $\|f_{ext}\|/(2\mu)$, where μ is the coefficient of friction – the contacts can support the tangential forces needed to resist the external force. ○

The grasp matrix $G(q_0)$ is surjective in the following generic grasps (see Exercises). In 2-D grasps, $G(q_0)$ is surjective for $k \geq 2$ frictional contacts, unless all finger contacts happen to be coincident. In 3-D grasps, $G(q_0)$ is surjective for $k \geq 3$ frictional hard-point contacts, unless all finger contacts lie on a common line. In 3-D grasps having soft-point contacts, $G(q_0)$ is surjective for $k \geq 2$ contacts unless all finger contacts happen to be coincident. Since the grasp matrix is generically surjective in frictional grasps, only Condition (ii) of Theorem 12.1 must be verified at a given grasp. In other words, the wrench resistant grasps basically correspond to equilibrium grasps that can realize non-marginal internal grasp forces at the contacts.

The catalog of equilibrium grasps developed in Chapter 5 describes the robustness of the frictional grasps with respect to small finger placement errors. The wrench resistant grasps inherit this property, as illustrated in the following example.

Robustness of wrench resistant grasps: Figure 12.2 shows an object $\mathcal{B}$ that is held at a wrench resistant grasp via antipodal frictional contacts. Wrench resistance requires a surjective grasp matrix $G(q_0)$, as well as feasible internal grasp forces that lie in the interior of the friction cones. The surjectivity of $G(q_0)$ is a generic property of frictional equilibrium grasps. Hence, all finger placements that support non-marginal internal grasp forces form wrench resistant grasps. To demonstrate this robustness property, fix the left finger body O_1 at x_1. The right finger body O_2 can contact any point on the

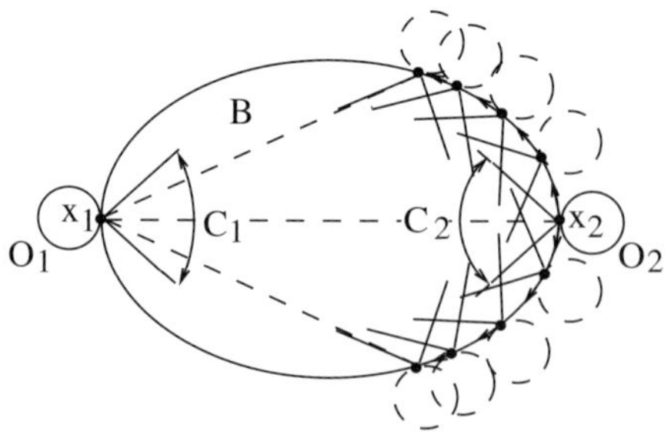

Figure 12.2 An initial wrench resistant grasp via antipodal frictional contacts. The grasp remains wrench resistant under fairly large finger placement errors in this example.

opposing boundary segment depicted in Figure 12.2, while still maintaining a wrench resistant grasp of the object B. □

The testing of wrench resistance can be efficiently performed using convex analysis tools that are discussed next.

12.2 Wrench Resistance as a Linear Matrix Inequality

The wrench resistance of many simple grasps can be resolved with simple analysis of the grasp matrix, based on Theorem 12.1. However, a computational procedure is generally required to assess the wrench resistance ability of arbitrary frictional grasps. This section shows how the problem of assessing wrench resistance can be converted into a *linear matrix inequality* (LMI) problem, which can be efficiently solved as a convex optimization problem. To simplify the LMI formulation, we will assume that the coefficients of friction at the contacts are slightly reduced from their true value. In this manner, every feasible equilibrium grasp of $\mathcal{B}$ forms a non-marginal equilibrium grasp (and hence is wrench resistant) under the true value of the coefficients of friction at the contacts.

12.2.1 Formulation of Friction Constraints in Terms of Positive Semi-Definite Matrices

The computational difficulty of determining the wrench resistance of frictional grasps arises from the cone-like nature of the Coulomb friction model. Fortunately, cone-like constraints can be formulated under the *convex optimization framework.* Consider a hard-point contact. The ith finger force can be expressed as $f_i = (f_i^s, f_i^t, f_i^n)$, using a reference frame at the contact point x_i. Under the Coulomb friction model, the finger forces are constrained by the friction cone inequalities

$$f_i^n \geq 0 \quad \text{and} \quad \sqrt{(f_i^s)^2 + (f_i^t)^2} \leq \mu_i f_i^n, \tag{12.2}$$

where μ_i is the coefficient of friction at x_i. The torque component about the contact normal is similarly bounded in terms of f_i^n under the soft-point contact model. These constraints can be equivalently expressed more generally as follows.

PROPOSITION 12.2 *Let $P(f_i)$ be the symmetric 3×3 matrix:*

$$P(f_i) = \begin{bmatrix} \mu_i f_i^n & 0 & f_i^s \\ 0 & \mu_i f_i^n & f_i^t \\ f_i^s & f_i^t & \mu_i f_i^n \end{bmatrix} \qquad f_i = (f_i^s, f_i^t, f_i^n).$$

The friction cone constraints of Eq. (12.2) can be expressed as the matrix inequality

$$P(f_i) \succeq 0, \tag{12.3}$$

where $P \succeq 0$ denotes that the matrix P is positive semi-definite.[2]

[2] A symmetric $n \times n$ matrix P is *positive semi-definite* if $v^T P v \geq 0$ for all $v \in \mathbb{R}^n$.

Proof: Since $P(f_i)$ is real symmetric, its eigenvalues are real and given by

$$\begin{aligned} \lambda_1 &= \mu_i f_i^n \\ \lambda_2 &= \mu_i f_i^n + \sqrt{(f_i^s)^2 + (f_i^t)^2} \\ \lambda_3 &= \mu_i f_i^n - \sqrt{(f_i^s)^2 + (f_i^t)^2}. \end{aligned}$$

The matrix $P(f_i)$ is positive semidefinite when $\lambda_1, \lambda_2, \lambda_3 \geq 0$. The requirement $\lambda_1 \geq 0$ implies that $f_i^n \geq 0$. Since $f_i^n \geq 0$, the eigenvalue λ_2 is positive semi-definite. Hence, $P(f_i) \succeq 0$ when $\lambda_3 \geq 0$, which implies the friction cone constraint: $\sqrt{(f_i^s)^2 + (f_i^t)^2} \leq \mu_i f_i^n$. □

Let us show how the matrix inequality constraint, $P(f_i) \succeq 0$, can be formulated as a linear matrix inequality. At a frictionless contact the matrix P reduces to a scalar, $P(f_i) = f_i^n$, and the friction cone constraint reduces to the inequality $f_i^n \geq 0$. At a frictional hard-point contact, the constraint matrix $P(f_i)$ is linear in the finger contact force components. Hence, it can be factored into a simple linear sum whose coefficients are constant symmetric matrices:

$$\begin{aligned} P(f_i) &= \begin{bmatrix} \mu_i f_i^n & 0 & f_i^s \\ 0 & \mu_i f_i^n & f_i^t \\ f_i^s & f_i^t & \mu_i f_i^n \end{bmatrix} \\ &= f_i^s \begin{bmatrix} 0 & 0 & 1 \\ 0 & 0 & 0 \\ 1 & 0 & 0 \end{bmatrix} + f_i^t \begin{bmatrix} 0 & 0 & 0 \\ 0 & 0 & 1 \\ 0 & 1 & 0 \end{bmatrix} + f_i^n \begin{bmatrix} \mu_i & 0 & 0 \\ 0 & \mu_i & 0 \\ 0 & 0 & \mu_i \end{bmatrix} \\ &= f_i^s S_s + f_i^t S_t + f_i^n S_n. \end{aligned}$$

The constraint matrix P for the soft-point contact model is similarly linear in the finger contact force and moment components (see Exercises). Hence, several practical rigid-body contact models can be put in the form

$$P(f_i) = \sum_{j=1}^{p} f_{ij} S_{ij} \succeq 0 \qquad f_i = (f_{i1}, \ldots, f_{ip}), \tag{12.4}$$

where p is the number of independent force and possibly moment components at each contact, and S_{ij} for $j = 1 \ldots p$ are constant symmetric $p \times p$ matrices associated with the components of f_i. Equation (12.4) forms a *linear matrix inequality,* defined as follows.

DEFINITION 12.2 (**LMI**) *A* **linear matrix inequality** *in the variable $\vec{z} \in \mathbb{R}^n$ takes the form*

$$Q(\vec{z}) = Q_0 + \sum_{i=1}^{n} z_i Q_i \succeq 0 \qquad \vec{z} = (z_1, \ldots, z_n) \tag{12.5}$$

where $Q_0, Q_1, \ldots, Q_n$ are constant symmetric $n \times n$ matrices.

The inequality $Q(\vec{z}) \succeq 0$ specifies a convex cone constraint based on the following arguments. Consider the vector space of real symmetric $n \times n$ matrices. Given two symmetric positive semi-definite matrices, Q_1 and Q_2, their sum $s_1Q_1 + s_2Q_2$ forms a symmetric positive semi-definite matrix for all $s_1, s_2 \geq 0$. The inequality $Q \succeq 0$ thus specifies a convex cone in the vector space of symmetric $n \times n$ matrices. Since $Q(\vec{z})$ is a linear parametrization, it retains the convex cone structure in the variable $\vec{z} \in \mathbb{R}^n$. The convex cone constraint $Q(\vec{z}) \succeq 0$ can represent a variety of linear constraints, quadratic constraints and matrix norm inequalities, as illustrated in the following example.

Example: Consider a matrix norm constraint of the form

$$\|A\vec{z} + \vec{a}\| \leq \vec{b} \cdot \vec{z} + c \qquad \vec{z} \in \mathbb{R}^n,$$

where A is a constant $n \times n$ matrix, $\vec{a}, \vec{b} \in \mathbb{R}^n$ are constant vectors and c is a constant scalar such that $\|\vec{a}\| \leq c$. This constraint can be expressed as the linear matrix inequality

$$\begin{bmatrix} (\vec{b} \cdot \vec{z} + c)I & A\vec{z} + \vec{a} \\ (A\vec{z} + \vec{a})^T & \vec{b} \cdot \vec{z} + c \end{bmatrix} \succeq 0,$$

where I is an $n \times n$ identity matrix. The matrix norm inequality, as well as the ith friction cone constraint $P(f_i) \succeq 0$, form second-order cone constraints. ○

From a computational point of view, there are two main types of LMI problems:

(i) **LMI feasibility problems** test for the existence of $\vec{z}$ such that $Q(\vec{z}) \succeq 0$.
(ii) **LMI optimization problems** minimize a convex or quasiconvex cost function, $\varphi(\vec{z})$, over all $\vec{z}$ such that $Q(\vec{z}) \succeq 0$.[3]

In addition to the wrench resistance problem that is discussed next, many problems in engineering can be formulated in terms of LMI. Polynomial time algorithms (e.g., interior point methods) have been developed for LMI problems, and efficient numerical tools are readily available for such problems.

12.2.2 Wrench Resistance as an LMI Feasibility Problem

Let us show how wrench resistance can be tested using LMI formulation. The composite friction cone constraint, $(f_1, \ldots, f_k) \in C_1 \times \cdots \times C_k$, can be represented in terms of a $kp \times kp$ block-diagonal matrix, whose diagonal blocks are determined by the friction cone constraints at the k contacts

$$\mathcal{P}(\vec{f}) = \mathsf{diag}\big(P(f_1), P(f_2), \ldots, P(f_k)\big) \succeq 0 \qquad \vec{f} = (f_1, \ldots, f_k) \in \mathbb{R}^{kp}. \tag{12.6}$$

Since the diagonal block matrices are linear in the finger force components, the matrix inequality of Eq. (12.6) can be written as the linear combination

$$\mathcal{P}(\vec{f}) = \sum_{i=1}^{k} \sum_{j=1}^{p} f_{ij} S_{ij} \succeq 0 \qquad \vec{f} = (f_1, \ldots, f_k) \in \mathbb{R}^{kp}, \tag{12.7}$$

[3] A function $\varphi(\vec{z}) : \mathcal{D} \to \mathbb{R}$ on a convex set $\mathcal{D} \subseteq \mathbb{R}^n$ is a *convex function* when its epigraph, $\{(\vec{z}, \zeta) : \zeta \geq \varphi(\vec{z})\}$, forms a convex set in $\mathbb{R}^{n+1}$. It is *quasiconvex* when its sublevel sets, $\{\vec{z} : \varphi(\vec{z}) \leq c\}$, are convex sets in $\mathbb{R}^n$.

where f_{ij} is the jth component of the contact force f_i, and each $\boldsymbol{S}_{ij}$ is a constant $kp \times kp$ symmetric matrix of the form

$$\boldsymbol{S}_{ij} = \mathrm{diag}(O, \ldots, O, S_{ij}, O, \ldots, O),$$

where S_{ij} is the $p \times p$ constant matrix associated with the force component f_{ij}, and O is a $p \times p$ zero matrix.

The wrench resistance test of Theorem 12.1 can now be formulated as an LMI feasibility problem. Condition (i) of Theorem 12.1 holds true for generic frictional grasps. Let $\mathcal{N}(G)$ denote the null space of the grasp matrix $G(q_0)$. Condition (ii) of Theorem 12.1 is equivalent to the existence of $\vec{f} \in \mathcal{N}(G)$ such that $\mathcal{P}(\vec{f}) \succeq 0$. This condition can be further adapted to the LMI framework by explicit parametrization of the null space. The grasp matrix $G(q_0)$ is $m \times kp$, and its null space is $(kp - m)$-dimensional in $\mathbb{R}^{kp}$. Let N be the matrix whose columns form a basis for the null space: $\mathcal{N}(G) = N\vec{v}$ for $\vec{v} \in \mathbb{R}^{kp-m}$. Condition (ii) of Theorem 12.1 can be stated as a single LMI, by parametrizing the matrix $\mathcal{P}(\vec{f})$ in Eq. (12.7) in terms of the null space basis vectors:

$$\mathcal{P}(\vec{v}) = \sum_{i=1}^{k} \sum_{j=1}^{p} (N\vec{v})_{ij} \boldsymbol{S}_{ij} \succeq 0 \qquad \vec{v} \in \mathbb{R}^{kp-m} \tag{12.8}$$

where $(N\vec{v})_{ij}$ is the lth element of $N\vec{v}$, having index $l = (i-1)k + j$ for $l = 1 \ldots kp$. This LMI formulation is summarized in the following theorem.

THEOREM 12.3 (**LMI wrench resistance test**) *A k-finger grasp of a rigid object $\mathcal{B}$ is* **wrench resistant** *iff (i) the grasp matrix $G(q_0)$ is surjective (the generic case), and (ii) there exists a non-zero vector $\vec{v} \in \mathbb{R}^{kp-m}$ such that $\mathcal{P}(\vec{v}) \succeq 0$ according to Eq. (12.8).*

Testing the wrench resistance of a k-finger grasp forms an LMI feasibility problem, as one only seeks the existence of at least one null space vector, $\vec{v} \neq \vec{0}$, such that $\mathcal{P}(\vec{v}) \succeq 0$. To solve this LMI feasibility problem, one can formulate the following auxiliary LMI optimization problem. First, note that the inequality $\mathcal{P}(\vec{v}) \succeq 0$ is satisfied iff there exists a negative semi-definite scalar, $\sigma \leq 0$, such that $\mathcal{P}(\vec{v}) + \sigma \cdot I \succeq 0$ (see Exercises). The LMI feasibility problem can therefore be stated as

$$\text{minimize} \quad \{\sigma\} \quad \text{subject to} \quad \mathcal{P}(\vec{v}) + \sigma \cdot I \succeq 0 \qquad \vec{v} \in \mathbb{R}^{kp-m}, \ \sigma \in \mathbb{R}.$$

The original LMI is feasible (and hence there exist internal grasp forces satisfying Condition (ii) of Theorem 12.1) if and only if the optimal value of the auxiliary problem is negative semi-definite. In this way, the powerful computational tools that have been developed for LMI optimization can be leveraged for testing wrench resistance of arbitrary grasps.

12.3 Grasp Force Optimization

The convex optimization framework is useful for more than testing wrench resistance of arbitrary k-finger grasps. This section formulates the grasp force optimization problem to show the practical utility of convex optimization techniques. Let a rigid object $\mathcal{B}$ be held at a k-finger grasp via a given set of frictional contacts against an external wrench $\boldsymbol{w}_{ext}$. The *grasp force optimization problem* seeks to select feasible finger forces at the contacts that resist $\boldsymbol{w}_{ext}$, while optimizing a cost function, $\varphi(\vec{f})$, of the finger contact forces:

$$\begin{array}{lll} \text{minimize} & \{ \varphi(\vec{f}) \} & \text{where } \vec{f} \in \mathbb{R}^{kp} \\ \text{subject to} & \vec{f} \in C_1 \times \cdots \times C_k & \text{and} \quad G(q_0)\vec{f} + \boldsymbol{w}_{ext} = \vec{0}\,. \end{array}$$

Grasp force optimization requires the existence of feasible finger forces that satisfy the constraints. The wrench resistant grasps satisfy this requirement, as such grasps can vary their internal grasp forces while resisting a specified $\boldsymbol{w}_{ext}$. When the cost function, $\varphi(\vec{f})$, forms a convex or quasiconvex function (see previous section), the grasp force optimization forms a *convex optimization problem.* That is, the minimization of a convex or quasiconvex cost function subject to equality and inequality constraints that define a convex set. Convex optimization problems possess a unique optimum solution (which possibly degenerates to a convex optimum solution set). These problems include many physically meaningful grasp optimization criteria, as we shall see in Chapter 13 on grasp quality measures. To put this formulation in a concrete context, consider the following examples.

- **The gentlest grasp criterion:** The potential for a robot hand to damage the grasped object is proportional to the magnitude of the normal force components at the contacts. The *gentlest grasp* seeks to minimize the worst-case normal force applied at the contacts. This objective can be expressed as the second-order cone optimization problem:

$$\begin{array}{lll} \text{minimize} & \max\{f_1^n, \ldots, f_k^n\} & \text{where } \vec{f} \in \mathbb{R}^{kp} \\ \text{subject to} & \mathcal{P}(\vec{f}) \succeq 0 \text{ and } G(q_0)\vec{f} + \boldsymbol{w}_{ext} = \vec{0}, & \end{array}$$

  where $\mathcal{P}(\vec{f}) \succeq 0$ models the friction cone constraints, f_i^n is the normal component of the finger force f_i and $\vec{f} = (f_1, \ldots, f_k)$. The cost function $\varphi(\vec{f}) = \max\{f_1^n, \ldots, f_k^n\}$ is convex, based on the fact that the pointwise maximum of convex functions is convex.
- **The robust grasp criterion:** The minimization of the normal force components may push the contact forces toward the boundary of the friction cones. Such a solution is *non-robust,* since it relies on realistically uncertain value of the coefficients of friction at the contacts. Alternatively, one can minimize the normal contact force components without relying too much on tangential frictional forces. This objective can be accomplished by adding a penalty term, or a *barrier function,* to the cost function in order to bias the finger contact forces toward the center of the friction cones, thus minimizing the grasp's reliance on friction. A natural barrier function in this case is $-\log(\det(\mathcal{P}(\vec{f})))$, where $\mathcal{P}$ is the positive semi-definite matrix modeling the friction cone constraints. The barrier function forms a convex function (see Exercises), whose value approaches infinity as the contact forces rotate toward the friction cone boundaries. It can therefore be

added to the cost function of the previous example, to form the second-order cone optimization problem:

$$\begin{aligned} &\text{minimize} && \max\{f_1^n,\ldots,f_k^n\} - \log\big(\det(\mathcal{P}(\vec{f}))\big) \;\; \text{where } \vec{f} \in \mathbb{R}^{kp} \\ &\text{subject to} && \mathcal{P}(\vec{f}) \succeq 0 \text{ and } G(q_0)\vec{f} + \boldsymbol{w}_{ext} = \vec{0}, \end{aligned}$$

where $(f_1^n,\ldots,f_k^n)$ are the normal finger force components and $\vec{f} = (f_1,\ldots,f_k)$.

12.4 Grasp Controllability

This section analyzes a k-finger grasp as a control system. The *grasped object system* consists of the grasped object $\mathcal{B}$, whose control inputs are feasible finger forces applied at the contacts. The grasp system is said to be *locally controllable* when feasible finger forces, $\vec{f}(t) \in C_1 \times \cdots \times C_k$, can steer the grasped object from a start configuration at $t = 0$ to any nearby configuration in finite time $t = T$. This section shows that wrench resistant grasps are locally controllable and then describes a classical control approach that stabilizes such grasps against external disturbances.

12.4.1 Grasped Object Dynamics

To model a k-finger grasp as a control system, we must develop the dynamical equation that governs the grasped object motions. Let the object $\mathcal{B}$ be held by *pointed fingertips.* As long as the pointed fingertips apply contact forces within the allowed friction cones, they will remain fixed on the object surface during local motions that maintain the grasp. Let us also include the effect of gravity in the object dynamics. The gravitational wrench that acts on $\mathcal{B}$ is simply the negated gradient, $-\nabla U(q)$, of the object's gravitational potential energy $U(q)$ (see Chapter 11).

The grasped object *state* consists of its c-space position and velocity, $(q,\dot{q}) \in \mathbb{R}^m \times \mathbb{R}^m$, where $m = 3$ in 2-D grasps and $m = 6$ in 3-D grasps. To derive the grasped object dynamics, consider the object's kinetic energy given by the quadratic form

$$\mathcal{K}(q,\dot{q}) = \tfrac{1}{2}\dot{q}^T M(q)\dot{q} \qquad (q,\dot{q}) \in \mathbb{R}^m \times \mathbb{R}^m,$$

where $M(q)$ is the object's $m \times m$ mass matrix, which is symmetric and positive definite. The grasped object motions under the influence of an external wrench, $\boldsymbol{w}_{ext}$, are determined by the *transfer of power* equation:

$$\frac{d}{dt} K\big(q(t),\dot{q}(t)\big) = \boldsymbol{w}_{ext} \cdot \dot{q} \qquad (q,\dot{q}) \in \mathbb{R}^m \times \mathbb{R}^m. \tag{12.9}$$

The inner product, $\boldsymbol{w}_{ext} \cdot \dot{q}$, represents the instantaneous work of the external wrench $\boldsymbol{w}_{ext}$ on the rigid body velocity $\dot{q}$. Application of the chain rule on the left side of Eq. (12.9), coupled with *D'alembert's principle* (the dynamic analogue to the principle of virtual work) gives the grasped object *Newton-Euler* dynamical equation:

$$M(q)\ddot{q} + \begin{pmatrix} \vec{0} \\ \omega \times J_B(q)\omega \end{pmatrix} = \boldsymbol{w}_{ext} \qquad \dot{q} = (v,\omega),\ \ddot{q} = \tfrac{d}{dt}\dot{q}(t) \tag{12.10}$$

where $J_B(q) = R(\theta) J_B R^T(\theta)$, such that $R(\theta)$ is the object's orientation matrix and J_B is the object's moments of inertia matrix (see Bibliographical Notes).

The grasped object dynamics described in Eq. (12.10) assumes pointed fingertips. One should be aware that when this assumption is relaxed net fingertip rolling becomes feasible, and Eq. (12.10) must be augmented with rolling constraints (see Bibliographical Notes). The external wrench that acts on $\mathcal{B}$, $\boldsymbol{w}_{ext}$, consists of two parts: the net wrench applied by the fingers, $G(q)\vec{f}$, and the gravitational wrench, $-\nabla U(q)$. Substituting $\boldsymbol{w}_{ext} = G(q)\vec{f} - \nabla U(q)$ and using the notation $h(q,\dot{q}) = (\vec{0}, \omega \times J_B(q)\omega)$ in Eq. (12.10) gives the dynamical equation:

$$M(q)\ddot{q} + h(q,\dot{q}) + \nabla U(q) = G(q)\vec{f}(t) \qquad \vec{f}(t) \in C_1 \times \cdots \times C_k. \tag{12.11}$$

Equation (12.11) represents a second-order nonlinear control system, whose control inputs consist of feasible finger contact forces, $\vec{f} \in \mathbb{R}^{kp}$ such that $\vec{f}(t) \in C_1 \times \cdots \times C_k$ for $t \geq 0$. One should be aware that the actuation of the finger mechanisms determines the type of control inputs that can be realized at the contacts. We assume that the contact forces $\vec{f}(t)$ are Lipschitz continuous in t. However, the notion of grasp controllability also holds when the contact forces $\vec{f}(t)$ are piecewise continuous with finite discontinuous jumps at discrete instants, which can model breakage and reestablishment of finger contacts with the object $\mathcal{B}$.

12.4.2 Local Controllability of Multi-Finger Grasps

For control analysis, it is useful to write the grasp system of Eq. (12.11) as a first-order control system. Define the state variables (z_1, z_2), where $z_1 = q$ and $z_2 = \dot{q}$. Since $M(z_1)$ is invertible, the grasped object is governed by the first-order control system:

$$\begin{aligned} \dot{z}_1 &= z_2 \\ \dot{z}_2 &= M^{-1}(z_1)\big(-h(z_1,z_2) - \nabla U(z_1) + G(z_1)\vec{f}(t)\big) \end{aligned} \qquad \vec{f}(t) \in C_1 \times \cdots \times C_k. \tag{12.12}$$

Equation (12.12) takes the form of an *affine control system*. In these systems, the control inputs, $u(t) \in \mathcal{U}$ such that $\mathcal{U} \subset \mathbb{R}^l$, enter in a linearly affine manner into the system dynamics:

$$\begin{pmatrix} \dot{z}_1 \\ \dot{z}_2 \end{pmatrix} = \begin{pmatrix} z_2 \\ \vec{a}(z_1,z_2) + \vec{B}(z_1,z_2)u(t) \end{pmatrix} \qquad u(t) \in \mathcal{U}. \tag{12.13}$$

An *equilibrium state* of the system (12.13) satisfies the conditions: $z_2 = \vec{0}$ and $\vec{a}(z_1,z_2) + \vec{B}(z_1,z_2)u(t) = \vec{0}$. Let $\mathcal{V}$ be an open neighborhood in the ambient state space, $(z_1,z_2) \in \mathbb{R}^m \times \mathbb{R}^m$, centered at an equilibrium state. The nonlinear system (12.13) is *locally controllable* at the equilibrium state if there exists a finite time, $T > 0$, such that starting at any initial state, $(z_1(0), z_2(0)) \in \mathcal{V}$, the system can be steered by feasible control inputs, $u(t) : [0,T] \rightarrow \mathcal{U}$, to the equilibrium state, such that $(z_1(t), z_2(t))$ remains in $\mathcal{V}$ for $0 \leq t \leq T$. Such systems are said to be *small-time locally controllable,* or STLC.[4]

[4] STLC additionally requires that the finite time T becomes arbitrarily small for suitably small neighborhoods about the equilibrium state.

A control system is said to be *fully actuated* when the number of independent control inputs equals or exceeds the number of state variables. The grasped object system is typically *not* fully actuated. For instance, when a 3-D rigid object $\mathcal{B}$ is held via three frictional point contacts, there are $p=3$ independent finger force components at each contact that give nine independent control inputs, while the system has $2m=12$ state variables. However, the grasped object system forms a mechanical control system, whose state consists of configuration variables and their velocities. A mechanical control system is said to be *fully configuration actuated* when the number of independent control inputs equals or exceeds the number of configuration variables. When a rigid object $\mathcal{B}$ is held at a wrench resistant grasp, the grasp system is *fully configuration actuated.* For instance, when a 3-D rigid object $\mathcal{B}$ is held via three frictional point contacts, there are nine independent control inputs and $m=6$ configuration variables.

The equilibrium states of the grasp system described in Eq. (12.12) consist of zero-velocity equilibrium grasps. When the object $\mathcal{B}$ is held at a wrench resistant equilibrium grasp, surjectivity of the grasp matrix ensures that the grasp is fully configuration actuated during local finger motions that maintain the grasp. It follows that wrench resistant grasps are *locally controllable,* as stated in the following theorem.

THEOREM 12.4 **(STLC of wrench resistant grasps)** *Let a rigid object $\mathcal{B}$ be held at a frictional k-finger equilibrium grasp, possibly under external influences such as gravity. If the grasp is wrench resistant, the system formed by the grasped object and feasible finger contact forces as control inputs is* **locally controllable** *at the equilibrium grasp.*

The proof of Theorem 12.4 is based on the following *linearized dynamics* of the grasped object system.

Linearized grasp dynamics: Consider a nonlinear control system of the form $\dot{z}=F(z,u)$. An equilibrium of this control system consists of an equilibrium state, z_0, and an equilibrium control input, u_0, such that $F(z_0,u_0)=\vec{0}$. The linearization of such a nonlinear control system about (z_0,u_0) is based on the first-order Taylor expansion:

$$F(z,u)=\left.\frac{\partial F(z,u)}{\partial z}\right|_{z_0,u_0}\cdot(z-z_0)+\left.\frac{\partial F(z,u)}{\partial u}\right|_{z_0,u_0}\cdot(u-u_0).$$

In the grasp system, $z_0=(q_0,\vec{0})$ while the finger forces at the equilibrium grasp, $u_0=\vec{f}_0$, provide the equilibrium control inputs. The linearization will use the state variable $\tilde{z}=z-z_0$, which represents local deviation from the object's zero-velocity equilibrium grasp, and the control input $\tilde{f}=\vec{f}-\vec{f}_0$, which represents local deviation from the equilibrium contact forces. The linearization of the grasp system (12.12) about (z_0,u_0) has the form (see Exercises)

$$\dot{\tilde{z}}(t)=\begin{bmatrix} O & I \\ A_c & O \end{bmatrix}\tilde{z}(t)+\begin{bmatrix} O \\ B_c \end{bmatrix}\tilde{f}(t) \qquad \tilde{z}(t)\in\mathbb{R}^{2m},\ \tilde{f}(t)\in\mathbb{R}^{kp}, \tag{12.14}$$

where I is an $m \times m$ identity matrix and O is an $m \times m$ zero matrix in the first summand and an $m \times kp$ zero matrix in the second summand. The $m \times m$ matrix A_c and the $m \times kp$ matrix B_c are given by

$$A_c = M^{-1}(q_0) DG(q_0) \vec{f}_0 \qquad \text{and} \qquad B_c = M^{-1}(q_0) G(q_0),$$

where $DG(q_0)$ is the derivative of the grasp matrix $G(q)$ at q_0, such that $DG(q_0)\vec{f}_0$ is the following $m \times m$ matrix. Let ρ_i^0 denote the vector from the object frame origin to the ith contact point at q_0, expressed in the world frame. Then $DG(q_0)\vec{f}_0$ has the form (see Exercises)

$$DG(q_0)\vec{f}_0 = \begin{bmatrix} O & O \\ O & Q \end{bmatrix} \qquad Q = \textstyle\sum_{i=1}^{k} \rho_i^0 (f_i^0)^T - (\rho_i^0 \cdot f_i^0) I, \tag{12.15}$$

where, in 3-D grasps, O is a 3×3 zero matrix. Note that in 3-D grasps, Q forms a *nonsymetric* 3×3 matrix.

Proof of theorem 12.4: First note that the finger force deviations in the linearized grasp system, $\tilde{f}(t) \in \mathbb{R}^{kp}$, are *not* restricted to lie within the friction cones, since wrench resistant grasps form non-marginal equilibrium grasps according to Condition (ii) of Theorem 12.1. From control theory, a linear control system of the form $\dot{\tilde{z}} = A\tilde{z} + B\tilde{u}$, such that $\tilde{z} \in \mathbb{R}^N$ and $\tilde{u} \in \mathbb{R}^l$, is *controllable* when the $N \times N \cdot l$ *controllability matrix*

$$\begin{bmatrix} B & AB & A^2B & \cdots & A^{N-1}B \end{bmatrix}$$

has full rank of N (see Bibliographical Notes). The controllability matrix of the linearized grasp system (12.14) attains the form

$$\begin{bmatrix} O & B_c & O & A_cB_c & \cdots & O & A_c^{N-1}B_c \\ B_c & O & A_cB_c & O & \cdots & A_c^{N-1}B_c & O \end{bmatrix} \qquad N = 2m,\ l = kp,$$

where O is an $m \times kp$ zero matrix. The controllability matrix has full rank of $N = 2m$ when B_c has full row rank of m. Since $B_c = M^{-1}(q_0)G(q_0)$ and $M(q_0)$ is invertible, the rank of B_c depends upon the rank of the grasp matrix, $G(q_0)$. The $m \times kp$ grasp matrix $G(q_0)$ is surjective and hence has full row rank of m at a wrench resistant grasp (Condition (i) of Theorem 12.1). Hence, the linear system (12.14) is controllable. Controllability of a linearized control system implies *local controllability* of the original nonlinear system at the equilibrium state (see Bibliographical Notes). The nonlinear grasp system (12.12) is thus locally controllable at any wrench resistant grasp. □

12.4.3 Exact Feedback Linearization

Local controllability only determines if the grasped object system *can* be locally controlled via a given set of contacts; it does not specify a feedback control law for the finger contact forces. In grasp mechanics, one seeks control inputs that stabilize an object held at an equilibrium grasp against external wrench disturbances. Such feedback controls command the finger contact forces as a function of sensor measurements of the object's deviations from the equilibrium grasp. The controlled grasp system is said to be *locally asymptotically stable* when the system converges back to the equilibrium state from any nearby perturbed state (which may include contact breakages) while staying in a local neighborhood of the equilibrium state.

The standard approach for the control of fully configuration-actuated mechanical systems, which include wrench resistant grasps, is *exact feedback linearization.* This method is also known as *computed torque* in robotics. When a grasp is wrench resistant, the finger contact forces can be chosen according to the feedback control law,

$$G(q)\vec{f}(t) = -K_p(q - q_0) - K_d(\dot{q} - \vec{0}) + h(q,\dot{q}) + \nabla U(q) \qquad \vec{f}(t) \in C_1 \times \cdots \times C_k, \tag{12.16}$$

where K_p and K_d are $m \times m$ symmetric positive definite matrices selected by the control law designer. The proportional term, $-K_p(q - q_0)$, can be interpreted as the negated gradient of an *artificial potential function,* $V(q) = \frac{1}{2}(q - q_0)^T K_p(q - q_0)$, induced on the grasped object instead of the gravitational potential energy that is cancelled by the term $\nabla U(q)$. Substituting the control law (12.16) into the nonlinear grasp system (12.11) gives the closed loop *error system:*

$$M(q)\ddot{e}(t) + K_d\dot{e}(t) + K_p e(t) = \vec{0} \qquad e(t) = q(t) - q_0,\ \dot{e}(t) = \dot{q}(t),\ \ddot{e}(t) = \ddot{q}(t). \tag{12.17}$$

When the error system is linearized at the equilibrium state, $(e, \dot{e}) = (\vec{0}, \vec{0})$, the linearized error dynamics takes the form of a second-order linear differential equation. Any selection of positive definite feedback gain matrices, K_p and K_d, will ensure global asymptotic stability of the linear error system and, consequently, the *local asymptotic stability* of the zero velocity equilibrium grasp (see Bibliographical Notes). That is, the controlled finger forces will guide the grasped object back to the original equilibrium grasp once the disturbance is removed.

Let us end with three comments. First, note that the error system specified in Eq. (12.17) is not fully linear, due to the mass matrix $M(q)$. Full feedback linearization requires the more demanding control law: $G(q)\vec{f}(t) = -M(q)(K_p(q - q_0) + K_d(\dot{q} - \vec{0})) + h(q, \dot{q}) + \nabla U(q)$. Second, the control approach generalizes to any control law of the form

$$G(q)\vec{f}(t) = -\nabla V(q) - K_d(\dot{q} - \vec{0}) + h(q,\dot{q}) + \nabla U(q),$$

where $V(q)$ is any artificial potential function satisfying $V(q_0) = 0$ and $V(q) > 0$ in a c-space neighborhood centered at q_0. That is, $V(q)$ must possess a strict local minimum at the equilibrium grasp. Moreover, the grasped object's total mechanical energy, $E(q, \dot{q}) = \frac{1}{2}\dot{q}^T M(q)\dot{q} + V(q)$, can be used to assess the basin of stability surrounding the equilibrium grasp. Finally, exact feedback linearization is straightforward only in theory. Its practical effectiveness depends upon reasonably accurate knowledge of the object's mass matrix, the object's geometric shape and the coefficients of friction at the contacts, as well as accurate sensors to measure the object's position and velocity deviations in the process of stabilizing the grasp.

Bibliographical Notes

The literature on *wrench resistant* or *force closure* grasps is discussed in the Bibliographical Notes of Chapter 7. The wrench resistance test of Theorem 12.1 was developed by Murray, Li and Sastry [1]. The analysis of multi-finger grasps using convex optimization techniques dates to Kerr and Roth [2] and Cheng and Orin [3]. These papers formulate force closure feasibility as a *linear programming problem,* based on polyhedral approximation of the friction cones. The formulation of the exact friction cone constraints as linear matrix inequalities started with Buss et al. [4]. Han, Trinkle and Li [5] subsequently recognized that many wrench resistance problems can be addressed within the LMI framework. The formulation of the grasp force optimization problem in terms of second-order cones is due to Lobo et al. [6].

The relationship between wrench resistance and grasp controllability was developed by Brook, Shoham and Dayan [7]. This relationship takes into consideration the grasped object dynamics, and Greenwood [8] is a good source for such rigid body dynamics. The second-order error system requires positive definite feedback gain matrices, K_p and K_d, to ensure asymptotic stability of the error system. The selection of such matrices is discussed by Shapiro [9]. Extending the grasped object dynamics to curved fingertips introduces contact rolling constraints that depend on the bodies' curvatures at the contacts [10]. See texts by Murray, Li and Sastry [1] and Lynch and Park [11] for techniques that incorporate such constraints into the grasped object dynamics.

The chapter established the local controllability of wrench resistant grasps using several facts from control theory. The controllability matrix is described in texts such as [12] and [13]. Local controllability of fully configuration actuated mechanical systems is discussed in Lewis and Murray [14]. Local controllability of nonlinear systems derived from the controllability of their linearized dynamics is discussed in Sussmann [15], which is the best overall source for STLC results. Once a feedback control law commands the finger contact forces, local asymptotic stability of the nonlinear system follows from asymptotic stability of the linearized system [16]. This property is based on the *algebraic Lyapunov equation,* which provides a Lyapunov function demonstrating the stability of the nonlinear system, as well as a crude estimate for the basin of attraction that surrounds a stable state of the nonlinear system.

Exercises

Section 12.1

Exercise 12.1: Explain why a *first-order immobilizing grasp* of a rigid object $\mathcal{B}$ is wrench resistant, even under frictionless contact conditions.

Exercise 12.2*: Do the second-order immobilizing grasps of a rigid object $\mathcal{B}$ form wrench resistant grasps? Consider whether wrench resistance accounts for second-order effects.

Exercise 12.3: Prove that the grasp matrix associated with frictional point contacts has full rank in (1) 2-D grasps involving two or more non-coincident contacts, and (2) 3-D grasps involving three or more non-collinear contacts.

Solution: First consider 2-D grasps. Let p_i and p_j denote the location of the two finger contacts. Assign the object frame origin at p_i, with its x-axis pointing from p_i to p_j. In this coordinate system, $p_i = (0,0)$ while $p_j = (x_j, 0)$. Hence, the portion of the grasp matrix associated with the two frictional contacts has the form:

$$G(q_0) = \begin{bmatrix} \cdots & G_i & \cdots & G_j & \cdots \end{bmatrix} \quad \text{such that} \quad \begin{bmatrix} G_i \; G_j \end{bmatrix} = \begin{bmatrix} 1 & 0 & 1 & 0 \\ 0 & 1 & 0 & 1 \\ 0 & 0 & 0 & x_j \end{bmatrix}$$

The sub-matrix $[G_i \; G_j]$ has full row rank of three as long as the contacts are non-coincident, $x_i \neq x_j$. Adding additional contacts to the grasp matrix will not change the rank of $G(q_0)$.

Next consider 3-D grasps. Let p_i and p_j be the location of two finger contacts. Assign the object frame origin at p_i, with its x-axis pointing from p_i to p_j. Orient the y and z axes to satisfy the right-hand rule. The basis vectors of the jth finger contact frame do not necessarily align with this reference frame. However, since the contact forces allowed under the frictional point contact model span all three Cartesian directions, without loss of generality one can assign a basis for these forces that aligns with the reference frame located at p_i. Under this choice of frames, $p_i = (0,0,0)$ while $p_j = (x_j, 0, 0)$. The portion of the grasp matrix associated with the two frictional contacts has the form:

$$G(q_0) = \begin{bmatrix} \cdots & G_i & \cdots & G_j & \cdots \end{bmatrix} = \begin{bmatrix} \cdots & 1 & 0 & 0 & \cdots & 1 & 0 & 0 & \cdots \\ \cdots & 0 & 1 & 0 & \cdots & 0 & 1 & 0 & \cdots \\ \cdots & 0 & 0 & 1 & \cdots & 0 & 0 & 1 & \cdots \\ \cdots & 0 & 0 & 0 & \cdots & 0 & 0 & 0 & \cdots \\ \cdots & 0 & 0 & 0 & \cdots & 0 & 0 & -x_j & \cdots \\ \cdots & 0 & 0 & 0 & \cdots & 0 & x_j & 0 & \cdots \end{bmatrix}.$$

The rank of the submatrix $[G_i \; G_j]$ is at most *five*. Under the frictional point contact model, two fingers are not able to generate on $\mathcal{B}$ net torque about the axis passing through the two contact points. Consider a third finger placed at the point $p_l = (x_l, y_l, z_l)$. The basis vectors of the contact forces at p_l can be aligned with the axes of the reference frame located at p_i. The grasp matrix has the form:

$$G(q_0) = \begin{bmatrix} \cdots & G_i & \cdots & G_j & \cdots & G_l & \cdots \end{bmatrix}$$
$$= \begin{bmatrix} \cdots & 1 & 0 & 0 & \cdots & 1 & 0 & 0 & \cdots & 1 & 0 & 0 & \cdots \\ \cdots & 0 & 1 & 0 & \cdots & 0 & 1 & 0 & \cdots & 0 & 1 & 0 & \cdots \\ \cdots & 0 & 0 & 1 & \cdots & 0 & 0 & 1 & \cdots & 0 & 0 & 1 & \cdots \\ \cdots & 0 & 0 & 0 & \cdots & 0 & 0 & 0 & \cdots & 0 & z_l & y_l & \cdots \\ \cdots & 0 & 0 & 0 & \cdots & 0 & 0 & -x_j & \cdots & z_l & 0 & -x_l & \cdots \\ \cdots & 0 & 0 & 0 & \cdots & 0 & x_j & 0 & \cdots & y_l & -x_l & 0 & \cdots \end{bmatrix}.$$

If the third finger contact lies on the line passing through p_i and p_j, then $y_l = z_l = 0$, and the grasp matrix is *not* full rank. Otherwise, the grasp matrix has full row rank of six. ○

Section 12.2

Exercise 12.4: One particular soft-point contact model is defined by the following contact force and moment constraints (see Bibliographical Notes of Chapter 4):

$$f_i^n \geq 0, \quad \frac{1}{\mu_i^2}\left((f_i^s)^2 + (f_i^t)^2\right) + \frac{1}{\gamma_i^2}(\tau_i^n)^2 \leq (f_i^n)^2, \tag{12.18}$$

where f_i^n is the normal force component, (f_i^s, f_i^t) are the tangential force components and τ_i^n is the torque about the contact normal. In addition to the Coulomb coefficient of friction μ_i, the coefficient γ_i describes the limit on the torques that can be supported by the contact. Show that Eq. (12.18) is equivalent to the linear matrix inequality:

$$P(\vec{f}_i, \tau_i^n) = \begin{bmatrix} f_i^n & 0 & 0 & \mu_i^{-1} f_i^s \\ 0 & f_i^n & 0 & \mu_i^{-1} f_i^t \\ 0 & 0 & f_i^n & \gamma_i^{-1} \tau_i^n \\ \mu_i^{-1} f_i^s & \mu_i^{-1} f_i^t & \gamma_i^{-1} \tau_i^n & f_i^n \end{bmatrix} \succeq 0. \tag{12.19}$$

Solution: The eigenvalues of the 4×4 matrix specified in Eq. (12.19) are the roots of the characteristic equation in s:

$$(s - f_i^n)^2\left(s^2 - 2s \cdot f_i^n + ((f_i^n)^2 - \psi^2)\right) = 0,$$

where $\psi^2 = ((f_i^s)^2 + (f_i^t)^2)/\mu_i^2 + (\tau_i^n)^2/\gamma_i^2$. For the eigenvalues to be positive semi-definite, two conditions must be met. First, $f_i^n \geq 0$, which is the first constraint specified in Eq. (12.18). Second, $(f_i^n)^2 - \left((\tau_i^n)^2/\gamma_i^2 + ((f_i^s)^2 + (f_i^t)^2)/\mu_i^2\right) \geq 0$, which is the second constraint specified in Eq. (12.18). ○

Exercise 12.5: Verify that the matrix norm constraint $\|A\vec{z} + \vec{a}\| \leq \vec{b} \cdot \vec{z} + c$ defines a convex cone in $\mathbb{R}^n$. The additional conditions $\|\vec{a}\| \leq c$ and $\vec{b} \cdot \vec{z} \geq 0$ might be needed here.

Exercise 12.6: Prove that the inequality $\mathcal{P}(\vec{v}) \succeq 0$ specified in Theorem 12.3 is satisfied iff there exists a negative semi-definite scalar, $\sigma \leq 0$, such that $\mathcal{P}(\vec{v}) + \sigma \cdot I \succeq 0$.

Section 12.3

Exercise 12.7*: Show that the barrier function associated with the robust grasp optimization problem, $-\log(\det(\mathcal{P}(\vec{f})))$, where the matrix $\mathcal{P}(\vec{f}) \succeq 0$ is specified in Eq. (12.7), forms a convex function of the finger force components.

Solution: The $kp \times kp$ matrix $\mathcal{P}(\vec{f})$ is block diagonal. Hence, it suffices to show that the ith block cost function, $-\log(\det(P(f_i)))$, is convex over the convex constraint $P(f_i) \succeq 0$. Let $X \in \mathbb{R}^{n \times n}$ parametrize the vector space of real symmetric $n \times n$ matrices.

To check that a function $f(x)$ is convex, it suffices to verify that $f\left(\frac{1}{2}(x_1 + x_2)\right) \leq \frac{1}{2}\left(f(x_1) + f(x_2)\right)$. Hence, to show that $-\log(\det(X))$ is convex, it suffices to verify the relation:

$$-\log\left(\det\left(\tfrac{1}{2}(X_1 + X_2)\right)\right) \leq -\tfrac{1}{2}\left(\log\left(\det(X_1)\right) + \log\left(\det(X_2)\right)\right) \quad X_1, X_2 \succeq 0. \tag{12.20}$$

Since the logarithm function is monotonically increasing, the inequality specified in Eq. (12.20) follows from the inequality

$$\det\left(\tfrac{1}{2}(X_1 + X_2)\right) \geq \left(\det(X_1 X_2)\right)^{1/2} \qquad X_1, X_2 \succeq 0, \tag{12.21}$$

where we used the identity $\det(X_1 X_2) = \det(X_1) \cdot \det(X_2)$. The inequality (12.21) is clearly satisfied when X_1 or X_2 possess zero eigenvalues. Let us therefore assume that X_1 and X_2 are strictly positive definite. Each matrix X_i can be written as $X_i = (X_i^{1/2})^T X_i^{1/2}$. The square-root matrix, $X_i^{1/2}$, has a well-defined inverse denoted $X_i^{-1/2}$. After some technical work, the inequality specified in Eq. (12.21) can be written as

$$\det\left(\tfrac{1}{2}(I + X^T X)\right) \geq \left(\det(X)\right)^{1/2} \qquad X = X_2^{1/2} X_1^{-1/2}.$$

The matrices X and $I + X^T X$ share the same eigenvectors. Hence, if $\sigma_i > 0$ is the ith eigenvalue of X, then $1 + \sigma_i^2$ is the corresponding eigenvalue of $I + X^T X$. Since $\frac{1}{2}(1 + \sigma_i^2) \geq \sqrt{\sigma_i}$ for $i = 1 \ldots n$, the product of the eigenvalues satisfies the inequality specified in Eq. (12.21). ○

Exercise 12.8: Instead of minimizing the worst-case normal force components, one can minimize the sum of the normal force components, $\varphi(\vec{f}) = \sum_{i=1}^{k} f_i^n$, or the squared norm of the finger forces, $\varphi(\vec{f}) = \sum_{i=1}^{k} ||f_i||^2$. Verify that both are convex functions and, hence, can be optimized as LMI problems.

Exercise 12.9: Verify that the cost function $\varphi(\vec{f}) = \max\{||f_1||, \ldots, ||f_k||\}$ is *quasiconvex* in the finger contact forces, and hence, can be optimized as an LMI problem.

Section 12.4

Exercise 12.10: Verify that at an equilibrium state of the grasp system described in Eq. (12.12), all higher derivatives of $q(t)$ such as the object's acceleration must be zero.

Exercise 12.11*: Obtain the linearized dynamics of the nonlinear grasp system described in Eq. (12.12) at (p_0, u_0), assuming pointed fingertips. (The derivative of $M(q)$ at $q = q_0$ does not appear in the linearization at $p_0 = (q_0, \vec{0})$.)

Exercise 12.12: Obtain the formula for the $m \times m$ matrix $DG(q_0)\vec{f}_0$ that appears in the linearized dynamics of the grasped object system, described in Eq. (12.15).

Solution: Let $\vec{f}_0 = (f_1^0, \ldots, f_k^0)$ denote the contact forces at the equilibrium grasp. Let us obtain the formula for $DG(q_0)\vec{f}_0$ using the *grasp map*, $\mathcal{L}_G : C_1 \times \cdots \times C_k \to T_{q_0}^* \mathbb{R}^m$, where $m = 6$ in the case of 3-D grasps. The grasp map associated with frictional hard-point contacts has the form (see Chapter 4)

$$\mathcal{L}_G(q_0)\vec{f}_0 = \begin{bmatrix} I & \cdots & I \\ [\rho_1^0 \times] & & [\rho_k^0 \times] \end{bmatrix} \begin{pmatrix} f_1^0 \\ \vdots \\ f_k^0 \end{pmatrix} = \sum_{i=1}^{k} \begin{pmatrix} f_i^0 \\ \rho_i^0 \times f_i^0 \end{pmatrix} \qquad q_0 = (d_0, \theta_0),$$

where $\rho_i^0 = R(\theta_0)b_i$ is the vector from $\mathcal{B}$'s frame origin to the ith contact point, expressed in the world frame. Consider the derivative of $\mathcal{L}_G(q(t))\vec{f}_0$ along a c-space path $q(t)$, such that $q(0) = q_0$ and $\dot{q}(0) = \dot{q}$. Using the chain rule,

$$\begin{aligned} \frac{d}{dt}\bigg|_{t=0} \mathcal{L}_G(q(t))\vec{f}_0 &= \sum_{i=1}^{k} \begin{pmatrix} \vec{0} \\ (\omega \times \rho_i^0) \times f_i^0 \end{pmatrix} \\ &= \sum_{i=1}^{k} \begin{bmatrix} O & O \\ O & \rho_i^0 (f_i^0)^T - (\rho_i^0 \cdot f_i^0) I \end{bmatrix} \begin{pmatrix} v \\ \omega \end{pmatrix} \qquad \dot{q} = (v, \omega), \end{aligned}$$

where we used the triple vector product identity: $(\omega \times \rho_i^0) \times f_i^0 = [\rho_i^0 (f_i^0)^T - (\rho_i^0 \cdot f_i^0) I]\omega$. In this expression, O is a 3×3 zero matrix and I is a 3×3 identity matrix. Based on the relation, $\frac{d}{dt}\big|_{t=0} \mathcal{L}_G(q(t))\vec{f}_0 = [D\mathcal{L}_G(q_0)\vec{f}_0]\dot{q}$, the expression for $D\mathcal{L}_G(q_0)\vec{f}_0$ is given by the 6×6 matrix:

$$D\mathcal{L}_G(q_0)\vec{f}_0 = \sum_{i=1}^{k} \begin{bmatrix} O & O \\ O & \rho_i^0 (f_i^0)^T - (\rho_i^0 \cdot f_i^0) I \end{bmatrix}.$$

○

Exercise 12.13*: Consider the control law, $G(q)\vec{f}(t) = -\nabla V(q) - K_d(\dot{q} - \vec{0}) + h(q, \dot{q}) + \nabla U(q)$, where $V(q)$ is any artificial potential function satisfying $V(q_0) = 0$ and $V(q) > 0$ in a local c-space neighborhood centered at q_0. Prove that this control law stabilizes the grasped object system at the equilibrium grasp. (The grasped object's total mechanical energy, $E(q, \dot{q}) = \frac{1}{2}\dot{q}^T M(q)\dot{q} + V(q)$, forms a Lyapunov function for the closed-loop grasp system.)

Exercise 12.14*: The term $h(q, \dot{q})$ represents centripetal and Coriolis torques in the grasped object dynamics. Can this term be removed from the nonlinear control law of Eq. (12.16) without affecting the local asymptotic stability of the equilibrium grasp?

References

[1] R. M. Murray, Z. Li and S. S. Sastry, *A Mathematical Introduction to Robotic Manipulation*. CRC Press, 1994.

[2] J. R. Kerr and B. Roth, "Analysis of multi-fingered hands," *International Journal of Robotics Research*, vol. 4, no. 4, pp. 3–17, 1986.

[3] F. Cheng and D. Orin, "Efficient algorithm for optimal force distribution – The compact-dual *LP* method," *IEEE Transactions on Robotics and Automation*, vol. 6, no. 2, pp. 25–32, 1990.

[4] M. Buss, H. Hashimoto and J. Moore, "Dextrous hand grasping force optimization," *IEEE Transactions on Robotics and Automation*, vol. 12, no. 3, pp. 406–418, 1996.

[5] L. Han, J. C. Trinkle and Z. X. Li, "Grasp analysis as linear matrix inequality problems," *IEEE Transactions on Robotics and Automation*, vol. 16, no. 6, pp. 663–674, 2000.

[6] M. S. Lobo, L. Vandenberghe, S. Boyd and H. Lebret, "Applications of second-order cone programming," *Linear Algebra and Its Applications*, vol. 284, pp. 193–228, 1998.

[7] N. Brook, M. Shoham and J. Dayan, "Controllability of grasps and manipulations in multi-fingered hands," *IEEE Transactions on Robotics and Automation*, vol. 14, no. 1, pp. 185–192, 1998.

[8] D. T. Greenwood, *Advanced Dynamics*. Cambridge University Press, 2006.

[9] A. Shapiro, "Stability of second-order asymmetric linear systems with application to robot grasping," *ASME Journal of Applied Mechanics*, vol. 72, no. 6, pp. 966–968, 2005.

[10] D. J. Montana, "Contact stability for two-fingered grasps," *IEEE Transactions on Robotics and Automation*, vol. 8, no. 4, pp. 421–230, 1992.

[11] K. M. Lynch and F. C. Park, *Modern Robotics: Mechanics, Planning, and Control*. Cambridge University Press, 2017.

[12] K. J. Astrom and R. M. Murray, *Feedback Systems: An Introduction for Scientists and Engineers*. Princeton University Press, 2008.

[13] G. F. Franklin, J. D. Powell and A. Emami-Naeini, *Feedback Control of Dynamic Systems*. Pearson, 2018.

[14] A. D. Lewis and R. M. Murray, "Configuration controllability of simple mechanical control systems," *SIAM Journal on Control and Optimization*, vol. 35, no. 3, pp. 766–790, 1997.

[15] H. J. Sussmann, "A general theorem on local controllability," *SIAM Journal on Control and Optimization*, vol. 25, no. 1, pp. 158–194, 1987.

[16] H. K. Khalil, *Nonlinear Systems*. Prentice Hall, 2002.

13 Grasp Quality Functions

One often needs a measure of grasp effectiveness during robot grasp planning and execution. A function that quantifies the effectiveness of a grasp is termed a *grasp quality function*.[1] Given a suitable grasp quality function, the process of grasp planning typically reduces to a constrained optimization problem. That is, the optimal grasp is the one that optimizes the quality function subject to grasp security constraints. Grasp planning based on machine learning techniques automatically approximate grasp quality functions derived from training examples. Grasp quality functions can additionally guide feedback control laws that command the robot hand joints during execution of a grasping task. Grasp quality functions thus form an important link between basic grasp mechanics principles and the processes of grasp planning and execution.

One would expect a useful grasp quality function to satisfy a few basic criteria. It ought to capture key physical parameters of interest in a clear and meaningful way. Hence, most grasp quality functions are formulated using parameters such as finger contact locations, finger contact forces and grasping task specification. A grasp quality function must be computable with a reasonable amount of effort so that fast online grasp planning can be practically realized. A grasp quality function should preferably be frame invariant. That is, the choice of units, coordinates, and reference frames should not affect the value of this function.

In designing or using a grasp quality function, one must keep in mind the important characteristics of the problem for which the function is being used. For example, many manufacturing processes require the design of fixtures to stably hold a workpiece during machining and assembly operations. Issues like maximum deflection of the fixtured workpiece under the expected manufacturing forces, or maximum net wrench experienced by the workpiece during manufacturing, are essential quantities to optimize in fixture design. Section 13.1 briefly overviews quality functions that measure rigid-body displacements or wrenches. In robot grasping systems, a typical robot hand must be able to securely grasp a variety of objects. Sections 13.2–13.4 describe grasp quality functions that guide the selection of finger placements on the object surface. Practical robot hands have limits on the magnitude of the finger forces that can be realized at the contacts. Alternatively, one may wish to minimize the contact force magnitudes

[1] The term *grasp metric* is also used instead of grasp quality function. However, a *metric* measures distance between points in a topological space and many grasp quality functions are *not* metrics in this sense.

when the grasped object may be easily damaged. Section 13.5 describes grasp quality functions that take into account the finger force magnitudes required to maintain the grasp. All of these functions measure the quality of a grasp in the presence of *generic disturbances* that may act on the grasped object. Section 13.6 describes quality functions formulated for grasping tasks where a dominant external wrench is expected to act on the grasped object.

13.1 Quality Functions Based on Rigid-Body Kinematics

This class of grasp quality functions is mainly concerned with workpiece displacements during manufacturing and assembly operations. The workpiece can often be modeled as a rigid object, $\mathcal{B}$, with a configuration $q \in \mathbb{R}^m$ where $m = 3$ or 6. The object velocity, $\dot{q} = (v, \omega) \in T_q\mathbb{R}^m$, can be used to approximate small object displacements. The Euclidean norm seems like a natural way to measure the size of such velocity vectors. But translations and rotations and hence their velocities have different physical units. In order to meaningfully combine instantaneous translations and rotations, one must choose a *scaling factor:* $\|\dot{q}\| = \sqrt{\|v\|^2 + \|\alpha \cdot \omega\|^2}$ for a rigid-body velocity $\dot{q} = (v, \omega)$. Unfortunately, such a function is *not* frame invariant (see Appendix II). The size of a rigid-body velocity vector depends upon the chosen units of measurement, clearly an undesirable property of any grasp quality function.

Similarly, one may wish to design a fixture arrangement that minimizes the worst-case net wrench, $w = (f, \tau) \in T_q^*\mathbb{R}^m$, experienced by the object $\mathcal{B}$ during manufacturing operations. Again, one may choose to use the Euclidean norm to measure the size of wrenches. The use of such a norm requires a *scaling factor* in order to relate forces and torques: $\|w\| = \sqrt{\|f\|^2 + \|\beta \cdot \tau^2\|}$ for a wrench $w = (f, \tau)$ that acts on $\mathcal{B}$. Like the velocity Euclidean norm, the wrench Euclidean norm is *not* frame invariant and hence needs to be cautiously applied in the computation of grasp quality functions.

Moving beyond simple Euclidean norms, one can more generally define *kinematic quality functions* as real-valued function of rigid-body configurations, velocities and wrenches that capture the behavior of a rigid object during manufacturing operations.[2]

DEFINITION 13.1 **(kinematic quality function)** *A rigid-body* **kinematic quality function** *associated with a grasped or fixtured rigid object $\mathcal{B}$ takes the general form*

$$Q(q_1, \ldots, q_{k_1};\ \dot{q}_1, \ldots, \dot{q}_{k_2};\ w_1, \ldots, w_{k_3}) : (\mathbb{R}^m)^{k_1} \times (T_q\mathbb{R}^m)^{k_2} \times (T_q^*\mathbb{R}^m)^{k_3} \to \mathbb{R}, \tag{13.1}$$

where $q_1, \ldots, q_{k_1} \in \mathbb{R}^m$ are rigid-body configurations, $\dot{q}_1, \ldots, \dot{q}_{k_2} \in T_q\mathbb{R}^m$ are rigid-body velocities, and $w_1, \ldots, w_{k_3} \in T_q^*\mathbb{R}^m$ are rigid-body wrenches, $m = 3$ or 6.

While a variety of functions can fit the form of Eq. (13.1), kinematic quality functions typically take one of the following forms. See Appendix I for a review of metrics and norms.

[2] As opposed to *kinematic* quality functions, *dynamic* quality functions additionally depend on rigid-body accelerations.

1. **Distance metrics:** positive semi-definite metric functions of the form $Q(q_1, q_2)$, which measure the displacement between two configurations of the object $\mathcal{B}$.
2. **Velocity and wrench norms:** positive semi-definite functions of the form $Q_q(\dot{q})$ or $Q_q(\boldsymbol{w})$, which measure the size of the object's rigid-body velocity or wrench.
3. **Inner products:** bilinear functions of the form $Q_q(\dot{q}_1, \dot{q}_2)$ or $Q_q(\dot{q}, \boldsymbol{w})$, which measure the relative angle or instantaneous work between their arguments.

The following examples describe quality functions that form a rigid-body distance metric and a rigid-body velocity norm.

Example: Let $\rho(b) : \mathcal{B} \to \mathbb{R}$ be a positive semi-definite function that might represent the mass density of a rigid object $\mathcal{B}$. When $\mathcal{B}$ is located at a configuration $q = (d, \theta)$, its body points are located at $X(q, b) = R(\theta)b + d$ for $b \in \mathcal{B}$, expressed in the fixed world frame. Given two rigid-body configurations, q_1 and q_2, a function that measures the relative displacement of $\mathcal{B}$ from q_1 to q_2 can be defined as

$$d(q_1, q_2) = \int_{\mathcal{B}} \rho(b) \cdot \|X(q_1, b) - X(q_2, b)\| dV. \tag{13.2}$$

The function $d(q_1, q_2)$ is a *frame-invariant* distance metric under any choice of norm, $\|\cdot\|$, in $\mathbb{R}^3$. This metric can be interpreted as the weighted average displacement of the object points from q_1 to q_2. ○

Example: When a rigid object $\mathcal{B}$ moves with velocity $\dot{q} = (v, \omega)$, its body points move along velocity vectors $\frac{d}{dt} X(q(t), b) = \omega \times (R(\theta)b) + v$ for $b \in \mathcal{B}$, expressed in the fixed world frame. Let $\Omega_{\mathcal{B}}$ be a fixed set of *feature points* on the object $\mathcal{B}$. The velocity dependent function,

$$\phi_q(\dot{q}) = \max_{b \in \Omega_{\mathcal{B}}} \{\|\omega \times (R(\theta)b) + v\|\} \qquad q = (d, \theta), \dot{q} = (v, \omega), \tag{13.3}$$

is a *frame-invariant* rigid-body velocity norm as long as $\Omega_{\mathcal{B}}$ contains three non-collinear points, under any choice of norm in $\mathbb{R}^3$. The function $\phi_q(\dot{q})$ represents the worst-case displacement of the object feature points along a small displacement of $\mathcal{B}$ approximated by $\dot{q} = (v, \omega)$. This norm can be used to select fixture arrangements that minimize workpiece deflection during manufacturing and assembly operations. ○

13.2 Quality Functions Based on the Grasp Matrix

Let a rigid object $\mathcal{B}$ be held at a k-finger equilibrium grasp, such that the world and object frames coincide at the grasp configuration q_0. Under this assumption, the wrench generated by a finger force f_i that acts on $\mathcal{B}$ at x_i is given by $\boldsymbol{w}_i = (f_i, x_i \times f_i)$. The *grasp matrix* $G(q_0)$ maps the vector of contact force components, $\vec{f} = (f_1, \ldots, f_k)$, to the net object wrench: $\boldsymbol{w} = G(q_0)\vec{f}$ (see Chapter 4). The grasp matrix is $m \times kp$, where $m = 3$ or 6, and p is the number of independent force and possibly moment components at each contact. For instance, in 3-D grasps governed by the frictional hard-point contact model, each finger force varies with $p=3$ independent parameters, and $G(q_0)$ forms the $6 \times 3k$ matrix

$$G(q_0)=\begin{bmatrix} G_1 & \cdots & G_k \end{bmatrix} \quad \text{where} \quad G_i=\begin{bmatrix} s_i & t_i & n_i \\ x_i \times s_i & x_i \times t_i & x_i \times n_i \end{bmatrix} \qquad i=1\ldots k, \tag{13.4}$$

where $\{s_i, t_i, n_i\}$ is a reference frame aligned with $\mathcal{B}$'s unit tangents and inward unit normal at x_i. Note that the contact forces must additionally satisfy inequality constraints associated with the Coulomb friction model, $f_i \in C_i$ for $i=1\ldots k$.

Quality functions based on the grasp matrix quantify how well the finger contacts are selected based on the *singular values* of $G(q_0)$. The singular value decomposition of $G(q_0)$ factors the $m \times kp$ grasp matrix into the product of three matrices

$$G(q_0) = U \cdot \Sigma \cdot V^T,$$

where U and V are $m \times m$ and $kp \times kp$ orthogonal matrices, and Σ is the $m \times kp$ block diagonal matrix:

$$\Sigma = \begin{bmatrix} \sigma_1 & 0 & 0 & \cdots & 0 & \cdots & 0 \\ 0 & \sigma_2 & 0 & \cdots & 0 & \cdots & 0 \\ 0 & 0 & \ddots & \cdots & 0 & \cdots & 0 \\ \vdots & \vdots & \vdots & \ddots & 0 & \cdots & 0 \\ 0 & 0 & \cdots & \cdots & \sigma_m & \cdots & 0 \end{bmatrix} \qquad \sigma_1, \ldots, \sigma_m \geq 0.$$

The *singular values* of $G(q_0)$ are defined as the diagaonal elements $\sigma_1, \ldots, \sigma_m$, where U and V are chosen such that $\sigma_1, \ldots, \sigma_m \geq 0$. The squared singular values of $G(q_0)$ are the eigenvalues of the $m \times m$ matrix $G(q_0)G^T(q_0)$.

Example: The singular values of $G(q_0)$ can be readily determined in frictionless 2-D grasps as follows. The grasp matrix associated with frictionless contacts located at $x_1, \ldots, x_k$ has the form:

$$G(q_0)=\begin{bmatrix} n_1 & \cdots & n_k \\ x_1 \times n_1 & \cdots & x_k \times n_k \end{bmatrix} \qquad x_i \times n_i = x_i^T J n_i,\ J=\begin{bmatrix} 0 & 1 \\ -1 & 0 \end{bmatrix}.$$

The eigenvalues of the 3×3 matrix,

$$G(q_0)G^T(q_0)=\begin{bmatrix} \sum_{i=1}^k n_i n_i^T & \sum_{i=1}^k (x_i \times n_i) n_i \\ \sum_{i=1}^k (x_i \times n_i) n_i^T & \sum_{i=1}^k (x_i \times n_i)^2 \end{bmatrix},$$

are the singular values of $G(q_0)$. Any set of finger contacts determines a distinguished point, $x_0 \in \mathbb{R}^2$, called the *center* of the grasp. When the world and object frames are set at this point, the matrix $G(q_0)G^T(q_0)$ attains the block diagonal form (see Exercises)

$$G(q_0)G^T(q_0) \cong \begin{bmatrix} \sum_{i=1}^k n_i n_i^T & \begin{matrix} 0 \\ 0 \end{matrix} \\ \begin{matrix} 0 & 0 \end{matrix} & \sum_{i=1}^k (x_i \times n_i)^2 \end{bmatrix},$$

where the contact points are expressed relative to the grasp center point. Under this selection of reference frames, the singular values of $G(q_0)$ correspond to the eigenvalues of the 2×2 matrix $\sum_{i=1}^{k} n_i n_i^T$ and the scalar $\sum_{i=1}^{k} (x_i \times n_i)^2$. ○

The singular values of $G(q_0)$ can be used to characterize the quality of a grasp in several ways. The following are three such classical quality functions.

Minimum singular value of the grasp matrix: When a rigid object $\mathcal{B}$ is held at a k-finger equilibrium grasp, a measure of *wrench resistance* of the grasp is the smallest singular value of the grasp matrix. If a given equilibrium grasp is *not* wrench resistant, the smallest singular value of $G(q_0)$ must be zero (see Exercises). The *minimum singular value* quality function is defined as

$$Q_{MSV}(G) = \min\{\sigma_1, \ldots, \sigma_m\}.$$

The function Q_{MSV} estimates the margin of security of a grasp when the primary consideration of grasp security is wrench resistance. The larger the magnitude of Q_{MSV}, the better the margin of security of this grasp. This quantity can also be interpreted as the grasp *mechanical advantage,* which measures the minimum amplification of the contact force magnitudes to net wrench magnitudes affecting the grasped object. However, one should be cautious when using the function Q_{MSV}, since its value depends on the choice of the object reference frame as illustrated in the following example.

Example: Figure 13.1 depicts a rectangular object $\mathcal{B}$ held by four disc fingers via frictionless contacts located at $x_1 = (4, -2)$, $x_2 = (3, -3)$, $x_3 = (-4, 2)$ and $x_4 = (-3, 3)$. The object is immobilized by the disc fingers and is therefore held in a wrench resistant grasp. Assume the fixed world frame is located at the object center, and consider two different object frames, $\mathcal{F}_1$ and $\mathcal{F}_2$, depicted in Figure 13.1. The grasp matrices associated with $\mathcal{F}_1$ and $\mathcal{F}_2$ are given by

$$G_{\mathcal{F}_1}(q_0) = \begin{bmatrix} -1 & 0 & 1 & 0 \\ 0 & 1 & 0 & -1 \\ -2 & 3 & -2 & 3 \end{bmatrix} \quad \text{and} \quad G_{\mathcal{F}_2}(q_0) = \begin{bmatrix} -1 & 0 & 1 & 0 \\ 0 & 1 & 0 & -1 \\ -2 & 7 & -2 & -1 \end{bmatrix}.$$

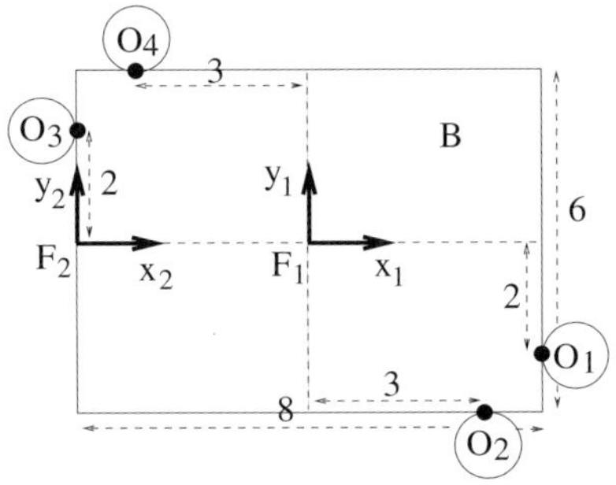

Figure 13.1 A frictionless four-finger grasp of a rigid object $\mathcal{B}$ with two choices of the object reference frame, $\mathcal{F}_1$ and $\mathcal{F}_2$.

To see how the choice of object frame affects the singular values of $G(q_0)$, consider the matrix products:

$$G_{\mathcal{F}_1}(q_0)G^T_{\mathcal{F}_1}(q_0) = \begin{bmatrix} 2 & 0 & 0 \\ 0 & 2 & 0 \\ 0 & 0 & 26 \end{bmatrix} \quad \text{and} \quad G_{\mathcal{F}_2}(q_0)G^T_{\mathcal{F}_2}(q_0) = \begin{bmatrix} 2 & 0 & 0 \\ 0 & 2 & 8 \\ 0 & 8 & 58 \end{bmatrix}.$$

When the frame $\mathcal{F}_1$ is selected, $Q_{MSV} = \sqrt{2}$. When the frame $\mathcal{F}_2$ is selected, $Q_{MSV} \cong 0.94$. Hence, when comparing alternative grasps using Q_{MSV}, the choice of object frame might influence the selection of the optimal grasp, which is clearly an undesirable property. ○

Volume of the grasp matrix ellipsoid: Consider a unit ball of finger contact force components, $\mathcal{D} = \{(f_1, \ldots, f_k) \in \mathbb{R}^{kp} : \sum_{i=1}^{k} \|f_i\|^2 \leq 1\}$. This unit ball is mapped by $G(q_0)$ to an m-dimensional ellipsoid in $\mathcal{B}$'s wrench space, termed the *grasp matrix ellipsoid.* The ellipsoid volume is the product of the principal axes half-lengths (multiplied by a numerical constant). The half-lengths of the grasp matrix ellipsoid are precisely the *singular values* of $G(q_0)$. Omitting the numerical constant, the volume of the grasp matrix ellipsoid is given by

$$\sigma_1 \cdot \sigma_2 \cdots \sigma_m = \sqrt{\det(\Sigma \cdot \Sigma^T)} = \sqrt{\det\left(G(q_0)G^T(q_0)\right)},$$

where we used the identities $V^T V = I$, $\det(AB) = \det(A) \cdot \det(B)$ and $|\det(U)| = 1$. The *wrench ellipsoidal volume* quality function is defined as the volume of the grasp matrix ellipsoid:

$$Q_{EV}(G) = \sqrt{\det\big(G(q_0)G^T(q_0)\big)}. \tag{13.5}$$

The function Q_{EV} attains zero value when the $m \times m$ matrix $G(q_0)G^T(q_0)$ becomes singular or, equivalently, when $G(q_0)$ loses its full rank. We know from Chapter 12 that wrench resistant grasps possess a full rank grasp matrix. Hence, Q_{EV} attains strictly positive values at such grasps. The function Q_{EV} estimates the *control authority* the grasping fingers have on the grasped object $\mathcal{B}$. As Q_{EV} attains higher values by varying the finger contact locations, small changes of the finger forces will be amplified to larger size control wrenches that can be affected on $\mathcal{B}$ while resisting external disturbances. Moreover, this quality function is *invariant* under different choices of the object reference frame (see Appendix II).

Isotropy of the grasp matrix ellipsoid: Let σ_{min} and σ_{max} denote the minimal and maximal singular values of the grasp matrix $G(q_0)$. The *grasp isotropy* is defined as the ratio

$$Q_{GI}(G) = \frac{\sigma_{min}}{\sigma_{max}} \qquad 0 \leq \sigma_{min} \leq \sigma_{max}.$$

The function Q_{GI} measures the *eccentricity* of the grasp matrix ellipsoid, a quantity that varies in the unit interval. To understand the meaning of Q_{GI}, suppose the grasped object $\mathcal{B}$ is subjected to an m-dimensional ball of disturbance wrenches in $\mathcal{B}$'s wrench space. When the grasp matrix is surjective, each of these wrenches can be resisted by finger contact forces of the form: $\vec{f}_p + \vec{f}_{null} \in \mathbb{R}^{kp}$, such that $\vec{f}_{null}$ lies in the null space of $G(q_0)$. The resistant contact forces, $\vec{f}_p$ such that $G(q_0)\vec{f}_p \neq \vec{0}$, span an m-dimensional linear subspace in $\mathbb{R}^{kp}$. When Q_{GI} approaches unity, the resistant contact forces needed to balance a ball of disturbance wrenches approximate a uniform ball of force components. The finger force magnitudes thus become more evenly balanced when resisting a ball of wrench disturbances that may act on $\mathcal{B}$. However, the value of the function Q_{GI} depends on the choice of the object reference frame.

The following example provides a numerical comparison of the three grasp matrix quality functions.

Example: Figure 13.2(a) shows a rigid box $\mathcal{B}$ of length $2r$ held under the influence of gravity by two fingers. The finger contacts are governed by the soft-point contact model, where each finger applies forces in the respective friction cone as well as torsional torque about the respective contact normal. Each contact force and torque varies with $p=4$ independent parameters, thus giving the 6×8 grasp matrix:

$$G(q_0) = \begin{bmatrix} 0 & 1 & 0 & 0 & 1 & 0 & 0 & 0 \\ 0 & 0 & 1 & 0 & 0 & 0 & -1 & 0 \\ 1 & 0 & 0 & 0 & 0 & 1 & 0 & 0 \\ -r & 0 & 0 & 0 & 0 & r & 0 & 0 \\ 0 & 0 & 0 & 1 & 0 & 0 & 0 & -1 \\ 0 & r & 0 & 0 & -r & 0 & 0 & 0 \end{bmatrix} \tag{13.6}$$

Note that two soft-point contacts suffice to ensure wrench resistance of this grasp. Figure 13.2(b) plots the three quality measures as a function of the object half-length r. The minimum singular value measure, $Q_{MSV}(G)$, initially captures the relatively small finger torques associated with small values of r. These torques increase with r until the minimum singular value of $G(q_0)$ becomes associated with the sideways finger forces (as well as torsional torques), which do *not* depend on r. The volume measure, $Q_{EV}(G)$,

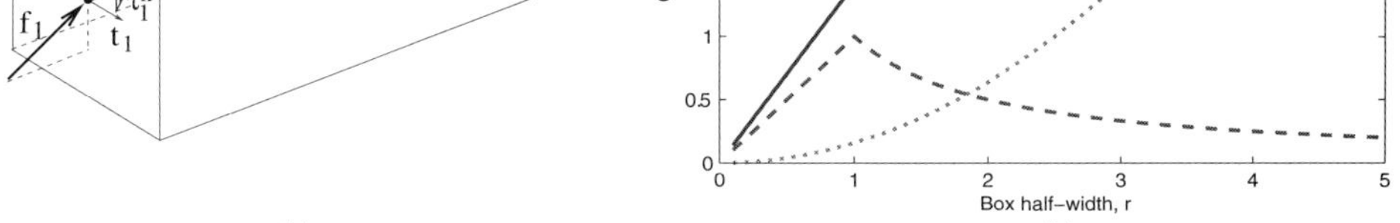

(a) (b)

Figure 13.2 (a) A two-finger grasp of a rigid box via soft-point contacts under gravity (the fingers are not shown). (b) Plot of the minimum singular value, $Q_{MSV}(G)$, the scaled grasp matrix ellipsoid volume, $Q_{EV}(G)$, and the isotropy index, $Q_{GI}(G)$, as a function of the object half-length r.

increases monotonically with r. This behavior reflects the fingers' ability to generate larger torques about the object x and z axes as r increases. The isotropy measure, $Q_{GI}(G)$, initially increases with r and attains a maximal value of unity at $r = 1$. At this object size, a unit ball of wrench disturbances acting on $\mathcal{B}$ can be resisted by a balanced ball of finger force components. As r increases beyond unity, σ_{min} becomes associated with the sideways finger forces (as well as torsional torques), while σ_{max} becomes associated with the finger torques about the object x and z axes. Since σ_{min} remains constant and σ_{max} increases linearly with r, $Q_{GI} = \sigma_{min}/\sigma_{max}$ decreases proportionally to $1/r$. This behavior indicates a decreased balance between the finger force magnitudes required to resist a unit ball of wrench disturbances that may act on $\mathcal{B}$ at this grasp. ○

13.3 Quality Functions Based on the Grasp Polygon

This section describes somewhat naive grasp quality functions which are based on the polygon (or polyhedron in 3-D grasps) formed by the finger contacts. Focusing on 2-D grasps, the *grasp polygon* is defined as the polygon whose vertices are located at the finger contact points, $x_1, \ldots, x_k$, ordered as they appear along the grasped object boundary (Figure 13.3).

Grasp polygon regularity index: It is intuitively appealing to choose grasps whose finger contacts spread out uniformly around the grasped object boundary. Evenly spread finger contacts will most likely evenly spread the contact force magnitudes required to maintain the equilibrium grasp. Based on this argument, the grasp polygon of a k-finger grasp should ideally form a regular k-gon. The grasp polygon quality is measured by how much the finger contacts deviate from the regular k-gon distribution. Let θ_i denote the internal angle of the grasp polygon at the vertex x_i, measured between the grasp polygon edges (Figure 13.3). A regular k-gon has identical interior angles of magnitude $\gamma = \pi - 2\pi/k$. The *grasp polygon regularity index* is defined as the normalized sum:

$$Q_{GPRI}(\theta_1, \ldots, \theta_k) = \frac{1}{\theta_{max}} \sum_{i=1}^{k} |\theta_i - \gamma|$$

where θ_{max} measures the deviation of the regular k-gon from the most ill-conditioned k-gon, formed when a k-gon collapses to a line segment with internal angles 0 or π.

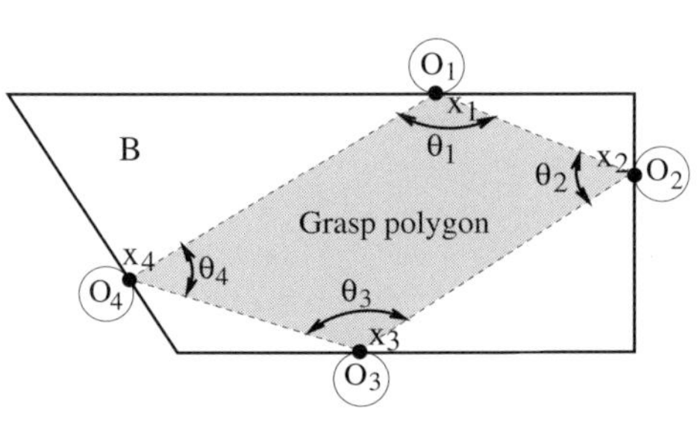

Figure 13.3 The grasp polygon of a quadrilateral object held by four disc fingers.

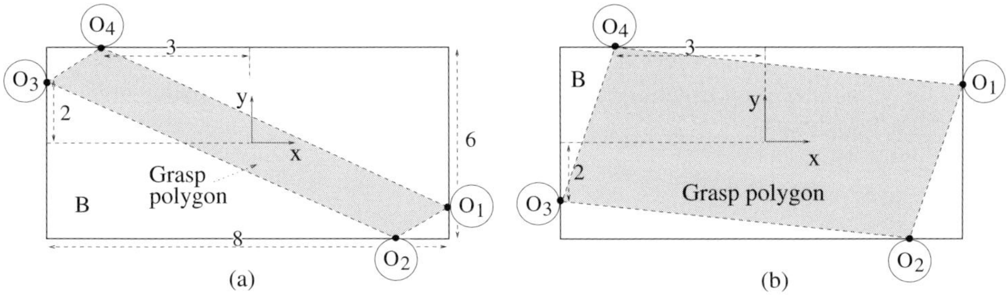

Figure 13.4 (a) The grasp polygon of a secure wrench resistant grasp with regularity index $Q_{GPRI} = 0.105$. (b) The grasp polygon of a non-secure grasp with regularity index $Q_{GPRI} = 0.035$. This grasp has a better finger spread but is *not* wrench resistant.

When a k-gon collapses to such a line segment, $\theta_1 = \theta_k = 0$ while $\theta_i = \pi$ for $i = 2 \ldots k-1$. Hence, $\theta_{max} = |0-\gamma|+|\pi-\gamma|+\cdots+|\pi-\gamma|+|0-\gamma| = (k-2)(\pi-\gamma)+2\gamma$. In particular, $Q_{GPRI} = 0$ at the optimal spread of the finger contacts, while $Q_{GPRI} = 1$ at the worst possible grasp regularity.

The grasp polygon regularity index emphasizes even spread of the finger contacts. However, it may be *poorly conditioned* in terms of grasp security as illustrated in the following example.

Example: Figure 13.4 shows two alternative grasps of a rectangular object $\mathcal{B}$ by four disc fingers. The grasp in Figure 13.4(a) is wrench resistant and hence secure. The grasp in Figure 13.4(b) is *not* wrench resistant and hence non-secure (see Exercises). The grasp polygons in both cases form parallelograms. The vertices of the first grasp polygon are located at $x_1 = (4, -2)$, $x_2 = (3, -3)$, $x_3 = (-4, 2)$ and $x_4 = (-3, 3)$. For this polygon $\theta_1 = 80.5°$ and $\theta_2 = 180° - \theta_1 = 99.5°$. Given that $\gamma = \pi/2$ and $\theta_{max} = 2\pi$, the regularity index of this grasp is $Q_{GPRI} = \frac{1}{360°}(|2\cdot(80.5°-90°)|+|2\cdot(99.5°-90°)|) = 0.105$. The vertices of the second grasp polygon are located at $x_1 = (4, 2)$, $x_2 = (3, -3)$, $x_3 = (-4, -2)$ and $x_4 = (-3, 3)$. For this polygon, $\theta_1 = 87°$ and $\theta_2 = 180° - \theta_1 = 93°$. The regularity index of this grasp, $Q_{GPRI} = \frac{1}{360°}(|2\cdot(87°-90°)| + |2\cdot(93°-90°)|) = 0.035$, reflects a better finger spread at this *non-secure* grasp. ○

Grasp polygon centroid: The geometric center of the grasp polygon (or grasp polyhedron in 3-D grasps) is termed the *grasp centroid.* When the grasp polygon has nonempty interior the position of its centroid, $\boldsymbol{x}_c$, is determined by the formula

$$\boldsymbol{x}_c = \frac{1}{A}\int x \cdot I_X(x)\,dx,$$

where A is the area of the grasp polygon, and $I_X(x)$ is an *indicator function* which attains unit value when x lies in the polygon and zero value elsewhere. The grasp centroid can be computed in terms of the grasp polygon vertices with the formula

$$\boldsymbol{x}_c = \frac{1}{6A}\sum_{i=1}^{k}(x_i \times x_{i+1})\cdot(x_i + x_{i+1}) \qquad A = \tfrac{1}{2}\textstyle\sum_{i=1}^{k} x_i \times x_{i+1},$$

where index addition is taken modulo k, $x_i \times x_{i+1} = x_i^T J x_{i+1}$ such that $J = \begin{bmatrix} 0 & 1 \\ -1 & 0 \end{bmatrix}$. When a rigid object $\mathcal{B}$ is held under the influence of gravity, the *grasp centroid displacement* quality function, Q_{GPCD}, is defined as the distance between $\boldsymbol{x}_c$ and $\mathcal{B}$'s center of mass at $\boldsymbol{x}_{cm}$

$$Q_{GPCD}(x_1, \ldots, x_k) = d(\boldsymbol{x}_c, \boldsymbol{x}_{cm}),$$

where $d(\boldsymbol{x}_c, \boldsymbol{x}_{cm})$ is any distance metric in $\mathbb{R}^2$. For instance, $d(\boldsymbol{x}_c, \boldsymbol{x}_{cm}) = \|\boldsymbol{x}_c - \boldsymbol{x}_{cm}\|$ under the Euclidean norm. The finger force magnitudes required to balance gravity are *reduced* when Q_{GPCD} approaches zero (see Exercises).

13.4 Quality Functions Based on Contact Point Locations

Let us recall the notion of *contact space* from Chapter 5. Which we will use in the context of 2-D grasps. The grasped object boundary is parametrized by a continuous function, $x(u): [0, L] \to \text{bdy}(\mathcal{B})$, where L is the object perimeter, and $x(0) = x(L)$. The ith finger contact is parametrized by $x_i = x(u_i)$. *Contact space* is defined as the space of k contact locations, $\mathcal{U} = (u_1, \ldots, u_k) \in \mathbb{R}^k$, with the associated periodicity rules. The vector $\vec{u} = (u_1, \ldots, u_k)$ thus determines a particular arrangement of the k finger contacts on the grasped object boundary.

Grasp quality functions based on contact point locations measure the location of a candidate k-contact arrangement within the following *wrench resistance regions.*

DEFINITION 13.2 *The* **wrench resistance regions** *of a rigid object $\mathcal{B}$ associated with k finger contacts are the connected sets in contact space $\mathcal{U}$ of all wrench resistant grasps of $\mathcal{B}$.*

Under frictional contact conditions wrench resistance is equivalent to feasibility of a *non-marginal equilibrium grasp* at the contacts (Chapter 12). The wrench resistance regions of 2-D grasps having $k \geq 2$ frictional contacts form k-dimensional sets in contact space $\mathcal{U}$. For instance, the wrench resistance regions associated with two frictional contacts form two-dimensional sets in the (u_1, u_2) plane (Figure 13.5(b)). In the case of frictionless contacts, wrench resistance is equivalent to *first-order immobilization* of the object $\mathcal{B}$, which requires $k \geq 4$ frictionless contacts in the case of 2-D grasps (Chapter 6). The wrench resistance regions associated with $k \geq 4$ frictionless contacts also form k-dimensional sets in contact space $\mathcal{U}$.

To obtain an intuitive feel for grasp quality functions measured in contact space, consider a polygonal object $\mathcal{B}$ held by *point fingers.* Under this assumption, the wrench resistance regions form polyhedral sets in contact space $\mathcal{U}$, as stated in the following lemma, whose proof appears in Appendix III.

LEMMA 13.1 (**wrench resistance regions**) *Let a polygonal rigid object $\mathcal{B}$ be held by $k \geq 2$ frictional point fingers or by $k \geq 4$ frictionless point fingers. The wrench resistance regions of $\mathcal{B}$ form* **k-dimensional polyhedral sets** *in contact space $\mathcal{U} \cong \mathbb{R}^k$.*

The wrench resistance regions are illustrated in the following example.

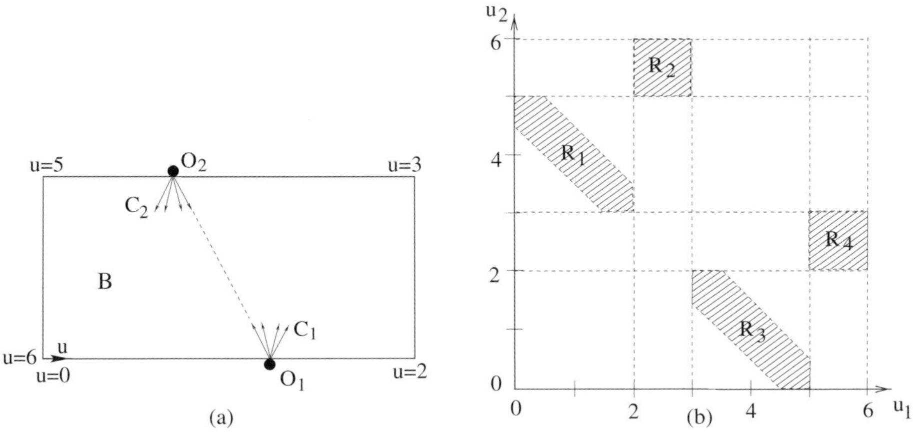

Figure 13.5 (a) A rectangular object $\mathcal{B}$ grasped by two frictional point fingers. (b) Contact space contains four symmetrically arranged wrench resistance regions $\mathcal{R}_1$, $\mathcal{R}_2$, $\mathcal{R}_3$ and $\mathcal{R}_4$.

Example: Figure 13.5(a) shows a rectangular object $\mathcal{B}$ whose vertices are located at $u = 0$, $u = 2$, $u = 3$, $u = 5$ and then back to the initial vertex at $u = 6$. The object is grasped by two frictional point fingers, with a uniform coefficient of friction $\mu = \tan(30^\circ) \cong 0.577$. Contact space forms the (u_1, u_2) rectangle shown in Figure 13.5(b), with the periodicity rules $u_1(0) = u_1(6)$ and $u_2(0) = u_2(6)$ (a topological torus). There are four wrench resistance regions in contact space, denoted $\mathcal{R}_1$, $\mathcal{R}_2$, $\mathcal{R}_3$, and $\mathcal{R}_4$. These regions are symmetrically arranged with respect to the diagonal, $u_1 = u_2$, since permutation of the finger placements does not affect the grasp's wrench resistance. While the wrench resistance regions correspond to particular edges of $\mathcal{B}$ in this example, these regions can generally span several contiguous edges along the grasped object boundary. ○

Practical robot hands are not able to precisely place their fingertips at specified points along a grasped object boundary. Hence, grasps that are *robust* with respect to small finger placement errors provide an important grasp quality measure. Contact space provides a convenient way to assess this type of robustness, and two such measures are discussed next.

Finger contact placement robustness quality function: One way to measure grasp robustness with respect to finger placement errors is based on the minimal distance between a given contact arrangement, $\vec{u} \in \mathcal{U}$, and the boundary of the wrench resistance region containing this point. Let $\mathcal{R}$ denote a wrench resistance region in contact space $\mathcal{U}$, and let $\mathsf{bdy}(\mathcal{R})$ denote its boundary. The *finger contact placement robustness* quality function, Q_{FCPR}, measures the minimum distance between $\vec{u}$ and $\mathsf{bdy}(\mathcal{R})$,

$$Q_{FCPR}(\vec{u}) = \min_{\vec{r} \in \mathsf{bdy}(\mathcal{R})} \{\|\vec{r} - \vec{u}\|\},$$

where $\|\cdot\|$ is any norm in $\mathbb{R}^k$. The quality function Q_{FCPR} is zero on the boundary of $\mathcal{R}$, which corresponds to grasps that are no longer wrench resistant. Higher values of Q_{FCPR} indicate better margins for finger placement errors, as illustrated in the following example.

Example: Consider the wrench resistance regions $\mathcal{R}_2$ and $\mathcal{R}_4$ depicted in Figure 13.5(b). The function $Q_{FCPR}(\vec{u})$ attains a maximal value of $\frac{1}{4}$ at the centers of these regions. The object $\mathcal{B}$ should therefore be held with an antipodal grasp, with the finger contacts located at the center points of the object left and right edges. Next consider the wrench resistance regions $\mathcal{R}_1$ and $\mathcal{R}_3$. The function Q_{FCPR} attains a maximal value of $\frac{1}{8}$ along these regions diagonal axis of symmetry. The corresponding finger contacts are antipodal and should be placed anywhere within unit-length segments centered on the lower and upper edges of the object $\mathcal{B}$. ○

Contact independent regions: Let $\mathcal{T}$ form a hyperrectangle region in contact space $\mathcal{U}$ aligned with the contact space axes, such that $\mathcal{T}$ lies in a wrench resistance region $\mathcal{R}$. The region $\mathcal{T}$ is *contact independent,* since the position of each finger contact within the corresponding object boundary segment is completely independent of the other finger contact positions. Consider the geometric center of $\mathcal{T}$. The half-lengths of $\mathcal{T}$ quantify how much the fingers can be independently misplaced from the geometric center of $\mathcal{T}$ while still maintaining wrench resistant grasps. This measure can be used to define the *contact independent region* quality function, Q_{CIR}, as the minimum size of $\mathcal{T}$,

$$Q_{CIR}(\mathcal{T}) = \min\{|I_1(\mathcal{T})|, \ldots, |I_k(\mathcal{T})|\} \qquad \mathcal{T} \subseteq \mathcal{R},$$

where $|I_i(\mathcal{T})|$ is the length of the interval formed by projecting $\mathcal{T}$ on the u_i-axis for $i = 1 \ldots k$. Larger values of $Q_{CIR}(\mathcal{T})$ indicate more robust grasps with respect to independent finger placement errors. One can thus identify which of the contact independent regions has the highest value of Q_{CIR} and then select the geometric center of this region as the optimal finger placement.

Example: Consider the two-finger grasp of the rectangular object $\mathcal{B}$ depicted in Figure 13.6(a). The wrench resistance regions are shown in Figure 13.6(b). The largest rectangular region contained in $\mathcal{R}_2$ and $\mathcal{R}_4$ occupies the entirety of these regions, $\mathcal{T}{=}\mathcal{R}_2$

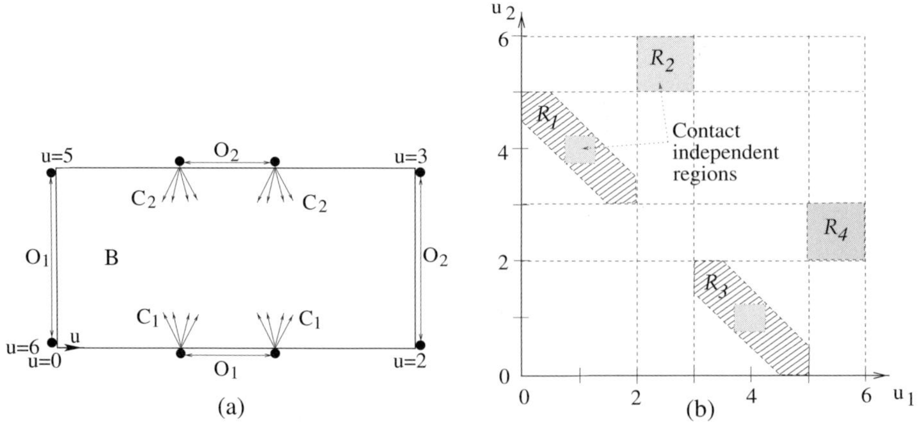

Figure 13.6 (a) A rectangular object $\mathcal{B}$ grasped by two frictional point fingers. (b) The largest contact independent regions are shown in solid gray within the wrench resistance regions.

and $\mathcal{T} = \mathcal{R}_4$, such that $Q_{CIR}(\mathcal{T}) = 1$ in each of these regions. The largest rectangular regions contained in $\mathcal{R}_1$ and $\mathcal{R}_3$, depicted in Figure 13.5(b), possess a lower value of $Q_{CIR}(\mathcal{T}) = 1/2$. The contact independent regions in $\mathcal{R}_2$ and $\mathcal{R}_4$ are thus more robust. These regions allow independent finger placement anywhere on the left and right edges of $\mathcal{B}$. The contact independent regions in $\mathcal{R}_1$ and $\mathcal{R}_3$ allow independent finger placement on shorter segments located at the middle of the lower and upper edges of $\mathcal{B}$. ○

13.5 Quality Functions Based on Contact Force Magnitudes

No matter what type of contact model holds at the finger contacts, the finger force magnitudes are constrained by the robot hand actuation mechanism (see Chapter 16). This practical consideration underlies grasp quality functions that measure the finger force magnitudes required to maintain a given grasp. A reasonable way to model practical limits on the finger force magnitudes would be to assume that each *normal* finger force component, f_i^n, is bounded within the unit interval:

$$0 \leq f_i^n \leq 1 \qquad i = 1 \dots k.$$

For frictional point contacts the tangential force components, (f_i^s, f_i^t), are consequently bounded as well,

$$\sqrt{(f_i^s)^2 + (f_i^t)^2} \leq \mu_i f_i^n \qquad i = 1 \dots k,$$

where μ_i is the coefficient of friction at x_i. This set of bounded finger forces is a truncation of the friction cone C_i by a tangent plane parallel shifted from the contact point x_i by a unit distance along the contact normal n_i. The bounded finger forces at x_i generate a *bounded* finger wrench cone in $\mathcal{B}$'s wrench space,

$$\overline{\mathcal{W}}_i = \left\{ \boldsymbol{w}_i = G_i f_i \; : \; 0 \leq f_i^n \leq 1 \quad \text{and} \quad \sqrt{(f_i^s)^2 + (f_i^t)^2} \leq \mu_i f_i^n \right\},$$

where G_i is the ith submatrix of the grasp matrix, $G(q_0) = [G_1 \cdots G_k]$. Quality functions based on contact force magnitudes are most naturally formulated in terms of the *convex hull* of the bounded finger wrench cones, defined as follows.

DEFINITION 13.3 (**bounded wrench cone $\overline{\mathcal{W}}$**) *Let a rigid object $\mathcal{B}$ be held at a k-finger grasp. The convex hull of the fingers' bounded wrench cones, $\overline{\mathcal{W}} = \text{co}\left(\overline{\mathcal{W}}_1, \dots, \overline{\mathcal{W}}_k\right)$, forms the* **bounded net wrench cone** *of the grasp.*

The bounded wrench cone $\overline{\mathcal{W}}$ consists of all net wrenches that can be affected on $\mathcal{B}$ by feasible finger forces whose normal components satisfy the inequality $\sum_{i=1}^{k} |f_i^n| \leq 1$. There are a number of ways to assess the quality of a grasp based on properties of the bounded net wrench cone $\overline{\mathcal{W}}$.

Largest inscribed ball quality function: At a wrench resistant grasp, $\overline{\mathcal{W}}$ contains an m-dimensional ball centered at $\mathcal{B}$'s wrench space origin, where $m = 3$ or 6. The smallest size wrench on the boundary of $\overline{\mathcal{W}}$ defines the quality of this grasp:

$$Q_{MINW}\left(\overline{\mathcal{W}}_1, \dots, \overline{\mathcal{W}}_k\right) = \min_{\boldsymbol{w} \in \text{bdy}(\overline{\mathcal{W}})} \{\|\boldsymbol{w}\|\}.$$

The function Q_{MINW} measures the radius of the largest m-ball centered at $\mathcal{B}$'s wrench space origin, such that the ball is inscribed in $\overline{\mathcal{W}}$. Its value represents the worst-case wrench size that can be resisted by bounded finger forces. Increasing the value of Q_{MINW} by varying the finger contact locations allows the fingers to better reject disturbances that may act on $\mathcal{B}$. However, a key issue is how to define the size or norm of a wrench $\boldsymbol{w}$ in a *frame invariant* manner. Unfortunately, this quality function is *not* frame invariant when the Euclidean norm is used to measure the size of $\boldsymbol{w}$. This important issue is illustrated in the following example.

Example: Figure 13.7(a) depicts an earlier example where a rectangular object $\mathcal{B}$ is held by four disc fingers via frictionless contacts located at $x_1 = (4, -2)$, $x_2 = (3, -3)$, $x_3 = (-4, 2)$ and $x_4 = (-3, 3)$. As noted earlier, the object $\mathcal{B}$ is held in a wrench resistant grasp. The fixed world frame is located at the object center, and the grasp matrices associated with two different object frames, $\mathcal{F}_1$ and $\mathcal{F}_2$, are given by

$$G_{\mathcal{F}_1}(q_0) = \begin{bmatrix} -1 & 0 & 1 & 0 \\ 0 & 1 & 0 & -1 \\ -2 & 3 & -2 & 3 \end{bmatrix} \quad \text{and} \quad G_{\mathcal{F}_2}(q_0) = \begin{bmatrix} -1 & 0 & 1 & 0 \\ 0 & 1 & 0 & -1 \\ -2 & 7 & -2 & -1 \end{bmatrix}.$$

The columns of $G_{\mathcal{F}_1}(q_0)$ and $G_{\mathcal{F}_2}(q_0)$ correspond to wrenches affected on $\mathcal{B}$ by finger forces that act along $\mathcal{B}$'s inward contact normals. Figures 13.7(b)–(c) show the tetrahedral convex hull, $\overline{\mathcal{W}}$, of these wrenches. Note that both tetrahedrons contain the origin in their interior, thus indiciating wrench resistance of the grasp. The choice of object frame clearly influences the shape of $\overline{\mathcal{W}}$. Figures 13.7(b)–(c) also show the largest inscribed ball centered at the origin and inscribed in each $\overline{\mathcal{W}}$. When the frame $\mathcal{F}_1$ is seleceted, $Q_{MINW} = 0.28$. When the frame $\mathcal{F}_2$ is selected, $Q_{MINW} = 0.42$. Hence,

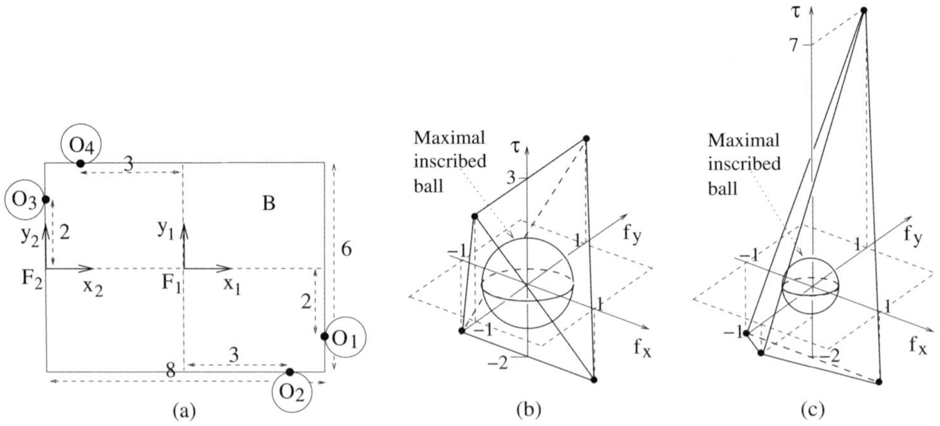

Figure 13.7 (a) A frictionless four-finger grasp of a rigid object $\mathcal{B}$ with two choices of the object reference frame, $\mathcal{F}_1$ and $\mathcal{F}_2$. (b)–(c) The bounded net wrench cone $\overline{\mathcal{W}}$ associated with $\mathcal{F}_1$ and $\mathcal{F}_2$, and the maximal inscribed ball centered at the respective wrench space origin. The inscribed ball in $\mathcal{F}_1$ is actually smaller than the inscribed ball in $\mathcal{F}_2$.

when comparing alternative grasps using Q_{MINW}, the choice of object frame might influence the selection of the optimal grasp. ○

Decoupled inscribed ball quality function: One way to handle the difficulty of comparing forces and torques is to decouple these components using a two-stage approach. First, determine the set of finger placements that optimize the grasp against pure force disturbances. This stage gives a set of candidate grasps. Next, select the best torque rejecting grasp among the force optimal grasps. Thus, consider the net forces that can be affected on $\mathcal{B}$ by bounded finger forces:

$$\overline{\mathcal{W}}_f = \{f = \Sigma_{i=1}^k f_i \,:\, 0 \leq f_i^n \leq 1 \quad \text{and} \quad \sqrt{(f_i^s)^2 + (f_i^t)^2} \leq \mu_i f_i^n \quad \text{for} \quad i = 1 \ldots k\}.$$

These forces can be thought of as the projection of the bounded net wrench cone, $\overline{\mathcal{W}}$, on the subspace of pure forces in $\mathcal{B}$'s wrench space. To measure the effectiveness of a candidate grasp, one computes the smallest size force on the boundary of $\overline{\mathcal{W}}_f$:

$$Q_{MinF} = \min_{f \in \text{bdy}(\overline{\mathcal{W}}_f)} \{\|f\|\}.$$

The function Q_{MinF} measures the worst-case force disturbance that can be resisted by bounded finger forces. Its value represents the largest disc in 2-D grasps or the largest ball in 3-D grasps that can be inscribed in $\overline{\mathcal{W}}_f$ with center at $\mathcal{B}$'s wrench space origin. Maximizing the function Q_{MinF} over all finger placements gives a set of force optimal grasps. Next, the net torques that can be affected on $\mathcal{B}$ are given by

$$\overline{\mathcal{W}}_\tau = \left\{\tau = \Sigma_{i=1}^k (x_i \times f_i) \,:\, 0 \leq f_i^n \leq 1 \quad \text{and} \quad \sqrt{(f_i^s)^2 + (f_i^t)^2} \leq \mu_i f_i^n \quad \text{for} \quad i = 1 \ldots k\right\},$$

where $x_1, \ldots, x_k$ are the finger contact locations at the force optimal grasps. The pure torque measure is defined as the smallest size torque on the boundary of $\overline{\mathcal{W}}_\tau$:

$$Q_{MinT} = \min_{\tau \in \text{bdy}(\overline{\mathcal{W}}_\tau)} \{\|\tau\|\}.$$

The function Q_{MinT} measures the worst-case torque disturbance that can be resisted by bounded finger forces. Maximizing the function Q_{MinT} over all force optimal grasps gives the optimal grasp. This two-stage approach is illustrated in the following example.

Example: Consider the ellipse $\mathcal{B}$ grasped by two disc fingers via frictional point contacts shown in Figure 13.8. The set of net forces, $\overline{\mathcal{W}}_f$, is the convex hull of the truncated friction cones placed at a common origin. When one slides the finger contacts along the ellipse boundary, $\overline{\mathcal{W}}_f$ changes its shape. The largest inscribed disc in $\overline{\mathcal{W}}_f$ is obtained at the two antipodal grasps of the ellipse (Figure 13.8(a)). Next, the projection of $\overline{\mathcal{W}}$ on the τ axis, $\overline{\mathcal{W}}_\tau$, forms a line segment. The net torques at the endpoints of $\overline{\mathcal{W}}_\tau$ corresponds to the two antipodal grasps. Up to a common multiplicative factor, the endpoint torques are proportional to the half-lengths of the ellipse's minor and major axes. The optimal pair (Q_{MinF}, Q_{MinT}) thus corresponds to the antipodal grasp along the ellipse major axis (Figure 13.8(b)). ○

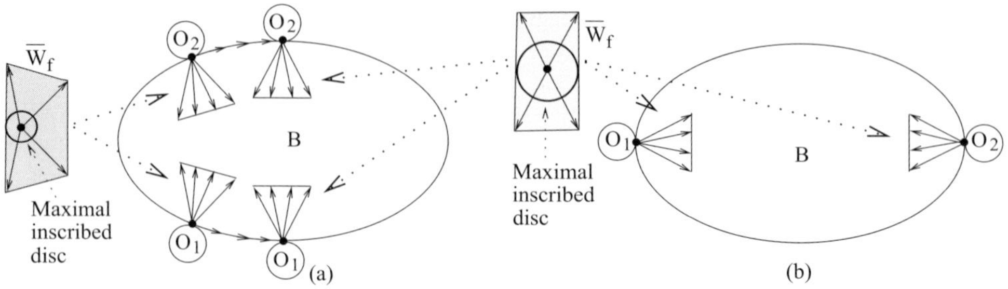

Figure 13.8 (a) Two finger placements with their pure force maximal discs are shown. (b) The optimal pair (Q_{MinF}, Q_{MinT}) gives the antipodal grasp along the ellipse major axis.

13.6 Finger Force Optimization Based on Task Specification

Many robot grasping tasks are influenced by a dominant external wrench, $\boldsymbol{w}_{ext}$, which is expected to act on the grasped object $\mathcal{B}$ during task execution. In order to hold the object stationary within the grasping fingers, the finger forces must satisfy the equilibrium grasp condition

$$G(q_0)\vec{f} + \boldsymbol{w}_{ext} = \vec{0} \qquad \vec{f} = (f_1, \ldots, f_k) \in \mathbb{R}^{kp},$$

where $G(q_0)$ is the $m \times kp$ grasp matrix associated with the finger contacts. The selection of the most suitable grasp for a given task can be formulated as a constrained optimization problem. The most general formulation seeks to compute the optimal grasp over different finger placements as well as different amounts of finger forces. Let us consider here the *grasp force optimization* problem, which seeks to select optimal finger forces at a fixed set of contacts. Let $Q(f_1, \ldots, f_k)$ be a grasp quality function that measures the finger contact forces. Assuming frictional point contacts, the grasp force optimization can be stated as the constrained minimization:

$$\begin{aligned} \min \quad & Q(f_1, \ldots, f_k) \\ \text{subject to} \quad & G(q_0)\vec{f} + \boldsymbol{w}_{ext} = \vec{0} \\ & f_i^n \geq 0 \quad \text{and} \quad \sqrt{(f_i^s)^2 + (f_i^t)^2} \leq \mu_i f_i^n \qquad i = 1 \ldots k. \end{aligned} \tag{13.7}$$

For a given set of contacts and task wrench, the equilibrium grasp constraints specify an affine hyperplane in the space of contact force components. We know from Chapter 12 that the friction cone constraints can be formulated as *linear matrix inequalities.* That is, there exist positive semi-definite matrices, $P(f_i)$ for $i = 1 \ldots k$, such that the friction cone constraints can be formulated in the form

$$P(f_i) = \begin{bmatrix} \mu_i f_i^n & 0 & f_i^s \\ 0 & \mu_i f_i^n & f_i^t \\ f_i^s & f_i^t & \mu_i f_i^n \end{bmatrix} \succeq 0 \qquad f_i = (f_i^s, f_i^t, f_i^n) \text{ for } i = 1 \ldots k,$$

where μ_i is the coefficient of friction at x_i. Note that $P(f_i) \succeq 0$ specifies a convex inequality in the space of finger force components. If $Q(f_1, \ldots, f_k)$ is convex or quasiconvex, Eq. (13.7) specifies a *convex optimization problem.* That is, the minimization of a convex or quasiconvex function subject to convex equality and inequality constraints. Such problems can be efficiently solved using convex optimization techniques.

The gentlest finger force magnitude quality function: No matter what type of contact model holds at the finger contacts, larger finger force magnitudes require greater mechanical effort within the finger mechanisms. Additionally, reducing the size of the finger force magnitudes may be necessary when grasping fragile or delicate objects. Hence, one often seeks to minimize the finger force magnitudes needed to maintain a specific task wrench. One approach seeks to *minimize* the worst-case finger force magnitude needed to balance the task wrench:

$$Q_{MAX}(f_1, \ldots, f_k) = \max\{\|f_1\|, \ldots, \|f_k\|\}.$$

Each finger force magnitude, $\|f_i\|$, is quasiconvex in the ith finger force components. The maximum of such functions forms a quasiconvex function. Hence, Q_{MAX} is quasiconvex in the finger force components.[3] The minimization of Q_{MAX} under the convex constraints specified in Eq. (13.7) thus forms a convex optimization problem. However, the minimization of Q_{MAX} tends to *decrease* the internal grasp forces at the contacts, resulting in finger forces that tend to align more closely with the friction cone edges. This undesirable behavior is illustrated in the following example.

Example: Figure 13.9 depicts two disc fingers attempting to lift a rigid object $\mathcal{B}$ against gravity via frictional point contacts located at x_1 and x_2. The task wrench is thus the gravitational force that acts on B. By symmetry of the grasp, $\|f_1\| = \|f_2\|$ at the minimum value of Q_{MAX}. The force part of the equilibrium equation, $f_1 + f_2 + f_g = \vec{0}$, fixes the vertical components of the finger forces. The horizontal finger force components that act along the x_1–x_2 line segment form the internal grasp forces. The minimum value of Q_{MAX} corresponds to finger forces f_1^* and f_2^* aligned with the upward edges of the

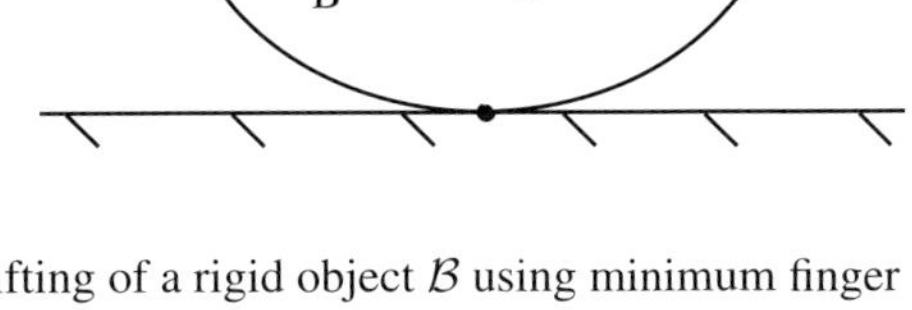

Figure 13.9 Two-finger lifting of a rigid object $\mathcal{B}$ using minimum finger force magnitudes.

[3] A function $\varphi(\vec{z}) : \mathcal{D} \to \mathbb{R}$ on a convex set $\mathcal{D} \subseteq \mathbb{R}^n$ is *quasiconvex* when its sublevel sets, $\{\vec{z} : \varphi(\vec{z}) \leq c\}$, are convex sets in $\mathbb{R}^n$. For instance, the function $\varphi(z_1, z_2) = \sqrt{z_1^2 + z_2^2}$ is quasiconvex.

friction cones C_1 and C_2 (Figure 13.9). This example shows that the gentlest grasp can be ill conditioned in terms of grasp security considerations. ○

The least friction reliant quality function: The amount of Coulomb friction at the contacts is often poorly known in advance and tends to fluctuate during grasping tasks. Hence, it may be practically desirable to minimize a grasp's reliance on precise knowledge of friction while the robot fingers attempt to balance a specific task wrench. To minimize the reliance on friction, the finger contact forces should align as closely as possible with the respective contact normals. Equivalently, the tangential force components should be minimized at the cost of larger normal force components at the finger contacts. When a finger contact force $f_i = (f_i^s, f_i^t, f_i^n)$ approaches the boundary of its friction cone, the value of the following nonnegative polynomial approaches zero.

$$\mu_i^2 (f_i^n)^2 - (f_i^s)^2 - (f_i^t)^2 \longrightarrow 0 \qquad f_i = (f_i^s, f_i^t, f_i^n) \in C_i,$$

where C_i is the friction cone at x_i. To obtain a quality function that depends only on the contact force direction, consider the quotient[4]:

$$\frac{\mu_i^2 (f_i^n)^2 - (f_i^s)^2 - (f_i^t)^2}{\mu_i^2 (f_i^n)^2} \qquad f_i = (f_i^s, f_i^t, f_i^n) \in C_i.$$

The quotient is zero when f_i aligns with the friction cone boundary, and increases monotonically to unity as f_i approaches the contact normal. Consequently, the negated logarithm

$$-\log\left(\frac{\mu_i^2 (f_i^n)^2 - (f_i^s)^2 - (f_i^t)^2}{\mu_i^2 (f_i^n)^2} \right) \qquad f_i = (f_i^s, f_i^t, f_i^n) \in C_i$$

attains values in the interval $[0, \infty]$ as f_i varies in C_i, with a value of zero when f_i aligns with the contact normal and ∞ when f_i aligns with the friction cone boundary. This fact can be used to define the *force normality* grasp quality function:

$$Q_{FN}(f_1, \ldots, f_k) = \max_{i=1\ldots k} \left\{ -\log\left(\frac{\mu_i^2 (f_i^n)^2 - (f_i^s)^2 - (f_i^t)^2}{\mu_i^2 (f_i^n)^2} \right) \right\} \qquad f_i \in C_i, i = 1 \ldots k.$$

The function Q_{FN} is convex in the finger force components (see Exercises). It attains non-negative values over the feasible finger forces, $f_i \in C_i$ for $i = 1 \ldots k$, and has a global minimum of zero when all finger forces align with the contact normals. Its value indicates the worst case non-normality across all k finger forces at the specified contacts. Lower values of Q_{FN} are associated with less friction reliant grasps, as illustrated in the following example.

Example: Figure 13.10 shows two disc fingers attempting to lift a rigid object $\mathcal{B}$ against gravity via frictional point contacts located at x_1 and x_2. The finger contacts are located

[4] The tangential force components in C_i can be written as $(f_i^s)^2 + (f_i^t)^2 = \sigma^2 \cdot (f_i^n)^2$ for $0 \le \sigma \le \mu_i$. The quotient thus becomes $(\mu_i^2 - \sigma^2)/\mu_i^2$, which varies in $[0, 1]$ as σ varies in $[0, \mu_i]$.

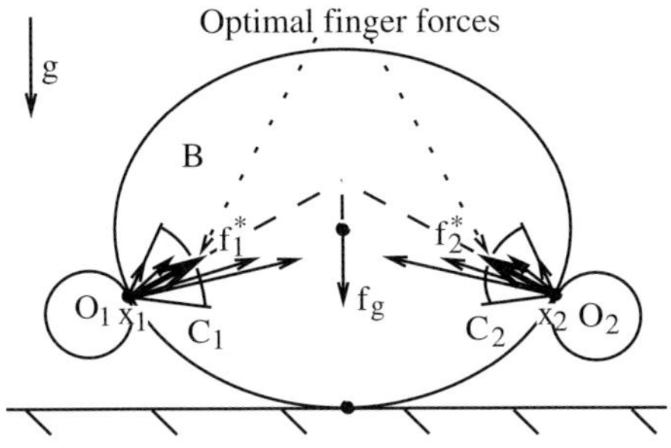

Figure 13.10 Two-finger lifting of a rigid object $\mathcal{B}$ using the least friction reliant finger forces.

on the object lower side where the friction cones contain upward-pointing contact normals. The function $Q_{FN}(f_1, f_2)$ consequently attains its global minimum along the finger forces f_1^* and f_2^* aligned with the contact normals at x_1 and x_2 (Figure 13.10). ○

The minimization of the quality function Q_{FN} seeks to align the finger forces along the contact normals. When the task wrench *cannot* be balanced by the contact normals, the minimization of Q_{FN} might yield finger forces with unbounded magnitudes (see Exercises). A practical approach to prevent this scenario would be to add upper bounds on the allowed normal force components. For instance, $0 \leq f_i^n \leq 1$ for $i = 1 \dots k$. The allowed contact force components now vary in a compact convex set, and the minimization of Q_{FN} would give bounded magnitude finger forces.

Bibliographical Notes

Grasp quality functions can be broadly divided into two types. The first type are quality functions that quantify desirable properties of *wrench resistant* grasps. Examples are quality functions based on singular values of the grasp matrix proposed by Li and Sastry [1], and the radius of the maximal ball inscribed in the grasp matrix ellipsoid proposed by Kirkpatrick, Mishra and Yap [2] and Ferrari and Canny [3]. However, these quality criteria depend on the choice of reference frames – a grasp that is optimal under one choice of reference frame may fail to be optimal under other reference frames. In order to avoid this problem, Markenscoff and Papadimitriou [4] as well as Mirtich and Canny [5] suggested to lexicographically optimize the grasp against pure force disturbances and then select among these grasps the one that best resists pure torque disturbances.

The second type of grasp quality functions quantify safety margins afforded by a given grasp. A primary example is the notion of *contact independent regions,* proposed by Nguyen [6] to ensure safety margin with respect to small finger placement errors. Another example is the least friction reliant grasp quality function proposed by Ji and Roth [7]. In related work, Nakamura, Nagai and Yoshikawa [8] considered the largest allowable *dynamic* perturbations to the contact point positions that will cause no slippage of the finger contacts on the grasped object surface. In the case of specific grasping tasks, Trinkle [9] formulated quality functions that minimize the worst-case finger force magnitude required for the given task.

Appendix I: Review of Distance Metrics and Norms

This appendix reviews standard notions of distance metrics and norms. Starting with distance metrics, let X be a set such as the configuration space of a rigid body $\mathcal{B}$.

DEFINITION 13.4 (**distance metric**) *A* **metric** *on X is a positive semi-definite function $d\colon X \times X \to \mathbb{R}$ with the following properties:*

1. $d(x_i, x_j) = d(x_j, x_i)$ (symmetry)
2. $d(x_i, x_k) \le d(x_i, x_j) + d(x_j, x_k)$ (triangle inequality)
3. $d(x_i, x_j) = 0$ *iff* $x_i = x_j$ (strict positivity).

The pair (X, d) forms a metric space.

When X represents the configuration space of a rigid body $\mathcal{B}$, a distance metric measures the distance between two rigid-body configurations, such as the function $d(q_1, q_2)$ described in Eq. (13.2).

Next consider the notion of a *norm*. Each configuration point of $\mathcal{B}$ possesses tangent and cotangent spaces with a vector space structure. The following norm quantifies the size of tangent and cotangent vectors in these vector spaces.

DEFINITION 13.5 (**vector norm**) *Let V be a finite dimensional real vector space. A* **vector norm** *is a real-valued positive semi-definite function $\|\cdot\|\colon V \to \mathbb{R}$, satisfying the following properties for all $u, v \in V$, and $a \in \mathbb{R}$:*

1. $\|a \cdot u\| = |a| \cdot \|u\|$ (homogeneity)
2. $\|u + v\| \le \|u\| + \|v\|$ (triangle inequality)
3. $\|u\| = 0$ *implies that* $u = 0$ (strict positivity).

The pair $(V, \|\cdot\|)$ forms a normed vector space.

Example: A vector norm can be used to induce a norm on matrices as follows. Let A be an $r \times l$ real matrix such that $l \le r$. The *matrix norm* of A is defined as

$$\|A\| = \max_{\|v\| \le 1} \|Av\| = \max_{\|v\| \le 1} v^T A^T A v \qquad v \in \mathbb{R}^l,$$

where $\|\cdot\|$ is any vector norm in $\mathbb{R}^l$. When $A = G^T(q_0)$, the matrix norm induced by the Euclidean vector norm equals the *maximal singular value* of the grasp matrix $G(q_0)$. ○

Norms on wrenches: Some grasp quality functions rely on covector norms that can be induced from vector norms as follows. Let V be a finite dimensional real vector space, and let V^* be its *dual vector space* – that is, the vector space of real-valued linear maps that act on V, denoted $\alpha : V \to \mathbb{R}$. If V is a normed vector space with elements $v \in V$, an *induced norm* on the dual vector space V^* can be obtained by

$$||\alpha|| = \max_{\|v\| \leq 1} \{|\alpha(v)|\} \qquad \alpha \in V^*,$$

where $|| \cdot ||$ is any norm in V.

Example: Let $q \in \mathbb{R}^m$ be a configuration of a rigid body $\mathcal{B}$. Let $\dot{q} \in T_q\mathbb{R}^m$ be a rigid-body velocity and $\boldsymbol{w} \in T_q^*\mathbb{R}^m$ a rigid-body wrench. The *kinetic energy* of $\mathcal{B}$ is given by the positive definite quadratic form $K = \frac{1}{2}\dot{q}^T M(q)\dot{q}$, where $M(q)$ is the $m \times m$ *mass matrix* of $\mathcal{B}$. The kinetic energy defines a rigid-body velocity norm, $\|\dot{q}\|_{rms}$, given by

$$\|\dot{q}\|_{rms} = \left(\dot{q}^T M(q)\dot{q}\right)^{\frac{1}{2}}. \tag{13.8}$$

A wrench norm, denoted $\|\boldsymbol{w}\|_{rms}$, can be induced from $\|\dot{q}\|_{rms}$ as follows. Consider the action of a wrench $\boldsymbol{w}$ on rigid-body velocities via the Euclidean inner product: $\boldsymbol{w}(\dot{q}) = \boldsymbol{w} \cdot \dot{q}$ for $\dot{q} \in T_q\mathbb{R}^m$. This action represents the instantaneous work done by the wrench $\boldsymbol{w}$ on $\mathcal{B}$ while it moves with velocity $\dot{q}$. It can be shown that the induced wrench norm has the form (see Exercises)

$$\|\boldsymbol{w}\|_{rms} = \max_{\|\dot{q}\|_{rms} \leq 1} \{|\boldsymbol{w} \cdot \dot{q}|\} = \left(\boldsymbol{w}^T M^{-1}(q)\boldsymbol{w}\right)^{\frac{1}{2}}.$$

○

Appendix II: Behavior of the Grasp Matrix under Coordinate Transformations

This appendix describes the coordinate transformation associated with different choices of reference frames and the behavior of the grasp matrix under such transformations. We focus on changes of the grasped object frame, $\mathcal{F}_B$, leaving the coordinate transformation associated with changes of the fixed world frame, $\mathcal{F}_W$, as an exercise.

Thus, consider a change of the object frame from $\mathcal{F}_B$ to $\mathcal{F}_B'$. This change is described by a pair (d_B, R_B), which describes the position and orientation of the new frame $\mathcal{F}_B'$ relative to the frame $\mathcal{F}_B$. Let $q = (d, \theta)$ and $\bar{q} = (\bar{d}, \bar{\theta})$ denote the object's c-space coordinates associated with the frames $\mathcal{F}_B$ and $\mathcal{F}_B'$, described relative to a fixed world frame. Let $b \in \mathcal{B}$ describe the position of the object points in $\mathcal{F}_B$, and let $\bar{b} \in \mathcal{B}$ describe the position of these points in $\mathcal{F}_B'$. The positions of each object point in the two reference frames, b and $\bar{b}$, are related by the rigid-body transformation: $b = R_B\bar{b} + d_B$. The position of each object point in the fixed world frame, x, must match for the two object frames. This constraint gives the relation:

$$x = R(\theta)b + d = R(\bar{\theta})\bar{b} + \bar{d} \qquad b = R_B\bar{b} + d_B.$$

It follows that the c-space coordinates, (d, θ) and $(\bar{d}, \bar{\theta})$, are related by the transformation

$$d = \bar{d} - R(\bar{\theta})R_B^T d_B \quad \text{and} \quad R(\theta) = R(\bar{\theta})R_B^T \tag{13.9}$$

where we used the identity $R_B R_B^T = I$. The coordinate transformation described in Eq. (13.9) forms a diffeomorphism, $q = h(\bar{q})$. Taking the derivative of both sides along a c-space trajectory $q(t) = h(\bar{q}(t))$, one obtains the velocity mapping: $\dot{q} = Dh(\bar{q})\dot{\bar{q}}$. The $m \times m$ Jacobian $Dh(\bar{q})$ thus maps the grasped object *velocities* from the tangent space $T_{\bar{q}}\mathbb{R}^m$ to the tangent space $T_q\mathbb{R}^m$, where $q = h(\bar{q})$ and $m = 3$ or 6. The transposed Jacobian, $Dh^T(\bar{q})$, maps the grasped object *wrenches* from the wrench space $T_q^*\mathbb{R}^m$ backward to the wrench space $T_{\bar{q}}^*\mathbb{R}^m$, where $\bar{q} = h^{-1}(q)$. The latter rule is based on the principle that the *instantaneous work* done by the two wrenches must match in the two c-space coordinate systems.

Assume that the object $\mathcal{B}$ is held at an equilibrium grasp configuration, $q_0 = (d_0, \theta_0)$, such that $R(\theta_0) = I$ at this grasp. That is, the world and object frames have parallel axes at the equilibrium grasp. The mapping of the object velocities from $T_{\bar{q}_0}\mathbb{R}^m$ to $T_{q_0}\mathbb{R}^m$ is given by the $m \times m$ matrix

$$\begin{pmatrix} v \\ \omega \end{pmatrix} = \begin{bmatrix} I & [d_B \times] \\ O & I \end{bmatrix} \begin{pmatrix} \bar{v} \\ \bar{\omega} \end{pmatrix} \qquad \dot{q} = (v, \omega),\ \dot{\bar{q}} = (\bar{v}, \bar{\omega}), \tag{13.10}$$

where I is a 3×3 identity matrix and O is a 3×3 zero matrix in the case of 3-D grasps. The transposed Jacobian that transforms the object wrenches is given by

$$\begin{pmatrix} \bar{f} \\ \bar{\tau} \end{pmatrix} = \begin{bmatrix} I & [d_B \times] \\ O & I \end{bmatrix}^T \begin{pmatrix} f \\ \tau \end{pmatrix} \qquad \boldsymbol{w} = (f, \tau),\ \bar{\boldsymbol{w}} = (\bar{f}, \bar{\tau}). \tag{13.11}$$

Note that the change of object frame orientation, R_B, does not appear in this transformation.

Let us next study how the *grasp matrix* changes under different choices of the object reference frame. Denote by $G(q_0)$ the grasp matrix associated with the object frame $\mathcal{F}_B$, and denote by $\bar{G}(\bar{q}_0)$ the grasp matrix associated with the object frame $\mathcal{F}_B'$. From Chapter 4, the two grasp matrices satisfy the relations $\boldsymbol{w} = G(q_0)\vec{f}$ and $\bar{\boldsymbol{w}} = \bar{G}(\bar{q}_0)\vec{f}$, where the finger contact forces components $\vec{f} = (f_1, \ldots, f_k)$ are described in the fixed world frame. Substituting $\boldsymbol{w} = G(q_0)\vec{f}$ in Eq. (13.11) gives

$$\begin{pmatrix} \bar{f} \\ \bar{\tau} \end{pmatrix} = \begin{bmatrix} I & [d_B \times] \\ O & I \end{bmatrix}^T G(q_0)\vec{f},$$

which leads to the grasp matrix transformation formula

$$\bar{G}(\bar{q}_0) = \begin{bmatrix} I & [d_B \times] \\ O & I \end{bmatrix}^T G(q_0). \tag{13.12}$$

Example: Let us verify that the volume of the grasp matrix ellipsoid, $\det(G(q_0)G^T(q_0))$, is invariant under different choices of the object reference frame. Using Formula (13.12):

$$\begin{aligned}\det\left(\bar{G}(\bar{q}_0)\bar{G}^T(\bar{q}_0)\right) &= \det\left(\begin{bmatrix} I & [d_B\times] \\ O & I \end{bmatrix}^T G(q_0)G^T(q_0)\begin{bmatrix} I & [d_B\times] \\ O & I \end{bmatrix}\right) \\ &= \det\left(\begin{bmatrix} I & [d_B\times] \\ O & I \end{bmatrix}^T\right)\cdot\det\left(G(q_0)G^T(q_0)\right) \\ &\quad\cdot\det\left(\begin{bmatrix} I & [d_B\times] \\ O & I \end{bmatrix}\right) \\ &= \det\left(G(q_0)G^T(q_0)\right),\end{aligned}$$

where we used the identity $\det(AB) = \det(A)\cdot\det(B)$ and the fact that the determinant of a triangular matrix is the product of its diagonal elements. Hence, the volume of the grasp matrix ellipsoid forms a frame-invariant quality function. ○

Appendix III: The Wrench Resistance Regions

This appendix contains a proof sketch of Lemma 13.1, concerning the wrench resistance regions in contact space $\mathcal{U}$.

Lemma 13.1 *Let a polygonal rigid object $\mathcal{B}$ be held by $k \geq 2$ frictional point fingers or by $k \geq 4$ frictionless point fingers. The wrench resistance regions of $\mathcal{B}$ form* ***k*-dimensional polyhedral sets** *in contact space $\mathcal{U}$.*

Proof sketch: Let f_i^+ and f_i^- denote unit magnitude forces collinear with the edges of the 2-D friction cone C_i.[5] Then $C_i = \{\lambda_{i1} f_i^- + \lambda_{i2} f_i^+ : \lambda_{i1}, \lambda_{i2} \geq 0\}$. The same positive sum applies to the net wrench cone that can be affected on the grasped object $\mathcal{B}$ by the k finger forces:

$$\mathcal{W} = \left\{ \sum_{i=1}^{k} \lambda_{i1} \begin{pmatrix} f_i^- \\ x_i \times f_i^- \end{pmatrix} + \lambda_{i2} \begin{pmatrix} f_i^+ \\ x_i \times f_i^+ \end{pmatrix} : \lambda_{i1}, \lambda_{i2} \geq 0 \quad \text{for} \quad i = 1 \ldots k \right\}.$$

A k-finger grasp is *wrench resistant* when $\mathcal{W} = \mathbb{R}^3$ or, equivalently, when $\mathcal{W}$ contains a three-dimensional ball centered at $\mathcal{B}$'s wrench space origin (see Chapter 12). First consider a point $\vec{u}$ in the interior of a wrench resistance region $\mathcal{R}$. Every pair of edge wrenches (associated with edges of the same or different friction cones) spans a two-dimensional sector based at $\mathcal{B}$'s wrench space origin. This sector lies in a plane V that passes through $\mathcal{B}$'s wrench space origin. Since $\vec{u} \in \mathcal{R}$ represents a wrench resistant grasp, there must exist edge wrenches on *both sides* of V. Otherwise, V would form a separating plane for $\mathcal{W}$, and then $\mathcal{W}$ could not contain a three-dimensional ball centered at $\mathcal{B}$'s wrench space origin.

[5] In the case of frictionless contacts, replace the forces $f_i^{\pm}$ by a single unit magnitude force collinear with the $\mathcal{B}$'s inward contact normal.

Next consider a point $\vec{u}_0$ that lies on the boundary of $\mathcal{R}$. The grasp corresponding to this point is no longer wrench resistant. Since $\mathcal{W}$ forms a convex cone, at least one of the wrench sectors defines a separating plane, denoted V_0, for $\mathcal{W}$ at the grasp corresponding to $\vec{u}_0$. At this grasp, all edge wrenches lie on the *same side* of V_0 in $\mathcal{B}$'s wrench space. Denote by $\boldsymbol{w}_{i_0}$ and $\boldsymbol{w}_{j_0}$ the edge wrenches whose sector lies in V_0. The wrench normal to the separating plane is given by $\boldsymbol{w}_{i_0} \times \boldsymbol{w}_{j_0}$. At the finger placement corresponding to $\vec{u}_0$, there exists a third edge wrench, $\boldsymbol{w}_{k_0}$ such that $k_0 \neq i_0, j_0$, which lies in the plane V_0. Hence, $(\boldsymbol{w}_{i_0} \times \boldsymbol{w}_{j_0}) \cdot \boldsymbol{w}_{k_0} = 0$. Now substitute $\boldsymbol{w}_i = (f_i^{\pm}, x_i \times f_i^{\pm})$ for the three wrenches. The unit forces $f_{i_0}, f_{j_0}, f_{k_0}$ remain constant along the edges of $\mathcal{B}$, while the contact points $x_{i_0}, x_{j_0}, x_{k_0}$ depend *linearly* on the contact parameters u_{i_0}, u_{j_0}, and u_{k_0}. Every boundary equation, $(\boldsymbol{w}_{i_0} \times \boldsymbol{w}_{j_0}) \cdot \boldsymbol{w}_{k_0} = 0$, thus forms a *linear constraint* in the contact parameters $u_1, \ldots, u_k$. The wrench resistance regions are consequently bounded by hyperplanes and hence form polyhedral sets in contact space $\mathcal{U}$. □

Exercises

Section 13.1

Exercise 13.1: Let $\rho(b) : \mathcal{B} \to \mathbb{R}$ be a nonnegative function that might represent the mass density of a rigid body $\mathcal{B}$. Show that the function

$$d(q_1, q_2) = \int_{\mathcal{B}} \rho(b) \cdot \|X(q_1, b) - X(q_2, b)\| dV$$

is a distance metric in $\mathcal{B}$'s c-space, where $|| \cdot ||$ is any vector norm in $\mathbb{R}^3$.

Exercise 13.2: Let $\Omega_{\mathcal{B}}$ be a set of feature point on a rigid body $\mathcal{B}$. Show that the function

$$\phi_q(\dot{q}) = \max_{b \in \Omega_{\mathcal{B}}} \{\|\omega \times (R(\theta)b) + v\|\} \qquad q = (d, \theta),\ \dot{q} = (v, \omega)$$

is a rigid-body velocity norm under the condition that the set of feature points $\Omega_{\mathcal{B}}$ contains three non-collinear points.

Exercise 13.3*: One must choose a *scaling factor* during the process of measuring the size of a rigid-body velocity vector: $||\dot{q}|| = \sqrt{||v||^2 + ||\alpha \cdot \omega||^2}$. Show that each choice of a different scaling factor, $\alpha \in \mathbb{R}$, is equivalent to the choice of a different object reference frame $\mathcal{F}_\alpha$.

Section 13.2

Exercise 13.4: Show that a surjective grasp matrix, $G(q_0)$, maps a unit ball of finger contact force components in $\mathbb{R}^{kp}$ to an m-dimensional ellipsoid in $\mathcal{B}$'s wrench space.

Hint: The m-dimensional ellipsoid consists of wrenches $\boldsymbol{w} \in T^*_{q_0}\mathbb{R}^m$ satisfying the inequality, $(\boldsymbol{w} - \boldsymbol{w}_c)P^{-1}(\boldsymbol{w} - \boldsymbol{w}_c) \leq 1$, where P is a positive definite $m \times m$ matrix. The eigenvectors of P define the ellipsoid's principal directions, and the eigenvalues of P are the squares of the principal axes half-lengths. ○

Exercise 13.5: Explain why the ellipsoid volume quality function, $Q_{EV} = \sqrt{\det(G(q_0)G^T(q_0))}$, attains positive values at wrench resistant grasps.

Exercise 13.6: Consider the four-finger grasp of a rectangular object shown in Figure 13.4(b). Explain why this grasp is *not* wrench resistant under the frictionless and frictional point contact models (assuming a sufficiently small coefficient of friction μ).

Exercise 13.7: Verify that the eigenvalues of the matrix $G(q_0)G^T(q_0)$ are the squared singular values of the grasp matrix $G(q_0)$.

Exercise 13.8: A 2-D rigid object $\mathcal{B}$ is held at a k-finger equilibrium grasp via frictionless contacts. Obtain a formula for the grasp's *center*, $x_0 \in \mathbb{R}^2$, at which the 3×3 matrix $G(q_0)G^T(q_0)$ attains the block diagonal form

$$G(q_0)G^T(q_0) \cong \begin{bmatrix} \sum_{i=1}^{k} n_i n_i^T & \begin{matrix} 0 \\ 0 \end{matrix} \\ 0 \quad 0 & \sum_{i=1}^{k}(x_i \times n_i)^2 \end{bmatrix}$$

where the contact points are expressed relative to the grasp center point.

Solution: For the 3×3 matrix $G(q_0)G^T(q_0)$ to become block diagonal, the point x_0 must satisfy the condition: $\sum_{i=1}^{k}((x_i - x_0) \times n_i)n_i = \vec{0}$. Or, equivalently,

$$\sum_{i=1}^{k}(x_i \times n_i)n_i = \sum_{i=1}^{k}(x_0 \times n_i)n_i = -\left[\sum_{i=1}^{k} n_i n_i^T\right] J x_0 \qquad J = \begin{bmatrix} 0 & 1 \\ -1 & 0 \end{bmatrix}.$$

Solving for x_0 gives

$$x_0 = J \cdot \left[\sum_{i=1}^{k} n_i n_i^T\right]^{-1} \sum_{i=1}^{k}(x_i \times n_i)n_i,$$

where we used the identity $J^2 = -I$. ○

Section 13.3

Exercise 13.9: A tool such as a hammer is held stationary by a k-finger robot hand under the influence of gravity. Let the finger contacts shift along the tool's handle while keeping a fixed grasp polygon. Explain how the value of the grasp polygon centroid displacement measure, Q_{GPCD}, influences the finger force magnitudes required to maintain the grasp.

Solution sketch: The torque balance equation is given by $\sum_{i=1}^{k} x_i \times f_i + \boldsymbol{x}_{cm} \times f_g = 0$, where $f_1, \ldots, f_k$ are the finger contact forces and f_g is the gravitational force that acts on $\mathcal{B}$ at $\boldsymbol{x}_{cm}$. Expressing the finger contact locations as $x_i = (x_i - \boldsymbol{x}_c) + \boldsymbol{x}_c$, the torque balance equation can be written as $\sum_{i=1}^{k}(x_i - \boldsymbol{x}_c) \times f_i - (\boldsymbol{x}_c - \boldsymbol{x}_{cm}) \times f_g = 0$, where we used the force balance equation $f_1 + \cdots + f_k + f_g = \vec{0}$. The vectors $x_i - \boldsymbol{x}_c$ remain constant as the grasp polygon shifts along the tool's handle. As the grasp polygon shifts *away* from the object's center of mass, the quantity $\|\boldsymbol{x}_c - \boldsymbol{x}_{cm}\|$ increases, and so must the finger force magnitudes $\|f_1\|, \ldots, \|f_k\|$ required to maintain the torque balance constraint. ○

Section 13.4

Exercise 13.10: Explain why the wrench resistance regions of a 2-D object $\mathcal{B}$ held via $k \geq 2$ frictional point contacts form k-dimensional subsets in contact space $\mathcal{U}$.

Exercise 13.11: Discuss the meaning of contact space $\mathcal{U}$ for a 2-D object $\mathcal{B}$ held by disc rather than point fingers (which may touch the object's convex vertices); then characterize the wrench resistance regions in contact space.

Exercise 13.12*: A rectangular object $\mathcal{B}$ is to be grasped by four *frictionless* point fingers. Determine the grasp that maximizes the contact placement robustness measure, Q_{FCPR}, in contact space $\mathcal{U}$. (Note that a four-finger placement at the center points of the object's edges represents a point on the *boundary* of the grasp's wrench resistance region.)

Exercise 13.13*: A rectangular object $\mathcal{B}$ is to be grasped by four *frictionless* point fingers. Determine the contact independent region with the maximal value of Q_{CIR}, as well as the finger placement corresponding to the geometric center of this region.

Section 13.5

Exercise 13.14: Show that the bounded net wrench cone at a k-finger grasp, $\overline{\mathcal{W}}$, consists of all net wrenches that can be affected on the grasped object by finger forces whose normal components satisfy the constraint $\sum_{i=1}^{k} |f_i^n| \leq 1$.

Exercise 13.15: The quality function Q_{MINW} measures the radius of the largest ball inscribed in the bounded net wrench cone $\overline{\mathcal{W}}$. Explain how this quality function scales when the normal finger force components vary in an interval $0 \leq f_i^n \leq \delta$ for some positive constant δ.

Exercise 13.16*: Figure 13.7 illustrates the dependency of the inscribed ball quality function, Q_{MINW}, on the choice of object reference frame. Show how different object frame choices can lead to different optimal grasps.

Exercise 13.17: An ellipse $\mathcal{B}$ is to be grasped by three disc fingers via *frictionless* contacts in a horizontal plane. Determine the optimal grasp based on the decoupled quality measure, (Q_{MinF}, Q_{MinT}).

Hint: The largest inscribed disc in the convex hull of the unit finger forces occurs when these forces form the vertices of an equilateral triangle. ○

Exercise 13.18: Under the conditions of the previous exercise, how the optimal grasp according the decoupled quality measure, (Q_{MinF}, Q_{MinT}), will change when $\mathcal{B}$ is held by three disc fingers via *frictional* contacts?

Exercise 13.19*: Consider the decoupled quality measure (Q_{MinF}, Q_{MinT}). Prove that the pure force measure, Q_{MinF}, is independent on the choice of the object frame. Prove that the pure torque measure, Q_{MinT}, is *order preserving* under different choices of object frames.

Exercise 13.20: Consider the least friction reliant grasp quality function, Q_{FN}. Show that in the case of 2-D grasps, the function: $-\log\big((\mu_i^2(f_i^n)^2 - (f_i^t)^2)/\mu_i^2(f_i^n)^2\big)$ is convex in the finger force components (f_i^t, f_i^n).

Hint: First show that the quotient is a convex function of (f_i^t, f_i^n) by checking its Hessian matrix. A plot of the function's level lines can help to verify convexity. ○

Section 13.6

Exercise 13.21: Consider the two-finger lifting of an object $\mathcal{B}$ depicted in Figure 13.9. Does the minimization of the least friction reliant quality function, Q_{FN}, forms a well-posed problem?

Exercise 13.22: Suggest how to modify the force optimization problem such that the quality function Q_{FN} would become well posed for all grasp arrangements, making sure that any additional constraints retain the convex nature of the problem.

Exercise 13.23*: Can the least friction reliant grasp be *precisely* attained with $Q_{FN} = 0$ when the task wrench, $\boldsymbol{w}_{ext}$, is balanced by $k \geq 4$ frictional point contacts in 2-D grasps or $k \geq 7$ frictional point contacts in 3-D grasps?

Appendix I

Exercise 13.24: Consider the rigid-body velocity norm $\|\dot{q}\|_{rms} = (\dot{q}^T M(q)\dot{q})^{\frac{1}{2}}$. Using the Euclidean inner product, $\boldsymbol{w}(\dot{q}) = \boldsymbol{w}\cdot\dot{q}$, prove that the wrench norm induced by $\|\dot{q}\|_{rms}$ is given by $\|\boldsymbol{w}\|_{rms} = (\boldsymbol{w}^T M^{-1}(q)\boldsymbol{w})^{\frac{1}{2}}$. Can you suggest a physical interpretation for $\|\dot{q}\|_{rms}$?

Appendix II

Exercise 13.25: Consider a change of the world reference frame from $\mathcal{F}_W$ to $\mathcal{F}'_W$, while the object reference frame remains fixed. The pair (d_W, R_W) describes the position and orientation of $\mathcal{F}'_W$ relative to $\mathcal{F}_W$. Show that the grasped object velocity transformation between these two frames takes the form

$$\begin{pmatrix} v \\ \omega \end{pmatrix} = \begin{bmatrix} R_W & R_W \\ O & R_W \end{bmatrix} \begin{pmatrix} \bar{v} \\ \bar{\omega} \end{pmatrix} \qquad \dot{q} = (v, \omega),\ \dot{\bar{q}} = (\bar{v}, \bar{\omega}),$$

where O is a 3×3 zero matrix in the case of 3-D grasps. Note that the change of the world frame origin, d_W, does not appear in this transformation.

References

[1] Z. Li and S. S. Sastry, "Task oriented optimal grasping by multifingered robot hands," *IEEE Transactions on Robotics and Automation*, vol. 4, no. 1, pp. 32–44, 1988.

[2] D. G. Kirpatrick, B. Mishra and C. K. Yap, "Quantitative steinitz's theorems with applications to multifingered grasping," in *20th ACM Symposium on Theory of Computing*, pp. 341–351, 1990.

[3] C. Ferrari and J. F. Canny, "Planning optimal grasps," in *IEEE International Conference on Robotics and Automation*, pp. 2290–2295, 1992.

[4] X. Markenscoff and C. H. Papadimitriou, "Optimum grip of a polygon," *International journal of Robotics Research*, vol. 8, no. 2, pp. 17–29, 1989.

[5] B. Mirtich and J. F. Canny, "Easily computable optimum grasps in 2-D and 3-D," in *IEEE International Conference on Robotics and Automation*, pp. 739–747, 1994.

[6] V.-D. Nguyen, "Constructing force-closure grasps," *International Journal of Robotics Research*, vol. 7, no. 3, pp. 3–16, 1988.

[7] Z. Ji and B. Roth, "Direct computation of grasping force for three-finger tip-prehension grasps," *Journal of Mechanics, Transmissions, and Automation in Design*, vol. 110, pp. 405–413, 1988.

[8] Y. Nakamura, Nagai and T. Yoshikawa, "Dynamics and stability in coordination of multiple robotic mechanisms," *International Journal of Robotics Research*, vol. 8, no. 2, pp. 44–61, 1989.

[9] J. C. Trinkle, "On the stability and instantaneous velocity of grasped frictionless objects," *IEEE Transactions on Robotics and Automation*, vol. 8, no. 5, pp. 560–572, 1992.

14 Hand-Supported Stances under Gravity – Part I

The next two chapters study the *equilibrium stances* of a rigid object supported by a robot hand under the influence of gravity. This chapter focuses on stances in two dimensions; the next chapter will consider stances in three dimensions. Hand-supported stances differ from the *wrench resistant* grasps discussed in Chapter 12. In a wrench resistant grasp, the robot fingers can resist any wrench that acts on the grasped object, including the gravitational wrench. In a hand-supported stance, the robot hand can resist only a specific set of wrenches that includes the gravitational wrench. This important distinction is illustrated with an example.

Example: Figure 14.1(a) depicts an object $\mathcal{B}$ held under the influence of gravity by two point fingers via frictional contacts. The fingers form a wrench resistant grasp that can actively oppose any external wrench that may act on $\mathcal{B}$, including the gravitational wrench. The situation changes when a robot hand passively supports a rigid object $\mathcal{B}$ against gravity in the manner depicted in Figure 14.1(b). While the object is *not* held in a wrench resistant grasp, the supporting contacts can oppose a set of external wrenches that may act on $\mathcal{B}$, which includes the gravitational wrench. In general, hand-supported stances are *not* as secure as wrench resistant grasps. However, hand-supported stances can safely perform a wide range of tasks such as carrying large nonprehensile objects and operating large hand held tools. ○

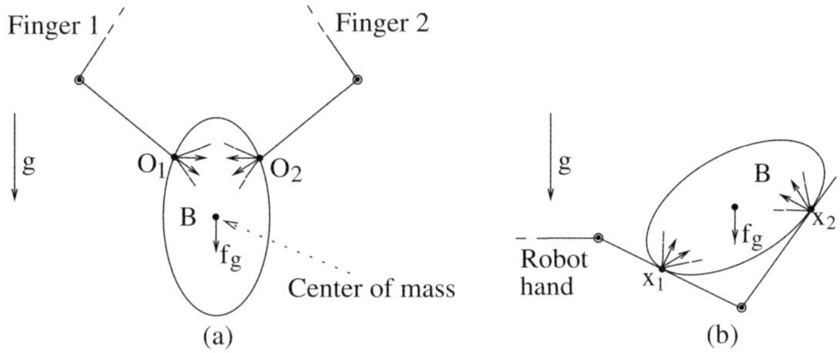

Figure 14.1 (a) In a wrench resistant grasp, the fingers can resist any external wrench that may act on $\mathcal{B}$. (b) In a hand-supported stance, the supporting contacts (palm and midfinger contacts allowed) can resist only a specific set of wrenches that may act on $\mathcal{B}$.

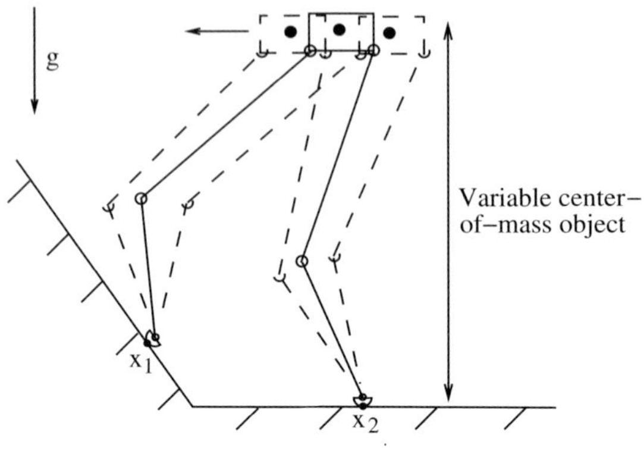

Figure 14.2 A two-legged robot shifting its center of mass during quasistatic locomotion. The legged robot equilibrium postures are equivalent to the equilibrium stances of a rigid body having a variable center of mass and supported by the same contacts.

This chapter introduces a notion of *local wrench resistance* that forms the limited security measure attainable by hand-supported stances.[1] Rather than study the hand postures that can support the object in static equilibrium, we will fix the robot hand and consider instead a rigid object that has a *variable center of mass* and is supported by the same contacts. The chapter studies the object's center-of-mass positions, termed the *stance equilibrium region,* at which the supporting contacts can balance the gravitational wrench that acts on the object supported by these contacts. Section 14.1 introduces the notion of local wrench resistant stances. Sections 14.2 and 14.3 discuss basic properties and the graphical construction of the stance equilibrium region. Section 14.4 studies a stronger safety requirement – that an equilibrium stance be maintained with respect to a *task-dependant neighborhood* of wrench disturbances that may act on the object supported by the robot hand.

Relation to legged robot locomotion: The modeling approach of this chapter is also useful for *legged robot locomotion.* As schematically depicted Figure 14.2, a rigid body with a variable center of mass forms a simplified model for a legged robot supported against gravity by a given set of foothold contacts. The stance equilibrium region describes the legged robot's safe center-of-mass positions, which can be used to plan the legged robot's safe quasistatic locomotion on uneven terrains. ○

14.1 Local Wrench-Resistant Stances

This section characterizes the equilibrium stances of a 2-D rigid object, $\mathcal{B}$, supported against gravity within a vertical plane by 2-D stationary rigid bodies $\mathcal{O}_1, \ldots, \mathcal{O}_k$, which represent the supporting robot hand. The object is supported via frictional point contacts

[1] The term *static stability* is often used to describe such equilibrium stances.

located at the points $x_1, \ldots, x_k$. Under the Coulomb friction model, the contact forces $f_1, \ldots, f_k$ must lie in their respective friction cones,

$$C_i = \left\{ f_i \in \mathbb{R}^2 : f_i \cdot n_i \geq 0 \text{ and } |f_i \cdot t_i| \leq \mu_i |f_i \cdot n_i| \right\} \qquad i = 1 \ldots k,$$

where μ_i is the coefficient of friction at x_i, and (t_i, n_i) are the unit tangent and $\mathcal{B}$'s inward unit normal at x_i. The position of the object's center of mass in the fixed world frame, denoted $\boldsymbol{x}_{cm}$, varies freely in the vertical plane. The gravitational force that acts on $\mathcal{B}$ at $\boldsymbol{x}_{cm}$, denoted f_g, acts along the vertical downward direction. Much like equilibrium grasps, a rigid object $\mathcal{B}$ is supported at a feasible equilibrium stance when feasible contact forces can balance the gravitational wrench that acts on $\mathcal{B}$ according to the following condition.

Equilibrium stance condition: *A 2-D rigid object $\mathcal{B}$ is supported by stationary rigid bodies $\mathcal{O}_1, \ldots, \mathcal{O}_k$ at a feasible* **equilibrium stance** *under the influence of gravity when*

$$\begin{pmatrix} f_1 \\ x_1 \times f_1 \end{pmatrix} + \cdots + \begin{pmatrix} f_k \\ x_k \times f_k \end{pmatrix} + \begin{pmatrix} f_g \\ \boldsymbol{x}_{cm} \times f_g \end{pmatrix} = \vec{0} \qquad f_i \in C_i,\ i = 1 \ldots k \quad (14.1)$$

where $\boldsymbol{w}_i = (f_i, x_i \times f_i)$ is the wrench generated by the ith contact force, $\boldsymbol{w}_g = (f_g, \boldsymbol{x}_{cm} \times f_g)$ is the gravitational wrench that acts on $\mathcal{B}$ and $x \times f = x^T J f$ where $J = \left[\begin{smallmatrix} 0 & 1 \\ -1 & 0 \end{smallmatrix}\right]$.

For a fixed set of supporting contacts and center-of-mass location, Eq. (14.1) specifies three linear constraints on the supporting contact forces, $(f_1, \ldots, f_k) \in \mathbb{R}^{2k}$. The feasible contact forces that solve Eq. (14.1), $f_i \in C_i$ for $i = 1 \ldots k$, form a *convex set* embedded in a $(2k - 3)$-dimensional affine subspace of $\mathbb{R}^{2k}$. For instance, at a two-contact stance, the equilibrium forces span a line segment (a convex set) in (f_1, f_2) space.

As illustrated in Figure 14.1(b), when a robot hand supports a rigid object against gravity via frictional contacts, the stance is typically *not* wrench resistant. However, the contacts can resist a local neighborhood of object wrenches that surrounds the gravitational wrench. This notion of *local wrench resistance* is defined as follows.

DEFINITION 14.1 (**local wrench resistance**) *Let a rigid object $\mathcal{B}$ be supported at a feasible equilibrium stance by stationary rigid bodies $\mathcal{O}_1, \ldots, \mathcal{O}_k$. The stance is* **locally wrench resistant** *if the contacts can resist with feasible forces all wrenches that act on $\mathcal{B}$ in a local neighborhood centered at the gravitational wrench, $\boldsymbol{w}_g$, in $\mathcal{B}$'s wrench space.*

Local wrench resistance forms the limited measure of security afforded by hand-supported stances. A practical robot hand is attached to a robot arm that is possibly mounted on a mobile robot platform. Local wrench resistance ensures that the object will remain in static equilibrium on the supporting robot hand, under all sufficiently small wrench disturbances generated by local movements of the robot arm and the mobile robot platform. Such disturbances can be lumped into a wrench neighborhood surrounding the gravitational wrench in $\mathcal{B}$'s wrench space. When the disturbances are small, local wrench resistance provides a useful safety criterion. A stronger measure of safety that ensures stance robustness against task-dependent neighborhoods of wrench disturbances is discussed later in this chapter.

Local wrench resistance is a generic property of hand-supported equilibrium stances. To justify this statement, consider the *net wrench cone* generated by the supporting contacts:

$$\mathcal{W} = \left\{ \sum_{i=1}^{k} \begin{pmatrix} f_i \\ x_i \times f_i \end{pmatrix} : f_i \in C_i \text{ for } i = 1 \ldots k \right\}.$$

The next lemma characterizes the net wrench cone, which will lead to the conclusion that local wrench resistance is a generic property (see Exercises for a proof).

LEMMA 14.1 **(net wrench cone)** *The net wrench cone forms a* **2-D convex sector** *at a single support stance and a* **3-D convex cone** *at stances supported by $k \geq 2$ non-coincident frictional contacts.*

Lemma 14.1 implies that local wrench resistance is a generic property of these stances, as stated in the following proposition. A *non-marginal equilibrium stance* is defined as a stance whose contact forces lie in the interior of the respective friction cones.

PROPOSITION 14.2 **(local wrench resistance)** *Let a 2-D rigid object $\mathcal{B}$ be supported against gravity by $k \geq 2$ stationary rigid bodies $\mathcal{O}_1, \ldots, \mathcal{O}_k$. If the object is supported at a non-marginal equilibrium stance (the generic case), the stance is* **locally wrench resistant.**

Proof sketch: Recall from Chapter 4 that the *grasp map, $\mathcal{L}_G$*, maps the feasible contact forces, $(f_1, \ldots, f_k) \in C_1 \times \cdots \times C_k$, to net object wrenches in $\mathcal{B}$'s wrench space. If the object $\mathcal{B}$ is supported at a feasible equilibrium stance, the image of the composite friction cone $C_1 \times \cdots \times C_k$ under $\mathcal{L}_G$ contains the negated gravitational wrench, $-\boldsymbol{w}_g$. The grasp map associated with $k \geq 2$ non-coincident frictional contacts is surjective (see Chapter 12). In this generic case, $\mathcal{L}_G$ maps the *interior* of $C_1 \times \cdots \times C_k$ in $\mathbb{R}^{2k}$ to the *interior* of $\mathcal{W}$ in $\mathcal{B}$'s wrench space $\mathbb{R}^3$. Based on this topological fact, when the contact forces lie in the interior of their respective friction cones, the net wrench affected on $\mathcal{B}$ by these forces lies in the *interior* of the net wrench cone $\mathcal{W}$. Since $\mathcal{W}$ is fully three-dimensional for $k \geq 2$ supporting contacts according to Lemma 14.1, the contacts can resist a local open neighborhood of object wrenches centered at the negated gravitational wrench, $-\boldsymbol{w}_g$, which gives local wrench resistance. □

14.2 The Feasible Equilibrium Region of 2-D Stances

A rigid object $\mathcal{B}$ with a variable center of mass, $\boldsymbol{x}_{cm}$, is supported by a given set of frictional contacts against gravity in the vertical plane. We would like to understand how far the object can shift its center of mass without losing its equilibrium stance. We would also like to understand which non-static motion modes will develop at the supporting contacts once the object loses its equilibrium stance. The stance equilibrium region is defined as follows.

DEFINITION 14.2 (**the region $\mathcal{ER}$**) *Let a 2-D rigid object $\mathcal{B}$ be supported against gravity by k frictional contacts. The* **stance equilibrium region,** $\mathcal{ER} \subseteq \mathbb{R}^2$*, is the set of all $\mathcal{B}$'s center-of-mass positions at which there exist feasible contact forces, $f_i \in C_i$ for $i = 1 \ldots k$, satisfying the equilibrium stance condition of Eq.* (14.1).

When the feasible equilibrium region of a k-contact stance is nonempty, it forms one of the following sets of vertical lines, collinear with the gravitational force direction.

LEMMA 14.3 (**shape of $\mathcal{ER}$**) *The stance equilibrium region, $\mathcal{ER} \subseteq \mathbb{R}^2$, forms a vertical line when $\mathcal{B}$ is supported by a single frictional contact, and a vertical strip, half plane, or the entire vertical plane when $\mathcal{B}$ is supported by $k \geq 2$ frictional contacts.*

Proof sketch: Let $\boldsymbol{e} = (0, 1)$ denote the vertical upward direction in $\mathbb{R}^2$, collinear with the gravitational force f_g. At any center-of-mass point $\boldsymbol{x}_{cm} \in \mathcal{ER}$, the gravitational wrench in Eq. (14.1) satisfies $(\boldsymbol{x}_{cm} + s\boldsymbol{e}) \times f_g = \boldsymbol{x}_{cm} \times f_g$ for all $s \in \mathbb{R}$. Hence, the entire vertical line $\{\boldsymbol{x}_{cm} + s \cdot \boldsymbol{e},\ s \in \mathbb{R}\}$ lies in $\mathcal{ER}$, which is thus a collection of vertical lines. To see that $\mathcal{ER}$ forms a connected set of vertical lines, consider two points $\boldsymbol{x}'_{cm}$ and $\boldsymbol{x}''_{cm}$ in $\mathcal{ER}$. Let $f'_i, f''_i \in C_i$ for $i = 1 \ldots k$ be the contact forces corresponding to the object's center of mass at $\boldsymbol{x}'_{cm}$ and $\boldsymbol{x}''_{cm}$. Any point on the line segment joining $\boldsymbol{x}'_{cm}$ and $\boldsymbol{x}''_{cm}$ can be written as $\boldsymbol{x}_{cm}(s) = s\boldsymbol{x}'_{cm} + (1-s)\boldsymbol{x}''_{cm}$ for some $0 \leq s \leq 1$. The corresponding contact forces, $f_i = sf'_i + (1 - s)f''_i$ lie in C_i for $i = 1 \ldots k$, and satisfy the equilibrium stance condition (14.1) with $\boldsymbol{x}_{cm} = \boldsymbol{x}_{cm}(s)$. Hence, $\mathcal{ER}$ forms a connected set of vertical lines (a single vertical line, a strip, a half plane, or the entire vertical plane) in $\mathbb{R}^2$. □

The feasible equilibrium region at a two-contact stance is illustrated in the following example.

Example: Figure 14.3(a) depicts a two-contact stance supported by rigid 35° slope segments, with coefficients of friction $\mu_1 = \mu_2 = 0.3$. Based on a graphical technique discussed in the next section, the region $\mathcal{ER}$ forms the vertical strip spanned by the intersection polygon $C_1 \cap C_2$. Figure 14.3(b) shows the region $\mathcal{ER}$ for the same stance – this time with much larger coefficients of friction, $\mu_1 = \mu_2 = 2.0$. The region $\mathcal{ER}$ now

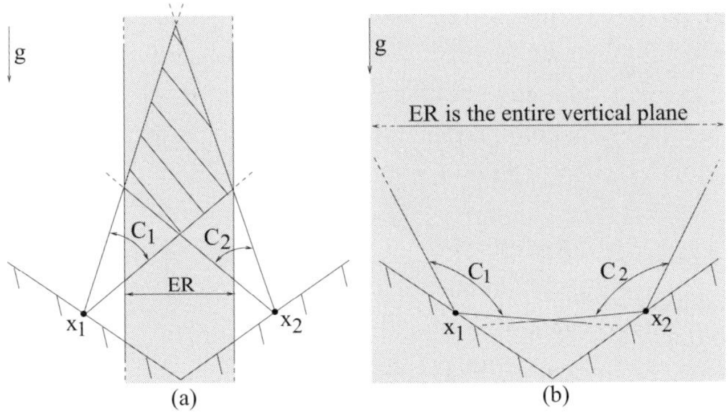

Figure 14.3 (a) The region $\mathcal{ER}$ of a two-contact stance with $\mu_1 = \mu_2 = 0.3$ forms the vertical strip spanned by the polygon $C_1 \cap C_2$. (b) The region $\mathcal{ER}$ of a two-contact stance with $\mu_1 = \mu_2 = 2.0$ occupies the entire vertical plane (the supported object $\mathcal{B}$ is not shown).

occupies the entire vertical plane. The line segment connecting the supporting contacts is contained in the friction cones C_1 and C_2, which implies full wrench resistance (Theorem 12.1). Such a contact arrangement is a wrench resistant grasp that is technically no longer a hand-supported stance. ○

When the region $\mathcal{ER}$ has a nonempty interior (the generic case), any placement of the object's center of mass in the interior of $\mathcal{ER}$ forms a local wrench resistant stance. This *robustness* with respect to small center-of-mass position errors is stated in the following proposition.

PROPOSITION 14.4 (**robustness of $\mathcal{ER}$**) *Let a 2-D rigid object $\mathcal{B}$ be supported against gravity by $k \geq 2$ frictional contacts. If the region $\mathcal{ER}$ has a nonempty interior (the generic case), all center-of-mass locations in the interior of $\mathcal{ER}$ form* **local wrench resistant** *stances.*

Proof sketch: Consider the two-contact stance depicted in Figure 14.4(a). As noted in Chapter 11, when $\mathcal{B}$'s center of mass, $\boldsymbol{x}_{cm}$, varies in $\mathbb{R}^2$ at the given stance, the gravitational wrench that acts on $\mathcal{B}$, $\boldsymbol{w}_g = (f_g, \boldsymbol{x}_{cm} \times f_g)$, spans an *affine line* in $\mathcal{B}$'s wrench space, denoted L. The line L has fixed force components and is depicted together with the net wrench cone $\mathcal{W}$ in Figure 14.4(b). Every point $\boldsymbol{x}_{cm} \in \mathcal{ER}$ determines a feasible equilibrium stance. In $\mathcal{B}$'s wrench space, every feasible equilibrium stance is associated with a particular point in the intersection $L \cap \mathcal{W}$. When the region $\mathcal{ER}$ has a nonempty interior, the line L intersects the interior of $\mathcal{W}$. Since $\mathcal{W}$ forms a 3-D cone in $\mathcal{B}$'s wrench space for $k \geq 2$ frictional contacts (Lemma 14.1), every point of L in the interior of $\mathcal{W}$ can be surrounded by a small open ball that lies in the interior of $\mathcal{W}$ (Figure 14.4(b)). Each of these balls forms a local wrench neighborhood that can be generated by feasible contact forces, which gives local wrench resistance. ○

Proposition 14.4 leads to the following practical insight concerning local wrench resistant stances. When a rigid object $\mathcal{B}$ is supported at a feasible equilibrium stance by $k \geq 2$ frictional contacts, the contacts will maintain the equilibrium stance against all

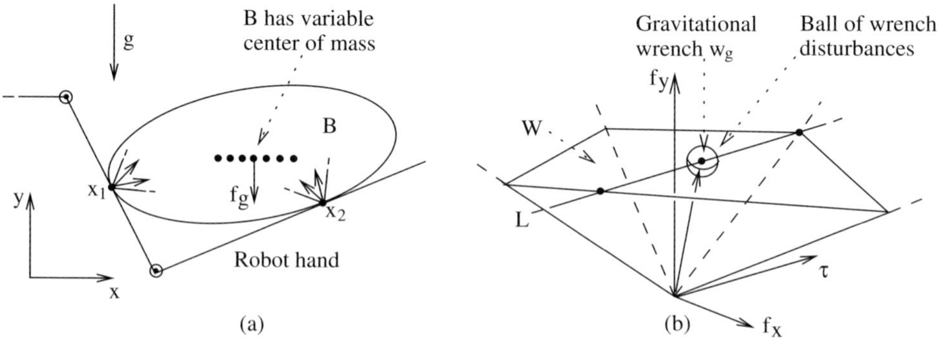

Figure 14.4 (a) A rigid object $\mathcal{B}$ with a variable center of mass supported by midfinger contacts x_1 and x_2. (b) The stance's net wrench cone, $\mathcal{W}$, and the affine line of gravitational wrenches, L, intersect along a line segment in $\mathcal{B}$'s wrench space.

sufficiently small external disturbances that may act on $\mathcal{B}$. That is, the object will *remain stationary* on the robot hand, while the contact forces change their direction and magnitude in response to small disturbances that act on $\mathcal{B}$. However, one is confronted at this stage with a fundamental limitation of the ideal rigid-body model, which can be summarized as follows.

Contact mode ambiguity: When a rigid object $\mathcal{B}$ is supported in static equilibrium by multiple frictional contacts, any combination of contact reaction forces can develop within the friction cones in response to a disturbance wrench that acts on $\mathcal{B}$. The ideal rigid-body model *cannot* predict exactly which reaction forces will develop within the Coulomb friction cones at the contacts. Local wrench resistance ensures that a nonempty set of these forces can maintain the equilibrium stance. However, other permissible contact reaction forces may initiate *nonstatic* motion modes at the contacts. For instance, the object may start rolling on one contact while breaking the other contacts in response to a disturbance wrench. This ambiguity reflects an *intrinsic limitation* of the ideal rigid-body model under frictional contact conditions (see Bibliography Notes). ○

The chapter makes the practical assumption that when the object's center of mass lies in the *interior* of the stance equilibrium region, the object will remain stationary under all sufficiently small disturbances.

14.3 Graphical Construction of the 2-D Stance Equilibrium Region

This section describes a graphical technique for computing the feasible equilibrium region of 2-D stances supported by any number of frictional contacts. The graphical technique is based on the following convex hull property.

PROPOSITION 14.5 (**convex hull property**) *Let $\mathcal{R}_{ij} \subseteq \mathbb{R}^2$ denote the feasible equilibrium region associated with two supporting contacts, x_i and x_j. The feasible equilibrium region associated with $k \geq 2$ frictional contacts is the* **convex hull** *of the pairwise stance equilibrium regions: $\mathcal{ER} = \text{conv}\{\mathcal{R}_{ij}$ for $1 \leq i, j \leq k\}$, where* conv *denotes convex hull.*

The proof of Proposition 14.5 appears in the chapter's appendix. The stance equilibrium region can thus be computed in terms of the pairwise equilibrium regions. Since these are all connected sets of vertical lines, their convex hull forms a single connected set of vertical lines, bounded by the *leftmost* and *rightmost* edges of the pairwise strips. Hence, we can focus on the graphical construction of the two-contact stance equilibrium region. The graphical construction requires the following mild assumption on the friction cones at the given stance.

DEFINITION 14.3 *A supporting contact is said to be* **upward pointing** *when all forces in its friction cone C_i possess a positive component along the vertical upward direction in $\mathbb{R}^2$.*

The stance equilibrium region associated with two upward-pointing contacts, $\mathcal{R}_{ij}$, can be obtained by the union and intersection of *five* vertical strips, each having a distinct geometric interpretation. The five strips are as follows. Let C_i^- denote the negative reflection of the friction cone C_i about its base point x_i, $C_i^- = \{-f_i : f_i \in C_i\}$. Let Π_{ij}^{++} denote the vertical strip whose lines pass through all points of the polygon $C_i \cap C_j$. For instance, the vertical strip depicted in Figure 14.3(a) is spanned by the polygon $C_1 \cap C_2$. Similarly, let Π_{ij}^{+-}, Π_{ij}^{-+}, and Π_{ij}^{--} denote the vertical strips spanned by the polygons $C_i \cap C_j^-$, $C_i^- \cap C_j$, and $C_i^- \cap C_j^-$. Note that some of these polygons and their associated strips may be empty. The fifth strip, denoted X_{ij}, is the vertical strip bounded by the contacts x_i and x_j. The graphical construction of the equilibrium region $\mathcal{R}_{ij}$ is summarized in the following theorem.

THEOREM 14.6 **(five-strip formula)** *The* **2-D feasible equilibrium region** *of a stance supported by two upward-pointing frictional contacts, x_i and x_j, forms a connected vertical strip according to the formula*

$$\mathcal{R}_{ij} = \left((\Pi_{ij}^{++} \cup \Pi_{ij}^{--}) \cap X_{ij}\right) \cup \left((\Pi_{ij}^{+-} \cup \Pi_{ij}^{-+}) \cap \bar{X}_{ij}\right), \tag{14.2}$$

where Π_{ij}^{++}, Π_{ij}^{+-}, Π_{ij}^{-+} and Π_{ij}^{--} are the vertical strips spanned by $C_i \cap C_j$, $C_i \cap C_j^-$, $C_i^- \cap C_j$ and $C_i^- \cap C_j^-$, X_{ij} is the vertical strip bounded by x_i and x_j, and $\bar{X}_{ij}$ is the complement of X_{ij} in $\mathbb{R}^2$.

Proof: The gravitational force f_g acts on $\mathcal{B}$ at $\boldsymbol{x}_{cm}$, while the contact forces f_i and f_j act on $\mathcal{B}$ at the supporting contacts x_i and x_j. The net torque generated by these forces must be zero at a feasible equilibrium stance. As discussed in Chapter 4, zero net torque requires that the lines underlying the forces f_i, f_j, and f_g intersect at a common point, $z \in \mathbb{R}^2$, which lies at infinity when all forces are parallel. Since f_i and f_j lie in their respective friction cones, C_i and C_j, the point z must lie within the intersection of the two double cones, $(C_i \cup C_i^-) \cap (C_j \cup C_j^-)$. Since $\boldsymbol{x}_{cm}$ and z lie along a common vertical line in $\mathbb{R}^2$, $\boldsymbol{x}_{cm}$ must lie in one of the infinite vertical strips spanned by the intersection $(C_i \cup C_i^-) \cap (C_j \cup C_j^-)$, which are the four strips Π_{ij}^{++}, Π_{ij}^{+-}, Π_{ij}^{-+}, and Π_{ij}^{--}.

Next consider the force part of the equilibrium stance equation, $f_i + f_j + f_g = \vec{0}$. Two *upward-pointing* contacts can always balance the gravitational force that acts on $\mathcal{B}$. Let us therefore focus on the horizontal component of the contacts forces at the equilibrium stance. Denote by u_i and u_j the unit vectors pointing from x_i and x_j to z. There are four cases to consider, depending on the location of z in the double cones $C_i \cup C_i^-$ and $C_j \cup C_j^-$.

Consider the case where $z \in C_i \cap C_j$. In this case $u_i \in C_i$ and $u_j \in C_j$. Hence $f_i = \lambda_i u_i$ and $f_j = \lambda_j u_j$ for some $\lambda_i, \lambda_j \geq 0$, which are not both zero. Let $h = (1,0)$ denote the horizontal unit direction in $\mathbb{R}^2$. Multiplying the equation $f_i + f_j + f_g = \vec{0}$ by h gives: $\lambda_i(u_i \cdot h) + \lambda_j(u_j \cdot h) = 0$. If λ_i and λ_j are strictly positive, the terms $u_i \cdot h$ and $u_j \cdot h$ must have *opposite* signs. The graphical interpretation of this fact is that z must lie *inside* the vertical strip X_{ij} bounded by x_i and x_j. If $\lambda_i > 0$ and $\lambda_j = 0$, the point z lies on the vertical line passing through x_i. If $\lambda_i = 0$ and $\lambda_j > 0$, the point z lies on the vertical line passing through x_j. Since $\boldsymbol{x}_{cm}$ and z lie on a common vertical line in $\mathbb{R}^2$, $\boldsymbol{x}_{cm}$ must lie in the intersection $\Pi_{ij}^{++} \cap X_{ij}$.

Next consider the case where $z \in C_i \cap C_j^-$. In this case, $u_i \in C_i$ while $u_j \in C_j^-$. Hence, $f_i = \lambda_i u_i$ and $f_j = \lambda_j u_j$ such that $\lambda_i \geq 0$ and $\lambda_j \leq 0$. The equation $\lambda_i(u_i \cdot h) + \lambda_j(u_j \cdot h) = 0$ now implies that $u_i \cdot h$ and $u_j \cdot h$ must have the same sign. The graphical interpretation of this fact is that z must lie *outside* the vertical strip X_{ij}. Thus, $\boldsymbol{x}_{cm}$ lies in the intersection $\Pi_{ij}^{+-} \cap \bar{X}_{ij}$. By applying similar arguments to the remaining two cases, $z \in C_i^- \cap C_j^-$ and $z \in C_i^- \cap C_j$, Formula (14.2) is obtained. □

Example: Consider the two-contact stance depicted in Figure 14.3(a). In this stance, only the polygon $C_1 \cap C_2$ is nonempty. According to the five-strip formula, $\mathcal{R}_{12} = \Pi_{12}^{++} \cap X_{12}$ at this stance. ○

The feasible equilibrium region of a two-contact stance contains the strip bounded by the contact points, $\mathcal{R}_{ij} \supseteq X_{ij}$, when the friction cones at x_i and x_j contain the vertical upward direction. At such stances, the region $\mathcal{R}_{ij}$ can spill outside the supporting contacts. For instance, the region $\mathcal{R}_{ij}$ of the stance depicted in Figure 14.1(b) spills to the right side of the supporting contact x_2 (see Exercises).

The feasible equilibrium region of 2-D stances associated with higher numbers of contacts can be computed based on the convex hull property described in Proposition 14.5. First, determine all the pairwise feasible equilibrium regions, $\mathcal{R}_{ij}$ for $1 \leq i, j \leq k$, using the five-strip formula for each pair of contacts. Since the region $\mathcal{ER}$ is the convex hull of the pairwise equilibrium strips, the *leftmost* and *rightmost* edges of the pairwise strips determine the left and right edges of the full equilibrium region $\mathcal{ER}$ (a single vertical strip). When a stance involves $k \geq 4$ supporting contacts, the stance equilibrium region is fully determined by at most *four* of the k contacts, as illustrated in the following example (see Exercises).

Example: Figure 14.5 depicts a six-contact stance in $\mathbb{R}^2$, without showing the object $\mathcal{B}$ supported by the contacts. All six contacts have a coefficient of friction $\mu = 0.4$. The

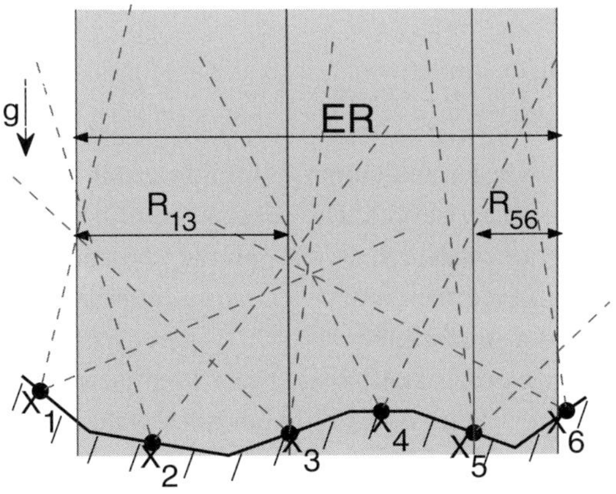

Figure 14.5 The feasible equilibrium region of this six-contact stance is the convex hull of the equilibrium strips $\mathcal{R}_{13}$ and $\mathcal{R}_{56}$ (the supported object is not shown).

leftmost edge of $\mathcal{ER}$ is the left edge of the equilibrium strip $\mathcal{R}_{13}$, while the rightmost edge of $\mathcal{ER}$ is the right edge of the equilibrium strip $\mathcal{R}_{56}$. In this example, $\mathcal{ER} = \text{conv}\{\mathcal{R}_{13}, \mathcal{R}_{56}\}$, which is fully determined by the contact points x_1, x_3, x_5, and x_6. ○

Development of non-static motion modes: When a 2-D rigid object $\mathcal{B}$ *slowly* shifts its center of mass while supported at an equilibrium stance, non-static motion modes will develop once $\mathcal{B}$'s center of mass crosses the boundary of the stance equilibrium region. In the context of hand-supported stances, the supported object will no longer maintain its static equilibrium on the supporting robot hand. To understand which non-static modes will develop at the contacts, consider the two vertical lines that form the boundary of the region $\mathcal{ER}$. These two lines correspond to two points in $\mathcal{B}$'s wrench space, where the affine line L of gravitational wrenches intersects the boundary of the net wrench cone $\mathcal{W}$ (Figure 14.4(b)). Each of these intersection points lies on a facet of $\mathcal{W}$, which forms a planar sector based at $\mathcal{B}$'s wrench space origin. The edges of this sector are generated by forces aligned with two friction cone edges, that may belong to the same or different friction cones. When an intersection point lies on a facet of $\mathcal{W}$ spanned by two edges that belong to the same friction cone, the object will start *rolling* at the corresponding contact while breaking away from the remaining $k-1$ contacts. When an intersection point lies on a facet of $\mathcal{W}$ spanned by two edges that belong to different friction cones, the object will start *sliding* at the two contacts while breaking away from the remaining $k-2$ contacts. ○

When the supporting contacts are not all upward pointing, the moment labeling technique of Chapter 4 can be used to determine the feasible equilibrium region of such stances (see Exercises).

14.4 Safety Margin on the 2-D Stance Equilibrium Region

The feasible equilibrium region of 2-D stances consists of infinitely high vertical lines. However, intuition suggests that any given stance must have a practical *upper limit* on the height of the object's safe center-of-mass positions. This concern is illustrated in the following example, which uses the analogous situation of a climber, whose shifting limbs are modeled as a rigid body having a variable center of mass.

Example: Consider the human climbing a tall ladder under the influence of gravity, depicted in Figure 14.6(a). The feasible equilibrium region for the combined climber–ladder center of mass, $\mathcal{ER}$, is the infinite vertical strip bounded by the ladder's ground contacts. However, the climber's moving limbs generate wrench disturbances. Intuition suggests that as the climber reaches higher elevations, it must *reduce* the disturbances generated by its moving limbs in order to prevent the ladder from tipping over. Equivalently, any pre-specified climbing pattern should impose a safe upper limit on the points of $\mathcal{ER}$. Based on graphical tools developed later in this section, the climber's safety region is schematically sketched in Figure 14.6(b). ○

This section characterizes the subset of $\mathcal{ER}$ that can balance a pre-specified neighborhood of disturbance wrenches that may act on $\mathcal{B}$ at a given stance. The convexity

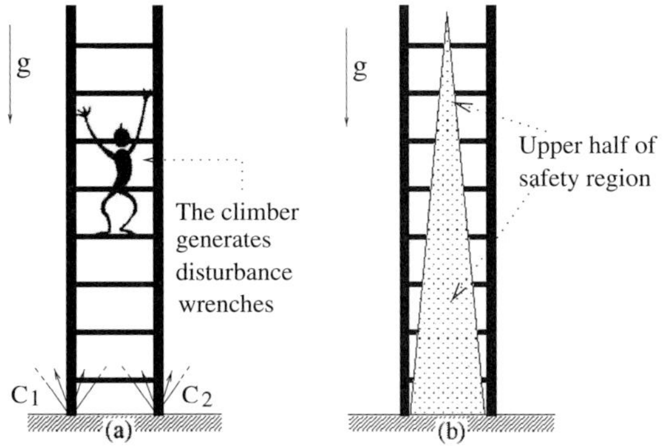

Figure 14.6 (a) A climbing human generates wrench disturbances. (b) A sketch of the climber–ladder center-of-mass safety region.

arguments of Section 14.3 still apply. Hence, we will focus on *two-contact stances* and comment on k-contact stances where appropriate. First consider the case where the object $\mathcal{B}$ is supported under the combined influence of a gravitational force f_g and a pre-specified disturbance wrench $\boldsymbol{w}_d = (f_d, \tau_d)$. The feasible equilibrium region associated with the disturbance wrench is defined as follows.

DEFINITION 14.4 (**the region $\mathcal{R}_{ij}(\boldsymbol{w}_{ext})$**) *Let a 2-D rigid object $\mathcal{B}$ be supported against gravity by frictional contacts x_i and x_j. The* **disturbance-safe equilibrium region,** *$\mathcal{R}_{ij}(\boldsymbol{w}_{ext}) \subseteq \mathbb{R}^2$, consists of all $\mathcal{B}$'s center-of-mass positions at which the supporting contacts can resist the net wrench $\boldsymbol{w}_{ext} = \boldsymbol{w}_g + \boldsymbol{w}_d$ which acts on $\mathcal{B}$.*

To graphically characterize the region $\mathcal{R}_{ij}(\boldsymbol{w}_{ext})$, first consider a wrench disturbance generated by a pure force, $\boldsymbol{w}_d = (f_d, 0)$. In this case, the net wrench affecting $\mathcal{B}$ can be represented as a pure force, $f_g + f_d$, acting along a line passing through $\mathcal{B}$'s center of mass. The region $\mathcal{R}_{ij}(\boldsymbol{w}_{ext})$ is constructed with the same five-strip formula of Theorem 14.6, except that now the five strips are *rotated* to match the direction of the net force $f_g + f_d$. When the wrench disturbance has a non-zero torque component, $\boldsymbol{w}_d = (f_d, \tau_d)$ such that $\tau_d \neq 0$, the net wrench acting on $\mathcal{B}$ is equivalent to a force $f_g + f_d$ that acts on $\mathcal{B}$ along a line parallel shifted by an offset of $\tau_d / \| f_g + f_d \|$ from $\mathcal{B}$'s center of mass.[2]

The following corollary to Theorem 14.6 describes the graphical construction of the region $\mathcal{R}_{ij}(\boldsymbol{w}_{ext})$.

COROLLARY 14.7 (**rotated five-strip formula**) *Let a 2-D rigid object $\mathcal{B}$ be supported by frictional contacts x_i and x_j under the combined influence of gravity, $\boldsymbol{w}_g = (f_g, \boldsymbol{x}_{cm} \times f_g)$, and a disturbance wrench, $\boldsymbol{w}_d = (f_d, \tau_d)$. Let Π_{ij}^{++}, Π_{ij}^{+-}, Π_{ij}^{-+}, and Π_{ij}^{--} be the strips spanned by the polygons $C_i \cap C_j$, $C_i^+ \cap C_j^-$, $C_i^- \cap C_j^+$, and $C_i^- \cap C_j^-$, and let X_{ij} be the strip bounded by the contacts, such that the five strips*

[2] This is *Poinsot's theorem* from Chapter 3: every wrench $\boldsymbol{w} = (f, \tau)$ that acts on $\mathcal{B}$ can be represented as a directed force line collinear with f, parallel shifted from $\mathcal{B}$'s frame origin by an offset determined by τ.

are parallel to $f_g + f_d$ and parallel shifted by an offset $-\tau_d/\|f_g + f_d\|$. The feasible equilibrium region $\mathcal{R}_{ij}(\boldsymbol{w}_{ext})$ is the **infinite strip** *parallel to $f_g + f_d$ given by*

$$\mathcal{R}_{ij}(\boldsymbol{w}_{ext}) = \Big((\Pi_{ij}^{++} \cup \Pi_{ij}^{--}) \cap X_{ij}\Big) \cup \Big((\Pi_{ij}^{+-} \cup \Pi_{ij}^{-+}) \cap \bar{X}_{ij}\Big),$$

where $\bar{X}_{ij}$ is the complement of X_{ij} in $\mathbb{R}^2$.

The proof of Corollary 14.7 is similar to the proof of Theorem 14.6 and is therefore omitted. When a rigid object $\mathcal{B}$ is supported by a higher number of contacts, start by computing the feasible equilibrium region associated with each pair of contacts, $\mathcal{R}_{ij}(\boldsymbol{w}_{ext})$ for $1 \le i, j \le k$, and then take the *convex hull* of these strips: $\mathcal{ER}(\boldsymbol{w}_{ext}) = \text{conv}\{\mathcal{R}_{ij}(\boldsymbol{w}_{ext})$ for $1 \le i, j \le k\}$. The graphical construction of the region $\mathcal{R}_{ij}(\boldsymbol{w}_{ext})$ is illustrated in the following example.

Example: First consider the unperturbed two-contact stance depicted in Figure 14.7(a). The feasible equilibrium region of this stance is the vertical strip $\mathcal{R}_{ij} = \Pi_{ij}^{++} \cap X_{ij}$. Next consider a disturbance wrench, $\boldsymbol{w}_d = (f_d, \tau_d)$, which acts on $\mathcal{B}$ as schematically depicted in Figure 14.7(b). The region $\mathcal{R}_{ij}(\boldsymbol{w}_{ext})$ can be constructed in two stages. First parallel shift the undisturbed equilibrium strip by an offset $-\tau_d/\|f_g + f_d\|$. Then rotate the strip as to match the direction of the net force $f_g + f_d$. Every center of mass of $\mathcal{B}$ that lies in the slanted strip of Figure 14.7(b) forms a feasible equilibrium stance under $\boldsymbol{w}_{ext} = \boldsymbol{w}_g + \boldsymbol{w}_d$. ○

The graphical construction of the region $\mathcal{R}_{ij}(\boldsymbol{w}_{ext})$ is next extended to the case where the wrench disturbances vary in a small neighborhood centered at the gravitational wrench. We have seen that two parameters influence the shape of $\mathcal{R}_{ij}(\boldsymbol{w}_{ext})$: the *direction* of the net force, $f_g + f_d$, and the *parallel shift parameter*, $\tau_d/\|f_g + f_d\|$. The wrench neighborhood will be similarly defined in terms of two parameters. Given a disturbance

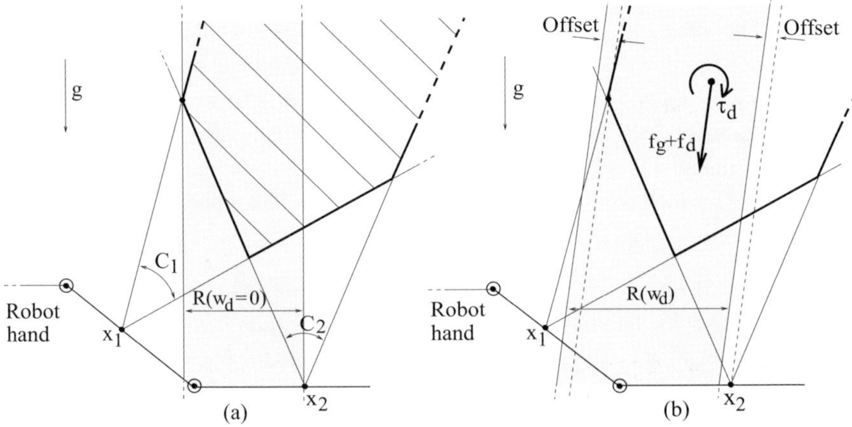

Figure 14.7 (a) The feasible equilibrium region $\mathcal{R}_{12}(\boldsymbol{w}_{ext})$ for (a) a zero-disturbance wrench, and (b) a disturbance wrench $\boldsymbol{w}_d = (f_d, \tau_d)$ (the supported object is not shown).

wrench $\boldsymbol{w}_d = (f_d^x, f_d^y, \tau_d)$, the two parameters are the horizontal force component, f_d^x, and the torque component, τ_d, of the disturbance wrench.

DEFINITION 14.5 **(local wrench neighborhood)** *Let a rigid object $\mathcal{B}$ be subjected to wrench disturbances $\left|f_d^x\right| \leq f_{max}$ and $|\tau_d| \leq \tau_{max}$. The wrench neighborhood determined by these disturbances, $\mathcal{W}_{ext}$, is defined as the* **convex hull** *of four wrenches,*

$$\mathcal{W}_{ext} = \textsf{conv}\left\{\boldsymbol{w}_g + \boldsymbol{w}_1, \boldsymbol{w}_g + \boldsymbol{w}_2, \boldsymbol{w}_g + \boldsymbol{w}_3, \boldsymbol{w}_g + \boldsymbol{w}_4\right\},$$

where $\boldsymbol{w}_g$ is the gravitational wrench that acts on $\mathcal{B}$, and the four disturbance wrenches are $\boldsymbol{w}_1 = (-f_{max}, 0, 0)$, $\boldsymbol{w}_2 = (f_{max}, 0, 0)$, $\boldsymbol{w}_3 = (0, 0, -\tau_{max})$, and $\boldsymbol{w}_4 = (0, 0, \tau_{max})$.

The wrench neighborhood is thus centered at the gravitational wrench, $\boldsymbol{w}_g$, and is spanned by wrenches having pure horizontal force or pure torque components. While $\mathcal{W}_{ext}$ forms a two-dimensional set of wrenches, it extends into a three-dimensional neighborhood of wrenches, based on the following homogeneity property. Let $(f_1, \ldots, f_k) \in \mathbb{R}^{2k}$ be the contact forces satisfying the equilibrium stance condition of Eq. (14.1) for a particular wrench $\boldsymbol{w}_{ext} \in \mathcal{W}_{ext}$. Then the contact forces $(s \cdot f_1, \ldots, s \cdot f_k)$ satisfy Eq. (14.1) for the wrenches: $\{s \cdot \boldsymbol{w}_{ext} : s \geq 0\}$. It follows that equilibrium feasibility with respect to $\mathcal{W}_{ext}$ implies equilibrium feasibility with respect to the three-dimensional neighborhood of wrenches: $\{s \cdot \boldsymbol{w}_{ext} : \boldsymbol{w}_{ext} \in \mathcal{W}_{ext}, s \geq 0\}$.

The feasible equilibrium region that can safely support the wrench neighborhood $\mathcal{W}_{ext}$ is defined next.

DEFINITION 14.6 **(the region $\mathcal{R}_{ij}(\mathcal{W}_{ext})$)** *Let a 2-D rigid object $\mathcal{B}$ be supported against gravity by frictional contacts x_i and x_j. The* **disturbance-safe equilibrium region,** *$\mathcal{R}_{ij}(\mathcal{W}_{ext})$, consists of all $\mathcal{B}$'s center-of-mass positions at which the supporting contacts can resist the entire wrench neighborhood $\mathcal{W}_{ext}$.*

At each center of mass location in the region $\mathcal{R}(\mathcal{W}_{ext})$, the contacts must be able to resist with feasible forces the entire wrench neighborhood $\mathcal{W}_{ext}$. Since $\mathcal{W}_{ext}$ is centered at the gravitational wrench, $\boldsymbol{w}_g$, the disturbance-safe equilibrium region $\mathcal{R}_{ij}(\mathcal{W}_{ext})$ lies within the stance equilibrium region $\mathcal{R}_{ij}$. The graphical construction of the disturbance-safe equilibrium region is described in the following proposition.

PROPOSITION 14.8 **(disturbance-safe region)** *Let a 2-D rigid object $\mathcal{B}$ be supported against gravity by frictional contacts x_i and x_j. The* **disturbance-safe equilibrium region** *associated with the wrench neighborhood $\mathcal{W}_{ext}$ is given by the intersection:*

$$\mathcal{R}_{ij}(\mathcal{W}_{ext}) = \mathcal{R}_{ij}(\boldsymbol{w}_1) \cap \mathcal{R}_{ij}(\boldsymbol{w}_2) \cap \mathcal{R}_{ij}(\boldsymbol{w}_3) \cap \mathcal{R}_{ij}(\boldsymbol{w}_4)$$

where $\mathcal{R}_{ij}(\boldsymbol{w}_1), \ldots, \mathcal{R}_{ij}(\boldsymbol{w}_4)$ are the feasible equilibrium regions associated with the wrenches $\boldsymbol{w}_1, \ldots, \boldsymbol{w}_4$ that define the neighborhood $\mathcal{W}_{ext}$.

The proof of Proposition 14.8 appears in the chapter's appendix. The feasible equilibrium regions $\mathcal{R}_{ij}(\boldsymbol{w}_1), \ldots, \mathcal{R}_{ij}(\boldsymbol{w}_4)$ form slanted strips in the vertical plane. These strips are oriented along the two directions $f_g - f_d$ and $f_g + f_d$, where $f_d = (f_{max}, 0)$. The disturbance-safe equilibrium region resulting from the intersection of the four strips forms a *bounded parallelogram,* as illustrated in the following example.

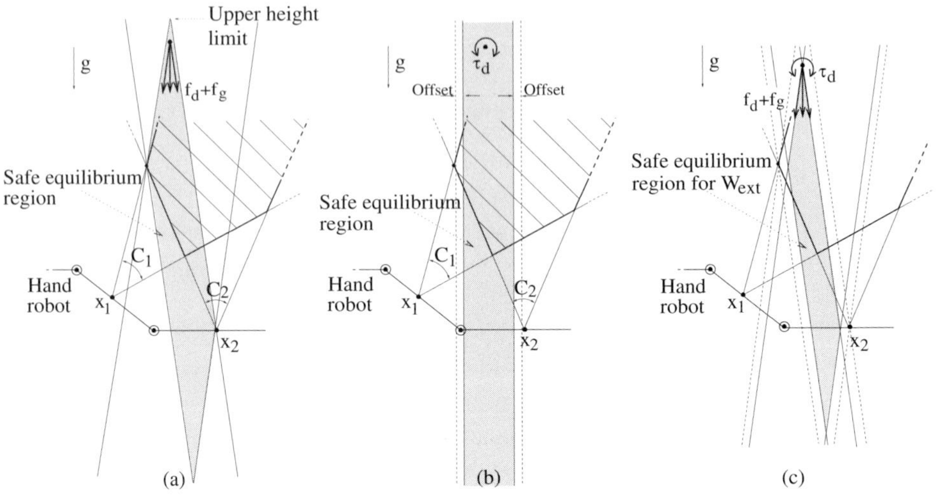

Figure 14.8 The disturbance-safe equilibrium region for (a) $|f_d^x| \leq f_{max}$, (b) $|\tau_d| \leq \tau_{max}$, and (c) the full wrench disturbance set $\mathcal{W}_{ext}$ (the supported object is not shown).

Example: Figure 14.8 depicts a two-contact stance with coefficients of friction $\mu_1 = \mu_2 = 0.5$ at x_1 and x_2 (the supported object is not shown). The point x_1 lies on a segment with 45° slope angle while x_2 is located on a horizontal segment. The *undisturbed* equilibrium region of this stance forms the vertical strip bounded by the two contacts, $\mathcal{R}_{ij} = X_{ij}$. First consider the disturbance-safe equilibrium region associated with pure horizontal force disturbances, satisfying $|f_d^x| \leq f_{max}$. This region is the intersection of two slanted strips associated with the horizontal force disturbances $f_d^x = \pm f_{max}$. The intersection gives the vertically oriented parallelogram depicted in Figure 14.8(a). It can be seen that the horizontal force disturbances, $|f_d^x| \leq f_{max}$, impose *height limit* on $\mathcal{B}$'s safe center-of-mass locations. Next consider pure torque disturbances, satisfying $|\tau_d| \leq \tau_{max}$. The disturbance-safe equilibrium region depicted in Figure 14.8(b) is obtained by parallel shifting the two bounding lines of the vertical strip $\mathcal{R}_{ij}$ *inward,* by $\tau_{max}/\|f_g\|$. The disturbance-safe equilibrium region associated with the full neighborhood, $\mathcal{R}_{ij}(\mathcal{W}_{ext})$, is the slanted parallelogram obtained by intersecting the four strips, depicted in Figure 14.8(c). ○

The disturbance-safe equilibrium region of two-contact stances thus forms a *bounded height* subset of the stance equilibrium region, $\mathcal{R}_{ij}(\mathcal{W}_{ext}) \subseteq \mathcal{R}_{ij}$. The disturbance-safe equilibrium region associated with a higher number of contacts, denoted $\mathcal{ER}(\mathcal{W}_{ext})$, can be computed in two stages. First, compute the four slanted strips $\mathcal{ER}(\boldsymbol{w}_1), \mathcal{ER}(\boldsymbol{w}_2), \mathcal{ER}(\boldsymbol{w}_3), \mathcal{ER}(\boldsymbol{w}_4)$ for the wrenches that determine the wrench neighborhood $\mathcal{W}_{ext}$, by taking the convex hull of the slanted strips associated with all contact pairs. Then take the intersection of the four slanted strips:

$$\mathcal{ER}(\mathcal{W}_{ext}) = \mathcal{ER}(\boldsymbol{w}_1) \cap \mathcal{ER}(\boldsymbol{w}_2) \cap \mathcal{ER}(\boldsymbol{w}_3) \cap \mathcal{ER}(\boldsymbol{w}_4)$$

which gives a *bounded height* parallelogram satisfying the relation $\mathcal{ER}(\mathcal{W}_{ext}) \subseteq \mathcal{ER}$.

Bibliographical Notes

Or and Rimon [1] describe the principles underlying the feasible equilibrium region of stances supported against gravity by multiple frictional contacts. While reference [1] focuses on legged robot postures over uneven terrains, the same principles apply to robot hands which form an uneven terrain when supporting rigid objects against gravity. Erdmann [2] as well as Akella, Ponce and coworkers [3] consider applications of hand-supported stances to object manipulation in 2-D gravitational environments.

The contact mode ambiguity associated with frictional contacts under the ideal rigid-body model has been originally observed on a single sliding contact. Using an example of a rigid rod sliding with friction on an inclined plane under gravity, Painlevé [4] observed that the rod's dynamics can exhibit *ambiguous* solutions as well as instants at which the dynamic solution becomes *inconsistent* with the ideal rigid-body model [5, 6]. Mason and Wang [7, 8] as well as Dupont [9] discuss the related issue of *dynamic jamming* during robot manipulation and legged locomotion.

In the context of hand-supported stances, contact mode ambiguity occurs when static equilibrium as well as non-static motion modes can simultaneously develop at the contacts in response to disturbance wrenches. Trinkle [10] discusses the notion of *strong stability* for hand-supported stances that do *not* suffer from any contact mode ambiguity. Techniques that can identify strongly stable hand-supported stances should be developed in order to ensure fully safe hand-supported stances during task execution.

Appendix: Proof Details

This appendix contains proofs of two propositions from Sections 14.3 and 14.4. The first proposition asserts that the feasible equilibrium region of k-contact stances can be computed as the convex hull of the pairwise feasible equilibrium regions.

Proposition 14.5 *Let $\mathcal{R}_{ij} \subseteq \mathbb{R}^2$ denote the feasible equilibrium region associated with two supporting contacts, x_i and x_j. The feasible equilibrium region associated with $k \geq 2$ frictional contacts is the* **convex hull** *of the pairwise stance equilibrium regions: $\mathcal{ER} = \textsf{conv}\{\mathcal{R}_{ij}$ for $1 \leq i, j \leq k\}$, where* $\textsf{conv}$ *denotes convex hull.*

Proof: The net wrench cone associated with $k \geq 2$ non-coincident frictional contacts, $\mathcal{W}$, is fully three-dimensional in $\mathcal{B}$'s wrench space. The boundary of $\mathcal{W}$ consists of planar sectors, each having a vertex at $\mathcal{B}$'s wrench space origin and two bounding edges. In particular, the net wrench cone associated with two frictional contacts x_i and x_j, denoted $\mathcal{W}_{ij}$, forms a 3-D cone bounded by four planar sectors.

Let us first show that $\textsf{conv}\{\mathcal{R}_{ij}$ *for* $1 \leq i, j \leq k\} \subseteq \mathcal{ER}$. The gravitational wrenches that can act on $\mathcal{B}$, obtained by varying $\boldsymbol{x}_{cm}$ in $\mathbb{R}^2$ at the given stance, span an affine line L in $\mathcal{B}$'s wrench space. When L intersects the interior of $\mathcal{W}_{ij}$, the intersection set $L \cap \mathcal{W}_{ij}$ forms a bounded line segment, a semi-infinite line segment, or the entire line L. Since $\mathcal{W}_{ij} \subset \mathcal{W}$, one concludes that $L \cap \mathcal{W}_{ij} \subseteq L \cap \mathcal{W}$ for each pair of contacts.

It follows that the convex hull of the segments $L \cap \mathcal{W}_{ij}$ associated with all contact pairs is a subset of $L \cap \mathcal{W}$:

$$\text{conv}\{L \cap \mathcal{W}_{ij} \text{ for } 1 \leq i, j \leq k\} \subseteq L \cap \mathcal{W}. \tag{14.3}$$

Each point of $L \cap \mathcal{W}_{ij}$ determines one vertical line of the stance equilibrium region $\mathcal{R}_{ij}$ in $\mathbb{R}^2$. Similarly, each point of $L \cap \mathcal{W}$ determines one vertical line of the full equilibrium region $\mathcal{ER}$. Based on this correspondence, it follows from Eq. (14.3) that $\text{conv}\{\mathcal{R}_{ij} \textit{ for } 1 \leq i, j \leq k\} \subseteq \mathcal{ER}$.

Let us next show that $\mathcal{ER} \subseteq \text{conv}\{\mathcal{R}_{ij} \textit{ for } 1 \leq i, j \leq k\}$. First consider the case where $L \cap \mathcal{W}$ forms a bounded line segment in $\mathcal{B}$'s wrench space. Each endpoint of $L \cap \mathcal{W}$ lies on a facet of $\mathcal{W}$. Each facet of $\mathcal{W}$ is a planar sector bounded by two edges. These edges are generated by contact forces aligned with two friction cone edges that may belong to distinct friction cones. Let the first endpoint of $L \cap \mathcal{W}$ lie on a facet spanned by wrenches generated by friction cone edges based at x_i and x_j. Let the other endpoint of $L \cap \mathcal{W}$ lie on a facet spanned by wrenches generated by friction cone edges based at x_k and x_l. The facet of $\mathcal{W}$ containing the first endpoint is also a facet of the wrench cone $\mathcal{W}_{ij}$ while the facet of $\mathcal{W}$ containing the other endpoint is also a facet of the wrench cone $\mathcal{W}_{kl}$. The convex combination of the two endpoints of $L \cap \mathcal{W}$ gives the entire line segment $L \cap \mathcal{W}$. One therefore concludes that

$$L \cap \mathcal{W} \subseteq \text{conv}\{L \cap \mathcal{W}_{ij}, L \cap \mathcal{W}_{kl}\}. \tag{14.4}$$

Based on the correspondence of points in $L \cap \mathcal{W}$ with vertical lines of $\mathcal{ER}$, one obtains from Eq. (14.4) the relation:

$$\mathcal{ER} \subseteq \text{conv}\{\mathcal{R}_{ij}, \mathcal{R}_{kl}\} \subseteq \text{conv}\{\mathcal{R}_{ij} \text{ for } 1 \leq i, j \leq k\}.$$

When $L \cap \mathcal{W}$ has only one endpoint, the line L intersects one of the two-contact wrench cones, say $\mathcal{W}_{ij}$, along a semi-infinite line segment. The single endpoint lies on a facet of $\mathcal{W}$ generated by two friction cone edges based at x_k and x_l, associate with the two-contact wrench cone $\mathcal{W}_{kl}$. One therefore obtains the relation $L \cap \mathcal{W} \subseteq \text{conv}\{L \cap \mathcal{W}_{ij}, L \cap \mathcal{W}_{kl}\}$. When L lies wholly inside $\mathcal{W}$, L must intersect a pair of two-contact wrench cones, $\mathcal{W}_{ij}$ and $\mathcal{W}_{kl}$, along semi-infinite line segments. This again leads to the conclusion that $L \cap \mathcal{W} \subseteq \text{conv}\{L \cap \mathcal{W}_{ij}, L \cap \mathcal{W}_{kl}\}$. Hence, $\mathcal{ER} \subseteq \text{conv}\{\mathcal{R}_{ij} \textit{ for } 1 \leq i, j \leq k\}$. □

The next proposition describes a graphical construction of the disturbance-safe equilibrium region associated with the wrench neighborhood $\mathcal{W}_{ext}$.

Proposition 14.8 *Let a 2-D rigid object $\mathcal{B}$ be supported against gravity by frictional contacts x_i and x_j. The* **disturbance-safe equilibrium region** *associated with the wrench neighborhood $\mathcal{W}_{ext}$ is given by the intersection*

$$\mathcal{R}_{ij}(\mathcal{W}_{ext}) = \mathcal{R}_{ij}(\mathbf{w}_1) \cap \mathcal{R}_{ij}(\mathbf{w}_2) \cap \mathcal{R}_{ij}(\mathbf{w}_3) \cap \mathcal{R}_{ij}(\mathbf{w}_4),$$

where $\mathcal{R}_{ij}(\mathbf{w}_1), \ldots, \mathcal{R}_{ij}(\mathbf{w}_4)$ are the feasible equilibrium regions associated with the wrenches $\mathbf{w}_1, \ldots, \mathbf{w}_4$ that define the neighborhood $\mathcal{W}_{ext}$.

Proof: Let us first show that $\mathcal{R}_{ij}(\boldsymbol{w}_1) \cap \mathcal{R}_{ij}(\boldsymbol{w}_2) \cap \mathcal{R}_{ij}(\boldsymbol{w}_3) \cap \mathcal{R}_{ij}(\boldsymbol{w}_4) \subseteq \mathcal{R}_{ij}(\mathcal{W}_{ext})$. When $\boldsymbol{x}_{cm} \in \mathcal{R}_{ij}(\boldsymbol{w}_1) \cap \mathcal{R}_{ij}(\boldsymbol{w}_2) \cap \mathcal{R}_{ij}(\boldsymbol{w}_3) \cap \mathcal{R}_{ij}(\boldsymbol{w}_4)$, an equilibrium stance is feasible with respect to each of the four wrenches $\boldsymbol{w}_1, \boldsymbol{w}_2, \boldsymbol{w}_3, \boldsymbol{w}_4$. Let $f_{l_1} \in C_i$ and $f_{l_2} \in C_j$ such that $l = 1,2,3,4$ denote the contact forces at the equilibrium stance associated with the four wrenches $\boldsymbol{w}_g + \boldsymbol{w}_1, \boldsymbol{w}_g + \boldsymbol{w}_2, \boldsymbol{w}_g + \boldsymbol{w}_3, \boldsymbol{w}_g + \boldsymbol{w}_4$:

$$\begin{pmatrix} f_{l_1} \\ x_i \times f_{l_1} \end{pmatrix} + \begin{pmatrix} f_{l_2} \\ x_j \times f_{l_2} \end{pmatrix} + \boldsymbol{w}_g + \boldsymbol{w}_l = \vec{0} \qquad l = 1,2,3,4. \tag{14.5}$$

Every wrench $\boldsymbol{w} \in \mathcal{W}_{ext}$ can be expressed as $\boldsymbol{w} = \boldsymbol{w}_g + \sum_{l=1}^{4} s_l \boldsymbol{w}_l$, such that $0 \le s_1, \ldots, s_4 \le 1$ and $\sum_{l=1}^{4} s_l = 1$. Summing the equilibrium stances of Eq. (14.5) with these coefficients gives

$$\sum_{p=1}^{4} s_l \cdot \left(\begin{pmatrix} f_{l_1} \\ x_i \times f_{l_1} \end{pmatrix} + \begin{pmatrix} f_{l_2} \\ x_j \times f_{l_2} \end{pmatrix} + \boldsymbol{w}_g + \boldsymbol{w}_l \right) = \vec{0}. \tag{14.6}$$

Consider the net contact forces $f_i = \sum_{l=1}^{4} s_l f_{l_1}$ and $f_j = \sum_{l=1}^{4} s_l f_{l_2}$. Since each friction cone is convex, $f_i \in C_i$ and $f_j \in C_j$. Substituting f_i, f_j, and $\boldsymbol{w} = \boldsymbol{w}_g + \sum_{l=1}^{4} s_l \boldsymbol{w}_l$ in Eq. (14.6) gives

$$\forall \boldsymbol{w} \in \mathcal{W}_{ext} \quad \begin{pmatrix} f_i \\ x_i \times f_i \end{pmatrix} + \begin{pmatrix} f_j \\ x_j \times f_j \end{pmatrix} + \boldsymbol{w} = \vec{0} \qquad f_i \in C_i,\ f_j \in C_j, \tag{14.7}$$

where we used the identity $s \cdot (x \times f) = x \times (s \cdot f)$ for $s \in \mathbb{R}$. Eq. (14.7) implies that for every center-of-mass position $\boldsymbol{x}_{cm} \in \mathcal{R}_{ij}(\boldsymbol{w}_1) \cap \mathcal{R}_{ij}(\boldsymbol{w}_2) \cap \mathcal{R}_{ij}(\boldsymbol{w}_3) \cap \mathcal{R}_{ij}(\boldsymbol{w}_4)$, the supporting contacts form a feasible equilibrium stance with respect to all wrenches $\boldsymbol{w} \in \mathcal{W}_{ext}$. Hence, $\boldsymbol{x}_{cm}$ lies in $\mathcal{R}_{ij}(\mathcal{W}_{ext})$.

To show that $\mathcal{R}_{ij}(\mathcal{W}_{ext}) \subseteq \mathcal{R}_{ij}(\boldsymbol{w}_1) \cap \mathcal{R}_{ij}(\boldsymbol{w}_2) \cap \mathcal{R}_{ij}(\boldsymbol{w}_3) \cap \mathcal{R}_{ij}(\boldsymbol{w}_4)$, note that, by construction, the wrenches $\boldsymbol{w}_g + \boldsymbol{w}_1, \boldsymbol{w}_g + \boldsymbol{w}_2, \boldsymbol{w}_g + \boldsymbol{w}_3, \boldsymbol{w}_g + \boldsymbol{w}_4$ lie in the wrench neighborhood $\mathcal{W}_{ext}$. Hence, at each center-of-mass position $\boldsymbol{x}_{cm} \in \mathcal{R}_{ij}(\mathcal{W}_{ext})$, the contacts are able to balance each of these four wrenches. Hence, $\boldsymbol{x}_{cm}$ lies in the individual pairwise regions $\mathcal{R}_{ij}(\boldsymbol{w}_l)$ for $l = 1,2,3,4$. ○

Exercises

Exercise 14.1: Explain why the two-finger grasp depicted in Figure 14.1(a) is fully wrench resistant, while the two-contact stance depicted in Figure 14.1(b) is only *locally* wrench resistant.

Section 14.1

Exercise 14.2: Explain why the contact forces solving the equilibrium stance equation (14.1) in 2-D stances span a convex set embedded in a $(2k-3)$-dimensional affine subspace of $\mathbb{R}^{2k}$.

Solution: The set of feasible contact reaction forces, $C_1 \times \cdots \times C_k$, is the product of convex sets and hence forms a convex set in $\mathbb{R}^{2k}$. The intersection of convex sets is a convex set. Now take the intersection of $C_1 \times \cdots \times C_k$ with the $(2k-3)$-dimensional affine subspace of contact forces solving Eq. (14.1). ○

Exercise 14.3: Verify that the net wrench cone at a 2-D stance, $\mathcal{W}$, forms a 2-D sector at a single frictional contact and a 3-D convex cone at stances supported by $k \geq 2$ non-coincident frictional contacts (which proves Lemma 14.1).

Hint: Denote by f_i^l and f_i^r the unit magnitude forces aligned with the edges of the friction cone C_i at x_i. Then $C_i = \{\lambda_1 f_i^l + \lambda_2 f_i^r : \lambda_1, \lambda_2 \geq 0\}$. Use this parametrization to represent the net wrench cone $\mathcal{W}$ as the image of the positive quadrant in (λ_1, λ_2)-space under the grasp map $\mathcal{L}_G$. ○

Section 14.2

Exercise 14.4: Explain why the feasible equilibrium region of a 3-D stance supported by two frictional contacts spans a vertical strip in $\mathbb{R}^3$ (which is the same as the feasible equilibrium strip of 2-D stances supported by two frictional contacts in the vertical plane).

Solution: At a 3-D stance supported by two frictional contacts, x_1 and x_2, the contact forces f_1 and f_2 generate zero moment about the line passing through x_i and x_2. Equilibrium stance requires that the gravitational wrench that acts on $\mathcal{B}$ will also generate zero moment about this line. Consequently, $\mathcal{B}$'s center of mass must lie in the vertical plane that contains the line passing through x_i and x_2. The stance equilibrium region thus forms a vertical strip, identical to the equilibrium strip of 2-D stances supported by two frictional contacts. ○

Section 14.3

Exercise 14.5: Use the five-strip formula to determine the stance equilibrium region associated with $k \geq 2$ contacts located on a horizontal line.

Solution: When all contacts lie on common horizontal line, only the polygons Π_{ij}^{++} and Π_{ij}^{--} are nonempty. The horizontal projections of Π_{ij}^{++} and Π_{ij}^{--} span the entire horizontal line. The region $\mathcal{ER}$ is simply the vertical strip bounded by the contacts, which forms a 2-D version of the *support polygon.* ○

Exercise 14.6: Explain why the 2-D feasible equilibrium region of a k-contact stance having $k \geq 4$ frictional contacts is fully determined by at most *four* of the k contacts (see proof of Proposition 14.5).

Exercise 14.7: In the two-contact stance of Figure 14.1(b), the contacts are located at $x_1 = (-1, 1)$ and $x_2 = (2, 2)$ relative to the intersection point of the supporting robot hand segments. The contact normals are oriented along 45° and 135° angles with respect to the horizontal x-axis. Determine the stance's feasible equilibrium region for the two cases of $\mu_1 = \mu_2 = 0.4$ and $\mu_1 = \mu_2 = 0.8$.

Exercise 14.8: Determine the range of the coefficient of friction, μ, for which the region $\mathcal{ER}_{ij}$ of the stance depicted in Figure 14.1(b) forms a semi-infinite vertical plane.

Exercise 14.9: A 2-D stance supported by frictional contacts x_i and x_j is defined as *tame* when the friction cones C_i and C_j lie *above* the line passing through x_i and x_j (a similar condition defines *tame* stances over 3-D terrains). Verify that such stances satisfy the relation $\mathcal{R}_{ij} \subseteq X_{ij}$.

Exercise 14.10: Under the tame stance condition, what additional condition ensures that the region $\mathcal{R}_{ij}$ coincides with the vertical strip X_{ij}?

Solution: Each friction cone should additionally contain the vertical upward direction. In this case $\Pi_{ij}^{++} = \Pi_{ij}^{--} = \mathbb{R}^2$ while $\Pi_{ij}^{+-} = \Pi_{ij}^{-+} = \emptyset$ in Formula (14.2). It follows that $\mathcal{R}_{ij}$ is the vertical strip bounded by the two contacts, $\mathcal{R}_{ij} = (\Pi_{ij}^{++} \cup \Pi_{ij}^{--}) \cap X_{ij} = X_{ij}$. Note that $\mathcal{R}_{ij} = X_{ij}$ on a flat horizontal terrain, since such a terrain is tame and contains the vertical upward direction at x_i and x_j. ○

Exercise 14.11: Determine which non-static motion modes will develop at the six-contact stance of Figure 14.5 when $\mathcal{B}$'s center of mass crosses the leftmost and rightmost bounding lines of $\mathcal{R}_{ij}$ (indicate which motion mode will develop at each of the six contacts). ○

Exercise 14.12*: Explain how the *moment labeling technique* of Chapter 4 can determine the feasible equilibrium region of 2-D stances whose supporting contacts are not all upward pointing.

Solution: Consider the gravitational force to be a force pushing on $\mathcal{B}$ within the vertical plane. Moment labeling depicts the collection of net object wrenches generated by the supporting contacts in terms of two signed polygons in $\mathbb{R}^2$, M^- and M^+. All lines passing between the two polygons along a direction determined by the sign of M^- and M^+ form feasible net object wrenches. When applied to stances under the influence of gravity, each of the two-contact stance equilibrium regions, $\mathcal{R}_{ij}$, forms a *vertical strip* bounded by M^- and M^+, which can be then used to construct the full region $\mathcal{ER}$ for k-contact stances based on the convex hull property. ○

Section 14.4

Exercise 14.13: In the two-contact stance of Figure 14.1(b), $\mathcal{B}$'s center of mass is located at the contact normals intersection point. In addition to a gravitational force of unit magnitude, a variable disturbance force f_d acts at $\mathcal{B}$'s center of mass. Assuming coefficients of friction $\mu_1 = \mu_2 = 0.4$ at the contacts, determine the maximal *magnitude* of f_d under which $\mathcal{B}$ remains at a feasible equilibrium stance.

Exercise 14.14: In the two-contact stance of Figure 14.1(b), $\mathcal{B}$'s center of mass is located at a height h above the contact normals intersection point. Plot a graph showing the maximal safe magnitude of f_d as a function of h for $-5 \leq h \leq 5$. Explain the meaning of this graph.

The following exercises support the proof of Lemma 14.1, concerning the dimension of the net wrench cone $\mathcal{W}$. Recall that L is the affine line of gravitational wrenches that act on $\mathcal{B}$.

Exercise 14.15*: A 2-D rigid object $\mathcal{B}$ is supported against gravity by two non-coincident frictional contacts in $\mathbb{R}^2$. Explain why the affine line L and the net wrench cone $\mathcal{W}$ intersect transversally in $\mathcal{B}$'s wrench space when the two contacts are *not* vertically aligned.

Exercise 14.16*: Assuming that $\mathcal{W}$ and L intersect transversally in $\mathcal{B}$'s wrench space, justify the dimension of the feasible equilibrium region specified in Lemma 14.1.

Exercise 14.17*: The disturbance-safe equilibrium region of a two-contact stance can be computed as the intersection: $\mathcal{R}_{ij}(\mathcal{W}_{ext}) = \mathcal{R}_{ij}(\boldsymbol{w}_1) \cap \mathcal{R}_{ij}(\boldsymbol{w}_2) \cap \mathcal{R}_{ij}(\boldsymbol{w}_3) \cap \mathcal{R}_{ij}(\boldsymbol{w}_4)$ (Proposition 14.8). Does this construction extend to any convex neighborhood of wrench disturbances which surrounds the gravitational wrench that acts on $\mathcal{B}$?

Application Exercises

Exercise 14.18: Figure 14.9 depicts a conceptual two-legged robot with flat footpads, supported by line contacts rather than point contacts. For a coefficient of friction $\mu = 1.0$ at the contacts and a 55° slope of the front segment, describe the 2-D feasible equilibrium region of this two-legged posture.

Hint: The contact reaction forces are continuously distributed along the supporting segments. These forces can be thought of as varying within identical friction cones based at all contacts. Under this interpretation, the contact reaction forces can be expressed as convex combinations of the contact reaction forces at the *endpoints* of these segments. ○

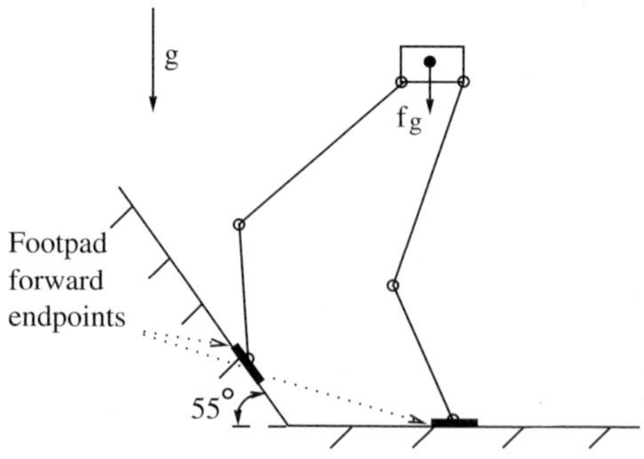

Figure 14.9 A conceptual two-legged robot with flat footpads supported by two terrain segments.

Exercise 14.19: The two-legged robot depicted in Figure 14.9 is supported by terrain segments with coefficient of friction $\mu = 0.5$ at the line segment contacts. Describe the feasible equilibrium strip of this two-legged posture.

Exercise 14.20: Describe the non-static motion modes that will develop at the contacts when the two-legged robot's center of mass *slowly* crosses the feasible equilibrium region boundaries.

Solution: When the two-legged robot retains a rigid-body shape with fixed joints values, non-static motion modes will develop at the endpoints of each footpad. The first mode involves *rolling* at one footpad endpoint while the other footpad breaks away from the terrain. In Figure 14.9, this motion mode will occur when the robot's center of mass, $\boldsymbol{x}_{cm}$, crosses the boundary line of $\mathcal{ER}$ that passes through the horizontal footpad's rear endpoint. At this instant, the horizontal footpad's rear endpoint will start rolling backward while the front footpad will break away from the terrain. The other motion mode involves *sliding* of two footpad endpoints. In Figure 14.9, this motion mode will occur when $\boldsymbol{x}_{cm}$ crosses the boundary line of $\mathcal{ER}$ that passes through the vertex of the polygon Π_{ij}^{++} associated with the front endpoints of the two footpads. At this instant, the front endpoints of both footpads will *slide* backward on the terrain while the rear part of each footpad will break away from the terrain. This motion mode incurs sliding of the full rear footpad on the terrain while the entire front footpad breaks away from the terrain. ○

Exercise 14.21: Figure 14.10 depicts a conceptual snake robot made of five rigid discs articulated by revolute joints and connected by rigid links. The discs are actuated in a manner that allows them to apply contact forces that push the snake robot along the terrain. For a coefficient of friction $\mu = 1.0$ at the contacts, determine the snake's feasible equilibrium region at the three climbing stages depicted in Figure 14.10.

Exercise 14.22: Assume that the snake robot mass is concentrated at the five discs that contact the terrain. Can the snake robot *slowly* climb the stair depicted in Figure 14.10 while continuously supported in static equilibrium against gravity?

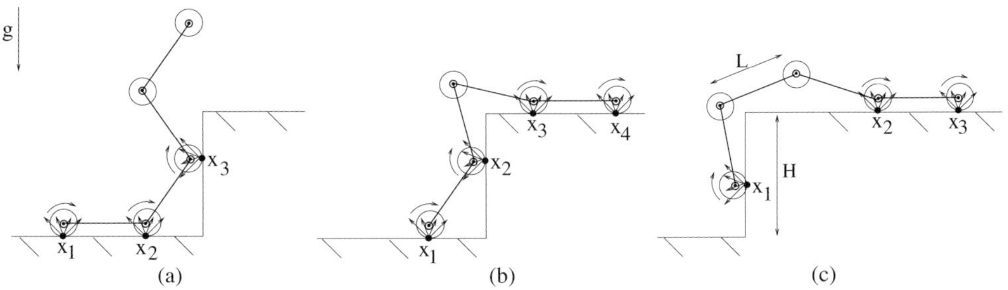

Figure 14.10 A conceptual snake robot attempting to climb a stair: (a) initial approach, (b) two-level support, and (c) tail lifting.

Exercise 14.23*: Determine the maximal stair height, H, that the snake robot can climb as a function of the rigid links' length, L (see Figure 14.10).

References

[1] Y. Or and E. Rimon, "Computation and graphical characterization of robust multiple-contact postures in two-dimensional gravitational environments," *International Journal of Robotics Research*, vol. 25, no. 11, pp. 1071–1086, 2006.

[2] M. A. Erdmann, "An exploration of nonprehensile two-palm manipulation," *International Journal of Robotics Research*, vol. 17, no. 5, pp. 485–503, 1998.

[3] S. J. Blind, C. C. McCullough, S. Akella and J. Ponce, "Manipulating parts with an array of pins: A method and a machine," *International Journal of Robotics Research*, vol. 20, no. 10, pp. 808–818, 2001.

[4] P. Painlevé, "Sur les lois du frottement de glissement," *Comptes Rendus de L'Acadamie des Sciences*, no. 121, pp. 112–115, 1895.

[5] F. Gnot and B. Brogliato, "New results on Painlevé paradoxes," *European Journal of Mechanics*, vol. A 18, no. 4, pp. 653–677, 1999.

[6] P. Lotstedt, "Coulomb friction in two-dimensional rigid body systems," *Zeitschrift fur Angewandte Mathematik und Mechanik*, vol. 61, pp. 605–615, 1981.

[7] M. T. Mason and Y. Wang, "On the inconsistency of rigid-body frictional planar mechanics," in *IEEE International Conference on Robotics and Automation*, pp. 524–528, 1988.

[8] Y. Wang and M. T. Mason, "Two-dimensional rigid body collisions with friction," *Journal of Applied Mechanics*, vol. 10, pp. 292–352, 1993.

[9] P. E. Dupont, "The effect of coulomb friction on the existence and uniqueness of the forward dynamics problem," *IEEE International Conference on Robotics and Automation*, pp. 1442–1447, 1992,

[10] J. S. Pang and J. C. Trinkle, "Stability characterizations of rigid body contact problems with Coulomb friction," *Zeitschrift fur Angewandte Mathematik und Mechanik*, vol. 80, no. 10, pp. 643–663, 2000.

15 Hand-Supported Stances under Gravity – Part II

This chapter studies the *equilibrium stances* of a rigid object supported by a robot hand under the influence of gravity in three dimensions. A limited level of security is achieved at such stances when the object is kept in static equilibrium that is robust with respect to small disturbances. The term *static stability* is often used to describe such equilibrium stances. For example, the two-finger robot hand depicted in Figure 15.1 supports a rigid object via three frictional contacts at a statically stable stance.

Much like the modeling approach taken in the previous chapter, we will fix the robot hand and treat the object as having a variable center of mass while supported by the robot hand. The object's center-of-mass positions at which the contacts can support the object form the *stance equilibrium region*. We will be particularly interested in *tame stances* that satisfy a generalized support polygon principle. This property makes their feasible equilibrium region reasonably intuitive to use as a margin of stability indicator.

The chapter first describes basic properties of frictional 3-D stances, which include the *local wrench resistance* property that forms the measure of security afforded by such stances. Section 15.2 introduces the class of *tame stances* and the generalized support polygon principle satisfied by such stances. Section 15.3 contains the core material of this chapter. Based on cellular decomposition of the supporting contacts' net wrench cone, this section develops an exact algebraic formulation for the feasible equilibrium region of general 3-D stances supported by frictional contacts. We will see that the horizontal cross section of this region forms a connected convex set that has five types

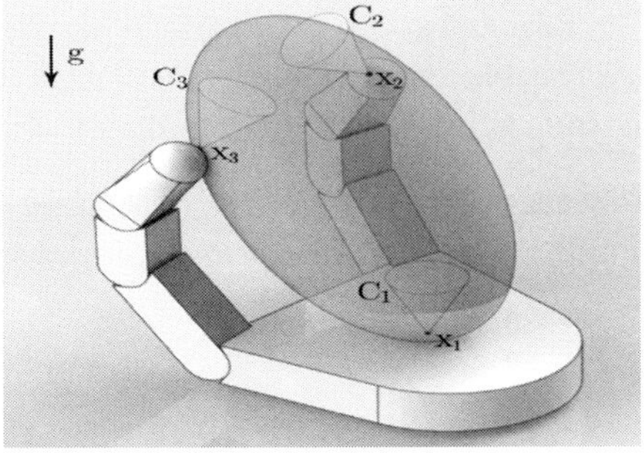

Figure 15.1 A two-finger robot hand supports a rigid object via frictional palm and fingertip contacts. The equilibrium stance is robust against sufficiently small disturbances.

of boundary curves. The chapter concludes with a discussion of the *non-static motion modes* that develop at the contacts once the object's center of mass crosses each of these boundary curves. At these events, the supported object loses its static stability and possibly escapes the robot hand.

Relation to legged robot locomotion: The modeling approach of this chapter is also useful for *quasistatic legged locomotion.* A legged robot such as a *humanoid robot* can be modeled as a rigid body having a variable center of mass and supported at the foothold contacts. The feasible equilibrium region provides a compact description of the legged robot's safe center-of-mass positions, which can be used to plan the legged robot's safe quasistatic locomotion on uneven terrains. ○

15.1 Basic Properties of 3-D Equilibrium Stances

This section describes basic properties of the frictional 3-D stances and defines the *stance equilibrium region* that is the focus of this chapter. The robot hand supports a rigid object $\mathcal{B}$ against gravity via frictional point contacts located at $x_1, \ldots, x_k$. Under the Coulomb friction model, the contact forces $f_1, \ldots, f_k$ must lie in their respective friction cones,

$$C_i = \left\{ f_i \in \mathbb{R}^3 : f_i \cdot n_i \geq 0 \text{ and } \sqrt{(f_i \cdot s_i)^2 + (f_i \cdot t_i)^2} \leq \mu_i |f_i \cdot n_i| \right\} \qquad i = 1 \ldots k, \tag{15.1}$$

where μ_i is the coefficient of friction at x_i, n_i is the object's inward unit normal at x_i, and (s_i, t_i) are mutually orthogonal unit tangents at x_i. The object's variable center-of-mass position in the fixed world frame is denoted $\boldsymbol{x}_{cm} \in \mathbb{R}^3$. The vertical upward direction in $\mathbb{R}^3$ is denoted $\boldsymbol{e} = (0, 0, 1)$. The horizontal directions in $\mathbb{R}^3$ define the horizontal projection matrix:

$$\mathbf{E} = \begin{pmatrix} 1 & 0 & 0 \\ 0 & 1 & 0 \end{pmatrix}.$$

The matrix $\mathbf{E}$ simply projects vectors in $\mathbb{R}^3$ to the horizontal plane. The horizontal projection of the contact points, contact forces and center-of-mass position are denoted $\tilde{x}_i = \mathbf{E}x_i$, $\tilde{f}_i = \mathbf{E}f_i$, and $\tilde{\boldsymbol{x}}_{cm} = \mathbf{E}\boldsymbol{x}_{cm}$. Much like equilibrium grasps, a rigid object $\mathcal{B}$ is supported at a feasible equilibrium stance when feasible contact forces can balance the gravitational wrench that acts on $\mathcal{B}$, according to the following condition.

Equilibrium stance condition: *A rigid object $\mathcal{B}$ is supported by k frictional contacts at a feasible* **equilibrium stance** *under the influence of gravity when*

$$\begin{pmatrix} f_1 \\ x_1 \times f_1 \end{pmatrix} + \cdots + \begin{pmatrix} f_k \\ x_k \times f_k \end{pmatrix} + \begin{pmatrix} f_g \\ \boldsymbol{x}_{cm} \times f_g \end{pmatrix} = \vec{0} \qquad f_i \in C_i,\ i = 1 \ldots k \tag{15.2}$$

where $\boldsymbol{w}_i = (f_i, x_i \times f_i)$ is the wrench generated by the ith contact force, and $\boldsymbol{w}_g = (f_g, \boldsymbol{x}_{cm} \times f_g)$ is the gravitational wrench that acts on $\mathcal{B}$. All wrenches are vectors in $\mathcal{B}$'s wrench space $T^*_{q_0}\mathbb{R}^6$, where q_0 is B's equilibrium stance configuration.

For a given set of supporting contacts and center-of-mass location, Eq. (15.2) specifies six linear constraints on the supporting contact forces, $(f_1, \ldots, f_k) \in \mathbb{R}^{3k}$. The feasible contact forces that solve Eq. (15.2) form a convex set embedded in a $(3k-3)$-dimensional affine subspace of $\mathbb{R}^{3k}$. As discussed in Chapter 14, when a robot hand supports a rigid object against gravity, the grasp is typically *not* wrench resistant. Instead, the robot hand can resist only a specific set of object wrenches that includes the gravitational wrench. This notion of *local wrench resistance* is repeated here in the context of 3-D stances.

Local wrench resistance: *Let a rigid object $\mathcal{B}$ be supported against gravity by k frictional contacts. The stance is* **locally wrench resistant** *when the contacts can resist with feasible forces a local neighborhood of wrenches centered at the gravitational wrench in $\mathcal{B}$'s wrench space.*

Local wrench resistance provides a limited safety criterion for hand-supported stances, as well as robustness with respect to small contact-placement errors of the object on the supporting robot hand. Local wrench resistance is a generic property of equilibrium stances supported by $k \geq 3$ frictional contacts. To justify this statement, consider the *net wrench cone* generated by the supporting contacts:

$$\mathcal{W} = \left\{ \sum_{i=1}^{k} \begin{pmatrix} f_i \\ x_i \times f_i \end{pmatrix} : f_i \in C_i \text{ for } i = 1 \ldots k \right\}.$$

The set $\mathcal{W}$ forms a convex cone based at $\mathcal{B}$'s wrench space origin, whose dimension is described in the following lemma (see Exercises for a proof). The supporting contacts are said to be *non-collinear* when they do not lie on a common line.

LEMMA 15.1 **(dimension of $\mathcal{W}$)** *The net wrench cone at a frictional 3-D stance forms a* **3-D cone** *for a single contact, a* **5-D cone** *for two non-coincident contacts and a* **full 6-D cone** *for $k \geq 3$ non-collinear contacts.*

The next proposition applies Lemma 15.1 to conclude that local wrench resistance is a generic property of frictional 3-D stances supported by $k \geq 3$ frictional contacts. A *non-marginal equilibrium stance* is defined as a stance whose contact forces lie in the interior of the respective friction cones.

PROPOSITION 15.2 **(local wrench resistance)** *Let a 3-D rigid object $\mathcal{B}$ be supported against gravity by $k \geq 3$ non-collinear frictional contacts. If the object is supported at a non-marginal equilibrium stance (the generic case), the stance is* **locally wrench resistant.**

The proof is similar to the one given for frictional 2-D stances (Proposition 14.2) and is therefore omitted. A robot hand is typically attached to a robot arm that is possibly mounted on a mobile robot platform. When the object $\mathcal{B}$ is supported at a local wrench resistant stance, it will remain stationary on the robot hand under all sufficiently small wrench disturbances generated by local movements of the robot arm and possibly the mobile robot platform.

Let us now fix the robot hand and consider instead the supported object $\mathcal{B}$ as a rigid body having a variable center of mass and supported by the contacts located on the robot hand. Our goal would be to characterize the object's center-of-mass positions at which the supporting contacts can resist the gravitational wrench that acts on $\mathcal{B}$. The *stance equilibrium region* is defined as follows.

DEFINITION 15.1 (**the region $\mathcal{ER}$**) *Let a 3-D rigid object $\mathcal{B}$ be supported against gravity by k frictional contacts. The* **stance equilibrium region,** $\mathcal{ER} \subseteq \mathbb{R}^3$*, is the set of all $\mathcal{B}$'s center-of-mass positions at which there exist feasible contact forces satisfying the equilibrium stance condition of Eq.* (15.2).

When the region $\mathcal{ER}$ is nonempty, it forms one of the following sets of vertical lines, collinear with the gravitational force direction.

PROPOSITION 15.3 (**shape of $\mathcal{ER}$**) *The stance equilibrium region, $\mathcal{ER} \subseteq \mathbb{R}^3$, is a* **vertical line** *when $\mathcal{B}$ is supported by a single contact, a* **vertical strip** *when $\mathcal{B}$ is supported by two non-coincident contacts and a* **vertical 3-D prism** *with a convex cross section when $\mathcal{B}$ is supported by $k \geq 3$ non-collinear frictional contacts.*[1]

Rather than giving a formal proof, consider the following intuitive arguments. The equilibrium stance condition described in Eq. (15.2) contains only the *horizontal coordinates* of $\mathcal{B}$'s center of mass, $\tilde{\boldsymbol{x}}_{cm}$. It follows that the region $\mathcal{ER}$ forms a collection of vertical lines in $\mathbb{R}^3$. For a single contact, $\mathcal{ER}$ is trivially the vertical line passing through the contact. For two contacts, the region $\mathcal{ER}$ forms a vertical strip, embedded in the vertical plane passing through the two contacts. For $k \geq 3$ contacts, the region $\mathcal{ER}$ forms a vertical 3-D cylindrical set with a convex cross section, based on the following argument. Let $\boldsymbol{x}'_{cm}$ and $\boldsymbol{x}''_{cm}$ be two center-of-mass points in $\mathcal{ER}$, and let $f'_i, f''_i \in C_i$ be the corresponding contact forces that maintain the equilibrium stance ($i = 1 \dots k$). The line segment joining $\boldsymbol{x}'_{cm}$ and $\boldsymbol{x}''_{cm}$ can be parametrized as $\boldsymbol{x}_{cm}(s) = s\boldsymbol{x}'_{cm} + (1-s)\boldsymbol{x}''_{cm}$ for $0 \leq s \leq 1$. The corresponding contact forces, $f_i = sf'_i + (1-s)f''_i$, lie in C_i and satisfy the equilibrium stance condition with $\mathcal{B}$'s center of mass located at $\boldsymbol{x}_{cm}(s)$. Hence, $\mathcal{ER}$ forms a connected convex set of vertical lines in $\mathbb{R}^3$.

It follows from Proposition 15.3 that the horizontal cross section of the stance equilibrium region, $\mathcal{ER} \subseteq \mathbb{R}^3$, forms a planar convex set. The computation of $\mathcal{ER}$ can therefore be reduced to the computation of its *horizontal cross section* in $\mathbb{R}^2$. From this point onward, the notation $\mathcal{ER}$ will refer to the horizontal cross section, $\mathcal{ER} \subseteq \mathbb{R}^2$, of the 3-D feasible equilibrium region.

15.2 The Tame Hand-Supported Stances

This section describes the class of *tame stances* and the generalized support polygon principle satisfied by these stances. Let us start with the classical *support polygon,* which

[1] When all k contacts lie on a common line in $\mathbb{R}^3$, the region $\mathcal{ER}$ degenerates into a two-dimensional vertical strip passing through this line.

is obtained by projecting the contact points to a common horizontal plane and then taking the convex hull of the projected contacts within the horizontal plane.

DEFINITION 15.2 *The* **support polygon** *of a k-contact stance is the horizontal polygon,* $\mathcal{P} = \text{conv}\{\tilde{x}_1, \ldots, \tilde{x}_k\} \subset \mathbb{R}^2$, *formed by the convex hull of the horizontal projection of the contacts.*

The support polygon of a k-contact stance forms a bounded convex set in $\mathbb{R}^2$, surrounded by a closed loop of line segments connecting pairs of projected contact pairs $(\tilde{x}_i, \tilde{x}_j)$ (not all contact points necessarily contribute to the boundary $\mathcal{P}$). The support polygon forms the feasible equilibrium region of stances located on flat horizontal terrains. However, a robot hand almost always forms an *uneven terrain* when supporting a rigid object against gravity. On such terrains, the relation between the stance equilibrium region, $\mathcal{ER} \subseteq \mathbb{R}^2$, and the support polygon $\mathcal{P}$ is rather complex. Fortunately, a large class of uneven terrains (which correspond to a large class of robot hand postures) satisfies the useful relation $\mathcal{ER} \subseteq \mathcal{P}$. This class of *tame stances* is defined in two stages. First consider tame stances supported by three contacts.

DEFINITION 15.3 **(tame three-contact stance)** *Let a 3-D rigid object* $\mathcal{B}$ *be supported by frictional contacts* x_1, x_2, x_3 *in* $\mathbb{R}^3$. *Let* Δ *be the plane passing through* x_1, x_2 *and* x_3. *Let* $\boldsymbol{n}$ *be the unit normal to* Δ *such that* $\boldsymbol{e} \cdot \boldsymbol{n} \geq 0$. *The three-contact stance is* **tame** *if the friction cones at the contacts lie strictly above the plane* Δ. *That is,* $f_i \cdot \boldsymbol{n} > 0$ *for all* $f_i \in C_i$, $i = 1, 2, 3$.

The generalization to k-contact stances requires that each contact triplet will form a tame stance, as specified in the following definition.

DEFINITION 15.4 **(tame K-contact stance)** *A k-contact stance is* **tame** *if each triplet of frictional contacts,* $x_i, x_j, x_p \in \mathbb{R}^3$ *such that* $1 \leq i, j, p \leq k$, *forms a tame three-contact stance.*

The tame stances can be equivalently characterized as follows. A k-contact stance is *tame* when each pair of contacts, x_i and x_j, determines a line in $\mathbb{R}^3$, such that the friction cones at all other contacts lie *above* this line with respect to the vertical upward direction $\boldsymbol{e}$. To describe the relation of the region $\mathcal{ER}$ to the support polygon $\mathcal{P}$, define a contact x_i as *quasi-flat* when its friction cone C_i contains the vertical upward direction $\boldsymbol{e}$. The following proposition describes the generalized support polygon principle satisfied by tame stances.

PROPOSITION 15.4 **(relation to support polygon)** *The feasible equilibrium region of a tame k-contact stance,* $\mathcal{ER} \subseteq \mathbb{R}^2$, *is* **fully contained** *in the support polygon,* $\mathcal{ER} \subseteq \mathcal{P}$. *The converse relationship,* $\mathcal{P} \subseteq \mathcal{ER}$, *holds for k-contact stances with quasi-flat contacts.*

Proof: First consider the relation $\mathcal{ER} \subseteq \mathcal{P}$. We have to show that every horizontal center-of-mass position $\tilde{\boldsymbol{x}}_{cm} \in \mathcal{ER}$ lies in the support polygon $\mathcal{P}$. The support polygon is bounded by a closed loop of linear segments connecting the horizontally projected contacts. Let $\tilde{x}_i$ and $\tilde{x}_j$ be two neighboring vertices along the support polygon's boundary, and let x_i and x_j be the corresponding contact points (which may lie at various heights). Consider the line l passing through x_i and x_j in $\mathbb{R}^3$. The contact forces f_i and f_j generate zero torque about l. The torque-balance part of Eq. (15.2) thus requires that the *net torque* generated about l by gravity and all other contact forces be zero. At a tame stance, each of the other contact forces satisfies $f_l \cdot \boldsymbol{n} \geq 0$ for all $f_p \in C_p$, where $\boldsymbol{n}$ is the upward-pointing normal to the plane Δ spanned by x_i, x_j and x_p. Torque balance about l implies that the object's center of mass, $\boldsymbol{x}_{cm} \in \mathbb{R}^3$, must lie within the half-space of $\mathbb{R}^3$ bounded by the vertical plane passing through the contacts x_i and x_j, such that the half-space contains the point x_p for all $p \neq i, j$. Repeating this argument for every segment (x_i, x_j) on the boundary of the support polygon $\mathcal{P}$, one concludes that $\boldsymbol{x}_{cm}$ must lie within the vertical prism spanned by the contacts, which implies that $\mathcal{ER} \subseteq \mathcal{P}$.

A quasi-flat contact can balance any object center-of-mass located on the vertical line passing through this contact. The vertical lines that pass through the quasi-flat contacts thus lie within the 3-D stance equilibrium region. The 3-D stance equilibrium region forms a convex set of vertical lines (Proposition 15.3). Hence, it contains the vertical prism spanned by the quasi-flat contacts, which gives the relation $\mathcal{P} \subseteq \mathcal{ER}$ for stances with quasi-flat contacts. □

A supporting robot hand typically contains tame stances as long as the terrain formed by the robot hand is not too steep. However, tame stances typically contain a mix of quasi-flat and non-quasi-flat contacts, as illustrated in the next example.

Example: Figure 15.2 depicts a two-finger robot hand supporting a rigid object $\mathcal{B}$ via four frictional contacts (the supported object is not shown). Two contacts are located at the fingertips while the other two contacts are located on the hand's palm. In Figure 15.2(a), the fingers are stretched outward and the object is supported by quasi-flat contacts. Hence, $\mathcal{ER} = \mathcal{P}$ at this stance. Next, consider the gradual closing

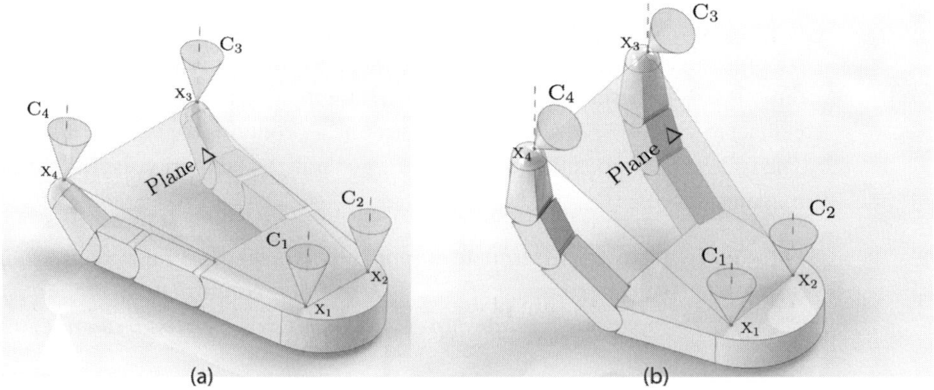

Figure 15.2 A two-finger robot hand supports a rigid object via four frictional contacts (the supported object is not shown). (a) The stance is tame, and all contacts are quasi-flat. Hence, $\mathcal{ER} = \mathcal{P}$ at this stance. (b) The stance is still tame but no longer quasi-flat at the fingertip contacts. Hence, $\mathcal{ER} \subset \mathcal{P}$ at this stance.

of the two fingers inward toward the hand's palm, accompanied by slight shift of the fingertip contacts. In Figure 15.2(b), the object is still supported by a tame stance, but the fingertip contacts are no longer quasi-flat. The feasible equilibrium region of this stance lies somewhere within the support polygon, $\mathcal{ER} \subset \mathcal{P}$, which highlights the need to understand exactly where the region $\mathcal{ER}$ lies. ○

Tame stances possess the useful property that their feasible equilibrium region lies within the classical support polygon, $\mathcal{ER} \subseteq \mathcal{P}$. However, the selection of *safe* object placements on a robot hand, and understanding the margin of stability away from non-static motion modes, require explicit characterization of the region $\mathcal{ER}$. This leads us to the core material of this chapter.

15.3 A Scheme for Computing the Stance Equilibrium Region

This section generalizes the scheme of Chapter 11, which identified the feasible equilibrium region of *frictionless* stances, to a scheme that computes the feasible equilibrium region of *frictional* stances. The scheme is based on a geometric interpretation of the equilibrium stance condition of Eq. (15.2), which can be written in the form:

$$\begin{pmatrix} f_1 \\ x_1 \times f_1 \end{pmatrix} + \cdots + \begin{pmatrix} f_k \\ x_k \times f_k \end{pmatrix} = -\begin{pmatrix} f_g \\ \boldsymbol{x}_{cm} \times f_g \end{pmatrix} \qquad f_i \in C_i,\ i = 1 \ldots k. \quad (15.3)$$

Recall that $\mathcal{B}$'s *wrench space* forms a six-dimensional vector space, parametrized by force and torque coordinates $(f, \tau) \in T^*_{q_0}\mathbb{R}^6$. The left side of Eq. (15.3) represents all net wrenches that can be affected on $\mathcal{B}$ by feasible forces at the contacts. It represents the *net wrench cone,* $\mathcal{W}$, which forms a 6-D convex cone based at $\mathcal{B}$'s wrench space origin for $k \geq 3$ frictional contacts (Lemma 15.1).

The right side of Eq. (15.3) represents all the gravitational wrenches that can act on $\mathcal{B}$ by varying the object's center-of-mass location at the given stance. The gravitational force that acts on $\mathcal{B}$ is constant, while the gravitational torque that acts on $\mathcal{B}$ can be written in the form:

$$\boldsymbol{x}_{cm} \times f_g = \left(\begin{pmatrix} \tilde{\boldsymbol{x}}_{cm} \\ 0 \end{pmatrix} + \begin{pmatrix} \vec{0} \\ \boldsymbol{x}_{cm} \cdot \boldsymbol{e} \end{pmatrix} \right) \times f_g = \begin{pmatrix} \tilde{\boldsymbol{x}}_{cm} \\ 0 \end{pmatrix} \times f_g \qquad \tilde{\boldsymbol{x}}_{cm} \in \mathbb{R}^2.$$

Since the gravitational force f_g is constant, the right side of Eq. (15.3) depends only on the horizontal coordinates of $\mathcal{B}$'s center of mass, $\tilde{\boldsymbol{x}}_{cm} \in \mathbb{R}^2$, which span the following *affine plane* in $\mathcal{B}$'s wrench space.

DEFINITION 15.5 (**affine plane L**) *The collection of gravitational wrenches that can act on a variable-center-of-mass object $\mathcal{B}$ at a given stance span the affine plane,*

$$L = \left\{ \begin{pmatrix} f_g \\ \begin{pmatrix} \tilde{\boldsymbol{x}}_{cm} \\ 0 \end{pmatrix} \times f_g \end{pmatrix} \in T^*_{q_0}\mathbb{R}^6 \ :\ \tilde{\boldsymbol{x}}_{cm} \in \mathbb{R}^2 \right\}, \qquad (15.4)$$

where f_g is a constant force while $\tilde{\boldsymbol{x}}_{cm} = \mathbf{E}\boldsymbol{x}_{cm}$ varies in $\mathbb{R}^2$.

Any wrench $(f,\tau)\in L$ corresponds to a *unique* horizontal center-of-mass position, $\tilde{\boldsymbol{x}}_{cm}$, of the supported object $\mathcal{B}$. To obtain a formula for $\tilde{\boldsymbol{x}}_{cm}$ as a function of $(f,\tau)\in L$, assume that $\|f_g\|=1$. This choice of force units has no influence on the generality of the results. Then $f_g=-\boldsymbol{e}$ and $\boldsymbol{x}_{cm}\times f_g=\boldsymbol{x}_{cm}\times(-\boldsymbol{e})$. Using the triple vector product identity: $\boldsymbol{e}\times(\boldsymbol{x}_{cm}\times\boldsymbol{e})=(\boldsymbol{x}_{cm}\cdot\boldsymbol{e})\boldsymbol{e}-\boldsymbol{x}_{cm}$, one obtains the expression for $\tilde{\boldsymbol{x}}_{cm}$:

$$\tilde{\boldsymbol{x}}_{cm}=-\mathbf{E}(\boldsymbol{e}\times\tau) \qquad \begin{pmatrix}-\boldsymbol{e}\\ \tau\end{pmatrix}\in L \text{ and } \mathbf{E}=\begin{pmatrix}1&0&0\\0&1&0\end{pmatrix}. \tag{15.5}$$

Based on the geometric interpretation of the equilibrium stance condition, the following theorem specifies a scheme for computing the horizontal cross section of the stance equilibrium region in terms of the net wrench cone $\mathcal{W}$ and the affine plane L.

THEOREM 15.5 (**computation of $\mathcal{ER}$**) *The horizontal cross section of a k-contact* **stance equilibrium region,** *$\mathcal{ER}\subseteq\mathbb{R}^2$, is the image of $\mathcal{W}\cap L$ under the linear map $\tilde{\boldsymbol{x}}_{cm}=-\mathbf{E}(\boldsymbol{e}\times\tau)$,*

$$\mathcal{ER}=\left\{\tilde{\boldsymbol{x}}_{cm}\in\mathbb{R}^2 : \tilde{\boldsymbol{x}}_{cm}=-\mathbf{E}(\boldsymbol{e}\times\tau) \text{ such that } \begin{pmatrix}-\boldsymbol{e}\\ \tau\end{pmatrix}\in\mathcal{W}\cap L\right\}, \tag{15.6}$$

where the forces are scaled such that $\|f_g\|=1$, $\mathcal{W}$ is the net wrench cone at the stance, and L is the affine plane of gravitational wrenches that can act on $\mathcal{B}$ at the given stance. The **stance equilibrium region boundary,** *$\mathsf{bdy}(\mathcal{ER})$, lies within the image of $\mathsf{bdy}(\mathcal{W})\cap L$*

$$\mathsf{bdy}(\mathcal{ER})\subseteq\left\{\tilde{\boldsymbol{x}}_{cm}\in\mathbb{R}^2 : \tilde{\boldsymbol{x}}_{cm}=-\mathbf{E}(\boldsymbol{e}\times\tau) \text{ such that } \begin{pmatrix}-\boldsymbol{e}\\ \tau\end{pmatrix}\in\mathsf{bdy}(\mathcal{W})\cap L\right\}, \tag{15.7}$$

where $\mathsf{bdy}(\mathcal{W})$ is the boundary of the net wrench cone $\mathcal{W}$.

Proof: Based on the geometric interpretation of Eq. (15.3), every wrench $(f,\tau)\in\mathcal{W}\cap L$ corresponds to a unique horizontal location of $\mathcal{B}$'s center of mass, $\tilde{\boldsymbol{x}}_{cm}=-\mathbf{E}(\boldsymbol{e}\times\tau)$. This gives Formula (15.6) for the region $\mathcal{ER}$. Next consider the boundary of $\mathcal{ER}$. The linear map $\tilde{\boldsymbol{x}}_{cm}=-\mathbf{E}(\boldsymbol{e}\times\tau)$ maps the affine plane L onto the horizontal plane $\mathbb{R}^2$. Being full rank, it maps the interior of $\mathcal{W}\cap L$ within the affine plane L to the interior of $\mathcal{ER}$ in $\mathbb{R}^2$. This implies that only the *boundary* of $\mathcal{W}\cap L$ within the affine plane L is possibly mapped to the boundary of $\mathcal{ER}$. The boundary of $\mathcal{W}\cap L$ within L consists of boundary points of $\mathcal{W}$ in the ambient wrench space. This gives the set $\mathsf{bdy}(\mathcal{W})\cap L$ specified in Eq. (15.7). Note that $\mathsf{bdy}(\mathcal{W})\cap L$ forms a collection of curves that are mapped to a collection of curves in $\mathcal{ER}$, such that certain segments of these curves form the outer boundary of the region $\mathcal{ER}$. □

Theorem 15.5 outlines a scheme for computing the feasible equilibrium region of k-contact stances. Based on this scheme, Sections 15.4 and 15.5 will characterize the cellular structure of the net wrench cone boundary, bdy($\mathcal{W}$). Section 15.6 will characterize the intersection bdy($\mathcal{W}$) $\cap$ L and map this set to the boundary of the stance equilibrium region, bdy($\mathcal{ER}$) in $\mathbb{R}^2$.

Polygonal approximation of $\mathcal{ER}$: A polygonal inner approximation for bdy($\mathcal{ER}$) can be efficiently computed as follows. First construct an inner approximation of the friction cones $C_1, \ldots, C_k$ by planar facets whose number is a user-specified parameter. This gives an inner approximation of the composite friction cone, $C_1 \times \cdots \times C_k$, by a convex polyhedral cone in $\mathbb{R}^{3k}$. The intersection set, bdy($\mathcal{W}$) $\cap$ L, forms a *convex polytope* in the composite space of contact force components and the object's horizontal center-of-mass position, $(f_1, \ldots, f_k, \tilde{\boldsymbol{x}}_{cm}) \in \mathbb{R}^{3k+2}$. The projection of this polytope to the horizontal plane can be computed as a series of linear programming problems, which gives a polygonal inner approximation for bdy($\mathcal{ER}$) in $\mathbb{R}^2$. However, this numerical technique does not offer any intuitive or analytic understanding of the stance equilibrium region. ○

15.4 The Boundary of the Net Wrench Cone $\mathcal{W}$

This section characterizes the boundary of the net wrench cone, bdy($\mathcal{W}$), at a k-contact stance. As a preliminary step, let us divide the contact forces of the composite friction cone, $(f_1, \ldots, f_k) \in C_1 \times \cdots \times C_k$, into *contact force cells*. The contact force cells that possibly contribute facets to the boundary of $\mathcal{W}$ will then be identified.

15.4.1 Cellular Decomposition of the Composite Friction Cone

The cellular decomposition of the composite friction cone, $C_1 \times \cdots \times C_k$, will be based on the interpretation of the contacts' net wrench cone, $\mathcal{W}$, as the image of the contact forces $(f_1, \ldots, f_k)$ under the *grasp map* (see Chapter 4):

$$\mathcal{L}_G(f_1, \ldots, f_k) = \begin{pmatrix} f_1 \\ x_1 \times f_1 \end{pmatrix} + \cdots + \begin{pmatrix} f_k \\ x_k \times f_k \end{pmatrix} \qquad (f_1, \ldots, f_k) \in C_1 \times \cdots \times C_k . \tag{15.8}$$

The set $C_1 \times \cdots \times C_k$ forms a *stratified set* in the space of contact forces $(f_1, \ldots, f_k)$. That is, a set that can be decomposed into disjoint *cells,* each being a manifold without boundary. The cells' dimension ranges between zero (a single-point manifold), and $3k$ (an open set in the ambient space of contacts forces). In order to describe how $C_1 \times \cdots \times C_k$ decomposes into cells, consider the decomposition of each friction cone into three subsets, as depicted in Figure 15.3(a):

$$C_i = O_i \cup I_i \cup S_i \qquad i = 1 \ldots k,$$

where $O_i = \{0\}$ is the cone's vertex point, I_i is the cone's interior and S_i is the cone's boundary surface excluding its vertex point. Note that I_i and S_i contain only non-zero

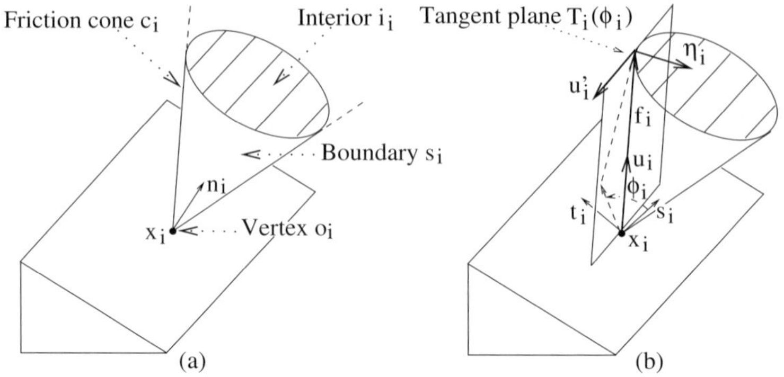

Figure 15.3 (a) Partition of the friction cone C_i into its vertex O_i, punctured surface S_i and interior I_i. (b) The tangent plane $T_i(\phi_i)$ touches the friction cone boundary S_i along a ray that starts at x_i along the contact force $f_i = \lambda_i \boldsymbol{u}_i(\phi_i)$.

forces. Using this notation, the set $C_1 \times \cdots \times C_k$ can be decomposed into 3^k cells of the form

$$C_1 \times \cdots \times C_k = (O_1 \cup I_1 \cup S_1) \times \cdots \times (O_k \cup I_k \cup S_k). \tag{15.9}$$

Each cell of $C_1 \times \cdots \times C_k$ represents a choice of one component from $\{O_i, I_i, S_i\}$ for $i = 1 \ldots k$. The dimension of any particular cell equals the *sum* of the dimensions of its component cells. For instance, the cell $O_1 \times S_2 \times S_3$ is four-dimensional, since O_1 is zero-dimensional while S_2 and S_3 are two-dimensional surfaces. Also note that $O_1 \times \cdots \times O_k$ is a zero-dimensional cell (a single-point manifold), while $I_1 \times \cdots \times I_k$ is a $3k$-dimensional cell (an open set in the ambient space of contact forces).

15.4.2 Contact Force Cells that Contribute Boundary Facets to $\mathcal{W}$

The net wrench cone is fully six-dimensional at 3-D stances supported by $k \geq 3$ non-collinear frictional contacts (Lemma 15.1). Hence its boundary, bdy($\mathcal{W}$), forms a five-dimensional set in $\mathcal{B}$'s wrench space. This set can be stratified into cells whose dimension ranges between zero and five. Since bdy($\mathcal{W}$) is five-dimensional, it is *fully captured* by the union of its five-dimensional cells, as cells of lower dimension can be obtained by intersecting two or more five-dimensional boundary cells. Let us therefore characterize the five-dimensional cells, termed *facets,* of bdy($\mathcal{W}$). The criterion that will help us identify which contact force cells contribute facets to bdy($\mathcal{W}$) is based on the Jacobian matrix of the grasp map. Since matrix rank is central to this section, let us recall how it is defined.

matrix rank: *Let A be an $l \times r$ real matrix and let $a_i \in \mathbb{R}^l$ denote the ith column of A. The* **range space** *of A, $\mathcal{R}(A)$, is the linear subspace of $\mathbb{R}^l$ spanned by the columns of A:*

$$\mathcal{R}(A) = \left\{ v \in \mathbb{R}^l \; : \; v = s_1 a_1 + \cdots + s_r a_r; \; s_1, \ldots, s_r \in \mathbb{R} \right\}.$$

The **rank** *of* A, $\text{rank}(A)$, *is the dimension of* A*'s range space. The matrix* A *has* **full rank** *when* $\text{rank}(A) = \min\{l, r\}$.

LEMMA 15.6 (**rank criterion**) *Let* $\mathcal{K}$ *be a cell of the composite friction cone* $C_1 \times \cdots \times C_k$. *If the image of* $\mathcal{K}$ *under the grasp map* $\mathcal{L}_G$ *contains a five-dimensional facet of* $\text{bdy}(\mathcal{W})$, *it must be the image of* **critical contact forces** *in* $\mathcal{K}$, *satisfying the rank deficiency condition*

$$\text{rank}(D\mathcal{L}_G) = 5, \tag{15.10}$$

where $D\mathcal{L}_G$ *is the Jacobian of the restriction of the grasp map* $\mathcal{L}_G$ *to the cell* $\mathcal{K}$.

Proof sketch: In the five-dimensional cells of $C_1 \times \cdots \times C_k$, the condition $\text{rank}(D\mathcal{L}_G) = 5$ means that $D\mathcal{L}_G$ has full rank. All of these cells are mapped by $\mathcal{L}_G$ to five-dimensional sets in $\mathcal{B}$'s wrench space that can possibly form five-dimensional facets of $\text{bdy}(\mathcal{W})$. Next consider the higher-dimensional cells of $C_1 \times \cdots \times C_k$. When a cell $\mathcal{K}$ has dimension six or higher, points where the Jacobian has full rank, $\text{rank}(D\mathcal{L}_G) = 6$, are mapped by $\mathcal{L}_G$ into the *interior* of $\mathcal{W}$ according to the *submersion theorem.*[2] Hence, only contact forces that satisfy the rank deficiency condition $\text{rank}(D\mathcal{L}_G) \leq 5$ are possibly mapped to $\text{bdy}(\mathcal{W})$. □

Lemma 15.6 can be used to identify which cells of $C_1 \times \cdots \times C_k$ can possibly contribute five-dimensional facets to $\text{bdy}(\mathcal{W})$. To apply Lemma 15.6, one must choose a parametrization for each d-dimensional cell $\mathcal{K}$ using d parameters, formulate the restriction of the grasp map $\mathcal{L}_G$ to $\mathcal{K}$ in terms of the d parameters, and then compute the $6 \times d$ Jacobian matrix $D\mathcal{L}_G$ and find conditions for its rank deficiency. A cell $\mathcal{K}$ can be assigned d parameters according to its S_i and I_i components as follows. Contact forces that lie in the ith friction cone interior, I_i, are simply parametrized by their Cartesian coordinates in the fixed world frame, $f_i \in \mathbb{R}^3$. Contact forces that lie in the ith friction cone boundary, S_i, are parametrized by $(\lambda_i, \phi_i) \in \mathbb{R}^+ \times \mathbb{R}$, where $\lambda_i > 0$ is the force magnitude and ϕ_i is the force angle, measured by projecting f_i on the plane spanned by the unit tangents at the contact point x_i (Figure 15.3(b)). The forces $f_i \in S_i$ are thus parametrized by

$$f_i(\lambda_i, \phi_i) = \lambda_i \boldsymbol{u}_i(\phi_i) \quad \text{such that} \quad \boldsymbol{u}_i(\phi_i) = \mu_i \cos(\phi_i) s_i + \mu_i \sin(\phi_i) t_i + n_i, \tag{15.11}$$

where μ_i is the coefficient of friction at x_i, (s_i, t_i) are the unit tangents at x_i and n_i is $\mathcal{B}$'s inward unit normal at x_i. The unit vector $\boldsymbol{u}_i(\phi_i)$ parametrizes a circle, formed by intersecting S_i with a plane orthogonal to n_i, such that the plane lies at a unit distance from x_i. Since $\boldsymbol{u}_i(\phi_i)$ has constant magnitude, $\boldsymbol{u}_i'(\phi_i) = -\mu_i \sin(\phi_i) s_i + \mu_i \cos(\phi_i) t_i$ is orthogonal to $\boldsymbol{u}_i(\phi_i)$. The pair $\{\boldsymbol{u}_i(\phi_i), \boldsymbol{u}_i'(\phi_i)\}$ thus spans the *tangent plane* to S_i at the point $f_i \in S_i$, denoted $T_i(\phi_i)$ (Figure 15.3(b)). Note that $T_i(\phi_i)$ is tangent to S_i along a ray collinear with the contact force $f_i = \lambda_i \boldsymbol{u}_i(\phi_i)$. Finally, the vector $\boldsymbol{\eta}_i(\phi_i) = \boldsymbol{u}_i(\phi_i) \times \boldsymbol{u}_i'(\phi_i) = -\mu_i \cos(\phi_i) s_i - \mu_i \sin(\phi_i) t_i - \mu_i^2 n_i$ is orthogonal to S_i at the point $f_i = \lambda_i \boldsymbol{u}_i(\phi_i)$ (Figure 15.3(b)).

[2] Let $f : \mathbb{R}^n \to \mathbb{R}^m$ be a differentiable function, $y = f(x)$ and $n \geq m$. If the Jacobian $Df(x)$ has full rank at x, f maps an open neighborhood of x in $\mathbb{R}^n$ onto an open neighborhood of y in $\mathbb{R}^m$.

The cells of the composite friction cone $C_1 \times \cdots \times C_k$ are next partitioned into *cell classes.*

DEFINITION 15.6 (**cell classes**) *Each* **cell class** *of the composite friction cone* $C_1 \times \cdots \times C_k = (O_1 \cup I_1 \cup S_1) \times \cdots \times (O_k \cup I_k \cup S_k)$ *is associated with* n_S *choices of* S_i *components,* n_I *choices of* I_i *components and* $k - n_S - n_I$ *choices of* O_i *components.*

Each cell class of $C_1 \times \cdots \times C_k$ can be represented by a word of k letters from the alphabet $\{S, I, O\}$, having n_S letters S, n_I letters I and $k - n_S - n_I$ letters O. For notational simplicity, we will use the convention that contact forces with indices $1 \ldots n_S$ lie in the boundary components S_i, contact forces with indices $n_S{+}1 \ldots n_S{+}n_I$ lie in the interior components I_i, while the remaining contact forces with indices $n_S{+}n_I{+}1 \ldots k$ are zero forces. This arbitrary choice of contact force indices will represent all other index permutations at any given cell class. For instance, for $k = 3$ contacts, the cell class SIO represents six cells: $S_1 \times I_2 \times O_3$, $S_1 \times O_2 \times I_3$, $O_1 \times S_2 \times I_3$, $O_1 \times I_2 \times S_3$, $I_1 \times O_2 \times S_3$, and $I_1 \times S_2 \times O_3$. The letter O represents zero contact force and, hence, will be omitted. For instance, the cell class SIO will be denoted as SI. Note that all cells of a particular class have the same dimension, given by $2n_S + 3n_I$. For instance, for $k = 3$ contacts, all cell classes of dimension five or higher are SI, II, SSS, SSI, SII, and III.

The following theorem specifies which cell classes of the composite friction cone $C_1 \times \cdots \times C_k$ can possibly contribute five-dimensional facets to $\mathsf{bdy}(\mathcal{W})$.

THEOREM 15.7 (**cell classes of** $\mathsf{bdy}(\mathcal{W})$) *Let a rigid object* $\mathcal{B}$ *be contacted by* $k \geq 3$ *non-collinear frictional contacts. The cell classes of* $C_1 \times \cdots \times C_k$ *that possibly contribute five-dimensional* **boundary facets** *to the net wrench cone* $\mathcal{W}$ *are* SI/II, SSI, SSS, $SSSS$ *and* $SSSSS$, *provided that the friction cones lie in generic positions relative to the contacts.*

The proof of Theorem 15.7 appears in the chapter's appendix. The proof verifies that the Jacobian of the grasp map, $D\mathcal{L}_G$, has *full rank* in each cell class of $C_1 \times \cdots \times C_k$ that is *not* listed in Theorem 15.7. When $D\mathcal{L}_G$ has full rank in a cell $\mathcal{K}$ of dimension six or higher, the cell is mapped into the interior of $\mathcal{W}$ in $\mathcal{B}$'s wrench space. For instance, consider the III and SII cell classes that are *not* listed in Theorem 15.7. In an III cell, $D\mathcal{L}_G$ has full rank in the entire cell when the contacts do not lie on a common line in $\mathbb{R}^3$. The entire cell is therefore mapped into the interior of $\mathcal{W}$. In an SII cell, $D\mathcal{L}_G$ becomes rank deficient only when the friction cone at the S contact is tangent to the line passing through the I contacts. This condition never occurs at tame stances and is highly non-generic at all other stances. The entire cell is therefore generically mapped into the interior of $\mathcal{W}$. Similar arguments hold for all cell classes that are *not* listed in Theorem 15.7.

Let us summarize the list of contact force cells that possibly contribute facets to $\mathsf{bdy}(\mathcal{W})$. At a three-contact stance, $\mathsf{bdy}(\mathcal{W})$ consists of SI/II, SSI, and SSS facets. At a four-contact stance, $\mathsf{bdy}(\mathcal{W})$ can additionally have $SSSS$ facets. At stances supported by $k \geq 5$ contacts, $\mathsf{bdy}(\mathcal{W})$ can have all five types of facets. Finally, no new types of boundary facets appear at stances supported by higher numbers of contacts. Our next task is to identify which contact forces in these cell classes are actually mapped to boundary facets of $\mathcal{W}$.

15.5 Critical Contact Forces That Contribute Boundary Facets to $\mathcal{W}$

This section characterizes the specific contact forces in the cell classes SI/II, SSI, SSS, $SSSS$, and $SSSSS$ that satisfy the rank deficiency condition, $\text{rank}(D\mathcal{L}_G)=5$. These *critical contact forces* are mapped by the grasp map $\mathcal{L}_G$ to facets of $\text{bdy}(\mathcal{W})$, provided that they satisfy an auxiliary condition discussed at the end of this section. The critical contact forces are characterized in the following proposition, whose proof appears in the chapter's appendix.

PROPOSITION 15.8 (**critical contact forces**) *For each cell class listed in Theorem 15.7, the* **critical contact forces** *that are possibly mapped to* $\text{bdy}(\mathcal{W})$ *satisfy the conditions:*

1. **Cell classes SI and II:** *The entire cells consist of critical contact forces.*
2. **Cell class SSI:** *A cell of this class is parametrized by (λ_1,ϕ_1), (λ_2,ϕ_2) and $f_3\in\mathbb{R}^3$. The critical contact force angles (ϕ_1,ϕ_2) satisfy two scalar equations,*

$$\eta_1(\phi_1)\cdot(x_1-x_3)=0 \quad \textit{and} \quad \eta_2(\phi_2)\cdot(x_2-x_3)=0, \tag{15.12}$$

where $\eta_i(\phi_i)$ is the normal to the friction cone tangent plane $T_i(\phi_i)$ for $i=1,2$.

3. **Cell class SSS:** *A cell of this class is parametrized by (λ_i,ϕ_i) for $i=1,2,3$. Let $(\bar{s},\bar{t},\bar{n})$ be a reference frame for the plane Δ spanned by x_1,x_2,x_3. The critical contact force angles (ϕ_1,ϕ_2,ϕ_3) satisfy the scalar equation*

$$\det\begin{bmatrix} \bar{s}\cdot\nu_1(\phi_1) & \bar{s}\cdot\nu_2(\phi_2) & \bar{s}\cdot\nu_3(\phi_3)\\ \bar{t}\cdot\nu_1(\phi_1) & \bar{t}\cdot\nu_2(\phi_2) & \bar{t}\cdot\nu_3(\phi_3)\\ \bar{n}\cdot(x_1\times\nu_1(\phi_1)) & \bar{n}\cdot(x_2\times\nu_2(\phi_2)) & \bar{n}\cdot(x_3\times\nu_3(\phi_3)) \end{bmatrix}=0, \tag{15.13}$$

where $\nu_i(\phi_i)=\bar{n}\times\eta_i(\phi_i)$, $\eta_i(\phi_i)$ is the normal to $T_i(\phi_i)$ for $i=1,2,3$.

4. **Cell class $SSSS$:** *A cell of this class is parametrized by (λ_i,ϕ_i) for $i=1\ldots4$. Let $(\bar{s}_i,\bar{t}_i,\bar{n}_i)$ be a reference frame for the plane spanned by the contacts x_i,x_{i+1},x_{i+2}, where index addition is taken modulo 4. The critical contact force angles $(\phi_1,\phi_2,\phi_3,\phi_4)$ satisfy three scalar equations,*

$$\det\begin{bmatrix} \bar{s}_i\cdot\nu_i(\phi_i) & \bar{s}_i\cdot\nu_{i+1}(\phi_{i+1}) & \bar{s}_i\cdot\nu_{i+2}(\phi_{i+2})\\ \bar{t}_i\cdot\nu_i(\phi_i) & \bar{t}_i\cdot\nu_{i+1}(\phi_{i+1}) & \bar{t}_i\cdot\nu_{i+2}(\phi_{i+2})\\ \bar{n}_i\cdot(x_i\times\nu_i(\phi_i)) & \bar{n}_i\cdot(x_{i+1}\times\nu_{i+1}(\phi_{i+1})) & \bar{n}_i\cdot(x_{i+2}\times\nu_{i+2}(\phi_{i+2})) \end{bmatrix}$$
$$=0 \quad i=1,2,3, \tag{15.14}$$

where $\nu_i(\phi_i)=\bar{n}_i\times\eta_i(\phi_i)$, $\eta_i(\phi_i)$ is the normal to $T_i(\phi_i)$ for $i=1\ldots4$.

5. **Cell class $SSSSS$:** *A cell of this class is parametrized by (λ_i,ϕ_i) for $i=1\ldots5$. Let $(\bar{s}_i,\bar{t}_i,\bar{n}_i)$ be a reference frame for the plane spanned by the contacts*

x_i, x_{i+1}, x_{i+2}, where index addition is taken modulo 5. The critical contact force angles $(\phi_1, \ldots, \phi_5)$ satisfy five scalar equations,

$$\det \begin{bmatrix} \bar{s}_i \cdot \boldsymbol{v}_i(\phi_i) & \bar{s}_i \cdot \boldsymbol{v}_{i+1}(\phi_{i+1}) & \bar{s}_i \cdot \boldsymbol{v}_{i+2}(\phi_{i+2}) \\ \bar{t}_i \cdot \boldsymbol{v}_i(\phi_i) & \bar{t}_i \cdot \boldsymbol{v}_{i+1}(\phi_{i+1}) & \bar{t}_i \cdot \boldsymbol{v}_{i+2}(\phi_{i+2}) \\ \bar{\boldsymbol{n}}_i \cdot (x_i \times \boldsymbol{v}_i(\phi_i)) & \bar{\boldsymbol{n}}_i \cdot (x_{i+1} \times \boldsymbol{v}_{i+1}(\phi_{i+1})) & \bar{\boldsymbol{n}}_i \cdot (x_{i+2} \times \boldsymbol{v}_{i+2}(\phi_{i+2})) \end{bmatrix}$$

$$= 0 \quad i = 1 \ldots 5, \tag{15.15}$$

where $\boldsymbol{v}_i(\phi_i) = \bar{\boldsymbol{n}}_i \times \boldsymbol{\eta}_i(\phi_i)$, $\boldsymbol{\eta}_i(\phi_i)$ is the normal to $T_i(\phi_i)$ for $i = 1 \ldots 5$.

Let us provide some intuitive insight for the critical contact forces in each cell class. In the SI and II cell classes, the critical contact forces lie in the vertical plane passing through the SI/II contacts. In the SSI cell class, Condition (15.12) yields up to four discrete solutions for the critical contact force angles (ϕ_1^*, ϕ_2^*). For each (ϕ_1^*, ϕ_2^*), the friction cone tangent planes, $T_1(\phi_1^*)$ and $T_2(\phi_2^*)$, intersect along a line that passes through the I contact point (Figure 15.4(a)). For each (ϕ_1^*, ϕ_2^*), the critical contact force magnitudes $\lambda_1, \lambda_2 > 0$ and the critical contact force f_3 vary freely within the SSI cell. These critical contact forces are mapped to a five-dimensional *hyperplane facet* of $\mathsf{bdy}(\mathcal{W})$.

In the SSS cell class, Condition (15.13) determines a two-dimensional solution set of critical contact force angles (ϕ_1, ϕ_2, ϕ_3), while the critical contact force magnitudes $\lambda_1, \lambda_2, \lambda_3 > 0$ vary freely in the SSS cell. Based on line geometry interpretation of the condition $\mathsf{rank}(D\mathcal{L}_G) = 5$, at each solution (ϕ_1, ϕ_2, ϕ_3) the friction cone tangent planes, $T_1(\phi_1)$, $T_2(\phi_2)$ and $T_3(\phi_3)$ intersect at a common point $z \in \Delta$, where Δ is the plane spanned by x_1, x_2 and x_3 (Figure 15.4(b)). The critical contact forces in an SSS cell are mapped to a five-dimensional *nonlinear facet* of $\mathsf{bdy}(\mathcal{W})$.

In the $SSSS$ cell class, Condition (15.14) specifies three scalar constraints in the contacts force angles $(\phi_1, \phi_2, \phi_3, \phi_4)$. These constraints determine a one-dimensional solution set in $(\phi_1, \phi_2, \phi_3, \phi_4)$ space, while the critical contact force magnitudes $\lambda_1, \lambda_2, \lambda_3, \lambda_4 > 0$ vary freely in the $SSSS$ cell. The geometric characterization of the critical contacts forces is similar to the one described for the SSS cells. These critical contact forces are also mapped to a five-dimensional *nonlinear facet* of $\mathsf{bdy}(\mathcal{W})$. In the $SSSSS$ cell class, Condition (15.15) specifies five scalar constraints in the contact force angles $(\phi_1, \ldots, \phi_5)$. These constraints determine a finite number of discrete solutions, $(\phi_1^*, \ldots, \phi_5^*)$, while the critical contact force magnitudes $\lambda_1, \ldots, \lambda_5 > 0$ vary freely in the $SSSSS$ cell. For each $(\phi_1^*, \ldots, \phi_5^*)$, the critical contact forces are mapped to a five-dimensional *hyperplane facet* of $\mathsf{bdy}(\mathcal{W})$.

The critical contact forces are illustrated in the following example.

Example: Figure 15.4 shows two views of a three-contact stance with coefficient of friction $\mu = 0.4$ at the contacts (the supported object is not shown). The contact points are located at $x_1 = (8, 0, 1)$, $x_2 = (4, 4\sqrt{3}, 1.2)$ and $x_3 = (0, 0, 1)$, expressed in the world frame. The contact normal directions are $n_i = (\sin\alpha_i \sin\beta_i, -\cos\alpha_i \sin\beta_i, \cos\beta_i)$, such that $(\alpha_1, \beta_1) = (70°, -20°)$, $(\alpha_2, \beta_2) = (-15°, 34°)$ and $(\alpha_3, \beta_3) = (-70°, -10°)$. A triplet of critical contact forces (f_1, f_2, f_3) in the cell $S_1 S_2 I_3$ is plotted in

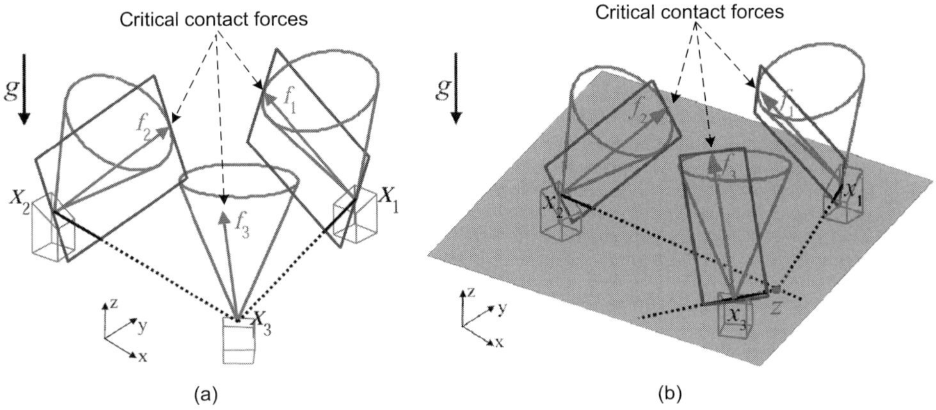

Figure 15.4 A three-contact stance with coefficient of friction $\mu = 0.4$. (a) A triplet of SSI critical contact forces, where the friction cone tangent planes at x_1 and x_2 intersect at the contact point x_3. (b) A triplet of SSS critical contact forces, where the friction cone tangent planes intersect at a common point, $z \in \Delta$, where Δ is the plane spanned by x_1, x_2, and x_3.

Figure 15.4(a). The critical forces f_1 and f_2 lie on the boundary of their friction cones, such that their friction cone tangent planes pass through the contact point x_3. Criticality of these contact forces can be viewed by the fact that infinitesimal changes in f_1 and f_2 along their friction cone tangent planes, accompanied by infinitesimal changes in $f_3 \in I_3$, will cause *no change* in the contacts' net torque about the intersection line of the two tangent planes.

A triplet of critical contact forces (f_1, f_2, f_3) in the cell $S_1S_2S_3$ is plotted in Figure 15.4(b). The three friction cone tangent planes intersect at a common point, $z \in \Delta$, where Δ is the plane spanned by x_1, x_2 and x_3. Criticality of these contact forces can be explained as follows. The line passing through z and x_i lies in the friction cone tangent plane $T_i(\phi_i)$ for $i = 1, 2, 3$. Hence, small changes in f_i along its friction cone tangent plane, $T_i(\phi_i)$, can be partitioned into two mutually orthogonal components. Changes along the vector $z - x_i$ and changes along the orthogonal complement $(z - x_i) \times \eta_i$, where η_i is normal to the tangent plane $T_i(\phi_i)$. The changes in f_1, f_2, f_3 along their z-pointing components generate a two-dimensional subspace of net wrenches, V_1, consisting of pure forces that lie in the plane Δ. The changes in f_1, f_2, f_3 along their complementary components generate a three-dimensional subspace, V_2, of net wrenches. Criticality can be viewed by the fact that all small changes of the contact forces along their friction cone tangent planes generate a five-dimensional subspace of net wrenches, $V_1 + V_2$, while $\mathcal{B}$'s wrench space is a six-dimensional vector space. In contrast, at any choice of *non-critical* contact forces, small changes of the contact forces along their friction cone tangent planes generate a full six-dimensional neighborhood of net wrenches in $\mathcal{B}$'s wrench space. ○

The conditions that define the critical contact forces in Proposition 15.8 are only *necessary* for finding the five-dimensional facets of $\text{bdy}(\mathcal{W})$. That is, critical contact forces in a cell $\mathcal{K}$ can be mapped by $\mathcal{L}_G$ to the interior rather than the boundary of $\mathcal{W}$. The next lemma specifies a *sign condition* that will be used to identify which of the critical contact forces are mapped to the actual boundary of $\mathcal{W}$.

LEMMA 15.9 (**sign condition**) *Let $\vec{f} = (f_1, \ldots, f_k)$ be critical contact forces in a cell $\mathcal{K}$ of $C_1 \times \cdots \times C_k$. The net object wrench $\boldsymbol{w} = \mathcal{L}_G(\vec{f})$ lies in* bdy($\mathcal{W}$) *iff there exists a five-dimensional separating hyperplane H that passes through $\boldsymbol{w}$, such that all wrenches $\boldsymbol{w} \in \mathcal{W}$ lie on the* **same side** *of H in $\mathcal{B}$'s wrench space. That is, there exists a sign $\sigma \in \{-1, +1\}$ satisfying the condition*

$$\sigma \cdot (\boldsymbol{w} \cdot \boldsymbol{\eta}_H) \geq 0 \qquad \text{for all } \boldsymbol{w} \in \mathcal{W} \tag{15.16}$$

where $\boldsymbol{\eta}_H \in \mathbb{R}^6$ is orthogonal to the hyperplane H.[3]

When a cell $\mathcal{K}$ contributes a five-dimensional facet to bdy($\mathcal{W}$), the critical contact forces generically form a smooth five-dimensional manifold within this cell. The image of this manifold under the grasp map $\mathcal{L}_G$ forms the five-dimensional facet of bdy($\mathcal{W}$). Based on this insight, let $\vec{f} = (f_1, \ldots, f_k)$ be particular combination of critical contact forces in $\mathcal{K}$. The *range space* of the Jacobian $D\mathcal{L}_G(\vec{f})$ forms a five-dimensional hyperplane, H, which is tangent to the facet of bdy($\mathcal{W}$) at the wrench $\boldsymbol{w} = \mathcal{L}_G(\vec{f})$. Hence, one simply checks the separation condition of Eq. (15.16) using the *left kernel* of $D\mathcal{L}_G(\vec{f})$ as the hyperplane's normal $\boldsymbol{\eta}_H$. That is, $\boldsymbol{\eta}_H$ is determined by the condition $\boldsymbol{\eta}_H^T D\mathcal{L}_G(\vec{f}) = (\vec{0})^T$.

The final proposition adapts the separation condition of Lemma 15.9 into a condition that identifies the boundary facets of $\mathcal{W}$. The proof of the proposition appears in the chapter's appendix.

PROPOSITION 15.10 (**separation condition**) *For each cell class listed in Theorem 15.7, the critical contact forces are mapped by the grasp map $\mathcal{L}_G(\vec{f})$ to an* **actual boundary facet** *of the net wrench cone $\mathcal{W}$ iff there exists a sign $\sigma \in \{-1, +1\}$ satisfying the inequality*

$$\sigma \cdot \left(\boldsymbol{w}_i \cdot \boldsymbol{\eta}_H\right) \geq 0 \qquad \text{for all } \boldsymbol{w}_i = (f_i, x_i \times f_i) \text{ such that } f_i \in C_i \text{ for } i = 1 \ldots k, \tag{15.17}$$

where $\boldsymbol{\eta}_H$ has the following form in each cell class.

1. **Cell classes SI/II:** *Set the origin at the contact point x_1. Then $\boldsymbol{\eta}_H = (\vec{0}, x_2 - x_1)$, where x_1 and x_2 are the active contacts in these cell classes.*
2. **Cell class SSI:** *The critical contact forces (f_1, f_2) have fixed angles (ϕ_1^*, ϕ_2^*). Set the origin at the contact point x_3. Then $\boldsymbol{\eta}_H = (\vec{0}, \boldsymbol{v}_{12}(\phi_1^*, \phi_2^*))$, where $\boldsymbol{v}_{12}(\phi_1^*, \phi_2^*) = \boldsymbol{\eta}_1(\phi_1^*) \times \boldsymbol{\eta}_2(\phi_2^*)$ is collinear with the line $T_1(\phi_1^*) \cap T_2(\phi_2^*)$, which passes through x_3.*
3. **Cell classes SSS, $SSSS$, $SSSSS$:** *Let (ϕ_1, ϕ_2, ϕ_3) be critical contact force angles. Set the origin at the intersection point of the friction cone tangent planes, $z \in \Delta$, where Δ is the plane spanned by x_1, x_2 and x_3. Then $\boldsymbol{\eta}_H = (\bar{\boldsymbol{n}}, \bar{\boldsymbol{\tau}})$, where*

[3] The vector η_H should be interpreted as a *tangent vector* in $\mathcal{B}$'s c-space. The inner product $\boldsymbol{w} \cdot \eta_H$ represents *instantaneous work* done on $\mathcal{B}$ by the wrench $\boldsymbol{w}$ along the object instantaneous motion $\dot{q} = \eta_H$.

$\bar{n}(\phi_1, \phi_2, \phi_3)$ is the unit normal to Δ, and $\bar{\tau}(\phi_1, \phi_2, \phi_3)$ is given by

$$\bar{\tau}(\phi_1, \phi_2, \phi_3) = [\eta_1(\phi_1)\, \eta_2(\phi_2)\, \eta_3(\phi_3)]^{-1} \boldsymbol{c}$$

where $\eta_i(\phi_i)$ is the normal to the tangent plane $T_i(\phi_i)$ $(i = 1, 2, 3)$, and $\mathbf{c} = (c_1, c_2, c_3)$ are the direction cosines of the friction cone tangent planes $T_1(\phi_1)$, $T_2(\phi_2)$, $T_3(\phi_3)$ with respect to Δ.

The boundary separation condition, $\sigma \cdot (w_i \cdot \eta_H) \geq 0$, checks the wrenches generated by the individual friction cones at the contacts. This simpler test relies on the structure of the net wrench cone $\mathcal{W}$ as the *sum* of wrenches generated by the individual friction cones. In the SI/II cell classes, one only needs to check the sign of $w_i \cdot \eta_H$ using the contact normals. This test ensures that the friction cones at the inactive contacts generate the same moment sign about the line passing through the active SI/II contacts. In the SSI cell class, one first checks the sign of $w_i \cdot \eta_H$ $(i = 1, 2)$, using the contact normals at the SS contacts. This test is followed by numerical evaluation of $w_i(\phi_i) \cdot \eta_H$, where $w_i(\phi_i)$ is the wrench generated by full rotation of the unit force $u_i(\phi_i) \in S_i$ at each of the inactive contacts. This two-part test ensures that the friction cones at the inactive contacts generate the same moment sign about the line $T_1(\phi_1^*) \cap T_2(\phi_2^*)$, which passes through the I contact point. In the SSS, $SSSS$ and $SSSSS$ cell classes, one numerically evaluates the sign of $w_i(\phi_i) \cdot \eta_H$ along full rotation of the unit force $u_i(\phi_i) \in S_i$ at each of the k contacts. This test ensures that all wrenches generated by the k friction cones lie in the same half-space of $\mathcal{B}$'s wrench space, bounded by the hyperplane that passes through the wrench $w = \mathcal{L}_G(\vec{f})$ with normal η_H.

15.6 The Stance Equilibrium Region Boundary Curves

This section provides a full description of the boundary curves surrounding the feasible equilibrium region, $\mathcal{ER} \subseteq \mathbb{R}^2$, of generic k-contact stances under the influence of gravity. An example will then show how these boundary curves appear at a tame stance as a function of a variable coefficient of friction at the contacts.

According to the scheme of Theorem 15.5, the intersection of $\mathsf{bdy}(\mathcal{W})$ with the affine plane of gravitational wrenches, L, maps to the boundary of $\mathcal{ER}$ in $\mathbb{R}^2$. The intersection $\mathsf{bdy}(\mathcal{W}) \cap L$ represents the two sides of the equilibrium stance condition specified in Eq. (15.3). This equation can be decomposed into three parts: (i) force balance equation, (ii) torque balance about the vertical direction and (iii) torque balance about the horizontal directions:

$$\begin{array}{ll} (i) & f_1 + \cdots + f_k = \boldsymbol{e} \\ (ii) & \boldsymbol{e} \cdot (x_1 \times f_1 + \cdots + x_k \times f_k) = 0 \\ (iii) & \mathbf{E}(x_1 \times f_1 + \cdots + x_k \times f_k) = \mathbf{E}(\boldsymbol{x}_{cm} \times \boldsymbol{e}) \end{array} \tag{15.18}$$

where the forces are scaled such that $\|f_g\| = 1$, e is the vertical upward direction in $\mathbb{R}^3$, and $\mathbf{E}$ is the horizontal projection matrix. The position of $\mathcal{B}$'s center of mass, x_{cm}, appears only in part (iii) of Eq. (15.18). The intersection $\text{bdy}(\mathcal{W}) \cap L$ will therefore be computed using parts (i) and (ii) of Eq. (15.18), then mapped to the horizontal position of $\mathcal{B}$'s center of mass, $\tilde{x}_{cm}$, by the linear map specified in Eq. (15.5): $\tilde{x}_{cm} = -\mathbf{E}(e \times \tau)$. That is, we know from part (iii) of Eq. (15.18) that $\tau = x_1 \times f_1 + \cdots + x_k \times f_k$. Substituting for τ in $\tilde{x}_{cm} = -\mathbf{E}(e \times \tau)$ gives the mapping of $\text{bdy}(\mathcal{W}) \cap L$ to the horizontal plane

$$\tilde{x}_{cm} = -\mathbf{E}\big(e \times (x_1 \times f_1 + \cdots + x_k \times f_k)\big),$$

where $(f_1, \ldots, f_k)$ are *critical contact forces* in the cell classes SI/II, SSI, SSS, $SSSS$ and $SSSSS$. The boundary of $\mathcal{W}$ consists of five types of facets associated with these cell classes. Hence, the mapping of $\text{bdy}(\mathcal{W}) \cap L$ to the horizontal plane gives *five types* of boundary curves of the region $\mathcal{ER}$, which are summarized in the following theorem. The theorem uses specific contact force indices, where the S_i contacts receive the lowest indices, then the I_i contacts and, last, the O_i contacts. However, this specific assignment of indices represents all possible index permutations.

THEOREM 15.11 (**boundary curves of $\mathcal{ER}$**) *The horizontal cross-section of the feasible equilibrium region of a k-contact stance, $\mathcal{ER} \subseteq \mathbb{R}^2$, forms a connected convex set whose boundary consists of up to* **five types** *of curves: SI/II line segments, SSI line segments, SSS convex arcs, $SSSS$ convex arcs and $SSSSS$ line segments, provided that the friction cones lie in generic positions relative to the contacts.*

The proof of Theorem 15.11 appears in the chapter's appendix. Let us describe how one determines each boundary curve of $\mathcal{ER}$, either graphically or by a numerical scheme. An SI/II boundary segment is associated with two non-zero contact forces, $(f_1, f_2) \in S_1 \times I_2$ or $(f_1, f_2) \in I_1 \times I_2$, that satisfy the separation condition described in Proposition 15.10. This segment can be graphically constructed using the *five-strip formula* described in Chapter 14. An SSI boundary segment is associated with three non-zero contact forces, $(f_1, f_2, f_3) \in S_1 \times S_2 \times I_3$, such that (f_1, f_2) have fixed angles, (ϕ_1^*, ϕ_2^*). A solution (ϕ_1^*, ϕ_2^*) that satisfies the separation condition described in Proposition 15.10 forms a linear boundary segment of $\mathcal{ER}$. This segment lies on the horizontal projection of the intersection line of the friction cone tangent planes, $T_1(\phi_1^*) \cap T_2(\phi_2^*)$, which passes through the I contact point (Figure 15.4).

Next consider the computation of a nonlinear SSS boundary curve of $\mathcal{ER}$. This curve is associated with three non-zero contact forces, $(f_1, f_2, f_3) \in S_1 \times S_2 \times S_3$, whose angles (ϕ_1, ϕ_2, ϕ_3) satisfy the scalar constraint of Eq. (15.13). The intersection $\text{bdy}(\mathcal{W}) \cap L$ according to parts (i) and (ii) of Eq. (15.18) introduces an additional scalar constraint:

$$\det\begin{bmatrix} \mathbf{E}u_1(\phi_1) & \mathbf{E}u_2(\phi_2) & \mathbf{E}u_3(\phi_3) \\ e \cdot (x_1 \times u_1(\phi_1)) & e \cdot (x_2 \times u_2(\phi_2)) & e \cdot (x_3 \times u_3(\phi_3)) \end{bmatrix} = 0 \qquad (15.19)$$

The angles (ϕ_1, ϕ_2, ϕ_3) that satisfy the two scalar constraints, Eq. (15.13) and Eq. (15.19), generically form a curve in (ϕ_1, ϕ_2, ϕ_3) space. The portion of the solution curve that satisfies the separation condition described in Proposition 15.10 maps to a convex arc on the boundary of $\mathcal{ER}$.[4] To compute the SSS boundary curve, at each point along the (ϕ_1, ϕ_2, ϕ_3) solution curve, obtain the force magnitudes $(\lambda_1, \lambda_2, \lambda_3)$ by solving the 3×3 linear system:

$$\lambda_1 \boldsymbol{u}_1(\phi_1) + \lambda_2 \boldsymbol{u}_2(\phi_2) + \lambda_3 \boldsymbol{u}_3(\phi_3) = \boldsymbol{e}.$$

Every solution satisfying $\lambda_1, \lambda_2, \lambda_3 \geq 0$ contributes a point $\tilde{\boldsymbol{x}}_{cm}$ to the boundary of $\mathcal{ER}$, located at $\tilde{\boldsymbol{x}}_{cm} = -\mathbf{E}\big(\boldsymbol{e} \times (\sum_{i=1}^{3} \lambda_i(\boldsymbol{\phi}) x_i \times \boldsymbol{u}_i(\phi_i))\big)$, where $\boldsymbol{\phi} = (\phi_1, \phi_2, \phi_3)$. A similar computation scheme holds for the nonlinear $SSSS$ boundary curve of $\mathcal{ER}$ (see Exercises).

Finally, consider the computation of a linear $SSSSS$ boundary segment of $\mathcal{ER}$. This segment is associated with five non-zero contact forces, $(f_1, \ldots, f_5) \in S_1 \times \ldots \times S_5$, whose angles form discrete solutions. A discrete solution $(\phi_1^*, \ldots, \phi_5^*)$ which satisfies the separation condition described in Proposition 15.10 contributes a hyperplane facet to $\mathsf{bdy}(\mathcal{W})$. This facet is embedded in a specific separating hyperplane, H, of the net wrench cone $\mathcal{W}$. The intersection $\mathsf{bdy}(\mathcal{W}) \cap L$ according to parts (i) and (ii) of Eq. (15.18) is equivalent to the intersection of H with the affine plane L. The intersection $H \cap L$ gives an affine line in $\mathcal{B}$'s wrench space, which maps to the line containing the linear $SSSSS$ boundary segment of $\mathcal{ER}$ in $\mathbb{R}^2$.

The remainder of this section illustrates the computation of the region $\mathcal{ER}$ for the three-contact stance depicted in Figure 15.5. This stance is *tame* as long as the coefficient of friction satisfies the inequality $\mu \leq 0.7$ at the contacts. Tame stances satisfy the relation $\mathcal{ER} \subseteq \mathcal{P}$, and we will study how the region $\mathcal{ER}$ changes within the support polygon $\mathcal{P}$ as a function of μ.

Computational example: The three-contact stance depicted in Figure 15.5(a) shows the localities of the supporting contacts without showing the supported object. When $\mu = 0.7$, the stance is tame and all three contacts are *quasi-flat*. Hence, $\mathcal{ER} = \mathcal{P}$ in Figure 15.5(a). When $\mu = 0.5$, the contact points x_1 and x_2 are still quasi-flat. Hence, their horizontal projections, $\tilde{x}_1$ and $\tilde{x}_2$, still lie in $\mathcal{ER}$ (Figure 15.5(b)). The convexity of $\mathcal{ER}$ implies that the entire line segment $\tilde{x}_1$–$\tilde{x}_2$ lies in $\mathcal{ER}$. Moreover, $\tilde{x}_1$–$\tilde{x}_2$ lies on the boundary of $\mathcal{ER}$, since it lies on the boundary of the support polygon $\mathcal{P}$ and $\mathcal{ER} \subseteq \mathcal{P}$ for tame stances. However, the contact point x_3 is no longer quasi-flat, as seen in top view in Figure 15.5(b). Now $\tilde{x}_3$ lies *outside* $\mathcal{ER}$, and only portions of the segments $\tilde{x}_1$–$\tilde{x}_3$ and $\tilde{x}_2$–$\tilde{x}_3$, ending at the points $\tilde{p}_1$ and $\tilde{p}_2$, lie on the boundary of $\mathcal{ER}$. The line segments $\tilde{x}_1$–$\tilde{p}_1$ and $\tilde{x}_2$–$\tilde{p}_2$ correspond to the vertical equilibrium strips associated with the contacts pairs (x_1, x_3) and (x_2, x_3) (see Chapter 14). Note that the boundary of $\mathcal{ER}$ has a missing piece between $\tilde{p}_1$ and $\tilde{p}_2$, which will be studied later.

Let $\tilde{C}_i$ denote the horizontal projection of the friction cone C_i, which forms a sector based at the projected contact point $\tilde{x}_i$ in $\mathbb{R}^2$. When the coefficient of friction is decreased to $\mu = 0.4$, the projected friction cone $\tilde{C}_3$ ceases to contain the projected contact point $\tilde{x}_2$ (Figure 15.5(c)). Hence, x_2 and x_3 alone cannot balance any gravitational load that acts on $\mathcal{B}$. Their vertical equilibrium strip becomes empty, and the entire segment $\tilde{x}_2$–$\tilde{x}_3$, except its endpoint $\tilde{x}_2$, lies *outside* the region $\mathcal{ER}$ (Figure 15.5(c)).

[4] The solution curve forms a single or multiple loops in (ϕ_1, ϕ_2, ϕ_3) space. Each of these loops maps to a loop inside $\mathcal{ER}$, whose outer segment forms a convex boundary arc of $\mathcal{ER}$.

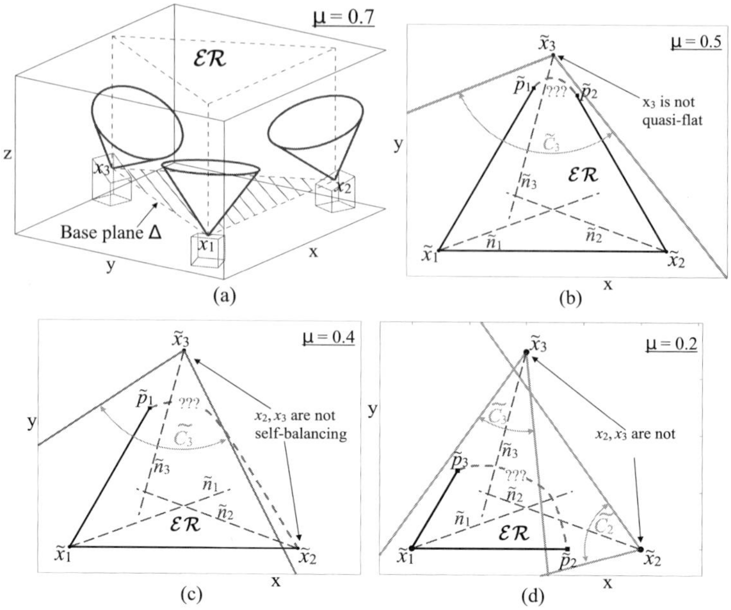

Figure 15.5 (a) A three-contact stance that is tame and has quasi-flat contacts for $\mu = 0.7$. (b) Top view of the stance for $\mu = 0.5$, where the projected contact point $\tilde{x}_3$ lies outside $\mathcal{ER}$. (c) Top view of the stance for $\mu = 0.4$, where the entire segment $\tilde{x}_2$–$\tilde{x}_3$ except its endpoint $\tilde{x}_2$ lies outside $\mathcal{ER}$. (d) Top view for $\mu = 0.2$, where both points $\tilde{x}_2$ and $\tilde{x}_3$ lie outside $\mathcal{ER}$.

The contacts x_1 and x_3 still contribute the segment $\tilde{x}_1$–$\tilde{p}_1$ to the boundary of $\mathcal{ER}$, but there is a missing piece between $\tilde{p}_1$ and $\tilde{x}_2$ (Figure 15.5(c)). Finally, let us decrease the coefficient of friction to $\mu = 0.2$. As shown in Figure 15.5(d), now only the contact point x_1 is quasi-flat. The projected friction cones $\tilde{C}_2$ and $\tilde{C}_3$ still contain the projected contact point $\tilde{x}_1$. Hence, the boundary of $\mathcal{ER}$ consists of portions of the segments $\tilde{x}_1$–$\tilde{x}_2$ and $\tilde{x}_1$–$\tilde{x}_3$, ending at the points $\tilde{p}_2$ and $\tilde{p}_3$, together with a missing piece connecting $\tilde{p}_2$ and $\tilde{p}_3$ (Figure 15.5(d)).

Let us next apply the tools developed in this chapter to complete the missing pieces of the stance equilibrium region. First consider the region $\mathcal{ER}$ for $\mu = 0.5$, where we left a missing boundary piece between $\tilde{p}_1$ and $\tilde{p}_2$ (Figure 15.5(b)). The missing piece is the nonlinear SSS curve depicted in Figure 15.6(a), which forms a *convex arc* along the boundary of $\mathcal{ER}$. The region $\mathcal{ER}$ forms a connected convex set, bounded by portions of the support polygon edges that are SS/II segments and the SSS boundary curve. When $\mu = 0.4$, we left a missing boundary piece between $\tilde{p}_2$ and $\tilde{x}_3$ (Figure 15.5(c)). This piece consists of an SSS curve and a linear SSI segment depicted as dashed line in Figure 15.6(b). Note that both segments lie in the interior of the support polygon. Next consider the region $\mathcal{ER}$ for $\mu = 0.2$, where we left a missing boundary piece between $\tilde{p}_2$ and $\tilde{p}_3$ (Figure 15.5(d)). This piece consists of the SSS boundary curve depicted in Figure 15.6(c), which forms a *convex arc* along the boundary of $\mathcal{ER}$. Finally, when $\mu = 0.1$ the entire region $\mathcal{ER}$ lies strictly inside the support polygon (Figure 15.6(d)). Its boundary consists of a nonlinear SSS curve and a linear SSI segment depicted as dashed line in Figure 15.6(d). ○

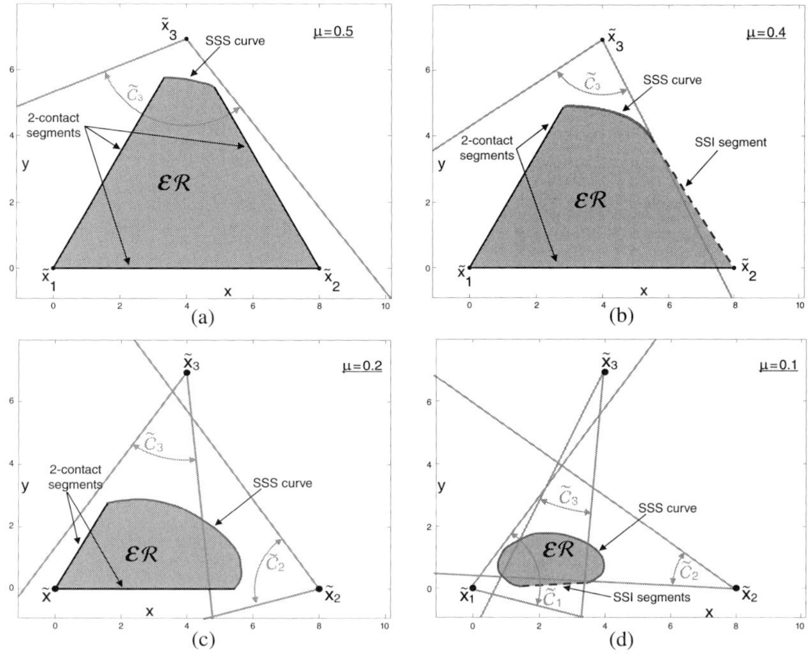

Figure 15.6 The region $\mathcal{ER}$ (shaded) for the tame three-contact stance of Figure 15.5. (a) At $\mu = 0.5$, a nonlinear SSS boundary curve appears. (b) At $\mu = 0.4$, a linear SSI segment (dashed) joins the nonlinear SSS boundary curve. (c) At $\mu = 0.2$, a nonlinear SSS curve forms the missing boundary piece of $\mathcal{ER}$. (d) At $\mu = 0.1$, the entire region $\mathcal{ER}$ lies in the interior of the support polygon. In all cases, $\mathcal{ER} \subseteq \mathcal{P}$ forms a connected convex set.

The example illustrates the intricate structure of the stance equilibrium region associated with frictional contacts on uneven terrains. Broadly speaking, when a robot hand supports a rigid object via quasi-flat contacts, the relation $\mathcal{P} \subseteq \mathcal{ER}$ stated in Proposition 15.4 justifies the use of the support polygon as a subset of *locally wrench resistant* and, hence, safe stances. However, hand-supported stances typically contain a mix of quasi-flat and non-quasi-flat contacts. Such stances are *tame* as long as the terrain formed by the robot hand is not too steep and, hence, satisfies the relation $\mathcal{ER} \subseteq \mathcal{P}$. The region $\mathcal{ER}$ thus forms a strict and sometimes much smaller subset of the support polygon $\mathcal{P}$. Hence, one must generally use the true region $\mathcal{ER}$ to ensure *local wrench resistance* and, hence, safety of hand-supported stances.

15.7 Onset of Non-Static Motion Modes at the Contacts

Consider the situation where a robot hand supports a rigid object in static equilibrium under the influence of gravity. As the robot hand starts to *slowly* change its orientation while keeping a fixed shape, the object will remain stationary on the supporting contacts as long as its center of mass remains in the feasible equilibrium region of the stance. When the object's horizontal center-of-mass position, $\tilde{x}_{cm}$, crosses the boundary of the stance equilibrium region, static equilibrium can no longer be maintained and non-static motion modes will develop at the contacts. These motion modes depend on the type of boundary curve of the region $\mathcal{ER}$ being crossed by $\mathcal{B}$'s center of mass.

Thus, consider a boundary curve of $\mathcal{ER}$ associated with critical contact forces in a cell $\mathcal{K}$ of the composite friction cone $C_1 \times \cdots \times C_k$. The contact forces $(f_1, \ldots, f_k) \in \mathcal{K}$ are associated with n_S choices of S_i components, n_I choices of I_i components and $k - n_S - n_I$ choices of O_i components. An S_i component represents a contact force that lies on the friction cone's surface, an I_i component represents a contact force that lies in the friction cone's interior, while an O_i component represents a zero contact force. The non-static motion modes that will develop at the contacts depend on the O_i, S_i and I_i components of $\mathcal{K}$. When a cell $\mathcal{K}$ contains an O_i component, the object will *break contact* at x_i, and the contact force f_i will vanish at the instant of contact separation. When a cell $\mathcal{K}$ contains an S_i component, the object will start *sliding* at x_i, and the contact force f_i will remain on the ith friction cone surface for some finite time interval. When a cell $\mathcal{K}$ contains an I_i component, the object will start *rolling* at x_i, and the contact force f_i will remain in the ith friction cone interior for some finite time interval.

Based on this insight, any crossing of the equilibrium region boundary curves incurs one of the following *five types* of motion modes at the contacts. First consider the case where the object's center of mass crosses an SI/II boundary segment of $\mathcal{ER}$. In this case, the object will start rolling about the line connecting the two I contacts in a tip-over motion or start sliding at the S contact and rolling at the I contact, such that all other contacts separate during these motions. When the object's center of mass crosses an SSI boundary segment of $\mathcal{ER}$, one expects the onset of rolling at the I contact accompanied by slippage at the two S contacts, while all other contacts separate during this motion. When the object's center of mass crosses an SSS, $SSSS$, or $SSSSS$ boundary curve of $\mathcal{ER}$, one expects the onset of slippage at the three, four or five S contacts, while all other contacts separate during these motions.

Remark: One should keep in mind that the ideal rigid-body model *cannot* predict exactly which contact reaction forces will develop within the Coulomb friction cones at the contacts. When a rigid body is supported in static equilibrium by multiple frictional contacts, *local wrench resistance* ensures that a nonempty set of feasible contact forces will be able to maintain the equilibrium stance against sufficiently small disturbances. This chapter made the implicit assumption that at least for *tame stances,* when the object's center of mass lies in the interior of the stance equilibrium region, the object will remain stationary under sufficiently small disturbances. ∘

Bibliographical Notes

Or and Rimon [1, 2] describe the analysis principles of the feasible equilibrium region associated with multiple frictional contacts. Reference [1] studies stances supported by three frictional contacts. These are the simplest stances that can be used in practical applications, since we have seen that $\mathcal{W}$ is fully six-dimensional at such stances. The feasible equilibrium region of three-contact stances is bounded by SI/SS, SSI, and SSS curves. Reference [2] considers stances supported by a higher number of contacts, whose feasible equilibrium region is surrounded by *all five types* of boundary curves.

The feasible equilibrium region of stances supported by $k \geq 4$ frictional contacts can be obtained by computing the feasible equilibrium region of each *4-tuple* of the k contacts, $\mathcal{ER}_{j_1 j_2 j_3 j_4}$, then taking the *convex hull* of these regions:

$$\mathcal{ER} = \text{conv}\{\mathcal{ER}_{j_1 j_2 j_3 j_4} : 1 \leq j_1, j_2, j_3, j_4 \leq k\} \qquad k \geq 4. \tag{15.20}$$

This property can be explained as follows. Each endpoint of a linear $SSSSS$ boundary segment of $\mathcal{ER}$ occurs at a center-of-mass location where the magnitude of one of the five contact forces associated with the $SSSSS$ segment reduces to zero. Each endpoint is therefore an endpoint of an adjacent $SSSS$ boundary curve of $\mathcal{ER}$. By convexity of $\mathcal{ER}$, the linear $SSSSS$ segments can be obtained by connecting the endpoints of the $SSSS$ curves by straight line segments, which is equivalent to the convex hull taken in Eq. (15.20).

The chapter is useful well beyond hand-supported stances. The understanding of the net wrench cone associated with multiple frictional contacts is fundamental in any *grasping or manipulation application* where an object can only be pushed by the robot hand's fingers. Another important application concerns *legged robot locomotion*. When a legged robot walks in a quasistatic manner on uneven terrain, the feasible equilibrium region determined by the robot footholds can be used to select statically stable postures, provided that the robot keeps its center of mass in a *bounded height* subset that takes into account a safety margin against disturbance wrenches. The disturbance-safe equilibrium region of 3-D stances is analogous to the disturbance-safe equilibrium region of 2-D stances (Chapter 14). It is the intersection of slanted 3-D prisms in R^3, which forms a bounded height subset of the stance's 3-D feasible equilibrium region [5].

A final note concerns the zero moment point or *ZMP principle* [3]. This principle lumps the net reaction wrench exerted by flat terrain on a flat robot footpad into a single force, which acts at a point p located within the footpad's area, together with pure moment about the footpad's contact normal at p. The ZMP principle is used to synthesize safe biped robot locomotion on flat terrains [4]. However, the ZMP principle is no longer useful when the legged robot is supported by an uneven terrain, which is also the terrain formed by robot hands.

Appendix: Proofs and Technical Details

This appendix contains proofs of Theorem 15.7, Proposition 15.8 and Proposition 15.10 from Section 15.4, and a proof of Theorem 15.11 from Section 15.6. The proofs rely on the following facts from *line geometry* (see Chapters 3 and 5). A directed line l that passes through a point $p \in \mathbb{R}^3$ along a unit direction $v \in \mathbb{R}^3$ is represented in *Plücker coordinates* by the vector $(v, p \times v) \in \mathbb{R}^6$. The linear subspaces in Plücker coordinates represent well-known line collections. Among these subspaces, we will need the following *flat and solid pencils*.

DEFINITION 15.7 *A* **flat pencil** *is a two-dimensional linear subspace in* $\mathbb{R}^6$, *spanned by the Plücker coordinates of two lines* $(v_1, p \times v_1)$ *and* $(v_2, p \times v_2)$ *that intersect at* p. *A* **solid pencil** *is a three-dimensional linear subspace in* $\mathbb{R}^6$, *spanned by the Plücker coordinates of three lines* $(v_1, p \times v_1)$, $(v_2, p \times v_2)$ *and* $(v_3, p \times v_3)$ *that intersect at* p.

A *flat pencil* represents a collection of lines that pass through a common point $p \in \mathbb{R}^3$, such that the lines are embedded in a plane spanned by the directions $v_1, v_2 \in \mathbb{R}^3$. A *solid pencil* represents a collection of lines that pass through a common point $p \in \mathbb{R}^3$, along all directions $v \in \mathbb{R}^3$. To see how these pencils appear in hand-supported stances, recall that each friction cone is partitioned into three subsets: $C_i = O_i \cup I_i \cup S_i$. A *flat pencil* appears in the Jacobian of an individual wrench, $\boldsymbol{w}_i$, generated by contact forces that vary in the friction cone boundary, $f_i \in S_i$. In this case, $f_i(\lambda_i, \phi_i) = \lambda_i \boldsymbol{u}_i(\phi_i)$, and the Jacobian of $\boldsymbol{w}_i = (f_i, x_i \times f_i)$ with respect to (λ_i, ϕ_i) is the 6×2 matrix:

$$D\boldsymbol{w}_i(\lambda_i, \phi_i) = \begin{bmatrix} \boldsymbol{u}_i(\phi_i) & \lambda_i \boldsymbol{u}_i'(\phi_i) \\ x_i \times \boldsymbol{u}_i(\phi_i) & \lambda_i x_i \times \boldsymbol{u}_i'(\phi_i) \end{bmatrix}. \tag{15.21}$$

The pair $(\boldsymbol{u}_i(\phi_i), \boldsymbol{u}_i'(\phi_i))$ spans the tangent plane $T_i(\phi_i)$ to the friction cone boundary S_i. The columns of $D\boldsymbol{w}_i$ in Eq. (15.21) span a flat pencil based at the contact point x_i, whose lines are embedded in $T_i(\phi_i)$. A *solid pencil* appears in the Jacobian of an individual wrench, $\boldsymbol{w}_i$, generated by contact forces that vary in the friction cone interior, $f_i \in I_i$. In this case, f_i is parametrized by its Cartesian coordinates, $f_i \in \mathbb{R}^3$, and the Jacobian of $\boldsymbol{w}_i = (f_i, x_i \times f_i)$ with respect to f_i is the 6×3 matrix

$$D\boldsymbol{w}_i(f_i) = \begin{bmatrix} I \\ [x_i \times] \end{bmatrix}, \tag{15.22}$$

where I is a 3×3 identity matrix, and $[x_i \times]$ is the 3×3 matrix that acts as cross product on vectors, $[x_i \times]v = x_i \times v$ for $v \in \mathbb{R}^3$ (this notation will be used throughout the appendix). The columns of $D\boldsymbol{w}_i$ in Eq. (15.22) span a solid pencil based at the contact point x_i.

The next lemma provides useful facts concerning linear dependency of pencils (see Bibliographical Notes).

LEMMA 15.12 **(linearly dependent pencils)** *Let* $p_1, p_2, p_3 \in \mathbb{R}^3$ *be three non-collinear points. The following relations hold for pencils based at these points.*

1. *Three flat pencils based at* p_1, p_2 *and* p_3 *are* **linearly dependent** *iff they intersect at a common point located on the plane spanned by the three points.*[5]
2. *Two flat pencils based at* p_1 *and* p_2 *and a solid pencil based at* p_3 *are* **linearly dependent** *iff the two flat pencils pass through the point* p_3.
3. *A flat pencil based at* p_1 *and two solid pencils based at* p_2 *and* p_3 *are* **linearly dependent** *iff the flat pencil at* p_1 *coincides with the plane spanned by the three points.*

[5] Three flat pencils that contain three parallel lines in $\mathbb{R}^3$ are considered to be intersecting at infinity.

Lemma 15.12 refers to linear dependency of vectors in $\mathbb{R}^6$ that represent the Plücker coordinates of directed lines in $\mathbb{R}^3$. Let us next recall that the null space of an $l \times r$ matrix of A, denoted $\mathcal{N}(A)$, is the linear subspace of $\mathbb{R}^r$ mapped by A to the origin: $\mathcal{N}(A)=\{v\in\mathbb{R}^r : Av=\vec{0}\}$. The *left kernel* of A is the linear subspace of $\mathbb{R}^l$ mapped by A^T to the origin: $\mathcal{N}(A^T)=\{v\in\mathbb{R}^l : Av=\vec{0}\}$.

The following theorem lists the cell classes of $C_1 \times \cdots \times C_k$ that can possibly contribute five-dimensional facets to the boundary of the net wrench cone $\mathcal{W}$.

Theorem 15.7 *Let a rigid object $\mathcal{B}$ be contacted by $k \geq 3$ non-collinear frictional contacts. The cell classes of $C_1 \times \cdots \times C_k$ that possibly contribute five-dimensional* **boundary facets** *to the net wrench cone $\mathcal{W}$ are SI/II, SSI, SSS, $SSSS$ and $SSSSS$, provided that the friction cones lie in generic positions relative to the contacts.*

Proof: Based on the *submersion theorem* discussed later, in Chapter 17, a cell $\mathcal{K}$ of $C_1 \times \cdots \times C_k$ is mapped by the grasp map $\mathcal{L}_G$ into the *interior* of $\mathcal{W}$ when the Jacobian matrix, $D\mathcal{L}_G$, has full row rank in this cell. Let us therefore establish that $D\mathcal{L}_G$ has full row rank in each cell class of dimension six or higher that is *not* listed in the theorem.

I. Cell classes with three non-zero contact forces: We have to show that $D\mathcal{L}_G$ has full row rank in the III and SII cell classes. In a cell $\mathcal{K}$ of the III class, the contact forces are parametrized by $(f_1, f_2, f_3)\in\mathbb{R}^9$. Using these parameters, the restriction of the grasp map $\mathcal{L}_G$ to the cell $\mathcal{K}$ forms a linear map, $\mathcal{L}_{\mathcal{K}} : \mathbb{R}^9 \to \mathbb{R}^6$, whose Jacobian is the 6×9 matrix,

$$D\mathcal{L}_{\mathcal{K}} = \begin{bmatrix} I & I & I \\ [x_1\times] & [x_2\times] & [x_3\times] \end{bmatrix} \cong \begin{bmatrix} O & O & I \\ [(x_1-x_3\times] & [(x_2-x_3)\times] & [x_3\times] \end{bmatrix},$$

where I and O are the identity and zero 3×3 matrices (this notation will be used throughout the proof). The Jacobian has a nontrivial left kernel only when $x_2 - x_1$ and $x_3 - x_1$ form collinear vectors in $\mathbb{R}^3$, which occurs only when the three contacts are collinear. Hence, $D\mathcal{L}_{\mathcal{K}}$ has full row rank in any III cell associated with three non-collinear contacts.

Next consider a cell $\mathcal{K}$ of the SII class, which has the form $\mathcal{K} = S_1 \times I_2 \times I_3$. The contact force $f_1 \in S_1$ is parametrized by (λ_1, ϕ_1), while the contact forces $(f_2, f_3)\in I_2 \times I_3$ are parametrized by $\mathbb{R}^6$. Using these parameters, the restriction of the grasp map $\mathcal{L}_G$ to the cell $\mathcal{K}$ forms a linear map, $\mathcal{L}_{\mathcal{K}} : \mathbb{R}^8 \to \mathbb{R}^6$, whose Jacobian is the 6×8 matrix,

$$D\mathcal{L}_{\mathcal{K}} = \begin{bmatrix} \boldsymbol{u}_1(\phi_1) & \lambda_1 \boldsymbol{u}_1'(\phi_1) & I & I \\ x_1 \times \boldsymbol{u}_1(\phi_1) & \lambda_1 x_1 \times \boldsymbol{u}_1'(\phi_1) & [x_2\times] & [x_3\times] \end{bmatrix}.$$

The two left columns of $D\mathcal{L}_{\mathcal{K}}$ are associated with the contact point x_1. These columns span a *flat pencil* tangent to S_1. The remaining six columns of $D\mathcal{L}_{\mathcal{K}}$ are associated with the contact points x_2 and x_3. These columns span two *solid pencils* based at x_2 and x_3. Let Δ be the plane spanned by x_1, x_2 and x_3. Based on Lemma 15.12, one flat pencil at x_1 and two solid pencils at x_2 and x_3 are linearly dependent as vectors in $\mathbb{R}^6$ only when the flat pencil at x_1 coincides with the plane Δ. Hence, $D\mathcal{L}_{\mathcal{K}}$ is rank deficient only when the friction cone C_1 at x_1 is *tangent* to Δ. At a *tame stance*, the friction cone C_1 lies strictly above Δ. At all other stances, such tangency requires non-generic alignment of the friction cone C_1 with the plane Δ.

II. **Cell classes with four non-zero contact forces:** We have to show that $D\mathcal{L}_G$ has full row rank in the cell classes $SSSI$, $SSII$, $SIII$ and IIII. In the $SIII$ and IIII cell classes, $D\mathcal{L}_{\mathcal{K}}$ has full row rank in the generic case where the I contacts do not lie on a common line in $\mathbb{R}^3$. A cell of the $SSII$ class has the form, $\mathcal{K} = S_1 \times S_2 \times I_3 \times I_4$. The restriction of the grasp map $\mathcal{L}_G$ to the cell $\mathcal{K}$ forms a linear map, $\mathcal{L}_{\mathcal{K}} : \mathbb{R}^{10} \to \mathbb{R}^6$, whose Jacobian is the 6×10 matrix,

$$D\mathcal{L}_{\mathcal{K}} = \begin{bmatrix} \boldsymbol{u}_1(\phi_1) & \lambda_1 \boldsymbol{u}_1'(\phi_1) & \boldsymbol{u}_2(\phi_2) & \lambda_2 \boldsymbol{u}_2'(\phi_2) & I & I \\ x_1 \times \boldsymbol{u}_1(\phi_1) & \lambda_1 x_1 \times \boldsymbol{u}_1'(\phi_1) & x_2 \times \boldsymbol{u}_2(\phi_2) & \lambda_2 x_2 \times \boldsymbol{u}_2'(\phi_2) & [x_3 \times] & [x_4 \times] \end{bmatrix}.$$

The columns of $D\mathcal{L}_{\mathcal{K}}$ associated with each SII triplet of the four contacts are linearly independent at a *tame stance* and linearly independent at all other generic stances. Hence, $D\mathcal{L}_{\mathcal{K}}$ has full row rank in the $SSII$ cell class. A cell $\mathcal{K}$ of the $SSSI$ class has the form: $\mathcal{K} = S_1 \times S_2 \times S_3 \times I_4$. The Jacobian $D\mathcal{L}_{\mathcal{K}}$ is the 6×9 matrix,

$$D\mathcal{L}_{\mathcal{K}} = \begin{bmatrix} \boldsymbol{u}_1(\phi_1) & \lambda_1 \boldsymbol{u}_1'(\phi_1) & \boldsymbol{u}_2(\phi_2) & \lambda_2 \boldsymbol{u}_2'(\phi_2) & \boldsymbol{u}_3(\phi_3) & \lambda_3 \boldsymbol{u}_3'(\phi_3) & I \\ x_1 \times \boldsymbol{u}_1(\phi_1) & \lambda_1 x_1 \times \boldsymbol{u}_1'(\phi_1) & x_2 \times \boldsymbol{u}_2(\phi_2) & \lambda_2 x_2 \times \boldsymbol{u}_2'(\phi_2) & x_3 \times \boldsymbol{u}_3(\phi_3) & \lambda_3 x_3 \times \boldsymbol{u}_3'(\phi_3) & [x_4 \times] \end{bmatrix}.$$

If $D\mathcal{L}_{\mathcal{K}}$ is rank deficient, $\text{rank}(D\mathcal{L}_{\mathcal{K}}) = 5$, it has a four-dimensional null space in $\mathbb{R}^9$. Based on the null space dimension, the column subsets associated with *four* contact triplets must simultaneously lose their full row rank. Thus consider the column subsets associated with the contact triplets $S_1 \times S_2 \times S_3$, $S_1 \times S_2 \times I_4$, $S_1 \times S_3 \times I_4$ and $S_2 \times S_3 \times I_4$. Let Δ be the plane spanned by x_1, x_2 and x_3. According to Lemma 15.12, linear dependency of the SSS columns requires that their friction cone tangent planes intersect at a common point, $z \in \Delta$. Lemma 15.12 also states that linear dependency of each SSI column subset requires that the tangent planes to the friction cones at the two S contacts pass through the I contact at x_4. These two facts imply that $\text{rank}(D\mathcal{L}_{\mathcal{K}}) = 5$ only when the tangent planes to the friction cones at x_1, x_2 and x_3 pass through both z and x_4. When the contact point x_4 lies outside Δ (the generic case on uneven terrains), the three tangent planes must intersect along a common line passing through z and x_4. The case where x_4 lies on Δ can be treated as the limit where x_4 approaches Δ from the outside and still requires that the three tangent planes intersect along a common line passing through the point $z = x_4$. The latter condition is a highly non-generic, and $D\mathcal{L}_{\mathcal{K}}$, therefore, has full row rank in the $SSSI$ cell class.

III. **Cell classes with five non-zero contact forces:** We have to show that $D\mathcal{L}_G$ has full row rank in the cell classes $SSSSI$, $SSSII$, $\ldots$, IIIII. In a cell $\mathcal{K}$ containing at least three I contacts, $D\mathcal{L}_{\mathcal{K}}$ has full row rank when the I contacts are not collinear. In the cell class $SSSII$, the columns of $D\mathcal{L}_{\mathcal{K}}$ associated with any SII

triplet are linearly independent at a *tame stance* and linearly independent at all other generic stances. Hence, $D\mathcal{L}_{\mathcal{K}}$ has full row rank in the $SSSII$ cell class. In the cell class $SSSSI$, the columns of $D\mathcal{L}_{\mathcal{K}}$ associated with any $SSSI$ quadruplet have been shown to have full row rank. Hence, $D\mathcal{L}_{\mathcal{K}}$ has full row rank in this cell class.

IV. **Cell classes with six non-zero contact forces:** All of these cell classes, except $SSSSSS$, contain previously considered cell classes where $D\mathcal{L}_G$ has full row rank. In a cell $\mathcal{K}$ of the $SSSSSS$ class, the Jacobian $D\mathcal{L}_{\mathcal{K}}$ forms a 6×12 matrix. If $\text{rank}(D\mathcal{L}_{\mathcal{K}}) = 5$ rather than full row rank, $D\mathcal{L}_{\mathcal{K}}$ has a seven-dimensional null space in $\mathbb{R}^{12}$. Based on the null space dimension, the column subsets associated with *seven* SSS contact triplets must simultaneously lose their full row rank. This condition is captured by seven scalar constraints, associated with rank deficiency of the seven SSS contact triplets. It can be verified that the seven constraints meet transversally in $(\phi_1, \ldots, \phi_6)$ space when the six contacts do not lie on a common plane in $\mathbb{R}^3$ (which is generic in hand-supported stances). As the number of constraints exceeds the $(\phi_1, \ldots, \phi_6)$ space dimension of six, the solution set where $\text{rank}(D\mathcal{L}_{\mathcal{K}}) = 5$ is generically empty. Hence, $D\mathcal{L}_{\mathcal{K}}$ has full row rank in the $SSSSSS$ cell class. When all six contacts lie on a common plane, the seven SSS contact triplets specify only four independent constraints. In this special case, $\text{rank}(D\mathcal{L}_{\mathcal{K}}) = 5$ is captured by three additional scalar constraints in $(\phi_1, \ldots, \phi_6)$ space, which leads to the same conclusion that $D\mathcal{L}_{\mathcal{K}}$ has full row rank in the $SSSSSS$ cell class.

V. **Cell classes with more than six non-zero contact forces:** These cell classes involve more than six active contacts. They contain previously considered cell classes, where $D\mathcal{L}_G$ was shown to have full row rank. Hence, $D\mathcal{L}_G$ has full row rank in all cell classes associated with more than six non-zero contact forces. □

The next proposition characterizes the *critical contact forces* in each cell class listed by Theorem 15.7. The critical contact forces are possibly mapped to the boundary of the net wrench cone $\mathcal{W}$, while all other contact forces are mapped to the interior of $\mathcal{W}$.

Proposition 15.8 *For each cell class listed in Theorem 15.7, the* **critical contact forces** *that are possibly mapped to* $\text{bdy}(\mathcal{W})$ *satisfy the conditions:*

1. **Cell classes SI and II:** *The entire cells consist of critical contact forces.*
2. **Cell class SSI:** *A cell of this class is parametrized by (λ_1, ϕ_1), (λ_2, ϕ_2) and $f_3 \in \mathbb{R}^3$. The critical contact force angles (ϕ_1, ϕ_2) satisfy two scalar equations,*

$$\eta_1(\phi_1) \cdot (x_1 - x_3) = 0 \quad \textit{and} \quad \eta_2(\phi_2) \cdot (x_2 - x_3) = 0, \tag{15.23}$$

where $\eta_i(\phi_i)$ is the normal to the friction cone tangent plane $T_i(\phi_i)$ for $i = 1, 2$.

3. **Cell class SSS:** *A cell of this class is parametrized by (λ_i, ϕ_i) for $i = 1, 2, 3$. Let $(\bar{s}, \bar{t}, \bar{n})$ be a reference frame for the plane Δ spanned by x_1, x_2, x_3. The critical*

contact force angles (ϕ_1, ϕ_2, ϕ_3) *satisfy the scalar equation,*

$$\det\begin{bmatrix} \bar{s}\cdot \boldsymbol{v}_1(\phi_1) & \bar{s}\cdot \boldsymbol{v}_2(\phi_2) & \bar{s}\cdot \boldsymbol{v}_3(\phi_3) \\ \bar{t}\cdot \boldsymbol{v}_1(\phi_1) & \bar{t}\cdot \boldsymbol{v}_2(\phi_2) & \bar{t}\cdot \boldsymbol{v}_3(\phi_3) \\ \bar{\boldsymbol{n}}\cdot (x_1\times \boldsymbol{v}_1(\phi_1)) & \bar{\boldsymbol{n}}\cdot (x_2\times \boldsymbol{v}_2(\phi_2)) & \bar{\boldsymbol{n}}\cdot (x_3\times \boldsymbol{v}_3(\phi_3)) \end{bmatrix} = 0, \quad (15.24)$$

where $\boldsymbol{v}_i(\phi_i) = \bar{\boldsymbol{n}} \times \boldsymbol{\eta}_i(\phi_i)$, $\boldsymbol{\eta}_i(\phi_i)$ *is the normal to* $T_i(\phi_i)$ *for* $i = 1,2,3$.

4. **Cell class** $SSSS$**:** *A cell of this class is parametrized by* (λ_i, ϕ_i) *for* $i = 1\ldots 4$. *Let* $(\bar{s}_i, \bar{t}_i, \bar{\boldsymbol{n}}_i)$ *be a reference frame for the plane spanned by the contacts* x_i, x_{i+1}, x_{i+2}, *where index addition is taken modulo 4. The critical contact force angles* $(\phi_1, \phi_2, \phi_3, \phi_4)$ *satisfy three scalar equations,*

$$\det\begin{bmatrix} \bar{s}_i\cdot \boldsymbol{v}_i(\phi_i) & \bar{s}_i\cdot \boldsymbol{v}_{i+1}(\phi_{i+1}) & \bar{s}_i\cdot \boldsymbol{v}_{i+2}(\phi_{i+2}) \\ \bar{t}_i\cdot \boldsymbol{v}_i(\phi_i) & \bar{t}_i\cdot \boldsymbol{v}_{i+1}(\phi_{i+1}) & \bar{t}_i\cdot \boldsymbol{v}_{i+2}(\phi_{i+2}) \\ \bar{\boldsymbol{n}}_i\cdot (x_i\times \boldsymbol{v}_i(\phi_i)) & \bar{\boldsymbol{n}}_i\cdot (x_{i+1}\times \boldsymbol{v}_{i+1}(\phi_{i+1})) & \bar{\boldsymbol{n}}_i\cdot (x_{i+2}\times \boldsymbol{v}_{i+2}(\phi_{i+2})) \end{bmatrix}$$
$$= 0 \quad i = 1,2,3, \quad (15.25)$$

where $\boldsymbol{v}_i(\phi_i) = \bar{\boldsymbol{n}}_i \times \boldsymbol{\eta}_i(\phi_i)$, $\boldsymbol{\eta}_i(\phi_i)$ *is the normal to* $T_i(\phi_i)$ *for* $i = 1\ldots 4$.

5. **Cell class** $SSSSS$**:** *A cell of this class is parametrized by* (λ_i, ϕ_i) *for* $i = 1\ldots 5$. *Let* $(\bar{s}_i, \bar{t}_i, \bar{\boldsymbol{n}}_i)$ *be a reference frame for the plane spanned by the contacts* x_i, x_{i+1}, x_{i+2}, *where index addition is taken modulo 5. The critical contact force angles* $(\phi_1, \ldots, \phi_5)$ *satisfy five scalar equations,*

$$\det\begin{bmatrix} \bar{s}_i\cdot \boldsymbol{v}_i(\phi_i) & \bar{s}_i\cdot \boldsymbol{v}_{i+1}(\phi_{i+1}) & \bar{s}_i\cdot \boldsymbol{v}_{i+2}(\phi_{i+2}) \\ \bar{t}_i\cdot \boldsymbol{v}_i(\phi_i) & \bar{t}_i\cdot \boldsymbol{v}_{i+1}(\phi_{i+1}) & \bar{t}_i\cdot \boldsymbol{v}_{i+2}(\phi_{i+2}) \\ \bar{\boldsymbol{n}}_i\cdot (x_i\times \boldsymbol{v}_i(\phi_i)) & \bar{\boldsymbol{n}}_i\cdot (x_{i+1}\times \boldsymbol{v}_{i+1}(\phi_{i+1})) & \bar{\boldsymbol{n}}_i\cdot (x_{i+2}\times \boldsymbol{v}_{i+2}(\phi_{i+2})) \end{bmatrix}$$
$$= 0 \quad i = 1\ldots 5, \quad (15.26)$$

where $\boldsymbol{v}_i(\phi_i) = \bar{\boldsymbol{n}}_i \times \boldsymbol{\eta}_i(\phi_i)$, $\boldsymbol{\eta}_i(\phi_i)$ *is the normal to* $T_i(\phi_i)$ *for* $i = 1\ldots 5$.

Proof: Using the grasp map $\mathcal{L}_G$, we have to characterize the critical contact forces satisfying the rank deficiency condition, $\mathrm{rank}(D\mathcal{L}_G) = 5$, in each cell class listed by Theorem 15.7.

I. **Cell classes** SI **and** II**:** All cells of the SI class form five-dimensional sets. Hence, the rank of $D\mathcal{L}_G$ is, at most, five in these cells. A cell $\mathcal{K}$ of the II class is parametrized by $(f_1, f_2) \in \mathbb{R}^6$. Using these parameters, the restriction of $\mathcal{L}_G$ to $\mathcal{K}$, $\mathcal{L}_{\mathcal{K}} : \mathbb{R}^6 \to \mathbb{R}^6$, has the 6×6 Jacobian matrix,

$$D\mathcal{L}_{\mathcal{K}} = \begin{bmatrix} I & I \\ [x_1\times] & [x_2\times] \end{bmatrix} \cong \begin{bmatrix} I & O \\ [x_1\times] & [(x_2 - x_1)\times] \end{bmatrix},$$

where I is the 3×3 identity matrix and O is a 3×3 zero matrix. Clearly, $\mathrm{rank}(D\mathcal{L}_{\mathcal{K}}) = 5$ on the entire II cell.

II. **Cell class** SSI**:** A cell $\mathcal{K}$ of this class has the form $\mathcal{K} = S_1 \times S_2 \times I_3$. Its contact forces (f_1, f_2) are parametrized by (λ_i, ϕ_i) for $i = 1,2$, while f_3 is parametrized by $\mathbb{R}^3$. Using these parameters, the restriction of $\mathcal{L}_G$ to $\mathcal{K}$, $\mathcal{L}_{\mathcal{K}}:\mathbb{R}^7\to\mathbb{R}^6$, has the 6×7 Jacobian matrix,

$$D\mathcal{L}_{\mathcal{K}}=\begin{bmatrix} \boldsymbol{u}_1(\phi_1) & \lambda_1\boldsymbol{u}_1'(\phi_1) & \boldsymbol{u}_2(\phi_2) & \lambda_2\boldsymbol{u}_2'(\phi_2) & I \\ x_1\times\boldsymbol{u}_1(\phi_1) & \lambda_1 x_1\times\boldsymbol{u}_1'(\phi_1) & x_2\times\boldsymbol{u}_2(\phi_2) & \lambda_2 x_2\times\boldsymbol{u}_2'(\phi_2) & [x_3\times] \end{bmatrix}. \tag{15.27}$$

The four columns of $D\mathcal{L}_{\mathcal{K}}$ associated with x_1 and x_2 span two *flat pencils,* which form the tangent planes $T_1(\phi_1)$ and $T_2(\phi_2)$. The columns associated with x_3 span a *solid pencil* based at x_3. According to Lemma 15.12, linear dependency of these pencils as vectors in $\mathbb{R}^6$ requires that the two flat pencils based at x_1 and x_2 pass through the contact point x_3. This condition gives the two scalar equations listed as Condition (15.23).

III. **Cell class SSS:** A cell $\mathcal{K}$ of this class has the form $\mathcal{K}=S_1\times S_2\times S_3$. Its contact forces (f_1,f_2,f_3) are parametrized by (λ_i,ϕ_i) for $i=1,2,3$. Using these parameters, the restriction of $\mathcal{L}_G$ to $\mathcal{K}$, $\mathcal{L}_{\mathcal{K}}:\mathbb{R}^6\to\mathbb{R}^6$ has the 6×6 Jacobian matrix,

$$D\mathcal{L}_{\mathcal{K}}=\begin{bmatrix} \boldsymbol{u}_1(\phi_1) & \lambda_1\boldsymbol{u}_1'(\phi_1) & \boldsymbol{u}_2(\phi_2) & \lambda_2\boldsymbol{u}_2'(\phi_2) & \boldsymbol{u}_3(\phi_3) & \lambda_3\boldsymbol{u}_3'(\phi_3) \\ x_1\times\boldsymbol{u}_1(\phi_1) & \lambda_1 x_1\times\boldsymbol{u}_1'(\phi_1) & x_2\times\boldsymbol{u}_2(\phi_2) & \lambda_2 x_2\times\boldsymbol{u}_2'(\phi_2) & x_3\times\boldsymbol{u}_3(\phi_3) & \lambda_3 x_3\times\boldsymbol{u}_3'(\phi_3) \end{bmatrix}.$$

The columns of $D\mathcal{L}_{\mathcal{K}}$ span three *flat pencils,* which form the friction cone tangent planes $T_1(\phi_1)$, $T_2(\phi_2)$ and $T_3(\phi_3)$. According to Lemma 15.12, these flat pencils are linearly dependent when they intersect at a common point, z, located on the plane Δ spanned by the three contacts. Let $\bar{\boldsymbol{n}}$ denote the unit normal to Δ, and let l_i be the intersection line $l_i=\Delta\cap T_i(\phi_i)$ for $i=1,2,3$. The Plücker coordinates of l_i are given by $(\boldsymbol{v}_i, x_i\times\boldsymbol{v}_i)\in\mathbb{R}^6$, where $\boldsymbol{v}_i(\phi_i)=\bar{\boldsymbol{n}}\times\boldsymbol{\eta}_i(\phi_i)$ is the direction of l_i. Let $(\bar{\boldsymbol{s}},\bar{\boldsymbol{t}},\bar{\boldsymbol{n}})$ be a reference frame for Δ. A standard result from line geometry states that three coplanar lines intersect at a common point, $z\in\Delta$, iff their Plücker coordinates with respect to this plane, $(\bar{\boldsymbol{s}}\cdot\boldsymbol{v}_i,\bar{\boldsymbol{t}}\cdot\boldsymbol{v}_i,\bar{\boldsymbol{n}}\cdot(x_i\times\boldsymbol{v}_i))\in\mathbb{R}^3$, form linearly dependent vectors in $\mathbb{R}^3$:

$$\det\begin{bmatrix} \bar{\boldsymbol{s}}\cdot\boldsymbol{v}_1(\phi_1) & \bar{\boldsymbol{s}}\cdot\boldsymbol{v}_2(\phi_2) & \bar{\boldsymbol{s}}\cdot\boldsymbol{v}_3(\phi_3) \\ \bar{\boldsymbol{t}}\cdot\boldsymbol{v}_1(\phi_1) & \bar{\boldsymbol{t}}\cdot\boldsymbol{v}_2(\phi_2) & \bar{\boldsymbol{t}}\cdot\boldsymbol{v}_3(\phi_3) \\ \bar{\boldsymbol{n}}\cdot(x_1\times\boldsymbol{v}_1(\phi_1)) & \bar{\boldsymbol{n}}\cdot(x_2\times\boldsymbol{v}_2(\phi_2)) & \bar{\boldsymbol{n}}\cdot(x_3\times\boldsymbol{v}_3(\phi_3)) \end{bmatrix}=0,$$

which gives Condition (15.24).

IV. **Cell classes $SSSS$ and $SSSSS$:** A cell $\mathcal{K}$ of the $SSSS$ class has the form $\mathcal{K}=S_1\times S_2\times S_3\times S_4$. Its contact forces (f_1,f_2,f_3,f_4) are parametrized by (λ_i,ϕ_i) for $i=1,2,3,4$. Using these parameters, the restriction of $\mathcal{L}_G$ to $\mathcal{K}$, $\mathcal{L}_{\mathcal{K}}:\mathbb{R}^8\to\mathbb{R}^6$ has a 6×8 Jacobian matrix. The columns of $D\mathcal{L}_{\mathcal{K}}$ form four *flat pencils,* which form the friction cone tangent planes at the contacts. When $\mathrm{rank}(D\mathcal{L}_{\mathcal{K}})=5$, the Jacobian matrix has a three-dimensional null space in $\mathbb{R}^8$. The null space dimension requires that *three* six-column subsets of $D\mathcal{L}_{\mathcal{K}}$, associated with three SSS contact triplets, will simultaneously lose their full row rank. The zero determinant of these contact triplets gives the three scalar equations specified as Condition (15.25). Finally, a cell of the $SSSSS$ class has the form $\mathcal{K}=S_1\times\cdots\times S_5$. Its contact forces $(f_1,\ldots,f_5)$ are parametrized by (λ_i,ϕ_i) for $i=1\ldots5$. Using these parameters, the restriction of $\mathcal{L}_G$ to $\mathcal{K}$, $\mathcal{L}_{\mathcal{K}}:\mathbb{R}^{10}\to\mathbb{R}^6$

has a 6×10 Jacobian matrix. When $\text{rank}(D\mathcal{L}_K) = 5$, the Jacobian matrix has a five-dimensional null space in $\mathbb{R}^{10}$. The null space dimension requires that *five* six-column subsets of $D\mathcal{L}_K$, associated with five SSS contact triplets, will simultaneously lose their full row rank. The zero determinant of these contact triplets gives the five scalar equations specified as Condition (15.26). □

Proposition 15.8 identifies the critical contact forces in a given cell class. However, it does not ensure that these contact forces are mapped to the actual boundary of the net wrench cone $\mathcal{W}$. The next proposition specifies sign conditions, which identify in each cell class the critical contact force that contribute to the actual boundary of $\mathcal{W}$.

Proposition 15.10 *For each cell class listed in Theorem 15.7, the critical contact forces are mapped by the grasp map $\mathcal{L}_G(\vec{f})$ to an* **actual boundary facet** *of the net wrench cone $\mathcal{W}$ iff there exists a sign $\sigma \in \{-1, +1\}$ satisfying the inequality*

$$\sigma \cdot \left(\boldsymbol{w}_i \cdot \boldsymbol{\eta}_H\right) \geq 0 \qquad \text{for all } \boldsymbol{w}_i = (f_i, x_i \times f_i) \text{ such that } f_i \in C_i \text{ for } i = 1 \ldots k, \quad (15.28)$$

where $\boldsymbol{\eta}_H$ has the following form in each cell class.

1. **Cell classes SI/II:** *Set the origin at the contact point x_1. Then $\boldsymbol{\eta}_H = (\vec{0}, x_2 - x_1)$, where x_1 and x_2 are the active contacts in these cell classes.*
2. **Cell class SSI:** *The critical contact forces (f_1, f_2) have fixed angles (ϕ_1^*, ϕ_2^*). Set the origin at the contact point x_3. Then $\boldsymbol{\eta}_H = (\vec{0}, \boldsymbol{v}_{12}(\phi_1^*, \phi_2^*))$, where $\boldsymbol{v}_{12}(\phi_1^*, \phi_2^*) = \boldsymbol{\eta}_1(\phi_1^*) \times \boldsymbol{\eta}_2(\phi_2^*)$ is collinear with the line $T_1(\phi_1^*) \cap T_2(\phi_2^*)$, which passes through x_3.*
3. **Cell classes SSS, $SSSS$, $SSSSS$:** *Let (ϕ_1, ϕ_2, ϕ_3) be critical contact force angles. Set the origin at the intersection point of the friction cone tangent planes, $z \in \Delta$ where Δ is the plane spanned by x_1, x_2 and x_3. Then $\boldsymbol{\eta}_H = (\bar{\boldsymbol{n}}, \bar{\boldsymbol{\tau}})$, where $\bar{\boldsymbol{n}}(\phi_1, \phi_2, \phi_3)$ is the unit normal to Δ, and $\bar{\boldsymbol{\tau}}(\phi_1, \phi_2, \phi_3)$ is given by*

$$\bar{\boldsymbol{\tau}}(\phi_1, \phi_2, \phi_3) = [\boldsymbol{\eta}_1(\phi_1)\, \boldsymbol{\eta}_2(\phi_2)\, \boldsymbol{\eta}_3(\phi_3)]^{-1} \boldsymbol{c},$$

where $\boldsymbol{\eta}_i(\phi_i)$ is the normal to the tangent plane $T_i(\phi_i)$ $(i = 1, 2, 3)$, and $\boldsymbol{c} = (c_1, c_2, c_3)$ are the direction cosines of the friction cone tangent planes $T_1(\phi_1)$, $T_2(\phi_2)$, $T_3(\phi_3)$ with respect to Δ.

Proof: The separating hyperplane normal, $\boldsymbol{\eta}_H$, is collinear with the *left kernel* of the Jacobian matrix $D\mathcal{L}_K$. Hence, we have to identify the left kernel of $D\mathcal{L}_K$ in each cell class listed by Theorem 15.7, where $\mathcal{L}_K$ denotes the restriction of the grasp map $\mathcal{L}_G$ to a cell $\mathcal{K}$. We will use the Jacobian matrices described in the proof of Proposition 15.8.

I. **Cell classes SI and II:** Setting the origin at x_1, $D\mathcal{L}_K$ is 6×5 in an SI cell and 6×6 in an II cell,

$$D\mathcal{L}_K = \begin{bmatrix} \boldsymbol{u}_1(\phi_1) & \lambda_1 \boldsymbol{u}_1'(\phi_1) & I \\ \vec{0} & \vec{0} & [(x_2 - x_1)\times] \end{bmatrix} \quad \text{and} \quad D\mathcal{L}_K = \begin{bmatrix} I & I \\ O & [(x_2 - x_1)\times] \end{bmatrix},$$

where I and and O are 3×3 identity and zero matrices (this notation will be used throughout the proof). In either cell class, the left kernel of $D\mathcal{L}_K$ is spanned by $\eta_H = (\vec{0}, x_2 - x_1)$.

II. **Cell class SSI:** Setting the origin at the I contact point, x_3, $D\mathcal{L}_K$ is the 6×7 matrix,

$$D\mathcal{L}_K = \begin{bmatrix} u_1(\phi_1) & \lambda_1 u_1'(\phi_1) & u_2(\phi_2) & \lambda_2 u_2'(\phi_2) & I \\ (x_1 - x_3) \times u_1(\phi_1) & \lambda_1 (x_1 - x_3) \times u_1'(\phi_1) & (x_2 - x_3) \times u_2(\phi_2) & \lambda_2 (x_2 - x_3) \times u_2'(\phi_2) & O \end{bmatrix}.$$

The last three columns of $D\mathcal{L}_K$ imply that $\eta_H = (\vec{0}, u)$, where u is yet to be determined. The tangent plane $T_i(\phi_i)$ passes through x_3 according to Lemma 15.12 ($i = 1, 2$). Hence, the vector $x_i - x_3$ lies in $T_i(\phi_i)$, while the vectors $(x_i - x_3) \times u_i(\phi_i)$ and $(x_i - x_3) \times u_i'(\phi_i)$ that appear in $D\mathcal{L}_K$ are *orthogonal* to the tangent plane $T_i(\phi_i)$ ($i = 1, 2$). The left kernel of $D\mathcal{L}_K$ is therefore spanned by the vector $\eta_H = (\vec{0}, v_{12})$, where $v_{12} = \eta_1(\phi_1) \times \eta_2(\phi_2)$ is collinear with the intersection line $T_1(\phi_1) \cap T_2(\phi_2)$, which passes through x_3.

III. **Cell class SSS:** The Jacobian $D\mathcal{L}_K$ is the 6×6 matrix,

$$D\mathcal{L}_K = \begin{bmatrix} u_1(\phi_1) & \lambda_1 u_1'(\phi_1) & u_2(\phi_2) & \lambda_2 u_2'(\phi_2) & u_3(\phi_3) & \lambda_3 u_3'(\phi_3) \\ x_1 \times u_1(\phi_1) & \lambda_1 x_1 \times u_1'(\phi_1) & x_2 \times u_2(\phi_2) & \lambda_2 x_2 \times u_2'(\phi_2) & x_3 \times u_3(\phi_3) & \lambda_3 x_3 \times u_3'(\phi_3) \end{bmatrix}.$$

Each pair $\{u_i(\phi_i), u_i'(\phi_i)\}$ spans the friction cone tangent plane $T_i(\phi_i)$ ($i = 1, 2, 3$). The tangent planes $T_1(\phi_1)$, $T_2(\phi_2)$ and $T_3(\phi_3)$ intersect at $z \in \Delta$ according to Lemma 15.12, where Δ is the plane spanned by x_1, x_2 and x_3. Hence, a suitable linear combination of $u_i(\phi_i)$ and $u_i'(\phi_i)$ gives the vector $x_i - z$ ($i = 1, 2, 3$). Recall that $\eta_i(\phi_i)$ is normal to the tangent plane $T_i(\phi_i)$. Another linear combination of $u_i(\phi_i)$ and $u_i'(\phi_i)$ gives the vector $(x_i - z) \times \eta_i(\phi_i)$, which is orthogonal to $x_i - z$ within the tangent plane $T_i(\phi_i)$ ($i = 1, 2, 3$). Selecting the origin at z, the two linear combinations can be used to represent the columns of $D\mathcal{L}_K$ as

$$D\mathcal{L}_K \cong \begin{bmatrix} x_1 - z & (x_1 - z) \times \eta_1(\phi_1) & x_2 - z & (x_2 - z) \times \eta_2(\phi_2) & x_3 - z & (x_3 - z) \times \eta_3(\phi_3) \\ \vec{0} & \|x_1 - z\| \eta_1(\phi_1) & \vec{0} & \|x_2 - z\| \eta_2(\phi_2) & \vec{0} & \|x_3 - z\| \eta_3(\phi_3) \end{bmatrix}.$$

The left kernel of $D\mathcal{L}_K$ is spanned by the vector $(\bar{n}, \bar{\tau})$, where $\bar{n}$ is the unit normal to the plane Δ, and $\bar{\tau} = \left[\eta_1(\phi_1)\, \eta_2(\phi_3)\, \eta_3(\phi_3)\right]^{-1} c$ such that $c = (c_1, c_2, c_3)$ are the direction cosines of the tangent planes $T_1(\phi_1)$, $T_2(\phi_2)$, $T_3(\phi_3)$ with respect to the plane Δ: $c_i = \bar{n} \cdot ((x_i - z) \times \eta_i))/(\|x_i - z\| \cdot \|\eta_i\|)$ for $i = 1, 2, 3$.

IV. **Cell classes $SSSS$ and $SSSSS$:** The Jacobian matrix $D\mathcal{L}_K$ is 6×8 in an $SSSS$ cell, while $D\mathcal{L}_K$ is 6×10 in an $SSSSS$ cell. The columns of $D\mathcal{L}_K$ associated with any SSS triplet must be linearly dependent at the critical contact forces. Hence, we may compute the left kernel of $D\mathcal{L}_K$ using the six columns associated with the contacts x_1, x_2 and x_3. This gives the same left kernel, $\eta_H = (\bar{n}, \bar{\tau})$, computed for the SSS cell class. □

Finally, consider Theorem 15.11, which describes the five types of boundary curves that surround the stance equilibrium region, $\mathcal{ER} \subseteq \mathbb{R}^2$.

Theorem 15.11 *The horizontal cross section of the feasible equilibrium region of a k-contact stance, $\mathcal{ER} \subseteq \mathbb{R}^2$, forms a connected convex set whose boundary consists of up to* **five types** *of curves: SI/II line segments, SSI line segments, SSS convex arcs, $SSSS$ convex arcs and $SSSSS$ line segments, provided that the friction cones lie in generic positions relative to the contacts.*

Proof: For each cell class listed by Theorem 15.7, we have to characterize the set $\mathsf{bdy}(\mathcal{W}) \cap L$ in $\mathcal{B}$'s wrench space and then map this set by the linear map $\tilde{\boldsymbol{x}}_{cm} = -\mathbf{E}(\boldsymbol{e} \times \tau)$ to obtain the curves that contains the boundary curves of the region $\mathcal{ER}$ in $\mathbb{R}^2$.

I. **Cell classes SI and II:** The SI and II cell classes are associated with two non-zero contact forces. These forces can possibly balance the gravitational force only when $\mathcal{B}$'s center of mass lies in the vertical plane containing the contact points x_1 and x_2. The horizontal cross section of this vertical plane forms a line that passes through the projected contacts $\tilde{x}_1$ and $\tilde{x}_2$, which contains the linear SI/II boundary segment of $\mathcal{ER}$ in $\mathbb{R}^2$.

II. **Cell class SSI:** The non-zero contact forces are parametrized by $f_i = \lambda_i \boldsymbol{u}_i(\phi_i)$ for $i = 1, 2$ and $f_3 \in \mathbb{R}^3$. The critical contact forces (f_1, f_2) have a discrete number of solution angles, (ϕ_1^*, ϕ_2^*). These angles are held fixed while the magnitudes $\lambda_1, \lambda_2 > 0$ and f_3 vary freely within the cell. The intersection $\mathsf{bdy}(\mathcal{W}) \cap L$ can be computed by solving Parts (i) and (ii) of Eq. (15.18) in the unknowns $(\lambda_1, \lambda_2, f_3)$:

$$\begin{aligned} &\lambda_1 \boldsymbol{u}_1(\phi_1^*) + \lambda_2 \boldsymbol{u}_2(\phi_2^*) + f_3 = \boldsymbol{e} \\ &\boldsymbol{e} \cdot (\lambda_1 x_1 \times \boldsymbol{u}_1(\phi_1^*) + \lambda_2 x_2 \times \boldsymbol{u}_1(\phi_1^*) + x_3 \times f_3) = 0. \end{aligned} \tag{15.29}$$

The solution of Eq. (15.29) forms an *affine line* in $(\lambda_1, \lambda_2, f_3)$ space. This affine line is mapped to the horizontal plane according to part (iii) of Eq. (15.18),

$$\tilde{\boldsymbol{x}}_{cm} = -\mathbf{E}\big(\boldsymbol{e} \times (\lambda_1 x_1 \times \boldsymbol{u}_1(\phi_1^*) + \lambda_2 x_2 \times \boldsymbol{u}_2(\phi_2^*) + \lambda_3 x_3 \times f_3)\big),$$

which gives the line containing the linear SSI boundary segment of $\mathcal{ER}$ in $\mathbb{R}^2$.

III. **Cell class SSS:** The non-zero contact forces are parametrized by $f_i = \lambda_i \boldsymbol{u}_i(\phi_i)$ for $i = 1, 2, 3$. The critical contact forces satisfy the scalar constraint (15.13) in the angles (ϕ_1, ϕ_2, ϕ_3), while the magnitudes $\lambda_1, \lambda_2, \lambda_3 > 0$ vary freely within the cell. The intersection $\mathsf{bdy}(\mathcal{W}) \cap L$ can be computed by solving parts (i) and (ii) of Eq. (15.18) in the unknowns $(\lambda_1, \lambda_2, \lambda_3)$:

$$\begin{aligned} &\lambda_1 \mathbf{E}\boldsymbol{u}_1(\phi_1) + \lambda_2 \mathbf{E}\boldsymbol{u}_2(\phi_2) + \lambda_3 \mathbf{E}\boldsymbol{u}_3(\phi_3) = \vec{0} \\ &\boldsymbol{e} \cdot \big(\lambda_1 x_1 \times \boldsymbol{u}_1(\phi_1) + \lambda_2 x_2 \times \boldsymbol{u}_2(\phi_2) + \lambda_3 x_3 \times \boldsymbol{u}_3(\phi_3)\big) = 0. \end{aligned} \tag{15.30}$$

The system specified by Eq. (15.30) in $(\lambda_1, \lambda_2, \lambda_3)$ has a nonempty solution set when the determinant of the corresponding 3×3 matrix is zero:

$$\det \begin{bmatrix} \mathbf{E}\boldsymbol{u}_1(\phi_1) & \mathbf{E}\boldsymbol{u}_2(\phi_2) & \mathbf{E}\boldsymbol{u}_3(\phi_3) \\ \boldsymbol{e} \cdot (x_1 \times \boldsymbol{u}_1(\phi_1)) & \boldsymbol{e} \cdot (x_2 \times \boldsymbol{u}_2(\phi_2)) & \boldsymbol{e} \cdot (x_3 \times \boldsymbol{u}_3(\phi_3)) \end{bmatrix} = 0. \tag{15.31}$$

This gives a second scalar constraint in the angles (ϕ_1, ϕ_2, ϕ_3). The solution of both constraints, Eq. (15.13) and Eq. (15.31), forms a curve in (ϕ_1, ϕ_2, ϕ_3)

space. The contact force magnitudes corresponding to each point on this curve are determined by part (i) of the equilibrium stance equation: $\lambda_1 \boldsymbol{u}_1(\phi_1)+\lambda_2 \boldsymbol{u}_2(\phi_2)+\lambda_3 \boldsymbol{u}_3(\phi_3)=\boldsymbol{e}$. The resulting contact forces are mapped to the horizontal plane according to part (iii) of Eq. (15.18): $\tilde{\boldsymbol{x}}_{cm}(\boldsymbol{\phi})=-\mathbf{E}(\boldsymbol{e}\times\sum_{i=1}^{3}\lambda_i(\boldsymbol{\phi})\,x_i\times\boldsymbol{u}_i(\phi_i))$, where $\boldsymbol{\phi}=(\phi_1,\phi_2,\phi_3)$. As $\boldsymbol{\phi}$ varies along the nonlinear solution curve in (ϕ_1,ϕ_2,ϕ_3) space, its image $\tilde{\boldsymbol{x}}_{cm}(\boldsymbol{\phi})$ contains the nonlinear SSS boundary segment of $\mathcal{ER}$ in $\mathbb{R}^2$.

IV. **Cell class $SSSS$:** The non-zero contact forces are parametrized by $f_i=\lambda_i\boldsymbol{u}_i(\phi_i)$ for $i=1,2,3,4$. The critical contact forces form a nonlinear curve in $(\phi_1,\phi_2,\phi_3,\phi_4)$ space. The force magnitudes corresponding to each solution $(\phi_1,\phi_2,\phi_3,\phi_4)$ are determined by the 4×4 linear system in $(\lambda_1,\lambda_2,\lambda_3,\lambda_4)$, consisting of parts (i) and (ii) of Eq. (15.18):

$$\lambda_1\boldsymbol{u}_1(\phi_1)+\lambda_2\boldsymbol{u}_2(\phi_2)+\lambda_3\boldsymbol{u}_3(\phi_3)+\lambda_4\boldsymbol{u}_4(\phi_4)=\boldsymbol{e}$$

$$\boldsymbol{e}\cdot\big(\lambda_1 x_1\times\boldsymbol{u}_1(\phi_1)+\lambda_2 x_2\times\boldsymbol{u}_2(\phi_2)+\lambda_3 x_3\times\boldsymbol{u}_3(\phi_3)+\lambda_4 x_4\times\boldsymbol{u}_4(\phi_4)\big)=0. \tag{15.32}$$

The resulting contact forces are mapped to $\mathbb{R}^2$ according to part (iii) of Eq. (15.18): $\tilde{\boldsymbol{x}}_{cm}(\boldsymbol{\phi})=-\mathbf{E}(\boldsymbol{e}\times\sum_{i=1}^{4}\lambda_i(\boldsymbol{\phi})\,x_i\times\boldsymbol{u}_i(\phi_i))$, where $\boldsymbol{\phi}=(\phi_1,\phi_2,\phi_3,\phi_4)$. As $\boldsymbol{\phi}$ varies along the nonlinear solution curve in $(\phi_1,\phi_2,\phi_3,\phi_4)$ space, its image $\tilde{\boldsymbol{x}}_{cm}(\boldsymbol{\phi})$ contains the nonlinear $SSSS$ boundary segment of $\mathcal{ER}$ in $\mathbb{R}^2$.

V. **Cell class $SSSSS$:** The non-zero contact forces are parametrized by $f_i=\lambda_i\boldsymbol{u}_i(\phi_i)$ for $i=1\ldots5$. The critical contact forces possess a discrete number of solution angles, $(\phi_1^*,\ldots,\phi_5^*)$. The force magnitudes associated with each discrete solution are obtained by solving the 4×5 linear system in $(\lambda_1,\ldots,\lambda_5)$, consisting of parts (i) and (ii) of Eq. (15.18):

$$\begin{aligned}&\lambda_1\boldsymbol{u}_1(\phi_1^*)+\cdots+\lambda_5\boldsymbol{u}_5(\phi_5^*)=\boldsymbol{e}\\&\boldsymbol{e}\cdot\big(\lambda_1 x_1\times\boldsymbol{u}_1(\phi_1^*)+\cdots+\lambda_5 x_5\times\boldsymbol{u}_5(\phi_5^*)\big)=0.\end{aligned} \tag{15.33}$$

The solution of Eq. (15.33) forms an *affine line* in $(\lambda_1,\ldots,\lambda_5)$ space. This affine line is mapped to the horizontal plane according to part (iii) of Eq. (15.18): $\tilde{\boldsymbol{x}}_{cm}(\boldsymbol{\lambda})=-\mathbf{E}(\boldsymbol{e}\times\sum_{i=1}^{5}\lambda_i(\phi^*)\,x_i\times\boldsymbol{u}_i(\phi_i^*))$, where $\boldsymbol{\lambda}=(\lambda_1,\ldots,\lambda_5)$. As $\boldsymbol{\lambda}$ varies along the affine line in $(\lambda_1,\ldots,\lambda_5)$ space, its image $\tilde{\boldsymbol{x}}_{cm}(\boldsymbol{\lambda})$ contains the linear $SSSSS$ boundary segment of $\mathcal{ER}$ in $\mathbb{R}^2$. □

Exercises

Section 15.1

Exercise 15.1: Verify that the net wrench cone at any k-contact stance, $\mathcal{W}$, forms a *convex cone* based at $\mathcal{B}$'s wrench space origin.

Solution: To verify this statement, we must verify that for any two wrenches $\boldsymbol{w}_1,\boldsymbol{w}_2\in\mathcal{W}$, the linear combination $s_1\boldsymbol{w}_1+s_2\boldsymbol{w}_2$ lies in $\mathcal{W}$ for all $s_1,s_2\geq0$. Let f_i' and f_i'' denote the contact forces at x_i associated with the wrenches $\boldsymbol{w}_1$ and $\boldsymbol{w}_2$. The linear combination $s_1\boldsymbol{w}_1 + s_2\boldsymbol{w}_2$ can be written as

$$s_1 \cdot \boldsymbol{w}_1 + s_2 \cdot \boldsymbol{w}_2 = s_1 \cdot \sum_{i=1}^{k} \begin{pmatrix} f_i' \\ x_i \times f_i' \end{pmatrix} + s_2 \cdot \sum_{i=1}^{k} \begin{pmatrix} f_i'' \\ x_i \times f_i'' \end{pmatrix}.$$

The coefficients $s_1, s_2 \geq 0$ can be interpreted as modulating the contact force magnitudes. Hence, $s_1 \cdot \boldsymbol{w}_1 + s_2 \cdot \boldsymbol{w}_2 = \sum_{i=1}^{k}(f_i, x_i \times f_i)$, where $f_i = s_1 f_i' + s_2 f_i''$ for $i = 1 \dots k$. Since each friction cone C_i is a convex set, each force f_i lies in the respective friction cone C_i, which establishes that $\mathcal{W}$ forms a convex cone in $\mathcal{B}$'s wrench space. ○

Exercise 15.2: Explain why the feasible equilibrium region of 3-D a stance supported by two contacts spans a vertical strip in $\mathbb{R}^3$.

Solution: At a 3-D stance supported by two contacts, x_1 and x_2, the net wrench cone $\mathcal{W}$ is five-dimensional in $\mathcal{B}$'s wrench space, since f_1 and f_2 cannot generate torques about the line passing through x_1 and x_2. It can be verified that $\mathcal{W}$ and L are transversal in $\mathcal{B}$'s wrench space as long as the two contacts are not vertically aligned in $\mathbb{R}^3$. Based on the intersection formula mentioned in the proof of Proposition 15.3, the intersection $\mathcal{W} \cap L$ is one-dimensional, since $\dim(\mathcal{W}) + \dim(L) - m = 5 + 2 - 6 = 1$, where m is $\mathcal{B}$'s wrench space dimension. When $\mathcal{W} \cap L$ is a bounded line segment, $\mathcal{R}_{ij}$ is a vertical strip in $\mathbb{R}^3$ embedded in the vertical plane passing through x_1 and x_2. When $\mathcal{W} \cap L$ is a semi-infinite line segment or a full line, $\mathcal{R}_{ij}$ is a semi-infinite vertical plane or a full vertical plane in $\mathbb{R}^3$, again passing through the two contacts. ○

Exercise 15.3*: Verify that the net wrench cone, $\mathcal{W}$, forms a 3-D cone at a single frictional contact stance, a 5-D cone at stances supported by two non-coincident frictional contacts, a full 6-D cone at stances supported by $k \geq 3$ non-collinear contacts (which proves Lemma 15.1).

Section 15.2

Exercise 15.4: Consider a k-contact stance whose contacts lie at the same height in $\mathbb{R}^3$, such that all friction cones contain the vertical upward direction $\boldsymbol{e}$. Explain why the horizontal cross section of the stance's feasible equilibrium region coincides with support polygon, $\mathcal{ER} = \mathcal{P}$.

Exercise 15.5*: A 3-D rigid object $\mathcal{B}$ is supported by $k \geq 3$ frictional contacts against gravity. Explain why $\mathcal{W}$ and L intersect transversally in $\mathcal{B}$'s wrench space when the gravitational wrench can be balanced by at least three contact reaction forces lying strictly in their friction cone interiors.

Section 15.3

Exercise 15.6: A polygonal *inner approximation* of the region $\mathcal{ER} \subseteq \mathbb{R}^2$ can be efficiently computed based on inner approximation of the 3-D friction cones by planar facets. Can a suitable approximation of the 3-D friction cones be used to compute a polygonal *outer approximation* of the feasible equilibrium region, $\mathcal{ER}$, as a series of linear programming problems?

Section 15.4

Exercise 15.7: Consider the critical contact forces in an SSI cell. Explain why Condition (15.12) of Proposition 15.8 implies that the friction cone tangent planes, $T_1(\phi_1^*)$ and $T_2(\phi_2^*)$, must intersect along a line passing through the contact point x_3.

Exercise 15.8: Explain why the critical contact forces in an SSI cell, $(f_1, f_2, f_3) \in S_1 \times S_2 \times I_3$, can have up to four discrete angles, (ϕ_1^*, ϕ_2^*), of the critical contact forces f_1 and f_2.

Exercise 15.9: Verify that the critical contact forces in an SSI cell are mapped by the grasp map $\mathcal{L}_G$ into a five-dimensional hyperplane in $\mathcal{B}$'s wrench space.

Hint: consider the moment generated by the critical contact forces about the line $T_1(\phi_1^*) \cap T_2(\phi_2^*)$.

Exercise 15.10: Consider the critical contact forces in an SSS cell. Explain why Condition (15.13) of Proposition 15.8 implies that the friction cone tangent planes $T_1(\phi_1)$, $T_2(\phi_2)$ and $T_3(\phi_3)$ must intersect at a common point, z, located on the plane Δ spanned by x_1, x_2 and x_3.

Solution sketch: An SSS cell has the form: $\mathcal{K} = S_1 \times S_2 \times S_3$, and its contact forces are parametrized by $f_i = \lambda_i \boldsymbol{u}_i(\phi_i)$ for $i = 1, 2, 3$. Using these parameters, the restriction of $\mathcal{L}_G$ to $\mathcal{K}$, $\mathcal{L}_K : \mathbb{R}^6 \to \mathbb{R}^6$ has the 6×6 Jacobian matrix:

$$D\mathcal{L}_K = \begin{bmatrix} \boldsymbol{u}_1(\phi_1) & \lambda_1 \boldsymbol{u}_1'(\phi_1) & \boldsymbol{u}_2(\phi_2) & \lambda_2 \boldsymbol{u}_2'(\phi_2) & \boldsymbol{u}_3(\phi_3) & \lambda_3 \boldsymbol{u}_3'(\phi_3) \\ x_1 \times \boldsymbol{u}_1(\phi_1) & \lambda_1 x_1 \times \boldsymbol{u}_1'(\phi_1) & x_2 \times \boldsymbol{u}_2(\phi_2) & \lambda_2 x_2 \times \boldsymbol{u}_2'(\phi_2) & x_3 \times \boldsymbol{u}_3(\phi_3) & \lambda_3 x_3 \times \boldsymbol{u}_3'(\phi_3) \end{bmatrix}.$$

As described in the chapter's appendix, the columns of $D\mathcal{L}_K$ span three *flat pencils* whose lines form the friction cone tangent planes $T_1(\phi_1)$, $T_2(\phi_2)$ and $T_3(\phi_3)$. According to Lemma 15.12, these flat pencils are linearly dependent when they intersect at a common point, z, located on the plane Δ spanned by the three contacts. Let $\bar{\boldsymbol{n}}$ denote the unit normal to Δ, and let l_i be the intersection line $l_i = \Delta \cap T_i(\phi_i)$ $(i = 1, 2, 3)$. The Plücker coordinates of l_i are given by $(\boldsymbol{v}_i, x_i \times \boldsymbol{v}_i) \in \mathbb{R}^6$, where $\boldsymbol{v}_i(\phi_i) = \bar{\boldsymbol{n}} \times \boldsymbol{\eta}_i(\phi_i)$ is the ith line direction. Let $(\bar{\boldsymbol{s}}, \bar{\boldsymbol{t}}, \bar{\boldsymbol{n}})$ be a reference frame for Δ. A standard result from line geometry states that three coplanar lines intersect at a common point, $z \in \Delta$, iff their Plücker coordinates with respect to this plane, $(\bar{\boldsymbol{s}} \cdot \boldsymbol{v}_i, \bar{\boldsymbol{t}} \cdot \boldsymbol{v}_i, \bar{\boldsymbol{n}} \cdot (x_i \times \boldsymbol{v}_i)) \in \mathbb{R}^3$, form linearly dependent vectors in $\mathbb{R}^3$:

$$\det \begin{bmatrix} \bar{\boldsymbol{s}} \cdot \boldsymbol{v}_1(\phi_1) & \bar{\boldsymbol{s}} \cdot \boldsymbol{v}_2(\phi_2) & \bar{\boldsymbol{s}} \cdot \boldsymbol{v}_3(\phi_3) \\ \bar{\boldsymbol{t}} \cdot \boldsymbol{v}_1(\phi_1) & \bar{\boldsymbol{t}} \cdot \boldsymbol{v}_2(\phi_2) & \bar{\boldsymbol{t}} \cdot \boldsymbol{v}_3(\phi_3) \\ \bar{\boldsymbol{n}} \cdot (x_1 \times \boldsymbol{v}_1(\phi_1)) & \bar{\boldsymbol{n}} \cdot (x_2 \times \boldsymbol{v}_2(\phi_2)) & \bar{\boldsymbol{n}} \cdot (x_3 \times \boldsymbol{v}_3(\phi_3)) \end{bmatrix} = 0.$$

This scalar equation gives Condition (15.13) specified in Proposition 15.8. ○

Exercise 15.11: Consider the critical contact forces in a cell $\mathcal{K}$ of the $SSSS$ class, described by Condition (15.14) of Proposition 15.8. Obtain a geometric characterization of these critical contact forces.

Hint: Consider the friction cone tangent planes associated with each triplet of the four contacts.

Exercise 15.12*: Consider the critical contact forces in a cell $\mathcal{K}$ of the $SSSS$ class. Explain how Condition (15.14) changes when all four contacts lie in a common plane in $\mathbb{R}^3$.

Hint: Condition (15.14) specifies only two independent constraints. However, the critical contact forces satisfy, in this case, an additional scalar equation and still form a one-dimensional set of angles $(\phi_1, \phi_2, \phi_3, \phi_4)$. ○

Exercise 15.13*: A similar special case occurs in an $SSSSS$ cell when all five contacts lie on a common plane in $\mathbb{R}^3$. Explain how Condition (15.15) for the critical contact forces changes in this case.

Hint: Condition (15.15) specifies only four independent constraints. However, the critical contact forces satisfy three additional scalar equations and still form a discrete solution set of angles $(\phi_1^*, \ldots, \phi_5^*)$. ○

Exercise 15.14: Verify that the critical contact forces vary with five free parameters in each cell class that possibly contributes a five-dimensional facet to $\text{bdy}(\mathcal{W})$. (Hence, these forces generically form a smooth five-dimensional manifold in the respective contact force cell.)

Exercise 15.15: Consider the computation of the separation condition of Eq. (15.17) in the cell classes SSS, $SSSS$ and $SSSSS$. Let $\boldsymbol{w}_i(\phi_i)$ be the wrench generated by full rotation of the unit force $\boldsymbol{u}_i(\phi_i) \in S_i$. Verify that the sign of $\boldsymbol{w}_i(\phi_i) \cdot \boldsymbol{\eta}_H$ can be reduced to checking the extremum values of a trigonometric function $a_i + b_i \cos\phi_i + c_i \sin\phi_i$ (which can be computed in closed form).

Section 15.5

Exercise 15.16: A linear SSI boundary segment of the region $\mathcal{ER}$ is associated with fixed angles (ϕ_1^*, ϕ_2^*) of the contact forces $(f_1, f_2) \in S_1 \times S_2$. Explain why this segment lies on the horizontal projection of the intersection line of the friction cone tangent planes, $T_1(\phi_1^*) \cap T_2(\phi_2^*)$.

Solution: Denote by l the intersection line $T_1(\phi_1^*) \cap T_2(\phi_2^*)$. Since l passes through the contact point x_3, the three contact force lines intersect the line l. The three contact forces thus generate zero torque about l. The gravitational force that acts on $\mathcal{B}$ must generate zero torque about l to maintain torque balance at the equilibrium stance. Hence, $\mathcal{B}$'s center of mass, $\boldsymbol{x}_{cm}$, must lie in the vertical plane containing this line. The horizontal cross section of this vertical plane contains the linear SSI boundary segment of $\mathcal{ER}$. ○

Exercise 15.17: The linear boundary segments of the region $\mathcal{ER}$ originate from the SI/SS, SSI and $SSSSS$ cell classes. Verify that the facets of bdy($\mathcal{W}$) associated with these cell classes are embedded in five-dimensional hyperplanes in $\mathcal{B}$'s wrench space.

Exercise 15.18: Verify that at a *tame* k-contact stance, a linear boundary segment of $\mathcal{ER}$ lies on an edge of the support polygon $\mathcal{P}$ iff the two contacts that determine the edge of $\mathcal{P}$ possess a nonempty vertical equilibrium strip.

Exercise 15.19: Describe a scheme for computing a nonlinear $SSSS$ boundary curve of the region $\mathcal{ER}$ in $\mathbb{R}^2$.

Solution: An $SSSS$ boundary curve of $\mathcal{ER}$ is associated with four non-zero contact forces, $(f_1, f_2, f_3, f_4) \in S_1 \times S_2 \times S_3 \times S_4$, whose angles $(\phi_1, \phi_2, \phi_3, \phi_4)$ satisfy the three constraints specified in Eq. (15.14). The angles that satisfy these constraints generically form a nonlinear curve in this space. The portion of the solution curve that satisfies the separation condition described in Proposition 15.10 maps to a *convex arc* on the boundary of $\mathcal{ER}$. To compute this arc, for each solution $(\phi_1, \phi_2, \phi_3, \phi_4)$, obtain the force magnitudes $(\lambda_1, \lambda_2, \lambda_3, \lambda_4)$ by solving the 4×4 linear system determined by parts (i) and (ii) of Eq. (15.18):

$$\begin{aligned}&\lambda_1 \boldsymbol{u}_1(\phi_1) + \lambda_2 \boldsymbol{u}_2(\phi_2) + \lambda_3 \boldsymbol{u}_3(\phi_3) + \lambda_4 \boldsymbol{u}_4(\phi_4) = \boldsymbol{e}\\ &\boldsymbol{e} \cdot \big(\lambda_1 x_1 \times \boldsymbol{u}_1(\phi_1) + \lambda_2 x_2 \times \boldsymbol{u}_2(\phi_2) + \lambda_3 x_3 \times \boldsymbol{u}_3(\phi_3) + \lambda_4 x_4 \times \boldsymbol{u}_4(\phi_4)\big) = 0.\end{aligned}$$

Every solution satisfying $\lambda_1, \ldots, \lambda_4 \geq 0$ contributes a point $\tilde{\boldsymbol{x}}_{cm}$ on the boundary of $\mathcal{ER}$, located at $\tilde{\boldsymbol{x}}_{cm} = -\mathbf{E}(\boldsymbol{e} \times \sum_{i=1}^{4} \lambda_i(\boldsymbol{\phi})\, x_i \times \boldsymbol{u}_i(\phi_i))$, where $\boldsymbol{\phi} = (\phi_1, \phi_2, \phi_3, \phi_4)$. ○

Exercise 15.20: Let $\eta_H = (\bar{\boldsymbol{n}}, \bar{\tau})$ be the normal to an $SSSSS$ facet of bdy($\mathcal{W}$), expressed with respect to an origin located at the point z described in Proposition 15.10. Verify that the line containing a linear $SSSSS$ boundary segment of $\mathcal{ER}$ satisfies the implicit equation $(\tilde{x}_{cm} - z) \cdot \tilde{\tau} = \tilde{\boldsymbol{n}}$, where $\tilde{\tau} = E\bar{\tau}$ is the projection of $\bar{\tau}$ on the horizontal plane, while $\tilde{\boldsymbol{n}} = \bar{\boldsymbol{n}} \cdot \boldsymbol{e}$ is the projection of $\bar{\boldsymbol{n}}$ along the vertical direction.

Exercise 15.21*: At a frictional three-contact stance, the region $\mathcal{ER}$ is bounded by SI/II and SSI linear segments and nonlinear SSS curves. Explain why each SSS curve must join SI/II or SSI segments at its endpoints (except when the boundary of $\mathcal{ER}$ consists of a single nonlinear SSS closed curve).

Section 15.6

Exercise 15.22: Consider the two non-static motion modes that can develop at 2-D stances, SS and I, when $\mathcal{B}$'s center of mass crosses a vertical boundary line of the stance equilibrium strip (see Chapter 14). Can these motion modes develop when $\mathcal{B}$'s center of mass crosses any of the boundary curves of the region $\mathcal{ER}$ in 3-D stances?

Solution sketch: The crossing of an II boundary segment incurs rolling of the object $\mathcal{B}$ about the line passing through the two I contacts. This 3-D motion mode corresponds to the I motion mode in 2-D stances, where the object $\mathcal{B}$ starts rolling at one contact while all other contacts separate during this motion. The *endpoints* of an SI/II boundary segment in 3-D stances correspond to vertical lines that bound the feasible equilibrium strip associated with two active contacts. Such endpoints may correspond to contact forces that lie on the friction cone surfaces at the two active contacts. When $\mathcal{B}$'s center of mass crosses such endpoints, the object will slip on the two active contacts, thus matching the SS motion mode at the 2-D stances.

Connection to ZMP

Exercise 15.23*: A 3-D rigid object $\mathcal{B}$ with a fixed center of mass is supported under the influence of gravity by a given set of frictional contacts at a feasible equilibrium stance. The contact forces that solve the equilibrium stance equation (15.2) are *not* uniquely determined. Show that the *vertical components* of these forces, aligned with gravity, are uniquely determined by the position of $\mathcal{B}$'s center of mass.

References

[1] Y. Or and E. Rimon, "Analytic characterization of a class of three-contact frictional equilibrium postures in three-dimensional gravitational environments," *International Journal of Robotics Research*, vol. 29, pp. 3–22, 2010.

[2] Y. Or and E. Rimon, "Characterization of frictional multi-contact equilibrium stances and postures on 3-D uneven tame terrains," *International Journal of Robotics Research*, vol. 36, no. 1, pp. 105–108, 2017.

[3] M. Vukobratovic and B. Borovac, "Zero moment point – thirty five years of its life," *International Journal of Humanoid Robotics*, vol. 1, no. 1, pp. 157–173, 2004.

[4] S. Kajita, H. Hirukawa, K. Harada and K. Yokoi, *Introduction to Humanoid Robots*. Springer: Springer Tracts in Advanced Robotics, 2013.

[5] H. Audren and A. Kheddar, "3-D robust stability polyhedron in multicontact," *IEEE Transactions on Robotics*, vol. 34, no. 2, pp. 388–403, 2018.

Part IV

Grasping Mechanisms

16 The Kinematics and Mechanics of Grasping Mechanisms

The previous chapters included only the *fingertips* in the grasp analysis and synthesis, with the grasping mechanism abstracted away to a device that could generate fingertip motions and forces as necessary. This advanced chapter considers the *combined system* of the robot hand mechanism and grasped object. The chapter first develops two basic relationships that govern the combined hand mechanism and grasped object system. The kinematic relation between velocities of the hand's joints and motions of the fingertips, which impart in turn rigid-body velocities to the grasped object, and the dual relation between torques or forces generated at the hand's joints and the net wrenches applied to the grasped object. Using these kinematic relations, the chapter describes *four types* of fingertip contact forces that can possibly be generated by any robot hand mechanism. Based on this understanding of the fingertip forces, we will be able to extend the notion of *wrench resistance* to include the effect of the full hand mechanism on this measure of grasp security.

16.1 The Relation between Finger-Joint Velocities and the Grasped Object Rigid-Body Velocity

A typical robot hand consists of multiple articulated fingers attached to a common base that forms the hand's *palm* (Figure 16.1). The hand is assumed to grasp a rigid object $\mathcal{B}$ through fingertip contacts, with the hand's palm possibly providing an additional passive contact. To derive the relation between the finger-joint velocities and the grasped object rigid-body velocity imparted through the fingertips, we must introduce notation for the finger joints and define reference frames for the palm and finger mechanisms.

The robot hand consists of k serial chain finger mechanisms. The ith finger is articulated by n_i joints, which can be either rotational or prismatic, denoted by the vector $\boldsymbol{\phi}_i = (\phi_{i,1}, \dots, \phi_{i,n_i})$. The composite vector $\Phi = (\boldsymbol{\phi}_1, \dots, \boldsymbol{\phi}_k) \in \mathbb{R}^N$ denotes all joint variables of the k-finger hand, where $N = \sum_{i=1}^{k} n_i$ is the total number of joints. The hand's palm is assigned a *palm frame,* denoted $\mathcal{F}_P$ (Figure 16.1). The ith finger is assigned a *base frame,* $\mathcal{F}_{K_i}$, which is fixed to the palm at the base of the ith finger (Figure 16.1). Each finger ends with a rigid fingertip, O_i, which is assigned a *fingertip frame,* $\mathcal{F}_{O_i}$, attached to the fingertip at the contact point x_i (Figure 16.1). The grasped object $\mathcal{B}$ is assigned a *central object frame,* $\mathcal{F}_B$, as well as *object contact frames* $\mathcal{F}_{C_1}, \dots, \mathcal{F}_{C_k}$

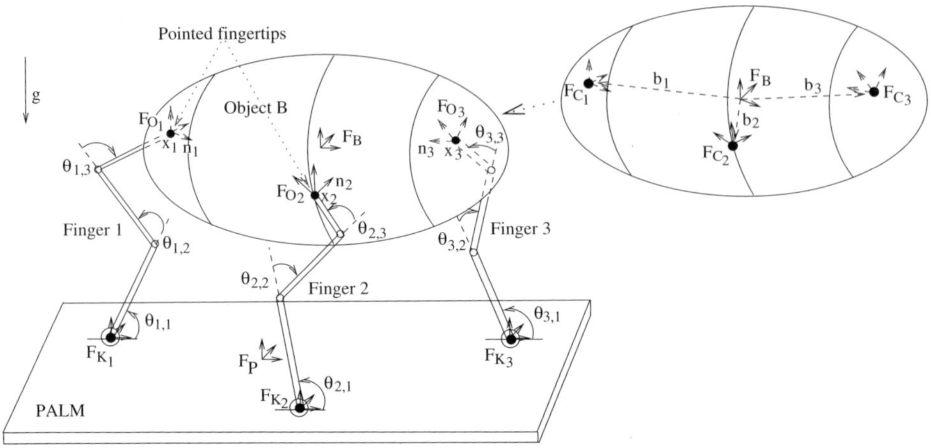

Figure 16.1 Key variables and reference frames for a three-finger robot hand grasping a rigid object $\mathcal{B}$. The palm and finger mechanisms are shown schematically; each finger ends with a pointed fingertip.

attached to $\mathcal{B}$ at the contact points. The ith contact frame, $\mathcal{F}_{C_i}$, is attached to $\mathcal{B}$ at the object contact point b_i, with its z-axis aligned with $\mathcal{B}$'s inward unit normal at this point (Figure 16.1).

When a robot hand holds a rigid object $\mathcal{B}$ via frictional contacts, the fingertips are able to roll without slippage during local finger motions that maintain the grasp. This chapter assumes that each finger ends with a *pointed fingertip* (Figure 16.1). When the finger mechanisms locally move while maintaining frictional contacts with the grasped object, the pointed fingertips will maintain fixed contacts on the object surface. Note that such fingertip contacts effectively form mechanical joints that allow relative angular motion between the fingertips and grasped object (see Exercises).

Keeping the palm fixed, let the finger mechanisms locally move while maintaining their contacts with the grasped object. During such local motions, certain components of the fingertip and object contact frame velocities must match at the contacts in order to maintain the grasp. These *contact compatibility constraints* clearly depend on the contact model that is assumed to hold at the contacts. This chapter will consider the frictionless and frictional hard-point contact models, leaving the case of soft-point contact model as an exercise.

First consider the contact compatibility constraints associated with frictionless contacts. In order to maintain fingertip contacts with the object surface during local motion of the finger mechanisms, the *normal* velocity of the fingertip and object contact frames must match at the contacts. To formulate this constraint, let $Y_i(\phi_i)$ denote the position of the ith fingertip contact point in the fixed palm frame. Let $X_{b_i}(q)$ denote the position of the ith object contact point, b_i, in the fixed palm frame. From Chapter 2, we know that $X_{b_i}(q)$ is given by

$$X_{b_i}(q) = R(\theta)b_i + d \qquad q = (d, \theta).$$

That is, $X_{b_i}(q)$ gives the position of a fixed object point b_i in the palm frame as a function of the grasped-object configuration q. The *contact compatibility constraints* that must be maintained during local motions of the finger mechanisms are given by

$$n_i^T(q)\frac{d}{dt}Y_i\big(\phi_i(t)\big) = n_i^T(q)\frac{d}{dt}X_{b_i}\big(q(t)\big) \qquad i = 1\ldots k, \tag{16.1}$$

where $n_i(q)$ is $\mathcal{B}$'s inward unit normal at x_i, expressed in the palm frame. Equation (16.1) models the fact that contact compatibility requires matching velocities along the object surface normals at the finger contacts.

In the case of frictional contacts, assume that the object is held in a non-marginal equilibrium grasp, with contact forces that point into the interior of the respective friction cones. In such grasps, friction effects ensure that the fingertips will not slide along the object surface during sufficiently local motion of the fingers. Hence, both the *normal and tangential* velocities of the fingertip and object contact frames must match at the contacts. However, the relative angular velocities of the fingertip and object contact frames are *not* constrained. To formulate the contact compatibility constraints, denote by $\{s_i(q), t_i(q), n_i(q)\}$ the unit tangents and normal axes of the ith object contact frame, $\mathcal{F}_{C_i}$, which depend on the object's configuration q. The *contact compatibility constraints* that must be maintained during local motion of the fingers are given by

$$[s_i(q)\, t_i(q)\, n_i(q)]^T \frac{d}{dt}Y_i\big(\phi_i(t)\big) = [s_i(q)\, t_i(q)\, n_i(q)]^T \frac{d}{dt}X_{b_i}\big(q(t)\big) \qquad i = 1\ldots k. \tag{16.2}$$

To turn Equations (16.1) and (16.2) into a concise relation between the finger-joint velocities and the grasped object rigid-body velocity, we need expressions for $\frac{d}{dt}Y_i(\phi_i(t))$ and $\frac{d}{dt}X_{b_i}(q(t))$.

First consider the velocity of the ith pointed fingertip, $Y_i(\phi_i)$, taken along a finger-joint path $\phi_i(t)$. The position of this point in the palm frame can be expressed as

$$Y_i(\phi_i) = Z_{PK_i} + R_{PK_i} Y_{K_i}(\phi_i),$$

where the pair (Z_{PK_i}, R_{PK_i}) describes the position and orientation of the ith finger base frame in the palm frame. Since Z_{PK_i} and R_{PK_i} are constants, application of the chain rule gives

$$\frac{d}{dt}Y_i\big(\phi_i(t)\big) = R_{PK_i} J_i(\phi_i)\,\dot{\phi}_i, \tag{16.3}$$

where $J_i(\phi_i)$ is the $3 \times n_i$ *Jacobian matrix* of the ith finger mechanism. This Jacobian maps the ith finger-joint velocities to the linear velocity of its pointed fingertip in the base frame $\mathcal{F}_{K_i}$. A quick review of this standard Jacobian appears in Appendix I. The linear velocity of the pointed fingertip is subsequently mapped to the palm frame by the rotation matrix R_{PK_i}.

Next, consider the velocity of the ith object contact point, taken along a c-space path $q(t)$ of the grasped object $\mathcal{B}$. Applying the chain rule to $X_{b_i}(q(t))$ while keeping the point b_i fixed on the object surface gives

$$\frac{d}{dt} X_{b_i}(q(t)) = DX_{b_i}(q)\dot{q}, \tag{16.4}$$

where $DX_{b_i}(q)$ is the Jacobian matrix of $X_{b_i}(q)$, and $\dot{q}$ is the rigid-body velocity of $\mathcal{B}$. From Chapter 3, we know that the Jacobian of $X_{b_i}(q)$ is the 3×6 matrix,

$$DX_{b_i}(q) = [I \;\; -[R(\theta)b_i \times]] \qquad q = (d, \theta),$$

where I is a 3×3 identity matrix, $R(\theta)$ is the object 3×3 orientation matrix and $[R(\theta)b_i \times]$ is a 3×3 skew symmetric matrix.

To obtain a concise relation between the finger-joint velocities and the grasped object rigid-body velocity, define the ith *contact frame basis* as the single-column matrix $W_i(q) = [n_i(q)]$ in the case of frictionless contacts, and as the three-column matrix $W_i(q) = [s_i(q)\, t_i(q)\, n_i(q)]$ in the case of frictional contacts. Substituting for the velocities $\frac{d}{dt} Y_i(\phi_i(t))$ and $\frac{d}{dt} X_{b_i}(q(t))$ in the contact compatibility constraints described in Eq. (16.1) and Eq. (16.2), we obtain the relation between the finger-joint velocities and the grasped object rigid-body velocity:

$$W_i^T(q) R_{PK_i} J_i(\phi_i)\, \dot{\phi}_i = W_i^T(q)[I \;\; -[R(\theta)b_i \times]]\dot{q} \qquad i = 1 \ldots k. \tag{16.5}$$

The right side of Eq. (16.5) is next expressed in terms of the $6 \times kp$ *grasp matrix,* $G = [G_1 \cdots G_k]$, where p is the number of independent force components at each contact point (see Chapter 4). At a frictionless contact $p = 1$ and each submatrix G_i is given by

$$G_i(q) = \begin{bmatrix} n_i \\ R(\theta)b_i \times n_i \end{bmatrix} \qquad q = (d, \theta).$$

At a frictional contact $p = 3$ and each submatrix G_i is given by

$$G_i(q) = \begin{bmatrix} s_i & t_i & n_i \\ R(\theta)b_i \times s_i & R(\theta)b_i \times t_i & R(\theta)b_i \times n_i \end{bmatrix} \qquad q = (d, \theta).$$

It follows that each submatrix G_i satisfies the relation:

$$G_i^T(q) = W_i^T(q) DX_{b_i}(q) = W_i^T(q)[I \;\; -[R(\theta)b_i \times]].$$

The constraints that govern the finger joints and grasped object velocities during local motions that maintain the grasp thus take the form

$$\begin{bmatrix} W_1^T(q)R_{PK_1}J_1(\phi_1) & O & \cdots & O \\ \vdots & & \ddots & O \\ O & O & \cdots & W_k^T(q)R_{PK_k}J_k(\phi_k) \end{bmatrix} \begin{bmatrix} \dot{\phi}_1 \\ \vdots \\ \dot{\phi}_k \end{bmatrix} = \begin{bmatrix} G_1^T(q) \\ \vdots \\ G_k^T(q) \end{bmatrix} \dot{q} = G^T(q)\dot{q}, \tag{16.6}$$

where each O is a $p \times n_i$ matrix of zeroes ($i = 1 \ldots k$). The matrix on the left side of Eq. (16.6) forms the *hand Jacobian,* as stated in the following definition.

DEFINITION 16.1 **(hand Jacobian)** *Let a k-finger robot hand positioned at joint values $\Phi = (\phi_1, \ldots, \phi_k) \in \mathbb{R}^N$ hold a rigid object $\mathcal{B}$ at a configuration q. Let p be the* *number of independent force components at each contact point. The* **hand Jacobian,** *denoted $J_H(\Phi, q)$, is the $kp \times N$ matrix*

$$J_H(\Phi,q) = \begin{bmatrix} W_1^T(q)R_{PK_1}J_1(\phi_1) & O & \cdots & O \\ \vdots & & \ddots & O \\ O & O & \cdots & W_k^T(q)R_{PK_k}J_k(\phi_k) \end{bmatrix} \quad \Phi = (\phi_1, \ldots, \phi_k),$$

where $W_i(q)$ forms the ith contact frame basis, R_{PK_i} transforms velocities from the ith finger base frame to the palm frame, and $J_i(\phi_i)$ is the ith finger Jacobian.

Each diagonal block of the hand Jacobian consists of two parts. The first part, $R_{PK_i}J_i(\phi_i)$, characterizes the ith finger ability to instantaneously move its pointed fingertip in physical space while the hand's palm remains stationary. This part depends on the kinematic structure of the ith finger mechanism. The second part, $W_i^T(q)$, describes the ith object contact frame basis as a function of the object's configuration q. The hand Jacobian is illustrated in the following example.

Example: Figure 16.2 shows a 2-D version of a parallel jaw gripper. The gripper forms a simple two-finger hand, actuated by two prismatic joints with linear displacements ϕ_1 and ϕ_2, which occur along a common axis. Each finger ends with a flat fingertip or *jaw.*[1] A rigid object $\mathcal{B}$ is held at a configuration q, such that the object and palm frame axes are parallel at the given grasp. When the gripper maintains frictional point contacts with the object $\mathcal{B}$, there are $p = 2$ independent force components at each contact. The hand Jacobian thus forms the 4×2 matrix (see Exercises):

$$J_H(\Phi,q) = \begin{bmatrix} 0 & 0 \\ 1 & 0 \\ 0 & 0 \\ 0 & 1 \end{bmatrix} \qquad \phi = (\phi_1, \phi_2).$$

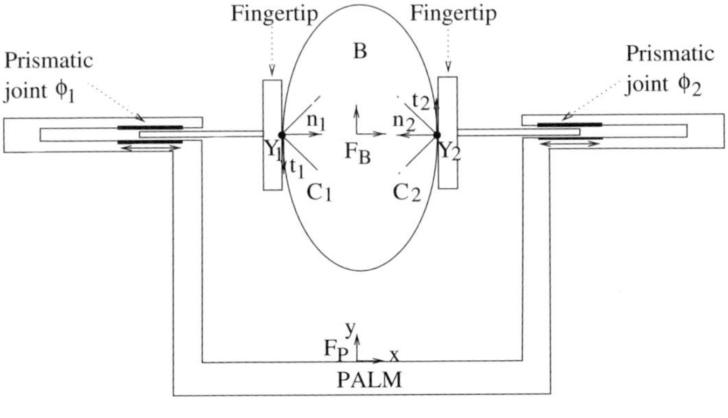

Figure 16.2 A parallel jaw gripper holding a rigid object $\mathcal{B}$ via antipodal frictional contacts.

[1] The gripper's mechanical structure prevents any rolling of $\mathcal{B}$ relative to the fingertips during local motion of the gripper's joints. The pointed fingertips assumption can therefore be relaxed in this example.

The hand Jacobian maps the joint velocities, $(\dot{\phi}_1, \dot{\phi}_2)$, to the fingertip contact point velocities, $(\dot{Y}_1, \dot{Y}_2)$, according to the rule:

$$\begin{pmatrix} \dot{Y}_1 \\ \dot{Y}_2 \end{pmatrix} = J_H(\Phi, q)\begin{pmatrix} \dot{\phi}_1 \\ \dot{\phi}_2 \end{pmatrix} = \begin{bmatrix} 0 & 0 \\ 1 & 0 \\ 0 & 0 \\ 0 & 1 \end{bmatrix}\begin{pmatrix} \dot{\phi}_1 \\ \dot{\phi}_1 \end{pmatrix}.$$

Note that the fingertip contact-point velocities $\dot{Y}_1$ and $\dot{Y}_2$ are expressed with respect to the object's contact frame bases, $\{t_1(q), n_1(q)\}$ and $\{t_2(q), n_2(q)\}$. Thus, $\frac{d}{dt} Y_1 = \dot{\phi}_1 n_1(q)$ and $\frac{d}{dt} Y_2 = \dot{\phi}_2 n_2(q)$ at this grasp. ○

The hand Jacobian forms a block-diagonal matrix whose blocks correspond to the individual finger mechanisms that maintain the grasp. When a grasp involves an additional palm contact, the palm contact has no diagonal block in $J_H(\Phi, q)$, as illustrated in the following example.

A palm and finger example: In Figure 16.3, a single finger actuated by two revolute joints, $\phi_{2,1}$ and $\phi_{2,2}$, acts against the hand's palm. The grasping mechanism holds a rigid object $\mathcal{B}$ via antipodal frictional contacts, such that the object and palm frame axes are parallel at the given grasp. The 4×2 hand Jacobian of this grasp at the indicated angles is given by (see Exercises)

$$J_H(\Phi, q) = \begin{bmatrix} 0 & 0 \\ 0 & 0 \\ 0 & -\sqrt{3}/4 \\ 1 & 1/4 \end{bmatrix}. \tag{16.7}$$

The upper two rows of $J_H(\Phi, q)$ are associated with the palm contact, while the lower two rows form the 2×2 diagonal block associated with the finger mechanism. Since the palm is a simple unarticulated rigid body, its contact point with the object $\mathcal{B}$ remains stationary during local motion of the finger mechanism that maintains the grasp. When an object $\mathcal{B}$ is held by several finger mechanisms against a palm contact, the upper two

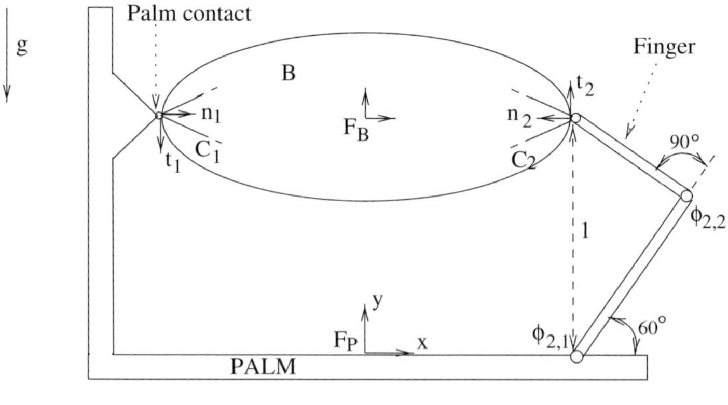

Figure 16.3 A rigid object $\mathcal{B}$ held by a finger and palm via antipodal frictional contacts.

rows of $J_H(\Phi, q)$ form a block of zeroes, while the lower part of $J_H(\Phi, q)$ is block diagonal. ○

When the hand Jacobian is incorporated into the contact compatibility constraints, we obtain the following concise relation between the finger joints and grasped object velocities.

THEOREM 16.1 (**hand-object velocities**) *The* **contact compatibility constraints** *that govern the finger joint velocities, $\dot{\Phi}$, and the grasped object rigid-body velocity, $\dot{q}$, during local motions of the fingers that maintain the grasp are given by*

$$J_H(\Phi, q)\dot{\Phi} = G^T(q)\dot{q}, \tag{16.8}$$

where $J_H(\Phi, q)$ is the hand Jacobian and $G(q)$ is the grasp matrix.

The transposed grasp matrix on the right side of Eq. (16.8), $G^T(q)$, performs two operations. First, it maps the grasped object rigid-body velocity to the object's contact-point velocities expressed in the palm frame. Next, it projects these velocities on the subspaces that must satisfy the compatibility constraints at the contacts. The right side of Eq. (16.8) thus represents the compatible components of the grasped-object velocities at the contacts. The hand Jacobian on the left side of Eq. (16.8), $J_H(\Phi, q)$, also performs two operations. First, it maps the finger-joint velocities to the fingertip contact-point velocities expressed in the palm frame. These velocities are then projected on the subspaces that must satisfy the compatibility constraints at the contacts. The left side of Eq. (16.8) thus represents the compatible components of the fingertip velocities during local motions that maintain the grasp.

Remark: When the fingers are *not* idealized as having pointed fingertips, the contact compatibility constraints must include the effect of fingertip rolling in the case of frictional contacts. As the fingertips roll on the object surface during local motions that maintain the grasp, the contact point locations vary in a manner that depends on the bodies' surface curvature at the contacts (see Bibliographical Notes). ○

16.2 The Relation between Finger-Joint Torques and Grasped Object Wrenches

This section describes the dual relation between the torques or forces applied at the finger joints and the net wrench affected on the grasped object $\mathcal{B}$ through the fingertip contacts. Assume all joints of the robot hand are actuated. Every rotational joint generates control torque about its rotation axis, while every prismatic joint generates control force along its sliding axis. For simplicity, the term *joint torque* will mean either joint torque or force. The torques generated by the joints of the ith finger mechanism are denoted by the vector $\boldsymbol{\tau}_i = (\tau_{i,1}, \ldots, \tau_{i,n_i})$. The composite vector of joint torques of all k fingers is denoted $\mathcal{T} = (\boldsymbol{\tau}_1, \ldots, \boldsymbol{\tau}_k)$. Note that $\mathcal{T}$ acts as a covector on the finger-joint velocities, hence $\mathcal{T} \in T^*_{\Phi}\mathbb{R}^N$.

To derive the dual relation, we will use the *virtual work principle.* When a robot hand holds with stationary fingers a rigid object $\mathcal{B}$ at an equilibrium grasp, the instantaneous work performed by the finger-joint actuators is fully transmitted to the grasped object. The instantaneous work performed by the finger-joint actuators is simply the sum of the power generated at each joint:

$$\mathcal{T} \cdot \dot{\Phi} = \sum_{i=1}^{k} \sum_{j=1}^{n_i} \tau_{i,j} \cdot \dot{\Phi}_{i,j}.$$

The inner product, $\mathcal{T} \cdot \dot{\Phi}$, represents the action of the covector $\mathcal{T}$ on the finger-joint velocities at the equilibrium grasp. The power transferred to the grasped object through the fingertip contacts satisfies a similar relation. Denote by $\boldsymbol{w}$ the net wrench applied by the fingers on the object $\mathcal{B}$. The power transferred to $\mathcal{B}$ through the fingertip contacts is given by the inner product $\boldsymbol{w} \cdot \dot{q}$, where $\dot{q}$ is the object's rigid-body velocity. This inner product represents the action of the covector $\boldsymbol{w}$ on the rigid-body velocities of $\mathcal{B}$ at the equilibrium grasp.

In order to apply the principle of virtual work to the hand mechanism, assume that the mechanical elements of the hand are perfectly rigid so that elastic deformation energy cannot be stored in the links or joints of the grasping mechanism. Under this assumption, the power generated by the finger-joint actuators must equal the power transferred to the grasped object through the fingertip contacts. Let $\boldsymbol{f}_i$ denote the ith fingertip contact force that acts on $\mathcal{B}$ at the given grasp, expressed in the palm frame. The transfer of power takes the form:

$$\mathcal{T} \cdot \dot{\Phi} = \sum_{i=1}^{k} \boldsymbol{f}_i \cdot \frac{d}{dt} Y\big(\phi_i(t)\big) \quad \dot{\Phi} = (\dot{\Phi}_1, \ldots, \dot{\Phi}_k). \tag{16.9}$$

Similarly, assume perfectly rigid fingertips and a perfectly rigid object $\mathcal{B}$, so that elastic energy cannot be stored by material deformations of the fingertips or grasped object. Under this assumption, the power generated by the fingertip contact forces must equal the power transferred to the grasped object:

$$\boldsymbol{w} \cdot \dot{q} = \sum_{i=1}^{k} \boldsymbol{f}_i \cdot \frac{d}{dt} X_{b_i}\big(q(t)\big). \tag{16.10}$$

The contact forces are next expressed in terms of the object contact frame bases. Using the ith contact frame basis $W_i(q) = [n_i(q)]$ at a frictionless contact and $W_i(q) = [s_i(q)\, t_i(q)\, n_i(q)]$ at a frictional contact, $\boldsymbol{f}_i = W_i(q) f_i$ for $f_i \in \mathbb{R}^p$. Let $\vec{f} = (f_1, \ldots, f_k) \in \mathbb{R}^{kp}$ denote the composite vector of all fingertip force components. The following proposition describes how $\vec{f}$ relates to the finger-joint torques, $\mathcal{T}$, as well as to the object net wrench $\boldsymbol{w}$.

PROPOSITION 16.2 **(joint torques and object wrench)** *Let a k-finger robot hand hold with stationary fingers a rigid object $\mathcal{B}$ at an equilibrium grasp. The finger-joint torques, $\mathcal{T}$, and the net object wrench applied by the robot hand, $\boldsymbol{w}$, satisfy the relations,*

$$\mathcal{T} = J_H^T(\Phi, q)\vec{f} \quad \textit{and} \quad \boldsymbol{w} = G(q)\vec{f} \qquad \vec{f} = (f_1, \ldots, f_k), \tag{16.11}$$

where $\vec{f} \in \mathbb{R}^{kp}$ are the fingertip contact force components, $J_H(\Phi, q)$ is the $kp \times N$ hand Jacobian and $G(q)$ is the $6 \times kp$ grasp matrix at the equilibrium grasp.

Proof: First consider the power generated by the finger joints. Substituting for the pointed fingertip velocities according to Eq. (16.3) gives

$$\mathcal{T} \cdot \dot{\Phi} = \sum_{i=1}^{k} \boldsymbol{f}_i \cdot R_{PK_i} J_i(\phi_i)\,\dot{\phi}_i$$

$$= \sum_{i=1}^{k} f_i^T W_i^T(q) R_{PK_i} J_i(\phi_i)\,\dot{\phi}_i \qquad \dot{\Phi} = (\dot{\phi}_1, \ldots, \dot{\phi}_k) \in T_{\Phi}\mathbb{R}^N.$$

Since $\mathcal{T}$ acts as a covector on the tangent space of finger-joint velocities, $T_{\Phi}\mathbb{R}^N$, one concludes that $\mathcal{T} = J_H^T(\Phi, q)\vec{f}$. Similarly, substituting for the object's contact point velocities according to Eq. (16.4) gives

$$\boldsymbol{w} \cdot \dot{q} = \sum_{i=1}^{k} \boldsymbol{f}_i \cdot DX_{b_i}(q)\dot{q} = \sum_{i=1}^{k} f_i^T G_i^T(q)\dot{q} \qquad \dot{q} \in T_q\mathbb{R}^6,$$

where $DX_{b_i}(q) = G_i^T(q)$ is the ith submatrix of the transposed grasp matrix $G^T(q)$. Since $\boldsymbol{w}$ acts as a covector on the tangent space of $\mathcal{B}$'s rigid-body velocities, $T_q\mathbb{R}^6$, one concludes that $\boldsymbol{w} = G(q)\vec{f}$. This relation was in fact used to define the grasp matrix in Chapter 4. □

Proposition 16.2 specifies that the fingertip forces map to finger-joint torques according to the rule $\mathcal{T} = J_H^T(\Phi, q)\vec{f}$. The *same* fingertip forces generate net object wrench according to the rule $\boldsymbol{w} = G(q)\vec{f}$. It is worth noting that the transposed hand Jacobian, $J_H^T(\Phi, q)$, as well as the grasp matrix, $G(q)$, act on the *same* vector space of fingertip force components. Proposition 16.2 can be used to obtain a direct relation between the finger-joint torques and the net object wrench by solving the linear system $\boldsymbol{w} = G(q)\vec{f}$ for $\vec{f}$. When the fingers apply net wrench $\boldsymbol{w}$ on the object $\mathcal{B}$, the solution for $\vec{f}$ takes the general form

$$\vec{f} = G^{\dagger}(q)\boldsymbol{w} + \vec{f}_{null}, \tag{16.12}$$

where $\vec{f}_{null}$ is any vector of internal grasp forces from the null space of $G(q)$ (see Chapter 4). The matrix $G^{\dagger}(q)$ is the *right pseudo-inverse* of $G(q)$. Briefly, $G^{\dagger}(q)$ has the same dimensions as $G^T(q)$. When $G(q)$ has full row rank (see Chapter 12), its right pseudo-inverse is given by the formula $G^{\dagger} = G^T(GG^T)^{-1}$. In this case $G^{\dagger}(q)$ forms the *right inverse* of $G(q)$, since it satisfies the relation $G(q)G^{\dagger}(q) = I$. The following example illustrates the role of $G^{\dagger}(q)$ in a given grasp.

Example: Figure 16.4 shows two fingers holding a rigid ellipse with major axis of length $2L$ via antipodal frictional contacts. Each finger mechanism is articulated by two

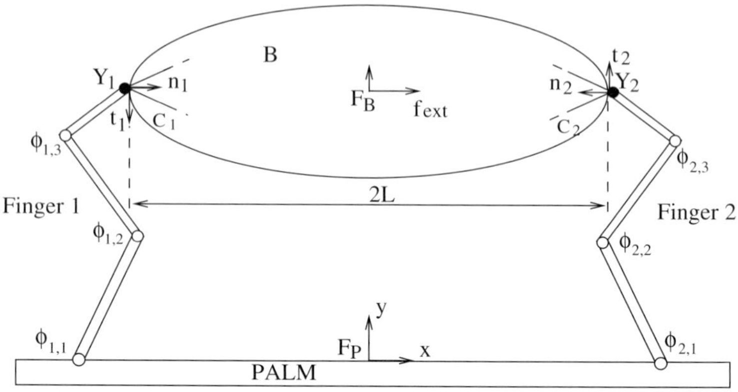

Figure 16.4 A rigid object $\mathcal{B}$ held by two finger mechanisms via antipodal frictional contacts in a horizontal plane without gravity. See text for computation of the grasp matrix right inverse, $G^{\dagger}(q)$, of this grasp.

revolute joints. The object and palm frames axes are parallel at the given grasp. The 3×4 grasp matrix is given by (see Exercises):

$$G(q) = \begin{bmatrix} 0 & 1 & 0 & -1 \\ -1 & 0 & 1 & 0 \\ L & 0 & L & 0 \end{bmatrix}.$$

Since $G(q)$ has full row rank, its right inverse is the 4×3 matrix:

$$G^{\dagger}(q) = G^T(GG^T)^{-1} = \begin{bmatrix} 0 & -1 & L \\ 1 & 0 & 0 \\ 0 & 1 & L \\ -1 & 0 & 0 \end{bmatrix} \begin{bmatrix} 2 & 0 & 0 \\ 0 & 2 & 0 \\ 0 & 0 & 2L^2 \end{bmatrix}^{-1} = \frac{1}{2} \begin{bmatrix} 0 & -1 & 1/L \\ 1 & 0 & 0 \\ 0 & 1 & 1/L \\ -1 & 0 & 0 \end{bmatrix}. \tag{16.13}$$

Note that $G(q)G^{\dagger}(q) = I$, where I is a 3×3 identity matrix. The null space of $G(q)$ consists of internal grasp forces expressed relative to the contact frame bases $\{t_1, n_1\}$ and $\{t_2, n_2\}$. The fingertip contact forces thus satisfy the relation,

$$\vec{f} = G^{\dagger}(q)\boldsymbol{w} + \vec{f}_{null} = \frac{1}{2} \begin{bmatrix} 0 & -1 & 1/L \\ 1 & 0 & 0 \\ 0 & 1 & 1/L \\ -1 & 0 & 0 \end{bmatrix} \boldsymbol{w} + \begin{pmatrix} 0 \\ s \\ 0 \\ s \end{pmatrix} \qquad s \in \mathbb{R}, \tag{16.14}$$

where $\vec{f}$ consists of force components expressed relative to the object contact frame bases. The fingertip contact forces that solve Eq. (16.14) must additionally lie in their respective friction cones. For instance, let a pure horizontal force, f_{ext}, act at the center of the object $\mathcal{B}$. In order to resist this force, the fingers must generate an opposing wrench, $\boldsymbol{w} = (-f_{ext}, 0)$, on the object $\mathcal{B}$. The fingertip contact forces are thus given by

$$\vec{f} = \begin{pmatrix} f_1 \\ f_2 \end{pmatrix} = G^{\dagger}(q)w + \vec{f}_{null} = \begin{pmatrix} 0 \\ -\frac{1}{2}f^x_{ext} + s \\ 0 \\ \frac{1}{2}f^x_{ext} + s \end{pmatrix} \qquad f_{ext} = (f^x_{ext}, 0),\ s \geq \frac{1}{2}|f^x_{ext}|.$$

The inequality $s \geq |f^x_{ext}|/2$ specifies the magnitude of the internal grasp forces that ensure feasibility of the fingertip forces, $f_1 \in C_1$ and $f_2 \in C_2$. ○

The concise relation between the finger-joint torques and the net object wrench is obtained by substituting the general solution for $\vec{f}$ into the equation $\mathcal{T} = J_H^T(\Phi, q)\vec{f}$, which leads to the following theorem.

THEOREM 16.3 (**dual hand-object relation**) *Let a k-finger robot hand hold a rigid object $\mathcal{B}$ at an equilibrium grasp, possibly under the influence of an external wrench such as gravity. The* **finger joint torques,** $\mathcal{T}$, *and the* **net object wrench,** w, *are related by the equation,*

$$\mathcal{T} = J_H^T(\Phi, q)\big(G^{\dagger}(q)w + \vec{f}_{null}\big) \qquad \mathcal{T} \in \mathbb{R}^N, \tag{16.15}$$

where $J_H(\Phi, q)$ is the hand Jacobian, $G^{\dagger}(q)$ is the right pseudo-inverse of the grasp matrix $G(q)$ and $\vec{f}_{null}$ is any vector of internal grasp forces in the null space of $G(q)$.

Theorem 16.3 describes the relation satisfied by three physical processes that underly the *combined system* of the grasping mechanism and grasped object. The finger-joint torques, the net object wrench and the internal grasp forces. The next section analyzes the physical meaning of this three-part relation.

16.3 The Four Types of Hand Mechanism Grasp Forces

A hand mechanism can be used to impart motions to the grasped object or to apply wrenches on the object through fingertip forces. This section analyzes the hand's ability to apply object wrenches. The next chapter will consider the hand's ability to locally manipulate an object. The analysis of this section will use linear algebraic properties of the hand Jacobian, $J_H(\Phi, q)$, and the grasp matrix $G(q)$. To this end, let us recall some basic definitions and facts concerning matrices.

Matrix range and null space: *Let A be an $l \times r$ real matrix. Let $a_i \in \mathbb{R}^l$ denote the ith column of A. The* **range space** *of A, denoted $\mathcal{R}(A)$, is the linear subspace of $\mathbb{R}^l$ spanned by linear combinations of the columns of A:*

$$\mathcal{R}(A) = \Big\{ v \in \mathbb{R}^l \ : \ v = s_1 a_1 + \cdots + s_r a_r;\ s_1, \ldots, s_r \in \mathbb{R} \Big\}.$$

The **null space** *of A, denoted $\mathcal{N}(A)$, is the linear subspace of $\mathbb{R}^r$ mapped by A to the origin:*

$$\mathcal{N}(A) = \Big\{ v \in \mathbb{R}^r \ : \ Av = \vec{0} \Big\}.$$

The null space of A is orthogonal to the rows of A. Hence, $\mathcal{N}(A)$ and $\mathcal{R}(A^T)$ form orthogonal complements in R^r. Next, recall the definition of matrix rank.

Matrix rank: *Let A be an $l \times r$ real matrix. The* **rank** *of A, denoted* $\text{rank}(A)$*, is the dimension of its range space:* $\text{rank}(A) = \dim(\mathcal{R}(A))$*. In general,* $\text{rank}(A) \leq \min\{l, r\}$*. The matrix A has* **full rank** *when* $\text{rank}(A) = \min\{l, r\}$*. The matrix A has* **full row rank** *when $l \leq r$ and* $\text{rank}(A) = l$*.*

The rank of a matrix satisfies the identity $\text{rank}(A) = \text{rank}(A^T)$. Full row rank means that the columns of A span the vector space $\mathbb{R}^l$, and in this case A represents a surjective linear map. The following classical theorem relates the rank of A to the dimension of its null space.

Rank nullity theorem: *Let A be an $l \times r$ matrix with real entries. Then*

$$\dim(\mathcal{N}(A)) + \dim(\mathcal{R}(A)) = r \tag{16.16}$$

where $\dim(\mathcal{N}(A))$ *and* $\dim(\mathcal{R}(A))$ *are the dimensions of the null and range spaces of A.*

Next recall the notion of direct sum of two linear subspaces.

Direct sum: The *sum* of two linear subspaces, U and V, is the linear subspace $U + V = \{u + v : u \in U \text{ and } v \in V\}$. It forms a *direct sum*, denoted $U \oplus V$, when $U \cap V = \{0\}$.

We also need formulas for the orthogonal projections of vectors in $\mathbb{R}^r$ to the null space $\mathcal{N}(A)$, as well as to its orthogonal complement that forms the range space of A^T.

Orthogonal projections: *Let A be an $l \times r$ real matrix. which possesses full row rank and a right inverse $A^\dagger$. The* **orthogonal projection** *of vectors in $\mathbb{R}^r$ to the null space of A, denoted $L_{\mathcal{N}(A)}$, has the form:*

$$L_{\mathcal{N}(A)}(v) = [I - A^\dagger A]v \qquad v \in \mathbb{R}^r.$$

The **orthogonal projection** *of vectors in $\mathbb{R}^r$ to the orthogonal complement of $\mathcal{N}(A)$ is given by $L_{\mathcal{R}(A^T)}(v) = [A^\dagger A]v$ for $v \in \mathbb{R}^r$.*

The linear map $\mathcal{L}_{\mathcal{N}(A)}$ forms an orthogonal projection, since the vector $\mathcal{L}_{\mathcal{N}(A)}(v)$ lies in the null space of A while the difference, $v - \mathcal{L}_{\mathcal{N}(A)}(v) = [A^\dagger A]v$, is orthogonal to $\mathcal{N}(A)$.

In order to identify the different types of forces that can appear at the fingertip contacts, let the robot hand hold with stationary fingers a rigid object $\mathcal{B}$ at an equilibrium grasp, possibly under the influence of an external wrench such as gravity. Let $\boldsymbol{w}$ be the net object wrench applied by the fingers in response to the external wrench. Consider the relation between the finger-joint torques and the net object wrench, specified in Eq. (16.15) of Theorem 16.3:

$$\tau = J_H^T(\Phi, q)(G^\dagger(q)\boldsymbol{w} + \vec{f}_{null}) = J_H^T(\Phi, q)(\vec{f} + \vec{f}_{null}) \qquad \vec{f}_{null} \in \mathcal{N}(G(q)). \tag{16.17}$$

Note that the range space of $G^\dagger(q)$ coincides with the range space of $G^T(q)$. Hence, the notation $\mathcal{R}(G^T(q))$ will be used for the range space of $G^\dagger(q)$. The following four types of fingertip contact forces may exist in Eq. (16.17):

- Resistant grasp forces, $\vec{f} \in \mathcal{R}(G^T(q))$, of two types:
  1. Contact forces $\vec{f} = G^{\dagger}(q)\boldsymbol{w}$ that lie in the null space of the transposed hand Jacobian, $\mathcal{N}\big(J_H^T(\Phi,q)\big)$.
  2. Contact forces $\vec{f} = G^{\dagger}(q)\boldsymbol{w}$ that lie in the orthogonal projection of $\mathcal{N}\big(J_H^T(\Phi,q)\big)^{\perp}$ on the range space $\mathcal{R}(G^T(q))$.
- Internal grasp forces, $\vec{f}_{null} \in \mathcal{N}(G(q))$, of two types:
  1. Contact forces $\vec{f}_{null}$ that lie in the null space of the transposed hand Jacobian, $\mathcal{N}\big(J_H^T(\Phi,q)\big)$.
  2. Contact forces $\vec{f}_{null}$ that lie in the orthogonal projection of $\mathcal{N}\big(J_H^T(\Phi,q)\big)^{\perp}$ on the null space, $\mathcal{N}(G(q))$.

Each of the four contact force types spans a particular linear subspace of $\mathbb{R}^{kp}$ and has a different physical interpretation which is next described. The first two types form *resistant grasp forces.* These forces generate net object wrenches needed in order to resist external influences that may act on $\mathcal{B}$. The *actively controlled* resistant contact forces are defined as follows.

DEFINITION 16.2 (**actively controlled forces**) *Let a k-finger robot hand hold a rigid object $\mathcal{B}$ at an equilibrium grasp characterized by the pair $(J_H(\Phi,q), G(q))$. The grasp is* **actively controlled** *if the net object wrench generated by the hand's fingertip contact forces, $\boldsymbol{w}$, satisfies the condition:*

$$\vec{f} = G^{\dagger}(q)\boldsymbol{w} \quad \text{such that } \vec{f} = L_{\mathcal{R}(G^T)}(\vec{g}) \text{ for some } \vec{g} \in \mathcal{N}(J_H^T(\Phi,q))^{\perp}.$$

Any such vector $\vec{f}$ represents **actively controlled** *fingertip contact forces.*

The actively controlled contact forces are generated by proper coordination of the finger-joint torques at the given grasp. These contact forces span a linear subspace of $\mathbb{R}^{kp}$, denoted U_{act}, which is the image of the subspace $\mathcal{N}(J_H^T(\Phi,q))^{\perp}$ under the orthogonal projection $L_{\mathcal{R}(G^T)}$:

$$U_{act} = L_{\mathcal{R}(G^T)}\big(\mathcal{N}(J_H^T(\Phi,q))^{\perp}\big).$$

The complementary type of resistant contact forces are generated in a passive manner by rigid parts of the hand mechanism. These forces are defined as follows.

DEFINITION 16.3 (**structurally dependent forces**) *Let a k-finger robot hand hold a rigid object $\mathcal{B}$ at an equilibrium grasp characterized by the pair $(J_H(\Phi,q), G(q))$. The grasp is* **structurally dependent** *if the net object wrench generated by the hand's fingertip contact forces, $\boldsymbol{w}$, satisfies the condition:*

$$\vec{f} = G^{\dagger}(q)\boldsymbol{w} \qquad \text{such that } \vec{f} \in \mathcal{N}(J_H^T(\Phi,q)).$$

Any such vector $\vec{f}$ represents **structurally dependent** *fingertip contact forces.*

The structurally dependent contact forces span a linear subspace of $\mathbb{R}^{kp}$, denoted U_{sd}, which is given by the intersection of two linear subspaces:

$$U_{sd} = \mathcal{R}\big(G^T(q)\big) \cap \mathcal{N}\big(J_H^T(\Phi,q)\big).$$

When a rigid object $\mathcal{B}$ is held in a structurally dependent grasp, there exist resistant contact forces, $\vec{f} \in U_{sd}$, which satisfy the relation $\mathcal{T} = J_H^T(\Phi, q)\vec{f} = \vec{0}$. Such contact forces resist external wrenches that may act on $\mathcal{B}$ without any need for torques at the finger joints. Structurally dependent contact forces appear when a robot hand uses its palm as an additional passive contact. When a grasp involves fully actuated finger mechanisms, structurally dependent contact forces appear when the fingers are positioned at singular configurations, as illustrated in the following example.

Example: Figure 16.5 shows a rigid object $\mathcal{B}$ supported against gravity by two finger mechanisms via frictional contacts. The gravitational force f_g must be balanced by feasible fingertip forces at the equilibrium grasp. When the finger mechanisms are folded as shown in Figure 16.5, the contact forces are generated in a passive manner by rigid parts of the finger mechanisms. These are structurally dependent contact forces, since the vector of contact force components, $\vec{f} = (f_1, f_2)$, satisfies the relation $\mathcal{T} = J_H^T(\Phi, q)\vec{f} = \vec{0}$, which means that the finger-joint actuators need not generate any torques in order to maintain the equilibrium grasp. Note that some amount of joint torques is required in order to *stabilize* the grasp against external disturbances that may act on $\mathcal{B}$ (see Exercises). ○

The subspaces U_{act} and U_{sd} represent *pure types* of resistant grasp forces. These two types are complementary and together span the entire set of resistant contact forces, which forms the range space of $G^T(q)$. This property is stated in the following lemma (see Appendix II for a proof).

LEMMA 16.4 (**resistant force decomposition**) *The subspaces of actively controlled and structurally dependent resistant contact forces at a k-finger grasp, U_{act} and U_{sd}, are mutually orthogonal in $\mathbb{R}^{kp}$ and satisfy the relation $U_{act} \oplus U_{sd} = \mathcal{R}(G^T(q))$.*

To summarize, the resistant contact forces that can be generated by any grasping mechanism arise from two different physical sources. One generated actively by the finger-joint actuators, the other generated passively by mechanical parts of the hand

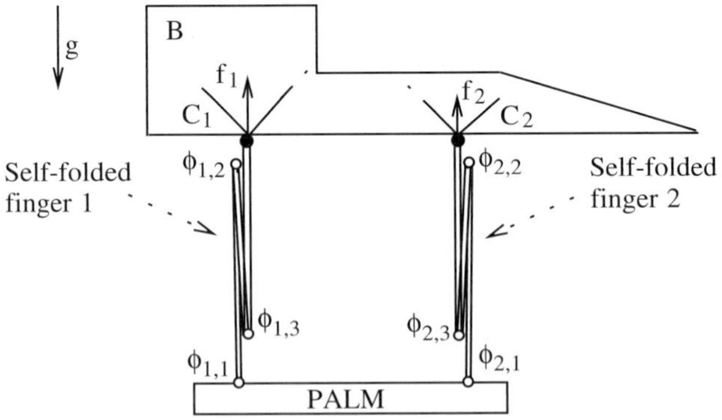

Figure 16.5 A rigid object $\mathcal{B}$ supported by two self-folded finger mechanisms in a structurally dependent grasp. The fingers are able to resist gravity without any need for joint torques.

mechanism. Based on Lemma 16.4, every vector of resistant grasp forces forms a unique combination of actively controlled and structurally dependent forces.

The next two types of contact forces are *internal grasp forces.* These forces assist in maintaining the equilibrium grasp of the object $\mathcal{B}$, but they do not generate any net wrench on $\mathcal{B}$. The actively controlled internal grasp forces are defined as follows.

DEFINITION 16.4 (**active internal forces**) *Let a k-finger robot hand hold a rigid object $\mathcal{B}$ at an equilibrium grasp characterized by the pair $(J_H(\Phi,q), G(q))$. The* **actively controlled** *internal grasp forces are contact forces $\vec{f}_{null} \in \mathcal{N}(G(q))$ that lie in the orthogonal complement of the null space of $J_H^T(\Phi,q)$.*

The actively controlled internal grasp forces can be directly generated and adjusted by the finger-joint actuators. These forces span a linear subspace of $\mathbb{R}^{kp}$, denoted N_{act}, which is the image of the subspace $\mathcal{N}(J_H^T(\Phi,q))^{\perp}$ under the orthogonal projection $L_{\mathcal{N}(G)}$:

$$N_{act} = \mathcal{L}_{\mathcal{N}(G)}\left(\mathcal{N}(J_H^T(\Phi,q))^{\perp}\right).$$

The complementary type of internal grasp forces are generated in a passive manner by rigid parts of the hand mechanism. These forces are defined as follows.

DEFINITION 16.5 (**preloaded internal forces**) *Let a k-finger robot hand hold a rigid object $\mathcal{B}$ at an equilibrium grasp characterized by the pair $(J_H(\Phi,q), G(q))$. If there exist internal grasp forces, $\vec{f}_{null} \in \mathcal{N}(G(q))$, which lie in the null space of $J_H^T(\Phi,q)$, they form* **preloaded** *internal grasp forces.*

The preloaded internal grasp forces span a linear subspace of $\mathbb{R}^{kp}$, denoted N_{pre}, which is given by the intersection of two linear subspaces:

$$N_{pre} = \mathcal{N}\big(G(q)\big) \cap \mathcal{N}\big(J_H^T(\Phi,q)\big).$$

These forces lie in the null space of $J_H^T(\Phi,q)$ and, hence, *cannot* be actively generated by the finger-joint actuators. Instead, such forces must have been preloaded by some other physical processes. For instance, fixturing systems typically require initial preloading of a workpiece when the fixture is first established. Once preloaded internal grasp forces have been established in a given grasping or fixturing system, these forces *cannot* be modified by the finger-joint actuators, as illustrated in the following example.

Example: Figure 16.6 shows a nonstandard parallel jaw gripper actuated by two prismatic joints, ϕ_1 and ϕ_2, which slide along axes *parallel* to the gripper jaws. The gripper is initially preloaded by a horizontal slider mounted on the hand's palm. Once the slider has been preloaded, the gripper applies equal magnitude horizontal forces on $\mathcal{B}$. These preloaded internal grasp forces lie in the null space of the transposed hand Jacobian, $J_H(\Phi,q)$ where $\Phi = (\phi_1, \phi_2)$. The gripper joints are *not* involved in the generation of these forces, and once the initial grasp has been established, the gripper joints cannot change the internal grasp forces applied to $\mathcal{B}$. ○

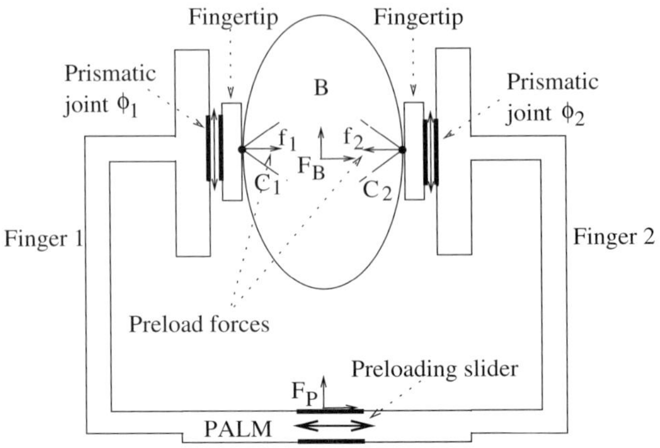

Figure 16.6 A nonstandard parallel jaw gripper holding a rigid object $\mathcal{B}$ via antipodal frictional contacts. The grasp is preloaded by a horizontal slider mounted on the hand's palm.

The subspaces N_{act} and N_{pre} represent *pure types* of internal grasp forces. These two types are complementary and together span the entire null space of the grasp matrix, as stated in the following lemma.

LEMMA 16.5 **(internal force decomposition)** *The subspaces of active and preloaded internal grasp forces, N_{act} and N_{pre}, are mutually orthogonal and satisfy the property: $N_{act} \oplus N_{pre} = \mathcal{N}(G(q))$.*

The proof of Lemma 16.5, which is similar to the proof of Lemma 16.4, is left as an exercise. To summarize, the internal grasp forces that can be generated by any grasping mechanism consist of two types of forces. One generated actively by the hand's joint actuators, the other generated by external loading and transmitted through rigid parts of the hand mechanism. A final insight into the physical meaning of the four types of contact forces is based on the following definition.

DEFINITION 16.6 **(active–passive subspaces)** *The* **active and passive** *contact forces of a k-finger grasping mechanism are the linear subspaces of $\mathbb{R}^{kp}$ given by*

$$\mathcal{A} = U_{act} + N_{act} \quad \textit{and} \quad \mathcal{P} = U_{sd} + N_{pre},$$

where U_{act} and U_{sd} are the subspaces of resistant grasp forces, while N_{act} and N_{pre} are the subspaces of internal grasp forces at the given grasp.

The subspaces of active and passive contact forces, $\mathcal{A}$ and $\mathcal{P}$, collectively span the entire set of contact forces that can be realized by a grasping mechanism at any given grasp. This property is stated in the following theorem.

THEOREM 16.6 **(active–passive contact forces)** *Let a k-finger robot hand hold a rigid object $\mathcal{B}$ at an equilibrium grasp. The active and passive subspaces are mutually disjoint and together span the entire space of the hand's contact forces: $\mathcal{A} \oplus \mathcal{P} = \mathbb{R}^{kp}$.*

Proof sketch: Assume generic grasps that possess a surjective grasp matrix (see Chapter 12). The null space of $G(q)$ and the range space of $G^T(q)$ are orthogonal complements in $\mathbb{R}^{kp}$. Hence, $\mathcal{N}(G(q)) \oplus \mathcal{R}(G^T(q)) = \mathbb{R}^{kp}$. According to Lemma 16.4, $\mathcal{R}(G^T(q)) = U_{act} \oplus U_{sd}$. According to Lemma 16.5, $\mathcal{N}(G(q)) = N_{act} \oplus N_{pre}$. The direct sum is associative. Hence, $\mathcal{N}(G(q)) \oplus \mathcal{R}(G^T(q)) = (U_{act} \oplus N_{act}) \oplus (U_{sd} \oplus N_{pre}) = \mathcal{A} \oplus \mathcal{P} = \mathbb{R}^{kp}$. □

Theorem 16.6 provides the following insight concerning the contact forces that can be realized by any grasping mechanism. The actively controlled resistant contact forces and the actively controlled internal grasp forces collectively form the *active contact forces* of the hand mechanism at a given grasp. These forces can be controlled and adjusted by the finger-joint actuators. The structurally dependent resistant contact forces and the preloaded internal grasp forces collectively form the *passive contact forces* of the hand mechanism at a given grasp. These forces arise from physical sources that do not involve the finger-joint actuators, nor can they be affected by the finger-joint actuators within the rigid-body model. Since $\mathcal{A} \oplus \mathcal{P} = \mathbb{R}^{kp}$, the hand's contact forces form *unique combinations* of active and passive contact forces at any given grasp.

16.4 Effect of the Robot Hand on Wrench-Resistant Grasps

Earlier chapters of the book described the notion of *wrench resistance* in terms of rigid fingertips while ignoring the effect of the robot hand on such grasps. This notion is now termed wrench resistance at the fingertip contacts. A k-finger grasp is *wrench resistant at the fingertip contacts* when every external wrench that may act on the grasped object, $\boldsymbol{w}_{ext}$, can be resisted by some combination of feasible contact forces:

$$G(q)\vec{f} + \boldsymbol{w}_{ext} = \vec{0} \qquad \vec{f} \in C_1 \times \cdots \times C_k$$

where $\vec{f} = (f_1, \ldots, f_k)$ are the hand's contact force components, and $C_1, \ldots, C_k$ are the generalized friction cones at the contacts. In Chapter 12, we saw that a grasp is wrench resistant at the fingertip contacts when the grasp matrix $G(q)$ is surjective, such that there exist internal grasp forces, $\vec{f}_{null}$, which lie in the interior of the respective friction cones. Based on the division of a hand's contact forces into four different subspaces, this section extends the notion of *wrench resistance* to include the hand's ability to realize such secure grasps. First, recall two basic facts concerning matrices.

Injective linear maps: Let A be an $l \times r$ real matrix, which acts on a subspace $V \subseteq \mathbb{R}^r$. The linear map represented by A is **injective** or **one-to-one** if each vector $v \in V$ is mapped to a unique vector in $\mathbb{R}^l$. Equivalently, the linear map is injective if A has only a trivial null space in V.

Full column rank: *Let A be an $l \times r$ real matrix. The matrix A has* **full rank** *when* $\text{rank}(A) = \min\{l, r\}$ *and* **full column rank** *when $r \leq l$ and* $\text{rank}(A) = r$.

We will study two types of wrench resistance that involve the robot hand mechanism. The first type involves fully active finger-joint actuators, as specified in the following definition.

DEFINITION 16.7 (**active wrench resistance**) *A k-finger robot hand holds a rigid object $\mathcal{B}$ at an* **actively wrench resistant grasp** *if every external wrench that acts on $\mathcal{B}$ can be resisted by a feasible combination of actively controlled contact forces, $\vec{f} \in \mathcal{A}$ such that $\vec{f} \in C_1 \times \cdots \times C_k$, where $\mathcal{A}$ is the subspace of active contact forces.*

Active wrench resistance requires fully actuated finger mechanisms. That is, each finger must be able to actively generate the full range of contact forces supported by the friction cones at the respective contacts. In order to develop a test for a hand mechanism to be actively wrench resistant, first note that the grasp must be wrench resistant at the fingertip contacts. In particular, the contacts must form a feasible equilibrium grasp of the object $\mathcal{B}$ in the absence of any external influences. Next consider the grasping mechanism. When an external wrench $\boldsymbol{w}_{ext}$ acts on the grasped object $\mathcal{B}$, the fingers must generate an opposing net wrench, $\boldsymbol{w} = -\boldsymbol{w}_{ext}$, with feasible contact forces of the form $\vec{f} = G^{\dagger}(q)\boldsymbol{w} + \vec{f}_{null}$. The vector $G^{\dagger}(q)\boldsymbol{w}$ is actively generated by the finger-joint actuators when it has no structurally dependent force components. Similarly, the vector $\vec{f}_{null}$ is actively controlled by the finger-joint actuators when it has no preloaded internal force components. These requirements lead to the following test for actively wrench resistant grasps.

THEOREM 16.7 (**active wrench resistance test**) *Let a k-finger robot hand hold a rigid object $\mathcal{B}$ at a grasp characterized by the pair $(J_H(\Phi, q), G(q))$. The grasp is* **actively wrench resistant** *when it satisfies the conditions:*

(i) The grasp matrix $G(q)$ satisfies the wrench resistance conditions at the fingertip contacts.

(ii) The transposed hand Jacobian, $J_H^T(\Phi, q)$, is injective on the space of contact forces $\mathbb{R}^{kp}$.

Proof: When a grasp satisfies Condition (i), every external wrench that acts on $\mathcal{B}$ can be opposed by some combination of feasible contact forces:

$$\vec{f} = G^{\dagger}(q)(-\boldsymbol{w}_{ext}) + \vec{f}_{null} \qquad \vec{f} \in C_1 \times \cdots \times C_k. \tag{16.18}$$

First consider the vector $G^{\dagger}(q)(-\boldsymbol{w}_{ext})$ in Eq. (16.18). Since $J_H^T(\Phi, q)$ is injective on the space of contact forces, $\mathbb{R}^{kp}$, it is injective on the range space of $G^{\dagger}(q)$. An injective linear map has a trivial null space. The subspace of structurally dependent contact forces, U_{sd}, is thus empty at this grasp. Since $\mathcal{R}(G^{\dagger}(q)) = U_{act} \oplus U_{sd}$ according to Lemma 16.4, the vector $\vec{f} = G^{\dagger}(q)(-\boldsymbol{w}_{ext})$ lies in U_{act} and is therefore actively generated by finger-joint torques, $\mathcal{T} = J_H^T(\Phi, q)G^{\dagger}(q)(-\boldsymbol{w}_{ext}) \neq \vec{0}$.

Next consider the vector of internal grasp forces, $\vec{f}_{null}$, in Eq. (16.18). Since $J_H^T(\Phi, q)$ is injective on the space of contact forces, $\mathbb{R}^{kp}$, it is injective on the null space of $G(q)$. The subspace of preloaded internal grasp forces, N_{pre}, is thus empty at this grasp. Since $\mathcal{N}(G(q)) = N_{act} \oplus N_{pre}$ according to Lemma 16.5, $\vec{f}_{null}$ lies in N_{act} and is therefore actively generated by finger-joint torques, $\mathcal{T} = J_H^T(\Phi,q)\vec{f}_{null} \neq \vec{0}$. The combined vector $\vec{f} = G^{\dagger}(q)(-\mathbf{w}_{ext}) + \vec{f}_{null}$ thus lies in the active contact forces subspace, $\mathcal{A} = U_{act} + N_{act}$. □

Theorem 16.7 can be used to identify a large class of robot hands that can realize actively wrench resistant grasps. The transposed hand Jacobian, $J_H^T(\Phi,q)$, is injective when its null space, $\mathcal{N}(J_H^T(\Phi,q))$, is trivial at the given grasp. The $N \times kp$ matrix $J_H^T(\Phi,q)$ satisfies the rank nullity theorem, $\dim(\mathcal{N}(J_H^T)) + \dim(\mathcal{R}(J_H^T)) = kp$. Hence, $J_H^T(\Phi,q)$ has a trivial null space when $\text{rank}(J_H^T(\Phi,q)) = kp$. The matrix $J_H^T(\Phi,q)$ consists of $n_i \times p$ diagonal blocks according to Definition 16.1. When the fingers are positioned at non-singular configurations (the generic case), the rank of $J_H^T(\Phi,q)$ is given by

$$\text{rank}\big(J_H^T(\Phi,q)\big) = \sum_{i=1}^{k} \min\{n_i, p\}.$$

It follows that $J_H^T(\Phi,q)$ has a trivial null space when each finger is articulated by $n_i \geq p$ joints, provided that the fingers do not form degenerate serial chain mechanisms.[2] For instance, a robot hand that grasps 3-D objects via frictional hard-point contacts has $p = 3$ independent force components at each contact. In general, at least three such contacts are required for wrench resistance at the fingertips. Hence, any robot hand having at least three fingers with $n_i \geq 3$ actuated joints in each finger is *actively wrench resistant* at any non-marginal equilibrium grasp. A planar version of such a hand is illustrated in the following example.

Example: The two-finger hand depicted in Figure 16.7 holds a rigid object $\mathcal{B}$ via antipodal frictional contacts. The fingers are actuated by rotational joints $(\phi_{1,1}, \phi_{1,2})$ and

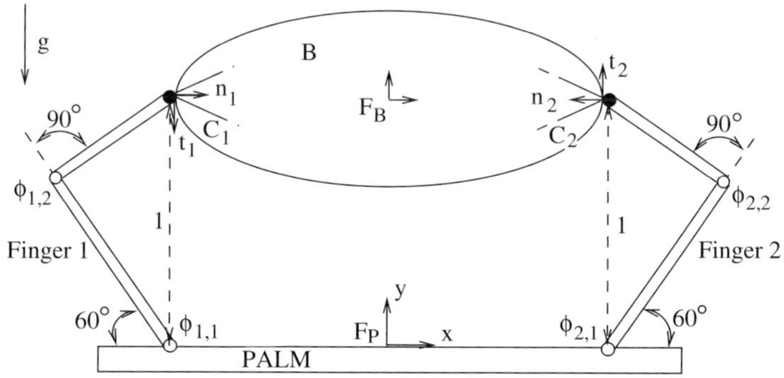

Figure 16.7 A rigid object $\mathcal{B}$ held by two finger mechanisms via antipodal frictional contacts. The grasp is actively wrench resistant, which means that any external wrench such as gravity can be actively resisted by the finger-joint actuators.

[2] A 3-D finger mechanism having three revolute joints with parallel axes is an example of a *degenerate* serial chain mechanism. Its pointed fingertip can locally move only a two-dimensional subset of $\mathbb{R}^3$.

$(\phi_{2,1}, \phi_{2,2})$. For the joint values depicted in Figure 16.7, the hand Jacobian forms the 4×4 block-diagonal matrix (see Exercises):

$$J_H(\Phi, q) = \begin{bmatrix} 0 & -\sqrt{3}/4 & 0 & 0 \\ -1 & -1/4 & 0 & 0 \\ 0 & 0 & 0 & -\sqrt{3}/4 \\ 0 & 0 & 1 & 1/4 \end{bmatrix}.$$

The grasp is wrench resistant at the fingertip contacts as required by Condition (i) of Theorem 16.7. Since $J_H^T(\Phi, q)$ has full rank at the given grasp, it is injective on the space of the hand's contact forces, $\vec{f} = (f_1, f_2)$, which gives Condition (ii) of Theorem 16.7. The hand thus maintains an actively wrench resistant grasp. Physically, any external wrench that acts on $\mathcal{B}$, such as gravity, can be actively resisted with feasible contact forces generated by the finger-joint actuators. ○

Additional insight into the behavior of actively wrench resistant grasps is based on the property that every controlled combination of finger-joint torques exerts a unique net wrench on the grasped object $\mathcal{B}$. This property is stated in the following lemma (see Exercises for a proof).

LEMMA 16.8 (**finger joints to grasped-object action**) *Let a k-finger robot hand hold a rigid object $\mathcal{B}$ at an actively wrench resistant grasp. Each controlled vector of finger-joint torques, $\mathcal{T} \in \mathbb{R}^N$, exerts a* **unique net wrench** *on $\mathcal{B}$, according to the formula,*

$$w = G(q)\left[J_H^T(\Phi, q)\right]^{\dagger} \mathcal{T}, \tag{16.19}$$

where $G(q)$ is the grasp matrix and $[J_H^T]^{\dagger}$ is the left inverse of $J_H^T(\Phi, q)$.

The left inverse of $J_H^T(\Phi, q)$ is well defined since the transposed hand Jacobian is injective according to Theorem 16.7. The left inverse is given by $[J_H^T]^{\dagger} = [J_H J_H^T]^{-1} J_H$. It has the same dimensions as J_H and satisfies the relation $(J_H^T)^{\dagger} J_H^T = I$, where I is an $N \times N$ identity matrix. A robot hand that maintains an actively wrench resistant grasp can thus affect a precise wrench profile on the grasped object during manipulation tasks such as part insertion. This property justifies the commonly used term *precision grasps* for actively wrench resistant grasps.

A complementary notion of wrench resistance allows rigid parts of the hand mechanism to passively generate some resistant contact forces at a given grasp. This notion of *structurally dependent wrench resistance* can lead to simpler hand designs and is defined as follows.

DEFINITION 16.8 (**structural wrench resistance**) *A k-finger robot hand holds a rigid object $\mathcal{B}$ at a* **structurally dependent wrench resistant grasp** *if every external wrench that acts on $\mathcal{B}$ can be resisted by a feasible combination of actively controlled and structurally dependent fingertip forces, $\vec{f} \in \mathcal{A} + \mathcal{P}$, such that $\vec{f} \in C_1 \times \cdots \times C_k$.*

Structural wrench resistance still requires wrench resistance at the fingertip contacts. However, some resistant contact forces, $\vec{f} \in U_{act} \oplus U_{sd}$, can be generated in a passive manner by rigid parts of the hand mechanism. Such contact forces lie in the null space of the transposed hand Jacobian and, therefore, do not require active application of finger-joint torques. However, the null space of $J_H^T(\Phi, q)$ should *not* contain any internal grasp forces, since such forces cannot be modulated by the finger-joint actuators in response to external wrenches that may act on $\mathcal{B}$. These considerations lead to the following test for structurally dependent wrench resistant grasps.

THEOREM 16.9 (**structurally dependent wrench resistance test**) *Let a k-finger robot hand hold a rigid object $\mathcal{B}$ at a grasp characterized by the pair $(J_H(\Phi, q), G(q))$. The grasp is* **structurally dependent wrench resistant** *when it satisfies the conditions:*

(i) The grasp matrix $G(q)$ satisfies the wrench resistance conditions at the fingertip contacts.
(ii) The transposed hand Jacobian, $J_H^T(\Phi, q)$, has a nonempty null space.
(iii) The null space of $J_H^T(\Phi, q)$ does not contain any internal grasp forces.

Proof: Since the grasp is wrench resistant at the fingertip contacts, every $\boldsymbol{w}_{ext}$ that acts on $\mathcal{B}$ can be resisted by some combination of feasible contact forces:

$$\vec{f} = G^{\dagger}(q)(-\boldsymbol{w}_{ext}) + \vec{f}_{null} \qquad \vec{f} \in C_1 \times \cdots \times C_k. \tag{16.20}$$

First consider the vector $G^{\dagger}(q)(-\boldsymbol{w}_{ext})$ in Eq. (16.20). The range space of $G^{\dagger}(q)$ satisfies the relation $\mathcal{R}(G^{\dagger}(q)) = U_{act} \oplus U_{sd}$ according to Lemma 16.4. The vector $G^{\dagger}(q)(-\boldsymbol{w}_{ext})$ thus lies in $U_{act} \oplus U_{sd}$, which implies that $G^{\dagger}(q)(-\boldsymbol{w}_{ext})$ consists of structurally dependent as well as actively controlled contact forces.

Next consider the vector of internal grasp forces, $\vec{f}_{null}$, in Eq. (16.20). Since the null space of $J_H^T(\Phi, q)$ does not contain any internal grasp forces, the subspace of preloaded internal grasp forces, N_{pre}, is empty at this grasp. Since $\mathcal{N}(G(q)) = N_{act} \oplus N_{pre}$ according to Lemma 16.5, $\vec{f}_{null}$ lies in N_{act}, which means that it is actively generated by the finger-joint actuators, with $\mathcal{T} = J_H^T(\Phi, q)\vec{f}_{null} \neq \vec{0}$. The hand mechanism can thus oppose every external wrench that acts on $\mathcal{B}$ with feasible combinations of actively controlled and structurally dependent contact forces. □

In contrast with fully active wrench resistant grasps, structurally dependent wrench resistant grasps do not need fully actuated fingers, which translates to simpler hand designs. A 2-D version of such a simpler hand is illustrated in the following example.

Example: The parallel jaw gripper depicted in Figure 16.8 holds a rigid ellipse $\mathcal{B}$ of width $2W$ via antipodal frictional contacts. The ellipse is located at a configuration q, such that its frame axes are parallel to the palm frame axes. The grasp is wrench resistant at the fingertip contacts. The 3×4 grasp matrix and its null space are given by

$$G(q) = \begin{bmatrix} 0 & 1 & 0 & -1 \\ -1 & 0 & 1 & 0 \\ W & 0 & W & 0 \end{bmatrix} \qquad \text{and} \qquad \mathcal{N}\big(G(q)\big) = \text{span}\left\{ \begin{pmatrix} 0 \\ 1 \\ 0 \\ 1 \end{pmatrix} \right\},$$

where the vector $\vec{f}_{null} = (0, 1, 0, 1)$ is expressed with respect to the contact frame bases $\{t_1, n_1\}$ and $\{t_2, n_2\}$. Since the gripper is actuated by two prismatic joints, its hand Jaco-

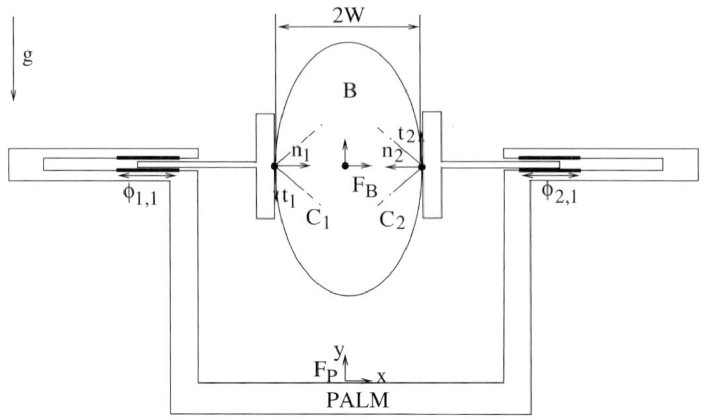

Figure 16.8 A parallel jaw gripper holding a rigid object $\mathcal{B}$ via antipodal frictional contacts. The grasp is structurally dependent wrench resistant, which means that external wrenches such as gravity are resisted by feasible combinations of actively controlled and structurally dependent contact forces.

bian forms a 4×2 matrix. The transposed hand Jacobian and its non empty null space are given by

$$J_H^T(\Phi, q) = \begin{bmatrix} 0 & 1 & 0 & 0 \\ 0 & 0 & 0 & 1 \end{bmatrix} \quad \text{and} \quad \mathcal{N}\big(J_H^T(\Phi, q)\big) = \text{span}\left\{ \begin{pmatrix} 1 \\ 0 \\ 0 \\ 0 \end{pmatrix}, \begin{pmatrix} 0 \\ 0 \\ 1 \\ 0 \end{pmatrix} \right\}.$$

The null space of $G(q)$ consists of horizontal contact forces, while the null space of $J_H^T(\Phi, q)$ consists of vertical contact forces that act along the tangents t_1 and t_2 at the contacts. It follows that $J_H^T(\Phi, q)$ is injective on the null space $\mathcal{N}(G(q))$. Condition (iii) of Theorem 16.9 is thus satisfied, and the gripper holds the object in a structurally dependent wrench resistant grasp. By proper adjustment of its internal grasp forces, the gripper can resist any external wrench that may act on $\mathcal{B}$ with feasible combinations of actively controlled and structurally dependent contact forces. ○

When a robot hand realizes a structurally dependent wrench resistant grasp, it can no longer exert unique net wrenches on the object $\mathcal{B}$. Instead, each controlled vector of finger-joint torques, $\mathcal{T}$, can resist a set of external object disturbances. This property can be seen in the parallel jaw gripper the previous example. For any application of internal grasp forces by the gripper joints, a set of structurally dependent contact forces can develop within the allowed friction cones, which in turn allows the gripper to resist a local neighborhood of external wrenches that may act on $\mathcal{B}$. This property underlies the robustness of industrial grippers. Additional insight into structural wrench resistance can be obtained by studying the following finger-and-palm hand.

Example: The hand depicted in Figure 16.9 consists of a single finger – actuated by two revolute joints, $\phi_{2,1}$ and $\phi_{2,2}$ – which acts against a rigid palm. The hand grasps a rigid

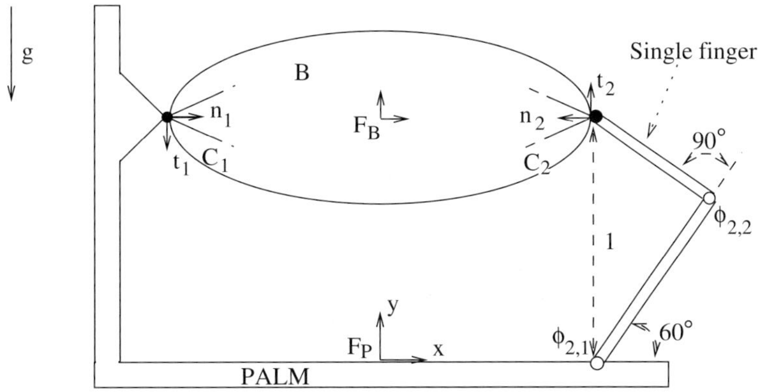

Figure 16.9 A rigid object $\mathcal{B}$ held by a finger-and-palm hand via antipodal frictional contacts. The grasp is structurally dependent wrench resistant.

object $\mathcal{B}$ via antipodal frictional contacts. At the indicated joint values, the transposed hand Jacobian and its null space are given by

$$J_H^T(\Phi,q) = \begin{bmatrix} 0 & 0 & 0 & 1 \\ 0 & 0 & -\sqrt{3}/4 & 1/4 \end{bmatrix} \quad \text{and} \quad \mathcal{N}\big(J_H^T(\Phi,q)\big) = \text{span}\left\{ \begin{pmatrix} 1 \\ 0 \\ 0 \\ 0 \end{pmatrix}, \begin{pmatrix} 0 \\ 1 \\ 0 \\ 0 \end{pmatrix} \right\}.$$

The grasp is wrench resistant at the fingertip contacts, which is Condition (i) of Theorem 16.9. The internal grasp forces are spanned by the vector $\vec{f}_{null} = (0,1,0,1)$, which represents equal magnitude horizontal contact forces. The null space of $J_H^T(\Phi,q)$ is nonempty at this grasp, which is Condition (ii) of Theorem 16.9. Moreover, the null space of $J_H^T(\Phi,q)$ does not contain any internal grasp forces, which is Condition (iii) of Theorem 16.9. The palm-and-finger hand thus maintains a structurally dependent wrench resistant grasp of the object $\mathcal{B}$.

Let us determine the hand's pure structurally dependent contact forces. The single finger can actively exert any contact force consistent with the friction cone at the contact. It follows that the pure structurally dependent contact forces consist only of feasible palm contact forces, $f_1 \in C_1$ such that $f_2 = \vec{0}$. These forces resist external forces that push the object $\mathcal{B}$ against the palm contact point. All other external wrenches that act on $\mathcal{B}$ can be resisted by feasible combinations of actively generated and structurally dependent contact forces. ○

Connection to immobilizing grasps: When a k-finger robot hand holds a rigid object $\mathcal{B}$ via *frictionless contacts,* active wrench resistance consists of two requirements according to Theorem 16.7. The fingertips must hold the object at a *first-order immobilizing grasp,* which ensures wrench resistance at the fingertip contacts. An injective transposed hand Jacobian then ensures that the hand mechanism will be able to actively resist any external wrench that may act on the immobilized object $\mathcal{B}$. However, wrench resistance forms a *first-order notion.* We know that curvature effects can achieve *second-order immobilization* using a smaller number of fingers. Hand mechanisms should thus be able to maintain second-order immobilizing grasps, such that the finger-joint actuators can actively resist any disturbance wrench that may act on the immobilized object $\mathcal{B}$. This notion of *second-order active wrench resistance* is still an open problem in grasp mechanics. ○

Bibliographical Notes

The contact forces that can be generated by any hand mechanism at the fingertips were first analyzed by Bicchi [1]. Bicchi analyzed the contact forces that can arise in compliant finger mechanisms and showed that the space of contact forces can be decomposed into three canonical subspaces. Melchiorri [2] subsequently showed that compliant finger mechanisms contain four canonical contact force subspaces. Readers who wish to further explore this topic are invited to consult the chapter on robot hands in the *Handbook of Robotics* [3]. The same canonical contact force subspaces appear in the robot hands considered in this chapter. The division of a robot hand's contact forces into actively generated and passively generated forces is due to Yoshikawa [4].

The chapter assumed *pointed fingertips* in order to simplify the analysis. Under this assumption, the combined system of the robot hand and grasped object can be interpreted as a parallel chain mechanism, which can lead to further insights. Readers who wish to explore this interpretation are invited to consult the chapter on parallel robots in the *Handbook of Robotics* [5]. When the fingers are *not* idealized as having pointed fingertips, the contact compatibility constraints must be extended to include the effect of rolling at the fingertip contacts. See texts by Mason [6] and Murray, Li and Sastry [7].

The chapter also assumed freely rotating finger joints. However, practical joints can have mechanical limits on their rotation, or they may not be *back drivable* (external torques that act on the joint cannot move the joint). Such considerations add another layer of structural dependency, this time at the finger-joint level. Haas-Heger et al. [8] discuss this topic in hand mechanisms.

Appendix I: The Jacobian of a Single Finger Mechanism

This appendix summarizes a well-known formula for the Jacobian of a serial chain finger mechanism. Recall that $\boldsymbol{\phi}_i = (\phi_{i,1}, \ldots, \phi_{i,n_i})$ denotes the ith finger-joint variables. Also recall that $\mathcal{F}_{K_i}$ is the ith finger base frame fixed on the hand's palm, while $\mathcal{F}_{O_i}$ is the ith fingertip frame located at the contact point with $\mathcal{B}$. The configuration of $\mathcal{F}_{O_i}$ expressed in the finger base frame $\mathcal{F}_{K_i}$ is parametrized by $\mathbb{R}^m$, where $m = 6$ in the case of 3-D finger mechanisms. The finger *full Jacobian,* denoted $J(\boldsymbol{\phi}_i)$, is the $m \times n_i$ matrix that maps the ith finger-joint velocities to the ith fingertip contact frame velocity,

$$\begin{pmatrix} v_i \\ \omega_i \end{pmatrix} = [J(\boldsymbol{\phi}_i)]\dot{\boldsymbol{\phi}}_i,$$

where (v_i, ω_i) is the linear and angular velocity of the ith fingertip contact frame, expressed in $\mathcal{F}_{K_i}$. This chapter assumed pointed fingertips and, hence, used only the *upper part* of $J(\boldsymbol{\phi}_i)$ to obtain the hand Jacobian.

The formula for $J(\boldsymbol{\phi}_i)$ is next written for its individual columns. Let $J(\boldsymbol{\phi}_i) = [\vec{J}_{i,1} \cdots \vec{J}_{i,n_i}]$. We need two quantities that depend on $\boldsymbol{\phi}_i$. Let $p_{i,j}$ denote the position of the ith fingertip frame origin with respect to the axis of the joint $\phi_{i,j}$, described in $\mathcal{F}_{K_i}$. Let $z_{i,j}$ denote a unit vector collinear with the axis of the joint $\phi_{i,j}$, described in $\mathcal{F}_{K_i}$. The unit vector $z_{i,j}$ is collinear with the joint rotation axis at a revolute joint and collinear with the joint sliding axis at a prismatic joint. Using this notation, the jth column of the Jacobian matrix is given by

$$\vec{J}_{i,j}(\boldsymbol{\phi}_i) = \begin{cases} \begin{pmatrix} -p_{i,j}(\boldsymbol{\phi}_i) \times z_{i,j}(\boldsymbol{\phi}_i) \\ z_{i,j}(\boldsymbol{\phi}_i) \end{pmatrix} & \phi_{i,j} \text{ revolute joint} \\ \\ \begin{pmatrix} z_{i,j}(\boldsymbol{\phi}_i) \\ \vec{0} \end{pmatrix} & \phi_{i,j} \text{ prismatic joint} \end{cases} \tag{16.21}$$

where $z_{i,j} = (0, 0, 1)$ in the case of a rotational joint of a planar finger mechanism.

Example: In the planar two-finger hand depicted in Figure 16.5, each finger is articulated by three rotational joints, $\boldsymbol{\phi}_i = (\phi_{i,1}, \phi_{i,2}, \phi_{i,3})$ for $i = 1, 2$. Let the ith finger links have lengths l_{i1}, l_{i2} and l_{i3}. Using Formula (16.21), the 3×3 Jacobian matrix of the ith finger mechanism is given by

$$J(\boldsymbol{\phi}_i) = \begin{bmatrix} -l_{i1}\sin(\phi_{i,1}) & -l_{i2}\sin(\phi_{i,1} + \phi_{i,2}) & -l_{i3}\sin(\phi_{i,1} + \phi_{i,2} + \phi_{i,3}) \\ l_{i1}\cos(\phi_{i,1}) & l_{i2}\cos(\phi_{i,1} + \phi_{i,2}) & l_{i3}\cos(\phi_{i,1} + \phi_{i,2} + \phi_{i,3}) \\ 1 & 1 & 1 \end{bmatrix} \quad i = 1, 2.$$

Note that $J(\boldsymbol{\phi}_i)$ loses its full rank when the three links of the ith finger are self-folded or fully stretched. ○

Appendix II: Resistant Contact Force Decomposition

This appendix contains a proof of Lemma 16.4, which describes the decomposition of the hand's *resistant contact forces* in terms of the subspaces U_{act} and U_{sd}.

Lemma 16.4 *The subspaces of actively controlled and structurally dependent resistant contact forces at a k-finger grasp, U_{act} and U_{sd}, are mutually orthogonal in $\mathbb{R}^{kp}$ and satisfy the relation, $U_{act} \oplus U_{sd} = \mathcal{R}(G^T(q))$.*

Proof sketch: Assume generic grasps whose $m \times kp$ grasp matrix, G, is surjective (see Chapter 12). In this case, the range space of G^T forms an m-dimensional linear subspace of $\mathbb{R}^{kp}$. First, consider the orthogonality of U_{act} and U_{sd}. Any vector $\vec{f} \in U_{act}$ is given by $\vec{f} = L_{\mathcal{R}(G^T)}(\vec{g}) = G^{\dagger}G\vec{g}$ for some $\vec{g} \in \mathcal{N}(J_H^T)^{\perp}$. It can be equivalently written as $\vec{f} = \vec{g} - [I - G^{\dagger}G]\vec{g}$. Let $\vec{h} = [I - G^{\dagger}G]\vec{g}$. Since $GG^{\dagger} = I$, one obtains that $G\vec{h} = \vec{0}$. Hence, $\vec{h}$ lies in the null space of G or, equivalently, $\vec{h} \in \mathcal{R}(G^T)^{\perp}$. Thus, $\vec{f} = \vec{g} - \vec{h}$, such that $\vec{g} \in \mathcal{N}(J_H^T)^\perp$ and $\vec{h} \in \mathcal{R}(G^T)^\perp$. Since $U_{sd} = \mathcal{R}(G^T) \cap \mathcal{N}(J_H^T)$, any vector in U_{sd} is orthogonal to *both vectors* $\vec{g}$ (since $\vec{g} \in \mathcal{N}(J_H^T)^\perp$) and $\vec{h}$ (since $\vec{h} \in \mathcal{R}(G^T)^\perp$). The subspaces U_{act} and U_{sd} are thus mutually orthogonal in $\mathbb{R}^{kp}$.

The direct sum $U_{act} \oplus U_{sd}$ spans the entire range space of G^T when $\dim(U_{act}) + \dim(U_{sd}) = m$. There are two cases to consider. In the first case, the dimension of $\mathcal{N}(J_H^T)^\perp$ is at least m. In this case, $L_{\mathcal{R}(G^T)}$ maps this subspace onto the range space of G^T, so that $\mathcal{R}(G^T) = U_{act}$ and $U_{sd} = \{0\}$. In the second case, the dimension of $\mathcal{N}(J_H^T)^\perp$ is less than m. The mapping $L_{\mathcal{R}(G^T)}$ generically preserves the dimension of its domain space. Hence, $\dim(U_{act}) = \dim(\mathcal{N}(J_H^T)^\perp) < m$. By construction, $U_{sd} = \mathcal{R}(G^T) \cap \mathcal{N}(J_H^T)$. Generically, $\mathcal{R}(G^T)$ and $\mathcal{N}(J_H^T)$ intersect in a transversal manner, and in this case, $\dim(U_{sd}) = \dim(\mathcal{R}(G^T)) + \dim(\mathcal{N}(J_H^T)) - kp$. Substituting $\dim(\mathcal{R}(G^T)) = m$ and $\dim(\mathcal{N}(J_H^T)) = kp - \dim(U_{act})$ gives

$$\dim(U_{sd}) = m + \big(kp - \dim(U_{act})\big) - kp = m - \dim(U_{act}).$$

Hence, $\dim(U_{act}) + \dim(U_{sd}) = m$, which implies that $U_{act} \oplus U_{sd} = \mathcal{R}(G^T)$. □

Exercises

Section 16.1

Exercise 16.1: A k-finger robot hand holds a 3-D rigid object with pointed fingertips via frictional hard-point contacts. Can the contacts be interpreted as standard mechanical joints during local finger motions that maintain the grasp?

Solution: When pointed fingertips hold a rigid object via frictional contacts, sufficiently local motions of the fingers that maintain the grasp do not incur contact slippage. The finger contacts can be interpreted as *spherical joints,* with three degrees of freedom at each contact. Note that the combined system of the hand mechanism and grasped object forms a *parallel chain* mechanism (see Bibliographical Notes). ○

Exercise 16.2: A k-finger robot hand holds a 3-D rigid object with pointed fingertips via frictional hard-point contacts. Determine the minimum number of fingers and the minimum number of joints in each finger that allow the fingers to locally move the object with full six degrees of freedom in $\mathbb{R}^3$ (assume identical finger mechanisms).

Solution: The combined hand and grasped object can be analyzed as a parallel chain mechanism. Such chains must possess at least three identical fingers and at least three joints in each finger to ensure that the combined system of the hand and grasped object possesses six degrees of freedom. Such hands can impart arbitrary local motions to the grasped object. Note that the classical *Salisbury hand* has three identical fingers with three joints in each finger. ○

Exercise 16.3: A robot hand holds a 3-D polyhedral rigid object with *flat fingertips* along the object facets via frictional *soft-point contacts.* Determine the minimum number of fingers and the minimum number of joints in each finger that allow the grasped object to locally move with full six degrees of freedom in $\mathbb{R}^3$ (assume identical finger mechanisms).

Exercise 16.4*: Extend the relations described in Equations (16.1) and (16.2) to fingertip contacts governed by the *soft-point contact* model, where the fingers can additionally apply torque about the respective contact normals.

Exercise 16.5: Obtain the hand Jacobian for the parallel jaw gripper in the grasp of Figure 16.2.

Exercise 16.6: Obtain the hand Jacobian for the palm-and-finger hand in the grasp of Figure 16.3.

Section 16.2

Exercise 16.7: Extend the relations described in Proposition 16.2 to fingertip contacts governed by the *soft-point contact* model.

Exercise 16.8: Obtain the grasp matrix, $G(q)$, for the two-finger grasp depicted in Figure 16.4.

Exercise 16.9: In Figure 16.4, a gravitational force acts at the center of $\mathcal{B}$. Determine the feasible fingertip forces, $f_1 \in C_1$ and $f_2 \in C_2$, which maintain the equilibrium grasp under the influence of gravity.

Section 16.3

Exercise 16.10: Let A be an $l \times r$ matrix with real entries. Justify the rank nullity theorem, $\dim(\mathcal{N}(A)) + \dim(\mathcal{R}(A)) = r$.

Hint: $\mathcal{R}(A^T)$ and its orthogonal complement in $\mathbb{R}^r$ satisfy the relation: $\mathcal{R}(A^T) \oplus \mathcal{R}(A^T)^{\perp} = \mathbb{R}^r$, where $\oplus$ denotes direct sum of linear subspaces. ○

Exercise 16.11: Verify that the linear map $\mathcal{L}_{\mathcal{N}(A)}(v) = [I - A^{\dagger}A]v$ maps vectors $v \in \mathbb{R}^r$ to the null space of A, such that $v - \mathcal{L}_{\mathcal{N}(A)}(v) = A^{\dagger}Av$ is orthogonal to the null space of A.

Solution: The first condition means that $A \cdot \mathcal{L}_{\mathcal{N}(A)}(v) = A[I - A^{\dagger}A]v = \vec{0}$ for all $v \in \mathbb{R}^r$. Since $A^{\dagger} = A^T(AA^T)^{-1}$, $A[I - A^{\dagger}A] = A - A = O$, where O is the zero $l \times r$ matrix. Thus, $A \cdot \mathcal{L}_{\mathcal{N}(A)}(v) = A[I - A^{\dagger}A]v = \vec{0}$ for all $v \in \mathbb{R}^r$. If $v - \mathcal{L}_{\mathcal{N}(A)}(v) = A^{\dagger}Av$ is orthogonal to the null space of A, it must lie in the range space of A^T. Since $A^{\dagger}$ has the same range space as A^T, the vector $A^{\dagger}Av$ lies in $\mathcal{R}(A^T)$ for all $v \in \mathbb{R}^r$. Hence, $v - \mathcal{L}_{\mathcal{N}(A)}(v)$ is orthogonal to the null space of A for all $v \in \mathbb{R}^r$. ○

Exercise 16.12: A serial-chain finger mechanism is positioned at a *singular configuration* when the finger Jacobian, $J_i(\vec{\phi}_i)$, loses its full rank. Explain why structurally dependent fingertip contact forces, $\vec{f} \in U_{sd}$ that satisfy $\mathcal{T} = J_H^T(\Phi, q)\vec{f} = \vec{0}$, exist only when some fingers of the robot hand are positioned at singular configurations. (Each finger is articulated by $n_i \geq 2$ joints in 2-D grasps, and by $n_i \geq 3$ joints in 3-D grasps.)

Exercise 16.13: In the structurally dependent grasp depicted in Figure 16.5, each finger forms a serial chain with three revolute joints and rigid links of unit length. As discussed in Appendix I, the Jacobian of each finger mechanism forms the 3×3 matrix:

$$J(\boldsymbol{\phi}_i) = \begin{bmatrix} -\sin(\phi_{i,1}) & -\sin(\phi_{i,1}+\phi_{i,2}) & -\sin(\phi_{i,1}+\phi_{i,2}+\phi_{i,3}) \\ \cos(\phi_{i,1}) & \cos(\phi_{i,1}+\phi_{i,2}) & \cos(\phi_{i,1}+\phi_{i,2}+\phi_{i,3}) \\ 1 & 1 & 1 \end{bmatrix} \quad i = 1, 2.$$

Write the hand Jacobian at $\phi_{i,1} = 90°$, $\phi_{i,1} + \phi_{i,2} = 90°$, and $\phi_{i,1} + \phi_{i,2} + \phi_{i,3} = 90°$ $(i = 1, 2)$. Verify that the fingertip contact forces, (f_1, f_2), lie in the null space of $J_H^T(\Phi, q)$.

Exercise 16.14: A k-finger robot hand holds a rigid object $\mathcal{B}$ at an equilibrium grasp with feasible internal grasp forces. Explain why the grasp must possess *actively controlled* internal forces when the hand Jacobian has a trivial null space.

Exercise 16.15: Provide a proof sketch of Lemma 16.5, that the subspaces of internal grasp forces N_{act} and N_{pre} are mutually orthogonal and satisfy the property: $N_{act} \oplus N_{pre} = \mathcal{N}(G(q))$. (Assume a surjective grasp matrix $G(q)$.)

Hint: First, verify that the subspaces N_{act} and N_{pre} are mutually orthogonal within the null space of the grasp matrix $G(q)$. The assumption that $G(q)$ is surjective leads to the conclusion that $\dim(N_{act}) + \dim(N_{pre}) = kp - m$, which is the dimension of the null space of $G(q)$. ○

Section 16.4

Exercise 16.16: Obtain the hand Jacobian for the two-finger hand depicted in the grasp of Figure 16.7.

Exercise 16.17: Consider the active wrench resistant grasp depicted in Figure 16.7. Determine the finger-joint torques, $\mathcal{T} = (\tau_{1,1}, \tau_{1,2}, \tau_{2,1}, \tau_{2,2})$, that resist a gravitational force f_g that acts at the grasped-object center.

Exercise 16.18*: Theorem 16.7 specifies *sufficient conditions* for a robot hand to be actively wrench resistant at a given grasp. Determine the *necessary conditions* for such grasps.

Exercise 16.19: Prove Lemma 16.8, that at an actively wrench resistant grasp every combination of finger-joint torques exerts a *unique* net wrench on the grasped object $\mathcal{B}$.

Solution: Suppose a particular $\mathcal{T} \neq \vec{0}$ resists two different external wrenches, $\boldsymbol{w}_{ext,1}$ and $\boldsymbol{w}_{ext,2}$. Based on Eq. (16.12), the contact forces that resist these wrenches, $\vec{f}_1$ and $\vec{f}_2$, satisfy the relations

$$\mathcal{T}=J_H^T(\Phi,q)\vec{f}_1 \;=\; J_H^T(\Phi,q)\big(G^{\dagger}(q)(-\boldsymbol{w}_{ext,1})+\vec{f}_{null,1}\big) \qquad \vec{f}_1\in C_1\times\cdots\times C_k$$

and

$$\mathcal{T}=J_H^T(\Phi,q)\vec{f}_2 \;=\; J_H^T(\Phi,q)\big(G^{\dagger}(q)(-\boldsymbol{w}_{ext,2})+\vec{f}_{null,2}\big) \qquad \vec{f}_2\in C_1\times\cdots\times C_k.$$

Subtracting the second equation from the first gives

$$J_H^T(\Phi,q)\big(-G^{\dagger}(q)(\boldsymbol{w}_{ext,1}-\boldsymbol{w}_{ext,2})+(\vec{f}_{null,1}-\vec{f}_{null,2})\big)=\vec{0}, \tag{16.22}$$

where $\vec{0}\in\mathbb{R}^N$. Since $J_H^T(\Phi,q)$ is injective on the space of finger contact forces, $-G^{\dagger}(q)(\boldsymbol{w}_{ext,1}-\boldsymbol{w}_{ext,2})+(\vec{f}_{null,1}-\vec{f}_{null,2})=\vec{0}$, where $\vec{0}$ is now the origin of contact force space $\mathbb{R}^{kp}$. The vector $\vec{f}_{null,1}-\vec{f}_{null,2}$ lies in the null space of $G(q)$, while the vector $G^{\dagger}(q)(\boldsymbol{w}_{ext,1}-\boldsymbol{w}_{ext,2})$ lies in the range space of $G^{\dagger}(q)$. Since $\text{range}(G^{\dagger})=\text{range}(G^T)$, the range space of $G^{\dagger}(q)$ is orthogonal to the null space of $G(q)$ in $\mathbb{R}^{kp}$. Hence, each summand of Eq. (16.22) must be individually equal to zero:

$$G^{\dagger}(q)(\boldsymbol{w}_{ext,1}-\boldsymbol{w}_{ext,2})=\vec{0} \quad\text{and}\quad \vec{f}_{null,1}-\vec{f}_{null,2}=\vec{0}. \tag{16.23}$$

From Chapter 12, wrench resistance at the fingertips requires that the object $\mathcal{B}$ be held in a non-marginal equilibrium grasp, such that the grasp matrix $G(q)$ has full row rank at the given grasp. When $G(q)$ has full row rank, the transposed grasp matrix $G^T(q)$ is injective by the rank nullity theorem. Hence, $G^{\dagger}(q)=G^T(GG^T)^{-1}$ is injective, and, consequently, $\boldsymbol{w}_{ext,1}=\boldsymbol{w}_{ext,2}$ in Eq. (16.23). This proves that $\mathcal{T}$ resists a unique external wrench that acts on $\mathcal{B}$. ○

Exercise 16.20: Consider the structurally dependent wrench resistant grasp depicted in Figure 16.8. Explain how the gripper resists with feasible combinations of structural and actively controlled contact forces any vertical force that may act at the grasped-object center.

Exercise 16.21: Consider the structurally dependent wrench resistant grasp depicted in Figure 16.8. Compute the gripper's joint forces required to resist a gravitational force that acts at the grasped-object center.

Exercise 16.22: Consider the palm-and-finger example depicted in Figure 16.9. Let the external wrenches that act on $\mathcal{B}$ be parametrized as $\boldsymbol{w}_{ext}=(f_x,f_y,\tau)$. Obtain the contact forces $\vec{f}=(f_1,f_2)$ that resist these external wrenches. Then obtain the set of external wrenches that can be resisted by pure structurally dependant resistant contact forces.

Solution: Using Formula (16.13) for $G^{\dagger}(q)$, the contact forces that can possibly resist the external wrench, $\vec{f}=G^{\dagger}(q)(-\boldsymbol{w}_{ext})+\vec{f}_{null}$, take the form

$$\vec{f}=\begin{pmatrix} f_1\\ f_2\end{pmatrix}=-\tfrac{1}{2}\begin{bmatrix} 0 & -1 & 1/L\\ 1 & 0 & 0\\ 0 & 1 & 1/L\\ -1 & 0 & 0\end{bmatrix}\begin{pmatrix} f_x\\ f_y\\ \tau\end{pmatrix}+\begin{pmatrix} 0\\ s\\ 0\\ s\end{pmatrix}=\begin{pmatrix} \frac{1}{2}(f_y-\frac{\tau}{L})\\ -\frac{1}{2}f_x+s\\ -\frac{1}{2}(f_y+\tau/L)\\ \frac{1}{2}f_x+s\end{pmatrix} \qquad s\in\mathbb{R},$$

where s parametrizes the internal grasp forces. The requirement $f_2=\vec{0}$ gives the constraints $f_y+\tau/L=0$ and $f_x/2+s=0$. Thus, $s=-f_x/2$, and the requirement $f_1\in C_1$

takes the form of a leftward force at the palm contact, $f_x \leq 0$ such that $|\frac{1}{2}(f_y - \tau/L)| \leq \mu|f_x|$, where μ is the coefficient of friction at the palm contact. ○

Exercise 16.23: A 2-D parallel jaw gripper is actuated by a *single* prismatic joint that acts against a rigid palm. Can the gripper realize a structurally dependent wrench resistant grasp when it holds a rigid object via antipodal frictional contacts?

References

[1] A. Bicchi, "On the problem of decomposing grasp and manipulation forces in multiple whole-limb manipulation," *International Journal of Robotics and Autonomous Systems*, vol. 13, no. 4, pp. 127–147, 1994.

[2] C. Melchiorri, "Multiple whole-limb manipulation: An analaysis in the force domain," *Robotics and Autonomous Systems*, vol. 20, pp. 15–38, 1997.

[3] C. Melchiorri and M. Kaneko, "Robot hands," in *Springer Handbook of Robotics*, Springer Verlag, 463–480, 2016.

[4] T. Yoshikawa, "Passive and active closure by constraining mechanisms," *Journal of Dynamic Systems, Measurement, and Control*, vol. 121, no. (3–4), pp. 418–424, 1999.

[5] J.-P. Merlet, C. Gosselin and T. Huang, "Parallel mechanisms and robots," in *Springer Handbook of Robotics*, Springer Verlag, 443–462, 2016.

[6] M. T. Mason, *Mechanics of Robotic Manipulation*. MIT Press, 2001.

[7] R. M. Murray, Li and S. S. Sastry, *A Mathematical Introduction to Robotic Manipulation*. CRC Press, 1994.

[8] M. Haas-Heger, Iyengar and M. Ciocarlie, "Passive reaction analysis for grasp stability," *IEEE Transactions on Automation Science and Engineering*, vol. 15, no. 3, pp. 955–966, 2018.

17 Grasp Manipulability

Grasp manipulability describes the ability of a robot hand to impart arbitrary local motions to a grasped object through the fingertip contacts. Grasp manipulability forms an important link between basic grasp mechanics principles and the ability of a robot hand to locally manipulate objects while maintaining the grasp. The notion of grasp manipulability is illustrated with the three-finger hand depicted in Figure 17.1. Each finger is articulated by three revolute joints and ends with a pointed fingertip. When this hand holds a rigid object $\mathcal{B}$ via frictional contacts, we will be able to assert that the object $\mathcal{B}$ is held at a locally manipulable grasp.

The chapter first studies the ability of a robot hand to impart arbitrary *instantaneous motions* to the grasped object. Based on instantaneous manipulability, the chapter describes how to assess the robot hand's ability to impart arbitrary *local motions* to the grasped object at a given grasp. The chapter ends with a short comment on the connection between local manipulability and controllability of multi-finger grasps.

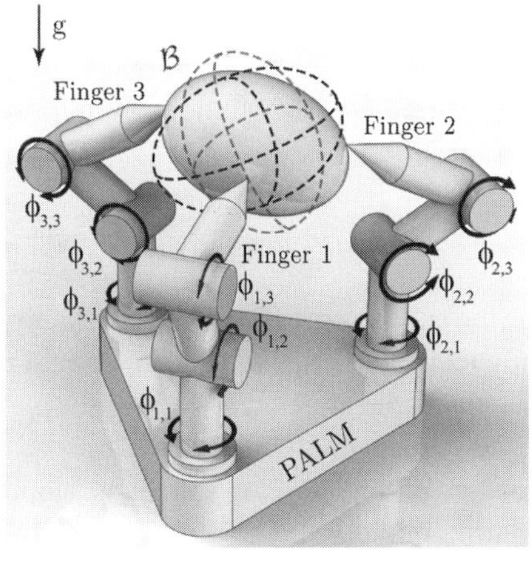

Figure 17.1 A three-finger hand with pointed fingertips holding a rigid object via frictional contacts (the friction cones are not shown). The object is locally manipulable at this grasp.

17.1 Instantaneous Manipulability

This section studies the ability of a robot hand to impart arbitrary instantaneous motions to the grasped object $\mathcal{B}$. We use the notation and assumptions introduced in the previous chapter. In particular, each finger mechanism is assumed to end with a *pointed fingertip*, so that no net rolling occurs at the contacts during local finger motions that maintain the grasp. The chapter relies on the *contact compatibility constraints* developed in the previous chapter. These constraints describe the kinematic relationships between the finger joint velocities and the grasped object rigid-body velocity during local finger motions that maintain the grasp. The contact compatibility constraints are based on the *hand Jacobian,* whose definition is repeated here. Recall that $\Phi = (\phi_1, \ldots, \phi_k) \in \mathbb{R}^N$ denotes the composite vector of all joint variables of the k-finger robot hand (Figure 17.1).

Hand Jacobian: *Let a k-finger hand positioned at joint values ϕ hold a rigid object $\mathcal{B}$ at a configuration q. The hand possesses N joints with p independent force components at each fingertip contact. The* **hand Jacobian,** *$J_H(\Phi, q)$, is the kp $\times$ N matrix:*

$$J_H(\Phi,q) = \begin{bmatrix} W_1^T(q)R_{PK_1}J_1(\phi_1) & 0 & \cdots & 0 \\ \vdots & & \ddots & 0 \\ 0 & 0 & \cdots & W_k^T(q)R_{PK_k}J_k(\phi_k) \end{bmatrix} \qquad \Phi = (\phi_1, \ldots, \phi_k),$$

where the columns of $W_i(q)$ form the ith object contact frame basis, R_{PK_i} transforms velocities from the ith finger base frame to the palm frame, and $J_i(\phi_i)$ is the ith finger Jacobian.

When a k-finger hand is positioned at joint values $\Phi \in \mathbb{R}^N$, the joint velocities span the vector space $T_\Phi \mathbb{R}^N$. Similarly, when a rigid object $\mathcal{B}$ is held at a configuration $q \in \mathbb{R}^m$, its rigid-body velocities span the vector space $T_q \mathbb{R}^m$, where $m = 3$ in 2-D grasps and $m = 6$ in 3-D grasps. According to Theorem 16.1, the *contact compatibility constraints* that govern the velocities $\dot{\Phi}$ and $\dot{q}$ during local finger motions that maintain the grasp are given by

$$J_H(\Phi, q)\dot{\Phi} = G^T(q)\dot{q} \qquad \dot{\Phi} \in T_\Phi \mathbb{R}^N,\ \dot{q} \in T_q \mathbb{R}^m, \tag{17.1}$$

where $G(q)$ is the $m \times kp$ grasp matrix. Every pair $(\dot{\Phi}, \dot{q})$ satisfying Eq. (17.1) represents a kinematically realizable instantaneous motion of the hand mechanism and the grasped object that maintains the grasp.

When every instantaneous motion of the grasped object, $\dot{q} \in T_q \mathbb{R}^m$, satisfies Eq. (17.1) with a unique set of finger joint velocities, application of finger joint velocities as kinematic control inputs will impart the desired velocity to the grasped object in a manner that maintains the grasp. This notion of *instantaneous manipulability* is defined as follows.

DEFINITION 17.1 **(instantaneous manipulability)** *A k-finger grasp of a rigid object $\mathcal{B}$ is* **instantaneously manipulable** *if every velocity of $\mathcal{B}$, $\dot{q} \in T_q \mathbb{R}^m$, can be realized by a* *unique set of finger joint velocities,* $\dot{\Phi} \in T_{\Phi}\mathbb{R}^N$*, which satisfies the contact compatibility constraints.*

The instantaneous manipulability of a grasp can be established from properties of the hand Jacobian $J_H(\Phi, q)$ and the grasp matrix $G(q)$, as stated in the following theorem. Recall that the *range space* of a matrix A, $\mathcal{R}(A)$, is the linear space spanned by the columns of A.

THEOREM 17.1 **(instantaneous manipulability test)** *Let a k-finger robot hand hold a rigid object* $\mathcal{B}$ *at a grasp characterized by the pair* $(J_H(\Phi, q), G(q))$*. The following conditions are necessary for* **instantaneous manipulability** *of the grasp:*

(i) *The transposed grasp matrix* $G^T(q)$ *forms an injective map on the vector space of* $\mathcal{B}$*'s rigid-body velocities* $T_q\mathbb{R}^m$*.*

(ii) *The range space of* $G^T(q)$ *and the range space of the hand Jacobian* $J_H(\Phi, q)$ *satisfy the relation:* $\mathcal{R}(G^T(q)) \subseteq \mathcal{R}(J_H(\Phi, q))$*.*

The two conditions are sufficient for **instantaneous manipulability** *when the object is held at a preloaded equilibrium grasp, possibly under an external influence such as gravity.*

Proof sketch: Instantaneous manipulability requires that for each velocity of the grasped object $\mathcal{B}$, $\dot{q} \in T_q\mathbb{R}^m$, there would exist a vector of finger joint velocities, $\dot{\Phi} \in T_{\Phi}\mathbb{R}^N$, that solves Eq. (17.1). For a given $\dot{q} \in T_q\mathbb{R}^m$, Eq. (17.1) has the form $Ax = b$, where $A = J_H(\Phi, q)$, $x = \dot{\Phi}$ and $b = G^T(q)\dot{q}$. A solution x exists iff $b \in \mathcal{R}(A)$. As $\dot{q}$ varies in the vector space $T_q\mathbb{R}^m$, the vectors $G^T(q)\dot{q}$ span the entire range space of $G^T(q)$. Condition (ii), $\mathcal{R}(G^T(q)) \subseteq \mathcal{R}(J_H(\Phi, q))$, is therefore necessary and sufficient for the existence of $\dot{\Phi}$ that solves Eq. (17.1).

Instantaneous manipulability additionally requires that for each velocity of the grasped object $\mathcal{B}$, $\dot{q} \in T_q\mathbb{R}^m$, there would exist a unique set of finger joint velocities that solves Eq. (17.1). If the uniqueness property does *not* hold at a given grasp, there exists a vector of finger joint velocities, $\dot{\Phi} \in T_{\Phi}\mathbb{R}^N$, which solves Eq. (17.1) for two different velocities of the grasped object $\mathcal{B}$: $J_H(\Phi, q)\dot{\Phi} = G^T(q)\dot{q}_1$ and $J_H(\Phi, q)\dot{\Phi} = G^T(q)\dot{q}_2$. It follows that $G^T(q)(\dot{q}_1 - \dot{q}_2) = \vec{0}$, which implies that $G^T(q)$ has a nontrivial null space. Condition (i) is thus necessary for instantaneous manipulability.

When $G^T(q)$ forms an injective map on the vector space of $\mathcal{B}$'s rigid-body velocities, $T_q\mathbb{R}^m$, it has a well-defined *left inverse* given by $(G^T)^{\dagger} = [GG^T]^{-1}G$. The matrix $(G^T)^{\dagger}$ has the same dimensions as G and satisfies the identity $(G^T)^{\dagger}G^T = I$, where I is the $m \times m$ identity matrix. The contact compatibility constraints of Eq. (17.1) can therefore be solved for $\dot{q}$ in terms of $\dot{\Phi}$:

$$\dot{q} = \left(G^T(q)\right)^{\dagger} J_H(\Phi, q)\dot{\Phi}. \tag{17.2}$$

Conditions (i) and (ii) imply that the $m \times N$ matrix $(G^T)^{\dagger} J_H$ in Eq. (17.2) is surjective (see Exercises). Every rigid-body velocity $\dot{q} \in T_q\mathbb{R}^m$ can therefore be realized with a unique affine subspace of finger joint velocities in $T_{\Phi}\mathbb{R}^N$, which implies instantaneous manipulability of the grasp. □

The requirement that $G^T(q)$ be injective on the vector space of $\mathcal{B}$'s rigid-body velocities ensures that each instantaneous motion of $\mathcal{B}$ induces a *unique combination* of contact point velocities on the object surface, expressed according to their contact frame basis components. The *alternative* situation occurs when $G^T(q)$ possesses a nontrivial null space. For instance, when a 3-D rigid object $\mathcal{B}$ is held via two frictional hard-point contacts, instantaneous rotations of $\mathcal{B}$ about the axis passing through the two contacts induce zero-velocity components at both contact points. Such instantaneous rotations of $\mathcal{B}$ clearly cannot be realized by the grasping fingers through the fingertip contacts.

Theorem 17.1 can be used to identify a large class of hand mechanisms that can realize instantaneously manipulable grasps. Condition (i) of Theorem 17.1 requires that the transposed grasp matrix, $G^T(q)$, be injective. Equivalently, the $m \times kp$ grasp matrix $G(q)$ must possess full row rank of m. The number of fingers, k, should therefore satisfy the lower bound $kp \geq m$. In the case of 3-D grasps with frictional hard-point contacts, $G(q)$ has full row rank when the object $\mathcal{B}$ is held by $k \geq 3$ fingers, provided that the fingertip contacts do not lie on a common line. Higher numbers of fingers are required when the object is held via frictionless contacts (see Exercises). Next consider the number of joints in each finger mechanism. The Jacobian of a k-finger hand forms the block diagonal matrix:

$$J_H(\Phi,q) = \begin{bmatrix} H_1(\phi_1,q) & 0 & \cdots & 0 \\ \vdots & & \ddots & 0 \\ 0 & 0 & \cdots & H_k(\phi_k,q) \end{bmatrix} \qquad \Phi = (\phi_1,\ldots,\phi_k).$$

The ith diagonal block, $H_i(\phi_i,q)$, has dimensions $p \times n_i$ where n_i is the number of joints in the ith finger. The grasp matrix can be written in the form $G(q) = [G_1(q) \cdots G_k(q)]$, where each submatrix $G_i(q)$ is $m \times p$. The contact compatibility constraints of Eq. (17.1) can thus be written in the form:

$$H_i(\phi_i,q)\dot{\phi}_i = G_i^T(q)\dot{q} \qquad i = 1 \ldots k.$$

It follows that Condition (ii) of Theorem 17.1 can be written as $\mathcal{R}(G_i^T(q)) \subseteq \mathcal{R}(H_i(\phi_i,q))$ for $i = 1 \ldots k$. Since $p \leq m$ under all contact models, each submtarix $G_i^T(q)$ has full row rank of p, and its range space is therefore p-dimensional. Hence, $\mathcal{R}(G_i^T(q)) \subseteq \mathcal{R}(H_i(\phi_i,q))$ when each diagonal block of the hand Jacobian, $H_i(\phi_i,q)$, is surjective. When the ith finger mechanism has a non-degenerate structure (see Chapter 16), $H_i(\phi_i,q)$ has full rank of $\min\{p,n_i\}$. To be surjective, each diagonal block must have full row rank, which leads to the conclusion that each finger mechanism must possess at least p joints to ensure instantaneous manipulability of the grasp. For instance, in robot hands that grasp 3-D objects via frictional hard-point contacts, there are $p = 3$ independent force components at each contact. The robot hand should therefore possess at least three fingers each articulated by $n_i \geq 3$ joints, as illustrated in Figure 17.1. The next example shows that an instantaneously manipulable grasp need not be wrench resistant.

Example: The hand depicted in Figure 17.2 consists of two identical fingers, each articulated by two revolute joints. The fingers support a rigid object $\mathcal{B}$ against gravity

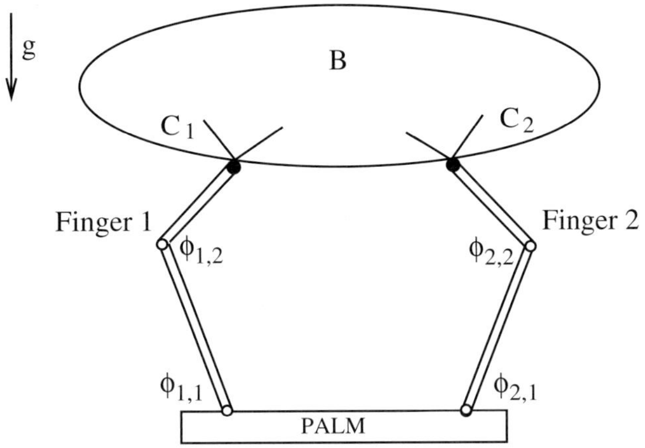

Figure 17.2 A rigid object supported against gravity by two fingers via frictional contacts. The object is instantaneously manipulable at this stance, which is *not* wrench resistant.

via frictional contacts. The 3×4 grasp matrix $G(q)$ has full row rank. Hence, $G^T(q)$ is injective on the vector space of $\mathcal{B}$'s rigid-body velocities, which is Condition (i) of Theorem 17.1. Intuitively, every instantaneous motion of $\mathcal{B}$ induces a unique combination of contact point velocities on the object boundary. Next, the range space of the submatrices $G_1^T(q)$ and $G_2^T(q)$ is two-dimensional at each contact. The range space of the diagonal blocks of the hand Jacobian, $H_1(\phi_1, q)$ and $H_2(\phi_2, q)$, is similarly two-dimensional at each contact. Condition (ii) of Theorem 17.1 is thus satisfied with equality: $\mathcal{R}(G_i^T(q)) = \mathcal{R}(H_i(\phi_i, q))$ for $i = 1, 2$. Since the object is supported at a preloaded equilibrium stance against gravity, it is instantaneously manipulable. Note that the contacts do *not* support internal grasp forces, which shows that an instantaneously manipulable grasp need *not* be wrench resistant. ○

The next example shows that instantaneous manipulability requires at least p joints in each finger mechanism, where p is the number of independent force components at each contact.

Example: The hand depicted in Figure 17.3 consists of nonidentical fingers. The left finger is articulated by a single revolute joint while the right finger is articulated by two revolute joints. The hand holds a rigid ellipse $\mathcal{B}$ of major axis length $2L$ via antipodal frictional contacts. The transposed grasp matrix and its left inverse are given by

$$G^T(q) = \begin{bmatrix} 0 & -1 & L \\ 1 & 0 & 0 \\ 0 & 1 & L \\ -1 & 0 & 0 \end{bmatrix} \quad \text{and} \quad \left(G^T(q)\right)^{\dagger} = \tfrac{1}{2} \begin{bmatrix} 0 & 1 & 0 & -1 \\ -1 & 0 & 1 & 0 \\ 1/L & 0 & 1/L & 0 \end{bmatrix}.$$

Condition (i) of Theorem 17.1 is satisfied, since $G^T(q)$ is injective on the vector space of $\mathcal{B}$'s rigid-body velocities. Next consider the hand Jacobian at the finger joint values $\phi_{1,1} = 90°$ and $(\phi_{2,1}, \phi_{2,2}) = (60°, 90°)$. Neglecting a small offset δ and assuming the

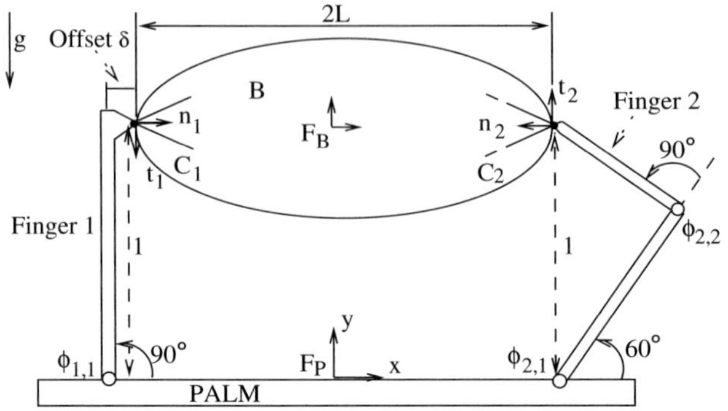

Figure 17.3 A rigid object held by two fingers via antipodal frictional contacts. The left finger has a deficient range space of fingertip velocities and consequently the object is *not* instantaneously manipulable at this grasp.

unit lengths indicated in Figure 17.3, the hand Jacobian forms the 4×3 block diagonal matrix (the upper-left block is 2×1, the lower right block is 2×2):

$$J_H(\Phi, q) = \begin{bmatrix} 0 & 0 & 0 \\ -1 & 0 & 0 \\ 0 & 0 & -\sqrt{3}/4 \\ 0 & 1 & 1/4 \end{bmatrix}$$

Condition (ii) of Theorem 17.1 is not satisfied, and the grasp is *not* instantaneously manipulable. To verify this statement, consider the grasped-object velocities that can be realized at this grasp. Using Eq. (17.2) with velocity coordinates $\dot{q} = (v_x, v_y, \omega)$, these grasped-object velocities are given by

$$\begin{pmatrix} v_x \\ v_y \\ \omega \end{pmatrix} = \tfrac{1}{2}(\dot{\phi}_{1,1} + \dot{\phi}_{2,1}) \begin{pmatrix} -1 \\ 0 \\ 0 \end{pmatrix} - \tfrac{1}{2}\dot{\phi}_{2,2} \begin{pmatrix} 1/4 \\ \sqrt{3}/4 \\ \sqrt{3}/4L \end{pmatrix}.$$

We see that the hand can realize only a subset of the grasped object rigid-body velocities. For instance, pure instantaneous rotation of $\mathcal{B}$ about the object frame origin, $\dot{q} = (0, 0, \omega)$, *cannot* be realized by any combination of finger joint velocities at this grasp. ○

17.2 Local Manipulability

This section studies the ability of a robot hand to impart arbitrary local motions to the grasped object at a given grasp. The notion of *local manipulability* is defined as follows.

DEFINITION 17.2 **(manipulable grasp)** *A k-finger robot hand holds a rigid object $\mathcal{B}$ in a* **locally manipulable** *grasp if every local motion of $\mathcal{B}$ can be realized by a unique set of finger joint trajectories that maintain the grasp.*

In order to develop a test for local manipulability, we need the classical *submersion theorem* from analysis. Let us summarize this theorem in terms of the spaces that will be used to characterize locally manipulable grasps.

Submersion: *Let $f:\mathbb{R}^N \to \mathbb{R}^m$ be a $C^{(1)}$ function such that $N \geq m$.*[1] *The function f is a* **submersion** *at $\Phi \in \mathbb{R}^N$ if the Jacobian matrix, $Df(\Phi): T_\Phi\mathbb{R}^N \to T_{f(\Phi)}\mathbb{R}^m$, is surjective.*

When a function f forms a submersion, it satisfies the following property (see Bibliographical Notes).

Submersion theorem: *If $f:\mathbb{R}^N \to \mathbb{R}^m$ is a submersion at Φ, it maps an open neighborhood of Φ in $\mathbb{R}^N$* **onto** *an open neighborhood of $f(\Phi)$ in $\mathbb{R}^m$.*

The following theorem asserts that instantaneous manipulability implies local manipulability of the grasp.

THEOREM 17.2 (**local manipulability test**) *If a k-finger robot hand holds a rigid object $\mathcal{B}$ at an instantaneously manipulable grasp, the grasp is* **locally manipulable.** *Conversely, a k-finger grasp that is not instantaneously manipulable is generically not locally manipulable.*

The proof of Theorem 17.2 that appears in the chapter's appendix is based on the *contact compatibility constraints* of Chapter 16. Recall that $Y_i(\phi_i)$ describes the position of the ith pointed fingertip in the palm frame, while $X_{b_i}(q)$ describes the position of the ith contact point on the object surface in the palm frame. The *local version* of the contact compatibility constraints takes the form

$$F(\Phi,q)=\begin{pmatrix} W_1^T(q)\big(Y_1(\phi_1)-X_{b_1}(q)\big) \\ \vdots \\ W_k^T(q)\big(Y_k(\phi_k)-X_{b_k}(q)\big)\end{pmatrix}=\vec{0} \qquad \Phi=(\phi_1,\ldots,\phi_k)\in\mathbb{R}^N,\ q\in\mathbb{R}^m. \tag{17.3}$$

where the zero vector has kp components. Eq. (17.3) specifies an implicit equation of the form $F(\Phi,q)=\vec{0}$. Let $\Phi_0=(\phi_{01},\ldots,\phi_{0k})$ be the finger joint values and let q_0 be the object configuration at the given grasp, so that $Y_i(\phi_{0i})=X_{b_i}(q_0)$ for $i=1\ldots k$. Condition (i) of Theorem 17.1 is used in the proof to establish the existence of a function, $q=f(\Phi)$, which satisfies the equation $F(\Phi,f(\Phi))=\vec{0}$ in an open neighborhood of Φ_0 in $\mathbb{R}^N$. Implicit differentiation of this equation gives the Jacobian matrix of $f(\Phi)$ at Φ_0:

$$Df(\Phi_0)=(G^T(q_0))^{\dagger}J_H(\Phi_0,q_0) \qquad q_0=f(\Phi_0),$$

where $G(q_0)$ is the grasp matrix, $(G^T(q_0))^{\dagger}$ is the left inverse of $G'(q_0)$, and $J_H(\Phi_0,q_0)$ is the hand Jacobian at the given grasp. From the proof of Theorem 17.1, we know that the Jacobian $Df(\Phi_0)$ is surjective. Hence, by the *submersion theorem,* the function

[1] A $C^{(1)}$ function is differentiable with a continuous first derivative.

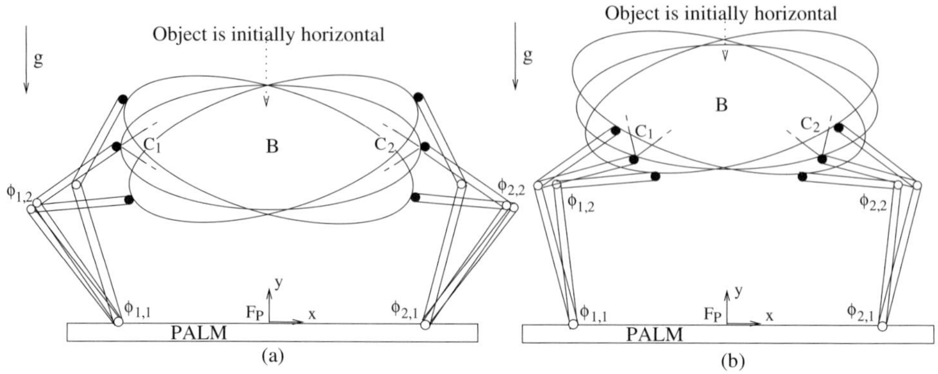

Figure 17.4 (a) Local motions of an object held at a locally manipulable grasp that is actively wrench resistant are shown. (b) Similar local motions can be realized when the object is supported against gravity at a locally manipulable stance that is only locally wrench resistant.

$f(\Phi)$ maps an open neighborhood centered at Φ_0 in $\mathbb{R}^N$ onto an open neighborhood centered at q_0 in $\mathbb{R}^m$, which implies local manipulability of the grasp.

The following example illustrates two types of locally manipulable grasps.

Example: The hand depicted in Figure 17.4 consists of two identical fingers, each articulated by two revolute joints. First consider the frictional equilibrium grasp shown in Figure 17.4(a). The object is instantaneously manipulable and hence locally manipulable at this grasp. This grasp is *actively wrench resistant* (see Chapter 16). The fingers can thus adjust their joint torques to ensure that the object is kept at an equilibrium grasp during such local motions. Now let the two fingers support the object $\mathcal{B}$ against gravity, as shown in Figure 17.4(b). The object is instantaneously manipulable and hence locally manipulable at this stance. While this stance is not wrench resistant, the supporting hand satisfies the *local wrench resistance* property described Chapter 14, which allows the hand to maintain the equilibrium stance during local motions of the object $\mathcal{B}$. ○

When a grasp *lacks* instantaneous manipulability, the robot hand is generically unable to impart arbitrary local motions to the grasped object. This property is illustrated in the following example.

Example: The hand depicted in Figure 17.5 consists of three identical fingers, each articulated by two revolute joints. The fingers end with pointed fingertips that hold a 3-D rigid object $\mathcal{B}$ via frictional contacts. There are $p = 3$ independent force components at each contact. Instantaneous manipulability thus requires at least three joints in each finger. However, each finger has only two joints and can realize only a two-dimensional set of fingertip velocities at its contact point with the object $\mathcal{B}$. The grasp is consequently *not* instantaneously manipulable, which generically implies lack of local manipulability. As can be seen in Figure 17.5, a vertical up-and-down motion of $\mathcal{B}$ *cannot* be realized by this hand. ○

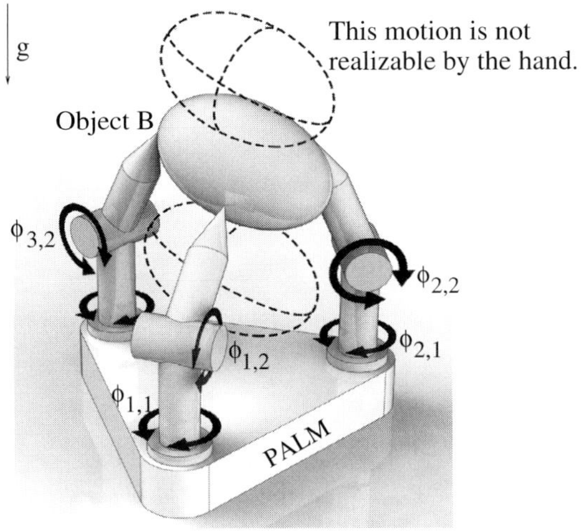

Figure 17.5 A rigid object held by three fingers with pointed fingertips via frictional contacts (the friction cones are not shown). The grasp is *not* locally manipulable, as can be seen by the hand's inability to locally move the object in a vertical up-and-down motion.

This section established that instantaneous manipulability generically implies local manipulability of a grasp. When the instantaneous manipulability test succeeds, the hand can impart arbitrary local motions to the grasped object in a manner consistent with the contact model at the fingertip contacts. When the test fails, the hand generically *cannot* impart arbitrary local motions to the grasped object.

Connection to local controllability: We know from Chapter 12 that a grasp which is wrench resistant at the fingertip contacts is locally controllable (STLC). This means that the grasped object can be steered to any nearby object configuration by feasible contact forces that form the control inputs. Similarly, when an object is supported at a non-marginal equilibrium stance against gravity, the grasp is locally controllable. When the full hand mechanism is taken into account, an actively wrench resistant grasp is locally controllable with finger joint torques as control inputs. Thus, we see parallel lines between *local controllability* of a grasp and *local manipulability* of the same grasp when interpreted as a kinematic control system, with contact point velocities or finger joint velocities as control inputs. This connection indicates that robot grasps form *kinematic control systems* (see Bibliographical Notes). ○

Bibliographical Notes

The notion of grasp manipulability was introduced by Yoshikawa [1]. Full analysis of robot hand manipulability appears in Li, Hsu and Sastry [2], as well as in Murray's PhD thesis [3]. The topic of grasp manipulability is also discussed by Mason [4] and Murray, Li and Sastry [5]. These texts also consider the effect of net fingertip rolling in locally manipulable grasps.

Kinematic control systems capture a wide variety of controlled mechanical systems. In these systems, controlled trajectories in the system's position-and-velocity state space can be replicated by kinematically controlled trajectories in the system's configuration space. Such systems are characterized by Lewis [6] and further discussed by Murphy and Burdick [7]. Finally, a good source for the *submersion theorem* from analysis is the text by Guillemin and Pollack [8].

Appendix: Proof of the Local Manipulability Theorem

This appendix contains a proof of the forward part of Theorem 17.2, that instantaneous manipulability of a grasp implies local manipulability of this grasp.

Theorem 17.2 *If a k-finger robot hand holds a rigid object $\mathcal{B}$ at an instantaneously manipulable grasp, the grasp is* **locally manipulable.**

Proof: Recall that $Y_i(\phi_i)$ denotes the position of the ith pointed fingertip in the palm frame, while $X_{b_i}(q)$ denotes the position of the ith contact point on the object surface in the palm frame. When the hand executes local motions that maintain the grasp, the contact compatibility constraints take the local form

$$\begin{pmatrix} W_1^T(q)\big(Y_1(\phi_1) - X_{b_1}(q)\big) \\ \vdots \\ W_k^T(q)\big(Y_k(\phi_k) - X_{b_k}(q)\big) \end{pmatrix} = \vec{0}, \tag{17.4}$$

where $\vec{0}$ has kp components, while the columns of each matrix $W_i(q)$ form the ith contact frame basis. Eq. (17.4) specifies an implicit equation of the form $F(\Phi,q)=\vec{0}$, where $\Phi=(\phi_1,\ldots,\phi_k)\in\mathbb{R}^N$ and $q\in\mathbb{R}^m$.

Let the combined system of hand mechanism and grasped object be positioned at (Φ_0,q_0) so that $F(\Phi_0,q_0)=\vec{0}$. If the constraints specified by Eq. (17.4) allow local manipulability, there must exist a *submersion* at Φ_0, $q=f(\Phi)$ such that $q_0=f(\Phi_0)$, from the fingers' joint space to the grasped object's c-space. The function $F(\Phi,q)$ is clearly $C^{(1)}$ in Φ and q. Moreover, it can be verified that $\vec{0}$ is a *regular value* of $F(\Phi,q)$, and, consequently, the solution set of $F(\Phi,q)=\vec{0}$ forms a smooth manifold in the composite vector space $\mathbb{R}^N\times\mathbb{R}^m$.[2]

To verify the existence of $q=f(\Phi)$, consider the partial derivative of $F(\Phi,q)$ with respect to q:

$$\frac{\partial}{\partial q}F(\Phi,q) = \begin{pmatrix} W_1^T(q)DX_{b_1}(q) \\ \vdots \\ W_k^T(q)DX_{b_k}(q) \end{pmatrix},$$

where terms proportional to $Y_i(\phi_i) - X_{b_i}(q)$ were omitted since these terms vanish at (Φ_0,q_0). From Chapter 3, each $DX_{b_i}(q)$ satisfies the formula $W_i^T(q)DX_{b_i}(q) =$

[2] When the Jacobian $DF(\Phi,q)$ is surjective at all points (Φ,q) such that $F(\Phi,q)=\vec{0}$, the zero vector forms a *regular value* of F. In this case, $F(\Phi,q)=\vec{0}$ specifies a smooth manifold in $\mathbb{R}^N\times\mathbb{R}^m$.

$G_i^T(q)$, where $G_i(q)$ is the submatrix of $G(q)$ associated with the ith contact. Hence, $\frac{\partial}{\partial q}F(\Phi_0,q_0)=G^T(q_0)$. The grasp is assumed to be instantaneously manipulable. Hence, by Condition (i) of Theorem 17.1, $G^T(q_0)$ is injective on the vector space $T_{q_0}\mathbb{R}^m$. It follows that $G^T(q_0)$ has a well-defined left inverse given by $(G^T)^{\dagger}=[GG^T]^{-1}G$. The derivative $\frac{\partial}{\partial q}F(\Phi_0,q_0)$ is thus invertible. Hence, by the *implicit function theorem,* there exists a $C^{(1)}$ function, $q=f(\Phi)$ such that $q_0=f(\Phi_0)$, which locally satisfies the contact compatibility constraints of Eq. (17.4).

To complete the proof, we have to show that the function $f(\Phi)$ maps an open neighborhood of Φ_0 in $\mathbb{R}^N$ onto an open neighborhood of q_0 in $\mathbb{R}^m$. Implicit differentiation of $F(\Phi,f(\Phi))=\vec{0}$ with respect to the independent variable Φ gives

$$\frac{d}{d\Phi}F\big(\Phi,f(\Phi)\big)=\frac{\partial}{\partial\Phi}F(\Phi,q)+\frac{\partial}{\partial q}F(\Phi,q)\cdot\frac{d}{d\Phi}f(\Phi)=\vec{0}\qquad q=f(\Phi).$$

Application of this formula to $F(\Phi,q)$ specified in Eq. (17.4) gives

$$\begin{pmatrix} W_1^T(q)\big(\frac{d}{d\phi_1}Y_1(\phi_1)-DX_{b_1}(q)\frac{d}{d\Phi}f(\Phi)\big) \\ \vdots \\ W_k^T(q)\big(\frac{d}{d\phi_k}Y_k(\phi_k)-DX_{b_k}(q)\frac{d}{d\Phi}f(\Phi)\big)\end{pmatrix}=\vec{0}. \tag{17.5}$$

Note that we omitted terms proportional to $Y_i(\phi_i)-X_{b_i}(q)$ that vanish at (Φ_0,q_0). From Section 16.1, we know that $W_i^T(q)\frac{d}{d\phi_i}Y_i(\phi_i)=H_i(\phi_i,q)$, where $H_i(\phi_i,q)$ is the ith diagonal block of the hand Jacobian, $J_H(\Phi,q)$. Let the $m\times N$ Jacobian matrix $Df(\Phi_0)$ denote the derivative $\frac{d}{d\Phi}f(\Phi)$ at Φ_0. Using the relation $G_i^T(q_0)=W_i^T(q)DX_{b_i}(q)$ $(i=1\ldots k)$, Eq. (17.5) can be written in the concise form

$$J_H(\Phi_0,q_0)-G^T(q_0)\cdot Df(\Phi_0)=\vec{0}, \tag{17.6}$$

where $G(q_0)=[G_1\cdots G_k]$. Multiplying both sides of Eq. (17.6) by $(G^T(q_0))^{\dagger}$ gives

$$Df(\Phi_0)=(G^T(q_0))^{\dagger}J_H(\Phi_0,q_0)\qquad q_0=f(\Phi_0),$$

where we used the identity $(G^T(q_0))^{\dagger}G^T(q_0)=I$. By Condition (i) of Theorem 17.1, the grasp matrix $G(q_0)$ is surjective. The left inverse matrix, $(G^T(q_0))^{\dagger}$, has the same range space as $G(q_0)$. Hence, $(G^T(q_0))^{\dagger}$ is also surjective. By Condition (ii) of Theorem 17.1, $\mathcal{R}(G^T(q_0))\subseteq\mathcal{R}(J_H(\Phi_0,q_0))$. Hence, the full Jacobian matrix, $Df(\Phi_0)=(G^T(q_0))^{\dagger}J_H(\Phi_0,q_0)$, is surjective as required by the submersion theorem. By the *submersion theorem,* the function $f(\Phi)$ maps an open neighborhood of Φ_0 in $\mathbb{R}^N$ onto an open neighborhood of q_0 in $\mathbb{R}^m$, which implies local manipulability of the grasp. □

Exercises

Section 17.1

Exercise 17.1: Can the notion of instantaneous manipulability be applied to the central body of a legged robot supported against gravity via foothold contacts on uneven terrain?

Exercise 17.2: Theorem 17.1 specifies two necessary conditions for instantaneous manipulability of a grasp. Verify that the $m \times N$ matrix $(G^T(q))^{\dagger} J_H(\Phi, q)$ is surjective under these conditions.

Exercise 17.3: A 3-D rigid object $\mathcal{B}$ is held by two fingers via frictional hard-point contacts. The fingers are *not* able to rotate the object about the axis that passes through the contacts. Can such rotations be realized by suitable motion of the hand's palm while the object $\mathcal{B}$ is held at an equilibrium grasp?

Exercise 17.4: A robot hand holds a rigid object $\mathcal{B}$ via *frictionless* point contacts. Determine the minimum number of fingers and the minimum number of joints in each finger that ensure instantaneous manipulability of the grasp.

Solution: To ensure instantaneous manipulability, the object $\mathcal{B}$ must be held at a frictionless equilibrium grasp. Let us next consider the two necessary conditions of Theorem 17.1. A frictionless contact has a single force component. The transposed grasp matrix, $G^T(q)$, forms an injective map on the vector space of $\mathcal{B}$'s rigid-body velocities when the grasp matrix $G(q)$ is surjective. In the case of frictionless equilibrium grasps, this condition is satisfied when the object $\mathcal{B}$ is held in a *first-order immobilizing* grasp. Next, each submatrix $G_i^T(q)$ forms a single-row matrix in the case of frictionless contacts. Hence, the condition $\mathcal{R}(G_i^T(q)) \subseteq \mathcal{R}(H_i(\phi_i, q))$ holds true when each finger mechanism has at least one joint. The minimum number of fingers is, thus, $m+1=4$ in 2-D grasps and $m+1=7$ in 3-D grasps, with at least one joint in each finger. ○

Exercise 17.5: Figure 17.2 depicts a rigid object $\mathcal{B}$ supported against gravity by a two fingers via frictional contacts. Characterize the external forces and torques that can act on $\mathcal{B}$ and be actively resisted by finger joint torques through the supporting contacts. (The graphical technique of Chapter 14 can be useful here.)

Section 17.2

Exercise 17.6: Compare the two locally manipulable grasps depicted in Figure 17.4 in terms of their robustness against external disturbances that may act on the grasped object $\mathcal{B}$.

Exercise 17.7: Specify the joint torques, $\mathcal{T}(t)$, that must by applied at the finger joints of the hand depicted in Figure 17.4, such that the object is kept at an equilibrium grasp during local motions that maintain the grasp.

Solution: Denote by $\boldsymbol{w}_g$ the gravitational wrench that acts on $\mathcal{B}$. The finger joint torques are set by the formula

$$\mathcal{T} = J_H^T(\Phi, q)\big(G^{\dagger}(q)(-\boldsymbol{w}_g) + \vec{f}_{null}\big),$$

where $G^{\dagger} = G^T(GG^T)^{-1}$ is the *right inverse* of $G(q)$. The internal grasp forces, $\vec{f}_{null}$, must be selected such that the net contact forces, $\vec{f} = G^{\dagger}(q)(-\boldsymbol{w}_g) + \vec{f}_{null}$, lie in the respective friction cones. ○

Exercise 17.8: A three-finger robot hand with three revolute joints in each finger can realize locally manipulable grasps of 3-D objects via frictional contacts. When the hand holds a 3-D object with a fourth finger that has only two revolute joints, some local object motions may cause the fourth fingertip to *break contact*. Can local manipulability be extended to allow contact breakage and reestablishment during local motions of the grasped object?

Exercise 17.9: A robot hand consists of four identical fingers, each articulated by two revolute joints. The hand grasps 3-D objects via frictional pointed fingertips. Describe how this hand can impart arbitrary local motions to the grasped object, despite the fact that such hands do not satisfy the formal definition of locally manipulable grasps.

Exercise 17.10*: Prove that every locally manipulable grasp is locally controllable as a kinematic control system whose control inputs are the finger joint angles or displacements.

References

[1] T. Yoshikawa, "Manipulability of robotic mechanisms," *International Journal of Robotics Research*, vol. 4, no. 2, pp. 3–9, 1985.

[2] Z. Li, P. Hsu and S. Sastry, "Grasping and coordinated manipulation by a multi-fingered robot hand," *International Journal of Robotics Research*, vol. 8, no. 4, pp. 33–50, 1989.

[3] R. M. Murray, "Robotic control and nonholonomic motion planning," PhD thesis, Dept. of Electrical Engineering and Computer Science, University of California, Berkeley, 1990. Available online.

[4] M. T. Mason, *Mechanics of Robotic Manipulation*. MIT Press, 2001.

[5] R. M. Murray, Z. Li and S. S. Sastry, *A Mathematical Introduction to Robotic Manipulation*. CRC Press, 1994.

[6] A. D. Lewis, "When is a mechanical control system kinematic?" In *IEEE International Conf. on Decision and Control*, pp. 1162–1167, 1999.

[7] T. D. Murphey and J. W. Burdick, "The power dissipation method and kinematic reducibility of multiple-model robotic systems," *IEEE Transactions on Robotics*, vol. 22, no. 4, pp. 694–710, 2006.

[8] V. Guillemin and A. Pollack, *Differential Topology*. Prentice-Hall, 1974.

18 Hand Mechanism Compliance

The preceding chapters modeled all the mechanical components of a robotic grasp as ideal rigid bodies. These components include the fingertips, the links of the finger mechanisms and the grasped object. However, *compliance*, which is the property of a material undergoing elastic deformation in response to applied stresses, plays an important role in robotic grasps involving compliant joints, soft fingertips or grasped objects made of soft materials. Compliance effects can appear as material deformation at the contacts, as deflection of inner parts in the finger structures or as part of the control strategy of a hand mechanism. This chapter considers how to model basic compliance phenomena in grasping mechanisms that arise from compliance-based hand control strategies.

Section 18.1 reviews idealized linear spring models. Common robot hand control algorithms implement a springlike behavior at the mechanism joints, leading to an overall compliant behavior of the hand. Based on these simple spring models, Section 18.2 studies the *grasp stiffness matrix* of a robot hand that holds rigid objects with compliant finger joints. This matrix describes the relationship between small movements of the grasped object and the restoring wrench applied to the object by the compliant robot hand. Section 18.3 studies the decoupling properties of the grasp stiffness matrix and its potential role in *task programming* of grasp stiffness. Section 18.4 analyzes the local stability of compliant grasps and describes the *coin-snapping* phenomenon, which demonstrates a loss of grasp stability under increased finger loadings.

18.1 One-Dimensional Stiffness and Compliance

This section provides basic background concerning the compliance introduced by two types of springs. An idealized linear spring is depicted in Figure 18.1. One end of the spring is fixed while the other end is free to displace along the x-axis. When no

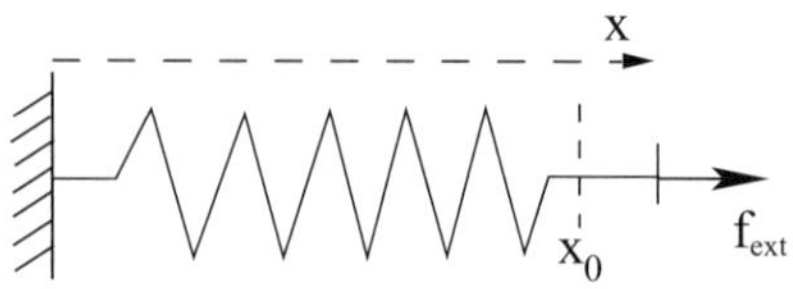

Figure 18.1 An idealized linear spring.

force is applied to the spring, its free end rests at an *unloaded* position, x_0. When an external force, f_{ext}, is applied to the free end of the spring along the negative x-axis direction, the spring compresses until it reaches an equilibrium position where the restoring force generated by the spring's compression balances the applied force. Under the idealized linear spring model, the spring follows *Hooke's law*, which states that the spring restoring force, f_R, is linear in the spring's free end displacement:

$$f_R(x) = -\kappa \cdot (x - x_0), \tag{18.1}$$

where $\kappa > 0$ is the *spring stiffness constant*. When an external force pulls on the spring, it extends under tension to an equilibrium position where the spring restoring force (pointing along the negative x-axis) balances the external force, according to Eq. (18.1). The spring constant is a function of the spring's material properties and its geometry. The spring is said to be *stiff* when κ is large and *soft* when κ is small.

The inverse of the stiffness constant, $c = 1/\kappa$, is the *spring compliance*. The value of the spring compliance determines the amount of spring deflection in response to an applied external force:

$$x - x_0 = c \cdot f_{ext}. \tag{18.2}$$

As the spring compresses or extends under an external load, energy is stored in the deformation of the spring material. If we assume that there are no losses during the deformation process (e.g., due to heat production during spring displacement or permanent plastic deformation of the spring material), then all of the work done by the external force to compress or extend the spring is stored in the spring as *potential energy.* When the external force on the spring is removed, the stored energy is released to perform useful work.

Under the linear spring idealization, the spring potential energy can be found by computing the total energy expended by the external force to compress or extend the spring from its unloaded length in a quasistatic fashion. At each instant during the spring deformation process, the spring pushes against the external force with a reaction force $f_R(x)$. The amount of potential energy, $U(x)$, stored in the spring is given by

$$U(x) = \int_{x_0}^{x} f_{ext}(y)dy = -\int_{x_0}^{x} f_R(y)dy = \int_{x_0}^{x} \kappa(y - x_0)\, dy = \tfrac{1}{2}\kappa(x - x_0)^2.$$

The potential energy can also be expressed in terms of the spring compliance. Substituting for $x - x_0$ according to Eq. (18.1) and $c = 1/\kappa$ into the expression for $U(x)$ gives

$$U(f_R) = \tfrac{1}{2}c \cdot f_R^2. \tag{18.3}$$

The following simple observations can be generalized to higher-dimensional analogs. First, the spring stiffness can be interpreted as the magnitude of the derivative of the restoring force with respect to changes in the spring position, Δx:

$$\lim_{\Delta t \to 0} \frac{f_R(x + \Delta x) - f_R(x)}{\Delta t} = \lim_{\Delta t \to 0} -\kappa \frac{\Delta x}{\Delta t} \quad \Rightarrow \quad \frac{d}{dt} f_R = -\kappa \frac{dx}{dt}.$$

This property will be generalized to define the *grasp stiffness matrix* in the next section.

Figure 18.2 shows two springs with different spring constants, κ_1 and κ_2, arranged in series and parallel combinations. Each of these combinations can be conceptually replaced by an *equivalent spring* with the following equivalent spring constants:

$$\kappa_{series} = \left(\kappa_1^{-1} + \kappa_2^{-1}\right)^{-1} \qquad \kappa_{parallel} = \kappa_1 + \kappa_2. \tag{18.4}$$

Note that the compliances, $c_1 = 1/\kappa_1$ and $c_2 = 1/\kappa_2$, of series springs are additive, while parallel compliances act like series stiffnesses:

$$c_{series} = c_1 + c_2 \qquad c_{parallel} = \left(c_1^{-1} + c_2^{-1}\right)^{-1}. \tag{18.5}$$

An idealized rotational spring is depicted in Figure 18.3(a). This type of spring can act around a revolute joint's axis as seen in Figure 18.3(b). The action of this spring is governed by a linear relationship between its rotational displacement, ϕ, and the restoring torque, τ_R:

$$\tau_R = -\kappa \cdot (\phi - \phi_0),$$

where ϕ is the angle of the distal end of the rotational spring and ϕ_0 is the unloaded rotational angle of the spring (Figure 18.3(a)).

Since idealized rotational springs respond linearly, the same laws regarding series and parallel combinations of rotational springs apply. In compliant multi-finger robot hands, increasing the number of fingers can only make the grasp *stiffer*, while increasing the number of spring-loaded joints in each finger mechanism can only make the grasp *softer*.

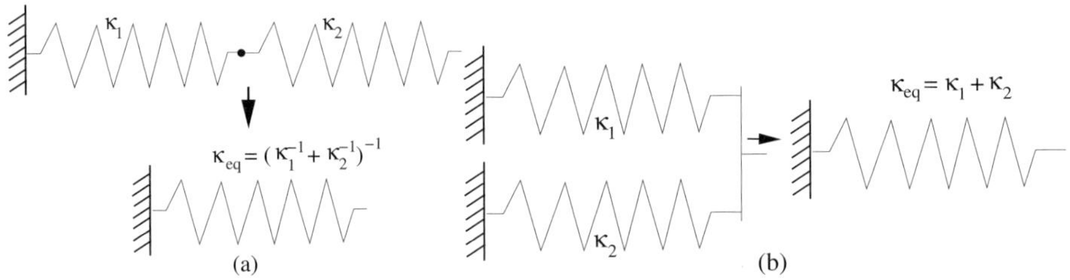

Figure 18.2 Idealized series and parallel linear spring arrangements and their equivalent springs. (a) Series springs equivalent. (b) Parallel springs equivalent.

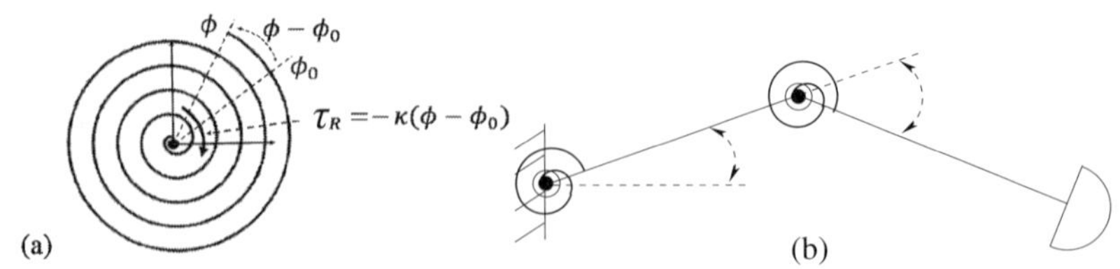

Figure 18.3 (a) Idealized geometry of a rotational spring governed by a linear stiffness relationship. (b) A planar robot finger with two compliant revolute joints.

18.2 The Effects of Joint Compliance on Grasp Stiffness

This section studies the modeling of compliant robotic grasps under the following assumptions. The finger mechanisms hold a rigid object $\mathcal{B}$ via *pointed fingertips* as schematically shown in Figure 18.4. While the finger links and fingertips are assumed to be perfectly rigid, the motion of each finger joint is influenced by a *compliant spring* of the type described in the previous section. As a finger joint displaces from its nominal position, a restoring torque or force is applied by the spring to resist the joint displacement. While it is not practically useful to build a hand mechanism with such joints, this situation arises in practice due to the nature of the feedback control strategies commonly used in robot grasping. In actively controlled hand mechanisms, feedback control algorithms apply joint torques or forces as to maintain each finger joint at a specified position. Very commonly, the joint-level control is implemented with a *proportional-integral-derivative* (PID) control algorithm. The integral term in PID control primarily cancels persistent steady-state errors in the desired joint positions. On short time scales its effect is negligible, and we can focus on how *proportional-derivative* (PD) control can lead to compliant hand behavior.

Recall from Chapter 16 that $\boldsymbol{\phi}_i = (\phi_{i,1}, \ldots, \phi_{i,n_i})$ denotes the joint values of the ith finger mechanism, where n_i is the number of joints in the ith finger. While the analysis applies to both revolute and prismatic joints, we will use joint angles and torques for both types of joints. Also recall that $\Phi = (\boldsymbol{\phi}_1, \ldots, \boldsymbol{\phi}_k) \in \mathbb{R}^N$ is the vector of all joints of the k-finger hand, where $N = \sum_{i=1}^{k} n_i$. Let $\phi^0_{i,j}$ denote a desired value of the jth joint angle in the ith finger. The joint torque applied by the feedback control algorithm to this joint typically takes the form:

$$\tau_{i,j}(t) = \tau^0_{i,j} - k_{i,j} \cdot (\phi_{i,j}(t) - \phi^0_{i,j}) - \eta_{i,j} \cdot \dot{\phi}_{i,j}(t), \tag{18.6}$$

where $\tau^0_{i,j}$ is the joint torque required to maintain the equilibrium grasp. The coefficients $k_{i,j}$ and $\eta_{i,j}$ are user-specified feedback gains for the joint. The *proportional* control term, $k_{i,j} \cdot (\phi_{i,j} - \phi^0_{i,j})$, behaves exactly like a rotational spring. A restoring torque is applied when the joint deviates from the nominal value $\phi^0_{i,j}$. The velocity feedback

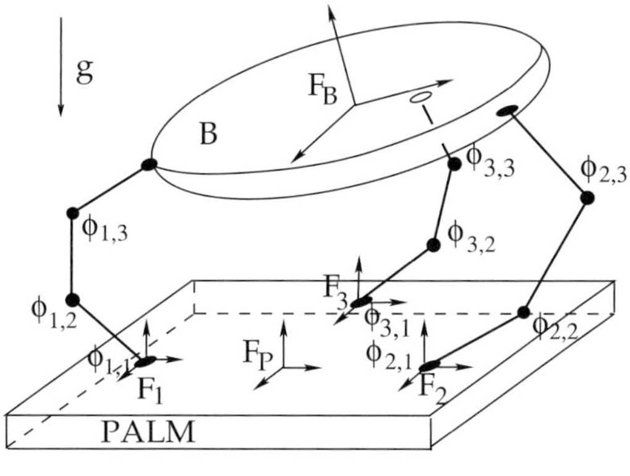

Figure 18.4 Geometry of a multi-finger hand mechanism.

term, $\eta_{i,j} \cdot \dot{\phi}_{i,j}$, effectively provides a form of damping during the joint movement, helping to stabilize the joint motion at the zero rotation velocity.

When the joints control system implements the PD control law just described, the ith finger *joint stiffness matrix* is the constant $n_i \times n_i$ matrix:

$$K_{\phi_i} = \begin{bmatrix} k_{i,1} & 0 & \cdots & 0 \\ 0 & k_{i,2} & \cdots & 0 \\ \vdots & \vdots & \ddots & \vdots \\ 0 & \cdots & \cdots & k_{i,n_i} \end{bmatrix}.$$

In order to realize specific grasping objectives, it may be desirable to *couple* the motions of the joints within each finger mechanism. For example, in the human hand, the movements of the distal finger bones are coupled with the motions of the intermediate bones via tendons. In such coupled joint movements, the ith finger joint stiffness matrix can have non zero off-diagonal terms. Coupling of joints within each finger (but not across finger mechanisms) will introduce a block diagonal structure to the composite *k-finger joint stiffness matrix,* which is the constant $N \times N$ matrix:

$$K_{\Phi} = \begin{bmatrix} K_{\phi_1} & 0 & \cdots & 0 \\ 0 & K_{\phi_2} & \cdots & 0 \\ \vdots & \vdots & \ddots & \vdots \\ 0 & \cdots & \cdots & K_{\phi_k} \end{bmatrix}.$$

We assume that the proportional gain matrices, K_{ϕ_i} for $i = 1 \ldots k$, are symmetric and full rank. The off-diagonal blocks of K_{Φ} could be fully populated if one wishes to allow for control coupling between the joints of *different* fingers, as might be needed to implement complex grasp stiffness behaviors.

Let us next introduce *grasp stiffness matrix,* which will serve as our model for compliant robotic grasps.

DEFINITION 18.1 **(grasp stiffness matrix)** *Let a k-finger robot hand whose joints are compliant hold a rigid object $\mathcal{B}$ at an equilibrium grasp (q_0, Φ_0). The $m \times m$* **grasp stiffness matrix**, *$K_G(q_0, \Phi_0)$, maps small object displacements approximated by rigid-body velocities to changes in the net object wrench:*[1]

$$\dot{w} = -[K_G(q_0, \Phi_0)]\dot{q}(t) \qquad \dot{q} \in T_{q_0}\mathbb{R}^m, \;\; \dot{w} \in T_{(q_0,\vec{0})}(T^*_{q_0}\mathbb{R}^m).$$

where $m = 3$ in 2-D grasps and $m = 6$ in 3-D grasps.

To derive an expression for the grasp stiffness matrix, we start with the potential energy of the system comprised of the compliant finger joints and the grasped object (note that $\nabla U(\Phi_0) = \vec{0}$ at equilibrium grasp):

$$U(\Phi) = U(\Phi_0) + \tfrac{1}{2}(\Phi - \Phi_0)^T K_{\Phi}\,(\Phi - \Phi_0) = U(\Phi_0) + \tfrac{1}{2}\sum_{i=1}^{k}(\phi_i - \phi_{i,0})^T K_{\phi_i}(\phi_i - \phi_{i,0})$$

[1] Small changes in the net object wrench are approximated by *wrench derivatives.* Formally, wrench derivatives lie in the tangent space of $\mathcal{B}$'s wrench space, $T_{(q_0,\vec{0})}(T^*_{q_0}\mathbb{R}^m)$, where $\vec{0}$ is the origin of $\mathcal{B}$'s wrench space.

where $\Phi_0 = (\phi_{1,0}, \ldots, \phi_{k,0})$ are the joint angles at the equilibrium grasp, which are also the desired joint angles in the PD control laws. The quadratic term in $U(\Phi)$ quantifies the amount of energy needed to deflect the joints away from Φ_0 via quasistatic movement of the finger joints. Our objective would be to express the grasp's potential energy in terms of the grasp stiffness matrix, $K_G(q_0, \Phi_0)$, and small object displacements approximated by $\dot{q}$:

$$U(\dot{q}) = U(\Phi_0) + \tfrac{1}{2}\dot{q}^T K_G(q_0, \Phi_0)\dot{q} \qquad \dot{q} \in T_{q_0}R^m.$$

One could obtain this expression by finding a relationship between small changes in Φ and small movements of the grasped object $\mathcal{B}$. Chapter 16 introduced a relationship between these key variables: $G^T(q)\dot{q} = J_H(\Phi, q)\dot{\Phi}$, where $G(q)$ is the grasp matrix and $J_H(\Phi, q)$ is the hand Jacobian. However, there is no simple solution for $\dot{q}$ in this relationship. Hence, the transformation of $U(\Phi)$ to $U(\dot{q})$ will be accomplished using a sequence of relationships that involve the finger joint torques, the finger contact forces applied to $\mathcal{B}$ by these torques, and, the resultant wrenches applied to the object $\mathcal{B}$.

Recall the notation introduced in Chapter 16 as reviewed in Figure 18.4. A reference frame $\mathcal{F}_P$ is attached to the hand's palm, while a reference frame $\mathcal{F}_{K_i}$ is attached to the ith finger base. A reference frame $\mathcal{F}_B$ is attached to the grasped object $\mathcal{B}$, while a contact frame $\mathcal{F}_{C_i}$ is attached to $\mathcal{B}$ at the ith contact point x_i, with its z-axis aligned with $\mathcal{B}$'s inward unit normal at this point. The vector b_i describes the position of the ith contact point in the object frame $\mathcal{F}_B$, while $x_i = X_{b_i}(q)$ describes this point in the palm frame $\mathcal{F}_P$. The vector $Y_i(\phi_i)$ describes the position of the ith pointed fingertip in the palm frame.

From Chapter 16, the ith finger Jacobian, $J_i(\phi_i)$, maps the ith finger joint velocities to the linear velocity of its pointed fingertip according to the formula:

$$\frac{d}{dt}Y_i(\phi_i) = R_{PK_i} J_i(\phi_i)\dot{\phi}_i,$$

where the constant matrix R_{PK_i} rotates linear velocities from the ith finger base frame to the palm frame. Recall that the contacts between the fingertips and the grasped object are subject to *contact compatibility constraints*, which restrict the relative motions between the fingertips and the object during local motions that maintain the grasp. In the case of frictionless point contacts, the contact compatibility constraints take the form:

$$n_i^T(q)\frac{d}{dt}Y_i(\phi_i) = n_i^T(q)\frac{d}{dt}X_{b_i}(q) \qquad i = 1 \ldots k,$$

where $n_i(q)$ is $\mathcal{B}$'s inward unit normal at the ith contact. In the case of frictional point contacts, the contact compatibility constraints take the form:

$$W_i^T(q)\frac{d}{dt}Y_i(\phi_i) = W_i^T(q)\frac{d}{dt}X_{b_i}(q) \qquad i = 1 \ldots k,$$

where $W_i(q) = [s_i(q)\ t_i(q)\ n_i(q)]$ consists of $\mathcal{B}$'s unit tangents and inward unit normal at the ith contact. We will use the notation $W_i(q) = [n_i(q)]$ for a frictionless contact and $W_i(q) = [s_i(q)\ t_i(q)\ n_i(q)]$ for a frictional contact.

From Chapter 16, the transposed hand Jacobian, $J_H^T(\Phi,q)$, maps the fingertip contact force components, $\vec{f} \in \mathbb{R}^{kp}$, to the finger joint torques, $\mathcal{T} \in \mathbb{R}^N$:

$$\mathcal{T} = [J_H^T(\Phi,q)]\ \vec{f} \tag{18.7}$$

where $J_H(\Phi,q)$ takes the form $J_H(\Phi,q) = \text{diag}\big(W_i^T(q) R_{PK_i} J_i(\phi_i)\big)$. We will use the shorter notation $H_i(\phi_i,q) = W_i^T(q) R_{PK_i} J_i(\phi_i)$ for the ith diagonal block of $J_H(\Phi,q)$ $(i = 1\ldots k)$.

The following proposition specifies an expression for the grasp stiffness matrix at an *unloaded* equilibrium grasp. The grasp stiffness matrix of preloaded grasps will be discussed in the next section (see coin-snapping phenomenon). Recall that the Jacobian of the object's ith contact point position, $X_{b_i}(q)$, takes the form $DX_{b_i}(q) = [I \;-\; [R_{PB}(\theta) b_i \times]]$, where $R_{PB}(\theta)$ describes the grasped object orientation in the palm frame. Note that $R_{PB}(\theta)$ is the same as the matrix $R(\theta)$ from previous chapters.

PROPOSITION 18.1 (**Grasp Stiffness Matrix**) *Let a k-finger robot hand contact a rigid object $\mathcal{B}$ with compliant finger joints at an unloaded equilibrium grasp (q_0, Φ_0). The $m \times m$ stiffness matrix of the ith finger, $K_{G,i}(q_0, \Phi_0)$, is given by*

$$K_{G,i}(q_0,\Phi_0) \;=\; DX_{b_i}^T(q_0) W_i(q_0) \big[H_i(\phi_{i,0}, q_0) K_{\phi_i}^{-1} H_i^T(\phi_{i,0}, q_0)\big]^{-1} W_i^T(q_0) DX_{b_i}(q_0),$$

where K_{ϕ_i} is the ith finger joint stiffness matrix and $H_i(\phi_{i,0}, q_0)$ is the ith diagonal block of the hand Jacobian. The **grasp stiffness matrix** *of the compliant hand-and-object system is the symmetric $m \times m$ matrix given by the parallel combination of the k finger stiffnesses:*

$$K_G(q_0,\Phi_0) = \sum_{i=1}^{k} K_{G,i}(q_0,\Phi_0).$$

where $m = 3$ in 2-D grasps and $m = 6$ in 3-D grasps

A proof of Proposition 18.1 appears in the chapter's appendix. The grasp stiffness matrix can be simplified in the case of frictional contacts. In these grasps, the ith contact frame matrix $W_i(q)$ is invertible, thus leading to the simplified grasp stiffness matrix:

$$K_G(q_0,\Phi_0) = \sum_{i=1}^{k} DX_{b_i}^T(q_0) \Big[R_{PK_i} J_i(\phi_{i,0}) K_{\phi_i}^{-1} J_i^T(\phi_{i,0}) R_{PK_i}^T \Big]^{-1} DX_{b_i}(q_0)\,.$$

For other contact models, recall that the contact basis matrices, $W_i(q)$ for $i = 1\ldots k$, possesses p columns, where p is the number of independent force and possibly moment components at the contact. Under these contact models, the grasp stiffness matrix formula requires the corresponding $p \times n_i$ Jacobian matrix of the ith fingertip-frame position and possibly orientation about the respective contact normal, where n_i is the number of joints in the ith finger mechanism.

The following step-by-step example illustrates the derivation of the grasp stiffness matrix for the case of frictional contacts.

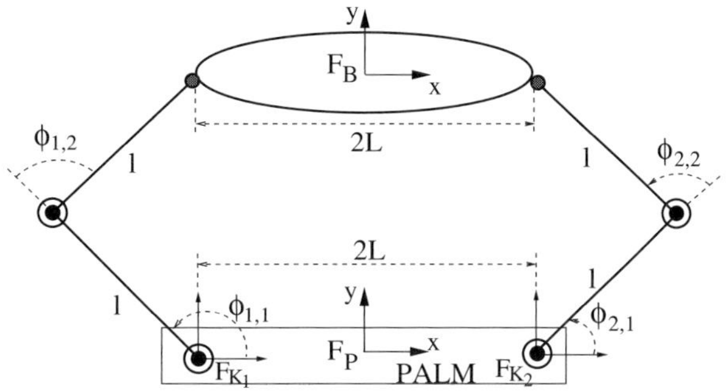

Figure 18.5 Top view of a two-finger hand with compliant finger joints.

Example: Consider the two-finger hand depicted in Figure 18.5. The hand holds with pointed fingertips a rigid ellipse of width $2L$ via frictional contacts. The object is held in a horizontal plane without gravity so that we can study the compliant behavior of the unloaded grasp. All finger link lengths are equal to l, and each finger is articulated by two revolute joints, $(\phi_{i,1}, \phi_{i,2})$ for $i = 1, 2$. The object frame origin lies at the ellipse's center, while the palm frame is located at the midpoint between the finger-base joints. The finger joint stiffnesses are symmetric so that $\kappa_{1,1} = \kappa_{2,1} = \kappa_1$ and $\kappa_{1,2} = \kappa_{2,2} = \kappa_2$, and there is no coupling between the joints. Let us compute the 3×3 grasp stiffness matrix associated with small displacements of the object when $(\phi_{1,1}, \phi_{1,2}) = (135°, -90°)$ and $(\phi_{2,1}, \phi_{2,2}) = (45°, 90°)$.

As discussed in Chapter 16, the 2×2 Jacobians of the finger mechanisms have the form:

$$J_i = \begin{bmatrix} -(l\sin\phi_{i,1} + l\sin(\phi_{i,1} + \phi_{i,2})) & -l\sin(\phi_{i,1} + \phi_{i,2}) \\ l\cos\phi_{i,1} + l\cos(\phi_{i,1} + \phi_{i,2}) & l\cos(\phi_{i,1} + \phi_{i,2}) \end{bmatrix} \quad i = 1, 2.$$

When evaluated at $(\phi_{1,1}, \phi_{1,2}) = (135°, -90°)$ and $(\phi_{2,1}, \phi_{2,2}) = (45°, 90°)$, the finger Jacobians take the form:

$$J_1 = \begin{bmatrix} -\sqrt{2}l & -\frac{\sqrt{2}}{2}l \\ 0 & \frac{\sqrt{2}}{2}l \end{bmatrix} \quad J_2 = \begin{bmatrix} -\sqrt{2}l & -\frac{\sqrt{2}}{2}l \\ 0 & -\frac{\sqrt{2}}{2}l \end{bmatrix}.$$

The contact basis matrices are: $W_1 = \left[\begin{smallmatrix} 0 & 1 \\ -1 & 0 \end{smallmatrix}\right]$ and $W_2 = \left[\begin{smallmatrix} 0 & -1 \\ 1 & 0 \end{smallmatrix}\right]$. The velocity transformations from the finger-base frames to the palm frame are the identity matrices, $R_{PK_1} = R_{PK_2} = I$. Using these expressions, the diagonal blocks of the hand Jacobian are

$$H_1 = W_1^T R_{PK_1} J_1 = \begin{bmatrix} 0 & -1 \\ 1 & 0 \end{bmatrix} \begin{bmatrix} 1 & 0 \\ 0 & 1 \end{bmatrix} \begin{bmatrix} -\sqrt{2}l & -\frac{\sqrt{2}}{2}l \\ 0 & \frac{\sqrt{2}}{2}l \end{bmatrix} = \begin{bmatrix} 0 & -\frac{\sqrt{2}}{2}l \\ -\sqrt{2}l & -\frac{\sqrt{2}}{2}l \end{bmatrix}$$

$$H_2 = W_2^T R_{PK_2} J_2 = \begin{bmatrix} 0 & 1 \\ -1 & 0 \end{bmatrix} \begin{bmatrix} 1 & 0 \\ 0 & 1 \end{bmatrix} \begin{bmatrix} -\sqrt{2}l & -\frac{\sqrt{2}}{2}l \\ 0 & -\frac{\sqrt{2}}{2}l \end{bmatrix} = \begin{bmatrix} 0 & 0 - \frac{\sqrt{2}}{2}l \\ \sqrt{2}l & \frac{\sqrt{2}}{2}l \end{bmatrix}.$$

The first-finger contact stiffness matrix (see proof of Proposition 18.1) is given by

$$[H_1 K_1^{-1} H_1^T]^{-1} = \left(\begin{bmatrix} 0 & -\frac{l}{\sqrt{2}} \\ -\sqrt{2}l & -\frac{l}{\sqrt{2}} \end{bmatrix} \begin{bmatrix} k_1^{-1} & 0 \\ 0 & k_2^{-1} \end{bmatrix} \begin{bmatrix} 0 & -\frac{l}{\sqrt{2}} \\ -\sqrt{2}l & -\frac{l}{\sqrt{2}} \end{bmatrix}^T \right)^{-1}$$
$$= \frac{1}{2l^2} \begin{bmatrix} 4k_2 + k_1 & -k_1 \\ -k_1 & k_1 \end{bmatrix}.$$

The second-finger contact stiffness matrix takes the form:

$$[H_2 K_2^{-1} H_2^T]^{-1} = \frac{1}{2l^2} \begin{bmatrix} 4k_2 + k_1 & k_1 \\ k_1 & k_1 \end{bmatrix}. \tag{18.8}$$

The Jacobians of the object's contact point positions, $DX_{b_1}(q_0)$ and $DX_{b_2}(q_0)$, are given by

$$DX_{b_1}(q_0) = \begin{bmatrix} 0 & -1 & -L \\ 1 & 0 & 0 \end{bmatrix} \qquad DX_{b_2}(q_0) = \begin{bmatrix} 0 & 1 & -L \\ -1 & 0 & 0 \end{bmatrix}.$$

Using Proposition 18.1, the contributions to the grasp stiffness matrix from the two fingers are

$$K_{G,1} = DX_{b_1}^T W_1 [H_1 K_1^{-1} H_1^T]^{-1} W_1^T DX_{b_1} = \frac{1}{2l^2} \begin{bmatrix} k_1 + 4k_2 & -k_1 & -Lk_1 \\ -k_1 & k_1 & Lk_1 \\ -Lk_1 & Lk_1 & L^2 k_1 \end{bmatrix}$$

and

$$K_{G,2} = DX_{b_2}^T W_2 [H_2 K_2^{-1} H_2^T]^{-1} W_2^T DX_{b_2} = \frac{1}{2l^2} \begin{bmatrix} k_1 + 4k_2 & k_1 & -Lk_1 \\ k_1 & k_1 & -Lk_1 \\ -Lk_1 & -Lk_1 & L^2 k_1 \end{bmatrix}.$$

The grasp stiffness matrix, $K_G(q_0, \Phi_0) = K_{G,1} + K_{G,2}$, is the symmetric 3×3 matrix:

$$K_G(q_0, \Phi_0) = K_{G,1} + K_{G,2} = \frac{1}{l^2} \begin{bmatrix} k_1 + 4k_2 & 0 & -Lk_1 \\ 0 & k_1 & 0 \\ -Lk_1 & 0 & L^2 k_1 \end{bmatrix}. \tag{18.9}$$

The form of the grasp stiffness matrix helps us interpret its compliant behavior. The 2×2 upper-left block in Eq. (18.9) is diagonal. Hence, small x-direction motions of the object do not generate a restoring force in the y-direction, and vice versa. The grasp stiffness matrix in Eq. (18.9) also contains a coupling term between the x and θ directions. This means that small object rotations will generate a reaction force in the x-direction in addition to a restoring torque. Likewise, small x-direction motions of the object will generate a reaction torque in addition to a restoring force along the object's x-direction. ○

18.3 The Grasp Center of Stiffness

The stiffness matrix of a compliant grasp captures many aspects of the robot hand behavior during manipulation tasks. This section studies grasp behavior that is relevant to understanding the interactions between a grasped object such as a tool and the environment. Let us first study how the structure of the grasp stiffness matrix changes under different choices of object reference frames. For notational convenience, we highlight the block structure of the grasp stiffness matrix as follows:

$$K_G(q_0, \Phi_0) = \begin{bmatrix} K_{tt} & K_{tr} \\ K_{tr}^T & K_{rr} \end{bmatrix}, \tag{18.10}$$

where the subscripts t and r refer to translations and rotations. Let us investigate what happens when the location of the object frame $\mathcal{F}_B$ changes to a new location $\mathcal{F}'_B$. Let (d_B, R_B) denote the position and orientation of $\mathcal{F}'_B$ relative to $\mathcal{F}_B$. The velocity of $\mathcal{F}'_B$ in the palm frame, $\dot{\bar{q}}$, is related to the velocity of $\mathcal{F}_B$ in the palm frame, $\dot{q}$, according to the transformation rule discussed in Appendix II of Chapter 13:

$$\dot{\bar{q}} = \begin{bmatrix} I & [d_B \times] \\ 0 & I \end{bmatrix} \dot{q} \, .$$

This change of coordinates results in a grasp stiffness matrix of the form:

$$\bar{K}_G(\bar{q}_0, \Phi_0) = \begin{bmatrix} I & [d_B \times] \\ 0 & I \end{bmatrix}^T K_G(q_0, \Phi_0) \begin{bmatrix} I & [d_B \times] \\ 0 & I \end{bmatrix} = \begin{bmatrix} \bar{K}_{tt} & \bar{K}_{tr} \\ \bar{K}_{tr}^T & \bar{K}_{rr} \end{bmatrix} \tag{18.11}$$

where

$$\begin{aligned} \bar{K}_{tt} &= K_{tt} \\ \bar{K}_{tr} &= K_{tt}[d_B \times] \; + \; K_{tr} \\ \bar{K}_{rr} &= K_{rr} + K_{tr}^T[d_B \times] + [d_B \times]^T K_{tr} + [d_B \times]^T K_{tt}[d_B \times]. \end{aligned}$$

This transformation shows that the grasp stiffness matrix is *not* frame invariant. Such coordinate transformations can be useful in *task programming* of grasp stiffness, as next described with an example.

Consider a robot hand that holds a tool such as a paint roller. To allow the robot to properly paint a wall, the grasped roller should be compliant in a direction normal to the wall so that it can adapt to small variations in the wall texture. The grasped tool should respond with moderate stiffness in translational directions tangent to the wall so that the roller can track a prescribed trajectory. The translational and rotational stiffnesses should be *decoupled* so that perturbations in the translational directions will not cause the roller to rotate away from the plane of the wall. Finally, rotational degrees of freedom of the hand should possess moderate stiffness so that the roller can comply to the wall's orientation. Thus, we would like to program the stiffness of the grasping robot hand to have a behavior that is synergistic with the given task. Task programming of grasp stiffness raises two related issues:

- How do changes in the object reference frame location modify the structure of the grasp stiffness matrix? In the paint roller example, the grasped roller stiffness should be approximately decoupled in a frame located at the roller's center.

- Can we select the joint stiffnesses and joint coupling factors in order to achieve task goals? In the paint roller example, can the finger joint stiffnesses of a given robot hand be selected so that the paint roller's stiffness can realize the previously described decoupling behavior?

These questions can be addressed in a systematic manner by looking at the issue of stiffness matrix *decoupling*, which is defined as follows.

DEFINITION 18.2 (**stiffness matrix decoupling**) *The grasp stiffness matrix is said to be* **decoupled** *if:*

(i) *Small translations of $\mathcal{B}$ result in pure restoring forces, without restoring torques.*
(ii) *Small rotations of $\mathcal{B}$ result in pure restoring torques, without restoring forces.*

The off-diagonal blocks of a decoupled grasp stiffness matrix are thus required to be zero. Does there exist a choice of object reference frame such that the grasp stiffness matrix is decoupled? Such a decoupling occurs if there exists a vector d_B such that

$$\bar{K}_{tr} = K_{tt}[d_B \times] + K_{tr} = O \tag{18.12}$$

where O is a 3×3 matrix of zeroes in 3-D grasps and a vector of two zeroes in 2-D grasps. The grasp *center of stiffness* is defined as the origin of the object frame that results in decoupling. When the grasped object displacements are defined with respect to an object frame that satisfies Eq. (18.12), the compliant hand–object system acts as if it is decoupled. Similarly, if there exists a choice of object frame that decouples the compliance matrix (the inverse of the grasp stiffness matrix), the origin of this frame defines the grasp *center of compliance*.

Let us consider the conditions under which it is possible to find a decoupling center of stiffness. We first consider 2-D grasps and then consider 3-D grasps.

Planar stiffness matrices: For 2-D compliant grasps, the 3×3 grasp stiffness matrix takes the form:

$$K_G(q_0, \Phi_0) = \begin{bmatrix} K_{tt} & K_{tr} \\ K_{tr}^T & K_{rr} \end{bmatrix} = \begin{bmatrix} k_{xx} & k_{xy} & k_{x\theta} \\ k_{xy} & k_{yy} & k_{y\theta} \\ k_{x\theta} & k_{y\theta} & k_{\theta\theta} \end{bmatrix}, \tag{18.13}$$

where $K_{tt} \in \mathbb{R}^{2\times 2}$, $K_{tr} \in \mathbb{R}^2$ and K_{rr} is a scalar. Applying the condition specified in Eq. (18.12) to the grasp stiffness matrix, we obtain two linear equations in the unknowns $d_B = (d_x, d_y)$:

$$K_{tt} \cdot \begin{pmatrix} -d_y \\ d_x \end{pmatrix} + K_{tr} = \begin{pmatrix} 0 \\ 0 \end{pmatrix} \qquad d_B = (d_x, d_y).$$

When the 2×2 matrix K_{tt} has full rank (the generic case), it is possible to find a unique object frame origin that results in a decoupled grasp stiffness matrix. The location of the

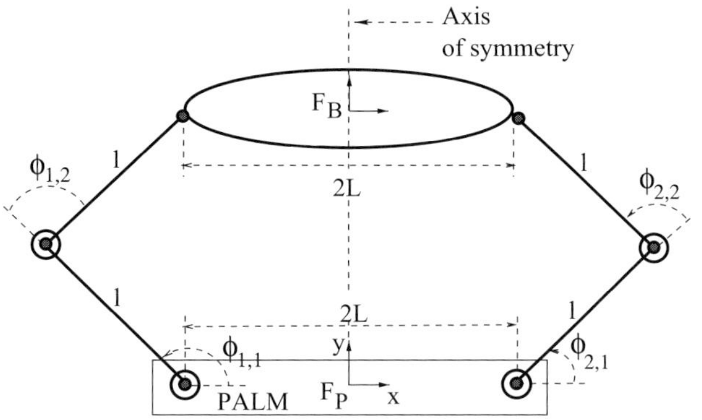

Figure 18.6 Top view of a two-finger hand with compliant joints.

decoupling frame $\mathcal{F}_B'$ origin relative to $\mathcal{F}_B$ is given by

$$d_x = \frac{k_{xy}k_{x\theta} - k_{xx}k_{y\theta}}{k_{xy}^2 - k_{xx}k_{yy}} \qquad d_y = \frac{k_{yy}k_{x\theta} - k_{xy}k_{y\theta}}{k_{xy}^2 - k_{xx}k_{yy}}. \tag{18.14}$$

These formulas show that one can potentially select the joint stiffnesses (which in turn dictate the grasp stiffness matrix via Proposition 18.1) to program a desired location of the grasp center of stiffness, as suggested by the following example.

Example: Consider the two-finger hand depicted Figure 18.5, which is reproduced in Figure 18.6. An expression for its 3×3 grasp stiffness matrix was given in Eq. (18.9). Let us find the center of stiffness of this grasp. Substituting the stiffness matrix specified in Eq. (18.9) into Eq. (18.14) gives the position of the grasp center of stiffness:

$$d_x = 0 \qquad d_y = L\frac{k_1}{k_1 + 4k_2}.$$

The center of stiffness of this grasp lies along the y-axis of the palm frame, which is the axis of the hand's left–right symmetry at the given grasp. The expression for d_y dictates the possible location of the grasp center of stiffness along the y-axis. Since L, k_1 and k_2 are all positive, $d_y > 0$. If $k_1 >> k_2$, then $d_y \to L$. If $k_2 >> k_1$, then $d_y \to 0$. Hence, by choosing the values of the joint stiffnesses, k_1 and k_2, the grasp center of stiffness can be programmed to lie anywhere along the y-axis within the interval $0 < y < L$. ○

Spatial stiffness matrices: The decoupling of the 6×6 grasp stiffness matrix in 3-D grasps is more delicate. Recall the decoupling requirement of Eq. (18.12): $\bar{K}_{tr} = K_{tt}[d_B\times] + K_{tr} = O$. When the 3×3 matrix K_{tt} has full rank (the generic case), one can invert K_{tt} to find the location of the decoupling object frame origin: $[d_B\times] = -K_{tt}^{-1}K_{tr}$. However, $[d_B\times]$ must be a 3×3 skew symmetric matrix. This naive solution does not guarantee a skew symmetric solution, except under very special conditions. While it is generally *not* possible to find an object frame displacement, d_B, which will drive the expression $\bar{K}_{tr} = K_{tt}[d_B\times] + K_{tr}$ to zero, as needed for decoupling, there exists a coordinate change which *simplifies* this term. The 3×3 matrix $\bar{K}_{tr}$ is in general non-symmetric. However, as shown in the following lemma, one can always choose an object frame origin that transforms the off-diagonal blocks $\bar{K}_{tr}$ and $\bar{K}_{tr}^T$ into symmetric matrices.

LEMMA 18.2 **(symmetric off-diagonal blocks)** *Generically, there exists a choice of object frame origin at which the* 3×3 *off-diagonal blocks of the grasp stiffness matrix,* $\bar{K}_G(\bar{q}_0, \Phi_0)$*, are symmetric:* $\bar{K}_{tr} = \bar{K}_{tr}^T$.

The proof of Lemma 18.2 appears in the chapter's appendix. To investigate further decoupling of the spatial grasp stiffness matrix, consider a change of the palm reference frame, $\mathcal{F}_P$, to a new location $\mathcal{F}'_P$. Let (d_P, R_P) denote the position and orientation of $\mathcal{F}'_P$ relative to $\mathcal{F}_P$. In the new palm frame, the 6×6 grasp stiffness matrix takes the form (see Exercises):

$$\bar{K}_G(\bar{q}_0, \Phi_0) = \begin{bmatrix} R_P^T K_{tt} R_P & R_P^T K_{tr} R_P \\ R_P^T K_{tr}^T R_P & R_P^T K_{rr} R_P \end{bmatrix}. \qquad (18.15)$$

Assume that the object reference frame is chosen to symmetrize K_{tr}. Since a symmetric matrix can be diagonalized by orthogonal matrices, Eq. (18.15) shows that a symmetric matrix K_{tr} can be further diagonalized by a change in the palm frame orientation, leading to the following decoupled form.

Canonical decoupled form: *Let a k-finger robot hand hold a rigid object* $\mathcal{B}$ *with compliant finger joints. There exists an object frame origin,* d_B*, such that* $\bar{K}_{tr}$ *forms a symmetric matrix. There additionally exists a palm frame orientation,* R_P*, such that* $R_P^T \bar{K}_{tr} R_P$ *forms a diagonal matrix. Under these choices of reference frames, the* 6×6 *grasp stiffness matrix takes the form:*

$$K_G(q_0, \Phi_0) = \begin{bmatrix} k_{xx} & k_{xy} & k_{xz} & k_{x,\phi_x} & 0 & 0 \\ k_{xy} & k_{yy} & k_{yz} & 0 & k_{y,\phi_y} & 0 \\ k_{xz} & k_{yz} & k_{zz} & 0 & 0 & k_{z,\phi_z} \\ k_{x,\phi_x} & 0 & 0 & k_{\phi_x,\phi_x} & k_{\phi_x,\phi_y} & k_{\phi_x,\phi_z} \\ 0 & k_{y,\phi_y} & 0 & k_{\phi_x,\phi_y} & k_{\phi_y,\phi_y} & k_{\phi_y,\phi_z} \\ 0 & 0 & k_{z,\phi_z} & k_{\phi_x,\phi_z} & k_{\phi_y,\phi_z} & k_{\phi_z,\phi_z} \end{bmatrix}, \qquad (18.16)$$

where ϕ_x*,* ϕ_y *and* ϕ_z *denote rotations about the* x*-,* y*- and* z*-axes of the palm frame.*

Consider the restoring wrench that results from pure object rotation about the palm frame x-axis in $\mathbb{R}^3$, when the grasp stiffness matrix takes the decoupled form of Eq. (18.16). The grasped object should experience restoring torques about all three axes, but the restoring forces are limited to the x-axis. Analogous behavior arises from object rotations around the palm frame y- and z-axes. When the grasped object is displaced along the palm frame x-axis, reaction forces will develop along all three directions, but the object will experience a restoring torque only about the x-axis. Analogous behavior arises from object displacements along the palm frame y- and z-axes.

Example: Recall the paint roller application described at the beginning of this section. The object frame associated with the decoupled form can be oriented with its z-axis collinear to the back-and-forth roller motions tangent to the wall. The object frame y-axis points into the wall, and its x-axis aligns with the roller's rotation axis. The robot hand gripping the roller must press the roller into the wall in order to reliably apply paint. Let us consider what happens with a generic robot hand and a generic choice of joint stiffnesses. If the grasp stiffness matrix is decoupled according to Eq. (18.16), when the robot hand pushes the roller against the wall, the roller will experience a rotation about the direction normal to the wall, which will misalign the roller away from its preferred z-direction motion. To minimize this unwanted behavior, one must choose hands with symmetric mechanism geometries and specialized joint stiffness values. These design choices allow for more complete grasp stiffness decoupling, as was seen in the example of Figure 18.6. ○

18.4 Stability of Compliant Grasps

When a grasped object initially held at an equilibrium grasp is perturbed, the compliant joints will generate reaction torques. These joint torques will generate net wrenches on the grasped object. A compliant grasp is said to be *locally asymptotically stable* if all perturbations to the grasped object result in restoring wrenches that guide the grasped object back to the equilibrium grasp once the disturbance is removed. From Section 18.2, we know that if the object is perturbed from its equilibrium grasp by a small amount, δq, the resulting wrench on the object due to the compliant finger joints is governed by the grasp stiffness matrix:

$$\delta \boldsymbol{w} \;=\; -[K_G(q_0, \Phi_0)]\delta q.$$

To move the grasped object through this small displacement against the finger forces, the perturbing agent must apply the *negated* wrench $-\delta \boldsymbol{w}$ to the object $\mathcal{B}$. The instantaneous work required to move the object through a small displacement δq is thus given by

$$(-\delta \boldsymbol{w}) \cdot \delta q \;=\; \delta q^T [K_G(q_0, \Phi_0)]\delta q.$$

This work is related to the potential energy of the compliant hand-and-object system. If the grasp stiffness matrix $K_G(q_0, \Phi_0)$ is positive definite, the unloaded equilibrium grasp lies at a *local minimum* of the grasp potential energy function. Hence, if a grasped object is perturbed within a compliant hand whose associated stiffness matrix is positive definite, positive work is required to displace the object along any direction.

Consider the case where the grasp stiffness matrix is positive definite. If there are no losses in the hand-and-object system, the object will oscillate indefinitely once the perturbation is released. In practical grasps, losses arise from damping introduced by the joint controls, such as the velocity feedback term in Eq. (18.6). Hence, one intuitively expects that as the stored energy is dissipated, the object will come to rest at its original equilibrium position, which gives local asymptotic stability of the grasp. We can make this observation formal by appealing to a classical result of Kelvin on the behavior of damped mechanical systems evolving under the influence of a potential field (see

Bibliographical Notes). The dynamics of the grasped object can be written as

$$\frac{d}{dt}\left[\frac{\partial}{\partial \dot{q}}\mathcal{K}(q,\dot{q})\right] - \frac{\partial}{\partial q}\mathcal{K}(q,\dot{q}) = \mathbf{w}(t), \tag{18.17}$$

where $\mathcal{K}(q,\dot{q})$ is the kinetic energy of the grasped object: $\mathcal{K}(q,\dot{q}) = \frac{1}{2}\dot{q}^T M(q)\dot{q}$, where $M(q)$ is the object $m \times m$ mass matrix. A system of the form (18.17) forms a *damped mechanical system* governed by a potential energy function when the external wrench that acts on $\mathcal{B}$ has the form: $\mathbf{w}(t) = -\nabla U(q) + f_d(q,\dot{q})$, where $U(q)$ is a potential energy function and $f_d(q,\dot{q})$ is a *dissipative vector field*. The vector field $f_d(q,\dot{q})$ is dissipative if it acts to reduce the total mechanical energy of the system, $E(q,\dot{q}) = U(q) + \mathcal{K}(q,\dot{q})$. That is, $\frac{d}{dt}E(q(t),\dot{q}(t)) < 0$ along system trajectories. In the compliant grasp case, $U(q)$ is the potential energy associated with the hand's joint stiffnesses.

The following theorem will be the basis for the grasp stability criterion.

Kelvin's stability theorem: *Consider a mechanical system whose dynamics takes the form of Eq. (18.17), with* $\mathbf{w}(t) = -\nabla U(q) + f_d(q,\dot{q})$*, where* $U(q)$ *is a potential energy function and* $f_d(q,\dot{q})$ *is a strictly* dissipative vector field. *The local minima of* $U(q)$ *with zero velocity are local attractors of the system trajectories.*

To apply Kelvin's stability theorem to compliant grasps, we must assume a source of damping, or energy loss, in the movements of the grasping mechanism. While small frictional processes in the finger joints dissipate energy, note that the PD control strategy of Eq. (18.6) contains a damping term $\tau_{i,j} = -\eta_{i,j}\dot{\phi}_{i,j}$. Since one or more finger joints must move when the grasped object moves, it is easy to show that the derivative term of the PD controls always leads to strict damping, as long as there is a damping term on every joint control law. These observations lead to the following stability theorem whose formal proof is omitted.

THEOREM 18.3 (**compliant grasp stability**) *Let a robot hand hold with pointed fingertips a rigid object* $\mathcal{B}$ *at an unloaded equilibrium grasp* (q_0, Φ_0)*. Each of the* n_i *joints is assumed be controlled by a PD control law in the form of Eq. (18.6). If the grasp stiffness matrix,* $K_G(q_0, \Phi_0)$*, is positive definite, the equilibrium grasp is* **locally asymptotically stable**.

The following example shows that local asymptotic stability is robust with respect to the choice of joint stiffnesses.

Example: Consider the two-finger compliant grasp depicted in Figure 18.5. Under what conditions is this grasp stable? The eigenvalues of the 3×3 grasp stiffness matrix are

$$\lambda_1 = \frac{k_1}{l^2}$$

$$\lambda_{2,3} = \frac{(k_1 + L^2 k_1 + 4k_2) \pm \sqrt{(k_1 + L^2 k_1 + 4k_2)^2 - 16L^2 k_1 k_2}}{2l^2}.$$

Since $k_1 > 0$ and $k_2 > 0$, all three eigenvalues are positive. Hence, the grasp depicted in Figure 18.5 is locally asymptotically stable under all joint stiffnesses $k_1 > 0$ and $k_2 > 0$. ○

Let us end this section with a discussion of the stability behavior of *preloaded* grasps. The grasp stiffness matrix is simply the Hessian of the potential energy of the system comprised of the compliant finger joints and the grasped object. However, new terms associated with the preloading contact forces give rise to intricate behavior that is demonstrated in the following example.

The coin-snapping phenomenon: The stability of a compliant grasp can behave in non-intuitive ways. In the coin-snapping phenomenon, an initially stable compliant grasp can abruptly lose its stability when the finger contact force magnitudes exceed a critical level. Due to this abrupt loss of grasp stability, the grasped object violently rotates in response to large moments generated by the finger contact forces at the instant of stability loss. This phenomenon can be formally analyzed by plotting the equilibrium solutions of the complaint grasp as a function of a scalar that modulates the finger contact force magnitudes. The stable equilibrium branch meets an unstable branch and ceases to exit at a *fold bifurcation,* or it merges with two unstable branches into an unstable branch at a *pitchfork bifurcation* (see Bibliographical Notes).

Even a simple planar hand can demonstrate this effect. Consider the situation where a compliant two-finger hand holds a disc-shaped object (the "coin") at antipodal points via frictional contacts. The grasping hand and object lie in a horizontal plane. Each finger mechanism consists of two mutually orthogonal prismatic joints. One joint axis aligns with the contact normal and has linear stiffness k_1; the other joint axis aligns with the tangent at the contact and has linear stiffness k_2. The grasp is *preloaded* along the contact normals. That is, the hand applies opposing normal forces, $f_{1,0} + f_{2,0} = \vec{0}$, which squeeze the disc along its diameter. We will investigate how the grasp stability varies as the preloading level increases.

Let $x_{0,1}$ and $x_{0,2}$ denote the location of the finger contacts at the equilibrium grasp. If the preloading finger force magnitudes are scaled by a common factor, $s \geq 1$, the force applied by each fingertip is given by

$$f_i(q) = -K_{G,i} \cdot (x_i(q) - x_{0,i}) + s \cdot f_{0,i} \qquad s \geq 1, i = 1,2$$

where $K_{G,1} = K_{G,2} = \text{diag}(k_1, k_2)$ are the finger stiffness matrices. The finger contact positions are given by $x_i(q) = R_{PB}(\theta)b_i + d$ where $q = (d, \theta)$ is the object's configuration. For small displacements relative to the equilibrium grasp configuration, the potential energy of this compliant grasp is given by

$$U(q) = \frac{1}{2}\sum_{i=1}^{2}(x_i(q) - x_{0,i})^T K_G, i(x_i(q) - x_{0,i}) - s f_{0,i} \cdot (x_i(q) - x_{0,i}) \qquad s \geq 1.$$

The first term in $U(q)$ is the potential energy due to compliant deflections of the finger joints as they respond to object motions. The second term in $U(q)$, $-s f_{0,i} \cdot (x_i(q) - x_{0,i})$, represents the energy needed to displace the object against the preloading forces. Note that it is positive in compressive grasps, where the finger forces act inward against the object $\mathcal{B}$. The *loaded* grasp stiffness matrix, $\widetilde{K}_G(q_0, \Phi_0)$, is simply the Hessian of the grasp's potential energy at the equilibrium grasp. This stiffness matrix is the sum of the unloaded grasp stiffness matrix and a term due to the preloading effects. Assuming the object frame is located at the center of the disc-shaped object, the grasp stiffness matrix is given by

$$\widetilde{K}_G(q_0, \Phi_0) = D^2U(q_0) = K_G(q_0, \Phi_0) + \sum_{i=1}^{2} \begin{bmatrix} O & & 0 \\ & & 0 \\ 0 & 0 & sf_{0,i} \cdot \rho_i(\theta_0) \end{bmatrix} \tag{18.18}$$

where O is a 2×2 matrix of zeroes, $\rho_i(\theta_0) = R_{PB}(\theta_0)b_i$ is the vector from $\mathcal{B}$'s frame origin to the ith contact point, and $K_G(q_0, \Phi_0)$ is the unloaded grasp stiffness matrix specified in Proposition 18.1. The terms $sf_{0,1} \cdot \rho_1(\theta_0)$ and $sf_{0,2} \cdot \rho_2(\theta_0)$ are both negative for compressive grasps, thus indicating their *destabilizing* effect with increased finger loading.

To see the coin-snapping effect, let us find an explicit formula for $\widetilde{K}_G(q_0, \Phi_0)$. The 2×2 diagonal blocks of the hand Jacobian are given by

$$H_1(\phi_{1,0}, q_0) = W_1^T R_{PK_1} J_1 = \begin{bmatrix} 0 & -1 \\ 1 & 0 \end{bmatrix} \qquad H_2(\phi_{2,0}, q_0) = W_2^T R_{PK_2} J_2 = \begin{bmatrix} 0 & 1 \\ 1 & 0 \end{bmatrix}.$$

Additionally,

$$DX_{b_1}(q_0) = \begin{bmatrix} 0 & 1 & 0 \\ -1 & 0 & -L \end{bmatrix} \quad \text{and} \quad DX_{b_2}(q_0) = \begin{bmatrix} 0 & -1 & 0 \\ -1 & 0 & L \end{bmatrix}.$$

The contributions to the unloaded grasp stiffness matrix from each finger are

$$K_{G,1} = DX_{b_1}^T(q_0) W_1 \left(H_1 K_1^{-1} H_1^T\right)^{-1} W_1^T DX_{b_1} = \begin{bmatrix} k_2 & 0 & Lk_2 \\ 0 & k_1 & 0 \\ Lk_2 & 0 & L^2k_2 \end{bmatrix}$$

and

$$K_{G,2} = DX_{b_2}^T(q_0) W_2 \left(H_2 K_2^{-1} H_2^T\right)^{-1} W_2^T DX_{b_2} = \begin{bmatrix} k_2 & 0 & -Lk_2 \\ 0 & k_1 & 0 \\ -Lk_2 & 0 & L^2k_2 \end{bmatrix}.$$

The unloaded grasp stiffness matrix is the sum of these terms:

$$K_G(q_0, \Phi_0) = K_{G,1} + K_{G,2} = 2 \begin{bmatrix} k_2 & 0 & 0 \\ 0 & k_1 & 0 \\ 0 & 0 & L^2k_2 \end{bmatrix}.$$

Let us next compute the preloading terms. At the equilibrium grasp: $f_{0,1} \cdot \rho_1(\theta_0) = -s\|f_{0,1}\|L$ and $f_{0,2} \cdot \rho_2(\theta_0) = -s\|f_{0,2}\|L$. Hence, the loaded grasp stiffness matrix takes the form:

$$\widetilde{K}_G(q_0, \Phi_0) = 2 \begin{bmatrix} k_2 & 0 & 0 \\ 0 & k_1 & 0 \\ 0 & 0 & L(k_2L - s\|f_0\|) \end{bmatrix} \qquad s \geq 1,$$

where $||f_0|| = ||f_{0,1}|| = ||f_{0,2}||$ are the finger force magnitudes at the initial grasp.

The first two eigenvalues of $\widetilde{K}_G(q_0, \Phi_0)$ are always positive since $k_1 > 0$ and $k_2 > 0$. In contrast, the sign of the third eigenvalue, $\lambda_3 = 2L(k_2 L - s||f_0||)$, depends on the squeezing forces magnitude, s. For low squeezing forces, λ_3 is positive and the grasp is stable. As s increases, the eigenvalue λ_3 eventually becomes negative at the critical loading level $s = k_2 L/||f_0||$. At this instant, the grasp becomes *unstable*. The eigenvector corresponding to the unstable eigenvalue represents pure rotation of the disc object about its center, which forms the grasp center of compliance. Once the grasp becomes unstable, the disc will violently rotate away from its equilibrium position and escape from the fingers. Readers can experience this unusual behavior by squeezing a moderate-sized or large coin between their thumb and forefinger until the coin snaps at a sufficiently high loading level. ○

Bibliographical Notes

Hanafus and Asada [1] and Salisbury [2] were the first to model grasp stiffness effects in robot hands governed by compliant finger joints. Cutkoskey, Kao and coworkers [3–5] carried out early work on the derivation of a robot hand's stiffness matrix, as well as studies on the properties of grasp stiffness matrices and their coordinate transformations.

Loncaric [6] and Ciblak and Lipkin [7, 8] analyzed the structure of spatial stiffness matrices. Selig [9] developed explicit expressions for the stiffness matrix of parallel spring systems. The realization of task specific spatial stiffness matrices using simple linear springs, which can be realized by "stiffness programming" of the finger joint PD control laws, is considered by Huang and Schimmels [10, 11]. Huang and Schimmels [12] and Patterson and Lipkin [13] also explore the *eigenscrews* of spatial stiffness matrices, which form the natural invariants of the spatial grasp stiffness matrix. Zefran and Kumar [14] used methods of differential geometry to show that the spatial grasp stiffness matrix depends upon a choice of affine connection.

Kelvin's stability result appears in the work of Koditschek [15]. For a different approach to grasp stability, see Shapiro et al. [16]. Analysis of the coin-snapping phenomena including its *bifurcation diagrams* can be found in Shapira et al. [17].

Appendix: Derivation of the Grasp Stiffness Matrix

This appendix contains a proof of Proposition 18.1, which derives an expression for the grasp stiffness matrix. The appendix also contains a proof of Lemma 18.2 on the decoupling properties of the grasp stiffness matrix.

Proposition 18.1 *Let a k-finger robot hand contact a rigid object $\mathcal{B}$ with compliant finger joints at an unloaded grasp (q_0, Φ_0). The $m \times m$ stiffness matrix of the ith finger, $K_{G,i}(q_0, \Phi_0)$, is given by*

$$K_{G,i}(q_0, \Phi_0) = DX^T_{b_i}(q_0) W_i(q_0) \left[H_i(\phi_{i,0}, q_0) K^{-1}_{\phi_i} H^T_i(\phi_{i,0}, q_0) \right]^{-1} W^T_i(q_0) DX_{b_i}(q_0),$$

where K_{ϕ_i} is ith finger joint stiffness matrix and $H_i(\phi_{i,0}, q_0)$ is the ith diagonal block of the hand Jacobian. The **grasp stiffness matrix** *of the compliant hand-and-object system is given by the parallel combination of the k finger stiffnesses:*

$$K_G(q_0, \Phi_0) = \sum_{i=1}^{k} K_{G,i}(q_0, \Phi_0).$$

Proof sketch: Let us ignore the constant term $U(\Phi_0)$ so that the grasp potential energy is given by $U(\Phi) = \frac{1}{2}(\Phi - \Phi_0)^T K_\Phi (\Phi - \Phi_0)$. We first convert the potential energy into an equivalent expression involving joint torques. Based on Eq. (18.3), the energy in each of the joint springs due to a torque $\tau_{i,j}$ about that joint axis can be expressed as a function of the joint compliance as $\frac{1}{2}\kappa^{-1}_{i,j}(\tau_{i,j} - \tau^0_{i,j})^2$. The total energy stored in the compliant joints of the k-finger hand can thus be expressed as

$$U(\mathcal{T}) = \frac{1}{2} \sum_{i=1}^{k} (\tau_i - \tau_{i,0})^T K^{-1}_{\phi_i} (\tau_i - \tau_{i,0}) = \frac{1}{2}(\mathcal{T} - \mathcal{T}_0)^T K^{-1}_\Phi (\mathcal{T} - \mathcal{T}_0), \quad (18.19)$$

where τ_i is the vector of ith finger joint torques, $\mathcal{T} = (\tau_1, \ldots, \tau_N)$ is the vector of all joint torques, and $\mathcal{T}_0$ are the joint torques at the equilibrium grasp.

Next we express the potential energy $U(\mathcal{T})$ in terms of small changes in the finger contact forces, $\vec{f} = (f_1, \ldots, f_k)$. Let us restrict our analysis to the situation where $\delta\mathcal{T} = \mathcal{T} - \mathcal{T}_0$ is a small change in $\mathcal{T}$ away from the equilibrium grasp torques. From Eq. (18.7), $\mathcal{T} = [J^T_H(\Phi, q)]\, \vec{f}$, we have that

$$\delta\mathcal{T} = \delta\Big([J^T_H(\Phi, q)\vec{f}] \Big) = J^T_H(\Phi_0, q_0)\delta\vec{f} + [DJ^T_H(\Phi_0, q_0)]\vec{f}_0,$$

where $\delta\vec{f} = (\delta f_1, \ldots, \delta f_k)$, $DJ^T_H(\Phi_0, q_0)$ is the total derivative of $J^T_H(\Phi, q)$ with respect to its arguments at (Φ_0, q_0), and $\vec{f}_0$ is the vector of contact forces at the equilibrium grasp. Substituting this expression for $\delta\mathcal{T}$ into Eq. (18.19) gives

$$\begin{aligned} U(\delta f_1, \ldots, \delta f_k) &= \tfrac{1}{2}\delta\mathcal{T}^T K^{-1}_\Phi \delta\mathcal{T} \\ &= \tfrac{1}{2}\sum_{i=1}^{k} \left(H^T_i(\phi_{i,0}, q_0)\delta f_i + DH^T_i(\phi_{i,0}, q_0) f_{i,0}\right)^T K^{-1}_{\phi_i} \left(H^T_i(\phi_{i,0}, q_0)\delta f_i \right. \\ &\qquad \left. + DH^T_i(\phi_{i,0}, q_0) f_{i,0}\right), \end{aligned}$$

where $H_i(\phi_{i,0}, q_0)$ is the ith diagonal block of the hand Jacobian. Typically non-zero joint torques are required in order to apply preloading contact forces at the equilibrium grasp. Here, we make the assumption that the grasp is *unloaded* so that $\vec{f_0} = \vec{0}$. Under this assumption, the potential energy stored in the compliant joints reduces to the form:

$$U(\delta f_1, \ldots, \delta f_k) = \frac{1}{2} \sum_{i=1}^{k} (\delta f_i)^T [H_i(\phi_{i,0}, q_0) K_{\phi_i}^{-1} H_i^T(\phi_{i,0}, q_0)] \delta f_i. \qquad (18.20)$$

Let us intuitively analyze Eq. (18.20) by physically interpreting the motions and forces underlying this expression. The product $H_i^T(\phi_{i,0}, q_0)\delta f_i$ gives the ith finger joint torques that are needed to support the change δf_i in the ith finger contact force. Since $K_{\phi_i}^{-1}$ is a matrix of joint compliances, its product with the joint torques, $K_{\phi_i}^{-1} H_i^T(\phi_{i,0}, q_0)\delta f_i$, consists of angular displacements of the ith finger joints. The product $H_i(\phi_{i,0}, q_0) K_{\phi_i}^{-1} H_i^T(\phi_{i,0}, q_0)\delta f_i$ represents small contact-point displacements. Such small displacements are approximated as contact point velocities, $(v_i, \ldots, v_k)$, that satisfy the local contact compatibility constraints:

$$v_i = W_i^T(q_0)\frac{d}{dt} X_{b_i}(q_0) = W_i^T(q_0) D X_{b_i}(q_0)\dot{q} \qquad i = 1 \ldots k.$$

Based on this interpretation, the gradient of the potential energy $U(\delta f_1, \ldots, \delta f_k)$ specified in Eq. (18.20) represents the *action* of the covectors $\delta f_1, \ldots, \delta f_k$ on contact point velocities, $v_1, \ldots, v_k$. Consequently, Eq. (18.20) implies that

$$v_i = \left[H_i(\phi_{i,0}, q_0) K_{\phi_i}^{-1} H_i^T(\phi_{i,0}, q_0)\right]\delta f_i \qquad i = 1 \ldots k.$$

This relationship between v_i and δf_i can be inverted:

$$\delta f_i = \left[H_i(\phi_{i,0}, q_0) K_{\phi_i}^{-1} H_i^T(\phi_{i,0}, q_0)\right]^{-1} v_i \qquad i = 1 \ldots k. \qquad (18.21)$$

Equation (18.21) describes the stiffness of the ith finger mechanism with respect to instantaneous displacements of its pointed fingertip under the contact compatibility constraints.[2]

The grasp potential energy can now be expressed in terms of the contact point velocities that satisfy the contact compatibility constraints. Substituting for δf_i according to Eq. (18.21) into Eq. (18.20), then making the identification $v_i = W_i^T(q_0) D X_{b_i}(q_0)\dot{q}$ for $i = 1 \ldots k$, gives the potential energy function:

$$U(\dot{q}) = \frac{1}{2}\dot{q}^T \sum_{i=1}^{k} D X_{b_i}^T(q_0)\Big(W_i(q_0)\big[H_i(\phi_{i,0}, q_0) K_{\phi_i}^{-1} H_i^T(\phi_{i,0}, q_0)\big]^{-1} W_i^T(q_0)\Big) D X_{b_i}(q_0)\dot{q},$$

which has the form $U(\dot{q}) = \frac{1}{2}\dot{q}^T K_G(q_0, \Phi_0)\dot{q}$. □

The following terminology and auxiliary lemma will be used in the characterization of the decoupling properties of the grasp stiffness matrix.

[2] The relationship specified in Eq. (18.21) is well defined only when the diagonal blocks of the hand Jacobian possess full row rank.

Skew symmetric matrices: A 3×3 skew symmetric matrix Z takes the form:

$$Z = \begin{bmatrix} 0 & -z_z & z_y \\ z_z & 0 & -z_x \\ -z_y & z_x & 0 \end{bmatrix}$$

The vector $z = (z_x, z_y, z_z)$ which corresponds to Z is denoted $z = Z^{\vee}$. Let Q be a 3×3 real matrix, and let $Q_{as} = \frac{1}{2}(Q - Q^T)$ be its skew symmetric part. The *axial vector* of Q is defined as $\text{axis}(Q) = (Q_{as})^{\vee}$.

LEMMA 18.4 *Let $z \in \mathbb{R}^3$, and let Z be a 3×3 skew-symmetric such that $Z = [z\times]$. Any 3×3 matrix A with real entries satisfies the identity:*

$$\text{axis}(Z \cdot A) = \tfrac{1}{2}(Z \cdot A - (Z \cdot A)^T)^{\vee} = \tfrac{1}{2}\big[\text{tr}(A)I - A\big]z$$

where $\text{tr}(A)$ is the trace of A, and I is the 3×3 identity matrix.

Proof: Let a_{ij} denote the entries of the matrix A, and let $z = (z_1, z_2, z_3)$. The product $Z \cdot A$, where $Z = [z\times]$ takes the form:

$$Z \cdot A = \begin{bmatrix} -a_{21}z_3 + a_{31}z_2 & -a_{22}z_3 + a_{32}z_2 & -a_{23}z_3 + a_{33}z_2 \\ a_{11}z_3 - a_{31}z_1 & a_{12}z_3 - a_{32}z_1 & a_{13}z_3 - a_{33}z_1 \\ -a_{11}z_2 + a_{21}z_1 & -a_{12}z_2 - a_{32}z_1 & a_{13}z_2 + a_{23}z_1 \end{bmatrix}.$$

We can use this product to find an expression for the axial vector of $Z \cdot A$:

$$(Z \cdot A)^{\vee}_{as} = \tfrac{1}{2}\big(Z \cdot A - (Z \cdot A)^T\big)^{\vee} = \tfrac{1}{2}\begin{pmatrix} (a_{22} + a_{33})z_1 - a_{12}z_2 - a_{13}z_3 \\ (a_{11} + a_{33})z_2 - a_{21}z_1 - a_{23}z_3 \\ (a_{11} + a_{22})z_3 - a_{31}z_1 - a_{32}z_3 \end{pmatrix}$$

$$= \tfrac{1}{2}\begin{bmatrix} a_{22} + a_{33} & -a_{12} & -a_{13} \\ -a_{21} & a_{11} + a_{33} & -a_{23} \\ -a_{31} & -a_{32} & a_{11} + a_{22} \end{bmatrix}\begin{pmatrix} z_1 \\ z_2 \\ z_3 \end{pmatrix} = \tfrac{1}{2}\big[\text{tr}(A)I - A\big]z\,.$$

□

Lemma 18.4 leads to the following decoupling property of the spatial grasp stiffness matrix.

Lemma 18.2 *Generically, there exists a choice of object frame origin at which the 3×3 off-diagonal blocks of the grasp stiffness matrix, $\bar{K}_G(\bar{q}_0, \Phi_0)$, are symmetric: $\bar{K}_{tr} = \bar{K}_{tr}^T$.*

Proof: Let us separate $\bar{K}_{tr}$ into its symmetric and skew symmetric parts:

$$\bar{K}_{tr} = (\bar{K}_{tr})_s + (\bar{K}_{tr})_{as} = \tfrac{1}{2}(\bar{K}_{tr} + \bar{K}_{tr}^T) + \tfrac{1}{2}(\bar{K}_{tr} - \bar{K}_{tr}^T)\,.$$

The skew symmetric part of $\bar{K}_{tr}$ takes the form:

$$(\bar{K}_{tr})_{as} = \tfrac{1}{2}(K_{tt}[d_B\times] - [d_B\times]^T K_{tt}^T) + \tfrac{1}{2}(K_{tr} - K_{tr}^T)\,.$$

Based on Lemma 18.4, the axial vector of $\bar{K}_{tr}$, $\bar{k}_{tr} \in \mathbb{R}^3$, is given by

$$\bar{k}_{tr} = \tfrac{1}{2}(K_{tt} - \text{tr}(K_{tt})I)d_B + \tfrac{1}{2}(K_{tr} - K_{tr}^T)^{\vee}\,. \tag{18.22}$$

If $\mathrm{tr}(K_{tt})$ is not in the spectrum of K_{tt} (the generic case), then Eq. (18.22) can be solved explicitly for the location of the object frame origin, d_B, that renders $\bar{K}_{tr}$ symmetric. Equivalently, the matrix $(\bar{K}_{tr})_{as}$ is zero when $d_B = -(K_{tt} - \mathrm{tr}(K_{tt})I)^{-1}(K_{tr} - K_{tr}^T)^{\vee}$. □

Exercises

Section 18.1

Exercise 18.1: Derive the equivalent spring relationships described in Eq. (18.4)–(18.5).

Exercise 18.2: Consider two spatial springs connected in series. Let K_1 be the spatial stiffness matrix of the first spring, and let K_1 be the spatial stiffness matrix of the second spring. Show that the equivalent stiffness of the series-connected springs is

$$K_{eq} = (K_1^{-1} + K_2^{-1})^{-1} .$$

Section 18.2

Exercise 18.3*: Consider the case where the k-finger joint stiffness matrix K_Φ is fully populated (i.e., it is not necessarily block diagonal). That is, coupling between joints across all k fingers is allowed. Show that the grasp stiffness matrix takes the form:

$$K_G(q_0, \Phi_0) = A_{CO}^T W(q_0)[W^T(q_0)J_H(\Phi_0, q_0)K_\Phi^{-1} J_H^T(\Phi_0, q_0)W(q_0)]^{-1} W^T(q_0)A_{CO},$$

where $W(q_0)$ is a block diagonal matrix having the form: $W(q_0) = \mathrm{diag}(W_i(q_0))$. The matrix A_{CO} has the form:

$$A_{CO} = \begin{bmatrix} T_{C_1 B} \\ \vdots \\ T_{C_k B} \end{bmatrix},$$

where $T_{C_i B}$ transforms rigid-body velocities from the object frame $\mathcal{F}_B$ to the object's ith contact frame, $\mathcal{F}_{C_i}$ for $i = 1 \ldots k$.

Exercise 18.4*: Obtain a formula for the grasp stiffness matrix in the case of the soft-point contact model (assume pointed fingertips).

Exercise 18.5*: Obtain a formula for the grasp stiffness matrix when the finger contacts are *loaded* at the grasp's equilibrium configuration (assume pointed fingertips).

Section 18.3

Exercise 18.6: Can the stiffness matrix of a planar compliant grasp be decoupled if the 2×2 matrix K_{tt} in Eq. (18.14) is *not* full rank?

Exercise 18.7: Show that if the 3×3 off-diagonal block $\bar{K}_{tr} = K_{tt}[d_B \times] + K_{tr} = O$, then the grasp stiffness matrix in Eq. (18.11) takes the form:

$$\bar{K}_G(\bar{q}_0, \Phi_0) = \begin{bmatrix} K_{tt} & O \\ O & (K_{rr} + [d_B \times]^T K_{tt}[d_B \times]) \end{bmatrix}$$

where O is a 3×3 matrix of zeroes.

Exercise 18.8: Consider a change of the palm frame, $\mathcal{F}_P$, to a new location $\mathcal{F}'_P$. Let (d_P, R_P) denote the position and orientation of $\mathcal{F}'_P$ relative to $\mathcal{F}_P$. Show that in the new palm frame, the 6×6 grasp stiffness matrix takes the form:

$$\bar{K}_G(q_0, \Phi_0) = \begin{bmatrix} R_P^T K_{tt} R_P & R_P^T K_{tr} R_P \\ R_P^T K_{tr}^T R_P & R_P^T K_{rr} R_P \end{bmatrix}.$$

Exercise 18.9: The *grasp compliance matrix* is defined as the inverse of the grasp stiffness matrix. In the case of 2-D grasps, can this matrix be decoupled into a block diagonal form by proper choice of the object frame origin?

Exercise 18.10: In the case of 2-D grasps, does the object frame origin that decouples the grasp stiffness matrix (the grasp's *center of stiffness*) coincide with the object frame origin that decouples the grasp compliance matrix (the grasp's *center of compliance*)?

References

[1] H. Hanafusa and H. Asada, "Stable prehension by a robot hand with elastic fingers," in *Proceedings of the 7th International Symposium Industrial Robots*, pp. 384–389, 1977.

[2] J. K. Salisbury, "Kinematic and force analysis of articulated hands," PhD thesis, Dept. of Mechanical Engineering, Stanford University, 1982.

[3] M. Cutkosky and I. Kao, "Computing and controlling the compliance of a robotic hand," *IEEE Transactions on Robotics and Automation*, vol. 5, no. 2, pp. 151–165, 1989.

[4] S.-F. Chen and I. Kao, "Conservative congruence transformation for joint and cartesian stiffness matrices of robotic hands and fingers," *International Journal on Robotics Research*, vol. 19, no. 9, pp. 835–847, 2000.

[5] I. Kao and C. Ngo, "Properties of the grasp stiffness matrix and conservative control strategies," *International Journal on Robotics Research*, vol. 18, no. 2, pp. 159–167, 1999.

[6] J. Loncaric, "Normal forms of stiffness and compliance matrices," *IEEE Journal on Robotics and Automation*, vol. 3, no. 6, pp. 567–572, 1987.

[7] N. Ciblak and H. Lipkin, "Asymmetric cartesian stiffness for the modeling of compliant robotic sytems," in *ASME Design Engineering Division*, vol. 72, pp. 197–204, 1997.

[8] N. Ciblak and H. Lipkin, "Design and analysis of remote center of compliance structures," *Journal of Robotic Systems*, vol. 20, no. 8, pp. 415–427, 2003.

[9] J. Selig, "The spatial stiffness matrix from simple stretches springs," in *Proceedings of the IEEE International Conference on Robotics and Automation*, pp. 3314–3319, 2000.

[10] S. Huang and J. Schimmels, "The bounds and realization of spatial stiffnesses achieved with simple springs connected in parallel," *IEEE Transactions on Robotics and Automation*, vol. 14, no. 3, pp. 466–475, 1998.

[11] S. Huang and J. Schimmels, "The bounds and realization of spatial compliances achieved with simple serial elastic mechanisms," *IEEE Transactions on Robotics and Automation*, vol. 16, no. 1, pp. 99–103, 2000.

[12] S. Huang and J. Schimmels, "The eigenscrew decomposition of spatial stiffness matrices," *IEEE Transactions on Robotics and Automation*, vol. 6, no. 2, pp. 146–156, 2000.

[13] T. Patterson and H. Lipkin, "Structure of robot compliance," *Journal of Mechanical Design*, vol. 115, no. 3, pp. 576–580, 2003.

[14] M. Zefron and V. Kumar, "Affine connections for the cartesian stiffness matrix," in *Proceedings of the IEEE International Conference on Robotics and Automation*, pp. 1376–1381, 1997.

[15] D. Koditschek, "The application of total energy as a lyapunov function for mechanical control systems," *Contemporary Mathematics*, vol. 97, pp. 131–157, 1989.

[16] A. Shapiro, E. Rimon and A. Ohev-Zion, "On the mechanics of natural compliance in frictional contacts and its effect on grasp stiffness and stability," *International Journal on Robotics Research*, vol. 32, no. 4, pp. 425–445, 2013.

[17] T. Shapira, E. Rimon and A. Shapiro, "Investigation of the coin snapping phenomenon in linearly compliant robot grasps," *IEEE Transactions on Robotics*, vol. 34, no. 3, pp. 794–804, 2018.

Appendix A Introduction to Non-Smooth Analysis

This appendix introduces some useful analysis tools for non-smooth functions. We will focus on the following class of Lipschitz continuous functions.

DEFINITION A.1 **(Lipschitz function)** *A function* $f(x) : \mathbb{R}^m \to \mathbb{R}$ *is* **Lipschitz continuous** *at x if there exists a local Lipschitz constant,* $c > 0$*, such that*

$$|f(x_1) - f(x_2)| \leq c \cdot \|x_1 - x_2\|$$

for all x_1 *and* x_2 *in a small open neighborhood of* x*. Lipschitz continuity is similarly defined for vector-valued functions.*

Lipschitz continuous functions lie between continuous and differentiable functions. According to *Rademacher's theorem,* every Lipschitz continuous function is differentiable almost everywhere in its domain. Among the piecewise differentiable functions, Lipschitz continuous functions possess a bounded derivative at the points of differentiability, and a bounded directional derivative at the points of non-differentiability. This property is illustrated in the following example.

Example: Figure A.1(a) shows the graph of the function $\mathsf{dst}(x, \mathcal{O}) = \min_{y \in \mathcal{O}}\{|x - y|\}$, where $\mathcal{O}$ is the semi-infinite interval: $\mathcal{O} = \{y \in \mathbb{R} : y \leq 0\}$. This function is piecewise differentiable and Lipschitz continuous. While this function is *not* differentiable at $x = 0$, it has bounded slopes on both sides of $x = 0$ and, hence, is Lipschitz continuous at this point. In contrast, the function $f(x)$ whose graph is shown in Figure A.1(b) has an

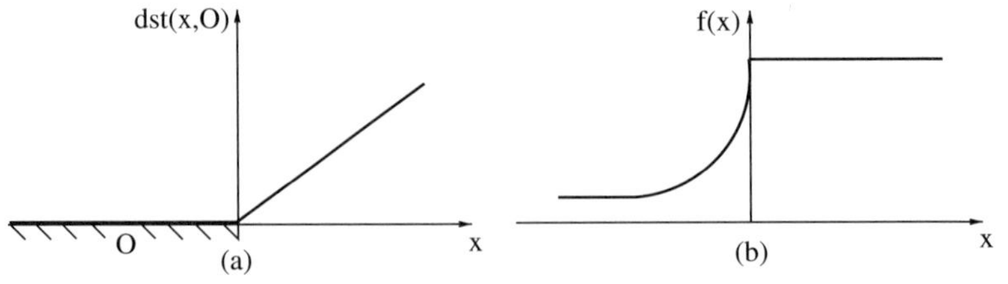

Figure A.1 (a) The function $\mathsf{dst}(x, \mathcal{O})$ is piecewise smooth and Lipschitz continuous. (b) The function $f(x)$ is piecewise smooth but *not* Lipschitz at $x = 0$.

infinite slope on the left side of $x = 0$. This function is piecewise differentiable but *not* Lipschitz at $x = 0$. ○

The distance of a point from a set: Consider the function $\mathsf{dst}(x, \mathcal{O}) : \mathbb{R}^m \to \mathbb{R}$, which measures the minimal distance of points $x \in \mathbb{R}^m$ from a closed set $\mathcal{O}$: $\mathsf{dst}(x, \mathcal{O}) = \min_{y \in \mathcal{O}}\{\|x - y\|\}$. Let us verify that $\mathsf{dst}(x, \mathcal{O})$ is Lipschitz continuous with a global Lipschitz constant $c = 1$. One must show that $|\mathsf{dst}(x_1, \mathcal{O}) - \mathsf{dst}(x_2, \mathcal{O})| \leq \|x_1 - x_2\|$ for all $x_1, x_2 \in \mathbb{R}^m$. Given any two points x_1 and x_2, assume that $\mathsf{dst}(x_1, \mathcal{O}) \geq \mathsf{dst}(x_2, \mathcal{O})$. Let $y_2 \in \mathcal{O}$ be a point that is closest to x_2, so that $\mathsf{dst}(x_2, \mathcal{O}) = \|x_2 - y_2\|$. The three points x_1, x_2 and y_2 satisfy the triangle inequality: $\|x_1 - y_2\| \leq \|x_1 - x_2\| + \|x_2 - y_2\|$. Since $\mathsf{dst}(x_1, \mathcal{O}) \leq \|x_1 - y_2\|$ and $\mathsf{dst}(x_2, \mathcal{O}) = \|x_2 - y_2\|$, the triangle inequality implies that $\mathsf{dst}(x_1, \mathcal{O}) \leq \|x_1 - x_2\| + \mathsf{dst}(x_2, \mathcal{O})$. Thus, $\mathsf{dst}(x_1, \mathcal{O}) - \mathsf{dst}(x_2, \mathcal{O}) \leq \|x_1 - x_2\|$, which implies that $\mathsf{dst}(x, \mathcal{O})$ is Lipschitz continuous with Lipschitz constant $c = 1$. ○

Some useful properties of the Lipschitz continuous functions are as follows.

1. Any differentiable function $f(x) : \mathbb{R}^m \to \mathbb{R}$ is Lipschitz continuous, with $c = \|\nabla f(x)\|$.
2. The composition of two Lipschitz continuous functions is Lipschitz continuous.

   To see this fact, let $g(y) : \mathbb{R}^n \to \mathbb{R}$ and $h(x) : \mathbb{R}^m \to \mathbb{R}^n$ be Lipschitz continuous functions with Lipschitz constants c_g and c_h. Let $f(x) = g(h(x))$. Then

   $$|f(x_1) - f(x_2)| = |g(h(x_1)) - g(h(x_2))| \leq c_g \|h(x_1) - h(x_2)\| \leq c_g c_h \|x_1 - x_2\|,$$

   thus proving that $f(x) = g(h(x))$ is Lipschitz continuous.
3. Let $F(x) = \min_{t \in \mathcal{T}}\{f_t(x)\}$ be the *pointwise minimum* over a parametrized family of Lipschitz continuous functions, where $\mathcal{T}$ is a discrete or a continuous set. If the functions $f_t(x)$ have a common upper bound on their Lipschitz constants, $F(x)$ is Lipschitz.

   To see this fact, consider two points x_1 and x_2 in a small open neighborhood of x. Let $F(x_1) = f_{t_1}(x_1)$, such that f_{t_1} has a Lipschitz constant c_{t_1}. Let $F(x_2) = f_{t_2}(x_2)$, such that f_{t_2} has a Lipschitz constant c_{t_2}. We have to show that $|F(x_1) - F(x_2)| = |f_{t_1}(x_1) - f_{t_2}(x_2)| \leq c\|x_1 - x_2\|$ for some $c > 0$. There are two cases to consider. If $f_{t_1}(x_1) \geq f_{t_2}(x_2)$, $|f_{t_1}(x_1) - f_{t_2}(x_2)| \leq |f_{t_2}(x_1) - f_{t_2}(x_2)| \leq c_{t_2}\|x_1 - x_2\|$ (since $f_{t_1}(x_1) \leq f_{t_2}(x_1)$). If $f_{t_1}(x_1) < f_{t_2}(x_2)$, $|f_{t_2}(x_2) - f_{t_1}(x_1)| \leq |f_{t_1}(x_2) - f_{t_1}(x_1)| \leq c_{t_1}\|x_1 - x_2\|$ (since $f_{t_2}(x_2) \leq f_{t_1}(x_2)$). Thus $|F(x_1) - F(x_2)| \leq \max\{c_{t_1}, c_{t_2}\} \cdot \|x_1 - x_2\|$. Hence, $F(x)$ is Lipschitz continuous with a Lipschitz constant $c = max\{c_t\}$ for $\mathcal{T}$.

The inter-body distance function: Let $\mathcal{B}(q)$ denote the set of points in $\mathbb{R}^m$ occupied by a rigid object $\mathcal{B}$, located at a configuration q. The minimum distance between $\mathcal{B}(q)$ and a fixed obstacle $\mathcal{O}$ is given by $\mathsf{dst}(B(q), \mathcal{O}) = \min_{x \in \mathcal{B}(q)}\{\mathsf{dst}(x, \mathcal{O})\}$. At every collision-free configuration, $q \in \mathcal{F}$, the minimum distance between $\mathcal{B}(q)$ and $\mathcal{O}$ is generically attained at a discrete set of points on the object boundary. That is, $\mathsf{dst}(B(q), \mathcal{O}) = \mathsf{dst}(x_1, \mathcal{O}) = \cdots = \mathsf{dst}(x_N, \mathcal{O})$, where $x_1, \ldots, x_N$ are the points on $\mathcal{B}$'s boundary that are closest to $\mathcal{O}$. The member functions $\mathsf{dst}(x_i, \mathcal{O})$ have a common Lipschitz constant,

$c=1$. Based on property (3), $\mathsf{dst}(B(q),\mathcal{O})$ is Lipschitz continuous in q with a Lipschitz constant $c=1$. ○

While Lipschitz continuous functions do not possess a gradient vector at the non-differentiable points, they possess a *generalized gradient* that can be used to analyze their behavior at these points. The generalized gradient is defined as follows.

DEFINITION A.2 (**generalized gradient**) *Let $f:\mathbb{R}^m \to \mathbb{R}$ be a Lipschitz continuous function. Let $x \in \mathbb{R}^m$ be surrounded by open sets $\mathcal{U}_1,\ldots,\mathcal{U}_N$ containing the point x at their common boundary point,*[1] *such that f is differentiable in each of these open sets. The* **generalized gradient** *of f at x, denoted $\partial f(x)$, is the convex combination:*

$$\partial f(x)=\sum_{i=1}^{N}\lambda_i\cdot \lim_{x_j\to x,\, x_j\in U_i}\nabla f(x_j) \qquad 0\le\lambda_1,\ldots,\lambda_N\le 1 \;\; \textstyle\sum_{i=1}^{N}\lambda_i=1.$$

The generalized gradient is thus a convex combination of limit vectors, $\lim_{x_j\to x}\nabla f(x_j)$, each taken along a sequence $\{x_j\}$ approaching the point x within one of the open sets $\mathcal{U}_1,\ldots,\mathcal{U}_N$. When $\nabla f(x)$ exists, it can be interpreted as a tangent vector in $T_x\mathbb{R}^m$. Similarly, the generalized gradient can be interpreted as a *convex set* of tangent vectors in $T_x\mathbb{R}^m$. The computation of $\partial f(x)$ is illustrated in the following example.

Example: Consider the absolute-value function: $f(x)=|x|$ for $x\in\mathbb{R}$. It is differentiable away from $x=0$ and Lipschitz continuous with $c=1$ at $x=0$ (since $\big||x_1|-|x_2|\big|\le |x_1-x_2|$). Let us compute the generalized gradient of f. Clearly, $f'(x)=-1$ for $x<0$ and $f'(x)=1$ for $x>0$. Based on Definition A.2, $\partial f(0)$ is the convex combination of -1 and 1, giving the interval $[-1,1]$. Hence, we obtain:

$$\partial f(x)=\begin{cases}-1 & x<0\\ [-1,1] & x=0\\ +1 & x>0\end{cases} \qquad \text{for } f(x)=|x|.$$

○

Some useful properties of the generalized gradient are as follows (see Bibliographical Notes).

(i) When $\partial f(x)$ is a single vector, the function f is differentiable at x and $\partial f(x)=\nabla f(x)$.

(ii) The generalized gradient satisfies the following *generalized chain rule*. Let $g(y):\mathbb{R}^n\to\mathbb{R}$ be Lipschitz continuous, and let $h(x):\mathbb{R}^m\to\mathbb{R}^n$ be differentiable. The function composition $f(x)=g(h(x))$ is Lipschitz continuous, and at points where the Jacobian $Dh(x)$ has full rank the generalized gradient of f has the form:

$$\partial f(x)=\partial g\big(h(x)\big)\cdot Dh(x)=\big\{w\cdot Dh(x)\,:\,w\in\partial g(h(x))\big\}.$$

(iii) The *generalized directional derivative* of f along a direction v is $\nabla f(x)\cdot v$ when f is differentiable at x, and $\partial f(x)\cdot v$ when f is non-differentiable but Lipschitz continuous at x.

[1] Technically, the closures of the sets $\mathcal{U}_1,\ldots,\mathcal{U}_N$ are required to *cover* a small open neighborhood centered at x, such that $x\in\mathsf{bdy}(\mathcal{U}_i)$ for $i=1\ldots N$.

To see this fact, consider a smooth path $\alpha(t):(-\epsilon,\epsilon)\to\mathbb{R}^m$, such that $\alpha(0)=x$ and $\dot{\alpha}(0)=v$. If $f(x)$ is Lipschitz continuous in x, $f(\alpha(t))$ is Lipschitz continuous in t. By the generalized chain rule, $\partial f(\alpha(0))=\partial f(\alpha(0))\cdot\dot{\alpha}(0)=\partial f(x)\cdot v$.

Let us next define the notion of an *extremum point* of Lipschitz continuous functions.

DEFINITION A.3 (**extremum point**) *A Lipschitz continuous function, $f:\mathbb{R}^m\to\mathbb{R}$, has an* **extremum point** *at x if every tangent direction, $v\in T_x\mathbb{R}^m$, satisfies one of the two conditions:*

1. *f has a zero generalized directional derivative along v, $w\cdot v=0$ for all $w\in\partial f(x)$.*
2. *f has a non-smooth local minimum or maximum along the line collinear with v at x.*

When f is differentiable at x, the extremum condition becomes $\nabla f(x)\cdot v=0$ for all $v\in T_x\mathbb{R}^m$. This gives the usual extremum condition $\nabla f(x)=\vec{0}$ for differentiable functions.

Example: The function $f(x,y)=|x|+ay^2$ is Lipschitz continuous. Its generalized gradient at the origin, $\partial f(0,0)=[-1,1]\times 0$, forms an interval aligned with the x axis. As depicted in Figure A.2(a), when $a>0$ this function has a non-smooth local minimum at $(0,0)$ along every direction transversal to the y-axis and a zero directional-derivative along the y-axis. The origin is thus an extremum point of f. ○

A useful property of the generalized gradient at an extremum point is stated in the following lemma.

LEMMA A.1 (**non-smooth extremum**) *If a Lipschitz continuous function $f:\mathbb{R}^m\to\mathbb{R}$ has an extremum point at x, then necessarily $\vec{0}\in\partial f(x)$.*

Proof sketch: Consider the case where f has a local maximum at x. Recall that x is surrounded by open sets, $\mathcal{U}_1,\ldots,\mathcal{U}_N$, such that f is differentiable in each of these sets. Let a smooth path $\alpha(t):(-\epsilon,0]\to\mathbb{R}^m$ approach x within the set $\mathcal{U}_i$, such that $\alpha(0)=x$ and $\dot{\alpha}(0)=v$. Since $f(\alpha(t))$ has a local maximum at $t=0$, $\frac{d}{dt}f(\alpha(t))\geq 0$ for

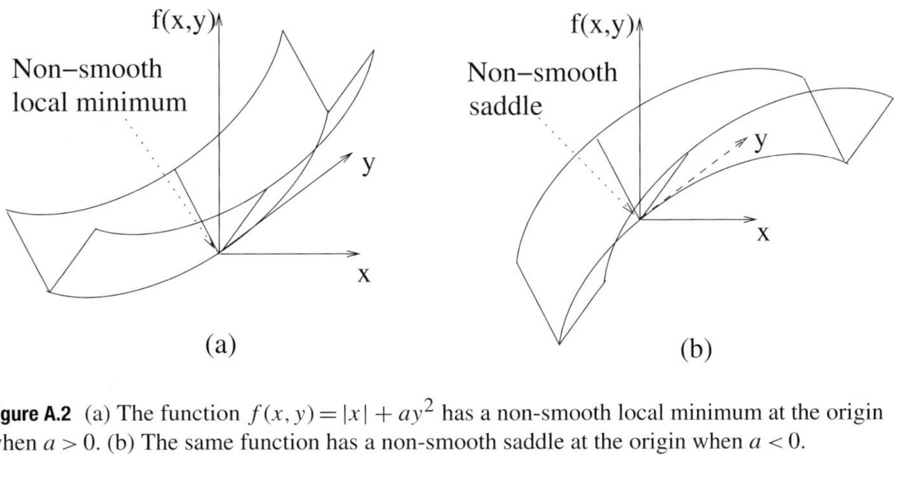

Figure A.2 (a) The function $f(x,y)=|x|+ay^2$ has a non-smooth local minimum at the origin when $a>0$. (b) The same function has a non-smooth saddle at the origin when $a<0$.

$t \in (-\epsilon, 0)$. It follows that $\lim_{t\to 0} \frac{d}{dt} f(\alpha(t)) \geq 0$. Based on the *generalized chain rule,* $\lim_{t\to 0} \frac{d}{dt} f(\alpha(t)) = w \cdot v \geq 0$ for some $w \in \partial f(x)$. Since x is surrounded by $\mathcal{U}_1, \ldots, \mathcal{U}_N$, we can repeat this argument for all directions $v \in T_x\mathbb{R}^m$. Hence, for every $v \in T_x\mathbb{R}^m$ there exists $w \in \partial f(x)$ such that $w \cdot v \geq 0$. This condition implies that $\partial f(x)$ (interpreted as a set in $T_x\mathbb{R}^m$) intersects *every* closed half-space of $T_x\mathbb{R}^m$ passing through the origin of this space. Since $\partial f(x)$ is a convex set, the latter condition can only occur when $\vec{0} \in \partial f(x)$. □

The following example shows that non-smooth extremum points can form local minima, saddles and local maxima (which can be viewed by reversing the following local minimum example).

Example: Consider the Lipschitz continuous function $f(x, y) = |x| + ay^2$ depicted in Figure A.2. Its generalized gradient at the origin is given by $\partial f(0,0) = [-1, 1] \times 0$. Since $\partial f(0,0)$ forms an open interval containing the origin, f satisfies the necessary condition of Lemma A.1. The origin forms a non-smooth extremum point, since f has a non-smooth minimum along the x-axis and a zero-directional derivative along the y-axis. When $a > 0$, the function f has a non-smooth local minimum at the origin (Figure A.2(a)). When $a < 0$, the function f has a non-smooth saddle at the origin (Figure A.2(b)). ○

The remainder of this appendix focuses on the *pointwise minimum function,* which serves to characterize the grasped object free motions when held by multiple finger bodies.

DEFINITION A.4 *The* **pointwise minimum function,** $f_{min} : \mathbb{R}^m \to \mathbb{R}$, *has the form,*

$$f_{min}(x) = \min\{f_1(x), \ldots, f_N(x)\},$$

where each member function $f_i(x) : \mathbb{R}^m \to \mathbb{R}$ *is Lipschitz continuous.*

The function $f_{min}(x)$ is Lipschitz continuous according to Property (iii) of these functions. Its generalized gradient, $\partial f_{min}(x)$, is determined by those functions that attain the minimum value at x. Let $\mathcal{I}(x)$ be the index set of the functions attaining the minimum value at x. That is, $f_{min}(x) = f_i(x)$ for $i \in \mathcal{I}(x)$. Assume, for simplicity, that these are differentiable functions. Based on Definition A.2, $\partial f_{min}(x)$ is the convex combination:

$$\partial f_{min}(x) = \sum_{i\in\mathcal{I}(x)} \lambda_i \nabla f_i(x) \qquad 0 \leq \lambda_i \leq 1 \text{ for } i \in \mathcal{I}(x),\ \textstyle\sum_{i\in\mathcal{I}(x)} \lambda_i = 1.$$

When an extremum point of $f_{min}(x)$ is associated with a single member function, $f_{min} = f_i(x)$, it is a differentiable extremum satisfying $\nabla f_i(x) = \vec{0}$. When an extremum point of $f_{min}(x)$ is associated with several member functions, $f_{min} = f_i(x)$ for $i \in \mathcal{I}(x)$, it is a non-smooth extremum satisfying the condition $\vec{0} \in \partial f_{min}(x)$.

Two useful properties of the non-smooth extremum points of $f_{min}(x)$ are as follows.

(a) The non-smooth extremum points of $f_{min}(x)$ are local maxima or saddles.

To see this fact, let $f_{min}(x)$ have a non-smooth extremum at x_0. In this case, $\partial f_{min}(x_0)$ is nontrivial and consequently, $\nabla f_i(x_0) \neq \vec{0}$ for some $i \in \mathcal{I}(x_0)$. It follows that f_i is strictly decreasing along the direction $-\nabla f_i(x_0)$. Since $f_{min}(x_0) = f_i(x_0)$, f_{min} is also strictly decreasing along this direction at x_0. Hence, x_0 *cannot* be a local minimum, and, therefore, it must be a local maximum or a saddle of f_{min}.

(b) A sufficient condition for $f_{min}(x)$ to have a strict local maximum at x_0 is that $\partial f_{min}(x_0)$ contains an open m-dimensional ball centered at the origin of $T_{x_0}\mathbb{R}^m$.

To see this fact, recall that $\partial f_{min}(x_0)$ is the convex combination of $\nabla f_i(x_0)$ for $i \in \mathcal{I}(x_0)$. By assumption, this convex set contains an open m-dimensional ball centered at the origin of $T_{x_0}\mathbb{R}^m$. Hence, for every direction $v \in T_{x_0}\mathbb{R}^m$, there exists some $i \in \mathcal{I}(x_0)$ such that $\nabla f_i(x_0)$ satisfies $\nabla f_i(x_0) \cdot v < 0$. It follows that f_i is strictly decreasing along v at x_0. Since $f_{min}(x_0) = f_i(x_0)$, f_{min} is also strictly decreasing along this direction. As this argument applies for every direction $v \in T_{x_0}\mathbb{R}^m$, f_{min} must have a strict local maximum at x_0.

Note that Property (b) forms a *first derivative test,* which gives a sufficient condition for a non-smooth local maximum of $f_{min}(x)$. When the member functions $f_1(x), \ldots, f_N(x)$ are twice differentiable, the local maxima of $f_{min}(x)$ are determined by the following *generalized Hessian matrix.*

DEFINITION A.5 (**generalized Hessian**) *Let $f_{min}(x)$ be the pointwise minimum of twice differentiable functions $f_1(x), \ldots, f_N(x)$. Let $w \in \partial f_{min}(x)$ be expressed as the convex combination:*

$$w = \sum_{i \in \mathcal{I}(x)} \lambda_i \nabla f_i(x) \qquad 0 \leq \lambda_i \leq 1 \text{ for } i \in \mathcal{I}(x),\ \textstyle\sum_{i \in \mathcal{I}(x)} \lambda_i = 1.$$

The **generalized Hessian** *of f_{min} at the composite point (x, w), denoted $\partial^2 f_{min}(x, w)$, is the $m \times m$ matrix,*

$$\partial^2 f_{min}(x, w) = \sum_{i \in \mathcal{I}(x)} \lambda_i D^2 f_i(x),$$

where $D^2 f_i(x)$ is the $m \times m$ Hessian matrix of f_i for $i \in \mathcal{I}(x)$.

Let $f_{min}(x)$ have an extremum point at x_0. The generalized gradient satisfies the condition $\vec{0} \in \partial f_{min}(x_0)$ according to Lemma A.1. Hence, the origin can be expressed as the convex combination:

$$\sum_{i \in \mathcal{I}(x_0)} \lambda_i \nabla f_i(x_0) = \vec{0} \qquad 0 \leq \lambda_i \leq 1 \text{ for } i \in \mathcal{I}(x_0),\ \textstyle\sum_{i \in \mathcal{I}(x_0)} \lambda_i = 1.$$

The generalized Hessian at $(x_0, \vec{0})$ provides the following local maximum test, which forms a key component of grasp immobilization theory.

THEOREM A.2 (**local maximum test**) *Let $f_{min} : \mathbb{R}^m \to \mathbb{R}$ have an extremum point at x_0, such that $\partial f_{min}(x_0)$ lies in a p-dimensional linear subspace V of $T_{x_0}\mathbb{R}^m$, where $0 \leq p \leq m$. Consider the quadratic form defined on the orthogonal complement subspace,*

$$\kappa(x_0, v) = v^T \left[\partial^2 f_{min}(x_0, \vec{0})\right] v \qquad v \in V^\perp,$$

where $\partial^2 f_{min}$ is the generalized Hessian of f_{min}. Then f_{min} has a **strict local maximum** *at x_0 under one of the following conditions:*

(i) *If the subspace V is fully m-dimensional, $\partial f_{min}(x_0)$ must contain a small m-dimensional ball centered at the origin of $T_{x_0}\mathbb{R}^m$.*

(ii) *If the subspace V has dimension $p < m$: (1) $\partial f_{min}(x_0)$ must contain a small p-dimensional ball centered at the origin in this subspace, and (2) $\kappa(x_0, v) < 0$ for all $v \in V^\perp$.*

Proof: Condition (i) is the same as Property (b) for a strict local maximum. Let us therefore prove Condition (ii). By assumption, the convex combination of $\nabla f_i(x_0)$ for $i \in \mathcal{I}(x_0)$ contains a p-dimensional ball centered at the origin of V. Hence, for any direction $v \notin V^\perp$ there exists $i \in \mathcal{I}(x_0)$ such that $\nabla f_i(x_0) \cdot v < 0$. Thus, f_i is strictly decreasing along v at x_0. Since $f_{min}(x_0) = f_i(x_0)$, the function f_{min} is also strictly decreasing along v at x_0. The function f_{min} thus has a strict local maximum along any direction $v \notin V^\perp$.

Next consider the behavior of f_{min} along directions $v \in V^\perp$. Let $\alpha(t) : (-\epsilon, \epsilon) \to \mathbb{R}^m$ be a smooth path such that $\alpha(0) = x_0$ and $\dot{\alpha}(0) = v \in V^\perp$. We have to show that $f_{min}(\alpha(t))$ is strictly decreasing along *all* such paths at $t = 0$. Up to second-order approximation, the collection of all paths tangent to $V^\perp$ at x is represented by the pairs $(\dot{\alpha}(0), \ddot{\alpha}(0))$, such that $\dot{\alpha}(0) = v \in V^\perp$ and $\ddot{\alpha}(0) = a \in \mathbb{R}^m$. These pairs can be written as $(v, a) \in V^\perp \times \mathbb{R}^m$. The second-order Taylor expansion of $f_i(x)$ along $\alpha(t)$ is given by

$$f_i(\alpha(t)) = f_i(x_0) + (\nabla f_i(x_0) \cdot v)t + \tfrac{1}{2}\left(v^T D^2 f_i(x_0) v + \nabla f_i(x_0) \cdot a\right) t^2 + o(t^3) \qquad t \in (-\epsilon, \epsilon),$$

where we substituted $\alpha(0) = x_0$, $\dot{\alpha}(0) = v$ and $\ddot{\alpha}(0) = a$. We may assume that $f_{min}(x_0) = 0$, so that $f_i(x_0) = 0$ for $i \in \mathcal{I}(x_0)$. Since $v \in V^\perp$ and $\nabla f_i(x_0) \in V$ for $i \in \mathcal{I}(x_0)$, the second-order approximation of $f_i(\alpha(t))$ has the form:

$$f_i(\alpha(t)) = \tfrac{1}{2}\left(v^T D^2 f_i(x_0) v + \nabla f_i(x_0) \cdot a\right) t^2 \qquad t \in (-\epsilon, \epsilon).$$

We now show that for every pair $(v, a) \in V^\perp \times \mathbb{R}^m$, there exists $i \in \mathcal{I}(x_0)$ such that $v^T D^2 f_i(x_0) v + \nabla f_i(x_0) \cdot a < 0$. This would imply that f_i is strictly decreasing along α at $t = 0$. Since $f_{min}(x_0) = f_i(x_0)$, this would imply that f_{min} is also strictly decreasing along α at $t = 0$.

By assumption, $\kappa(x_0, v) = v^T \sum_{i \in \mathcal{I}(x_0)} \lambda_i D^2 f_i(x_0) v < 0$ for all $v \in V^\perp$. Adding the zero term $\sum_{i \in \mathcal{I}(x_0)} \lambda_i (\nabla f_i(x_0) \cdot a)$ to the left side of this inequality gives

$$\sum_{i \in \mathcal{I}(x_0)} \lambda_i \left(v^T D^2 f_i(x_0) v + \nabla f_i(x_0) \cdot a\right) < 0 \qquad (v, a) \in V^\perp \times \mathbb{R}^m. \tag{A.1}$$

Since $\lambda_i \geq 0$ for $i \in \mathcal{I}(x_0)$, at least one of the summands in Eq. (A.1) must be negative. The function f_i associated with this summand has a strict local maximum along α at $t = 0$:

$$f_i(\alpha(t)) = \frac{1}{2}\left(v^T D^2 f_i(x_0)v + \nabla f_i(x_0) \cdot a\right) t^2 < 0 \qquad t \in (-\epsilon, \epsilon).$$

Since this argument holds for *all* pairs $(v, a) \in V^{\perp} \times \mathbb{R}^m$, f_{min} has a strict local maximum along all paths tangent to $V^{\perp}$ at x_0. Since f_{min} is also strictly decreasing along every direction $v \notin V^{\perp}$ at x_0, it has a strict local maximum at this point. □

The local maximum test of Theorem A.2 combines first and second-derivative tests. The existence of a p-dimensional ball contained in $\partial f_{min}(x_0)$ is a *first derivative test* within the subspace V. The negative definiteness of the quadratic form $\kappa(x_0, v)$ is a *second derivative test* within the orthogonal complement subspace $V^{\perp}$.

Bibliographical Notes

The generalized gradient has its origin with the notion of *subgradient* from convex analysis [1]. A comprehensive calculus for Lipschitz continuous functions, including the notion of the *generalized gradient* and the *generalized chain rule,* appears in Clarke's *Optimization and Nonsmooth Analysis* [2]. Clarke considers a generic type of *regular* Lipschitz continuous functions, which are the functions considered in this book.

The *generalized Hessian* of the pointwise minimum function, $\partial^2 f_{min}(x)$, is a modest extension of the non-smooth theory developed by the book authors. It allows rigorous development of second-order immobilization theory, which accounts for second-order penetration motions along the subspace of first-order free motions, as well as first-order penetration motions along the orthogonal complement subspace as specified by the local maximum text of Theorem A.2.

References

[1] R. T. Rockafellar, *The Theory of Subgradients and its Applications*. Heldermann Verlag, 1981.

[2] F. H. Clarke, *Optimization and Nonsmooth Analysis*. SIAM, 1990.

Appendix B Summary of Stratified Morse Theory

The free c-space of a rigid object $\mathcal{B}$ contacted by stationary rigid finger bodies $\mathcal{O}_1, \ldots, \mathcal{O}_k$ forms a *stratified set* $\mathcal{F}$ in $\mathbb{R}^m$, where $m = 3$ or 6. The strata of $\mathcal{F}$ are smooth manifolds of varying dimension,[1] each representing one possible contact combination of $\mathcal{B}$ with the finger bodies. Stratified Morse theory characterizes the critical points (local minima, saddles and maxima) of smooth scalar valued functions defined on stratified sets such as $\mathcal{F}$. This appendix summarizes the main results of this theory.

A *regularly stratified set* is a set $\mathcal{X} \subset \mathbb{R}^m$, partitioned into a finite union of disjoint smooth manifolds called *strata*, satisfying the Whitney condition. The dimension of the strata varies between zero (isolated point manifolds) and m (open subsets of the ambient space $\mathbb{R}^m$). The Whitney condition requires that the tangents of two neighboring strata meet nicely as follows. Let p be a point on the common boundary of two neighboring strata, $\mathcal{S}$ and $\mathcal{S}'$, such that p belongs to $\mathcal{S}$. Since p lies on the boundary of $\mathcal{S}'$, it can be approached along a sequence of points $\{p_i\}$ from the stratum $\mathcal{S}'$. The two strata satisfy the *regular stratification condition* when the tangent space $T_p\mathcal{S}$ is contained in the limit of tangent spaces, $\lim_{p_i \to p} T_{p_i}\mathcal{S}'$, taken along the sequence $\{p_i\}$. This condition is illustrated in the following example.

Example: Figure B.1(a) shows a set $\mathcal{X} \subset \mathbb{R}^3$ consisting of two intersecting planes, $\mathcal{X} = P_1 \cup P_2$. First consider a stratification of $\mathcal{X}$ into the plane P_1 and the two half planes of P_2 that lie on both sides of P_1, shown in Figure B.1(b). Let p be a point on the line $P_1 \cap P_2$. Since $p \in P_1$, the tangent plane $T_p P_1$ is *not* contained in the limit tangent plane $\lim_{p_i \to p} T_{p_i} P_2$. Hence, this is *not* a regular stratification of $\mathcal{X}$. Next consider the decomposition of $\mathcal{X}$ into four half planes and the line $P_1 \cap P_2$, shown in Figure B.1(c). The tangent line to $P_1 \cap P_2$ is contained in the limit tangent plane to each of the neighboring half planes. Hence, this is a regular stratification of $\mathcal{X}$. ○

Let us next introduce the notion of critical points on a stratified set. First, consider the definition of a critical point on a smooth manifold $\mathcal{M}$ in $\mathbb{R}^m$. Let $\tilde{\varphi}$ be a smooth real-valued function defined on $\mathbb{R}^m$, and let $\varphi : \mathcal{M} \to \mathbb{R}$ denote the restriction

[1] A *d-dimensional manifold* in $\mathbb{R}^m$ is a set $\mathcal{M}$ whose local neighborhoods can be parametrized by d independent parameters. The manifold locally looks like $\mathbb{R}^d$ at each point $p \in \mathcal{M}$, for a fixed dimension $0 \le d \le m$.

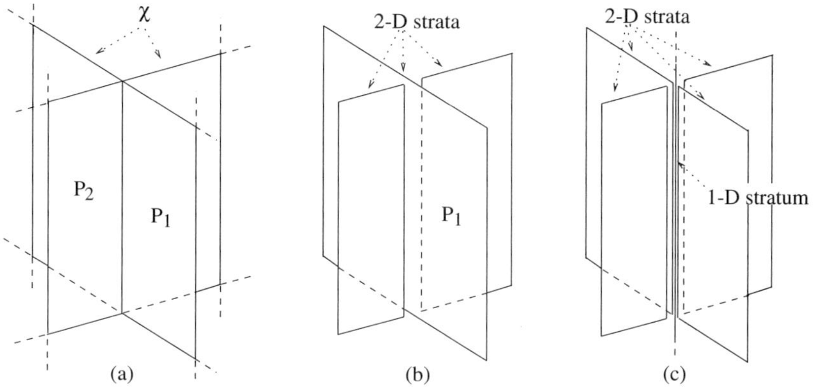

Figure B.1 (a) The set $\mathcal{X} = P_1 \cup P_2$. (b) A non-regular stratification of $\mathcal{X}$ into three 2-D strata. (c) A regular stratification of $\mathcal{X}$ into four 2-D strata and a single 1-D stratum.

of $\tilde{\varphi}$ to $\mathcal{M}$. A point $x \in \mathcal{M}$ is a *critical point* of φ if its derivative at x, denoted $D\varphi(x)$, vanishes along the tangent space $T_x\mathcal{M}$:

$$D\varphi(x) \cdot \dot{x} = 0 \qquad \text{for all } \dot{x} \in T_x\mathcal{M}.$$

Equivalently, the critical points of φ are points $x \in \mathcal{M}$ at which the gradient vector $\nabla\tilde{\varphi}(x)$ is orthogonal to the manifold $\mathcal{M}$ in the ambient space $\mathbb{R}^m$. A *critical value* of φ is the image of a critical point x, $c = \varphi(x)$. The following definition extends the notion of a critical point to stratified sets.

DEFINITION B.1 (**critical point**) *Let $\mathcal{X} \subset \mathbb{R}^m$ be a regularly stratified set, with $\varphi : \mathcal{X} \to \mathbb{R}$ the restriction of a smooth function $\tilde{\varphi}$ to $\mathcal{X}$. The* **critical points** *of φ in $\mathcal{X}$ are the* **union** *of the critical points obtained by restricting φ to the individual strata of $\mathcal{X}$.*

Note that every zero-dimensional stratum of $\mathcal{X}$ is automatically a critical point of φ. Next consider the notion of a Morse function on a smooth manifold $\mathcal{M}$. A smooth function $\varphi : \mathcal{M} \to \mathbb{R}$ is a *Morse function* when all its critical points are non-degenerate – i.e., when its Hessian matrix, $D^2\varphi(x)$, is non-singular at the critical points. The *Morse index* of φ at a non-degenerate critical point, denoted ν, is the number of negative eigenvalues of $D^2\varphi(x)$ at this point. In general, $0 \leq \nu \leq \dim(\mathcal{M})$, where $\dim(\mathcal{M})$ denotes the dimension of $\mathcal{M}$. At a local minimum, the eigenvalues are all positive and $\nu = 0$. At a saddle point, the eigenvalues have mixed signs and $0 < \nu < \dim(\mathcal{M})$. At a local maximum, the eigenvalues are all negative and $\nu = \dim(\mathcal{M})$.

Morse functions are generic and possess isolated critical points on the manifold $\mathcal{M}$. The following definition extends the notion of a Morse function to stratified sets.

DEFINITION B.2 (**Morse function**) *Let $\mathcal{X} \subset \mathbb{R}^m$ be a regularly stratified set, with $\varphi : \mathcal{X} \to \mathbb{R}$ the restriction of a smooth function $\tilde{\varphi}$ to $\mathcal{X}$. Then φ is a* **Morse function** *on $\mathcal{X}$ if* *(i)* *It is Morse on every stratum containing a critical point of* φ.

(ii) *The gradient vector* $\nabla\tilde{\varphi}$ *at every critical point is non-orthogonal to any of the neighboring strata meeting at this point.*

Morse functions are also generic on stratified sets. That is, almost all smooth functions on a regularly stratified set $\mathcal{X}$ are Morse functions. The *Morse index* of φ at a critical point, ν, is the number of negative eigenvalues of the Hessian $D^2\varphi(x)$ evaluated on the stratum containing the point x. For instance, $\nu=0$ when φ has a local minimum on the stratum containing the point x. Note, however, that φ need not possess a local minimum with respect to the neighboring strata that meet at this point. The Morse index of φ at a zero-dimensional stratum is defined as $\nu=0$. Condition (ii) is illustrated in the following example.

Example: Figure B.2 shows a stratified set $\mathcal{X}$ that consists of the surface of an upward-curving cylinder $\mathcal{S}_1$, a plane $\mathcal{S}_2$, and their intersection curve $\mathcal{S}_{12}$. The height function, $\varphi(x_1,x_2,x_3)=x_3$, has a critical point at $x_0\in\mathcal{S}_{12}$, which is a local minimum on $\mathcal{S}_{12}$. However, when x_0 is approached along a sequence of points from the stratum $\mathcal{S}_1$, the limit of the normals to $\mathcal{S}_1$ along these points is collinear with $\nabla\tilde{\varphi}(x_0)=(0,0,1)$ (Figure B.2). It follows that φ is *not* a Morse function at x_0. Note that φ becomes a Morse function for almost all small displacement of $\mathcal{S}_1$ or $\mathcal{S}_2$. ○

Stratified Morse theory studies the behavior of smooth real-valued functions on regularly stratified sets. The theory guarantees that as the value of a Morse function φ varies in the open interval between adjacent critical values of φ, the level sets $\mathcal{X}|_c=\{x\in\mathcal{X}:\varphi(x)=c\}$ are *topologically equivalent* (homeomorphic) to each other. The topological change in the level sets $\mathcal{X}|_c$ occurs locally at the critical points of φ. Let x_0 be such a critical point, with $c_0=\varphi(x_0)$. Stratified Morse theory characterizes the topological change at x_0 in terms of the behavior of φ on two complementary subsets of $\mathcal{X}$. The first set is the stratum of $\mathcal{X}$ that contains the critical point x_0, denoted $\mathcal{S}$. The other set, called the *normal slice* of $\mathcal{X}$ at x_0, is constructed in two stages. Let $D(x_0)$ be a small d-dimensional disc centered at x_0 that has the following two properties. The disc

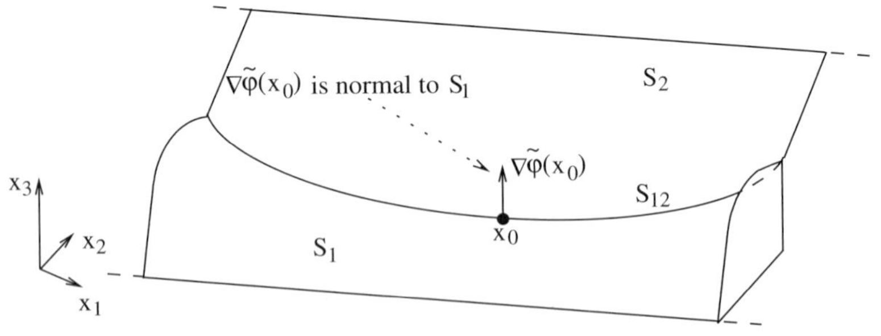

Figure B.2 The height function $\varphi(x_1,x_2,x_3)=x_3$ has a critical point at the point $x_0\in\mathcal{S}_{12}$ but is not Morse at this point since $\nabla\tilde{\varphi}(x_0)$ is normal to the neighboring stratum $\mathcal{S}_1$.

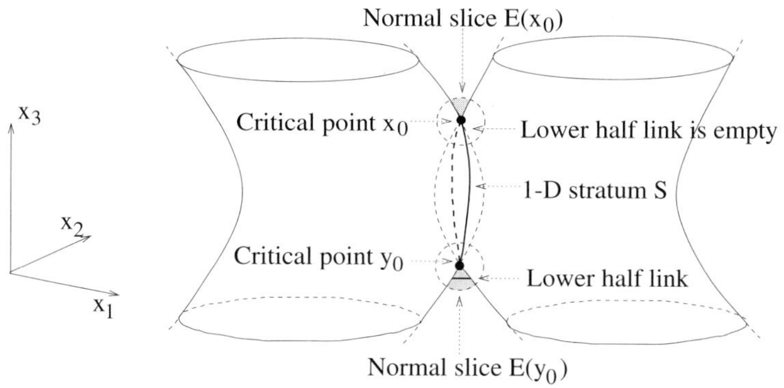

Figure B.3 An example showing the lower half link at the critical points x_0 and y_0. The function being used is $\varphi(x_1,x_2,x_3)=x_3$.

intersects the stratum $\mathcal{S}$ only at the point x_0, and the disc is transversal to $\mathcal{S}$ at x_0.[2] The latter requirement is satisfied if one chooses the disc to be normal to $\mathcal{S}$ at x_0, such that its dimension equals the dimension of the ambient space, m, minus the dimension of $\mathcal{S}$. In the second stage, one constructs the *normal slice* of $\mathcal{X}$ at x_0, denoted $E(x_0)$, as the intersection of $D(x_0)$ with the stratified set $\mathcal{X}$. Thus, $E(x_0)= D(x_0)\cap \mathcal{X}$.

Example: Figure B.3 depicts a stratified set $\mathcal{X}$ in $\mathbb{R}^3$, formed by removing from $\mathbb{R}^3$ the interior of two curved cylinders. The set $\mathcal{X}$ consists of the 3-D region surrounding the two cylinders, as well as the boundary of the cylinders' union. The height function, $\varphi(x_1,x_2,x_3)=x_3$, has two critical points, at x_0 and y_0. The stratum containing these points, $\mathcal{S}$, is a one-dimensional curve. The normal slice at each of these points is the intersection of a two-dimensional disc with $\mathcal{X}$. It is an upward-pointing sector at x_0, and a downward-pointing sector at y_0. ○

Let x_0 be a critical point of a Morse function $\varphi:\mathcal{X}\to\mathbb{R}$, such that x_0 lies in a stratum $\mathcal{S}$. The topological change in the level sets of φ at x_0 is characterized by its behavior on $\mathcal{S}$ and $E(x_0)$. On the stratum $\mathcal{S}$, the topological change is determined by the Morse index of φ at x_0. On the normal slice, $E(x_0)$, the topological change is determined by the *lower half link set,* denoted l^-, defined as the intersection of $E(x_0)$ with the level set $\mathcal{X}|_{c_0-\epsilon}=\{x\in\mathcal{X}:\varphi(x)=c_0-\epsilon\}$,

$$l^- = E(x_0)\cap \mathcal{X}|_{c_0-\epsilon}\,,$$

where $\epsilon>0$ is a small parameter and $c_0=\varphi(x_0)$. Stratified Morse theory guarantees that the topological nature of l^- does not change for all $\epsilon>0$ sufficiently small.

Example: Figure B.3 shows the stratified set $\mathcal{X}$ discussed in the previous example. The height function, $\varphi(x_1,x_2,x_3)=x_3$, has critical points at x_0 and y_0. The normal slice at x_0 has no points below this point, hence l^- is empty at this point. At the point y_0, the lower

[2] Two manifolds $\mathcal{M}_1$ and $\mathcal{M}_2$ in $\mathbb{R}^m$ *intersect transversally* at p if $T_p\mathcal{M}_1+T_p\mathcal{M}_2=T_p\mathbb{R}^m$.

half link is the intersection of a downward-pointing sector with a horizontal plane lying just below y_0. Hence, l^- is a line segment at this point. ○

The *Morse data* at a critical point consists of the Morse index ν and the lower half link set l^-. The main theorem of stratified Morse theory (or SMT) characterizes the topological change at the critical points in terms of the Morse data.

THEOREM B.1 **(main SMT theorem)** *Let $\varphi : \mathcal{X} \to \mathbb{R}$ be a Morse function on a regularly stratified set $\mathcal{X}$. Let x_0 be a critical point of φ with Morse index ν and lower half link l^-, with $c_0 = \varphi(x_0)$. The* **topological change** *in the level sets $\mathcal{X}|_c = \{x \in \mathcal{X} : \varphi(x) = c\}$ as c varies in the interval $[c_0 - \epsilon, c_0 + \epsilon]$ consists of taking a handle set,*

$$H = D^\nu \times \mathsf{cone}(l^-),$$

and gluing it to the level set $\mathcal{X}|_{c_0-\epsilon}$ along the gluing seam,

$$\Gamma = \mathsf{bdy}(D^\nu) \times \mathsf{cone}(l^-) \cup D^\nu \times l^-,$$

where D^ν is a closed ν-dimensional disc, and $\mathsf{cone}(l^-)$ is the cone with base set l^- and vertex at the critical point x_0.

The terms D^ν and $\mathsf{cone}(l^-)$ need some explanation. The closed ν-dimensional disc, D^ν, is a single point when $\nu = 0$ and a closed interval when $\nu = 1$. The boundary of D^ν, $\mathsf{bdy}(D^\nu)$, forms a $(\nu - 1)$-dimensional sphere. For instance, $\mathsf{bdy}(D^2)$ is a circle, $\mathsf{bdy}(D^1)$ consists of the two endpoints of D^1 and $\mathsf{bdy}(D^0)$ is empty. Given a lower half link l^- at a critical point x_0, the set $\mathsf{cone}(l^-)$ is defined as the collection of rays that emanate from x_0 and pass through all points of l^-. By definition, $\mathsf{cone}(l^-) = \{x_0\}$ when l^- is empty.

The remainder of this appendix applies Theorem B.1 to the characterization of two important types of critical points. The first type are the local minima of φ in $\mathcal{X}$.

COROLLARY B.2 **(local minimum)** *Let $\varphi : \mathcal{X} \to \mathbb{R}$ be a Morse function on a regularly stratified set $\mathcal{X} \subset \mathbb{R}^m$, and let $x_0 \in \mathcal{X}$ be a critical point of φ. Then φ has a* **local minimum** *at x_0 if and only if φ satisfies the two conditions,*

$$\nu = 0 \quad \text{and} \quad l^- = \emptyset,$$

where ν is the Morse index of φ at x_0 and l^- is the lower half link of φ at x_0.

Proof: First assume that x_0 is a local minimum of φ, with $c_0 = \varphi(x_0)$. In this case, the level set $\mathcal{X}|_{c_0-\epsilon} = \{x \in \mathcal{X} : \varphi(x) = c_0 - \epsilon\}$ must be empty in a sufficiently small neighborhood of x_0, where $\epsilon > 0$ is a small positive parameter. The lower half link l^- is a subset of this level set within the normal slice of $\mathcal{X}$ at x_0. Hence, $l^- = \emptyset$. Since φ has in particular a local minimum along the stratum containing x_0, $\nu = 0$. Thus, $l^- = \emptyset$ and $\nu = 0$.

Next assume that $\nu = 0$ and $l^- = \emptyset$. According to Theorem B.1, the topological change in the level sets $\mathcal{X}|_c$ at a critical point x_0 consists of taking a handle set, $H = D^\nu \times \mathsf{cone}(l^-)$, and gluing it to the level set $\mathcal{X}|_{c_0-\epsilon}$ along the gluing seam: $\Gamma = \mathsf{bdy}(D^\nu) \times \mathsf{cone}(l^-) \cup D^\nu \times l^-$. In our case, $\nu = 0$ and $l^- = \emptyset$. Hence, the handle set is $H = D^0 \times \{x_0\}$, which is topologically equivalent to the single-point set $H = \{x_0\}$. The gluing seam Γ is empty, since $\mathsf{bdy}(D^0)$ and l^- are both empty. Since Γ is empty, the handle set $H = \{x_0\}$ is disjoint from the sublevel set $\mathcal{X}|_{\le c_0-\epsilon} = \{x \in \mathcal{X} : \varphi(x) \le c_0 - \epsilon\}$ in a local neighborhood of $\mathbb{R}^m$ centered at x_0. Since $H = \{x_0\}$ and $\mathcal{X}|_{\le c_0-\epsilon}$ form locally disjoint closed sets in $\mathbb{R}^m$, they are locally separated from each other at x_0. Thus, a sufficiently small neighborhood about $H = \{x_0\}$ contains no points from the sublevel set $\mathcal{X}|_{\le c_0-\epsilon}$, and x_0 is consequently a local minimum of φ in $\mathcal{X}$. □

The condition $\nu = 0$ is the usual *second derivative* test ensuring that φ has a local minimum on the stratum containing x_0. The condition $l^- = \emptyset$ is a *first derivative test,* which verifies that φ has a local minimum with respect to the neighboring strata that meet at x_0.

The second type of critical point, a *join point,* is defined as follows.

DEFINITION B.3 **(join point)** *Let $\varphi : \mathcal{X} \to \mathbb{R}$ be a Morse function on a regularly stratified set $\mathcal{X} \subset \mathbb{R}^m$, and let $x_0 \in \mathcal{X}$ be a critical point of φ with $c_0 = \varphi(x_0)$. Then x_0 is a* **join point** *if two locally disjoint connected components of the level set $\mathcal{X}|_{c_0-\epsilon}$ meet at x_0 and become a single connected component in the level set $\mathcal{X}|_{c_0+\epsilon}$, where $\epsilon > 0$ is a small parameter.*

The following corollary of Theorem B.1 characterizes the Morse data at a join point.

COROLLARY B.3 **(join point)** *Let $\varphi : \mathcal{X} \to \mathbb{R}$ be a Morse function on a regularly stratified set $\mathcal{X} \subset \mathbb{R}^m$, and let $x_0 \in \mathcal{X}$ be a critical point of φ. Then x_0 is a* **join point** *if and only if φ satisfies one of the two conditions,*

$$\nu = 1 \quad \textit{and} \quad l^- \textit{ is empty} \tag{B.1}$$

or

$$\nu = 0 \quad \textit{and} \quad l^- \textit{ is disconnected,} \tag{B.2}$$

where ν is the Morse index of φ at x_0 and l^- the lower half link of φ at x_0.

Proof: At a join point, two locally distinct components of the level set $\mathcal{X}|_{c_0-\epsilon}$ meet and become a single component in the level set $\mathcal{X}|_{c_0+\epsilon}$. Hence, $\mathcal{X}|_{c_0-\epsilon}$ consists of two locally disjoint sets, while $\mathcal{X}|_{c_0+\epsilon}$ consists of a single locally connected set. It follows that the handle set H must be glued to $\mathcal{X}|_{c_0-\epsilon}$ along a *disconnected* gluing set Γ. The rest of the proof consists of determining the conditions under which Γ forms a disconnected set.

The gluing set Γ is the union of two sets, $\Gamma = \Gamma_1 \cup \Gamma_2$, such that $\Gamma_1 = \mathsf{bdy}(D^\nu) \times \mathsf{cone}(l^-)$ and $\Gamma_2 = D^\nu \times l^-$. Any two points in $\mathsf{bdy}(D^\nu)$ can be connected by a curve in D^ν. Similarly, any two points in l^- can be connected by a curve in $\mathsf{cone}(l^-)$. Hence, if Γ_1 and Γ_2 are both nonempty, their union $\Gamma = \Gamma_1 \cup \Gamma_2$ forms a connected set. Hence, $\Gamma = \Gamma_1 \cup \Gamma_2$ is disconnected only when Γ_1 is empty and Γ_2 is disconnected, or when Γ_1 is disconnected and Γ_2 is empty. The set $\Gamma_1 = \mathsf{bdy}(D^\nu) \times \mathsf{cone}(l^-)$ is empty iff $\nu = 0$, since $\mathsf{bdy}(D^0)$ is empty. The set $\Gamma_2 = D^\nu \times l^-$ is disconnected iff l^- is disconnected. This gives Condition (B.2). Similarly, the set $\Gamma_1 = \mathsf{bdy}(D^\nu) \times \mathsf{cone}(l^-)$ is disconnected iff $\nu = 1$, since $\text{bdy}(D^{\nu})$ forms a $(\nu - 1)$-dimensional sphere and only the zero-dimensional sphere is disconnected. But when $\nu = 1$, the only way in which $\Gamma_2 = D^{\nu} \times l^{-}$ can be empty is that l^{-} be empty. This gives Condition (B.1). □

The meaning of the condition $\nu = 1$ in Eq. (B.1) is as follows. Let $\mathcal{S}$ be the stratum containing the critical point x_0. When $\mathcal{S}$ is one-dimensional, $\nu = 1$ means that φ has a local maximum along $\mathcal{S}$ at x_0. When $\mathcal{S}$ has a higher dimension, $\nu = 1$ means that φ has a saddle point along $\mathcal{S}$ at x_0, with one negative eigenvalue and the rest positive eigenvalues. For instance, the point x_0 in Figure B.3 is a *join point* of the function $\varphi(x_1, x_2, x_3) = x_3$. Indeed, $l^{-} = \emptyset$ at x_0, while x_0 is a local maximum of φ along the one-dimensional stratum $\mathcal{S}$ containing this point.

Bibliographical Notes

Stratified Morse theory is described by Goresky and Macpherson [1]. Where Theorem B.1 is stated as the *main theorem* in Section 3.7 of this text. The local minimum test of Corollary B.2 can be obtained by other means. For instance, Trinkle et al. [2, 3] derived the first- and second-order local minimum criteria using linear complementarity approach.

References

[1] Goresky and Macpherson, *Stratified Morse Theory*. Springer-Verlag, 1980.

[2] J. C. Trinkle, A. O. Farahat and P. F. Stiller, *IEEE International Conference on Robotics and Automation*, 2815–2821, 1995.

[3] J. C. Trinkle, A. O. Farahat and P. F. Stiller, "First-order stability cells of active multi-rigid-body systems," *IEEE Transactions of Robotics and Automation*, vol. 11, no. 4, pp. 545–477, 1995.

Index